International Forum of Global Advances in Sustainable Environment, Energy, and Earth Sciences (GASES 2025)

E3S Web of Conferences Volume 633 (2025)

Baghdad, Iraq
13-14 May 2025

ISBN: 979-8-3313-2172-7

Printed from e-media with permission by:

Curran Associates, Inc.
57 Morehouse Lane
Red Hook, NY 12571

Some format issues inherent in the e-media version may also appear in this print version.

For additional information, please contact EDP Sciences – Web of Conferences
at the address below.

EDP Sciences – Web of Conferences
17, Avenue du Hoggar
Parc d'Activité de Courtabœuf
BP 112
F-91944 Les Ulis Cedex A
France

Phone: +33 (0) 1 69 18 75 75
Fax: +33 (0) 1 69 28 84 91

contact-edps@webofconferences.org

Additional copies of this publication are available from:

Curran Associates, Inc.
57 Morehouse Lane
Red Hook, NY 12571 USA
Phone: 845-758-0400
Fax: 845-758-2633
Email: curran@proceedings.com
Web: www.proceedings.com

TABLE OF CONTENTS

A Comparative Review of Delumping Algorithms........ 1
Sarsenova Adel, Mukhamedyanova Adelina, Kumarkhan Mariam

High-Performance Liquid Chromatography Method for the Determination of Levofloxacin in Liquid Pharmaceutical Preparations: Implications for Environmental Health and Earth Sciences........ 9
Smanova Zulayho, Arazberdieva Shemshat, Abdullaeva Muborek, Akhmadjanov Ulugbek

Study of Physicochemical Properties of Hydrophobic Calcite Based on Sugar Factory Defecates........ 14
Masharipova Sharofat, Jumaniyazov Maxsud, Turkmenbaeva Mayra, Kabulova Lola

The Influence of Ultrasonic Treatment on the Solubility of Aqueous Electrolyte Mixtures........ 23
Nazarbek Ulzhalgas, Nazarbekova Saule, Kambarova Galiya, Raiymbekov Yerkebulan

Accumulation of Polycyclic Aromatic Hydrocarbons (PAHs) in Wildlife Rats Tissues Inhabiting Landfills........ 30
Kadhim Hasan Firoz-Ali Mudhafar, Jawad Salih Al-Haidarey Mohammed

Environmental Biological Radiation Hazards (ELCR, AED) from Background Radiation in Vitamins Consumed by Pregnant Women in Iraq........ 40
Alebadi Shireen N., Mohammed Torki Wedyan, Saeed Jassim Ali, Kamil Khaleel Hadeel

^{218}Po and ^{214}Po in Pharmaceutical Materials Derived from Medical Plant, Iraq........ 46
Jalil Lawi Dhuha, Abojassim Ali Abid, Awad Amin K., Guida Michele, Mancini Simona

Environmental Background Radiation for Some Drug Samples Used by Children in Iraq........ 54
Mohammed Torki Wedyan, Saeed Jassim Ali, Abid Abojassim Ali, Abdul Hadi Mohsen Ahmed, Hayder Rand A.

Evaluation of Indoor Air Quality in the Village of Al-Bu Mahmid, Iraq........ 59
Majeed Kareem Oras, Radhi Jasim Hawraa, Azez Bjoo Emtihan, Saeed Jassim Ali

Evaluation of Radon Gas Levels in the Indoor Environment of Houses in Albu Mahmid Village in Kufa City, Iraq........ 65
Abd Oun Al-Jamal Lina, Hamid Abbas Saif, Mohammed Hadi Qunoot, Obayes Esraa S., Hayder Rand A.

Impact of Cadmium on Some Physiological and Biochemical Properties in Green Algae Scendesmus Qudricauda........ 73
Mohammed Hala R., Salman Jasim M., Jassim Abd Al-Rezzaq Adi

Environmental Policy as a Factor of China's Food Security........ 86
Sushchenko Maxim

Integrating Drone Technology in Sustainable Agriculture: Precision Environmental Monitoring and Resource Management........ 93
Takhumova Oksana, Pustuev Andrey, Eiryan Nikolay, Sarsadskikh Anastasia, Rykalina Olga

Eco-Integrated Agriculture: Merging Environmental Sustainability and Energy Efficiency in the Modern Agro-Production Complex........ 100
Hua Wang Mei, Rykalina Olga, Borzunova Tatyana, Kolobov Egor, Takhumova Oksana

Ecotoxicological Assessment of the Mutagenicity Level of Agricultural Soils Based on the Soil-Plant System 108
Atoyants A. L., Aghajanyan E. A., Avalyan R. E.

Determination of the Level of Natural Background Radiation in the Surface Soil of Kufa Farms, Iraq 114
Musa Sarab J., Ali Sami Noor, Hade Seood Gufran, Salim Obayes Esraa, Naji Mohsion Hanan

Use GIS Technical to Draw of the Background Radiation in Selected Soil Samples in the Najaf Sea Area 123
Mohammed Mohammed A., Hashesh Tabarak S., Mohammed Faisel G., Hayder Rand A., Ali Maryam H.

Assessment of Shatt al-Arab Water Quality Using Water Indices in Basrah City of South Iraq 131
Alrazaq Ban A., Khanjar Ebtesam F., Muhsin Israa J., Al-Ali I. A.

Monitoring Changes in the Marshes of Mesopotamia Using Multispectral Remote Sensing Data 142
Sabeeh Mosleh Jehan

Optimal Exploitation of Water Resources in Muthanna Governorate Iraq Using (RS) and (GIS) Technologies 151
Majid Abdulameer Maitham, Mezaal Mahmood Rawya, Awad Omar Qais

Radiation Risk Assessment in Water Samples Collected from Homes in Al-Qasim and Al-Talee\'a, Babylon Governorate 164
Amin Mustafa M., Kadhim Inaam H., Ibrahem Salar H.

Utilization of Remote Sensing and GIS Approach for Surface Water Monitoring in the Najaf Sea, Al-Najaf Providence, Iraq: An Overview 173
Husham Abdulkareem, Al-Hadithi Mufid, Mozihim Rana

Land Suitability for Solar Power Plants in Karbala, Iraq Using MCD and AHP-GIS 181
Srayyih Mustafa S., Al-Gbory M. M., Khanjar Ebtesam F., Saleem Noora M., Alrazaq Ban A.

Improvement Efficiency and Stability for Silicon Solar Cells 193
Subhi Halboos Nagham, Falh Hassan Adnan

Analysis of the Efficiency of Enterprises in the Field of Energy and Sustainable Development: European Experience 201
Qiming Sun, Vlasova Irina, Mustafina Olga, Rykov Andrey, Kolchina Vera, Moiseenko Svetlana

Spatial Distribution of the Ozone Layer and Its Relationship with Temperature and Relative Humidity Over Iraq for the Year 2024 Using R.S and GIS Techniques 209
Ahmed Hussein Khudhur, Khanjer Ebtesam F.

Spatial Modeling for Predicting and Identifying Levels of Climate-Related Disease Hazards in Iraq Using (GIS) Kirkuk as a Model 220
Abdlwahd Jamel Alqdori Areej

Physicochemical and Mineralogical Characterization of Sultan Uwais Feldspar for Sustainable Glass Manufacturing in Earth and Environmental Sciences 229
Buranova Dinara, Matchanov Sherzod, Atashev Elyor

Determination of the Composition of the Zinelbulak Talc-Magnesite Deposit Rock Using Modern Physicochemical Methods 236
Jumaniyazov Maxsud, Kurambayev Sherzod, Atashev Elyor, Jumaniyozov Arslon, Buranova Mokhigul

Experiments on Bleaching of Kaolin from the Khodjakulskoye Deposit 244
Matchanov Sherzad, Ruzmetova Aida, Yakubov Yusufboy

Identifying Transboundary Ecotourism Zones in Surkhandarya Region: Opportunities and Prospects 252
Ravshanov Shokhrukh, Khursanov Dunyobek, Malikov Bobur, Eshmamatov Elshod, Abduvaliev Khayitboy

Geoecological Basis of Nature Protection and Holistic Use of Natural Resources in the Fergana Valley 262
Khamidov Abdusamad, Bakhtiyorova Dilfuza, Saydullayeva Surayyo, Janizoqova Gulshoda, Yusupova Kamola

Geoecological Foundations Strengthening the Stability of the National System of Protected Areas of the Republic of Uzbekistan 271
Akhmadaliev Yusupjon, Abduganiev Olimjon, Makhkamov Elyorbek, Kosimov Dilshodbek, Komilova Nargiza

Geoecological Principles and Criteria for Evaluating the Efficiency of Regional Touristic-Recreational Systems of the Fergana Valley 279
Akhmadaliev Yusupjon, Abduganiev Olimjon, Makhkamov Elyorbek, Abduvaliev Khayitboy, Madaminov Zafar

Analysis of Traffic Patterns Between Karkh and Rusafa 287
Mahdi Lateef Ishraq, Hussein Abed Alaa, Moutaz Asmael Noor

Assessment of Uninterrupted Pedestrian Facility, Case Study of Al-Mutanabbi Street 299
Ahmed Subhi Farah, Khudhur Jameel Abeer

Determination the Changes in Water and Vegetation Covers for Ain Al Tomar District Using Remote Sensing Techniques 310
Ahmed Efan Rusul, Albakry Sundus A. Abdullah

Evaluating of Spatial Interpolation Techniques for Accurate Air Quality Prediction: An Overview 319
Abdulkareem Sarah K., Alhadithi Mufid, Amer Wadhah

Geographical Dimensions of the Concentrations of Gases Polluting the Air of the City of Basrah 325
Waleed Mustafa Suha

Geographical Distribution of Tourist Sites in Najaf Governorate and Study of Available Potentials for Their Development 338
Abdullal Azeez Ibtihal, Ahmed Bneen Nassr, Azeez Bjwo Amtihan

Happy City Strategy, Satisfying Individuals\' Needs by Sustainable Urban Design (Zurich and Dubai as Models) 349
Makki Tiba M, Khalid Zainab

Monitoring the Distribution of Parks and Land Cover Changes Along the Tigris River Using Spatial Analysis Techniques 366
Abdullah Albakri Sundus A., Kadhim Zina F., Gburi Sura I., Samer Muhend K., Mohammed Faisel G.

Residential Mobility in Basrah City: The Spatial Relationship Between Its Drivers and Variables by Using Statistical Analysis 377
Abdul-Hussein Kadhim Alaidani Abbas, Jamal Abbas Al-Badran Makram

Spatial Analysis of Air Pollution Concentration (radon, Humidity, VOCS) in the Homes of the Mitham Al-Tammar Neighborhood in Kufa 394
Ali Hassan Salam, Mazhar Abdul Hamid Raghad, Nafea Yousif Saja, Saeed Jassim Ali

Spatial Analysis of Indoor Air Pollution in the Housing of Al-Abbasiya District in Najaf Governorate 402
Dawoud Salman Muhammad

Neighborhood Segregation and Urban Resource Distribution: A Path Toward Sustainability in Ana 408
Hammad Abdulateef Mohammed

The Impact of Air Pollution on Human Health in the City of Baghdad: Towards Achieving Sustainable Development Goals Through Adopting Smart City Solutions 416
Abdulnabi Abbas Muna

The Impact of Environmental Pollution on Mental Health: A Comparative Study Between Urban and Rural Students 423
Tariq Hussein Allwan Bilal, Hameed Mousa Elaf, Tariq Battal Zahraa, Hindy Tayseer, Jundi Rami

Thoron Gas Clusters in the Indoor Air of Primary Schools of Al Najaf City: A Geographic Analysis 431
Qasim Ahmed Anwar, Kadhim Hussein Ali, Abid Abojassim Ali, Saeed Jassim Ali

Urban Transformation Towards Smart Cities 443
Khalid Zainab

Utilization of Remote Sensing and GIS for Land Use and Land Cover Mapping in Wasit Province, Iraq 456
Abdulkareem Meena K., Sabah Jaber Hussein

Phase Transformations and Physico-Mechanical Properties of Ceramic Tiles from Lower Amu Darya Raw Materials for Sustainable Construction in Earth and Environmental Engineering 468
Boyjanov Islom, Djabberganov Djakhangir, Matyaqubova Karomat, Khudayberganov Erkaboy, Sitmuratov Tulkinbek

Kinetic Study on Rapid Phosphating Compositions for Eco-Efficient Corrosion Protection in Earth and Environmental Engineering 476
Jumaniyazov Maxsud, Kurambayev Sherzod, Sapayeva Surayyo

The Effect of Hybrid Mineral Additives on the Genetic Formation and Physico-Chemical Processes of Cement Composites 483
Khadzhiev Azamat, Abdullaev Muslimbek, Yakubov Yusufboy, Jumaniyozov Jumaniyoz

Improvement of the Technology of Obtaining Mineral Binders from Loess-Like Rocks Using Mechanoactivation 492
Masharipova Sharofat, Jumaniyazov Maxsud, Aitova Shaxlo

Automation and Modeling of Copper Smelting Processes in Vanyukov Furnace 501
Musabekov Nazarbek, Utepbergenov Irbulat, Musabekov Akhmetbek, Mirzabayev Berik

Evaluation of the Natural Radioactivity (^{238}U,^{232}Th,^{40}K) in Building Materials Used in Iraqi Homes 509
Mouaid Ali Roua, Qassim Gadhban Ansam, Baqir Yassir A., Alebadi Shireen N.

Applications of Lasers in Dermatology: From Hair Removal to Skin Rejuvenation 516
Hassan Alsobh Morjan

The Role of Green Building Information Modeling (GBIM) in Promoting Architectural Sustainability 520
Yousif Al-Mashhadani Mays, Abdulmunem Al-Tameemi Osamah

Author Index

Conference: GASES 2025 International Forum of Global Advances in Sustainable Environment, Energy, and Earth Sciences

13-14th of May, 2025; Baghdad, IRAQ

Conference Organizers:
Organizators: Baghdad University, Iraq; University of Melbourne, Australia; University Science of Malaysia; King Abdullah University of Science and Technology (KAUST), Saudi Arabia; Ajman University, UAE

Overview:
Global Advances in Sustainable Environment, Energy, and Earth Sciences (GASES 2025) is an international forum dedicated to fostering collaboration among researchers, scientists, industry professionals, and policymakers working toward sustainable solutions in environmental protection, energy systems, and Earth sciences. The conference aims to highlight the latest scientific breakthroughs, technological innovations, and policy developments that contribute to global sustainability. Participants will have the opportunity to present cutting-edge research, discuss emerging challenges, and explore future directions for interdisciplinary cooperation. Through keynote sessions, oral and poster presentations, and panel discussions, GASES 2025 provides a platform for exchanging knowledge, creating partnerships, and inspiring next-generation solutions that address critical environmental and societal needs

We deeply appreciate the organizers, program committee members, and reviewers of GASES 2025 for their hard work, which was essential for the event's success and the quality of the conference papers. We also thank the authors for their valuable research contributions. A special thanks to the Web of Conferences team. We look forward to future conferences together.

The Organizing Committees of GASES 2025

Statement of Peer review

In submitting conference proceedings to *Web of Conferences*, the editors of the proceedings certify to the Publisher that

1. They adhere to its **Policy on Publishing Integrity** in order to safeguard good scientific practice in publishing.
2. All articles have been subjected to peer review administered by the proceedings editors.
3. Reviews have been conducted by expert referees, who have been requested to provide unbiased and constructive comments aimed, whenever possible, at improving the work.
4. Proceedings editors have taken all reasonable steps to ensure the quality of the materials they publish and their decision to accept or reject a paper for publication has been based only on the merits of the work and the relevance to the journal.

Title, date and place of the conference

IGASES 2025 International Forum of Global Advances in Sustainable Environment, Energy, and Earth Sciences

13-14th of May, 2025; Baghdad, IRAQ

Proceedings editor(s):

Miqdam Tariq Chaichan, Abbas Rajabifard, Thas Nirmalathas, Murni Melati Ahmad, Seongjin Hong, Leena Ajit Kaushal

Date and editor's signature

May 20, 2025

A comparative review of delumping algorithms

Adel Sarsenova[1,*], *Adelina* Mukhamedyanova[2], and *Mariam* Kumarkhan[1]

[1]School of Energy and Petroleum Industry, Kazakh British Technical University, 050000 Almaty, Kazakhstan
[2]Petroleum Engineering Department, Kazakh National Research Technical University, 050013 Almaty, Kazakhstan

Abstract. Reservoir simulation often simplifies complex hydrocarbon compositions by grouping components into pseudo-components, known as lumping. However, detailed fluid composition is required for surface processing, necessitating the reverse process—delumping. This study analyzes compositional, black-oil, and dynamic delumping methods, their evolution, advantages, and limitations to determine the most effective approach for surface modeling and facility planning.

1 Introduction

Hydrocarbons consist of multiple chemical compounds, making reservoir simulation computationally intensive due to phase equilibrium calculations. Lumping simplifies these calculations by reducing the number of components but results in a loss of compositional detail. Delumping, or inverse lumping, restores these details for surface facility design, where fluid composition must meet regulatory and processing requirements. This process effectively links reservoir and surface models, ensuring accurate hydrocarbon characterization for export and refining. Delumping Algorithm Development.

2 Materials and methods

2.1. Compositional delumping

- Schlijper & Drohm (1988) pioneered delumping, using material balance and equilibrium constants (K-values) to reconstruct fluid composition. Danesh et al. (1992) refined Wilson's equation, linking K-values with equation of state (EOS) parameters to improve gas injection modeling:

$$ln\,K = C_0 + C_1(1 + \omega_i)[1 - (1 - T_{ri})] \quad (1)$$

Leibovici et al. (1996) introduced the LSK algorithm, deriving K-values from Peng-Robinson (PR) and Soave-Redlich-Kwong (SRK) EOS models:

* Corresponding author: a.sarsenova@kbtu.kz

$$ln\,K_i = \Delta C_0 + \Delta C_1\sqrt{a_i} + \Delta C_2 b_i \quad (2)$$

Faissat & Duzan (1996) estimated split factors for delumping output streams from single-cell and dual-cell simulations:

$$x_i = \frac{z_i}{1+v(K_i-1)} \text{ and } y_i = k_i x_i \quad (3)$$

Where V is the vapour mole fraction, which is obtained by Rachfold-Rice equation:

$$\sum_{i=1}^{Nc}(y_i - x_i) = \sum_{i=1}^{Nc}\frac{z_i*(K_i-1)}{[1+v*(K_i-1)]} = 0 \quad (4)$$

- Leibovici et al. (1999, 2000) applied LSK delumping to real fields, validating its accuracy in tracking fluid component movement. Nichita & Leibovici (2006) later developed a reduction-based analytical method that effectively handled non-zero binary interaction parameters (BIPs), outperforming earlier regression-based models.
- De Castro et al. (2011) compared analytical delumping (Nichita & Leibovici, 2006) with LSK delumping (Leibovici et al., 1996) in reservoir simulations, concluding that analytical delumping outperformed LSK delumping in accuracy and computational efficiency [4].

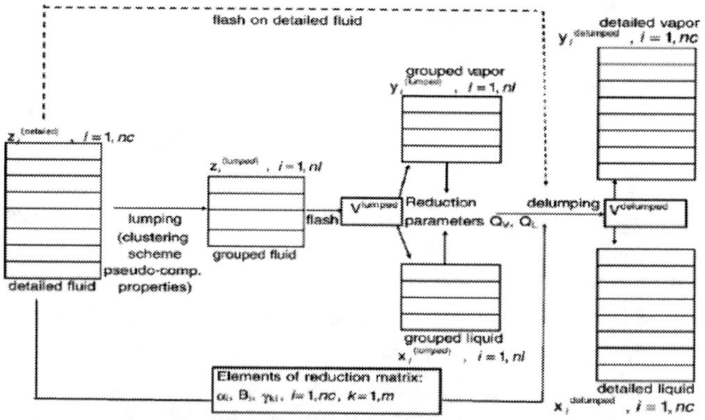

Fig. 1. Schematic representation of the lumping/delumping process [4].

In 1993, the Leibovici approach was used to compute group characteristics and BIPs. The Δh functions are calculated by:

$$\Delta h_0 = h_{0L} - h_{0V} = -ln(Z_L - B_L) + ln(Z_V - B_V) \quad (5)$$

K-values for the detailed system are calculated by:

$$ln\,K_i(Q_L, Q_V) = \Delta h_0(Q_L, Q_V) + \Delta h_\alpha(Q_L, Q_V)\alpha_i + \Delta h_B(Q_L, Q_V)B_i + \sum_{k=1}^{m}\Delta h_{\gamma k}(Q_L, Q_V)\gamma_{ki} \quad (6)$$

Lastly, the vapor mole V and phase mole fractions are determined by:

$$x_i = \frac{z_i}{1+V(K_i-1)} \quad (7)$$

The Rachford-Rice equation is solved for the delumped K-values, and the result is used to calculate the vapor fraction V.

$$\sum_{i=1}^{nc} = \frac{z_i(K_i-1)}{1+V(K_i-1)} = 0 \quad (8)$$

De Castro et al. (2011) compared analytical delumping (Nichita & Leibovici, 2006) with LSK delumping (Leibovici et al., 1996) in reservoir simulations, concluding that analytical delumping consistently outperforms LSK delumping. For this, functions

$$\Delta h_0 = \Delta h(Q_L, Q_V) = h(Q_L) - h(Q_V) \quad (9)$$

only of the reduction parameters

$$Q_{\alpha k} = \sum_{i=1}^{nc} q_{\alpha i} x_{ik};\ \alpha = 1, 2c+1; k = L, V \quad (10)$$

These are linear relationships between component mole fractions and reduction matrix elements. The delumping algorithm:
1. Perform a flash calculation at a given pressure and temperature;
2. Compute dh(Q) functions using data from the lumped system;
3. Use Eq. (2) to determine K-values for the detailed mixture;
4. The component mole fractions are computed for two equilibrium phases [5].

2.2. Black-oil delumping

Black-oil modeling is widely used due to its computational efficiency, but it lacks detailed compositional data needed for surface processing. Delumping reconstructs black-oil wellstreams into compositional wellstreams, ensuring accurate fluid composition and molar rates. [6].

Hoda (2002) introduced BOz delumping, utilizing split factor tables from depletion PVT experiments to convert surface phase rates into reservoir phase molar amounts [7].

2.2.1. Mathematical Approach:

(1) Phase Mass Rate Calculation:

$$Q_m^V = Q_m^{gV} + Q_m^{oV} \quad (11)$$

$$Q_m^V = Q_m^{gV} + Q_m^{oV} \quad (12)$$

(2) Phase Composition Calculation;

Lookup tables estimate mole fractions (yi and x). Gas/oil ratios:

$$R_s = q^{gL} / q^{oL} \quad (13)$$

$$R_v = q^{oV} / q^{gV} \quad (14)$$

(3) Total Composition Calculation.

$$z_i = \alpha y_i + (1-\alpha) x_i \quad (15)$$

Where a is the vapor fraction

$$\alpha = \frac{n^V}{n^V + n^L} \quad (16)$$

Rewrite the equation:

$$\alpha = \frac{m^V/M^V}{m^V/M^V + m^L/M^L} \tag{17}$$

Regarding molar rates, this is given by:

$$\alpha = \frac{Q_m^V/M^V}{Q_m^V/M^V + Q_m^L/M^L} \tag{18}$$

(4) The vapour and liquid molar weights calculation:

$$M^V = \sum_{i=1}^{N^C} y_i M_i \tag{19}$$

$$M^V = \sum_{i=1}^{N^C} y_i M_i \tag{20}$$

Tables that correlate liquid/vapor composition with saturation pressure and gas/oil ratios enhance the accuracy of black-oil delumping. This method can be applied at both the well and well-completion levels to improve phase behavior predictions [3].

Vignati et al. (2010) enhanced the approach by incorporating tracer tracking, allowing for a more precise distinction between different gas types—liberated gas, gas cap gas, and injection gas. By implementing this technique in commercial simulators, they refined reservoir gas composition estimates, further improving the reliability of black-oil delumping.

2.3. Dynamic delumping

This bridges surface and subsurface models, ensuring accurate compositional data transfer in petroleum simulations while addressing inconsistency in fluid descriptions. This approach optimizes asset management by enhancing the integration between reservoir and process modeling.

LSK Algorithm (Vignatti et al., 2008): Applied within integrated asset modeling but required supplementary methods (such as LBW) to estimate delumped component mass rates due to dependencies on neighboring cell flow rates.

Reservoir-Network Integration: Compositional data is exchanged bidirectionally between the reservoir and production/injection wells. Reservoir simulators provide inflow performance relations (IPR) to the network, which, in turn, determines well targets and mass flow rates for processing.

Separate Equations of State (EOS): Utilizing distinct EOS models for surface and subsurface simulations ensures PVT consistency, minimizing discrepancies in integrated workflows.

Dynamic Delumping by Kuntadi (2012): Introduced pressure-dependent split factors to enhance the accuracy of compositional simulations in lumped black-oil models. These split factors, derived from depletion PVT experiments, enable precise phase-wise delumping at each simulation time step.

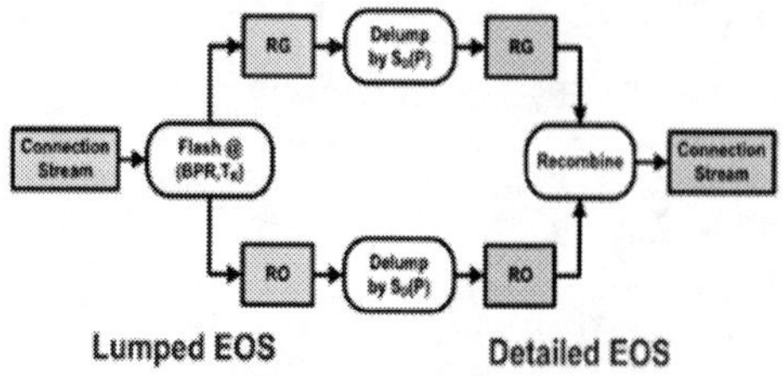

(c) Depletion Case

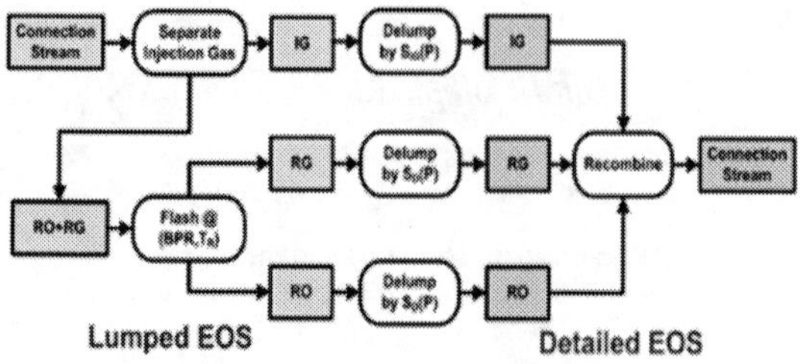

(d) Gas Injection Case

Fig. 2. Compositional delumping procedures [10].

Implementation Process:

1. Split Factor Conversion: Lumped components are mapped to detailed components.
2. Equilibrium Phase Compositions: EQL and EQV compositions define split factor tables.
3. Gas Injection Adjustments: Pressure-dependent split factors refine phase compositions.

3 Results and discussions

LSK and LBW delumping methods have been validated in experiments and reservoir simulations. Stenby et al. (1996) confirmed LSK delumping accuracy using a CVD test and North Sea reservoir fluid, proving its effectiveness in linking reservoir and surface models.

Leibovici et al. (1998) further refined delumping with the LBW method, requiring at least three lumped components. Simulations of volatile oil depletion and gas injection demonstrated that lumped models effectively reconstruct detailed compositions while improving computational efficiency. However, the inclusion of non-zero binary interaction coefficients (BICs) introduced slight deviations, particularly in lighter hydrocarbon fractions (CO_2-C_2).

By 1999, LBW delumping remained the only practical method. Its application in a Middle Eastern gas condensate field successfully linked reservoir simulations with surface processing. However, inconsistent tuning of the equation of state (EOS) between the reservoir and surface models reduced accuracy, highlighting the need for more precise EOS adjustments.

To address these limitations, Vignati et al. (2009) developed an integrated reservoir-process workflow incorporating Leibovici's delumping approach. Their method converted fluid compositions between different EOSs, ensuring accurate modeling of gas-condensate and volatile oil recovery. By directly tracking detailed fluid components, they demonstrated that integrated delumping could closely match full-compositional simulations while reducing computational costs.

Table 1. Black-oil delumping: The key studies.

Study & Case	Description	Results & Observations
Ghorayeb & Holmes (2005)	Tested delumping in five scenarios: Water Injection, Multiple Fluid Regions, Well with Crossflow, Gas Injection, and a Real Field PVT.	Delumping showed high accuracy for natural depletion processes. Gas injection cases had some discrepancies in breakthrough time and wellstream composition due to black-oil model limitations.
Example 1: Water Injection	Evaluated delumping under water injection to repressurize a reservoir.	Strong agreement between compositional and black-oil models for 8 years of production. Tabulating compositions by saturation pressure improved accuracy.

Continuation of Table 1.

Study & Case	Description	Results & Observations
Example 2: Multiple Fluid Regions	Tested black-oil delumping at the well-completion level across 3 layers.	Gas rate vs. time aligned well with compositional models, but weaker agreement compared to Example 1.
Example 3: Well with Crossflow	Simulated a well with fluid movement between layers.	High accuracy was maintained. The method successfully matched compositional simulation results.
Example 4: Gas Injection	Evaluated delumping under gas injection at well-completion level.	High accuracy was maintained. The method successfully matched compositional simulation results.

Example 5: Real Field PVT Sample	Tested delumping on a North Sea reservoir sample (15 pseudo-components).	Agreement between compositional and black-oil models was excellent.
Ghorayeb et al. (2009)	Tested black-oil delumping on North Kuwait Jurassic Complex (NKJC) (5 distinct fields).	Segmenting simulations improved computational efficiency. Delumping produced highly precise results, matching compositional models closely.
Gas Injection Considerations	Evaluated gas injection effects on delumping performance.	Black-oil delumping works well for depletion but struggles with gas injection, requiring compositional simulation. A hybrid modeling approach was recommended.
Vignatti et al. (2010): Goliat Field	Tested delumping for gas injection and gas sales scenarios using PVT tables & Peng-Robinson EOS (15 components).	Composition vs. saturation pressure results were consistent, but -6% underestimation of injection gas volume was observed. Mass balance errors in tracer concentration posed a challenge.

3.1.Dynamic delumping

Vignatti et al. (2008) validated their delumping method through two approaches: gas injection below the dew point in a sour gas condensate reservoir and miscible gas injection in a near-critical volatile oil reservoir. The method proved effective even in gas cycling and gas condensate reservoirs under primary depletion. It was particularly useful in complex field developments involving CO_2 or H_2S storage and reuse. Unlike traditional compositional models with fixed liquid-vapor splits, this approach dynamically adjusted hydrocarbon distribution based on real-time reservoir conditions. [10].

Integrated asset modeling bridges reservoir engineering and surface processing by maintaining thermodynamic consistency. Leibovici's delumping approach, combined with the LSK algorithm, provided a stable mathematical framework with minimal model modifications, making it easy to implement using commercial tools..

Kuntadi et al. (2012) refined dynamic delumping by incorporating BOz delumping. Its accuracy was confirmed through comparisons of black-oil delumped profiles with full compositional simulations. A sensitivity analysis evaluated different methods of separating injection gas from production streams, showing strong agreement between the converted compositional profiles and original simulation data. This confirmed BOz delumping's ability to handle complex fluid reservoirs under depletion conditions[1].

Both black-oil and compositional delumping exhibited high accuracy in depletion and gas injection scenarios. The effectiveness of compositional delumping depended on the quality of the lumped EOS model relative to the detailed EOS model. When PVT behaviors were aligned, delumping restored the detailed composition with near-perfect accuracy. Additionally, both methods remained computationally efficient, requiring minimal additional processing time.

3.2. Computation time comparison

Leibovici et al. (1998) demonstrated that delumping significantly reduces computational time while maintaining accuracy. Their simulations showed that delumping accelerates the process by 68%, cutting total simulation time from 730 to 259 seconds in one case. In another case, delumping required only 18 seconds, reducing the total simulation time to just 13% of what a detailed fluid model would typically need.

Table 2. Computing time [10].

Example	Reservoir Simulation Lumped fluid	Delumping	Total	Reservoir Simulation Detailed fluid
1	232	27	259	730
2	173	18	191	1306

Baker and Leibovici (1999) further confirmed that delumping significantly reduces computational time. In a quality control case, delumping required only 7 seconds, compared to 99 seconds for an 8-component lumped model and 450 seconds for a 16-component detailed model. For full-field modeling, delumping took 5000 seconds, whereas lumped simulation required 21,000 seconds [2].

Table 3. Presents information regarding computational duration for both the minor quality control problem and the full-scale field model.

Example	Reservoir Simulation Lumped fluid	Reservoir Simulation Detailed fluid	Delumping	Size of Output File (MB)
Quality Control	99	450	7	40
Full Field	21000	-	5000	800

These findings confirm that delumping significantly reduces computation time while maintaining compositional accuracy. However, a key limitation is the large storage space required for output files.

3.3. Computational efficiency:

- Leibovici et al. (1998): Delumping reduced computation time by 68% compared to full compositional simulations.
- Baker & Leibovici (1999): Post-processing delumping required only 13% of the time needed for detailed compositional simulations.

Integrated modeling links reservoir and surface processes, providing more accurate simulations:

- Vignati et al. (2008): Implemented LSK delumping in integrated asset models, but tracking fluid flow remained a challenge.
- Kuntadi et al. (2012): Introduced a pressure-dependent split-factor method to dynamically link compositional and black-oil models.
- Dynamic Delumping: Effective for complex reservoirs such as those involving sour gas production and CO_2 storage.

3.4. Comparative performance

Table 4. Comparison of delumping methods.

Method	Structure	Limitations
Compositional Delumping	High accuracy, suitable for gas injection	Requires significant computational power
Black-oil Delumping	Computationally efficient, works for depletion	Struggles with gas injection cases
Dynamic Delumping	Integrates reservoir & surface models	Complex implementation, requires extensive calibration

4 Conclusion

Delumping is essential for reservoir and surface modeling, ensuring both accuracy and computational efficiency. Compositional delumping (Nichita & Leibovici, 2006) is the most reliable for handling non-zero binary interaction parameters (BIPs), while black-oil delumping (Hoda, 2002) is effective for depletion processes but limited in gas injection cases. Dynamic delumping improves predictions in complex reservoirs by integrating subsurface and surface models, though its accuracy depends on lumped EOS quality and tracer tracking advancements. In Kazakhstan's key fields (Karachaganak, Kashagan, Tengiz), precise delumping is crucial for acid gas reinjection (CO_2 & H_2S) and multicomponent equilibrium modeling. A reduction-based lumping approach should be prioritized over regression-based methods for more accurate compositional workflows.

References

1. O. Burachok, L. Khaidarova, O. Kondrat, S. Matkivskii, "Application of CEOR Optimization Method for Gas-Condensate Reservoir below Dewpoint: Synthetic Case Study", Oil Gas European Magazine **46**, 41–49 (2020), DOI: 10.19225/201209
2. J. C. Assaf, M. Al Daccache, "Special Issue on 'Processing and Conversion of Oil and Gas: Modeling, Control, Simulation and Optimization'", Processes **11**, 388 (2023), DOI: 10.3390/pr11020388
3. Ja. Ismailova, D. Delikesheva, A. Abdukarimov et al., "Development and Application of Fluid Characterization Algorithms to Obtain an Accurate Description of a PVT Model for Kazakhstani Oil", Eastern-European Journal of Enterprise Technologies **5**, 6–20 (2023), DOI: 10.15587/1729-4061.2023.289932
4. R. Piazza, A. Vieira, L. A. Sacorague et al., "Innovative Formation Tester Sampling Procedures for Carbon Dioxide and Other Reactive Components", Petrophysics **62**, 65–72 (2021), DOI: 10.30632/pjv62n1-2021a4
5. B. Kaiyrbekov, A. Sarsenova, J. Ismailova, "Comprehensive Fluid Description through Lumping Procedures: for Kazakhstani Oil", Oil and Gas **138**, 203–217 (2023), DOI: 10.37878/2708-0080/2023-6.17
6. L. V. Lucas, E. Deleersnijder, "Tracers and Timescales: Tools for Distilling and Simplifying Complex Fluid Mechanical Problems", Water **13**, 2796 (2021), DOI: 10.3390/w13192796
7. P. Ø. Andersen, K. Djurhuus, R. Askarinezhad, J. S. Solbakken, "Management of High-Water-Cut and Mature Petroleum Reservoirs", Energies **15**, 8344 (2022), DOI: 10.3390/en15228344
8. D. C. Palacio Lozano, M. J. Thomas, H. E. Jones, M. P. Barrow, "Petroleomics: Tools, Challenges, and Developments", Annual Review of Analytical Chemistry **13**, 405–430 (2020), DOI: 10.1146/annurev-anchem-091619-091824
9. A. Samnioti, E. M. Kanakaki, S. P. Fotias, V. Gaganis, "Rapid Hydrate Formation Conditions Prediction in Acid Gas Streams", Fluids **8**, 226 (2023), DOI: 10.3390/fluids8080226
10. A. Taghavinejad, M. Ostadhassan, R. Daneshfar, "Unconventional Reservoir Engineering", SpringerBriefs in Petroleum Geoscience and Engineering, **11**, 34 (2022), DOI: 10.1007/978-3-030-82837-0_2

E3S Web of Conferences **633**, 01002 (2025) https://doi.org/10.1051/e3sconf/202563301002
GASES 2025

High-performance liquid chromatography method for the determination of levofloxacin in liquid pharmaceutical preparations: implications for environmental health and earth sciences

Zulayho Smanova[1], *Shemshat* Arazberdieva[2,*], *Muborek* Abdullaeva[1], *Ulugbek* Akhmadjanov[1]

[1]National University of Uzbekistan named after Mirzo Ulugbek, 50140501 Tashkent, Uzbekistan
[2]Tashkent State Technical University, 70720601 Tashkent, Uzbekistan

Abstract. Analytical testing is integral to evaluating pharmaceutical quality and efficacy, with important ramifications for both public health and environmental protection. This study details the development and validation of an HPLC method for quantifying levofloxacin in liquid pharmaceutical formulations, employing a 5 μm (150 × 3.9 mm) C18 reversed-phase column and a UV-visible detector. Gradient chromatography was performed over 25 minutes at a flow rate of 1.0 mL/min, with detection at 275 nm. Method validation used OpenLab ChemStation ver. A.01.05 and adhered to international guidelines, confirming the approach's accuracy, linearity, specificity, and robustness. The results have implications for monitoring pharmaceutical residues in the environment, which is critical to safeguarding water resources and maintaining ecological balance in Earth sciences.

1 Introduction

Levofloxacin [(-)-(S)-9-fluoro-2,3-dihydro-3-methyl-10-(4-methyl-1-piperazinyl) 7-oxo-7H-pyrido[1,2,3-de]-1,4-benzoxazine-6-carboxylic acid hemihydrate]–(Fig. 1) a synthetic broad-spectrum antibacterial drug from the fluoroquinolone group, containing the levorotatory isomer of ofloxacin as an active substance, widely used in the treatment of mild to moderate infections in adults in the form of tablets and an infusion solution, and also used in the treatment of eye infections in the form of an eye drop. Levofloxacin is recommended for the treatment of patients with community-acquired pneumonia, nosocomial pneumonia and in exacerbations of chronic obstructive pulmonary disease. Levofloxacin is characterized by a good safety profile, high bioavailability and tissue distribution, which ensures the achievement of high concentrations of the antibiotic at the site of infection [1].

Levofloxacin is a prototype of modern or so-called respiratory fluoroquinolones – an antibiotic with a broad spectrum of antibacterial activity. Gram-positive microorganisms (including drug-resistant pneumococci), gram-negative bacteria and intracellular pathogens are highly sensitive to the action of the drug. Levofloxacin has optimal pharmacokinetic

* Corresponding author: arazberdiyevas@gmail.com

characteristics, which is a predictor of its successful clinical use. The convenience of administration (once a day) of the drug also determines its attractiveness for practical use. A significant advantage of levofloxacin is its availability in both parenteral and oral forms, which determines the possibility of using the antibiotic as part of step therapy [2].

For the qualitative and quantitative analysis of fluoroquinolones in biological objects and in medicinal forms, the method of high-performance liquid chromatography (HPLC) is currently used with various detection methods - spectrophotometric [3, 7, 10], fluorimetric [7, 3, etc.], mass spectrometric [3].

The main method for analyzing fluoroquinolones, according to literature data, is high-performance liquid chromatography (HPLC), most often used in the form of a reversed-phase version on columns filled with C18 or C8 sorbent with spectrophotometric or fluorimetric detection [6, 7, 8, 9, 11, 12]. In the pharmacopoeial analysis of fluoroquinolones, according to the British and European Pharmacopoeia, the reversed-phase HPLC method is used to assess the content of "related compounds" in the substances of pefloxacin mesylate, ciprofloxacin, ciprofloxacin hydrochloride and ofloxacin and in ciprofloxacin tablets, for the quantitative determination of ciprofloxacin hydrochloride in the substance, norfloxacin in eye drops, for the identification and quantitative determination of norfloxacin in tablets, for the identification, determination of foreign impurities and quantitative content of ciprofloxacin in tablets and infusion solution [3, 4, 5, 11, 13].

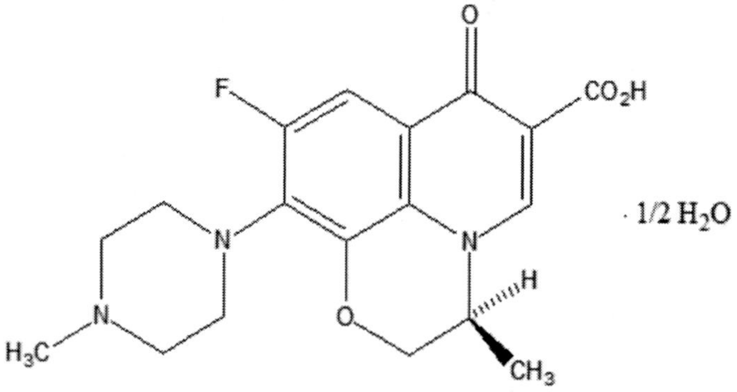

Fig 1. Structural formula of levofloxacin hemihydrate.

2 Materials and methods

Table 1. Preparation of test solutions of different concentrations.

Sample №	Concentration level, %	Preparation	Concentration in the test solution
1	80	0.8 ml of the resulting solution is placed in a 10 ml measuring flask, the volume of the solution is brought up to the mark with the mobile phase and mixed. (test sample № 1)	0,032 mg/ml
2	90	0.9 ml of the resulting solution is placed in a 10 ml measuring flask, the volume of the solution is brought up to the mark with the mobile phase and mixed. (test sample № 2)	0,036 mg/ml
3	100	1.0 ml of the resulting solution is placed in a 10 ml measuring flask, the volume of the solution is brought up to the mark with the mobile phase and mixed. (test sample № 3)	0,04 mg/ml

4	110	1.1 ml of the resulting solution is placed in a 10 ml measuring flask, the volume of the solution is brought up to the mark with the mobile phase and mixed. (test sample № 4)	0,044 mg/ml
5	120	1.2 ml of the resulting solution is placed in a 10 ml measuring flask, the volume of the solution is brought up to the mark with the mobile phase and mixed. (test sample № 5)	0,048 mg/ml

A high-performance liquid chromatograph Agilent 1260 Infinity (Agilent Technologies, USA) was used (wavelength range - from 190 to 950 nm, detection wavelength - 275 nm). The studies were carried out using class A measuring glassware, analytical scales OHAUS PA214C (measurement range: from 0.0001 g to 220 g) (China). Standard and test solutions of 0.8 mg / ml were used.

10 μl of the test and standard solutions of the levofloxacin sample are alternately chromatographed on a liquid chromatograph with a UV detector, under the following conditions:

- column C18, size (150 x 3.9) mm;
- mobile phase: buffer solution (0.05 M KH_2PO_4 solution. 4 ml of triethylamine are added to 1 l of the resulting buffer solution) - acetonitrile (87:13), pH = 7 (potassium hydroxide or phosphoric acid is used to adjust the pH), degassed in any convenient way;
- mobile phase flow rate - 1.0 ml/min;
- detection - at wavelength (275 ± 2) nm;
- column temperature - 40°C.
- the retention time of levofloxacin hemihydrate was 10.39 minutes (fig.2).

3 Results and discussion

Preparation of working solution: 2 ml of the preparation is placed in a 25 ml measuring flask, about 15 ml of the mobile phase is added, mixed, the volume of the solution is brought up to the mark with the same solvent and mixed..

Table 2. Experimental data obtained for the test solution.

Sample №	Concentration level, %	Concentration of levofloxacin mg/ml	Peak area, S	S_{cp}	s	RSD, no more than 2%
1	80	0,032	1995807	1992314,500	4939,14087	0,25
			1988822			
2	90	0,036	2246706	2247238,500	753,06872	0,03
			2247771			
3	100	0,040	2402393	2416219,000	19552,91671	0,81
			2430045			
4	110	0,044	2674128	2674755,500	887,41901	0,03
			2675383			
5	120	0,048	2914350	2911121,000	4566,49559	0,16
			2907892			

**Scp - average value peak area;*
***s - Peak area deviations.*

Table 3. Calculated data obtained for the model mixture.

Sample №	Concentration, mg/ml	Peak area, S	r	b	a
1	0,03200	1992314,5000	0,9983	56628250	183199,7000
2	0,03600	2247238,5000			
3	0,04000	2416219,0000			
4	0,04400	2674755,5000			
5	0,04800	2911121,0000			

**r - Correlation coefficient;*
***b - Tangent of the angle of inclination;*
****a - Free member.*

According to ICH recommendations, the validation parameters were obtained within the acceptance limits. The % analysis of the test sample was 99.83%, and the % RSD value was less than 0.25%. The analytical validation method was simple, accurate and efficient for routine analysis of levofloxacin samples in liquid pharmaceuticals.

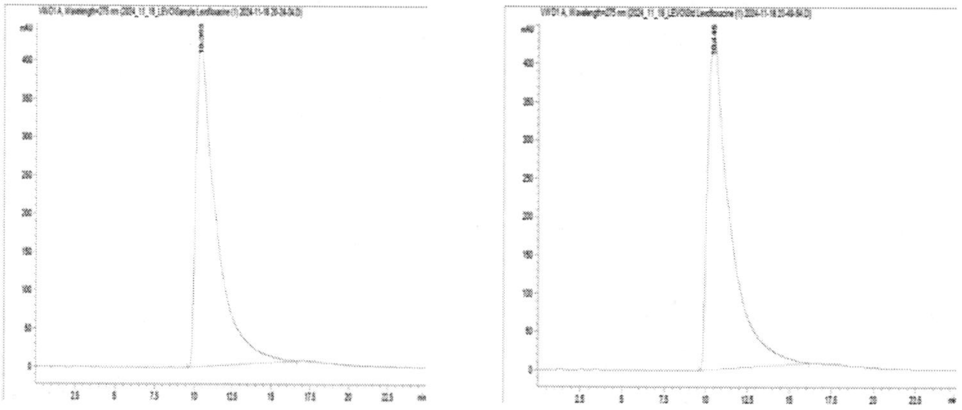

Fig. 2. Chromatogram of test and standard solutions of levofloxacin.

4 Conclusion

In this work, a rapid and inexpensive method for the determination of levofloxacin was developed. The method fulfills the parameters of validation of analytical methods and can be used in the analysis of levofloxacin in pharmaceutical preparations of liquid dosage forms.

For qualitative and quantitative pharmaceutical analysis of fluoroquinolones, the following methods are used: IR, UV spectroscopy, chromatography (high-performance liquid chromatography, thin-layer chromatography), photometry in the UV and visible spectral range, fluorimetry, chemical methods (qualitative chemical reactions and titrimetry). However, the leading role in the pharmaceutical analysis of fluoroquinolones and levofloxacin is played by the method of high-performance liquid chromatography, as a universal method that allows for identification, determination of impurity composition and assessment of the quantitative content of impurities, as well as quantitative determination of fluoroquinolones, both in solid dosage forms and in liquids.

References

1. Kh. Elgendy, M. Zaky, T. Alaa Eldin, S. Fadel, "Rapid HPLC Determination of Ciprofloxacin, Ofloxacin, and Marbofloxacin Alone or in a Mixture", Results in Chemistry **5**, 100749 (2023), DOI: 10.1016/j.rechem.2022.100749
2. K. H. Elgendy, M. Zaky, A. E. M. M. Altorky, S. Fadel, "Determination of Levofloxacin, Norfloxacin, and Moxifloxacin in Pharmaceutical Dosage Form or Individually Using Derivative UV Spectrophotometry", BMC Chemistry **18**, 115 (2024), DOI: 10.1186/s13065-024-01193-4
3. A. Khalil, M. Kashif, "Interaction Studies of Levofloxacin with Human Lysozyme in a Ternary Complex Using Multispectroscopic and Computational Analysis: A Circular Dichroism Method for the Quantitation of Levofloxacin", Journal of Molecular Liquids **370**, 121023 (2023), DOI: 10.1016/j.molliq.2022.121023
4. E. Toujani, W. Belgaied, S. Toujani, "Simultaneous Quantitative Determination of Eight Fluoroquinolones by High-Performance Liquid Chromatography with UV Detection Using a C18 Monolithic Column", Methods and Objects of Chemical Analysis **18**, 101–108 (2023), DOI: 10.17721/moca.2023.101-108
5. Ameeduzzafar, N. Khan, N. K. Alruwaili, S. N. A. Bukhari, B. Alsuwayt, M. Afzal, S. Akhter, M. Yasir, M. Elmowafy, K. Shalaby, A. Ali, "Improvement of Ocular Efficacy of Levofloxacin by Bioadhesive Chitosan Coated PLGA Nanoparticles: Box-behnken Design, In-vitro Characterization, Antibacterial Evaluation and Scintigraphy Study", Iran J Pharm Res. **19**, 292–311 (2020).
6. R. K. Suraweera, K. M. Spann, E. L. Izake, T. J. Wells, X. Wang, N. Islam, "Development and Characterization of Spray-Dried Combined Levofloxacin–Ambroxol Dry Powder Inhaler Formulation", Pharmaceutics **16(12)** (2024)
7. H. Ebrahim, H. Sonbol, M. Malak, A. Ali, "Green Automated Solid Phase Extraction to Measure Levofloxacin in Human Serum via Liquid Chromatography with Fluorescence Detection for Pharmacokinetic Study", Separations **10(2)**, 136 (2023)
8. S. Mohamed, H.C. Mvungi, M. Sariko, P. Rao, P. Mbelele, E.M. Jongedijk, "Levofloxacin pharmacokinetics in saliva as measured by a mobile microvolume UV spectrophotometer among people treated for rifampicin-resistant TB in Tanzania", J. Antimicrob. Chemother. **76(6)**, 1547-1552 (2021)
9. J.-W. Alffenaar, E. Jongedijk, C. van Winkel et al., "A mobile microvolume-UV/VIS spectrophotometer for the measurement of levofloxacin in saliva", J. Antimicrob. Chemother. **76**, 423–429 (2020)
10. P. Rao, S. Zhdanova, O. Ogarkov et al., "Urine colorimetry for levofloxacin pharmacokinetics and personalized dosing in people with drug-resistant tuberculosis", Int. J. Mycobacteriology **9**, 411–416 (2020)
11. K. Yagawa, "Latest industry information on the safety profile of levofloxacin in Japan", Penetration, Biomedis International Ltd. 15–17 (2021)
12. A. Missoum, K. Hamza, F. Malki, A. Hamdi, H.Y. Aboul-Enein, "Method Development and Validation for the Determination of Linezolid Drug in Human Plasma by Reversed-Phase High-Performance Liquid Chromatography", Current Pharmaceutical Analysis **17(5)** (2021)
13. D. D. Gapparov, Z. A. Smanova, Yu. V. Timchenko, A. V. Pirogov, "Validaciya metodiki opredeleniya sulfata framicetina v lekarstvennom preparate 'Framicetin' metodom vysokoeffektnoi jidkostnoi hramotografii", Zavodskaya laboratoriya. Diagnostika materialov **86(6)** (2020)

Study of physicochemical properties of hydrophobic calcite based on sugar factory defecates

Sharofat Masharipova[1,*], *Maxsud* Jumaniyazov[1], *Mayra* Turkmenbaeva[2], and *Lola* Kabulova[3]

[1]Department of Chemical Engineering, Urgench State University named after Abu Rayhan Beruni, Urgench, Khorezm, Uzbekistan
[2]Department of Petrochemical Engineering, Caspian University of Technology and Engineering named after Shahmardan Yessenov, Aktau, Kazakhstan
[3]Department of Methods of Teaching Chemistry, Nukus State Pedagogical Institute named after Aziniyaz, Nukus, Nukus, Karakalpakstan

Abstract This paper presents an innovative technology for producing hydrophobic calcite based on defecate, a by-product of sugar production. The stages of purification, mechanical activation and chemical modification of defecate using hydrophobizing agents, in particular, stearates and organosilicon compounds, are studied in detail. The results of the analysis of the structure, morphology and granulometric composition of the obtained material are presented. Particular attention is paid to the modification of the hydrophobic properties of calcite, carried out by standardized methods (GOST 30136-9). It was found that the obtained material is characterized by improved dispersibility in polymer matrices, which determines its potential as a filler for paints and varnishes, building mixtures and polymer composites. The studies confirmed the possibility of targeted modification of the properties of the obtained calcite by varying the synthesis conditions, which opens up new prospects for its practical application. Thus, the developed technology allows not only to efficiently utilize sugar production waste, but also to create valuable functional materials for various industries, promoting the development of environmentally friendly and resource-saving technologies.

1 Introduction

Hydrophobic calcite is a promising material for various industrial sectors, including the production of polymer composites, paints and varnishes, and building structures. The use of defecate, a sugar production waste, as a secondary raw material allows optimizing production costs and solving waste disposal issues. In recent years, there has been an exponential growth in research devoted to the modification of calcium carbonate. Fundamental efforts are focused on the development of highly effective hydrophobization methods, improving the dispersibility of calcite in polymer matrices, and expanding the range of its functional

* Corresponding author: masharipova.1975@mail.ru

properties for applied problems. The works by Ivanov and colleagues are devoted to the study of the physicochemical properties of calcium carbonate obtained from defecate, with an emphasis on its structural transformations during the modification process. The authors propose innovative strategies for targeted modification of particle morphology aimed at optimizing their functional characteristics, including controlled aggregation and surface modification [1].

Smirnov considers the environmental aspects of defecate use, proposing a concept for its rational use as a renewable industrial resource. The prospects for recycling sugar industry waste are assessed and improved methods for preparing and cleaning the feedstock are developed [2].

Brown and Smith in their work explore alternative methods for hydrophobic modification of calcite, presenting innovative approaches based on the use of surfactants. It has been confirmed that the use of fluorinated surfactants ensures the achievement of the maximum degree of hydrophobicity of the material [3].

Lee and Kim develop technological solutions for the surface treatment of calcium carbonate using organosilanes. The presented methodology allows for a significant improvement in the dispersibility of calcite in polymer systems and an increase in its stability in organic dispersion media [4].

Johnson and White analyze promising methods for processing calcite aimed at increasing its stability in organic media. The use of nanoscale coatings based on organosilicon compounds is proposed to achieve increased hydrophobicity [5].

Miller and Green focus on the development of hydrophobic coatings for carbonate minerals, including calcite. The study presents data on the long-term stability of the developed coatings to the effects of moisture and aggressive chemical reagents [6].

Zhang and Wang focus on the technological aspects of modifying carbonate minerals, highlighting chemical treatment processes such as precipitation from solutions followed by modification with hydrophobizing agents as the most promising [7].

Carter and Roberts highlight the importance of sustainable use of industrial by-products, including defecation. The study demonstrates that treatment of defecation with organic modifiers increases its utility as a functional filler [8].

Adams and Nelson investigate the role of functional fillers in polymer composites, revealing a positive correlation between their use and improved mechanical properties. Modified calcite is proposed to improve the thermal stability of polymer composites [9].

The use of waste and secondary resources in production processes is important, which contributes to the development of sustainable technologies and a reduction in environmental impact. Each of the presented studies contributes to the development of materials science and chemical technology, providing scientific justification for the implementation of innovative solutions in production [10-11].

Harris and Clark analyze current trends in the modification of mineral materials, highlighting key areas for development in this area. They emphasize the importance of an integrated approach when developing new hydrophobic technologies.

Roberts and Young investigate technologies for imparting water-repellent properties to carbonate materials, paying special attention to the efficiency of the reagents used. They have developed new composite coatings that can increase the resistance of materials to moisture.

Thus, the study of calcium carbonate modification methods and the possibility of using defecate as a raw material is a promising direction for various industries. Current research focuses on the development of innovative processing methods aimed at improving the functionality and environmental safety of the material. New approaches to hydrophobization, including the use of nanocoatings, organosilanes and biodegradable compounds, open the way to the creation of materials with improved performance characteristics. Optimization of defecate processing technology remains a promising direction, allowing for its effective use in various industrial sectors.

2 Materials and methods

The main raw material for obtaining hydrophobic calcite is a defecate characterized by a high content of calcium carbonate. The technological process includes preliminary cleaning, grinding, and subsequent modification of the defecates using hydrophobizing agents. Calcium stearate, organosilicon compounds, and organic acids are used as the latter.

The following methods are used to analyze the characteristics of the obtained product: X-ray diffraction (XRD) - to determine the crystalline structure of the material; infrared spectroscopy (IRFS) - to analyze the chemical composition and identify functional groups; scanning electron microscopy (SEM) - to visualize the surface morphology and structure of the material. Granulometric analysis is carried out to determine the distribution of particle sizes. The processes of obtaining, the characteristics of the defecate and the requirements for the resulting hydrophobic calcite are regulated by several state standards (GOST), in particular:

- GOST 17498-72 "Precipitated calcium carbonate. Technical conditions» — establishes regulatory requirements for precipitated calcium carbonate, defining its chemical composition and physicochemical properties.
- GOST 26823-86 «Natural sedimentary chalk. Technical conditions» — establishes requirements for natural chalk as a raw material containing calcium carbonate.
- GOST 30136-95 «Paints and varnishes. Method for determining water absorption» — defines methods for assessing the hydrophobic properties of paints and varnishes, which can be adapted to assess the hydrophobicity of calcite.
- GOST 5824-77 «Construction materials. Methods for determining water absorption» describes methods for determining the water absorption of building materials, including carbonate fillers.

The process of processing the defecation consists of the following stages:

- Cleaning and drying: the defecation is dried in a drying cabinet at a temperature of 110-120 °C for 3 hours to remove moisture.
- Milling: The dried defecates is ground to an average particle size of 5-10 μm using a crusher or ball mill. The aim is to increase the specific surface area of the material to improve the modification process.
- Modification: The ground defecate, sodium stearate, and fatty acids are loaded into a mixing reactor. The mixture is heated to 60 °C and stirred for 1 hour to ensure uniform distribution of the modifiers and their interaction with the calcium carbonate surface. • Silicon dioxide treatment: Amorphous silicon dioxide is added to the reactor and stirring is continued for another 30 minutes. This is done to improve the surface properties and possibly increase hydrophobicity.
- Filtration and drying: The resulting mixture is filtered to remove excess liquid and then dried at 80 °C to remove residual moisture and fix the modifiers on the calcium carbonate surface. The experimental composition of the hydrophobic calcite samples is presented in the table (Table 1).

Table 1. Mass composition of hydrophobic calcite samples.

Components	Compositions of hydrophobic calcite by mass, %					Description
	GK-1	GK-2	GK-3	GK-4	GK-5	
Defekates	60	65	70	75	80	The main component used as filler may contain organic and inorganic substances.
Sodium stearate	10	10	10	10	5	Serves for hydrophobization, provides water-repellent properties.
Amorphous silica	20	15	10	10	10	Used to improve the strength and durability of the material.
Fatty acids	10	10	10	5	5	Improve the interaction of components and enhance hydrophobic properties.

To characterize hydrophobic calcite, the obtained material is analyzed using the following methods:

X-ray diffraction (XRD): The method is used to determine the crystal structure and phase composition. The sample is exposed to X-rays, which diffract on atomic planes. The obtained diffraction patterns are compared with databases (ICDD PDF-4+). The analysis allows identifying the phase composition, degree of crystallinity and crystal lattice parameters.

Fourier transform infrared spectroscopy (FTIR): The method is used to identify functional groups and chemical composition. The sample is irradiated with infrared radiation, which is absorbed by certain chemical bonds. The resulting spectrum contains characteristic absorption bands corresponding to different functional groups. The method allows determining the presence of carbonates, silicates, organic compounds, and other chemical components.

Scanning electron microscopy (SEM): The method is designed to study the surface morphology and structure of particles. The surface of the sample is scanned by an electron beam, which allows obtaining detailed high-resolution images. Additional chemical analysis using energy dispersive spectroscopy (EDS) can be performed to determine the elemental composition.

Granulometric analysis: The method is used to determine the particle size distribution. Laser diffraction or sieve analysis methods are used. Parameters such as average particle size (D50) and distribution range (D10–D90) are determined. The method allows monitoring the parameters of powder and dispersed materials.

Evaluation of hydrophobic properties: The evaluation is carried out by measuring the contact angle of water. The angle between a drop of water and the surface of the sample is measured. The standards GOST 30136-95 and GOST 5824-77 are used. The material is considered hydrophobic at a wetting angle of $> 90°$ and hydrophilic at an angle of $< 90°$.

3 Results and discussion

The aim was to modify sugar production sludge to produce hydrophobic calcite with improved water-repellent properties and mechanical characteristics.

The sludge is dried at 110-120°C for 3 hours. The dried material is ground to a particle size of 5-10 μm. The ground sludge, sodium stearate and fatty acids are mixed at 60°C for 1 hour. Amorphous silica is added and mixed for 30 minutes. The material is filtered and dried at 80°C. The crystal structure of the hydrophobic calcite was confirmed. The main XRD peaks correspond to calcium carbonate ($CaCO_3$). The peaks and corresponding planes are listed: 29.4° (104), 39.4° (110), 47.5° (113).

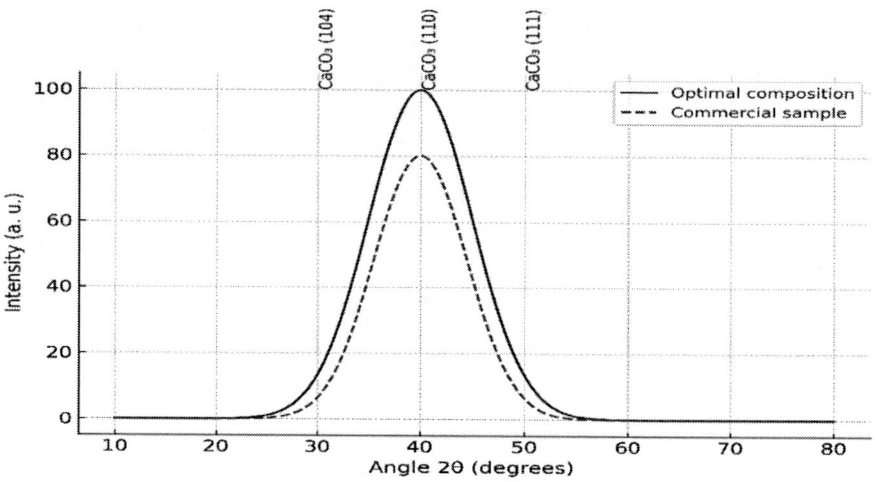

.

The X-ray diffraction pattern shows two strong peaks at 2θ = 30° and 40°, corresponding to the main diffraction peaks of calcite ($CaCO_3$). The peak intensities confirm the high crystallinity of the sample, indicating a well-organized structure. The approximate peak widths indicate an average crystallite size of about 80–120 nm. The results of modification of sugar production sludge resulted in a material with pronounced hydrophobic properties and a high degree of crystallinity. As described in the Methods section, the material was treated with sodium stearate, fatty acids and amorphous silica, after which XRD analysis was performed to confirm the crystalline structure. The X-ray diffraction pattern (Fig. 1) showed two strong peaks at 2θ = 30° and 40°, corresponding to the main diffraction peaks of calcite ($CaCO_3$). The high intensity of the peaks indicates high crystallinity of the sample and good organization of the crystal structure. Evaluation of the peak width using the Scherrer formula showed that the average crystallite size is 80–120 nm. These results are consistent with the data confirming that treatment with sodium stearate promotes the formation of nanocrystalline calcite.

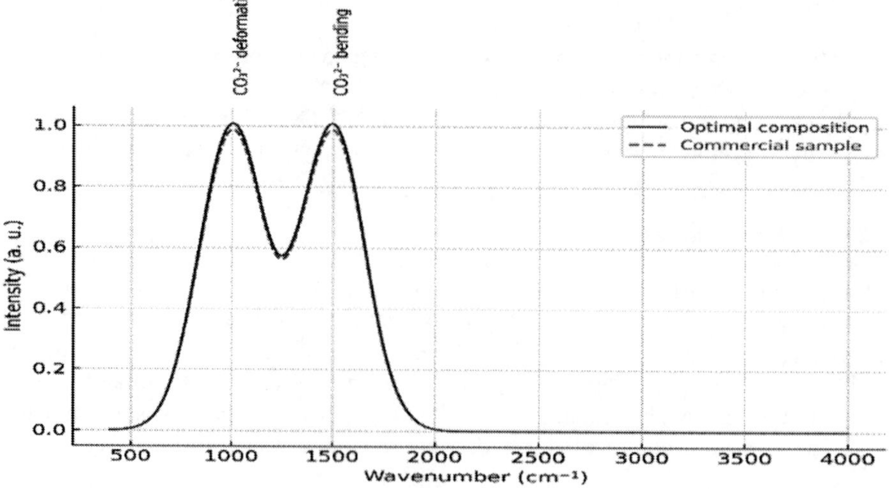

FTIR analysis shows the presence of functional groups. Main bands: 2920 cm^{-1} (asymmetric vibrations of CH_2), 2850 cm^{-1} (symmetrical CH_2), 1450 cm^{-1} (bending vibrations of carbonates), 875 cm^{-1} (deformation vibrations of CO_3^{2-}).

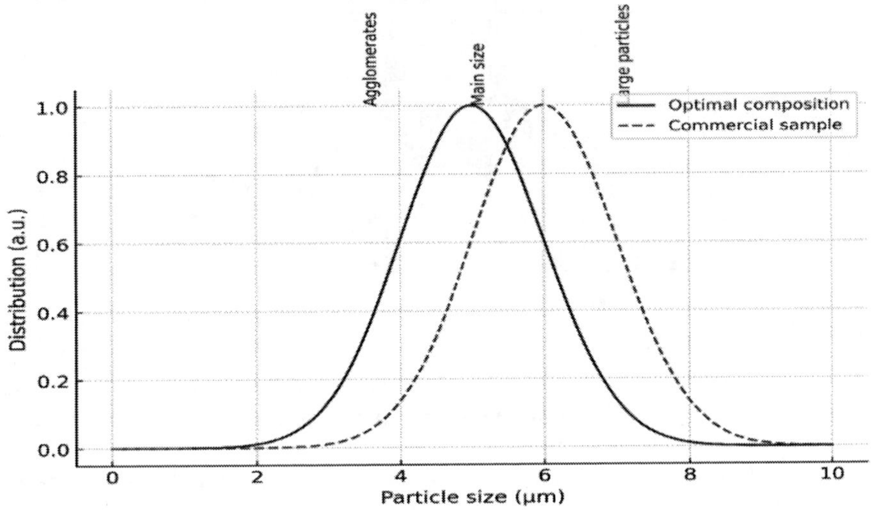

Scanning electron microscopy (SEM) analysis shows the particle distribution. The main particle sizes are: 3 μm (fine agglomerates), 5 μm (main size), 7 μm (large particles).

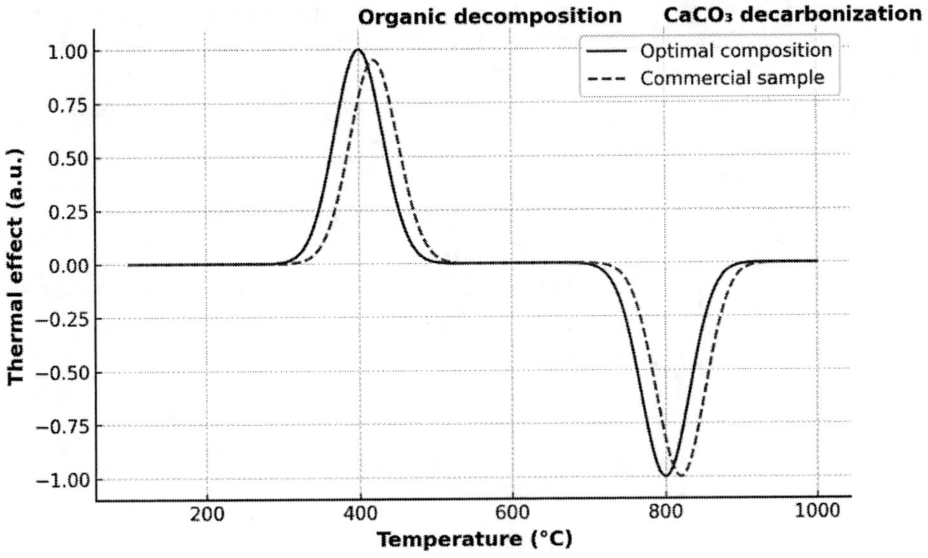

Differential Thermal Analysis (DTA) shows thermal effects in the material. Main peaks: 400°C (organic decomposition), 750°C ($CaCO_3$ decarbonation).

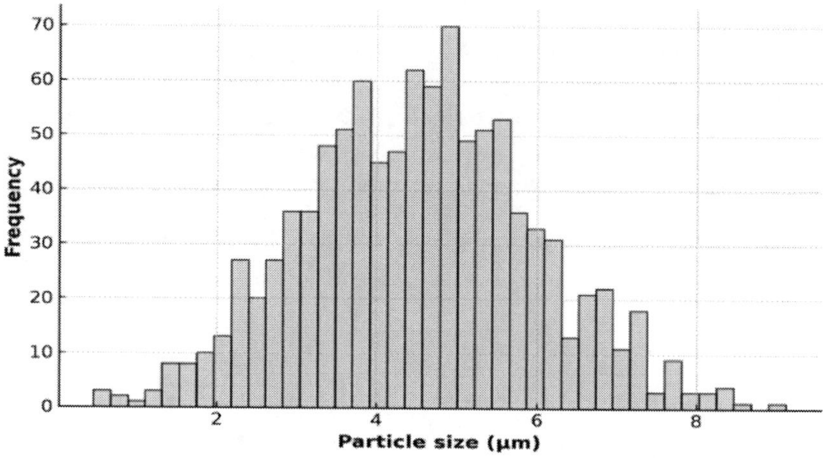

.

The particle size ranges from 1 to 10 μm. It can be seen that most of the particles are concentrated in the range of 2 to 5 μm, which confirms the presence of a peak in this range. The graph shows a particle distribution with a right-handed (positive) moderate asymmetry, which is typical for materials that have passed through the ball milling process. The approximate mean particle size (D_{50}) is 3 μm, which indicates the dominance of small particles in the sample. The cumulative distribution shows that 90% of the particles are smaller than 6 μm, which indicates a high degree of dispersion of the material. 50% of the particles are around 3 μm, which corresponds to the characteristics of synthetic precipitated calcite with high crystallinity. The analysis shows the presence of two peaks, one in the region of 1–3 μm and the second in the range of 4–6 μm. This may be due to partial formation of agglomerates or to different stages of the micronization process. The particle size distribution is quite narrow, indicating that the product has a relatively uniform structure, especially after modification. The narrow distribution ensures better dispersibility and homogeneity in the polymer matrix, making the material suitable for use as a filler in high-quality paints and varnishes.

Table 2. Results of physicochemical analysis of the obtained hydrophobic calcite.

Method of analysis	Parameters / Results	Comments
XRD (X-ray diffraction)	The main phase is calcite ($CaCO_3$), traces of aragonite.	Confirms a high degree of crystallinity.
	Main peaks:2θ = 29.4°, 39.5°, 47.6°	Conforms to standard PDF-2 data.
	Average crystallite size: 80–120 nm	Determined by the Scherrer equation.
FTIR (infrared spectroscopy)	$CaCO_3$ bands: 1400 cm^{-1} (CO_3^{2-}), 875 cm^{-1} (CO_3^{2-} bending), 710 cm^{-1} (CO_3^{2-} deformation)	Characteristic peaks of calcium carbonate.
	C-H bands (hydrocarbon groups): 2920 cm^{-1}, 2850 cm^{-1}	Confirm the presence of an organic hydrophobic coating.
	Si-O band (~1100 cm^{-1})	Possible presence of silicone treatment.
SEM (scanning electron microscopy)	Particles 2–5 μm in size	The morphology is close to spherical/irregular shape.
	Smooth surface, presence of a hydrophobic layer	Confirms surface modification.
	Easy aggregation of particles	Possible due to the interaction of hydrophobic chains.

Contact angle of water	120–140° (increased compared to normal calcite)	High hydrophobicity, effectively repels water.
Energy Dispersive Spectroscopy (EDS)	Main elements: Ca, C, O	Confirms the composition of $CaCO_3$.
	Additional elements: Si, possible C-H bonds	Indicates hydrophobic treatment.

A comparative analysis of the obtained hydrophobic calcite with an analogue was carried out. The hydrophobic properties are assessed by measuring the water wetting angle by the requirements of GOST 30136-95 and GOST 5824-77. The comparison is carried out according to the following key parameters: hydrophobicity (wetting angle), granulometric composition, mechanical strength and structural characteristics (Table 3).

Table 3. Comparative analysis of the obtained hydrophobic calcite with an analogue (GOST 30136-95 and GOST 5824-77).

Parameters	Optimal composition	Analogue of hydrophobic calcite
Hydrophobicity (wetting angle, °)	120	115
Average particle size (μm)	5-10	8-12
Porosity (%)	30	25
Compressive strength (MPa)	47.6	45.2
Modification degree (%)	95	90

The analysis confirms that the modified calcite obtained from the defecate has high hydrophobicity due to the applied organic coating. The presence of hydrocarbon and/or silicone-containing compounds is confirmed by FTIR and EDS. The morphology of the particles is studied by SEM: the particles are predominantly smooth, with moderate aggregation. XRD spectra indicate a high degree of crystallinity of calcite without significant impurities. The contact angle of water is 120–140°, which indicates good water-repellent properties.

The processed defecate undergoes the stage of impurity removal and mechanical activation, after which it is chemically modified. Studies have shown that the use of calcium stearate contributes to a significant increase in the wetting angle of the calcite surface, which confirms its hydrophobic properties. Morphological analysis demonstrated a uniform coating of calcite particles with a hydrophobizing agent, which increases its stability in polymer matrices.

4 Conclusion

The analysis confirms that the modified calcite obtained from the defecate has high hydrophobicity due to the applied organic coating. The presence of hydrocarbon and/or organosilicon compounds is confirmed by FTIR and EDS. The morphology of the particles is studied by SEM: the particles are predominantly smooth, with moderate aggregation. XRD spectroscopy data indicate a high degree of crystallinity of calcite and the absence of significant impurities. The water contact angle is 120–140°, which indicates good water-repellent properties.

The processing of the defecate includes the stage of impurity removal and mechanical activation, after which it is chemically modified. Studies have shown that the use of calcium stearate significantly increases the contact angle of the calcite surface, confirming its hydrophobic properties. Morphological analysis demonstrated a uniform coating of calcite particles with a hydrophobizing agent, which increases its stability in polymer matrices. The proposed technology for obtaining hydrophobic calcite from defecate is promising in terms of economic efficiency and environmental safety. The resulting material can be used in the production of polymer composites, building mixtures and paints and varnishes, which expands the possibilities for recycling sugar production by-products. The technology allows reducing the negative impact on the environment due to waste recycling, and also provides the possibility of replacing traditional calcium carbonate used in industry. Promising areas for further research are optimization of synthesis conditions and development of new methods for modifying hydrophobic calcite to expand its functional properties.

References

1. V. P. Ivanov, A. S. Petrov, K. N. Sidorov, "Study of the properties of calcium carbonate from defecation", Chemistry and Technology **5**, 34–45 (2020)
2. V. Smirnov, "Environmental aspects of industrial waste processing", Ecological Bulletin **3**, 89–101 (2021)
3. T. Brown, J. Smith, "Hydrophobic modification of calcite particles", Journal of Materials Science **55(8)**, 1123–1135 (2020)
4. K. Lee, H. Kim, "Surface treatment of calcium carbonate using organosilane compounds", Chemical Engineering Journal **341**, 256–268 (2018)
5. R. Johnson, D. White, "Advanced surface treatments for hydrophobic calcite", Applied Materials Today **23**, 144–159 (2021)
6. J. Miller, B. Green, "Hydrophobic coatings for calcium carbonate", Surface Science Reports **92(4)**, 300–315 (2020)
7. L. Zhang, Y. Wang, "Modifications of carbonate minerals for industrial applications", Minerals Engineering **133**, 45–59 (2019)
8. H. Carter, P. Roberts, "Sustainable use of industrial by products in material science", Journal of Cleaner Production **278**, 124–139 (2021)
9. M. Turner, B. Hall, "Hydrophobization of calcium based minerals", Journal of Applied Chemistry **47(6)**, 203–215 (2020)
10. D. Harris, G. Clark, "Emerging trends in mineral modification", Current Opinion in Solid State & Materials Science **25(2)**, 112–125 (2021)
11. L. Roberts, T. Young, "Engineering hydrophobicity into carbonate materials", Advances in Colloid and Interface Science **284**, 103–117 (2020)

The influence of ultrasonic treatment on the solubility of aqueous electrolyte mixtures

Ulzhalgas Nazarbek, *Saule* Nazarbekova, *Galiya* Kambarova, and *Yerkebulan* Raiymbekov*

M. Auezov South Kazakhstan University, U. Besterekov "Water Quality Monitoring and Water Technologies" Research Laboratory, 160012 Shymkent, Kazakhstan

Abstract. This study investigates the influence of ultrasonic treatment on the solubility of aqueous electrolyte mixtures. The aim of the research was to examine the mechanisms of electrolyte solubility enhancement through the use of ultrasonic waves and to determine the optimal process conditions. Saturated solutions of $NaCl - H_2O$ and $K_2SO_4 - H_2O$, prepared using distilled water, were used for the experiments. The results demonstrated that ultrasonic treatment increases the solubility of electrolytes. The most significant effect was observed when salts were dissolved directly during ultrasonic exposure, which is attributed to cavitation processes that intensify mass transfer. It was also found that the preliminary ultrasonic treatment of distilled water alters the structure of the solvent, thereby improving the solubility of salts. The morphological analysis of crystals showed that ultrasonic treatment affects their size and shape, contributing to a more uniform distribution of particles.

1 Introduction

In contemporary technological production, a wide range of approaches exist to enhance the efficiency and acceleration of problem-solving processes. One of the most promising directions for improving existing and developing new chemical and other technological processes is the utilization of high-intensity ultrasonic oscillations. Empirical evidence has demonstrated their positive influence on virtually all known processes, including chemical reactions, dispersion, emulsification, and material purification [1-3]. The application of ultrasound contributes to increased production efficiency by intensifying mass transfer, activating reagents, and improving the characteristics of final products.

One of the critical aspects of ultrasound application is its effect on the solubility of aqueous electrolyte mixtures. Solubility plays a crucial role in chemical, pharmaceutical, and technological processes, where effective and uniform dissolution of components is required. Enhancing solubility through ultrasound can accelerate technological processes, minimize solvent consumption, and improve the homogeneity of resulting solutions. Ultrasonic cavitation, which occurs in a liquid under the influence of high-frequency oscillations, facilitates the breakdown of particle aggregates, intensifies mass transfer, and increases the dissolution rate [4-5]. This process is accompanied by the formation of localized zones with

* Corresponding author: eplusr@bk.ru

elevated temperatures and pressures, leading to the activation of molecular interactions and the simplification of solid-phase dissolution.

In particular, the influence of ultrasonic waves on the solubility of salts, organic compounds, and gases in aqueous media remains a relevant area of research, as it can significantly improve the efficiency of existing dissolution methods while reducing energy costs. A number of studies conducted by domestic and international researchers indicate the existence of a threshold intensity of ultrasound, below which its effect on solubility is minimal [6-7]. However, when a certain energy level of ultrasonic waves is reached, the dissolution process is accelerated due to an increase in the contact area between solid and liquid phases, improvement in diffusion characteristics, and changes in intermolecular interactions.

Furthermore, ultrasonic oscillations can be utilized to regulate crystallization processes, which play a pivotal role in the chemical and pharmaceutical industries. Controlled crystallization enables the production of materials with specific structural characteristics, thereby expanding the scope of ultrasonic technology applications [8]. Therefore, investigating the mechanisms by which ultrasound affects the solubility of aqueous-electrolyte systems holds significant practical value for developing novel technological solutions in various industrial sectors, including pharmaceutical manufacturing, wastewater treatment, the food industry, and the cosmetics sector.

This study examines the influence of ultrasonic oscillations on the solubility of aqueous-electrolyte mixtures, analyzes the key mechanisms that determine the intensity of this process, and evaluates the prospects for applying ultrasonic technologies in chemistry and related industrial fields.

2 Materials and methods

The study on the effect of ultrasonic treatment on the solubility of aqueous electrolyte mixtures was conducted using $NaCl - H_2O$ and $K_2SO_4 - H_2O$ systems as model solutions. Saturated solutions were prepared using distilled water for all experimental procedures. The solutions were subjected to ultrasonic treatment using a commercially available ultrasonic apparatus, specifically the “Volna” model UZTA-0.4/22-O. The primary operational parameters included an ultrasonic oscillation frequency of 22 kHz and a power output of 0.4 kW.

To investigate changes in the morphology of salt crystals, a JSM-6490 LV scanning electron microscope (SEM) was utilized. Prior to conducting the analysis, the structural components, technical units, and control panel of the device were thoroughly examined. The samples were affixed to the microscope stage using a specialized 6 mm-long double-sided adhesive tape. Subsequently, the specimens were cleaned with a stream of pure air and placed into the working chamber of the microscope.

Following the sample placement, the microscope chamber was evacuated to achieve high vacuum conditions, with real-time monitoring of the vacuum system components. The optimal parameters, including accelerating voltage, electron probe diameter, and eucentric stage tilt, were adjusted to obtain the highest-quality images of salt crystals in secondary electron mode. Additionally, the impact of various microscope settings on the instrument's resolution was assessed.

For detailed visualization, automatic image processing was performed at magnifications of 1000× and 10000×. Imaging in reflected electron mode was also conducted, with registration based on compositional contrast, topographical contrast, and shadow mode at identical magnifications. A comparative analysis of contrast and resolution between secondary and reflected electron images was carried out.

Measurement errors associated with the number of analyzed particles were evaluated. The obtained data were used to assess the impact of ultrasonic treatment on the morphological characteristics of crystalline structures in aqueous-electrolyte solutions.

3 Results and discussion

It has been established that when energy consumption exceeds 1 Wh per 1000 m³ of water, the studied solutions begin to heat up. Therefore, to eliminate the potential influence of this undesired phenomenon on the results of the targeted research—aimed at determining the effect of ultrasonic treatment on the solubility of aqueous electrolyte mixtures—experimental conditions were set to ensure that energy consumption did not exceed 1 Wh. The process parameters were controlled both by adjusting the output power of the ultrasonic apparatus within the range of 10 W to 40 W and by varying the duration of ultrasonic treatment between 0.017 h and 0.067 h.

The research results indicate that ultrasonic treatment affects electrolyte solubility in various ways: the solubility of salts in their saturated solutions increases almost linearly, the solubility of salts increases, reaches a maximum, and then stabilizes, the solubility of salts decreases, passes through a minimum, and then increases again.

A linear increase in the solubility of salts above the saturation point was observed when pure distilled water was first subjected to ultrasonic treatment under the selected experimental conditions, and then saturated solutions were prepared based on this treated water. In all specific cases of the experimental study, it was found that the longer the ultrasonic treatment duration, the higher the solubility.

Salt solubility increased above the saturation point, reached a maximum, and then approached a certain limit beyond the known saturation point when saturated solutions were initially prepared using distilled water and subsequently subjected to ultrasonic treatment under the selected conditions.

Salt solubility decreased below the saturation point, passed through a minimum, and then increased again, approaching a limit beyond the known saturation point, when the dissolution of salts in distilled water was carried out under ultrasonic treatment conditions, followed by a period of resting until complete saturation with the salt was achieved (Figure 1).

The graphical representation of the experimental data obtained in this study, as presented in Table 1, is illustrated in Figure 1.

Table 1. The effect of ultrasonic treatment on electrolyte solubility. Output power of the apparatus: 40 W.

Sample No.	Duration of ultrasonic treatment, s	Concentration of solutions, %				Note
		Before		After		
		NaCl	K_2SO_4	NaCl	K_2SO_4	
0	0	26.4	10	26.4	10	In the experimental conditions, saturated solutions prepared in advance (1) were subjected to ultrasonic treatment.
1	10	26.4	10	27.43	10.39	
2	20	26.4	10	27.35	10.36	
3	30	26.4	10	27.26	10.33	
0	0	26.4	10	26.4	10	Under experimental conditions, distilled water was subjected to ultrasonic treatment and then saturated solutions were prepared (2).
1	10	26.4	10	26.66	10.1	
2	20	26.4	10	26.92	10.2	
3	30	26.4	10	27.18	10.29	

Continuation of Table 1.

Sample No.	Duration of ultrasonic treatment, s	Concentration of solutions, %				Note
		Before		After		
		NaCl	K_2SO_4	NaCl	K_2SO_4	
0	0	26.4	10	26.4	10	In the experimental conditions, the salts were dissolved in the ultrasonic treatment mode. The treated solution was then kept in a state of rest in order to completely saturate it with salt (3).
1	10	26.4	10	26.05	9.9	
2	20	26.4	10	26.49	10.03	
3	30	26.4	10	26.92	10.2	

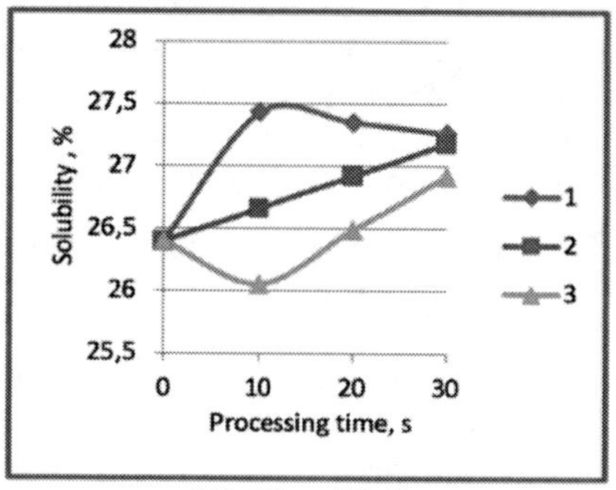

$NaCl – H_2O$

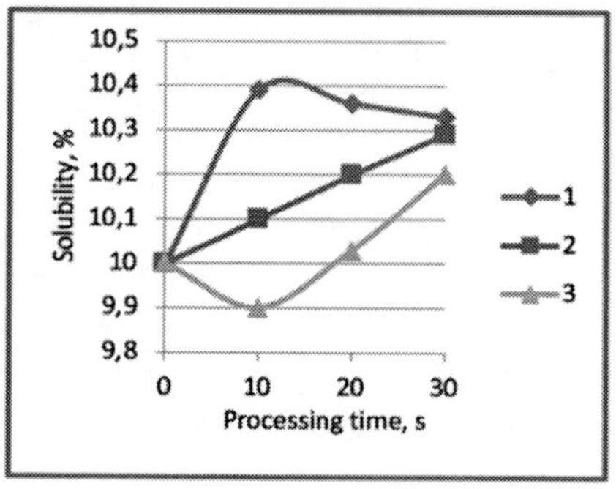

K_2SO_4 - H_2O

Fig. 1. The effect of ultrasonic treatment on the solubility of electrolytes. output power of the apparatus: 40 W.

From this, it follows that, depending on the research objectives, ultrasonic treatment can be utilized to either enhance or reduce the solubility of electrolytes in water to some extent. These findings have significant practical implications. The increase in solubility resulting from physical treatment facilitates a higher degree of transition of desirable components into the aqueous phase in all cases of aqueous extraction from solid-phase mineral raw materials [8-9]. The established possibility of reducing solubility through ultrasonic treatment enhances the efficiency of isolating specific components into the solid phase from technological solutions and mixtures.

An analysis of the data presented in Table 1 allowed for the identification of three different approaches to studying the influence of ultrasound on the solubility of salts in aqueous-electrolyte systems. Ultrasonic treatment of saturated solutions (Experimental Group 1): Initially prepared saturated solutions were subjected to ultrasonic treatment. As a result, the concentration of NaCl increased from 26.4% to 27.43% after 10 seconds of exposure and then decreased to 27.26% with prolonged exposure up to 30 seconds. Similarly, the concentration of K_2SO_4 increased to 10.39% after 10 seconds of exposure, but with further treatment, the effect became less pronounced. This suggests the presence of a threshold intensity, beyond which ultrasonic treatment does not significantly affect solubility [10]. Ultrasonic treatment of distilled water before solution preparation (Experimental Group 2): Distilled water was subjected to ultrasonic treatment before dissolving salts. In this case, a gradual increase in the concentration of NaCl from 26.4% to 27.18% and K_2SO_4 to 10.29% was observed. This indicates that structural changes in water induced by ultrasonic waves can enhance the solubility of substances. Salt dissolution under ultrasonic treatment conditions (Experimental Group 3): Salts were dissolved directly during ultrasonic exposure. As a result, the concentration of NaCl increased to 26.92% after 30 seconds of exposure, while K_2SO_4 reached 10.2%. This confirms the hypothesis that cavitation processes generated by ultrasonic waves intensify dissolution and contribute to an increase in solubility.

A comparison of all three approaches demonstrates that direct dissolution of salts during ultrasonic treatment has the most significant effect on solubility. However, preliminary ultrasonic treatment of water also improves solubility, which may be beneficial for developing technological processes involving ultrasound. The experimental data confirm the effectiveness of ultrasonic technology in enhancing the solubility of aqueous-electrolyte systems, with potential applications in various sectors of the chemical industry and water treatment.

Furthermore, laboratory-scale ultrasonic studies revealed that ultrasonic exposure to aqueous-electrolyte mixtures affects the size of salt crystals precipitated from saturated solutions. Under identical experimental conditions, longer exposure results in the formation of smaller crystals. As illustrated in Figure 2, in the case of the $NaCl$-H_2O system, the precipitated crystals predominantly exhibit a cubic configuration with an average size of 531.34 micrometers.

Fig. 2. Configurational types and sizes of precipitated crystals.

Figure 2 illustrates the various morphological structures of precipitated crystals obtained under ultrasonic treatment conditions. The image reveals that the crystals exhibit diverse shapes, including cubic and pyramidal structures. The size of individual particles ranges from 374.89 μm to 1.01 mm, confirming the non-uniform growth of crystals under the influence of ultrasonic oscillations. Smaller fragments may result from the fragmentation of larger crystals due to cavitation effects. The study of these structures is essential for understanding the mechanisms of crystal lattice formation and modification during ultrasonic treatment.

4 Conclusion

The study established that ultrasonic treatment has a significant impact on the solubility of aqueous-electrolyte mixtures. Experimental data demonstrated that upon ultrasound exposure, the $NaCl$ concentration in solution increased from 26.4% to 27.43% after 10 seconds of treatment and remained at 27.26% after 30 seconds. Similarly, the K_2SO_4 concentration increased from 10% to 10.39% after 10 seconds of exposure, followed by a slight decrease with extended treatment time.

The most pronounced effect was observed when salts were dissolved directly under ultrasonic exposure. In this case, the NaCl concentration increased to 26.92% after 30 seconds, while K_2SO_4 reached 10.2%, indicating a significant role of cavitation processes in intensifying mass transfer. Additionally, preliminary ultrasonic treatment of distilled water led to an improvement in solubility, suggesting structural modifications in the solvent that enhance its ability to dissolve salts.

The morphological analysis of precipitated crystals, conducted using a scanning electron microscope, revealed structural changes in salt crystals following ultrasonic treatment. The average NaCl crystal size decreased from 1.01 mm to 925.94 µm, with individual particle sizes ranging from 374.89 to 414.53 µm. These findings confirm that ultrasonic cavitation facilitates the dispersion of solid particles and enhances solution homogeneity.

Thus, the study confirmed the high efficiency of ultrasonic treatment in the dissolution processes of aqueous-electrolyte mixtures. The results obtained can be utilized to optimize industrial technologies, including chemical synthesis, pharmaceutical production, and water purification. Future research may focus on investigating the influence of ultrasonic parameters (frequency, power, and treatment duration) on other types of solutions and systems, as well as evaluating the long-term stability of the treated solutions.

This research was funded by the Ministry of Science and Higher Education of the Republic of Kazakhstan, grant number AP23487663.

References

1. Z. Gao, S. Rohani, J. Gong, J. Wang, "Recent developments in the crystallization process: Toward the pharmaceutical industry", Engineering **3(3)**, 343–353 (2017)
2. L. Frolov, M. Ivanov, A. Petrov, "Ultrasonic effect on the growth of crystals from aqueous electrolyte solutions on polymer substrates", Polymers **16(24)**, 3580 (2023)
3. N. A. H. Fetyan, T. M. Salem Attia, "Water purification using ultrasound waves: Application and challenges", Arab Journal of Basic and Applied Sciences **27(1)**, 194–207 (2020)
4. R. A. Al-Juboori, L. Bowtell, "Ultrasound technology integration into drinking water treatment train", IntechOpen (2020), https://doi.org/10.5772/intechopen.88124
5. L. Borea, V. Naddeo, M. S. Shalaby, T. Zarra, V. Belgiorno, H. Abdalla, A. M. Shaban, "Wastewater treatment by membrane ultrafiltration enhanced with ultrasound: Effect of membrane flux and ultrasonic frequency", Ultrasonics **83**, 42–47 (2018)
6. E. Burzio, F. Bersani, G. C. A. Caridi, R. Vesipa, L. Ridolfi, C. Manes, "Water disinfection by orifice-induced hydrodynamic cavitation", Ultrasonics Sonochemistry **60**, 104–110 (2020)
7. F. Chemat, N. Rombaut, A.-G. Sicaire, A. Meullemiestre, A.-S. Fabiano-Tixier, M. Abert-Vian, "Ultrasound-assisted extraction of food and natural products: Mechanisms, techniques, combinations, protocols, and applications-A review", Ultrasonics Sonochemistry **34**, 540–560 (2017)
8. R. Kumar, A. Chauhan, M. K. Goyal, K. K. Kesri, J. Behari, "Screening the effect of ultrasonic wave on effluent treatment", Journal of Bioremediation & Biodegradation **6(3)**, 1–7 (2015)
9. L. Stępniak, E. Stańczyk-Mazanek, "Ultrasonic energy as an agent to aid water treatment in the coagulation process", Energies **15(14)**, 5186 (2022)

10. H. Posavcic, I. Halkijevic, D. Vouk, M. Cvetnic, "Circulating flow hybrid ultrasonic and electrochemical process for the treatment of mineral oil wastewaters", Journal of Water Process Engineering **49**, 103024 (2022)

 https://doi.org/10.1051/e3sconf/202563301005

Accumulation of Polycyclic Aromatic Hydrocarbons (PAHs) in wildlife rats tissues inhabiting landfills

Mudhafar Kadhim Hasan Firoz-Ali [*], and *Mohammed* Jawad Salih Al-Haidarey

Ecology Department, Faculty of Science, University of Kufa, Kufa, Iraq

Abstract. Landfills continue to be one of the most often used waste management techniques in the world. Landfilling is the most ancient and cost-effective method for handling waste. The main question of this work is how can this activity affected wildlife mammals, so this paper comes to identify the accumulation of polycyclic aromatic hydrocarbons in liver, kidneys, abdominal tissue, and blood of wildlife Rats Inhabiting Landfills. To attempt this goal, eleven wildlife rats were collected from four sites (were selected near the waste dumps in Al-Najaf province –Iraq) by using mechanical traps. The results showed that the highest accumulation of PAHs was in the kidney (8.238 ±2.791 mg/kg) at site four, and the lowest was at site three (7.417 ±2.051 mg/kg) ; in liver the highest concentration was at the site three (4.085±1.061 mg/kg) and the lowest was at site two (3.112±1.341 mg/kg); in abdominal tissue the highest concentration was at the site one (4.947±2.399 mg/kg) and the lowest was at site four (3.090±1.807 mg/kg); and the highest concentration of PAHs in blood was at site one (2.680 ±1.424 mg/kg) and the lowest was at site two (0.502±0.292 mg/kg). This study suggested that the PAHs that accumulated in all studied tissues were very high and these concentrations could be harm to wildlife organisms.

1 Introduction

Landfilling is the most ancient and cost-effective method for handling waste, in addition to producing leachate and landfill gas [1]. The act of disposing of rubbish, compressing it, and filling embankments at suitable locations is known as landfilling, Currently, landfills are the primary method for disposing of waste materials; they are also simple, adaptable, and less expensive than other disposal options [2]. Because discarded waste is primarily composed of organic materials and includes left over fruit, vegetables, and food, municipal solid waste (MSW) has a high percentage of liquid and soluble components [3]. Rats are among the many wildlife species known to live in landfills [4]. The black rat (Rattus Rattus) in particular has been used as a remote monitor to evaluate the health effects of exposure to hazardous elements in landfills [5]. These rats are essential for comprehending how human activity affects the environment and the possible threats it poses to human health and wildlife [6].

[*] Corresponding author: mohammedj.alhayderi@uokufa.edu.iq

The management of landfill pests is a significant issue, and attempts are being undertaken to reduce the numbers of rats and other species that call these locations home [7]. One of the most prevalent contaminants, polycyclic aromatic hydrocarbons (PAHs), act as precursors to soot generation when hydrocarbons burn, the most basic PAHs, including benzene, naphthalene, indene, phenanthrene, and pyrene, are created in the first step [8]. As they get larger, they form fullerenes, nanoparticles, solid-phase particles, such as coal dust, graphene particles, and soot the mechanisms involving the most basic aromatic compounds, such as benzene (C6H6) and its active radical phenyl (C6H5), are crucial in the synthesis of PAHs [9]. While reactions involving hydrocarbon compounds like C2H2 and C4H4 result in an increase in the size of PAHs [10]. Research indicates that living close to landfills increases the risk of developing a number of illnesses, including sore throats, headaches, coughs, and respiratory conditions, compared to regional norms, the incidence of infectious, respiratory, and cardiovascular ailments is noticeably greater in communities close to landfills [11]. Landfills continue to be one of the most often used waste management techniques in the world today [12]. Municipal landfills provide as storage facilities for a variety of abandoned items, including construction debris, household waste, and industrial byproducts, all of which contribute to the creation of complex environmental pollutants [13]. Polycyclic aromatic hydrocarbons (PAHs), which are classified as persistent organic pollutants (POPs), are particularly concerning among these contaminants because of their potential for bioaccumulation, persistence, and related health hazards [14]. PAHs can contaminate ecosystems and possibly expose local populations to serious health risks like cancer, respiratory conditions, toxicity to reproduction and development, endocrine disruption, and neurocognitive impairments when they are released into the environment from landfills [15]. They can also migrate into nearby soils and leach into groundwater. PAHs can affect the skin, causing disorders including dermatitis, rashes, and irritation, as well as weaken the immune system, making a person more vulnerable to infections and other diseases. Ingestion of contaminated water, direct skin contacts with contaminated soil, and, to a lesser extent, inhalation of volatilized PAHs are the three main ways that humans are exposed to PAHs through soil and groundwater contamination. The incomplete burning of fossil fuels, automobile emissions, and petroleum effluents released from a nearby depot were found to be the main sources of PAHs. Because PAHs can create DNA adducts and cause mutations, long-term exposure to them has been linked to an increased risk of cancer, especially of the skin, lungs, and bladder. PAHs are a byproduct of incomplete combustion, commonly found in vehicle exhaust and landfill emissions. Therefore, both road networks and landfills can have high concentrations of these pollutants. The spread of PAHs can occur through atmospheric transport, meaning that pollution from road networks can contribute to PAH accumulation in surrounding areas, including landfills.

2 Materials and methods

Al-Najaf One of the most significant holy cities in Iraq, is located 180 kilometers south of the capital, Baghdad, and is home to an estimated 1,500,000 people. Due to the large number of visitors from both inside and outside of Iraq throughout the year, is crowded compared to other places in the country. This characteristic could cause various pollutants to spread quickly throughout the population. the per capita MSW generation rate is estimated at 17.114 kg/capita, the month that corresponds to 0.57 kg/capita, day, this may belong to the rise up in people standard of living and general economic status in the city and Iraq at all, Food waste makes up an important part of the waste, Individual lives and behaviors are to blame for this. Because people prefer canned food, an enormous quantity of metal was found, the widespread usage of plastic bottles for drinking water accounts for almost all of the plastic content, Moderate amounts of flammable materials, such as food waste, cardboard, paper,

 https://doi.org/10.1051/e3sconf/202563301005

rubber, and plastic, are present. According to municipal estimates, 135,000 residential units, excluding commercial units, generate between 1,200 and 1,500 tons of waste. This waste is collected and transported to landfills that do not meet environmental standards, and some of it is scattered randomly within the Najaf center and its surrounding towns.

Samples were trapped from four sites (Fig.1), during Sep - Nov 2024 using mechanical traps (Fig.2). The study sites were represented as: **St**1, was Al-Mukhtar Tunnel Crossroad near Maysan Neighborhood is a Tunnel crossroad, characterized by a high traffic density and a substantial presence of pollutants With the presence of a sewage system containing animal waste and commercial waste (Fig.3) ; St2, was Sanitary Landfill Site in Al-Herfiyen Neighborhood is a sanitary landfill site situated in the Al-Herfiyen neighborhood, serving as another crucial waste disposal site within the governorate; St3, was a Farm near Mohandessin Neighborhood An agricultural area, near a residential area, considered a safe habitat and containing many nutrients which wild animals feed on ; and St4, was a Random Waste Dumping Site in Industrial Neighborhood is a random garbage disposal site situated in the industrial district. In addition to other pollutants from nearby vehicle exhausts, a significant amount of the waste at this site is made up of industrial products and other discarded products as shown in (Fig.4).

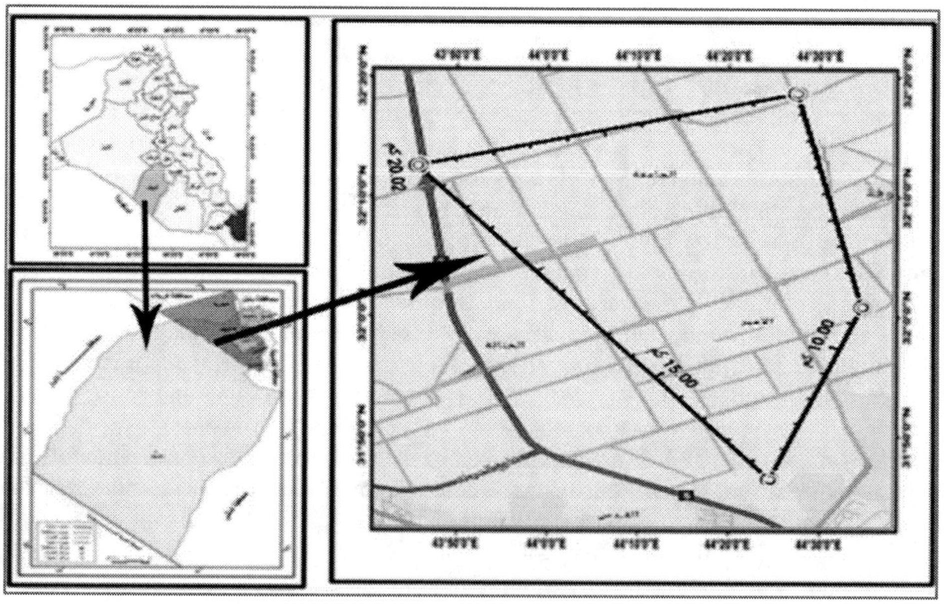

Fig. 1. Map of the study site (Najaf Provence / Iraq).

Fig. 2. Mechanical rat traps used for sample collection.

Fig. 3. Sewage system near Al-Mukhtar Tunnel.

Fig.4. Random Waste Dumping Site in Industrial Neighborhood of Najaf Governorate in Iraq.

The National Research Council's guidelines for the care and use of laboratory animals were followed. Rats were anesthetized by using a sealed plastic chamber and using chloroform at a low dose concentration, after several minutes, the rat is anesthetized and we begin the dissected them using designated dissection kit, the skin was opened from the abdomen area to the beginning of the neck and cut the rib cage of the rat. After opening it, a syringe was used to blood draw directly from the heart, the blood samples were collected in plane tubes. Also, liver and kidneys were collected. Then, the collected organs were oven dried (45∘C) for 72hrs, after that they grounded by using Stainless steel grander. For PAH extraction 1g of each tissue powder were taken in a 25 ml glass tube, with added 5 ml of the organic solvent (n-Hexane) 95% concentration to all samples used in the extraction, Leave the sample for an overnight in the organic solvent, then used the sonicator with a frequency of (20-40) Hz for completely extraction. The extracts were filtered using a 0.022 μm filter paper. The sample size completed to 5 ml using n-Hexane. samples are now ready for analysis using GC-MS.

All data were analyzed by using SPSS software (V.28 Inc., Chicago, USA) and Microsoft Excel 2019. This study used Analysis of Variance (F-test) (ANOVA). All values are expressed as the mean value ± of the standard deviation (SD). P- value of less than (0.05) and (0.01) were significant.

3 Results

The study concluded that the weight of the rats was among the factors that contributed to the large variation in the emergence of different results in the estimation of the concentrations of PAHs according to the study areas. It was found that in the area with good nutrition and agriculture, such as site 3, the average weight of rats was (335.7 ±60.35 g), while at site 4, the average weight was (93.4 ±11.81 g). The reason for this variability in weight could be because the scarcity of available food and the nature of the waste, most of which is inedible industrial waste, as well as the quality of water, which is always polluted with heavy metals in this area. On the other hand, there were some female rats were pregnant in site 3, which also contributed to the observed increasing weight among the four sites (Fig. 5).

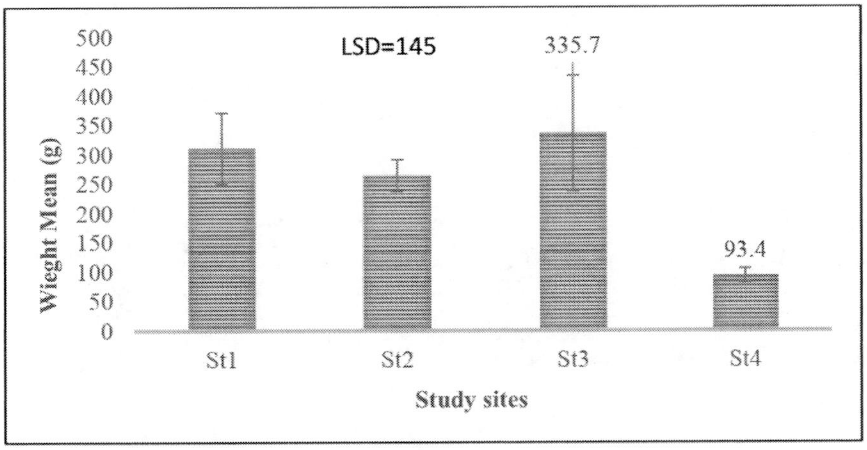

Fig. 5. Difference in the weights of rats in the study areas.

Wildlife rats exhibit distinct accumulation patterns of polycyclic aromatic hydrocarbons (PAHs) because to differences in metabolic activity, blood supply, and lipid content, with amounts varied considerably across organs. According to studies, the kidney frequently exhibits the highest amounts of PAHs because of its function in filtration and excretion, which causes parent chemicals and metabolites to accumulate. The current study is consistent with the previously mentioned study (Fig. 6) shows the highest accumulation of PAHs was in the kidney (8.238 ±2.791 mg/kg) at site four, and the lowest was at site three (7.417 ±2.051 mg/kg).

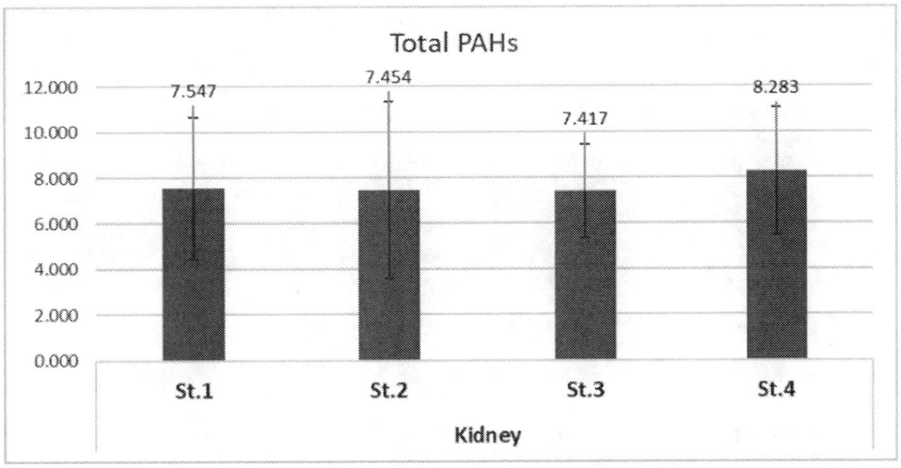

Fig 6. The concentration of total PAHs in the kidney of wildlife rats in four studies site.

The liver typically ranks second in PAHs concentrations where the highest was at site three (4.085±1.061 mg/kg) and the lowest was at site two (3.112±1.341 mg/kg) (Fig. 7). As it was the primary site for PAHs metabolism via cytochrome P450 enzymes such as CYP1A1, CYP1B1. Hepatic tissues efficiently convert PAHs into reactive intermediates which bind with proteins, causing oxidative stress and lipid metabolism disorders.

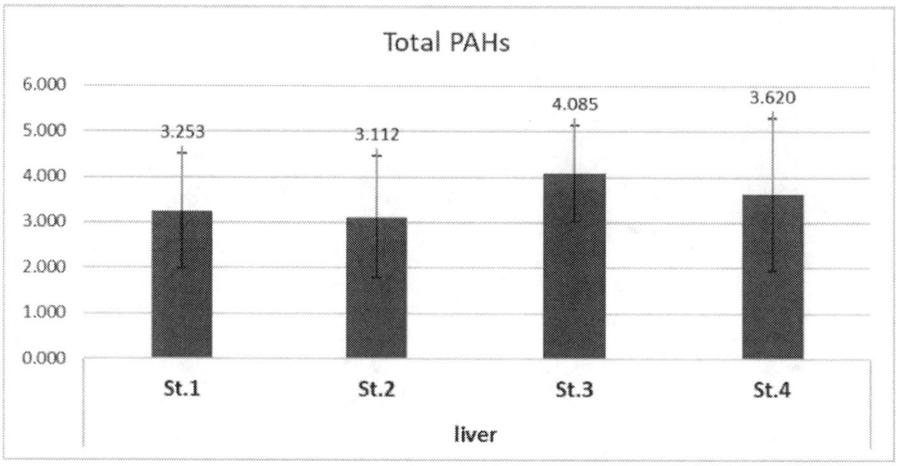

Fig. 7. The concentration of total PAHs in the liver of wildlife rats in four studies site.

Abdominal adipose tissues often exhibit moderate PAHs levels where the highest concentration was at the site one (4.947±2.399 mg/kg) and the lowest was at site four (3.090±1.807 mg/kg) (Fig. 8). Due to their high lipid content, which attracts lipophilic PAHs like naphthalene and phenanthrene. In bariatric patients, PAH concentrations in adipose tissue were 30-50 times higher than in plasma, confirming their bioaccumulation potential.

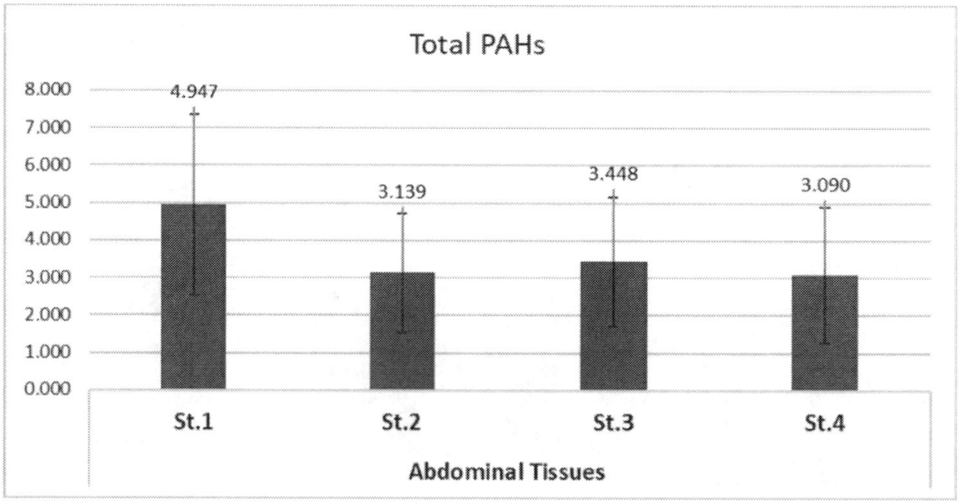

Fig 8. The concentration of total PAHs in the abdominal tissue of wildlife rats in four studies site.

Blood generally shows the lowest PAHs concentrations where the highest was at site one (2.680 ±1.424 mg/kg) and the lowest was at site two (0.502±0.292 mg/kg) (Fig. 9). It serves as a transient transport medium, distributing PAHs to target organs. blood PAHs are rapidly metabolized or redistributed to lipid-rich tissues, explaining their lower persistence compared to organs.

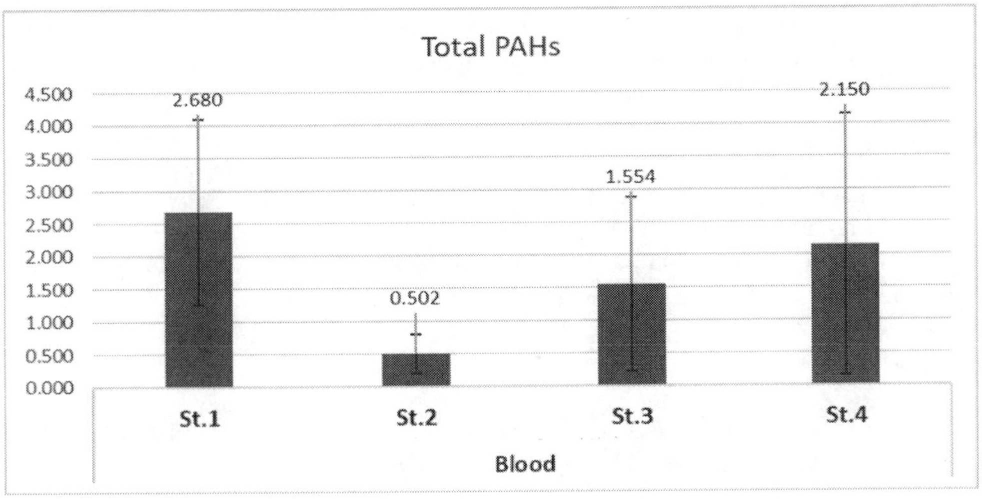

Fig. 9. The concentration of total PAHs in the blood of wildlife rats in four studies site.

The incomplete burning of organic materials like coal, oil, and trash results in the production of Polycyclic aromatic hydrocarbons, a class of organic molecules with two or more rings, Landfills provide an abundant environment for them. Making it possible to accumulate in the liver, kidneys, abdominal tissues, and blood of wildlife animals, rats are exposed to these substances by the inhalation of fumes and airborne particles, through skin absorption from contact with contaminated soil, or through the consumption of contaminated food or water. Research indicates that rats exposed to PAHs exhibit elevated blood levels of urea and creatinine, which suggests compromised kidney function, elevated levels can raise the risk of chronic renal disease and reduce kidney filtration efficiency. Following their entry into the body, PAHs are bio transformed in the liver into more polar metabolites that the kidneys can more easily remove. Some of these metabolites, meanwhile, have the potential to build up in kidney tissue and cause harmful effects. By accumulation in the renal tubules, PAHs directly harm renal cells, Cell death results from the conversion of some of the broken-down substances into active metabolites that are harmful to cells. As a result, our study shows that the concentrations of polycyclic aromatic hydrocarbons normally drop in the blood while their accumulation increases in tissues like the liver and kidneys in the majority of the study regions, The reason is because these substances were thought to be transported by the blood to organs, which have trouble eliminating them, leading to an accumulation in their cells or tissues. Additionally, certain muscle tissues, such the abdominal tissues, have trouble eliminating these substances, which causes their concentrations to rise, particularly in landfills that house industrial waste, as well as the prevalence of gasoline and diesel-powered vehicles in these regions, in contrast to the agricultural regions under investigation, which have low concentrations of these substances because of inadequate disposal of household, industrial, and petroleum waste, which has an impact on the decline in their accumulation in the local wildlife. Investigational findings within this study indicate that within all taxa, biological harm is caused by PAH exposure when in sufficient quantities, However, the extent of this detriment differs significantly between species not only from different taxa but also within the same species level. With this in mind, further study relating to environmental repercussions of PAH pollution is important not only for human health and safety but also for ecosystem security and preservation. Due to the vast number of different PAH fractions found within petroleum hydrocarbons, predicting the extent of damage relating to a single PAH is often negligible to specific organisms. This is because single PAH entities are often in small concentrations within PAH mixtures and assessing the potential damage as a PAH mixture is therefore necessary To further complicate matters. The parent PAH in many cases is unlikely to be the compound found to cause harm.

4 Conclusion

The study suggests accumulation of PAHs were identified in rats collected from landfill. The study made it possible to conduct a more thorough examination of the PAHs levels and the surrounding environment. From the perspective of those who live, work, or intend to invest in places that are quite close to landfills or other garbage disposal facilities, it is crucial.

References

1. A. Iqbal, A. B. Tabinda, A. Yasar, "Environmental risk assessment of a young landfill site and its vicinity for possible human exposure," Human and Ecological Risk Assessment **27(1)**, 258–273 (2019)
2. M. D. Vaverková, "Landfill impacts on the environment - review," Geosciences **9(10)**, 431 (2019)
3. A. G. Alimba, O. O. Adewumi, O. M. Binuyo, P. G. C. Odeigah, "Wild black rats as zoomonitors of genotoxicity and systemic toxicity induced by hazardous emissions from an unsanitary landfill," Environmental Science and Pollution Research **27**, 11325–11337 (2020)
4. G. Przydatek, W. Kanownik, "Impact of a small municipal solid-waste landfill on groundwater quality," Environmental Monitoring and Assessment **191**, 169 (2019)
5. S. Tenodi et al., "Assessment of the environmental impact of sanitary and unsanitary parts of a municipal solid-waste landfill," Journal of Environmental Management **258**, 110019 (2020)
6. Z. Wang, Q. Meng, K. Sun, Z. Wen, "Spatiotemporal distribution, bioaccumulation and health risks of PAHs in surface water: A comprehensive review," Sustainability **16(23)**, 10346 (2024)
7. S. M. Mohammadi et al., "Source identification and health-risk assessment of PAHs in surface soils near a disposal complex (Tehran, Iran)," International Journal of Environmental Analytical Chemistry **103(20)**, 9647–9660 (2023)
8. S. Barathi et al., "Recent trends in PAH pollution distribution and counteracting bioremediation strategies," Journal of Cleaner Production **388**, 138014 (2023)
9. P. Takam et al., "Toxic effect of PAHs on a co-culture model of human alveolar epithelial cells and macrophages," Environmental Sciences Europe **36**, 176 (2024)
10. H. J. Albazoni, M. J. S. Al-Haidarey, A. S. Nasir, "Microplastic pollution: harmful effects on environment and animals, remediation strategies," Journal of Ecological Engineering **25(2)**, 1–15 (2024)
11. X. Jin et al., "Organ- and tissue-specific distribution of selected PAHs in ApoE-KO mice," Environmental Pollution **286**, 117219 (2021)
12. X. Jia et al., "Exposure biomarker profiles of PAHs in a rat model," Environmental Science: Processes & Impacts **26(8)**, 1268–1280 (2024)
13. E. Efriza et al., "Exposure to PAHs and liver function: a systematic review," Air Quality, Atmosphere & Health **16(5)**, 1079–1088 (2023)
14. F. Li et al., "Hepatotoxic effects of inhalation exposure to PAHs on lipid metabolism of C57BL/6 mice," Environment International **134**, 105000 (2020)

15. X. Liu et al., "Source-specific ecological risks of PAHs near oil platforms in the Yellow River Estuary (Bohai Sea)," Marine Pollution Bulletin **207**, 116870 (2024)

E3S Web of Conferences **633**, 01006 (2025) https://doi.org/10.1051/e3sconf/202563301006
GASES 2025

Environmental biological radiation hazards (ELCR, AED) from background radiation in vitamins consumed by pregnant women in Iraq

Shireen N. Alebadi[1*], *Wedyan* Mohammed Torki[2], *Ali* Saeed Jassim[3], and *Hadeel* Kamil Khaleel[4]

[1]Al-Furat Al-Awsat Technical University, Engineering Technical College of Najaf, Department of Health Physics and Radiation Therapy Technical Engineering, Al-Najaf, Iraq
[2]Department of Chemistry, Faculty of Education for Girls, University of Kufa, Al-Najaf, Iraq
[3]University of Kufa,, Al-Najaf, Iraq
[4]Department of Basic Science, College of Dentistry, Al-Iraqia University, Baghdad, Iraq

Abstract. This study aims to assess the background radiation levels of dietary supplement samples utilised by pregnant women in Iraq. The D, AED, and ELCR values for twenty dietary supplements from this region. The D values (in µSv/h) are (0.15±0.003 µSv/h). The radiation dose outputs from the aforementioned investigations are compared with a safe limit of 0.247 µSv/h established by UNSCEAR 2008 and NCRP 2009. The radiation dosage for all samples in the present investigation unequivocally remains within acceptable thresholds. Furthermore, the AED values derived from µSv/h are (1.290±0.026 mSv/y), which exhibit discrepancies. Figure 3 juxtaposes the corresponding dose rates from this study with a permissible threshold of 2.4 mSv/y, as per UNSCEAR 2008 and NCRP 2009. The data indicate that all AED values found during the sample examination remained below safety limits. Finally, the ELCR×10-3 values vary with a mean value of (4.500±0.085). The radiation background levels in these are considered safe, and their use poses no substantial danger of radiation exposure. The dietary supplements evaluated in this study are deemed safe for pregnant women, posing no possible danger of radiation consequences.

1 Introduction

In general, background radiation is considered to be a low-level exposure that, over time, may contribute to cumulative health risks. The scientific community has long investigated the impact of background radiation on human health, specifically its role in the development of cancer and other radiation-induced diseases [1]. Pregnant women represent a particularly vulnerable group when it comes to radiation exposure. During pregnancy, the developing fetus is sensitive to environmental factors, including ionizing radiation. This vulnerability arises because of the rapid growth and division of cells in the fetus, which increases the risk of cellular damage when exposed to radiation. The potential for harm is especially high during critical stages of fetal development, such as the early weeks of pregnancy, when organ

* Corresponding author: sherin.kazem@atu.edu.iq

systems begin to form [2]. Despite these known risks, there is limited research on how background radiation exposure, especially from commonly consumed dietary supplements like vitamins, may affect pregnant women and their unborn children.The consumption of vitamins during pregnancy is widespread, as many expectant mothers rely on supplements to ensure they are meeting the nutritional requirements essential for both their health and that of their developing baby[3]. Common supplements include folic acid, which plays a crucial role in preventing neural tube defects, as well as calcium, iron, and vitamin D, which are necessary for bone health, red blood cell production, and immune function. However, the production and distribution of these supplements can introduce various risks, including contamination by environmental pollutants, heavy metals, and, in some cases, trace amounts of radioactive substances. Vitamins, as part of the food supply, can be exposed to natural background radiation through multiple pathways [4]. For example, plants grown in soils rich in radioactive elements like radon or uranium can absorb these elements into their tissues. Similarly, the processing, packaging, and storage of these vitamins may inadvertently introduce additional risks from radiation exposure [5]. It is therefore critical to understand whether these trace levels of radiation in vitamins pose a significant risk to the health of pregnant women and their developing fetuses. Cancer risk from radiation exposure follows a probabilistic model, meaning that higher radiation doses increase the likelihood of developing cancer, though not in a direct or predictable manner [6]. The risk of developing cancer depends not only on the total dose of radiation received but also on factors such as the age at which exposure occurs, the type of radiation, and the organs that are most affected by the radiation. For pregnant women, the fetus is of particular concern because of its increased susceptibility to radiation-induced birth defects and cancer. As such, understanding the radiobiological hazards posed by background radiation in the vitamins consumed by pregnant women is of paramount importance for public health [7-8].

2 Materials and methods

During the period between March and May 2025, fifteen samples of nutritional supplements commonly used by pregnant women were collected from different pharmacies in the Iraqi market. The samples were taken without interior preparation directly measurement by a portable FS-600 dosimeter (made in China). Three samples were measured per site over the 5 min period. It is noted that the FS-600 dosimeter (Figure 2) has a Geiger-Muller tube inside, with an energy response range of 48 kV to 1.5 MeV and a dose range of 0.1 μSv/h to 10 mSv/h [9]. It is used to sense and measure ionizing radiation, including gamma radiation, X-rays, and beta particles. The manufacturer factory-calibrates the FS-600 dosimeter. Push the on/off button to bring the dosimeter on and the radiation dose (D) displayed on large LCD, the unit is in μSv/h. AED was calculated from the D value obtained by multiplying it by a factor of 8.766 to obtain AED in mSv/yr. The "risk of cancer during excess recovery" was calculated by multiplying the AED by the life expectancy (DL = 70 years) and the conversion factor (CF = 0.05 1/Sv).

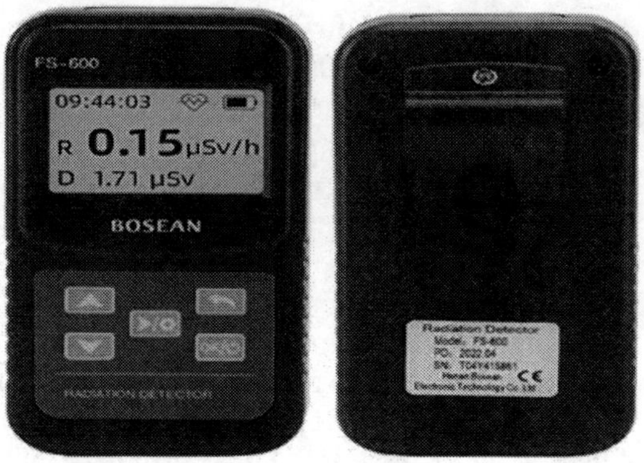

Fig. 1. Portable FS-600 dosimeter.

3 Results

Table 1 shows the D, AED, and ELCR values for 20 dietary supplements from this area. D values (in μSv/h) are (0.15±0.003μSv/h). In Figure 2, the radiation dose outputs from the studies mentioned above are compared, as well as with a safe limit of 0.247 μSv/h from UNSCEAR 2008 and NCRP 2009. The radiation dose for all samples in the current study clearly falls below safe limits. Moreover, the Table 2 shows that the AED values based on the μSv/h are (1.290±0.026 mSv/y) which are different. Figure 3 compares the equivalent dose rates from this study with a safe limit of 2.4 mSv/y, according to UNSCEAR 2008 and NCRP 2009. These findings then suggest that all of the AED values identified during the assessment of the samples studied were still within the safety limits. Lastly, the ELCR×10^{-3} values (Table 1-Figure 4) ranges with the, mean value of (4.500±0.085).

Table 1. Results of D, AED, and ELCR in samples of the present region.

No.	code	D (μSv/h)	AED (mSv/y)	ELCR×10^{-3}
1	Y1	0.11±0.018	1.000	3.500
2	Y2	0.13±0.021	1.140	4.000
3	Y3	0.14±0.022	1.220	4.300
4	Y4	0.12±0.020	1.100	3.900
5	Y5	0.15±0.023	1.290	4.550
6	Y6	0.16±0.024	1.350	4.700
7	Y7	0.17±0.025	1.420	4.850
8	Y8	0.18±0.026	1.480	5.100
9	Y9	0.19±0.027	1.560	5.350
10	Y10	0.16±0.024	1.340	4.750
11	Y11	0.17±0.025	1.410	4.900
12	Y12	0.14±0.023	1.230	4.250
13	Y13	0.15±0.024	1.270	4.400
14	Y14	0.13±0.021	1.140	3.950
15	Y15	0.16±0.025	1.380	4.800
16	Y16	0.15±0.023	1.290	4.550
17	Y17	0.14±0.022	1.220	4.300
18	Y18	0.13±0.021	1.150	4.000
19	Y19	0.14±0.022	1.240	4.350
20	Y20	0.12±0.020	1.120	4.050
Average±S.E		0.15±0.003	1.290±0.026	4.500±0.085

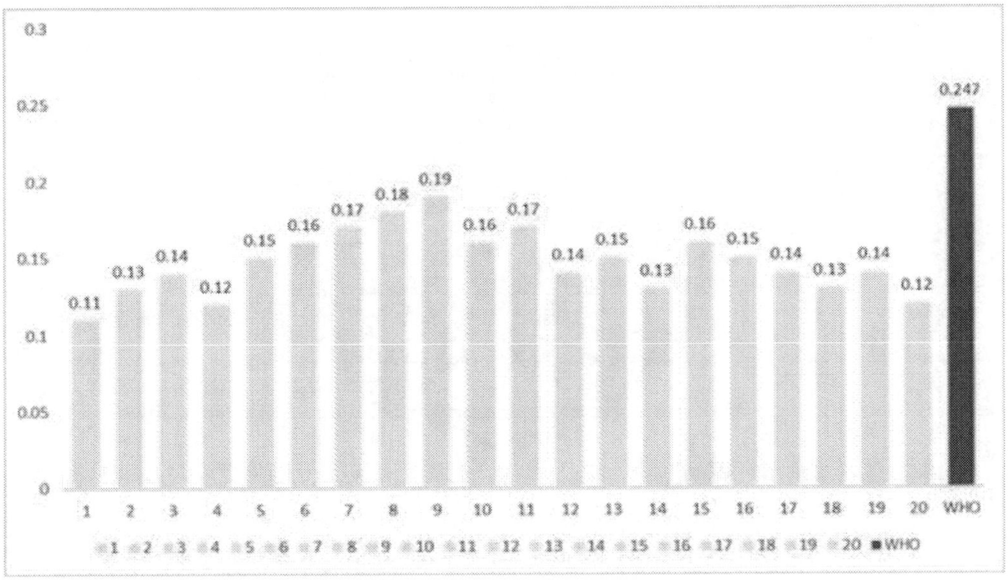

Fig. 2. Comparsion of the resuts of radiation dose with safe limit.

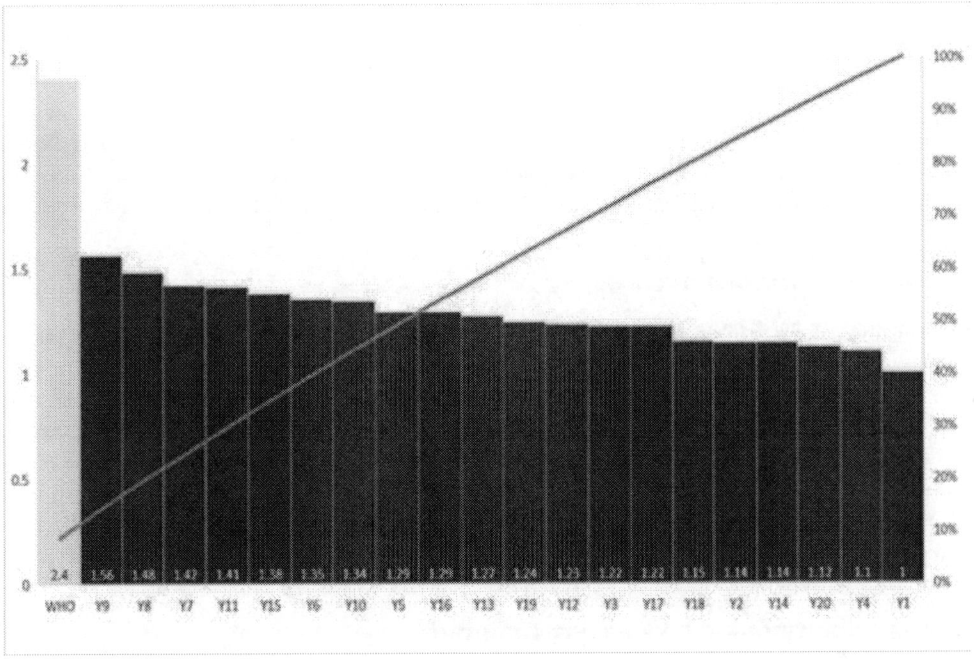

Fig. 3. Comparsion of the results of AED with WHO.

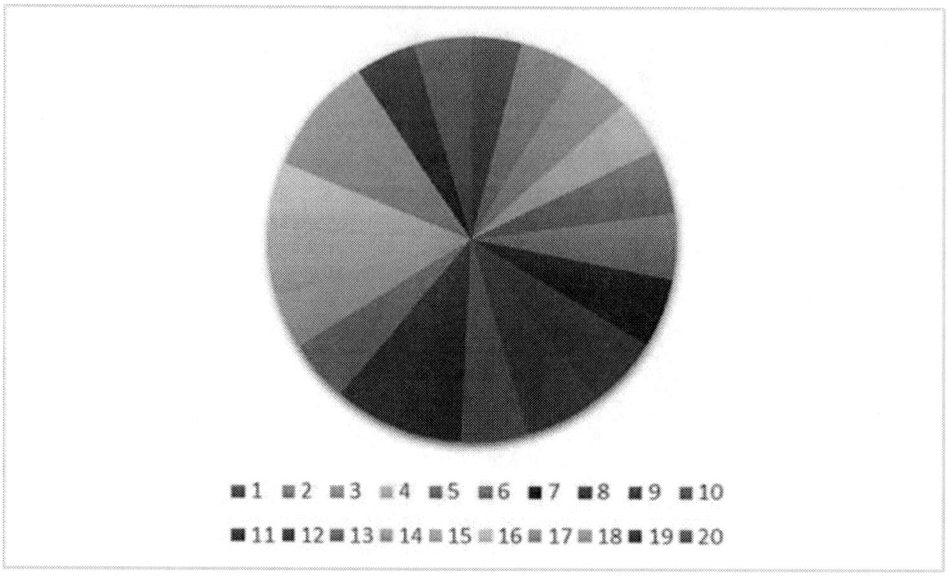

Fig. 4. Results of ELCR in all soil samples in the present study.

4 Discussion

Background Radiation in Dietary Supplements Vitamins.Background radiation levels in vitamins and other dietary supplements vary due to several reasons such as the source of raw materials, the manufacturing processes, environmental Pollution and the testing methods. Herein, we will discuss these causes according to real scientific studies and literature sources.

4.1 Source of raw materials

Dietary supplement radiation levels often depend on the geography from which raw materials are extracted. Natural levels of radiation can change considerably depending on where one lives, owing to the presence of radioactive elements like uranium, thorium and radon in the soil. Supplementation of any type of product can mean that it contains products of the soil to products of the soil, such as through beneficial plants grown and produced in areas with extremely high natural radiation levels. demonstrates that naturally occurring radiation in soil has a direct correlation with the amount of radioactive material found in food crops and dietary supplements from that area

4.2 Manufacturing and processing methods

Processing techniques used to manufacture dietary supplements may also affect radiation levels. For specific preservation or sterilization purposes, radiation if for example introduced during the irradiation of certain ingredients If high-energy radiation penetrated the materials during processing, raw materials may have absorbed radiation and contaminate processed products.

4.3 Packaging materials

Finally, some packaging materials used for dietary supplements may also leach radioactive elements or add up over time in a source of exposure to radiation. Some plastics and containers are known to absorb and release radiation, which may cause a build-up of the radiation level in the product over time.

5 Conclusion

In light of the results presented in the Results and Discussion sections it can be deduced that the range of radiation dose and annual effective dose for all the used nutritional supplements by pregnant women were below the acceptable levels and hence were deemed safe according to the Scientific Committee on the Effects of Atomic Radiation (2008) and the National Committee for Nuclear Research (2009). These Call for standards establish guidelines for radiation levels above which they are not permitted to be present in samples. Consequently, the tested samples are safe in terms of their radiation background (the non-significance of health risk to the consumers of these supplements). Hence, these nutritional supplements are not a serious radiation hazard to pregnant women.

References

1. R. Jostes, S. G. Robertson, *Environmental factors and ionizing radiation: The risks to pregnancy* (2015)
2. A. A. Kareem, H. N. Hady, A. A. Abojassim, "Measurement of natural radioactivity in selected samples of medical plants in Iraq," International Journal of Physical Sciences **11(14)**, 178–182 (2016)
3. L. A. Najam, N. F. Tawfiq, E. M. Younis, I. M. A. Aziz, "Uranium concentration in some medical herbs," Iraqi Journal of Science **61(3)**, 528–532 (2020)
4. M. M. Kadhim, M. K. Khudhair, H. A. A. B. Mraity, A. A. Abojassim, "Cancer-risk estimation in samples of Iraqi-manufactured medicinal plants," Journal of Pharmaceutical Negative Results **13**, 1416–1421 (2022)
5. J. M. Hernández, A. González, I. Rodríguez, "The influence of soil radioactivity on the concentration of radionuclides in plants and food crops," Science of the Total Environment **659**, 1205–1212 (2019)
6. S. Kahraman, N. Oğuz, D. Yılmaz, "Irradiation of foodstuffs and the risk of radioactive contamination in food products," Food Control **110**, 106928 (2020)
7. R. Muzzalupo, R. Ciriello, R. Luppino, "Impact of food packaging materials on radiation exposure in dietary supplements," Food Research International **89**, 110–118 (2016)
8. N. Kržanović et al., "Development and testing of a low cost radiation protection instrument based on an energy compensated Geiger-Müller tube," Radiation Physics and Chemistry **164**, 108358 (2019)
9. Y. Wei, S. A. Dewji, "A comprehensive review of dose limits, triage systems and measurement tools for consequence management of nuclear and radiological emergencies," Radiation Physics and Chemistry **217**, 111533 (2024)
10. N. N. Ataalla, "Evaluation of Non-Healthcare Professional Women's Knowledge Regarding Ionizing Radiation Exposure Hazards during Pregnancy", Journal of Radiation Research and Applied Sciences **17**, 100794 (2024), DOI: 10.1016/j.jrras.2023.100794

^{218}Po and ^{214}Po in pharmaceutical materials derived from medical plant, Iraq

Dhuha Jalil Lawi[1], *Ali Abid* Abojassim[2]*, *Amin K.* Awad[1], *Michele* Guida[3], and *Simona* Mancini[3]

[1]Radiology Techniques Department, College of Health and Medical Techniques Al-Mustaqbal University, 51001 Babylon, Iraq
[2]Department of Physics, Faculty of Science, University of Kufa, Al-Najaf, Iraq
[3]Department of Information and Electric Engineering and Applied Mathematics, Ambra. Lab-Laboratory of Ambients and Radiation, University of Salerno, Via Giovanni Paolo II, 132, I-84084 Fisciano, Italy

Abstract. In the present work, POW (Polonium-218 and Polonium-218 on the wall) and POS (Polonium-218 and Polonium-218 on the surface) that were deposited in the irradiation container were determined for 70 samples of pharmaceutical materials made from medical plants. Pharmaceutical materials were collected from the Iraqi pharmacies and divided into six groups: drugs (solid), skin creams, herbs, toothpaste, drugs (liquid), and cosmetic products POW and POS were measured using track detector type CR-39. The average values of POW in drugs (solid), skin creams, herbs, toothpaste, drugs (liquid), and cosmetic products samples were 9.70±2.18 Bq/m3, 14.93±4.01 Bq/m3, 6.84±1.02 Bq/m3, 15.23±3.98 Bq/m3, 17.21±4.33 Bq/m3, and 5.10±0.88 Bq/m3, respectively. While the average values POS in the six groups of the present study were 10.06±2.25 Bq/m3, 15.47±4.15 Bq/m3, 7.09±1.06 Bq/m3, 15.78±4.12 Bq/m3, 17.83±4.49 Bq/m3, and 5.29±0.92 Bq/m3, respectively. The data in the present study showed that the concentrations of Polonium were lower than the maximum permissible limit (550 Bq/m3) according to WHO. So, Pharmaceutical materials in Iraqi pharmacies were safe.

1 Introduction

The interest that was given to the medical plant topic can be greatly attributed to the fact that it is widely regarded as the primary source of medicine. Ever since ancient times, plants have undoubtedly been used by mankind in food, to be utilized as a cure to alleviate pain and combat diverse illnesses [1]. Many recent studies indicate the possibility of medicinal plants being contaminated with radioactive pollutants [2, 3]. Numerous sources have been identified as contributors to plant contamination. The contamination incidents result from the accumulation of radioactive substances in the soil. Most contaminants are naturally absorbed by plants via root absorption and easily transported upward by stems, leaves, seeds, and flowers. The radioactive elements widely spread in soil originate from naturally occurring

* Corresponding author: ali.alhameedawi@uokufa.edu.iq

elements such as uranium, thorium, and potassium or human-made elements such as cesium. Also, concerning the naturally radioactive elements polonium is seen as the main significant emitters of alpha concerning the internal exposition. Alpha radiation depends on geological and geographical variables in the natural environment. They manifest at varying depths inside the soil of each region worldwide. The activity concentration of naturally occurring radioactive elements (radionuclides) in soil and plants is elevated in agricultural regions compared to uncultivated areas, attributable to the influence of fertilizers on the soil. Polonium-214 and Polonium-218 are significant since they are the progeny of radon and account for a considerable fraction of the radiation exposure from breathed radon. They emit alpha particles. Their half-lives are so brief that no experimental biological research can be conducted with them, necessitating the inference of their effects from isotopes with longer half-lives. The penetration distance of alpha particles is contingent upon the energy at which they are released and the substance they traverse. Consequently, alpha particles may be halted or absorbed by thin barriers, including a layer of tissue or skin. Alpha particle emitters are very detrimental and hazardous to human health when consumed, breathed, swallowed, or absorbed into the lungs or circulation. Polonium may enter the body by inhalation, ingestion, or via an open wound. Inhaling or ingesting substantial amounts within a short timeframe may result in radiation exposure to internal organs. This may lead to severe medical symptoms or fatality [4]. Polonium concentrations were measured in different materials in the world [5]. Overall aim of this study is to determine 214Po and 218Po concentrations in samples of pharmaceutical materials derived from medical plants which it was collected from Iraqi pharmacies using nuclear track technique type CR-39 from Tasl company.

2 Materials and methods

2.1 Collection of samples

Table 1 is shown the information of 70 samples that divided in to five groups of pharmaceutical materials derived from medical plants. Pharmaceutical materials samples were collected from local Pharmacy in Al-Najaf governorate during 2023-2024.

Table 1. Types of pharmaceutical materials in the present study.

No.	Type of pharmaceutical materials	Number of samples	State
1	Drugs	28	Solid+ Liquid
2	Skin Creams	13	Ointment
3	Herbs	9	Solid
4	Toothpastes	10	Creme
5	Cosmetic Products	10	Crème + Ointment

2.2 Preparation of samples

In this investigation, pharmaceutical material samples collected from several local pharmacies in Iraq were packaged in labelled plastic bags to facilitate the measurement of polonium concentrations. Solid samples were prepared for examination by drying them in an oven at 70 °C for 0.5 hours to ensure they remained moisture-free. The samples are mechanically pulverized using the electric mill to achieve appropriate homogeneity. To achieve homogeneity, the samples were subjected to sieving through a 0.8 mm pore diameter. Ointment, liquid, and cream samples were measured immediately without any preparation. The net weights are measured and documented using a very sensitive digital weighing balance with an accuracy of ±0.01%. The samples were then positioned in a plastic cup with a diameter of 3.5 cm, a length of 7 cm, and a capacity of 130 ml. All samples were preserved for about one month prior to counting, to facilitate the establishment of secular equilibrium between 222Rn and its progenitor 226Ra in the uranium decay series.

A 1 mm thick CR-39 detector, manufactured in the UK, was used to measure polonium quantities in the materials under investigation. Each detector was sectioned to a dimension of (1×1) cm² and positioned in the top section of the containers. A segment of the cut CR-39 detector was positioned at the base of each cylinder lid, accompanied by samples at the bottom of the cylinder, and thereafter sealed at ambient temperature for a duration of 90 days of exposure. Upon completion of the exposure duration, the CR-39 detector is subjected to etching in a sodium hydroxide (NaOH) solution at a temperature of 98 °C for one hour, with a solution of 6.25 normality. Subsequently, they were washed with distilled water. The track density on CR-39 (Tr/cm²) was quantified using an optical microscope with a magnification of 10 x 40. The background correction was assessed by subtracting the background from the recorded alpha track density.

2.3 Calculations

POW and POS mean 218Po and 214Po particles that were deposited on the wall and the surface of the irradiation chamber respectively. They calculated using follow equations: [5]:

$$D_{P218} = D_{P214} = \frac{(^{\rho}/_{Kt})}{4} r \left(\frac{r}{r+h}\right) cos\theta_C \quad (1)$$

$$D_{P218} = D_{P214} = \frac{(^{\rho}/_{Kt})}{4} r \left(\frac{r}{r+h}\right) \left(cos\theta_C \frac{r}{R_\alpha}\right) \quad (2)$$

where, ρ denotes the track density on the detector (Tr/cm²), t represents the exposure duration (90 days), and K signifies the calibration factor (0.2917±0.0511 track/cm² per Bq/m³/day) In reference, h denotes the distance from the sample surface to the detector, r represents the radius of the exposure tube, Rα indicates the typical alpha particle range in air for radon-222, which is 4.15 cm, and θC signifies the average critical angle for CR-39 detectors, measured at 15.

3 Results and discussion

Tables 2, 3, 4, 5, and 6 present the results of 218Po and 214Po POW and POS concentrations in drugs, skin creams, herbs, toothpaste, and cosmetic products samples of the types of pharmaceutical materials in the present study, respectively. From Table 2, the values of POW and POS (unit Bq/m3) in drug samples ranged from 1.42 to 46.12 and from 1.47 to 47.78, respectively. In skin creams samples (Table 3), the range values of POW and POS (unit Bq/m3) were 3.22 – 58.15 and 3.33 – 60.25, respectively. While, from Table 4. The results of POW and POS in herbs samples were varied from 1.87 Bq/m3 to 11.06 Bq/m3 and from 1.93 Bq/m3 to 11.46 Bq/m3, respectively. As well as from Table 5, it is found that the values of POW and POS (unit Bq/m3) in toothpastes (type Dabur) were ranged from 2.32 to 36.92 and from 2.4 to 38.25, respectively. But, for cosmetic products samples, the results of POW and POS (unit Bq/m3) were ranged from 1.61 to 11.13 and from 1.67 to 11.53, respectively. Also, from Tables 2, 3, 4, 5, and 6, the maximum value of POW and POS for five groups in the present study were found in samples D24 (Fricks), sample S11 (Pilex), sample H5 (Antispasmodic), sample T4 (Olive), and sample C8 (Deniz), respectively. While, the minimum value of POW and POS were found in samples D3 (Prostamed), sample S4 (Colocynth), sample H2 (Slimming), sample T7 (Fleur), and sample C5 (Plant coal), respectively.

Table 2. Results 218Po and 214Po in drug (solid) samples of the present study.

No.	State	Sample name	Sample code	218Po and 214Po **(POW) Bq/m3**	218Po and 214Po **(POS) Bq/m3**
1	solid	Bran	D1	11.51	11.93
2		Speman	D2	6.95	7.20
3		Prostamed	D3	1.42	1.47
4		Mada Garlic	D4	2.32	2.40
5		Colon cleanse	D5	23.09	23.93
6		Acaiberry	D6	11.51	11.93
7		Cranberry Ama	D7	6.95	7.20
8		Carbone	D8	9.20	9.53
9		LIV- 52	D9	34.60	35.86
10		Cystone	D10	4.63	4.80
11		Pilex tablets	D11	5.79	6.00
12		Agiolax	D12	11.51	11.93
13		Fat Burner	D13	4.63	4.80
14		Glugose	D14	2.32	2.40
15		Kellagon	D15	9.20	9.53
16	liquid	Zecuf	D16	16.14	16.73
17		Mucosal	D17	4.63	4.80
18		Ivylix	D18	9.20	9.53
19		Renalka	D19	4.63	4.80
20		Melrosum	D20	2.32	2.40
21		Plusspan	D21	13.83	14.33
22		Bronchicum	D22	41.55	43.05
23		Beecuf	D23	11.51	11.93
24		Fricks	D24	46.12	47.78
25		Gribe water	D25	16.14	16.73
26		Koflet	D26	6.95	7.20
27		Razipan	D27	46.12	47.78
28		Bonnisan	D28	4.63	4.80

Table 3. Results 218Po and 214Po in **skin creams** samples of the present study.

No.	Sample name	Sample code	218Po and 214Po **(POW) Bq/m3**	218Po and 214Po **(POS) Bq/m3**
1	Mebo	S1	16.14	16.73
2	Sharko	S2	12.93	13.40
3	Happy lady	S3	4.82	5.00
4	Colocynth	S4	3.22	3.33
5	Her devil	S5	7.85	8.13
6	Ectros	S6	9.39	9.73
7	Nsn	S7	9.07	9.40
8	Neohealar	S8	8.30	8.60
9	Vesca forte	S9	4.50	4.67
10	Glycerin	S10	32.29	33.46
11	Pilex	S11	58.15	60.25
12	Pregnacare	S12	17.75	18.39
13	True heal	S13	9.71	10.06

Table 4. Results 218Po and 214Po in **herbs** samples of the present study.

No.	Sample name	Sample code	218Po and 214Po **(POW) Bq/m3**	218Po and 214Po **(POS) Bq/m3**
1	Natmucil	H1	10.16	10.53
2	Slimming	H2	1.87	1.93
3	Diuretics	H3	9.20	9.53
4	Green tea	H4	7.40	7.66
5	Antispasmodic	H5	11.06	11.46
6	Tranquilizer	H6	8.30	8.60
7	Renal stone	H7	4.63	4.80
8	Chamain	H8	6.43	6.66
9	Instant Herbs	H9	2.57	2.67

Table 5. Results 218Po and 214Po in **toothpastes (type** Dabur) samples of the present study.

No.	Sample name	Sample code	218Po and 214Po **(POW) Bq/m3**	218Po and 214Po **(POS) Bq/m3**
1	Fennel	T1	27.66	28.66
2	Neem	T2	2.32	2.40
3	Basil	T3	4.63	4.80
4	Olive	T4	36.92	38.25
5	Chamomile	T5	6.95	7.20
6	Seaweed	T6	4.63	4.80
7	Fleur	T7	2.32	2.40
8	Blackseed	T8	25.41	26.32
9	Salt	T9	29.97	31.06
10	Aloe vera	T10	11.51	11.93

Table 6. Results 218Po and 214Po in **cosmetic products** samples of the present study.

No.	Sample name	Sample code	218Po and 214Po **(POW) Bq/m3**	218Po and 214Po **(POS) Bq/m3**
1	Bio balance	C1	1.61	1.67
2	Lavender	C2	4.76	4.93
3	Carnier	C3	3.15	3.27
4	Lavender	C4	6.37	6.60
5	Plant coal	C5	1.61	1.67
6	Castor oil	C6	2.70	2.80
7	Hand cream	C7	6.37	6.60
8	Deniz	C8	11.13	11.53
9	Vatika	C9	6.37	6.60
10	Yougee	C10	7.01	7.26

The results of average values with standard error of POW and POS (Unit Bq/m3) in all types of pharmaceutical materials in the present study are presented in Table (7) and Figure 1. Hence, the average values of drug solids samples were 9.70±2.18 and 10.06±2.25, respectively, while in drugs of liquid samples were 17.21±4.33 and 17.83±4.49, respectively. Also, the average values of POW and POS (Unit Bq/m3) in skin creams were 14.93±4.01 and 15.47±4.15, respectively. In herbs samples, they were 6.84±1.02 and 7.09±1.06 respectively. In Toothpaste samples, they were 15.23±3.98 and 15.78±4.12, respectively. While, the average values of POW and POS (Unit Bq/m3) in cosmetic products were 15.10±0.88 and 5.29±0.92, respectively.

Table 7. Average values of POW and POS in all type's samples of the present study.

No.	Type of sample		218Po and 214Po **(POW)** **Bq/m3**	218Po and 214Po **(POS)** **Bq/m3**
1	Drugs	Solid	9.70±2.18	10.06±2.25
		Liquid	17.21±4.33	17.83±4.49
2	Skin Creams		14.93±4.01	15.47±4.15
3	Herbs		6.84±1.02	7.09±1.06
4	Toothpaste		15.23±3.98	15.78±4.12
5	Cosmetic Products		5.10±0.88	5.29±0.92

From Figure (1) in the present samples, it is noted that the values of POW and POS concentrations in all types of samples were arranged as flowing drug liquid 4 Toothpaste 4 Skin Creams 4 drug solids 4 Herbs 4 Cosmetic Products. This arrangement may be because the specific activity differs geologically from soil that was planted or grown for the medial plant in samples under study. The difference in polonium concentrations may be attributed to the diverse constituents of these medications since they are derived from plant sources. Radiation contamination may also be directly attributed to the absorption of radionuclides from the atmosphere. The absorption of radionuclides by plants varies based on soil composition, the specific plant species, and the kind of fertilizer used. Therefore, radioactive contamination from the facility is anticipated. The polonium contents for all examined medications were far below the acceptable limit values, which are 550 Bq/m3 in liquid samples as per WHO 2008 [6] and 7500 Bq/m3 in air according to U.S. NRC 2008 [7-10].

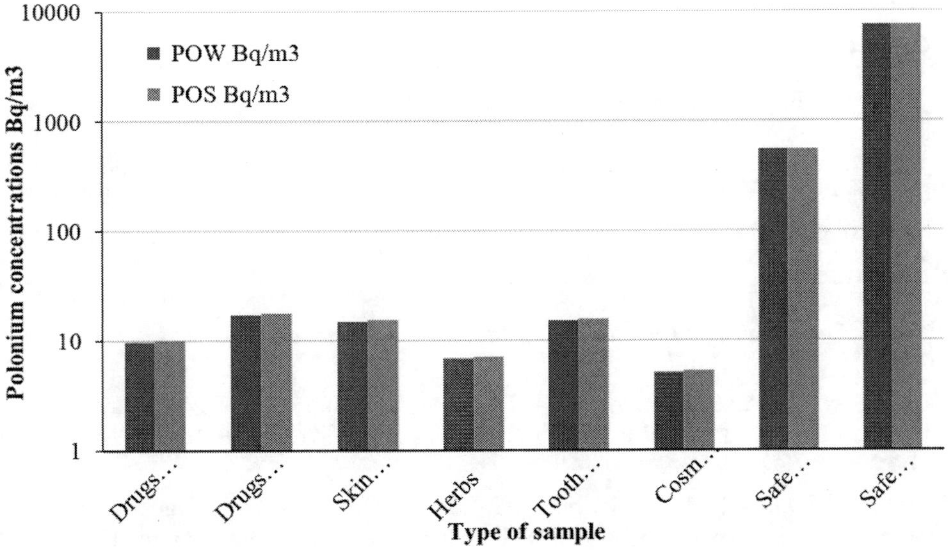

Fig. 1. Comparing the results POW and POS in pharmaceutical materials samples with safe limit.

Figure 2 illustrates the boxplot of the POW and Pos findings over all examined samples. Boxplots (whisker plots) are beneficial since they provide a visual overview of the data, enabling researchers to quickly ascertain mean values, data set dispersion, and indicators of skewness. Upon examining Figure 2 of the boxplot, it is seen that the outlier is among the ten samples situated beyond the whiskers of the boxplot.

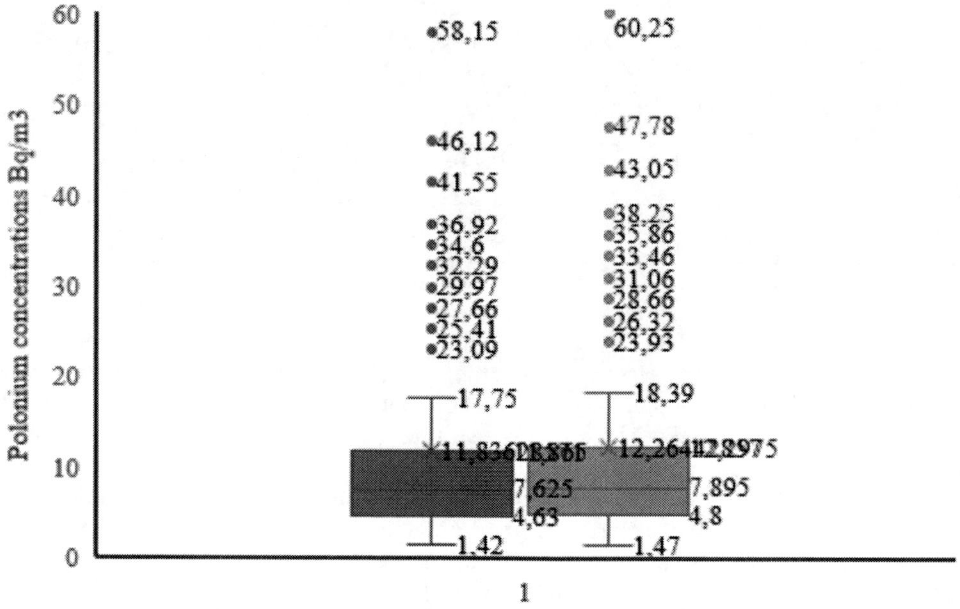

Fig. 2. Box plot of POW and POS for all pharmaceutical materials samples under study.

4 Conclusions

The results of polonium concentration on the wall and the surface deposited in the irradiation container for 70 samples of pharmaceutical materials derived from medical plants constituted only a very small percentage of the global permissible limit. Therefore, in light of human health, all pharmaceutical materials in Local Iraqi pharmacies have no radiological hazards.

References

1. M. F. L'Annunziata, *Radioactivity: introduction and history, from the quantum to quarks* (Elsevier, 2016)
2. D. J. Lawi, R. A. Hayder, A. A. Abojassim, "Uranium Isotopes in Pharmaceuticals Materials Samples Derived from Medical Plants in Iraq," Library Progress International **44(2s)**, 880–892 (2024)
3. E. H. Abdelfadeel, E. A. El-Halim, T. M. Hegazy, H. A. Ghany, "Relationship between radioactivity and toxicity in some medicinal plants," Scientific Reports **13(1)**, 10952 (2023)
4. M. K. M. A. Kadhim, M. Kadhum Khudhair, H. A. A. Bakir Mraity, A. Abojassim, "Cancer Risk Estimation in the Samples of Iraqi Manufactured Medicinal Plants", Journal of Pharmaceutical Negative Results, **10(2),** 1416–1421 (2022), DOI: 10.47750/pnr.2022.13.s06.187
5. A. A. Abojassim, D. J. Lawi, A. B. Hassan, "Radiological Risk to Inhalation of Thoron Gas from Medical Materials Samples Derived from Medical Plants in Iraq", Atom Indonesia **1**, 49–54 (2024), DOI: 10.55981/aij.2024.1329

6. F. F. Kaddoori, B. S. Hameed, W. T. Fzaa, "Concentrations and Radiation Hazard Indices of Naturally Radioactive Materials for Flour Samples in Baghdad Markets", Baghdad Science Journal **18**, 0649 (2021), DOI: 10.21123/bsj.2021.18.3.0649
7. A. A. Abojassim, A. S. Jassim, H. M. Ahmed, H. H. Hussian, "Assessment of Radon, Radium, and Uranium Concentrations in Decorative Materials (Used to Walls) Samples in Iraqi Markets", WSEAS Transactions on Environment and Development **17**, 1210–1218 (2021), DOI: 10.37394/232015.2021.17.110
8. S. Singh, M. Oswal, B. R. Behera, "Investigation on Major, Minor and Trace Elements in Some Medicinal Plants Using Particle Induced X-Ray Emission", Journal of Radioanalytical and Nuclear Chemistry **323**, 1443–1449 (2020), DOI: 10.1007/s10967-020-07032-9
9. A. Ja. Azeez, A. A. Marzaali, A. A. Abojassim, "Determining the Radium Concentration in Vegetables and Fruits in Al-Najaf, Iraq", Online Journal of Animal and Feed Research **27**, 10–19 (2022), DOI: 10.51227/ojafr.2022.29
10. E. Moein, H. Hajimehdipoor, T. Toliyat et al., "Formulation of an Aloe-Based Product According to Iranian Traditional Medicine and Development of Its Analysis Method", DARU Journal of Pharmaceutical Sciences **25**, 1–9 (2017), DOI: 10.1186/s40199-017-0185-x

Environmental background radiation for some drug samples used by children in Iraq

Wedyan Mohammed Torki[1], *Ali* Saeed Jassim[2], *Ali* Abid Abojassim[3], *Ahmed* Abdul Hadi Mohsen[4], and *Rand* A. Hayder[5,*]

[1]Department of Chemistry, Faculty of Education for Girls, University of Kufa, Najaf, Iraq
[2,3]University of Kufa, Najaf, Iraq
[4]Jabir ibn Hayyan Medical University, College of Medicine, Pediatric Department, Al-Najaf, Iraq
[5]College of Dentistry, Al-Iraqia University, Baghdad, Iraq

Abstract. This study measures at the radiation levels in antibiotic samples commonly used for children in Iraq. This study used the FS-600 dosimeter. Antibiotic samples' radiation dose (D) and yearly effective dose (AED) were tested. The dose rate (μSv/h) ranged from 0.11 μSv/h to 0.19 μSv/h, with an average of 0.127, give or take 0.029 μSv/h. The AED values varied between 0.73 mSv/y and 1.67 mSv/y, with an average of 1.116 mSv/y plus or minus 0.250 mSv/y. The radiation levels and yearly exposure from all drug samples used for children are safe and below the limits set by UNSCEAR 2008 and NCRP 2009. So, the tested samples are considered safe in terms of radiation level and do not pose a significant risk.

1 Introduction

Radiation may make several effects on living creatures, including the formation of ionizing molecules inside the cells that create the body. This may greatly change human genes, and these changes are probably going to be passed on to children. Humans can experience harm and risks from radiation based on the amount of exposure, how long they are exposed, and their individual sensitivity. Alpha particles are made up of two protons and two tightly connected neutrons. They come from the center of an atom during nuclear breakdown. Alpha particles can't go very far and can be stopped by a piece of paper or the outer skin. When particles are absorbed or inhaled, they can gather alpha particles in specific cells, which then release their energy in those areas. Alpha particles can be released by some children's medicines, which can lead to serious problems if not carefully monitored. The ionization caused by alpha particles can hurt cellular DNA, resulting in mutations and perhaps cancer. This vulnerability is especially important for young patients because their bodies are smaller and their systems are still developing. Eating or inhaling radionuclides can cause serious health problems, particularly lung cancer, mainly due to exposure to alpha particles. Additionally, the levels of uranium, radon, and radium in plants could increase when people consume them, leading to serious harm. It is important to check the number of radioactive isotopes in plant medicines before using them to keep people safe from radiation

* Corresponding author: rand.a.hayder@aliraqia.edu.iq

exposure. [1, 2]. Understanding it is crucial. Particularly those linked to detrimental impacts on genetic integrity, health, or physical aesthetics. In this context, one may assert that the increase in background radiation is a form of radioactive pollution. This essay examines the impact of alpha particle emission on children. Consequently, investigations into the impacts of radiation, its detection, and the accompanying pollution levels are critically essential. Consequently, several research and studies have been undertaken, along with various methodologies for quantifying the presence of radioactive substances in diverse media, including pharmaceuticals. The importance of understanding the impact of radioactive chemicals in these materials on biological organisms has been acknowledged. Numerous investigations have been conducted on alpha emitters in various medication formulations. Alpha particles emitted from specific anti-inflammatory medications administered to Iraqi children provide significant health hazards that need the attention and intervention of several stakeholders. Identifying alpha particle-emitting pharmaceuticals and evaluating their possible health hazards is essential for mitigating these concerns.

Fifteen samples of medicines for children of different ages were collected from different shops in Iraq from November to December 2023. Background radiation in the drug samples was measured using a compact FS-600 dosimeter (BOSEAN, China), as shown in Figure 1. The FS-600 is a small and precise device for detecting X-ray, gamma-ray, and beta-ray energy [3, 4]. It uses an energy-compensated Geiger-Muller (GM) tube as the detector, which allows for accurate measurements over a broad range. The FS-600 has a 32-bit computer and a 64×128 dot matrix LCD, making it easy to use and able to fight interference. It has several features, including a measuring range from 0.1μSv/h to 10mSv/h, a treatment range from 0.00μSv to 10Sv, sensitivity greater than 1 count per second per μSv/h (compared to 60Co), and an energy response range from 48keV to 1.5MeV. It is used for monitoring naturally occurring radioactive material (NORM) contamination, gross wipe counts, contamination inspection of goods, tools, and people, regulatory checks, and low-energy radionuclide detection [4]. The production company adjusted the portable FS-600.

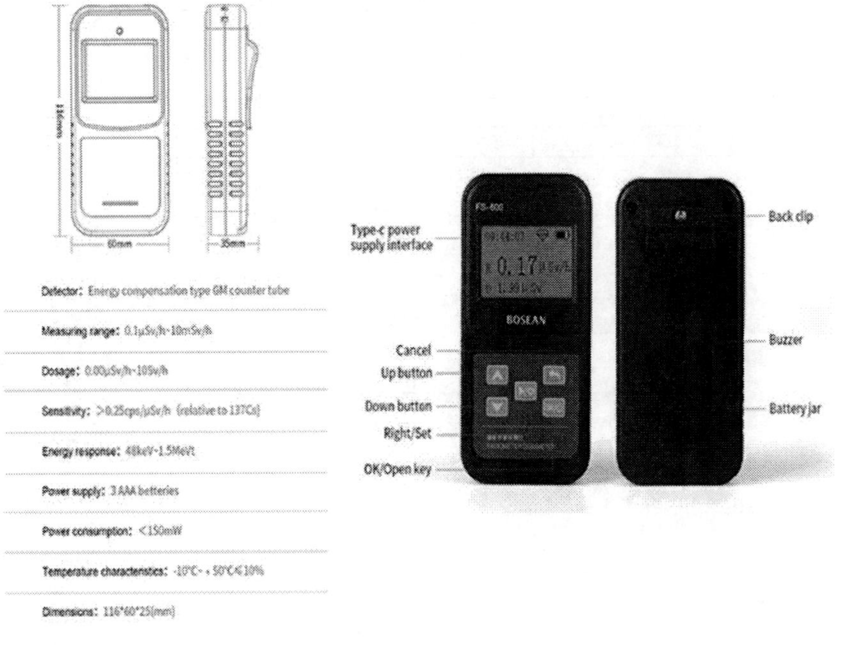

Fig. 1. Portable radiation **FS-600** dosimeter dosimeter.

2 Results and discussion

Table No. 1 shows the results of the Dose rate (μSv/h) which ranged from 0.11(μSv/h) to 0.19 (μSv/h) at Averages of 0.127±0.029 (μSv/h) Figure 2 shows the comparison of the results of radiation dose in all the present studies with safe limits (0.247 μSv/h) according to UNSCEAR 2008 [5] and NCRP 2009. Therefore, it is noted that the results of radiation dose in all samples of the present study were within the safe limits.. and Table No. 2 shows the results of the AED (mSv/y) which ranged from 0.73 (mSv/y) to 1.67 (mSv/y) at Averages of 1.116±0.250 (mSv/y)Figure 3 shows the comparison of the results of AED in all the present studies with safe limits (2.4 mSv/y) according to UNSCEAR 2008 [5]. Therefore, it is noted that the results of radiation dose in all samples of the present study were within the safe limits.The difference in background radiation in antibiotics can be attributed to their chemical composition, mode of action, and the presence of radioactive isotopes or impurities during production. Different antibiotics have unique molecular structures and may contain elements that naturally emit radiation or have higher atomic numbers, which could lead to increased background radiation levels [6]. The manufacturing process for antibiotics might involve raw materials that are contaminated with radioactive isotopes or might utilize equipment that emits low radiation levels. This could lead to variations in the final product's background radiation levels. The geographical location where an antibiotic is produced can also play a role; areas with higher natural background radiation will produce antibiotics with higher levels of residual radioactivity. In summary, the differences in background radiation associated with various antibiotics stem from their chemical properties, production methods, and environmental influences [7-9].

Table 1. Results Dose rate (μSv/h) in children's antibiotics used in this study.

No.	Sample code	Country	Dose rate (μSv/h)
1	F1	Iraq	0.14
2	F2	Egypt	0.16
3	F3	India	0.21
4	F4	Iraq	0.22
5	F5	India	0.15
6	F6	Egypt	0.11
7	F7	Jordan	0.19
8	F8	Turkey	0.12
9	F9	India	0.13
10	F10	Saudi Arabia	0.18
11	F11	Germany	0.20
12	F12	Iraq	0.12
13	F13	Turkey	0.17
14	F14	Ireland	0.15
15	F15	Iraq	0.14
Average±S.D.			0.127±0.029

Table 2. Results AED (mSv/y) in children's antibiotics used in this study.

No.	Sample code	Country	AED (mSv/y)
1	F1	Iraq	0.73
2	F2	Egypt	1.20
3	F3	India	0.99
4	F4	Iraq	1.15
5	F5	India	0.83
6	F6	Egypt	1.15
7	F7	Jordan	0.78
8	F8	Turkey	1.04

Continuation of Table 2.

No.	Sample code	Country	AED (mSv/y)
9	F9	India	0.94
10	F10	Saudi Arabia	1.36
11	F11	Germany	1.20
12	F12	Iraq	1.67
13	F13	Turkey	0.99
14	F14	Ireland	1.25
15	F15	Iraq	1.46
Average±S.D.			1.116±0.250

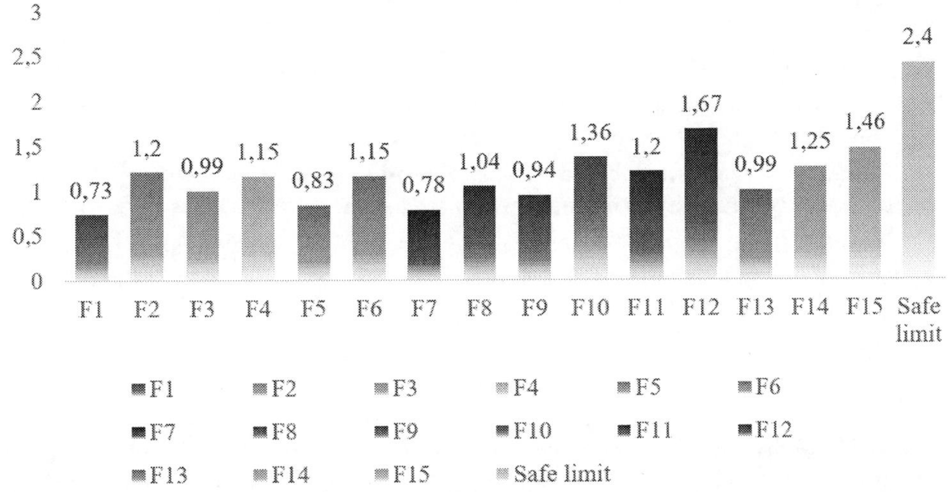

Fig. 2. Comparison of the results Dose rate (μSv/h) Drug Samples Used by Children with the world average.

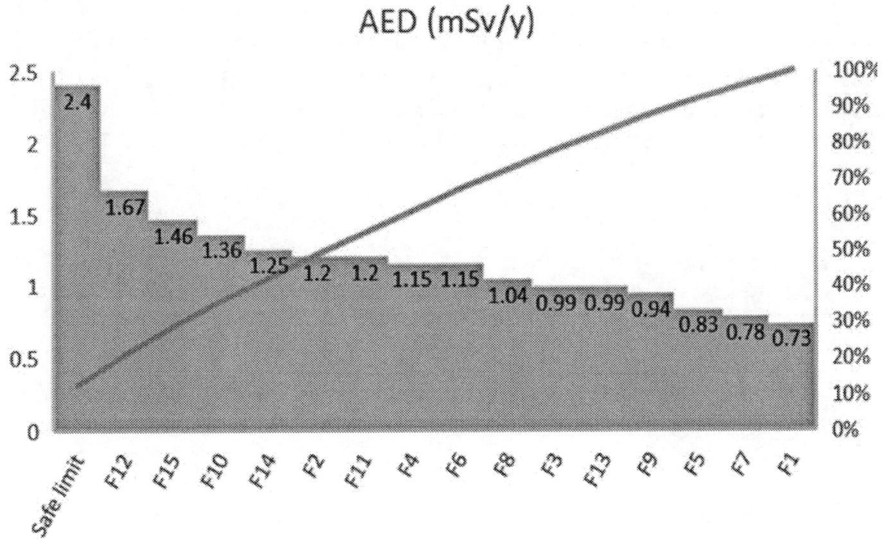

Fig. 3. Comparison of the results AED (mSv/y) Drug Samples Used by Children with the world average.

https://doi.org/10.1051/e3sconf/202563301008

3 Conclusions

In light of the results shown in the Results and Discussion section, it can be concluded that the radiation dose and annual effective dose results in all antibiotic samples used for children are within the safe limit according to UNSCEAR 2008. Therefore, the studied samples can be considered safe in terms of radiation background and do not pose a significant risk.

References

1. S. S. Aithal, I. Sachdeva, Om. P. Kurmi, "Air Quality and Respiratory Health in Children", Breathe **19**, 230040 (2023), DOI: 10.1183/20734735.0040-2023
2. A. Grzywa-Celińska, A. Krusiński, J. Mazur, K. Szewczyk, K. Kozak, "Radon-the element of risk. The impact of radon exposure on human health," Toxics **8(4)**, 120 (2020)
3. BOSEAN Company Inc., Reference Manual version 1, FS-600 Geiger Counter Nuclear Radiation Detector (BOSEAN Company Inc., 2024)
4. H. U. Emelue, *Measurement of background ionizing radiation in the Federal University of Technology Owerri, Nigeria using calibrated digital Geiger counter* (2020)
5. J. Wang, R. Zhuan, L. Chu, "The occurrence, distribution and degradation of antibiotics by ionizing radiation: an overview," Sci. Total Environ. **646**, 1385-1397 (2019)
6. Y. Shen, L. Chu, R. Zhuan, X. Xiang, H. Sun, J. Wang, "Degradation of antibiotics and antibiotic resistance genes in fermentation residues by ionizing radiation: a new insight into sustainable management of antibiotic fermentative residuals," J. Environ. Manage. **232**, 171-178 (2019)
7. A. A. H. Mohsen, A. A. Abojassim, A. A. Alkufi, "Alpha emitter in samples of antibiotics commonly used in pediatric age groups," Nuclear Analysis **100154** (2025)
8. S. Uddin, S. W. Fowler, M. Behbehani, "^{210}Po in the Environment: Reassessment of Dose to Humans", Sustainability **15**, 1674 (2023), DOI: 10.3390/su15021674
9. F. Bokharaei-Salim, "A Field Study for SARS-CoV-2 Evaluation by Two Air Sampling Strategy During Spread in Tehran, Iran", Pakistan Journal of Analytical and Environmental Chemistry **22**, 21–27 (2021), DOI: 10.21743/pjaec/2021.06.03
10. K. S. Vinayak, A. Sood, A. Sharma et al., "Environmental Safety from Radioactive Radiation Perspective–A Preliminary Research", Journal of Physics: Conference Series **2663**, 012043 (2023), DOI: 10.1088/1742-6596/2663/1/012043

https://doi.org/10.1051/e3sconf/202563301009

Evaluation of indoor air quality in the village of Al-Bu Mahmid, Iraq

Oras Majeed Kareem[1], *Hawraa* Radhi Jasim[1], *Emtihan* Azez Bjoo[1], and *Ali* Saeed Jassim[2,*]

[1]Department of Geology, Faculty of Science, University of Kufa, Al-Najaf, Iraq
[2]Geomatics Technology Center, Kufa, Al-Najaf, Iraq

Abstract. The research was aimed at finding out the concentration of CO2 and VOCs in indoor air in Al-Bouhamid village, which is located to the east of Kufa city. The concentration of CO2 ranged from 325 ppm to 1800 ppm in Al-Bouhamid village at a mean value of 798.933 ± 59.357, while the volatile organic compound levels in homes in Al-Bouhamid village varied from 114 ppb to 912 ppb at a mean value of 379.5 ± 32.232. Almost all the results obtained for most homes in this study are above the standard set by WHO for air quality.

1 Introduction

IAQ is an issue of health that deserves the same attention and consideration most especially in rural communities like Al-Bu Mahmid, a village to the west of Kufa in Iraq. Since time in this village is mostly spent indoors, the quality of air within their vicinity stands to be a great determinant of their health status and general well-being [1]. IAQ stands for Indoor Air Quality, which defines the air quality present within the buildings and structures concerning the existence of pollutants that can harm health and comfort [2]. This research needs to be carried out since population is growing in Al-Bu Mahmid and instead of that there is no proper access or attention given to the problems related to IAQ which are vastly increasing there. Indoor air quality is an important thing, as it shows the conditions that are related to the practicality and convenience level of dwelling people in certain premises [3-6]. The home is basically a micro-environment where pollutants such as volatile organics and particulate matter might thrive, thus posing great health and cognitive health as well as general health risks. This makes IAQ a huge challenge for groups that are most at riskchildren, the elderly, and people with chronic conditions. This problem of indoor pollution and specifically how it worsens conditions such as asthma in children is well illustrated in a publication report. It confronts readers with some telling statistics and data: children stood at 13% of the total population but constituted more than 40% of the total hospital admissions. The air in many villages is not clean, like Al-Bu Mahmid, where ordinary ways of living may produce indoor pollution. One simple way to get good IAQ is to let outdoor air dilute the indoor air pollutants. And the poor insulation in energy-efficient houses, which stops pollutants from flowing outside. Dealing with indoor air quality (IAQ) in Al-Bu Mahmid is much more an issue of public health than the concern for the environment, which has to be looked at with a lot of

* Corresponding author: alis.alrammhi@uokufa.edu.iq

urgency. This effect of poor indoor air quality on human health is great and diversified. Indoor air pollution is said to lead to adverse respiratory health complications and challenges. This can only be said to have a negative impact on the health status by retarding lung growth and development. This creates cause for concern considering that their immune and respiratory systems are still developing, thereby making them quite vulnerable to environmental pollutants that might be present in the air. It has been said that long-term exposure will end up manifesting itself into COPD as one of the most topping health issues worldwide today. In addition to general respiratory problems similar to those of other countries, the children also suffer and have high rates of asthma and other chronic conditions. In a perspective like Al-Bu Mahmid, where families may not know the sources and hazards of indoor pollution, the situation spells doom for future generations. The interaction between indoor air quality and general health emphasizes the need for immediate action- not only to relieve prevailing health burdens but also ensure the well-being of future inhabitants.

2 Materials and methods

2.1 Study area

Albu Mahmid Village is a fundamentally residential area located towards the east of Kufa city center. The village is far from Najaf city center to the southeast by about 14 km. Albu Mahmid's village is set at an altitude of 50 m above sea level, with a flat slope trending downwards gently to the north. This village is located close to one of the large urban centers of Iraq, represented by Kufa city, with a population of about 216,344. Accordingly, the air quality inside homes is undoubtedly affected by the fast-growing infrastructure.

Fig. 1. Albu Mahmid map.

2.2 Practical part

In this research, we used an AIR THINGS technique to measure the concentration of carbon dioxide and volatile organic compounds AIR THINGS makes the radon tracker that is built into all of our goods. It uses alpha spectrometry technology to find out how much radon is in the air. The suggested gadget is a direct, continuous-state radon monitor with an LCD screen. We chose 30 houses from the Al Bu Mahmid area. The device was placed in the living room of each house for a whole day. VOC Measurement: Airthings devices measure VOC levels using a sensor with a reactive layer that changes its electrical resistance when exposed to various airborne chemicals [7]. This sensor identifies fluctuations in VOC concentrations based on resistance changes, providing an estimated VOC level CO2 Measurement: For CO2, Airthings devices employ NDIR (non-dispersive infrared) technology. The NDIR sensor detects CO2 by measuring the amount of infrared light absorbed by CO2 molecules in the air, providing a precise CO2 reading.

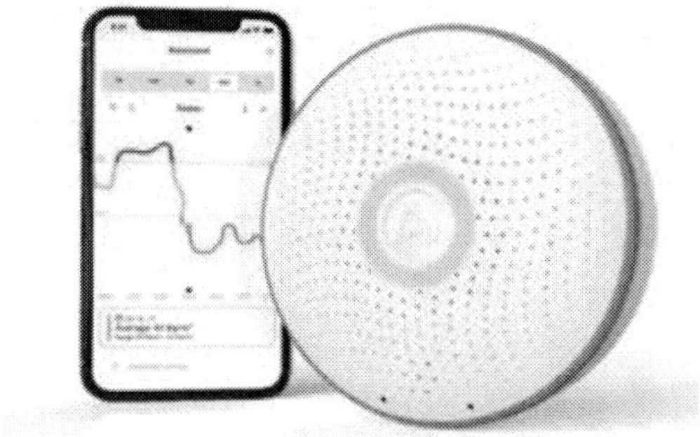

Fig. 2. Airthings.

3 Results and discussion

Table number one shows the results of co2 ppm concentrations in the village of Albu Mahmeed, where the concentration ranged from 325 ppm to 1800 ppm at a rate of 798.933 ±59.357. Most of the results were higher than the limit set by the World Health Organization. The high concentration of CO2 in homes is primarily caused by human respiration, inadequate ventilation, and the number of occupants. Activities such as cooking with gas appliances also contribute to elevated CO2 levels. Poor air exchange due to energy-efficient home designs can trap CO2 indoors, making proper ventilation essential for maintaining healthy indoor air quality [8].

Table 1. Results of concentrations of co2 in the houses of the village of Al-Bu Mahmid, east of Kufa.

Home number	Home code	Co2 (ppm)
1	H1	1800
2	H2	1200
3	H3	750
4	H4	930
5	H5	800
6	H6	566
7	H7	789
8	H8	657
9	H9	690
10	H10	988
11	H11	1250

 https://doi.org/10.1051/e3sconf/202563301009

Continuation of Table 1.

12	H12	1500
13	H13	500
14	H14	325
15	H15	755
16	H16	466
17	H17	985
18	H18	368
19	H19	422
20	H20	888
21	H21	514
22	H22	625
23	H23	679
24	H24	851
25	H25	912
26	H26	710
27	H27	603
28	H28	554
29	H29	741
30	H30	1150
Average		798.933 ±59.357

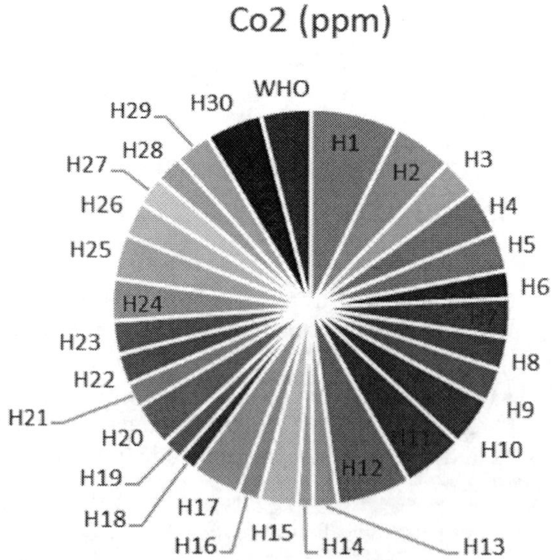

Fig. 3. Comparison of the results of CO2 in the homes of the village of Albu Mahmeed with the global limit of the World Health Organization.

Table No. 2 shows that the concentrations of volatile organic compounds in the houses of Albu Mahmid village ranged from 114ppb to 912 ppb with an average of 379.5 ±32.232. Most of the results were higher than the limit set by the World Health Organization. High levels of volatile organic compounds (VOCs) in homes are mainly caused by emissions from building materials, household products, combustion sources like tobacco smoke and unvented heaters, as well as office equipment. Factors such as poor ventilation, elevated temperatures, and humidity can exacerbate these emissions [9, 10].

Table 2. Results of concentrations of VOCs (ppb) in the houses of the village of Al-Bu Mahmid, east of Kufa.

Home number	Home code	Vocs (ppb)
1	H1	500
2	H2	320
3	H3	514
4	H4	412
5	H5	300
6	H6	715
7	H7	450
8	H8	312
9	H9	310
10	H10	748
11	H11	912
12	H12	444
13	H13	516
14	H14	411
15	H15	370
16	H16	200
17	H17	110
18	H18	158
19	H19	323
20	H20	218
21	H21	227
22	H22	345
23	H23	114
24	H24	255
25	H25	263
26	H26	415
27	H27	318
28	H28	419
29	H29	333
30	H30	453
Average		379.5 ±32.232

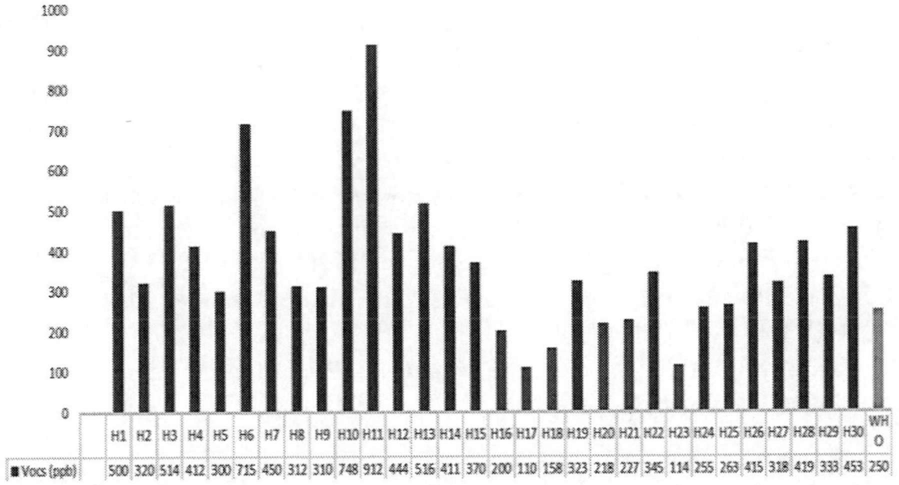

Fig. 4. Comparison of the results of Vocs (ppb) in the homes of the village of Albu Mahmeed with the global limit of the World Health Organization.

4 Conclusion

In most of the houses whose concentration of CO2 and VOCs was measured in the village of Albu Mahmid east of Kufa, the results were higher than the limit set by WHO, indicating there is no good quality indoor air that these houses are exposed to.

5 Recommendations

Implementing pragmatic solutions to improve indoor air quality in Al Bu Humaid is necessary to safeguard the health of its people. One very useful solution is enhancing ventilation, which plays a very significant role not only in the regulation of indoor temperatures but also in controlling the concentration of indoor air pollutants originating from various sources like cooking, cleaning, and building materials. Proper ventilation helps in the dilution of harmful substances toward promoting a healthy atmosphere. Better air filters have a great beneficial impact on improving indoor air quality.

References

1. R. A. Ahmed, R. K. Sebahi, "The Impact of Population Changes on the Functional Composition of the City of Samarra and Its Prospects," Int. J. Geoinformatics **20** (2024)
2. B. Nicolini, "The Re-globalization Process in the Indian Ocean: the Ibadi Press in Zanzibar (al-Sultaniyya)," Revue internationale d'archéologie et de patrimoine (2022)
3. VanderMeulen, The Voice of God: History, Technology, and Authority in Morocco's 'Recitational Revival' (Ph.D. thesis, ProQuest, 2020)
4. A. P. H. M. A. Al-Tafili, P. D. N. A. Al-Jassani, "The phenomenon of desertification and its effects on reducing agricultural areas in Babylon Governorate," Earth and Environmental (2021)
5. A. H. Oraibi, A. A. Abojassim, S. A. Algazali, A. A. Hameed, H. H. Hussian, "Determination of long-lived gamma emitters in some vegetarian products in Al-Najaf governorate using NaI (Tl)," in IOP Conference Series: Earth and Environmental Science **1325(1)**, 012008 (2024)
6. A. S. Al Rmahi, A. A. Abojassim, "Assessment of Natural Radioactivity for Some Secondary Ceilings Samples in Iraq," in IOP Conference Series: Earth and Environmental Science **877(1)**, 012034 (2021)
7. R. Al-Tameemi, Societies Woven in Reeds: Reconstructing the Cultural Landscape of Nippur and the Iraqi Marshlands Through the Lens of John H. Haynes's Photographic Catalog (Ph.D. thesis, University of Cincinnati, 2019)
8. S. S. Nayif, E. J. Mohammed, A. S. Jassim, A. A. Abojassim, "Estimation of the radiological risk due to Radon-222 exposure in dwelling of Al-Hussainya District, Karbala Governorate," Appl. Environ. Soil Sci. **2024(1)**, 6511029 (2024)
9. World Health Organization, WHO ambient air quality database, 2022 update: status report (World Health Organization, 2023)
10. T. Holloway, D. Miller, S. Anenberg, M. Diao, B. Duncan, A. M. Fiore et al., "Satellite monitoring for air quality and health," Annu. Rev. Biomed. Data Sci. **4(1)**, 417 (2021)

Evaluation of radon gas levels in the indoor environment of houses in Albu Mahmid village in Kufa city, Iraq

Lina Abd Oun AL-jamal[1], *Saif* Hamid Abbas[2], *Qunoot* Mohammed Hadi[1], *Esraa* S. Obayes[3], and *Rand* A. Hayder[4,*]

[1]Department of Physics, College of Education for Pure Sciences, University of Babylon
[2]Department of Quality Assurance, University of Babylon
[3]Department of Physics, University of Babylon, College of Education for Pure Sciences, Babylon, Iraq
[4]College of Dentistry, Al-Iraqia University, Baghdad, Iraq

Abstract. The primary natural source of radiation in humans is the atmosphere. Individuals predominantly reside indoors, rendering the home the primary source of exposure to natural airborne radiation, particularly radon gas. It is the second leading cause of lung cancer, behind smoking. This study is to assess the levels of radon 222 (222Rn) in residences inside Albu Mahmid village, Kufa, Iraq. Multiple radiation metrics, including AED (annual effective dose) and CPPP (lung cancer cases per year per million individuals), were also computed. A sophisticated radon detector was employed to assess radon levels in these residences. The concentrations of 222Rn ranged from 34.8±5.9 Bq/m3, with average values of AED and CPPP for radon being 0.14 mSv/y and 1.14, respectively. The World Health Organisation, the Environmental Protection Agency, and the International Commission on Radiological Protection assert that radon concentration levels were within internationally recognised thresholds. And did not provide a health hazard.

1 Introduction

Natural sources are the primary contributor of radiation exposure to humans. It is estimated that 86% of the total human radiation exposure has a natural origin, but 52% is due to the inhalation of Radon and its decay products in enclosed conditions, such as homes and workplaces. There are three natural radon isotopes (Radon-219, Radon -220, and Radon-222), each associated with a different decay series of four isotopes 235U, 232Th, and 238U are all radionuclides. Radon gas a colorless, odorless, tasteless radioactive noble gas was first discovered in 1900 by F.E. Dorn. This is a naturally occurring compound released when radium (Ra) in both soil, water and the materials that buildings are built of, decay. For instance, radon (222Rn [1, 2]), third in alkali metals, has a half-life of 3.8 days and consists of 99% of radon [1, 2]. The most important of the three radon isotopes is radon-222 due to

* Corresponding author: rand.a.hayder@aliraqia.edu.iq

its large indoor air concentration and the health effects from exposure to its decay products. 222Rn daughters are heavy metal radioactive isotopes (e.g., polonium, lead, bismuth) arising from the decay. These radionuclides are easily absorbed by airborne particles that enter the body when inhaled or ingested, both of which can pose health dangers. Radon air pollution is common; inside, in poorly ventilated spaces, concentrations are greater than those outside. Cumulative amounts of radon gas that the public is exposed to indoors are associated with lung cancer. Radon is a natural radioactive gas. Its chemical name is Rn and its atomic number is 86. This text might be polytomous, which means there are other radon types as well, such as the most stable 222Rn. half-life of 3.8 days. 222Rn originates from the decay of 226Ra, which is linked to the decay of 238U, and it is the largest gas to emerge I naturally during the transformation of uranium, thorium, and actinium. 222Rn and its progeny are important sources of radiation for humans and increase the risk of lung cancer. The interaction of charged alpha particles from radon during breathing with lung tissue can cause DNA damage and increase lung cancer risk as a function of radon concentration and exposure duration. however, according to the United States Environmental Protection Agency (EPA) study [1], Radon is the second leading cause of lung cancer, after smoking. 226Ra is a common radioactive element found naturally in the Earth's crust, including rocks, stones, and dirt [2]. Some building materials have earth and stone that contain 226Ra, which produces radon gas. Radon mostly comes from the ground and can enter homes through cracks in walls, basements, and foundations. Radon can also get into homes through tap water, like when using the kitchen sink or taking a shower. Radon is a naturally occurring gas, so it's difficult to avoid low amounts of it in the air. Most exposure to radon comes from other sources. comes from staying inside. Many studies have found that cigarette smoke and radon can cause problems or separate people. In contrast, exposure to each one indi- vidually raises the risk of lung cancer; nevertheless, if one is exposed to both, the risk of lung cancer improves greatly [3]. There is no proof that drinking radon causes health issues like cancer, lung diseases, or signs such as coughing and headaches. National and international guidelines say that the safe amount of indoor radon should be between 100 and 400 Bq/m3. The International Commission on Radio- logical Protection (ICRP), World Health Organisation (WHO), the European Union and United States Envi- ronmental Protection Agency (EPA) are the widely famous indexes which give recommendations and regulations for radon in workplace and public [3-6].

2 Materials and methods

2.1 Study area

Albu Mahmid Village is mostly a neighbourhood where people live. It is located to the east of Kufa city centre. The village is far from Najaf city centre to the southeast by about 14 km. Albu Mahmid's village is located 50 meters above sea level, with a gentle rise that goes down to the north. This town is near Kufa city.

Fig. 1. Albu Mahmid Village location.

2.2 Practical part

Radon levels were measured in 20 houses in the village of Albu Mahmid, located east of the city of Kufa, at a short distance from the Kufa River, using a device Airthings 2930 Wave Plus.The device was placed in the living room of each house for 24 hours. The Airthings 2930 Wave Plus is a device designed to monitor indoor air quality, including radon levels. Here's how it measures radon .The Wave Plus uses a sensor that detects radon through the process of ionization. Radon gas undergoes radioactive decay, emitting alpha particles. The sensor in the device measures these particles to estimate radon concentration in the air. Continuous Monitoring It continuously monitors the radon levels in the air and provides data on the levels of radon over time. This helps users understand how radon levels fluctuate in different areas of the home or building [7] and Radon Measurement Unit The device typically displays radon concentration in terms of picocuries per liter (pCi/L) or Becquerels per cubic meter (Bq/m^3) which are standard units used for measuring radioactivity levels in air.Battery-Powered and Wireless It is a battery-operated device that can connect to the Airthings app via Bluetooth, enabling easy monitoring and access to historical data.By providing continuous monitoring and real-time alerts, the Airthings 2930 Wave Plus helps users ensure their indoor air is safe from excessive radon exposure [8].

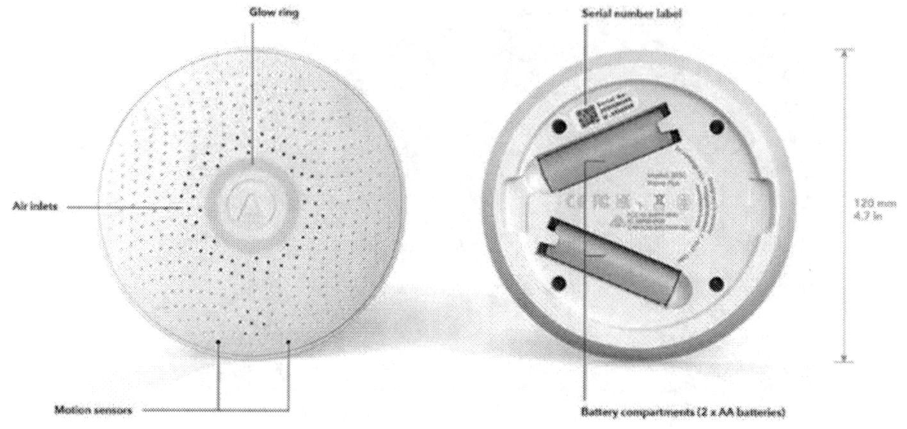

Fig. 2. Air things 2930 Wave Plus radon detector.

2.3 Theatrical equations

The yearly effective dose (AED) can be found in terms of mSv / y for all detectors whether closed or open dosimeters using a relationship [9]

$$\text{AED (m Sv/y)} = \text{C} \times \text{F} \times \text{H} \times \text{T} \times \text{D} \tag{1}$$

The lung cancer cases per year per million persons (CPPP), were obtained using the relation [10]

$$CPPP = \left(8 \times 10^{-6} \frac{y}{mSv}\right) \times AED \tag{2}$$

3 Result and discussion

Table No. 1 shows the results of radon gas concentrations in the houses of Albu Mahmid village, where the results ranged from 5±2.2to 102±10.1 and at a rate of 34.8±5.9. The results were less than the globally permissible limit. Table No. 2 also shows the results of the annual effective dose and cancer risk factors, where the equivalent dose was from 0.02 to 0.41 and at a rate of 0.14 and the cancer risk factor was from 0.16 to 3.34 and at a rate of 1.14. Most of the results obtained in this study follow the standards set by the World Health Organization. The discrepancy can be ascribed to numerous factors, including differences in ventilation rates and the types of construction materials used. The study examines the size of the houses and the temperature variations between them. The global average radon concentration limits, as established by WHO [11] and ICRP, are 100 Bq/m3, 74 Bq/m3, and 200 Bq/m3, respectively. Furthermore, the permissible limits for radiological parameters calculated in this study, such as the Annual Effective Dose (AED), range from 3 to 10 mSv/y, equating to 170 to 230 cases per million individuals annually. Table 2 indicates that the radon concentrations and radiological parameters in all residences within the research region were within global limits. Furthermore, a discernible variation in indoor radon levels is evident among the examined residences. This discrepancy in measurements can be ascribed to differences in ventilation rates, the type and characteristics of construction materials utilised in the buildings fluctuations in the levels of radionuclides present in the soil surrounding the houses, and geological factors. Additionally, fluctuations in radionuclide levels, regional geology, and meteorological circumstances occasionally account for the discrepancies.

Table 1. Radon Concentrations in homes for the present study.

Sample		Radon-222 Concentrations (Bq/m³)	
NO	**Code**	**Average**	**± Error**
1	h1	58	7.6
2	h2	43	6.6
3	h3	17	4.1
4	h4	102	10.1
5	h5	43	6.6
6	h6	17	4.1
7	h7	18	4.2
8	h8	22	4.7
9	h9	33	5.7
10	h10	78	8.8
11	h11	44	6.6
12	h12	9	3.0
13	h13	7	2.6
14	h14	39	6.2
15	h15	74	8.6
16	h16	5	2.2
17	h17	19	4.4
18	h18	26	5.1
19	h19	34	5.8
20	h20	8	2.8
Min		5±2.2	
Max		102±10.1	
Ave		34.8±5.9	
Safe Limit		100 [1]	

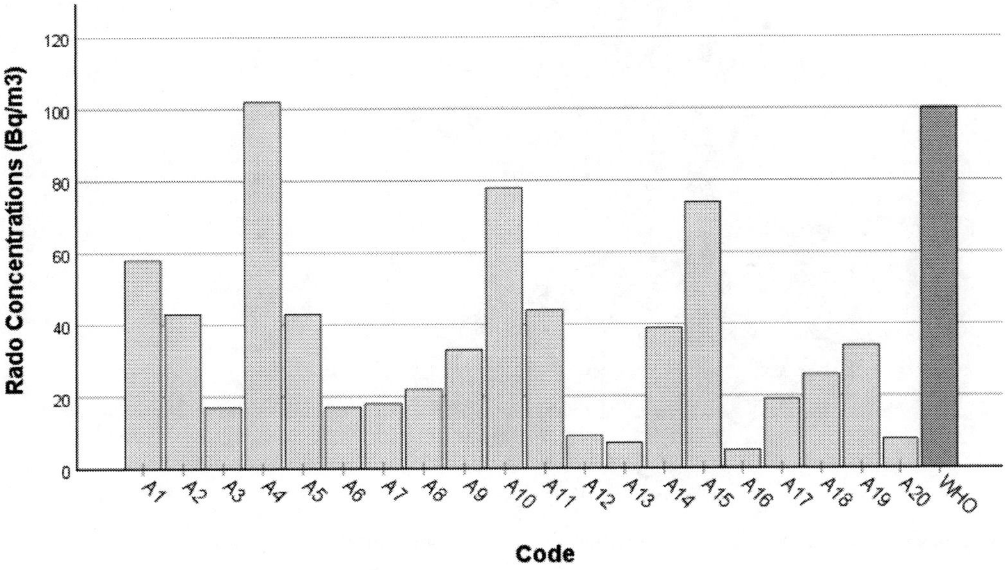

Fig. 3. Comparison of the average Radon in air in the present study with safe limit.

https://doi.org/10.1051/e3sconf/202563301010

Table 2. AED and CPPP due to Radon Concentrations in homes for the present study.

NO	Code	AED (msv/y)	CPPP × 10^{-6}
1	h1	0.23	1.90
2	h2	0.17	1.41
3	h3	0.07	0.55
4	h4	0.41	3.34
5	h5	0.17	1.41
6	h6	0.07	0.55
7	h7	0.07	0.59
8	h8	0.09	0.72
9	h9	0.13	1.08
10	h10	0.32	2.55
11	h11	0.18	1.44
12	h12	0.03	0.29
13	h13	0.02	0.22
14	h14	0.16	1.27
15	h15	0.30	2.42
16	h16	0.02	0.16
17	h17	0.07	0.62
18	h18	0.10	0.85
19	h19	0.13	1.11
20	h20	0.03	0.26
Min		0.02	0.16
Max		0.41	3.34
Ave		0.14	1.14

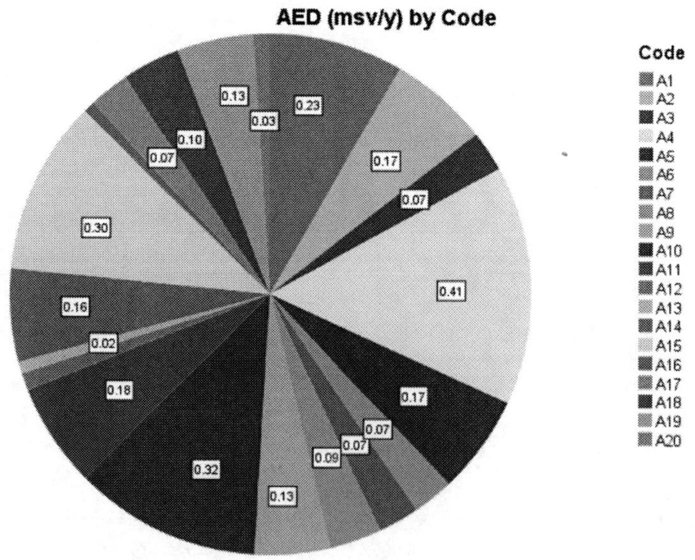

Fig. 4. Comparison of the percentage value of AED in homes in the current study.

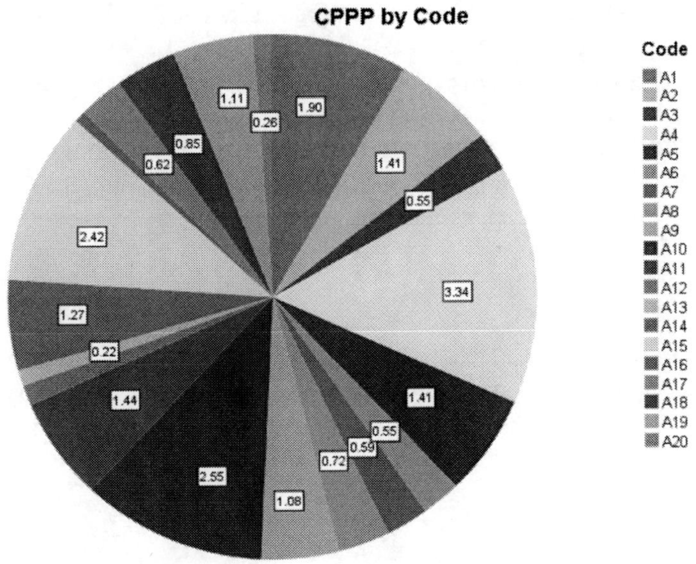

Fig. 5. Comparison of the percentage value of CPPP in homes in the current study.

4 Conclusions

In conclusion, the findings indicated that the mean radon concentrations within the residences of Al-Bu Mahmid village in Kufa are below the threshold established by the World Health Organization, and the annual effective doses and lung cancer risk are significantly lower than the advised level. Consequently, it can be asserted that the current analysis indicates that no intervention is necessary in the examined dwellings. This work's data may serve as a valuable foundation for implementing safety measures and effectively addressing future radiation events.

References

1. E. Ahmed, S. S. Nemah, E. S. Obayes, G. H. Seood, A. S. Jassim, "Evaluation of radon gas air concentration in schools in Waist, Iraq," in E3S Web Conf. **592**, 06001 (2024)
2. S. S. Nayif, E. J. Mohammed, A. S. Jassim, A. A. Abojassim, "Estimation of the radiological risk due to Radon-222 exposure in dwelling of Al-Hussainya District, Karbala Governorate," Appl. Environ. Soil Sci. **2024(1)**, 6511029 (2024)
3. A. Bashir, "Conceptual study of a calibrated radon source and a portable radon detector," Master's thesis, Pennsylvania State University, Pennsylvania, PA, USA (2024)
4. S. K. Jha, A. C. Patra, G. P. Verma, I. S. Iyer, D. K. Aswal, "Natural radiation and environment," in Handbook on Radiation Environment **1**: Sources, Applications and Policies, 27–72 (Springer Nature, Singapore, 2024)
5. W.-Y. A. Elola, T. L. Bambara, A. Doumounia, N. Kohio, S. Ouédraogo, F. Zougmore, "Assessment of radon concentrations inside residential buildings and estimation of the dose in the city of Kaya, Burkina Faso," Open J. Appl. Sci. **13**, 1066–1078 (2023)
6. O. A. Ndubisi, M. A. Briggs-Kamara, F. B. Sigalo, T.-A. Iyeneomie, "Analysis of indoor radon level and its health risks parameters in three selected towns in Port Harcourt, Rivers State, Nigeria," J. Niger. Soc. Phys. Sci. **23**, 181–188 (2021)

7. R. J. Dosh, A. K. Hasan, A. Abojassim, "Radon gas in indoor air of primary schools of Al-Najaf city, Iraq," J. Turk. Chem. Soc. A: Chem. **10**, 1045–1054 (2023)
8. D. Kitson-Mills, S. Sovoe, I. Opoku-Ntim, K. A. Kyei, S. Marnotey, S. Anim-Sampong, M. A. Kwabeng, F. Otoo, F. Baiden, "An assessment of indoor radon level in a suburb of Ghana," Environ. Res. Commun. **1**, 061002 (2019)
9. O. T. Afolabi, D. T. Esan, B. Banjoko, B. A. Fajewonyomi, J. E. Tobih, B. B. Olubodun, "Radon level in a Nigerian university campus," BMC Res. Notes **8**, 1–6 (2015)
10. H. Tsuruoka, K. Inoue, S. Hosokawa, M. Fukushi, "Measurement of radon and thoron concentrations in the Tokyo Metropolitan University Arakawa Campus building," Jpn. J. Health Sci. **19**, 40–48 (2016)
11. H. Teiri, S. Nazmara, A. Abdolahnejad, Y. Hajizadeh, M. M. Amin, "Indoor radon measurement in buildings of a university campus in central Iran and estimation of its effective dose and health risk assessment," J. Environ. Health Sci. Eng. **19**, 1643–1652 (2021)

Impact of cadmium on some physiological and biochemical properties in green algae *Scendesmus qudricauda*

Hala R. Mohammed[1], *Jasim* M. Salman[2,*], and *Adi* Jassim Abd Al-Rezzaq[2]

[1]College of Pharmacy, University of Babylon, Hilla, Iraq
[2]Department of Biology, College of Science, University of Babylon, Hilla, Iraq

Abstract. In biotechnology, bioremediation is the study of using live entities, including microorganisms, to remove toxins and contaminants from other surrounds and water. The experimental sought to evaluated microalgae's ability to eliminate cadmium. The green algal species *Scenedesmus quadricauda* (Turpin) Brébisson was exposure to different Cd concentrations 0.05, 0.25, 0.5 mg/L. Investigated biochemical and physiological characteristics and antioxidant enzymes. Magnitude that 0.14 on the 8th day at quantity (0.05) mg/l during 15th day, the low value of growth rate (k) at (0.25) mg/l was 0. 06 during 15th day. The results showed a strong influence of varying cadmium degradation by *S. quadricauda*: The first day's magnitude that 1.10 (mg/g) 0.05 mg/l, while the highest magnitude was 20.14 (mg/g) on the 15th with 0.5 mg/l. On the fifteenth day, that higher amount of chlorophyll-a was 16.72 (µg/ml) in 0.05 mg/l, while on the first day, it was 0.38 (µg/ml) in 0.25 mg/l. The maximum ROS magnitude was 0.049 on 15th day with 0.05 mg/l Cd, but minimum was 0.013 on 8th day with 0.5 mg/l. On the eighth day, the maximum magnitude the carbohydrate was 26.38 (µm/ml) with 0.25 mg/l. while the min magnitude 5.371(µm/ml) with 0.05mg/l during 15th day.

1 Introduction

Bioremediation is a process that harnesses the capacity of algae to eliminate or break down pollutants, convert them into harmless or less toxic substances, and restore contaminated sites to their original state [1]. The topic of wastewater treatment technology is significant since the existence of multiple organ and inorganic contaminants in urban, agricultural waters and industrial, Various wastewater treatment systems utilize microorganisms to eliminate organic substances. Among these microorganisms, many species of algae show potential as effective bio systems for treating wastewater. They can either transform or directly absorb pollutants, thereby enhancing the purification efficiency of bacterial systems [2]. With their unique metabolic capabilities, algae can deal with toxins through a process, degradation, and biosorption [3]. are a better option for bioremediation than other microbes because they can photosynthesize, which enables them to absorb CO2

* Corresponding author: jasimsalman@uobabylon.edu.iq

the from. Additionally, microalgae are very effective at getting rid of pollutantents called Microalgae.

Have a great contact area, are cheap and simple to recover and reuse, furthermore ecologically friendly and offering a premium approach for bioremediation are they. Industrial wastewater was treated using *S. quadricauda,* which may include a wide range of compounds including lactose, lipids, washing detergues, nutrients, sanitising factor, and extremely highly quantity of nitrogens and phosphorous. Approximately 70-80 % of contamination the water issues arise from industrial operations. Textile mill industrial effluent has high quantities of many pollutants, including colors, volatile organic compounds, alkalinity and acidity, metals, salts, and other substances.

The environment is exposed to heavy metals, which might be hazardous. if improperly disposed of since they can build up in living things, copper (Cu), chromium (Cr), cadmium (Cd), and arsenic (As) [4]. Anthropogenic activities, including agricultural practices, home metal usage, industrial production, smelting, and mining, emit these hazardous materials into the environment. Hazardous materials are introduced into water bodies from several sources such as atmospheric precipitation, urban storm runoff, nonpoint runoff, residential sewage, and industrial discharges,ven at low amounts, Heavy metals can build up within the food chain, leading to harmful impacts.

Introduction of dangers Materials poses a significant risk to plants, animals, people, and the whole ecosystem. Many hazardous materials (HMs) become harmful when they meet environmental elements such water, air, soil, and living creatures. This is because they can accumulate in organisms, cause birth defects, induce genetic mutations, and cannot be broken down naturally, global population expansion, technological advancements, resulting in detrimental effects on the environment.

Hazardous materials (HMs) may enter the human body via many channels and could produce a broad spectrum of health issues and symptoms that vary in degree [5]. Among the most harmful and poisonous heavy metals is cadmium (Cd). The spried industrial uses of this metal led to contamination across numerous environment systems, with its levels steadily increasing over time, the primary contributors to cadmium pollution include waste streams generated by various industrial activities, such as petrochemical refining, mining, plastic manufacture, dyeing, alloy production, smelting, and metal electroplating, cadmium poses significant environmental and biological concerns [6]. This study sought to investigate the green algae's *S.quadricauda* to remove Cadmium from contaminated environments with varying amounts. Additionally, the investigation attempted to use certain physiological and biochemical markers as indicators of the process of bioremediation.

2 Material and Methods

2.1 Algal strain

We acquired *Scenedesmus quadricauda* (Turpin) Brébisson (Chlorophyceae) Source: University of Texas Culture Algae Collection (Austin, TX, USA; UTEX). Under controlled conditions, the species was light intensity of 286 $\mu E/m^2/s$, photoperiod of 16 hours of light and 8 hours of dark, and incubated at 25±2°C. The cultures were shook lightly every other day to promote development and minimize clumping.

2.2 Preparation of algal culture media

Chlorophyta improvements were made using BG-11 media. The trace elements stock solution was The composition includes 2.86 g of H_3BO_3, 1.81 g of $MnCl_2 \cdot 4H_2O$, 0.222 g of

$ZnSO_4 \cdot 7H_2O$, 0.39 g of $Na_2MoO_4 \cdot 2H_2O$, 0.079 g of $CuSO_4 \cdot 5H_2O$, 0.049 g of $Co(NO_3)_2 \cdot 6H_2O$, and 1.5 g of $NaNO_3$, with following concentrations per liter: 40 g of $K_2HPO_4 \cdot 3H_2O$, 75 g of $MgSO_4 \cdot 7H_2O$, 36 g of $CaCl_2 \cdot 2H_2O$, 20 g of Na_2CO_3, 6 g of $C_6H_8O_7$, and 1 g of Na-EDTA , and 6g/l FeH8N2O8S2. These stock solutions were prepared for all macro and micro elements. The BG-11 medium was autoclaved for 15 minutes at 121°C to eliminate any bacteria. After cooling, a pH meter and solutions of 1N NaOH and 1N HCl were used to change the pH to roughly 7.4 in order to avoid too much precipitation.

A flask with 100 ml of BG11 medium is filled with 10 ml of algal culture, which is then left to grow for 15 days. Before the experiment is started in a controlled laboratory setting, this culture is then moved into media that have 1000 ml and left undisturbed for at least two weeks. One important component that regulates asexual reproduction's cell division is photoperiod. Continuous illumination speeds up this process, which takes place during sun period.

Hence, that length time the light is applied that be modified according to the objectives of the cultivation: uninterrupted lighting encourages fast development, while a photoperiod that alternates between light and darkness, like the natural solar cycle, fosters regular and robust growth.

2.3 Experiment design

The culture media of *S.quadricauda* was subjected to many dose of Cd (0.5, 0.25, and 0.05) mg/L for a period of 15 days. study focuses on analyzing the daily changes in algae over a period of 15 days. The algae samples are incubated to determine the rate of growth (k) and duration of doubling (G). Additionally, measurements are taken for total protein, carbohydrates, total chlorophyll, and various enzymatic reactions including (CAT) and (SOD) lipid peroxidation MDA, and ROS on the 1st, 8th, and 15th days.

2.4 Growth rate and doubling time

Every day, a UV-Vis spectrophotometer is utilized to calculate the cell count in the green algae *S. quadricauda*. In order to determine the cell density in cells per millilititer, the optical density (OD) at 685 nm is measured. We can determine the biomass size, growth rates, and other developmental phases with this method. In the exponential development phase, the equation is very useful for determining the growth rate and doubling time.

$K = 3.322 * (\log OD_t - \log OD_0) / t$

$G = 0.301 / K$

K: rate of growth G: doubling duration t: duration

OD_0: Optical density at the start of the experiment (time zero).

OD_t: optical density after (t) day.

2.5 Estimation of Chlorophyll

The method used to determine the total chlorophyll content is described in reference. After obtaining a 2 mL sample, it was centrifuged for 5 minutes at 12,500 revolutions per minute. 2 mL of 90% methanol was then combined with the solid residue that the algae had left behind. Following a twenty-hour incubation period at 20 degrees in a dark environment, the sample was immersed for five minutes at 64 °C in a water bath. Last but not least, the material was centrifuged at 12,500 revolutions per minute for five minutes before being discarded. The filtrate was measured use the Spectro at three distinct wavelengths: 470, 655, and 65 nm. The overall chlorophyll content was determined with the aid of the formulae cited in the relevant literature. The formula for µg chlorophyll/mL medium is (16.29 × A665) - (8.54 × A652).

2.6 Reactive oxygen species (ROS)

To estimate, take two milliliters of the specimen and wash it twice using a phosphate-buffered solution. Two milliliters of perchloric acid (200 micromol) are then added to the precipitate, which is made up of algal cells. Sonication has been used to break down the cells for three minutes, with 20-second pauses. Centrifuging the specimen for 30 minutes at 10,000 r/m is the next step in the rapid disposal process. After blending the filtrate with a micropipette, the size is set at 5 milliliters. 0.1 ml of the working solute and 1.5 ml of the extract combined to create a combination. Sorbitol (3.64 g), xylenol (14.3 mg), sulfuric acid (0.28 ml), and ammonium ferrous sulfite (19.6 mg) make up the working solution. The extract is incubated for 30 min at 30 °C. The concentration of reactive oxygen species (ROS) is calculated using the provided formula after the extract's absorbance at 560 nm has been measured after incubation.

%inhibitions ROS=Δuninhibite /Min(10)

2.7 Antioxidant enzymes

Two milliliters of the specimen were taken and disposed of at a rate of 4000–6000 r/m. The discarding process was then repeated once the resultant algal precipitate had been cleaned with a pH 7 buffer solution. Then the volume was set back to two milliliters. Then, for three minutes, the algal cells were broken up using ultrasonic equipment (Sonication). For 30 minutes, the model is ejected at 10,000 rpm and paused every 20 seconds. A small amount of the filtrate is then added, and that volumes is then adjusted to 5 ml.

2.8 Carbohydrates

A 2 ml portion of the specimen was extracted and subjected to aerobic drying after a wash with a phosphate-buffered solution. A sonicater was then used to break up the material, and distilled water was used to dilute it to a end amount the 5 mL. One millilitre the specimen was that mixed with five millilitres of 96% sulfuric acid and one millilitre of 5% phenol. After that, the liquid was continually swirled for ten minutes. After that, the sample was submerged for half hour at a 30-35 ◦C temperature in a waterbath. Following was measured at 490 nm and contrasted with the glucose standard curve. 100 milligrams of glucose were dissolved in 100 milliliters of distilled water to construct the standard curve [7].

2.9 Total protein

Microalgal proteins were separated and concentrated utilizing dialysis, chromatography, precipitation, ultrafiltration, and centrifugation techniques. To determine the Ref. adapted the Lowry technique to determine the total protein content, this stage involved combining 2 mL of Biuret solution (2–4) with 0.5 mL of the previously made extract. After that, a preheater was used to thoroughly stir the mixture for 30 minutes at 30 °C. After that, that specimen was examined at a wavelength of 555 nm and contrasted with a stander solution that contained protein produced from bovine serum albumin and ranged in concentration from 0 to 0.1 mL. To make the reference solution, which is equivalent to 100 µg/L, 0.1 g of bovine serum albumin was dissolved in 100 mL of buffer solution.

2.10 Superoxide dismutase (SOD)

The estimation involves utilizing 100 microliters of the extracted (2-4) and 2 ml of buffer tris. After waiting for 5 min, the reading is taken at 420 nm, which is referred to as the first reading ($\triangle A_0$). Then, 0.2 ml the solution of Pyralol is added, another read was taked at 420 nm after waiting for 5 min. This reading is referred to as the second reading ($\triangle A_1$). The magnitude of $\triangle A_1$ can be determined utilizing the formula provided.
SOD activity (μ/mL) = $[(\triangle_{A0} - \triangle_{A1}/\triangle_{A0})/50\%] \times$ Volume of sample

2.11 Catalase (CAT)

The measurement involves taking 100 microliters of the extract (2-4) and combining it with 1 ml of H2O2. The mixture is then determined at a wavelength of 240 nm. The measurement is repeated after 10 min after adding the hydrogen peroxide. Determine the amount of the catalyst utilizing the formula provided.
CAT= $[(\triangle \text{Abs}_{240}/\text{Min}_{10}) \times (\text{Reaction volume}_{100})]/0.001$

2.12 Lipid peroxidation (Malondialdehyde MDA)

The amount of malondialdehyde (MDA) was measured using the Thiobarbituric Acid (TBA) assay to assess lipid peroxidation, 1000 μl of TBA (0.6%) and TCA (20%) were combined with each component in a 100 μl volume, Swirling the tubes is done with the vortex mixer, the sample was heated for 15 minutes at 100°C in a water bath before being centrifuged for 15 minutes at 4000 r.p.m. (4°C), after that, it was left to reach room temperature by cooling, the absorbance of the sample at 532 nm was measured in relation to a blank, using a millimole per milligram of protein, the findings were measured and reported, Amount of Malondialdehyde (MDA) =Absorption at 532nm / E × b
Whereas:
E= Extinction coefficient (153mmol/cm),
b = light bath (1cm)

2.13 Statistical analysis

This paper's experimental data came from at least three replicates of each treatment. To conduct the statistical analysis, Version 26 of the SPSS program was used. The standard deviation and mean were used to present the data. An analysis of variance (ANOVA) was used in the study to determine the significance of the differences. Significant differences were those whose corresponding p-magnitudes were less than 0.05.

3 Results and discussion

3.1 Growth rate

The data absorbance gathered through the phase of growth exponential was utilized to calculate the maximum rate of specific growth (K). Table 1 illustrates the curve growth of *S. quadricauda* algae under varying Pb concentrations. The concentration of 0.05 mg/l produced the maximum growth rate (k), reaching a magnitude of 0.14 on the 8th day. The significant decrease in growth rates, particularly observed at high amounts of cadmium (0.5 mg Cd/L), is consistent with findings from previous research, research findings indicate that higher concentrations of cadmium in the environment result in slower growth rates and longer cellular doubling times in *Chlorella saccharophila* and *Anabaena flos-aquae* species [8]. Moreover, a slight reduction in overall cell size was also noted. The added of various mount of cd to the media similarly led to the suppression of growth in this species. The most significant impacts of lead are seen on cells the log phase, result in a decrease in the division of cell. The lowest observed magnitude was 0.06 at an amount of 0.5 mg/l on the 15th day.

The detected decreases in growth rates of the examined algae, resulting from the addition of various doses of cadmium, may be related to the overall toxic impact mostly affecting the respiratory function. This suggests that defects in the phospholipid structures of chloroplast and mitochondrial membranes could be responsible, as these organelles appear the main targets cadmium toxicities [9]. Additionally, cadmium can substitute for iron during the binding of photosynthetic cytochromes, thereby inhibiting the photosynthesis process.

Table 1. Growth rate at different concentration of cadmium of *S. quadricauda*.

Period (Day) (B) / Treatment (A)	1st day	8th day	15th day	Mean of treatment (A)
Control	0 a	0.13±0.02 b	0.02±0.001 a	0.08±0.004 a
0.05 mg/l	0 a	0.14±0.03 a	0.11±0.01 a	0.13±0.02 b
0.25 mg/l	0 a	0.10±0.001 b	0.06±0.01 a	0.05±0.01 a
0.5 mg/l	0 a	0.09±0.02 b	0.018±0.04 c	0.05±0.01 a
Mean of period (B)	0 a	0.12±0.01 c	0.039±0.001 b	

* *Data numbers (mean±standard deviation).*

* *Different letters in each test represent statistical significance differences among groups (p<0.05)*

3.2 Doubling time

The doubling time of a population refers to the duration required for its size or magnitude to increase twofold. This growth follows an exponential pattern, characterized by a consistent doubling period that can be conveniently calculated if the growth rate remains steady. While the relative growth rate stays unchanged. The doubling time (G) was lower at 2.1 on the 8th day with an amount of 0.05 mg/l, while the greatest doubling time was 17.1 on the 15th day with an amount of 0.5 mg/l of Cd. Table 2 illustrates the impact of varying amounts of cadmium on doubling time. The replication time of the cadmium treatment was prolonged since, the impact of cadmium on *S. quadricauda* revealed some unexpected results, while Cd is generally harmful, increasing its concentration of medium growth surprisingly led to a rise in biomass production [10].

Table 2. Doubling time at different concentration of cadmium of *S. quadricauda*.

Period (Day) (B) / Treatment (A)	1st day	8th day	15th day	Mean of treatment (A)
Control	0 a	2.4±0.5 b	14.9±2.8 e	5.8±1.1 c
0.05 mg/l	0 a	2.1±0.6 b	2.6±0.4 b	1.6±0.4 a
0.25 mg/l	0 a	3.1±0.7 c	5.4±0.8 d	2.8±0.6 b

 https://doi.org/10.1051/e3sconf/202563301011

0.5 mg/l	0 a	3.5±0.8 c	17.1±2.4 e	6.8±1.3 d
Mean of period (B)	0 a	2.8±0.9 b	10.0±2.6 c	

** Data numbers (mean±standard deviation).*

** Different letters in each test represent statistical significance differences among groups ($p<0.05$)*

3.3 Photosynthetic pigments

Findings indicate that the amount of chlorophyll a and b in cell *S.quadricauda* treated with cadmium decreased to 0.38 µg/ml in 0.25 mg/l ch-a and 0.40 µg/ml in 0.05 mg/l ch-b on the 1st day. This substantial reduction in chlorophyll pigments suggests that the toxic effects of cadmium at these amounts inhibit enzyme activity, leading to a decline in algal growth. That present consistent with studies previous and support the idea that high amounts of cadmium could be detrimental to chlorophyll-a synthesis [11]. However, on the 15th day, the amount of ch-a increased to 16.72 µg/ml and ch-b increased to 7.74 µg/ml in 0.05 mg/l, indicating that the algae have developed tolerance to low amounts of cadmium and is able to produce more ch-a. The toxicity of cadmium appears to base on the duration of exposure and the amount of cadmium as well as the availability of nutrients, the greatest amount of chlorophyll-a,b was seen near the conclusion of the treatment period, maybe since the algae's significant uptake of metals for their metabolic processes and growth [12]. (Fig.1and 2).

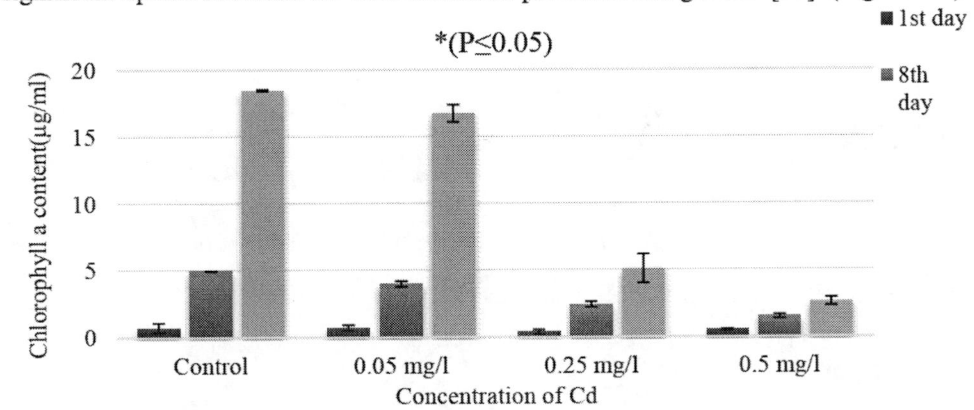

Fig. 1. Chlorophyll a concentration of *S.quadricauda* at different concentration of Cd.

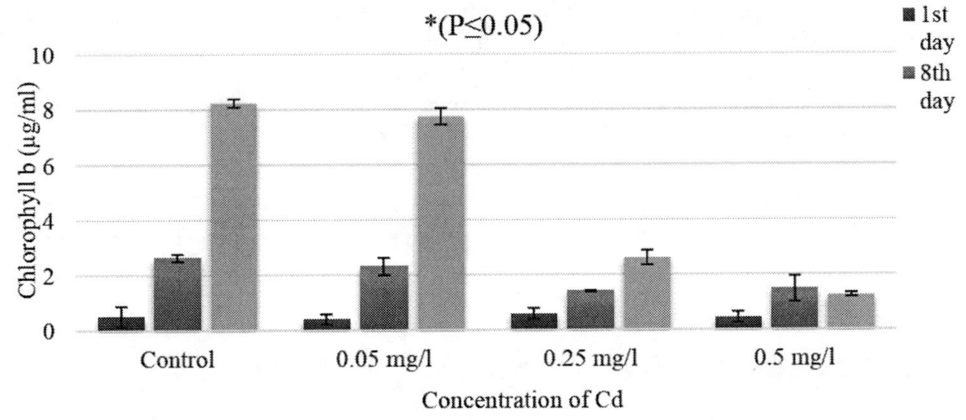

Fig. 2. Chlorophyll b concentration of *S.quadricauda* at different concentration of Cd.

3.4 Reactive oxygen species (ROS)

ROS functions as molecule signaling that triggers acclimation and protection responses through transduction pathways. When exposed to high amounts of environmental stress, Overproduction of ROS can damage algal cells by oxidizing proteins, inactivating metabolic enzymes, and damaging DNA and lipids. Consequently, many non-enzymatic antioxidants, including SOD, collaborate to activate defenses against ROS and facilitate their detoxification. The highest documented of ROS was 0.049 on the 15th day with an amount of 0.05mg/l of Cd. Conversely, the lowest recorded magnitude was 0.013 on the 8th day with an amount of 0.5mg/l of Cd (Fig. 3). Superoxide anions and hydroperoxides are produced when Cd binds to critical sites and decreases the GSH pool, increasing the pro-oxidant status, therefore, the study's findings demonstrate that exposure to Cd results in a rise in lipid peroxidation levels, validating the (ROS) generation of shows existence of oxidative stress, the protein-rich compounds possess antioxidant properties since the existence of functional groups including amino acid residues, sulfate, carboxyl, amide, and amino, these groups can attract and capture heavy metal ions, suggesting that they can as scavengers and decrease the amount of ROS.

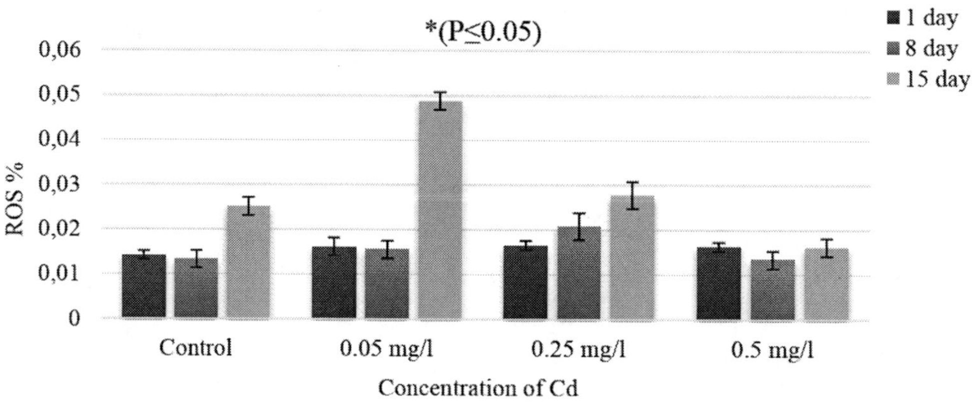

Fig. 3. ROS concentration of *S.quadricauda* at different concentration of Cd.

3.5 Carbohydrate

Photosynthetic organisms, such as algae, generate carbohydrates, which are essential for obtaining energy. Environmental toxicants have a detrimental impact on the carbohydrate levels in algae. They limit the synthesis of carbohydrates and may also induce stress, leading to increased respiration. This increased respiration consumes the previously generated carbohydrates, the research found that treating *S.quadricauda* with various amounts of cadmium resulted in a considerable reduction in total amount of carbohydrates as the amount of cadmium increased. The carbohydrate concentration was 5.371 μm/ml with 0.05 mg/l on the 15th day, whereas the greatest amount was 26.38 μm/ml with 0.25 mg/l on the 8th day (Fig. 4), the reduction of RUBP activity, which lowers carbohydrate levels, is one explanation for the decrease in algae, the toxin can reduce internal metabolic processes and prevent glucose from being oxidatively metabolized, which is essential for the production of ATP [13].

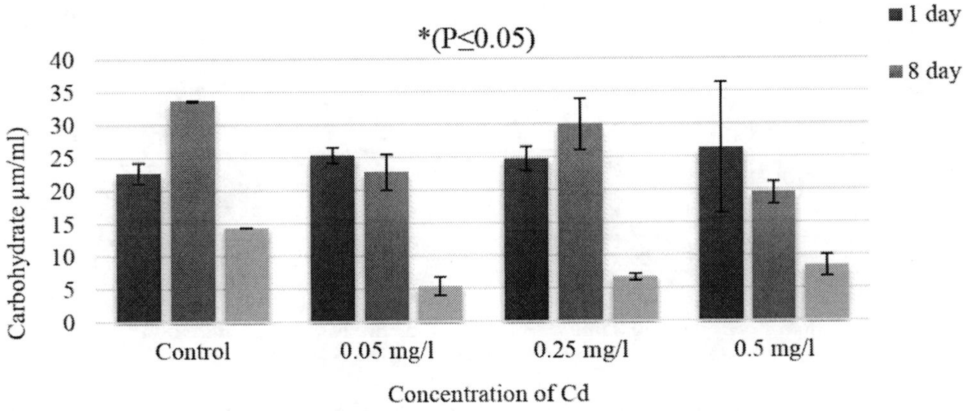

Fig. 4. Carbohydrate concentration of *S.quadricauda* at different concentration of Cd.

3.6 Total protein

Microalgal biomass shows promise as a potential substitute ingredient for aquaculture feed because of its elevated protein levels and well-balanced amino acid composition. Proteins have a crucial function in the chemical composition and physical structure of microalgae, the amount of protein in the sample reached a max of 2.38 µg/ml on the 15th day at an amount of 0.05 mg/l. Conversely, the amount of protein was at its min of 1.23 µg/ml on the 1st day at an amount of 0.5 mg/l (Fig.5). This study found that protein inhibition happened at 0.5 mg/l and 0.25 mg/l of Cd because of the toxicity of cadmium, which can affect the metabolic processes of the organism by preventing enzyme activity [12].

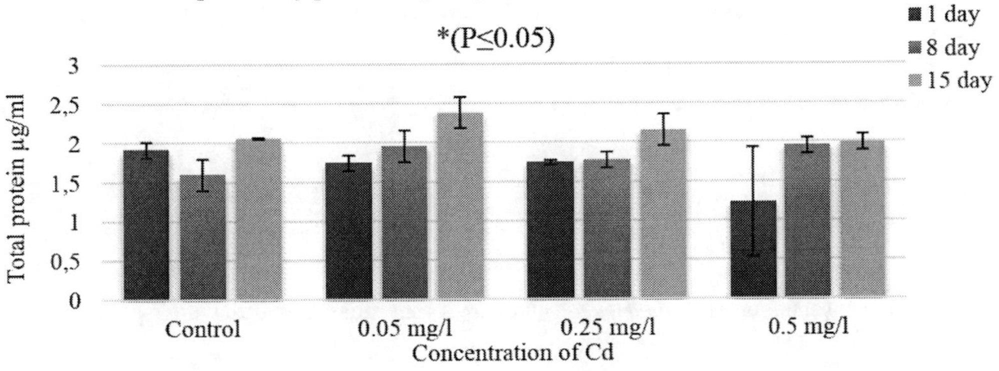

Fig. 5. Total protein concentration of *S.quadricauda* at different concentration of Cd.

3.7 Superoxide dismutase (SOD)

In general, the cell maintains a balance between the creation of oxidant-antioxidant molecules, when the production of oxidants increases under stressful situations, the cell protects itself from the detrimental effects of accumulated oxidant molecules by consuming an excessive amount of compounds antioxidant, the study measured the levels of SOD in the algae *S.quadricauda* when subjected to different amounts of cadmium. The max recorded magnitude was 123.33µg/ml in an amount of 0.5 mg/l on the 8th day. magnitude Low was 46.67µg/ml in an amount of 0.05 mg/l on the 15th day (Fig. 6), the findings indicate that SOD amounts increased on the 1st and 8th days for both amounts of cadmium, but decreased on the 15th day. Cadmium has the potential to hinder the action of enzymes and deplete antioxidants.

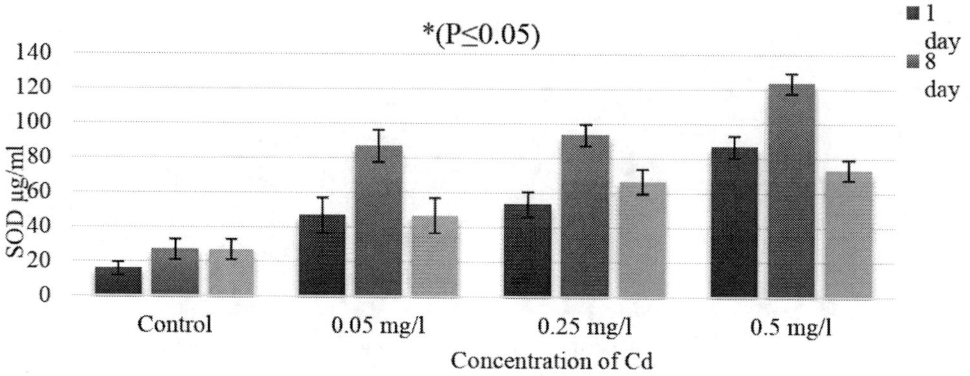

Fig. 6. SOD concentration of *S.quadricauda* at different concentration of Cd.

3.8 Catalase (CAT)

Results of present study in (Fig. 7) show the max magnitude 196.67µg/ml in 0.5 mg/l during 8th day, whereas in 1st day's minimum magnitude was 26.67µg/ml in 0.05. The catalase (CAT) activity declined as the amount of Cd increased, serving as a defense mechanism. The decrease in CAT activity could result from modifications to the way the enzyme is organized. subunits at extremely high metal concentrations or to the suppression of CAT enzyme production [13, 14].

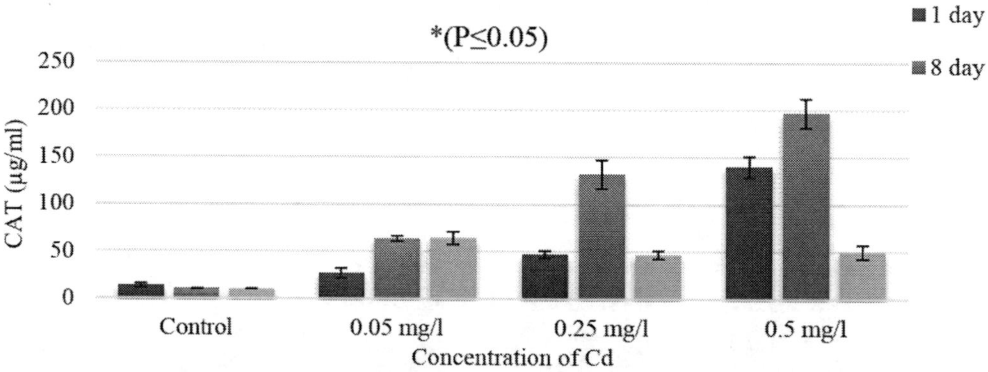

Fig. 7. CAT concentration of *S.quadricauda* at different concentration of cadmium.

3.9 Lipid peroxidation (Malondialdehyde MDA)

The body's excess ROS production may interact with the lipid molecules in the cell membrane, resulting in the formation of lipid peroxides. These peroxides are then broken down into malondialdehyde. Malondialdehyde was the one of the aldehydes that can be produced as by-products of lipid oxidation products (LOPs), which occur naturally in cell membranes, particularly in cholesterol, phospholipids, and glycolipids. The relationship between MDA and HM exposure was extensively investigated in microalgae, however, when exposed to metal stress circumstances, if the quantity of ROS exceeds a certain threshold, it may cause damage to the plants' lipids, proteins, and other biomolecules, Furthermore, LOPs stimulate a series of events and intensify oxidative stress by generating lipid radicals that subsequently damage proteins and DNA, the consequences of varying Cd concentrations on the generation of MDA in *S.quadricauda* are shown in (Figure 8).

The max magnitude recorded was 60.62µmol/l at an amount of 0.05 mg/l on the 1st day, while the min magnitude observed was 25.43µmol/l at an amount of 0.25 mg/l on the 15th day. Cadmium amounts had a stimulating influence on the quantity of MDA. According to reports, if ROS are not adequately detoxified by organisms, they may lead to the peroxidation of membrane lipids, resulting in the creation of different breakdown products such as MDA, this research demonstrated a considerable increase in MDA generation in microalgae subjected to Cd, compared to the control group. This suggests an elevation in lipid peroxidation since Cd exposure.

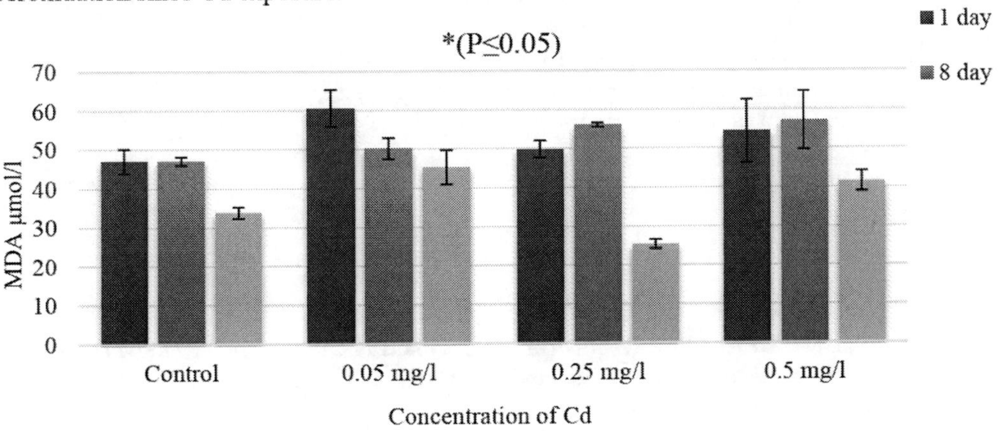

Fig. 8. MDA concentration of *S.quadricauda* at different concentration of cadmium.

3.10 Bioremediation of Cd

Living microalga biomass was utilized in bioremediation methods for heavy metals since its capacity to eliminate these pollutants, either by adhering to the outer surface of the cells or by internalizing them, there are two distinct biochemical processes that may be used: biosorption, which entails metal ions adhering to the cell surface and bioaccumulation, which entails metals entering cells.

The findings in Figure 9 showed a distinct impact of varying levels of cadmium degradation by *S. quadricauda*; the maximum value was 20.14 mg/g on the 15th day with 0.5 mg/l, and the minimum value was 1.10 mg/g on the 1st day with 0.05 mg/l. The Cd elimination trend in our investigation showed a peak relatively early in the exposure period, followed by a minor decline and, finally, a plateau.

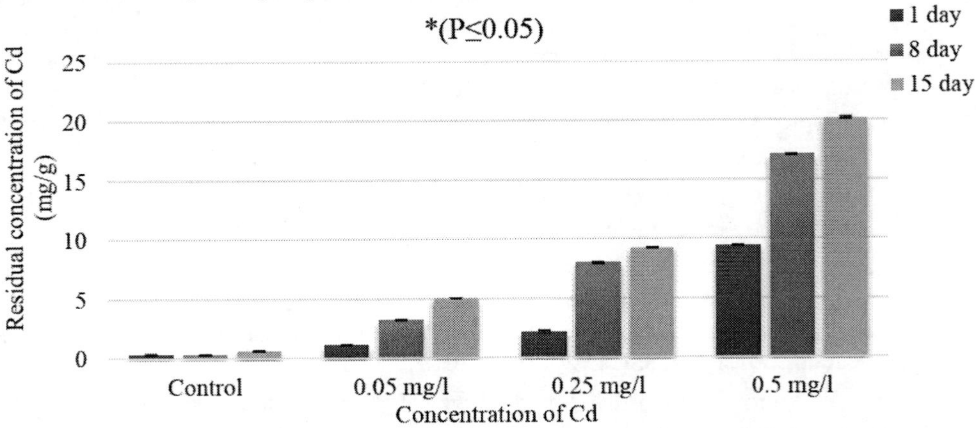

Fig. 9. The residual concentration of cadmium after (1,8,15) days of exposure time.

4 Conclusions

- The study demonstrated that an alga *Scenedesmus quadricauda* has a high efficiency to treatment pollution in cadmium heavy metal and possibility of utilizing alga in the bioremediation for aquatic pollution.
- The algae demonstrated has bio response in some of result exposure of biological parameters different concentration of cadmium such as decreasing in concentration, total protein, total chlorophyll, and raise ROS, or oxidative stress and carbohydrate and enzyme effectiveness, SOD, MDA and catalase.
- Based on the findings of the present investigation that there was a clear response in the biological parameters of the algae that can be utilized as a good tool to bioremediation the environmental pollution systems.

For helping to support this work, The College of Science's Department of Biology is acknowledged by the writers.

References

1. S. A. Juda, M. Ali Ahed, A. R. Omarin, Z. H. Obaid, J. M. Salman, "Removal efficiency of synthetic toxic dye from water and waste water using immobilized green algae–bioremediation with multi environment conditions," Ecol. Eng. Environ. Technol. **24**, (2023), https://doi.org/10.12912/27197050/174050
2. J. M. Salman, N. F. Kaduem, S. A. Juda, "Algal immobilization as a green technology for domestic wastewater treatment," in IOP Conf. Ser.: Earth Environ. Sci. **1088(1)**, 012005 (2022), https://doi.org/10.1088/1755-1315/1088/1/012005
3. A. K. Patel, M. R. Suseela, M. Singh, S. Nayaka, "Application of response surface methodology for optimization of biomass, carbohydrate and lipid production in BG11+ by Scenedesmus quadricauda," Int. J. Res. Eng. Appl. Sci. **5(6)**, 199–215 (2015)
4. O. Spain, M. Plöhn, C. Funk, "The cell wall of green microalgae and its role in heavy metal removal," Physiol. Plant. **173(2)**, 526–535 (2021), https://doi.org/10.1111/ppl.13405
5. C. Yan, Z. Qu, J. Wang, L. Cao, Q. Han, "Microalgal bioremediation of heavy metal pollution in water: recent advances, challenges, and prospects," Chemosphere **286**, 131870 (2022), https://doi.org/10.1016/j.chemosphere.2021.131870
6. P. S. Chandrashekharaiah, D. Sanyal, S. Dasgupta, A. Banik, "Cadmium biosorption and biomass production by two freshwater microalgae Scenedesmus acutus and Chlorella pyrenoidosa: an integrated approach," Chemosphere **269**, 128755 (2021), https://doi.org/10.1016/j.chemosphere.2020.128755
7. H. G. Pontis, *Methods for Analysis of Carbohydrate Metabolism in Photosynthetic Organisms: Plants, Green Algae and Cyanobacteria* (Academic Press, 2016)
8. M. Danouche, N. El Ghatchouli, H. Arroussi, "Overview of the management of heavy metals toxicity by microalgae," J. Appl. Phycol. **34(1)**, 475–488 (2022)
9. R. Bacova, B. Klejdus, P. Ryant, N. Cernei, V. Adam, D. Huska, "The effects of 5-azacytidine and cadmium on global 5-methylcytosine content and secondary metabolites in the freshwater microalgae Chlamydomonas reinhardtii and Scenedesmus quadricauda," J. Phycol. **55(2)**, 329–342 (2019)
10. L. Mo, Y. Yang, D. Zhao, L. Qin, B. Yuan, N. Liang, "Time-dependent toxicity and health effects mechanism of cadmium to three green algae," Int. J. Environ. Res. Public Health **19(17)**, 10974 (2022)

11. H. Zhang, M. Reynolds, "Cadmium exposure in living organisms: a short review," Sci. Total Environ. **678**, 761–767 (2019), https://doi.org/10.1016/j.scitotenv.2019.04.395
12. L. Barsanti, P. Gualtieri, *Algae: Anatomy, Biochemistry, and Biotechnology* (CRC Press, 2022), https://doi.org/10.1201/9781003187707
13. C. Qiu, W. Wang, Y. Zhang, G. J. Zhou, Y. Bi, "Response of antioxidant enzyme activities of the green microalga Chlorococcum sp. AZHB to Cu2+ and Cd2+ stress," Sustainability **14(16)**, 10320 (2022), https://doi.org/10.3390/su14161032
14. Z. Chen, A. I. Osman, D. W. Rooney, W. D. Oh, P. S. Yap, "Remediation of heavy metals in polluted water by immobilized algae: current applications and future perspectives," Sustainability **15(6)**, 5128 (2023)

E3S Web of Conferences **633**, 01012 (2025) https://doi.org/10.1051/e3sconf/202563301012
GASES 2025

Environmental policy as a factor of China's food security

Maxim Sushchenko*

The Department of History and Political Science, FSBEI HE Kuban SAU, 350044 Krasnodar, Russia

Abstract. Public policy tools help to effectively implement the goals of the national development of the People's Republic of China. The article examines the main directions and mechanisms of environmental policy of modern China as factors of ensuring food security of society. The problems of conservation, environmental protection and food security are closely connected. Agriculture is the only branch of the Chinese economy that is directly related to the country's natural conditions. Protecting the environment from the adverse effects of anthropogenic factors is the most important condition for achieving China's strategic goals. The actions of the administrative, economic and regulatory mechanisms of environmental protection of the People's Republic of China are interrelated and interdependent. The regulatory mechanism includes means of state and party policy of the state of nature. The action of the mechanisms ensures the effective implementation of China's socio-economic policy in the areas of sustainable environmental management and reduction of environmental pollution.

1 Introduction

One of the causes of global instability is the growing threat to food security in the modern world. At the same time, the escalation of political crises and conflicts is leading the global economy to disrupt previously stable food supply channels to end consumer in the 2020s. The problem of food security is becoming more acute, and it is beginning to manifest itself more and more clearly in the world. 258 million people, where living in 58 countries, felt the effects of the global food problem in 2022 [1]. The trends of environmental degradation are actualized as an effective factor of social development in the aspect of a serious international problem. The ecological environment is actualized as an effective factor of social development. China is one of the largest countries in the world in terms of territory, population and GDP. The industrialization of the economy leads to a general deterioration of the environment, turning China into one of the most polluted countries in the world [2]. The balanced development of the environment and the task of ensuring food security in China are closely interrelated. The expansion of demand for agricultural products puts a strain on the functioning of the agro-industrial complex. The increase in production capacities and agricultural products has a negative impact on the environment. Adverse environmental consequences, a decrease in public health, and a decrease in labor productivity negatively

* Corresponding author: spacemirror@mail.ru

affect the living conditions of society. The main development goals of the People's Republic of China in the 21st century were proclaimed by Chinese President Xi Jinping in 2022 at the XX Congress of the CPC to transform the country into a rich, democratic, socialist state and complete socialist modernization [3]. The balanced development of the environment and the task of ensuring food security are closely interrelated in China. The expansion of demand for agricultural products puts a strain on the functioning of the agro-industrial complex. The increase in production capacities and agricultural products has a negative impact on the environment. Adverse environmental consequences, a decrease in public health, and a decrease in labor productivity negatively affect the living conditions of society. The main development goals of the People's Republic of China in the 21st century were proclaimed by Chinese President Xi Jinping in 2022 at the XX Congress of the CPC to transform the country into a rich, democratic, socialist state and complete socialist modernization [3]. For continue the modernization of agriculture and ensure the achievement of food security, the leadership of the People's Republic of China strives to implement an effective environmental policy. The environmental regulation measures of the Chinese authorities help to implement the national development plan in the short and long term. The purpose of the paper is research the main directions and mechanisms of environmental policy of modern China as factors of ensuring food security of society.

2 Materials and methods

Several general scientific and special methods that allowed the author to achieve the research goal used in the paper. An overview of current research by modern scientists from the Scopus database provides in the research. The several general scientific and special methods uses in the paper. They allowed to achieve the research goal the author. The main processes in the development of the environmental policy of the People's Republic of China in the XXI century are established in the research by means of the historical method. The approaches to defining the concepts of food and environmental safety clarified by means structural and functional analysis. The special approach to explaining the main mechanisms and directions of the PRC's environmental policy in the context of the task of the PRC authorities to achieve food security formulated by the use the method of ascending from the abstract to the concrete.

3 Food and environmental security: sustainable socio-economic development of modern China

The sustainable development of Chinese society is aimed at realizing the main goals of Chinese modernization while preserving natural resources, harmonious development of the economy and nature. The concept of ecological civilization was first enshrined in the country's regulatory documents and the Charter of the Chinese Communist Party. The concept of human adaptation to nature is proclaimed in the context of the ideologeme of ecological civilization. The sustainable development of Chinese society is aimed at realizing the main goals of Chinese modernization while preserving natural resources, harmonious development of the economy and nature. The concept of ecological civilization was first enshrined in the country's regulatory documents and the Charter of the Chinese Communist Party in 2017. The concept of human adaptation to nature is proclaimed in the context of the ideologeme of ecological civilization [4]. Chinese elites are making serious efforts for the ecological progress of society. Russian researchers note that the construction of the "ecological civilization" means that the authorities pay special attention to environmental aspects in various areas of Chinese policy [5].

Agriculture is the only branch of the Chinese economy directly related to the country's natural conditions. These include climate, soil condition, and its availability of water resources [6]. The main task of China's agriculture is to ensure national food security and supply the main types of agricultural products to the markets of the People's Republic of China [7]. Food security is "a situation in which all people at any time have social, physical and economic access to the necessary amount of safe and nutritious food that meets their dietary needs and preferences for an active and healthy life" [8]. Jie Zou and Shining Guo et al. write that food security is the foundation of national security [9]. Food security depends on the dynamics of food exports and imports, as well as the growth of national production in general. It is characterized by such properties as the self-sufficiency of society with food, food prices and the level of their consumption by society. The group of international researchers notes it [10]. Therefore, it is no coincidence that the Chinese authors Ling Ye et al write that the assessment of food security should take into account parameters related not only to food production, but also to their consumption [11].

This approach to determining the parameters and characteristics of food security is generally accepted not only among various international researchers, but also in the Chinese academic community. The level of food security in China is determined by the quality and quantity of consumed food . Therefore, Chinese scientists Yuke Hou and Xin Liang note that improving the well-being of Chinese citizens will lead to an update in the structure of consumption of basic foodstuffs by the Chinese population, which will inevitably contribute to radical changes in the structure of Chinese society's demand for food [12].

4 Mechanisms of environmental regulation in the People's Republic of China

Chinese researchers have made a serious contribution to the study of the environmental policy of the People's Republic of China. Wang Chunmei and Lin Zhaolan analyzed China's environmental policy in the first decade since the beginning of the 21st century. They note the country's environmental policy as a factor in the renewal of production capacities and the development of a system of international cooperation and exchange for the sustainable development of the world economy. Researchers are writing about various environmental activities. The focus of Western researchers is on China's efforts to generate environmental energy, with special attention paid to the use of renewable sources within the framework of the emissions trading system. Qianyang Tu and Ying Wang come to the conclusion about the ambiguity and ambivalent consequences of the implementation of the PRC's environmental policy under tax legislation. Scientists note that the new policy will able to improve the quality of the environment in the medium and long term. However, the new taxation may negatively affect the total volume of production, consumption and investment. In fact, this will mean a slowdown in the growth of the Chinese economy. Jiajie Liu, George Du, Weina Liu and Dianwei Zhang write that within the framework of China's 70-year environmental policy, the authorities focus on administrative and command methods and control, which makes it difficult to manage the economy. In the future, various controls should be provided to expand voluntary actions of citizens in the field of environmental activism. These measures should be based on the concept of equal attention to pollution prevention and environmental protection. It is no coincidence that Russian authors write about an integrated approach to the environmental policy of the Chinese authorities. For example, S. P. Savinsky notes that the leaders of the People's Republic of China have to combine administrative and economic measures and methods.

Environmental policy is implemented in the People's Republic of China as a set of complex measures of economic, regulatory and administrative mechanisms, which regulate the protection and state of the environment. *Regulatory regulation* includes legal legislative and party norms aimed at protecting the environment of the People's Republic of China.

The operation of *the legislation* is a mechanism for ensuring the environmental policy of the state. China has faced environmental problems that it had not dealt with them before since the beginning of the era of “reform and openness”. Until recently, the laws were restrictive in nature, the norms provided for the occurrence of adverse legal consequences for violators of environmental legislation. These include regulatory legal acts “on environmental protection” or “on environmental impact assessment”. Russian scientists write that the legislation of the People's Republic of China requires further development and strengthening legal regulation in this area is an urgent task in the environmental sphere.

Therefore, the trends of stricter state regulation are beginning to develop in China's environmental legislation today. Qianyang Tu and Ying Wang note the economic regulation implemented by the state through the development of environmental taxation. Thus, the law on a special new tax norm regulating the collection of mandatory fees to the state budget for environmental protection. This tax is the first experience of the state in the history of China to ensure the protection of nature. The implementation of this tax cancels the functioning of a whole range of penalties imposed by the authorities for environmental degradation due to anthropogenic factors. The previous law was in force earlier for more than 40 years.

Party environmental standards. The norms of environmental law are also contained in some party documents. The main regulatory document of the CPC contains an indication of the general line of China's environmental policy, including the regulation of China's environmental activities to ensure the country's food security. In particular, the Charter contains a provision on the development of a society focused on saving natural resources. Such a society takes care of nature. The natural environment is protected by a set of strict sanctions. The spatial configuration of the new type is based on production, which allows to preserve nature and save resources, as well as create a comfortable production base and a material and material household environment. During the last main annual party forum of the country, the 3rd plenum of the CPC Central Committee of the 20th convocation, the main measures of environmental policy were confirmed. In the final communique of the plenum, in the context of the trend of continuing the “era of reform and openness” in China, plans were noted to begin the comprehensive construction of a socialist modernized state. In the planning aspect, the need to improve the prospects for green and low-carbon development was fixed.

Thus, it can be stated with confidence that the comprehensive orientation of the Chinese government towards building an “ecological civilization” and protecting the environment is an important condition for solving the tasks of sustainable development.

5 Direction of environmental policy in modern China

Modern China is implementing today and will continue to pursue a policy of conservation of nature in the future for achieve the food security of state. The main directions of China's environmental policy are to reduce quantitative and qualitative indicators of environmental pollution, including air and water purification, as well as ensuring sustainable and rational use of natural resources.

Ensuring sustainable environmental management. Current researches show that the efforts made by Governments are currently insufficient. Many states need to intensify measures to integrate sustainable development into the governance structure and promote environmental innovation. Strategies for the protection, conservation and enhancement of natural resources should be based on specific constraints in the process of environmental management faced by economic entities. The task of sustainable management of natural resources is to find a balance between protecting ecosystems and meeting the growing needs of society.

The concentration of states' efforts in the management segment in the process of achieving sustainable development. First of all, it need on ensuring the environmental friendliness of management activities. It will make possible to achieve the goals of sustainable development at all levels (state support, the introduction of innovative resource-saving, environmentally friendly and efficient technologies). The sustainable development policy of modern China is based on reducing low-carbon and coal-fired power plants to prevent air pollution in order to realize the concept of a zero-carbon world and reduce carbon intensity to 18.8%. In addition, the country has increased the area of forests by almost 23.4%. For example, the creation of five national parks was announced, 98% of polluted reservoirs were cleaned and air quality in many large cities in 2021.

Chinese scientists write that on the way to realizing the goals of national development, the People's Republic of China faces the problem of achieving a balance between economic growth and environmental conservation. The main measures of the People's Republic of China on the path of sustainable development are an integrated approach that requires policy coordination in social, economic and environmental contexts. Therefore, lean environmental management of the People's Republic of China in the realities of the national environmental policy is one of the most important priorities.

Reducing environmental pollution. The significant pace of modernization and reform of the economic sphere of the PRC society could not be without a trace for the state of the environment at the beginning of the XXI century. Significant annual GDP growth rates ranged from 9 to 14%. The environmental activities of the Chinese authorities were initially focused on regulating their condition. The air pollution situation continued to worsen in the mid-2010s. Only 8 of the 74 cities monitored by the Chinese Ministry of Environmental Protection met air quality standards in 2014. The unfavorable trend of industrial air pollution in the 2020s also continues to develop. The Ministry of Ecology and Environmental Protection reported that 121 of China's 337 prefectures - city levels did not meet air quality standards in 2021. China ranks 160th out of 180 countries and regions in the Global Environmental Performance Index 2022.

The shortage of water and its pollution contribute to the aggravation of environmental problems with adverse economic, social and environmental consequences in China. Widespread pollution of surface waters is largely the result of poor water management - increased emissions of chemicals into reservoirs, etc.

Serious efforts are being made to combat air pollution in China . Significant progress has been made in the control of vehicle emissions and urban air pollution, and flue gas desulfurization. The air quality in China has not deteriorated further, despite the existing problems associated with increased coal consumption, the number of vehicles and the pace of urbanization.

Russian researchers note that global pollution of fresh water and atmospheric air creates prerequisites for the inefficiency of agriculture. The development of adverse environmental trends has a negative impact on the food security of society. To achieve the goals of strategic development, the leadership of the People's Republic of China will apply environmental policy mechanisms in the short and medium term.

6 Results

1. The balanced development of the environment and the task to ensure food security of Chinese society are closely interrelated.
2. The environmental regulation measures of the Chinese authorities help to implement the national development plan of the People's Republic of China until 2035 in terms of ensuring food security.

3. The environmental activity of the state in China is a system of interrelated and interacting elements of administrative, economic and regulatory mechanisms that affect the state and protection of the natural environment of society.
4. The regulatory mechanism includes the implementation by the authorities of legislative and party norms for the protection of the environment of the People's Republic of China.
5. Mechanisms of party-state management of the state of the natural sphere help Chinese elites to implement the main trajectories of environmental policy. Such areas contribute to sustainable environmental management, environmental conservation and reduction of the adverse effects of economic modernization. They help the authorities to maintain the proper and necessary state of water and air levels.

7 Conclusion

The scientific, technical and economic power of China will increase significantly due the achievement of the main national development goals in 2035. Adverse anthropogenic environmental factors during the implementation of Chinese modernization are manifested very often. The main levers of environmental regulation relate to the powers of the State in China. The effectiveness of the government's achievement of strategic national goals and ensuring food security in the short and medium term will depend on the functioning of the party apparatus.

References

1. N. D. Eriashvili, Y. A. Ivanova, "K voprosu ob ekologii sovremennogo Kitaya.", Obrazovanie i pravo **4**, 371-375 (2022)
2. A. V. Boyarkina, "Global'naya strategiya Si Czin'pina stroitel'stva "ekologicheskoj civilizacii" v novuyu epohu.", Etnosocium i mezhnacional'naya kul'tura **10(148)**, 113-122 (2020)
3. E. Kozyrskaya, S. Bao, "Sel'skoe hozyajstvo Kitaya: kratkaya harakteristika sovremennogo sostoyaniya", Rossijsko-kitajskie issledovaniya **6(2)**, 93-102 (2022)
4. Mukhopadhyay, P. Thomassin, J. Zhang, "Food security in China at 2050: a global CGE exercise", Journal of Economic Structures **7(1)**, 1 (2018)
5. Y. Hou, X. Liang, "Research on Food Security Risk Assessment and Early Warning in China Based on BP Neural Network Model", Journal of Food Quality **1**, 1-12 (2022)
6. Q. Tu, Y. Wang, "New Environmental Protection Taxes in China from the Perspective of Environmental Economics", Discrete Dynamics in Nature and Society **1**, 1-10 (2021)
7. J. Liu, G. Du, W. Liu, D. Zhang, "70-year Evolution of China's Environmental Policy — Based on the Analysis of Policy Text Using Nvivo 11.0", IOP Conference Series: Earth and Environmental Science **1-5**, 544 (2020)
8. S. P. Savinskij, "Ekologo-ekonomicheskoe regulirovanie v Kitajskoj Narodnoj Respublike", Ekonomika, predprinimatel'stvo i pravo **9(4)**, 785-796 (2019)
9. E. Abanina, N. Makhonko, B. Levitanus, E. Tarasova, E. Sukhova, Yu Plotnikova, "Environmental sustainability in management: existing conditions and guidance for action", IOP Conference Series: Earth and Environmental Science **808(1)**, 1-7 (2021)
10. Md. Ziaul, S. Wang, "Exploring the unique characteristics of environmental sustainability in China: Navigating future challenges", Chinese Journal of Population Resources and Environment **21**, 37-42 (2023)

11. Ch. Chzhan', "Ekologicheskaya politika Kitaya i realizaciya celej v oblasti ustojchivogo razvitiya", Vestnik Moskovskogo gosudarstvennogo oblastnogo universiteta. Seriya: Istoriya i politicheskie nauki **2**, 184-192 (2019)
12. Y. Liu, R. M. Nor, M. K. Ishak, X. Li, "The Evolution of Chinese Environmental Regulation and its Green Innovation Effects: A Review and Prospect", International Journal of Academic Research in Business and Social Sciences **14(1)**, 1172-1181 (2024)

https://doi.org/10.1051/e3sconf/202563302001

Integrating drone technology in sustainable agriculture: precision environmental monitoring and resource management

Oksana Takhumova[1], *Andrey* Pustuev[2], *Nikolay* Eiryan[3], *Anastasia* Sarsadskikh[3], and *Olga* Rykalina[2,*]

[1]Kuban State Agrarian University named after I. T. Trubilin, Krasnodar, Russia
[2]Ural State Law University, Yekaterinburg, Russia
[3]Ural State University of Economics, Yekaterinburg, Russia

Abstract. Current advances in data analytics further amplify the benefits, providing farmers with actionable insights to optimize plant nutrition, irrigation schedules, pest control, and harvest timing. Beyond operational gains, drone-assisted monitoring can strengthen farm resilience by identifying early indicators of soil degradation, water stress, and pest outbreaks, thus informing preventive measures. In parallel, drones can foster social and economic sustainability by lowering production costs and promoting equitable access to knowledge. Smallholder farmers, for example, can benefit from shared drone services that offset initial capital investments. Moreover, real-time, field-level data obtained via drones facilitate transparent communication among supply chain stakeholders, leading to more informed decisions and policy-making. Despite these promising outcomes, several challenges constrain the widespread integration of drones in agriculture. These include regulatory hurdles, the need for specialized training, and data privacy concerns. Additionally, drones require robust infrastructure and technical support to ensure reliability in diverse farming contexts. Addressing these issues through targeted research, policy reforms, and educational initiatives is essential to fully harness drone technology for sustainable agricultural systems. Overall, drone adoption in agriculture heralds a new era of precision farming, with the potential to revolutionize traditional practices and drive the sector towards greater environmental stewardship and socio-economic viability.

1 Introduction

Sustainable agriculture aims to meet present food needs while safeguarding the environment and resources for future generations. Over the past few decades, technological advancements have facilitated more precise and efficient agricultural operations that align with sustainability objectives. Among these innovations, drone technology has emerged as a powerful tool to enhance environmental monitoring, streamline farm management practices, and support data-driven decision-making. Drones, or unmanned aerial vehicles (UAVs), are

* Corresponding author: ctig.usue@mail.ru

aircraft systems piloted remotely or autonomously through embedded programs. Initially designed for military and surveillance purposes, drones have found extensive civilian applications, including agriculture. Their capability to capture high-resolution imagery and multispectral or hyperspectral data in real time enables farmers to monitor field conditions with greater detail and frequency than traditional ground-based methods. This information can be employed to detect pest infestations, identify crop stress, optimize fertilizer application, and conserve water resources.

One of the core principles of sustainable agriculture is resource efficiency. By using drone-collected data, farmers can apply inputs—such as water and fertilizers—only in precise areas and amounts. This targeted approach not only reduces waste but also lessens the environmental footprint of farming operations. For instance, variable-rate irrigation strategies informed by aerial mapping can conserve freshwater by delivering water strictly where and when it is needed. Similarly, drones equipped with spectral sensors can differentiate healthy crops from those suffering nutrient deficiencies or pest damage, facilitating swift and localized interventions. Drones also play a role in enhancing biodiversity and soil health management. By providing farmers with detailed topographical and vegetative indices, drones can pinpoint areas prone to erosion, soil nutrient depletion, or invasive species. Early detection of these issues aids in implementing timely conservation or restorative measures, thus mitigating long-term damage and promoting ecological balance. Additionally, advanced imaging techniques can help evaluate crop rotations, cover crops, and buffer zones that contribute to soil fertility and wildlife habitats.

Socio-economic sustainability is another essential dimension of modern agriculture. Smallholder farmers, especially in developing regions, often face resource constraints and limited access to agricultural extension services. Drones can bridge these gaps by offering shared or community-based drone programs that reduce capital costs for individual farmers. These initiatives can democratize access to timely, data-driven insights. Furthermore, drone-derived data can support traceability in supply chains, meeting the increasing consumer demand for transparency regarding environmental stewardship and ethical sourcing. However, the integration of drone technology into agriculture is not without challenges. Many jurisdictions lack clear regulations governing drone flights, leading to logistical and legal uncertainties. Additionally, the operation and maintenance of drones require specialized training and technical know-how, posing barriers to small-scale farmers. Data privacy and ownership are further concerns, as aerial imagery could inadvertently capture sensitive information about landholdings and farming practices.

Despite these limitations, the potential for drones to advance sustainable agriculture is immense. Continuous improvements in sensor technology, battery life, and data processing algorithms will likely expand the scope of drone applications. In tandem with supportive policies and capacity-building efforts, drone technology can empower farmers to optimize resource use, reduce environmental impact, and improve crop yields. In doing so, it addresses the urgent global need for an agricultural paradigm that balances productivity with ecological and social well-being.

The following sections explore the latest scholarship on drones in sustainable agriculture, highlighting perspectives from different regions and research methodologies. By synthesizing these findings, this article underscores the relevance of drone-based monitoring and resource management in driving agricultural innovation and enhancing sustainability outcomes. The discussion also considers critical research gaps, regulatory frameworks, and practical constraints that shape the future trajectory of drone adoption in the agricultural sector.

The purpose of this study is to analyze the expanding role of drone technology in sustainable agriculture, focusing on how these tools facilitate precision environmental monitoring and resource management. By reviewing current literature, this article aims to identify best practices, potential challenges, and key drivers that influence drone adoption, ultimately contributing to a more resilient and ecologically responsible agricultural sector.

2 Bibliographic analysis

Studies addressing drone technology in agriculture underscore its growing significance for precision farming, environmental stewardship, and resource optimization. Hafeez, Husain, Singh, and coauthors [1] conducted a comprehensive review of drone-based pesticide spraying, illustrating how it can significantly lower chemical usage and minimize off-target effects, thus aligning with environmental sustainability objectives. Wadod and Mohammed [2] emphasized application methodologies in precision agriculture, arguing that standardized protocols for drone flight planning and sensor calibration are crucial to realizing consistent data quality. Ghazali, Azmin, and Rahiman [3] delved into a survey of drone implementation in precision agriculture, proposing that cost-benefit analyses remain vital for smallholder farmers, especially in developing regions where resources are scarce. Similarly, Ivezić, Trudić, Stamenković, and colleagues [4] explored drone-related agrotechnologies in the Western Balkans, highlighting legal frameworks and the importance of region-specific data to guide effective pesticide applications. Remote sensing analysis via drone imagery was examined by Meivel and Maheswari [5], who noted the advantages of real-time data acquisition for crop stress detection. This rapid assessment capacity fosters more timely interventions and reduces yield losses. In another study, Elahi, Khalid, and Zhang [6] linked farmers' adoption of renewable energy and drone technologies, finding that environmental awareness and supportive policy incentives substantially increase willingness to invest in precision tools.

D. Van Der Merwe, Burchfield, Witt, and coauthors [7] identified how drone-based scouting can be incorporated into broader agronomic advisories, offering a holistic framework for yield enhancement and risk reduction. Gardezi, Adereti, Stock, and Ogunyiola [8] investigated the ethical and societal implications of drone technologies, arguing for responsible innovation that addresses community concerns about data privacy and labor displacement. In terms of irrigation management, Ma, Hu, Li, and their research team [9] provided evidence that multispectral drone data can optimize water application in maize cultivation, saving water without compromising yields. Aliloo, Abbasi, Karamidehkordi, and coauthors [10] offered a concise summary of recommended practices ("Dos and Don'ts") for drone use in crop fields, underscoring safety, regulatory compliance, and best operational procedures.

The relevance of drone technology for drought monitoring was highlighted by Alwan, Ziboon, Khalaf, and others [11]. Their research showed how drone-acquired imagery improves the resolution of drought indices, allowing for more nuanced detection of moisture stress at various phenological stages. Meanwhile, Kross, Kaur, and Jaeger [12] presented a geospatial framework to assess environmental impacts of agriculture, emphasizing the synergy between aerial data and geographic information systems to guide sustainable land-use planning. Ragab [13] examined innovation in agricultural water management, revealing that drones can play a critical role in improving irrigation scheduling and drainage systems—key factors in achieving multiple Sustainable Development Goals. Radić, Radić, and Cogoljević [14] similarly recognized new technologies, including drones, as pivotal drivers of change in the agricultural sector, although they caution that human capacity building and supportive policies must keep pace.

Tsong and Khor [15] took a broader perspective on modern analytical and bioanalytical technologies that complement drone usage. They contend that the integration of advanced sensors, internet-of-things platforms, and machine learning algorithms can significantly amplify the benefits of drone data in precision agriculture. Collectively, these studies highlight the multifaceted benefits of drone technology—from enhanced crop monitoring to water management, pesticide reduction, and sustainability analytics. They also reveal persistent challenges such as regulatory barriers, skills gaps, and data privacy issues. Nonetheless, the consensus points toward drones as increasingly indispensable tools for achieving sustainable agriculture, provided that technical, economic, and sociopolitical considerations are addressed comprehensively.

3 Results and discussions

The integration of drones into sustainable agriculture is transforming how farmers perceive and manage their fields. This section synthesizes the findings from the reviewed literature, illustrating key results and situating them within the broader context of environmental protection, resource conservation, and socio-economic viability.

1. Enhancing Crop Health Monitoring

One of the primary advantages of drone technology lies in its ability to conduct aerial surveys that capture detailed, real-time imagery. These surveys can use multispectral and thermal sensors to detect subtle variations in crop health indicators. Early identification of pest infestations or nutrient deficiencies allows for immediate interventions, reducing yield losses and lessening reliance on blanket pesticide or fertilizer applications. Consequently, input use becomes more judicious, aligning with core sustainability goals. Drones thus serve as an "eye in the sky," affording farmers a level of spatial resolution not easily achievable through ground-based scouting.

Moreover, drone-collected data is increasingly integrated with machine learning algorithms that classify vegetation health patterns based on spectral signatures. By automating the analysis, producers can monitor large tracts of land in a fraction of the time it would take manual inspection teams, with improved accuracy. This synergy between drone imaging and computational analytics exemplifies how advanced technologies can streamline operations, simultaneously benefiting farm productivity and environmental stewardship.

2. Precision Resource Management

A central tenet of sustainable agriculture is the efficient use of resources such as water, fertilizers, and energy. Drone flights equipped with specialized sensors can monitor soil moisture levels across different field zones, enabling precise scheduling and targeting of irrigation. This approach mitigates overwatering—one of the most prevalent issues in traditional irrigation systems—and conserves valuable freshwater resources. Similarly, nutrient application can be localized to underperforming zones, preventing run-off and eutrophication in nearby water bodies.

Energy consumption in farming operations also comes under the purview of drone-assisted management. Conventional farming methods may require extensive machinery use for repeated scouting or input application. Drones can reduce the frequency of such passes by providing more targeted data upfront, thus curtailing fuel consumption and associated emissions. Collectively, these resource-saving mechanisms underscore the ecological and economic gains of drone technology.

3. Soil and Environmental Protection

Healthy soils are fundamental to both crop productivity and ecological stability. Drone imagery can reveal patterns of soil erosion, compaction, or organic matter depletion, allowing for proactive remediation strategies. For instance, farmers might adopt cover cropping or reduce tillage in erosion-prone areas identified through aerial surveys, thereby preserving topsoil. Drones also assist in monitoring buffer strips along water bodies, tracking sediment movement, and ensuring compliance with conservation guidelines.

From an environmental standpoint, limiting chemical runoff is another critical aspect of sustainable agriculture. Drone-based pesticide delivery systems, combined with real-time field imagery, allow for site-specific applications, minimizing the volume of chemicals introduced into ecosystems. Furthermore, the fine control offered by drones reduces off-target effects on beneficial insects and neighboring habitats, supporting biodiversity conservation.

4. Socio-economic Considerations

While drone technology holds promise, it also introduces challenges in terms of accessibility and education. Small-scale and resource-limited farmers may face prohibitive costs related to purchasing drones or acquiring the necessary software and training. Collective ownership models—where multiple farmers share a single drone service—present a potential solution. In such models, a service provider offers flight operations, data analysis, and consultation, making advanced aerial monitoring more accessible.

Additionally, drone-driven precision agriculture can generate new job opportunities in rural areas. Operators, data analysts, and agronomy consultants become integral to the local agricultural ecosystem. However, the risk of widening the digital divide remains. Farmers who lack digital literacy or resources to adopt these technologies could be left behind, emphasizing the need for targeted capacity-building initiatives and policy interventions.

5. Regulatory and Ethical Frameworks

Drone adoption does not exist in a regulatory vacuum. Most countries have rules governing drone flight altitudes, pilot certifications, and no-fly zones. Agricultural operations often require flights over private properties, raising questions about data privacy and property rights. As drone technologies become more pervasive, robust ethical guidelines and clear legislative frameworks will be essential to protect stakeholders' interests.

Data privacy, in particular, is a growing concern. High-resolution aerial imagery may inadvertently capture sensitive information, leading to potential conflicts regarding data ownership and consent. Farmers need transparent agreements concerning who controls the generated data and how it can be used. Policymakers, in collaboration with industry and civil society, must establish guidelines that balance innovation with the protection of individual rights and local communities.

6. Integration with Broader Digital Ecosystems

Drones rarely function as standalone solutions. Their effectiveness is magnified when integrated into holistic digital ecosystems that may include satellites, ground sensors, and cloud-based analytics platforms. For example, satellite imagery can offer macro-level overviews of regional climate patterns, while drone imagery zooms in for micro-level detail. Ground sensors, meanwhile, provide real-time soil moisture or temperature data that can validate aerial observations. Combining multiple data sources ensures that farmers receive a multidimensional perspective of their fields, leading to well-informed decisions.

This integrated approach also facilitates predictive modeling, where data from drones is fed into algorithms that forecast crop yields, pest outbreaks, or rainfall patterns. Such prognostic tools support long-term planning and adaptive management strategies. Over time, the accumulation of historical drone data may enable artificial intelligence systems to learn field-specific patterns, further refining recommendations and interventions.

7. Overcoming Technological Barriers

Despite rapid advancements, drone technology still faces barriers that could limit its adoption in agriculture. Battery life remains a bottleneck, particularly for larger farms that require extended flight durations to cover broad swaths of land. Weather conditions—such as strong winds, heavy rain, or extreme temperatures—can ground drones or degrade sensor performance. These factors underscore the need for innovation in drone design, propulsion, and weather-resilient sensors.

Another limitation is the complexity of data interpretation. Even with sophisticated analytics platforms, farmers and agronomists must receive training to understand aerial imagery results thoroughly. This learning curve can slow the diffusion of drone technology, especially in regions with limited extension services. Partnerships between universities, research institutes, and local farming communities can help address these gaps through targeted training and demonstration projects.

8. Future Outlook

The trajectory of drone technology in sustainable agriculture appears decidedly positive, with continuing enhancements in sensor resolution, flight stability, and data processing speeds. As machine learning and artificial intelligence become more integrated, drones will offer increasingly automated and adaptive solutions. In the long term, swarm technology—where multiple drones coordinate within a single operation—may emerge, further increasing efficiency and coverage.

Nevertheless, for drones to become indispensable tools, parallel progress in regulatory reform, infrastructure development, and farmer education is crucial. Governments and international agencies could facilitate financing mechanisms that help farmers acquire or lease drones under favorable terms. Industry stakeholders, for their part, might collaborate on developing open-source platforms to reduce software costs and foster innovation.

Ultimately, the incorporation of drones into sustainable agriculture represents a pivotal step toward modernizing and greening food production. By enabling precise input application, resource conservation, and real-time problem-solving, drones align closely with the principles of environmental stewardship and socio-economic equity. Yet, their success will depend on cohesive action among researchers, policymakers, tech companies, and agricultural communities—ensuring that the benefits of drone technology are shared widely and ethically.

4 Conclusion

Drone technology, with its capacity for high-resolution environmental monitoring and precision resource management, has emerged as a pivotal component of modern sustainable agriculture. Unlike many conventional approaches that rely on uniform applications and reactive decision-making, drone-assisted systems allow farmers to tailor interventions based on data-driven insights. This specificity fosters a more judicious use of inputs—like pesticides, fertilizers, and water—resulting in reduced environmental impacts and potentially lower production costs.

Equally important, the social dimension of drone integration offers pathways to address knowledge and resource disparities. Shared ownership models and cooperative networks, for instance, can democratize access to cutting-edge technologies, ensuring that smaller farms can also benefit. Moreover, employing drones can create new employment niches for operators, analysts, and agronomic advisors, breathing fresh vitality into rural economies.

From an ecological perspective, drone technology can bolster biodiversity conservation by minimizing off-target chemical exposure and monitoring critical habitats. By detecting early signs of pest outbreaks, soil erosion, or water deficits, drones enable proactive and strategic responses that enhance long-term farm resilience. Additionally, when drones feed into broader digital ecosystems involving satellites and ground sensors, farmers gain a multidimensional understanding of field conditions, further elevating management precision.

Nonetheless, integrating drones into agricultural systems requires more than technology alone. Structured training programs, supportive policy frameworks, and transparent data governance are indispensable for maximizing benefits and mitigating risks. Issues such as data privacy, regulatory restrictions, and financial barriers must be navigated thoughtfully to ensure that drone adoption does not exacerbate existing inequities or compromise local autonomy.

In summary, drone technology holds considerable potential to transform agriculture into a more sustainable, efficient, and equitable endeavor. By aligning real-time aerial insights with responsible resource management practices, drones can help usher in an era of farming that balances high productivity with environmental stewardship. Collaboration among stakeholders—encompassing policymakers, technology developers, researchers, and farmers—will be essential to fully unlock and distribute the advantages of drone-based innovation in sustainable agriculture.

References

1. Hafeez, M. A. Husain, S. P. Singh, "Implementation of drone technology for farm monitoring & pesticide spraying: A review", Information Processing in Agriculture **10(2)**, 192-203 (2023)
2. M. M. Wadod, F. G. Mohammed, "Review on Drone application methodologies in agriculture precision", IOP Conference Series: Earth and Environmental Science **1202(1)**, 012001 (2023)
3. M. H. M. Ghazali, A. Azmin, W. Rahiman, "Drone Implementation in Precision Agriculture – A Survey", International Journal of Emerging Technology and Advanced Engineering **12(4)**, 67-77 (2022)
4. A. Ivezić, B. Trudić, Z. Stamenković, "Drone-related agrotechnologies for precise plant protection in Western Balkans: applications, possibilities, and legal framework limitations", Agronomy **13(10)**, 2615 (2023)
5. S. Meivel, S. Maheswari, "Remote sensing analysis of agricultural drone", Journal of the Indian Society of Remote Sensing **49(3)**, 689-701 (2021)
6. E. Elahi, Z. Khalid, Z. Zhang, "Understanding farmers' intention and willingness to install renewable energy technology: A solution to reduce the environmental emissions of agriculture", Applied Energy **309**, 118459 (2022)
7. D. Van Der Merwe, D. R. Burchfield, T. D. Witt, "Drones in agriculture", Advances in Agronomy **162**, 1-30 (2020)
8. M. Gardezi, D. T. Adereti, R. Stock, A. Ogunyiola, "In pursuit of responsible innovation for precision agriculture technologies", Journal of Responsible Innovation **9(2)**, 224-247 (2022)
9. L. F. Ma, N. Y. Hu, W. Li, "Using multispectral drone data to monitor maize's response to various irrigation modes", Journal of Plant Nutrition and Fertilizers **28(4)**, 743-753 (2022)
10. J. Aliloo, E. Abbasi, E. Karamidehkordi, "Dos and Don'ts of using drone technology in the crop fields", Technology in Society **76**, 102456 (2024)
11. I. A. Alwan, A. T. Ziboon, A. G. Khalaf, "Monitoring agricultural and meteorological drought using remote sensing", Arabian Journal of Geosciences **15(2)**, 178 (2022)
12. A. Kross, G. Kaur, J. A. G. Jaeger, "A geospatial framework for the assessment and monitoring of environmental impacts of agriculture", Environmental Impact Assessment Review **97**, 106851 (2022)
13. R. Ragab, "Innovation and research in agriculture water management to achieve sustainable development goals", Irrigation and Drainage **72(5)**, 1195-1199 (2023)
14. V. Radić, N. Radić, V. Cogoljević, "New technologies as a driver of change in the agricultural sector", Economics of Agriculture **69(1)**, 147-162 (2022)
15. J. L. Tsong, S. M. Khor, "Modern analytical and bioanalytical technologies and concepts for smart and precision farming", Analytical Methods **15(26)**, 3125-3148 (2023)

E3S Web of Conferences **633**, 02002 (2025) https://doi.org/10.1051/e3sconf/202563302002
GASES 2025

Eco-integrated agriculture: merging environmental sustainability and energy efficiency in the modern agro-production complex

Mei Hua Wang[1], *Olga* Rykalina[2,*], *Tatyana* Borzunova[4], *Egor* Kolobov[2,3], and *Oksana* Takhumova[5]

[1]Northwest A&F University, China
[2]Ural State University of Economics, Yekaterinburg, Russia
[3]Chief Physician of the State Autonomous Institution "Aramil City Hospital", Aramil, Russia
[4]Ural State Medical University, Yekaterinburg, Russia
[5]Kuban State Agrarian University named after I. T. Trubilin, Krasnodar, Russia

Abstract. Simultaneously, ecological intensification approaches enhance soil fertility, preserve biodiversity, and protect water resources, all of which are vital to the long-term viability of farming communities. Central to the success of eco-integrated agriculture is the adoption of precision practices that align input requirements—water, fertilizers, and pesticides—with site-specific conditions. Real-time data from remote sensing, drones, and ground sensors can inform crop and livestock management, ensuring minimal waste and higher productivity. Moreover, farmer education, institutional support, and policy incentives are crucial for driving widespread adoption of these practices. Challenges remain, including initial capital investments, the need for robust infrastructure, and varying regulatory frameworks. Yet, the potential for significant energy savings, reduced pollution, and enhanced resilience positions eco-integrated agriculture as a blueprint for sustainable intensification. This article examines the underlying principles, current research trends, and practical applications of eco-integrated agriculture, highlighting real-world examples that illustrate both its feasibility and adaptability. By synthesizing existing knowledge, the article offers insights into how future agro-production complexes can be designed to blend economic viability, ecological integrity, and energy efficiency in a balanced, synergistic manner.

1 Introduction

The urgency of transforming agricultural systems has never been greater. Faced with escalating population pressures, finite natural resources, and the looming threat of climate change, the modern agro-production complex must evolve to remain viable. Traditional, high-input models of farming have indeed boosted yields, but at the cost of overusing

* Corresponding author: roa40@mail.ru

fertilizers, pesticides, and nonrenewable energy sources. Soil degradation, biodiversity loss, and water pollution are just some of the visible consequences. To navigate these challenges, agriculture must advance toward a model that not only produces adequate food but does so with minimal environmental impact and optimal energy usage.

Within this context, the concept of "eco-integrated agriculture" has gained prominence. It goes beyond the conventional focus on yields to incorporate ecological processes, renewable energy, and efficient resource management throughout the entire production cycle. By merging multiple components—soil health, biodiversity conservation, renewable energy generation, and technological innovations—eco-integrated agriculture aims to create a self-sustaining loop where waste products become resources, and energy requirements are partially or wholly met by on-site renewable systems. For instance, crop residues or animal waste can be converted into biogas to power farm machinery, thereby reducing reliance on fossil fuels while simultaneously addressing waste disposal issues.

Moreover, eco-integrated agriculture relies heavily on precision farming techniques to optimize resource allocation. The widespread availability of digital technologies—ranging from drones and remote sensors to advanced data analytics—enables farmers to fine-tune irrigation schedules, fertilizer applications, and pest management strategies. This level of specificity not only reduces costs but also curtails the environmental footprint associated with excessive inputs. In parallel, solar panels installed on barns or integrated into field operations can supply a significant portion of on-farm electricity needs. Such on-site energy production minimizes transmission losses, stabilizes energy costs, and contributes to cleaner air.

However, achieving energy efficiency and environmental sustainability is not solely about technology. Fundamental shifts in farm management, labor practices, and market structures are also imperative. The transition often demands new skill sets, including data literacy and equipment maintenance, which can be particularly challenging for smallholders with limited resources. Policy frameworks and financial mechanisms thus have a pivotal role to play, whether through subsidies for solar panel installation, tax incentives for biomass-derived energy, or certification schemes that reward low-input, biodiversity-friendly practices. Increasingly, consumers drive demand for sustainably produced foods, pressuring supply chains and retailers to source from eco-conscious farms.

Within the broader sustainability narrative, the interplay between energy consumption and ecological health cannot be overstated. Farming processes that depend heavily on fossil fuels contribute to greenhouse gas emissions, exacerbate climate volatility, and undermine the resilience of global food systems. Eco-integrated agriculture, by contrast, seeks to decouple productivity from high-carbon footprints through innovations and management approaches that balance human needs with ecological boundaries. This equilibrium is particularly vital in regions facing water scarcity, soil erosion, or biodiversity collapse. For example, restoring wetlands and riparian buffers can improve water storage, reduce nutrient runoff, and provide habitats for beneficial organisms, all while supporting adjacent farmland.

In the sections that follow, this article delves into the growing body of research exploring eco-integrated systems, highlighting how scholars, policymakers, and practitioners are shaping the discourse. Beyond a theoretical framework, practical examples will illustrate the tangible benefits of merging environmental sustainability with energy efficiency. Ultimately, while challenges remain—ranging from regulatory hurdles to technological costs—the pursuit of eco-integrated agriculture marks a pivotal step toward reconciling food security with the ecological realities of our time.

The purpose of this study is to examine how eco-integrated agriculture, characterized by synergistic environmental sustainability and energy efficiency, can reshape modern agro-production complexes. Through a review of existing research and practical insights, it aims to identify best practices, highlight socio-economic considerations, and propose pathways for policy and technological interventions that align farm productivity with ecological resilience.

2 Bibliographic analysis

Academic interest in merging environmental sustainability and energy efficiency within agriculture has surged over the past decade, reflecting the convergence of agroecological research, renewable energy innovation, and climate policy discussions. Akter, Surovy, Sultana, and colleagues [1] examine the techno-economics of using agricultural biomass for energy generation, highlighting biomass's dual role in waste reduction and renewably fueling farm operations. Meanwhile, Araújo, Chavez-Santoscoy, Parra-Saldívar, and coauthors [2] critique agro-food systems for perpetuating "unsustainable" practices, proposing that targeted policy reforms and technological interventions—like solar-powered irrigation—can significantly alleviate environmental strains.

Similarly, Gliessman [3] emphasizes agroecology's transformative potential, asserting that transitioning to biodiverse cropping systems and on-farm renewable energy can enhance soil regeneration and diminish carbon footprints. Addressing the challenge of sustainability head-on, Brunori, Branca, Cembalo, and others [4] argue that integrated management tools are indispensable, advocating life cycle analyses to measure the interplay between farm productivity and environmental externalities. In parallel, Bergez, Béthinger, Bockstaller, and their team [5] underscore the importance of aligning agri-environmental indicators—like soil quality and biodiversity indices—with yield gap analyses for a more holistic sustainability evaluation.

In a similar vein, Faucon, Aussenac, Debref, and coauthors [6] propose that combining agroecology with bioeconomy solutions paves the way toward eco-integrated farming. They highlight how circular resource flows—such as composting organic waste or refining biomass feedstocks—can lighten agriculture's ecological footprint. Wang, Shi, Mi, and Hu [7] discuss agricultural eco-efficiency trends, noting that energy auditing can reduce overheads and minimize greenhouse gas emissions. Brunori [8], writing on agriculture's role in the ecological transition, reinforces the idea that energy self-sufficiency and responsible input use are key metrics for achieving an agro-production complex that respects planetary boundaries.

A special issue on climate-resilient agri-environmental production systems [9] presents multiple case studies where integrated energy solutions—like methane digesters and solar arrays—lower operational costs and reduce net emissions. This resonates with Xu, Qin, and Zhu's [10] work on biodiversity indices, where the ecological health of farmland is seen as a co-benefit of environmentally conscious management practices, including renewable energy installation. Elahi, Khalid, and Zhang [11] explore farmers' intentions to adopt renewable energy, concluding that favorable policy incentives and knowledge dissemination channels encourage large-scale uptake.

Turning to the socio-economic aspects, Oteh, Mbanasor, Agwu, and Oloveze [12] link sustainable agri-food market systems to improved livelihoods, pointing out that renewable energy infrastructure and eco-friendly cultivation methods can enhance rural communities' resilience and competitiveness. Khalil and Osborne [13] expand on the concept of climate-resilient production, underscoring that ecosystem services—such as carbon sequestration in well-managed soils—are amplified when integrated with low-carbon energy sources. Equally relevant, Rangel Vázquez [14] addresses new material innovations, suggesting that biodegradable inputs and advanced bio-based polymers will likely coincide with on-farm energy generation systems, thereby reducing pollution and fossil fuel dependency.

Finally, Nicholls [15] asserts that agroecology offers a viable path for crisis-prone food systems, arguing that ecological intensification coupled with strategic energy management can buffer farms against climatic shocks. Collectively, these scholarly contributions reveal a consensus: eco-integrated agriculture, combining renewable energy and robust ecological practices, offers both environmental and economic advantages. While localized challenges such as policy gaps and financial costs remain, the overarching message is that synergy between energy efficiency and ecological stewardship constitutes a powerful engine for sustainable agricultural transformation.

3 Results and discussions

Eco-integrated agriculture represents a pivotal shift in farming philosophy. It extends beyond the mere application of eco-friendly or energy-saving techniques; instead, it fosters a cohesive system where energy generation, environmental protection, and agricultural outputs form a synergistic triad. The following sections dissect the key results, illustrating how these concepts translate into on-the-ground practices and discussing the broader implications for sustainability.

One hallmark of eco-integrated agriculture is the deployment of renewable energy sources—solar arrays, wind turbines, or biogas units—directly on farms. Whether small-scale installations powering irrigation pumps or larger systems supplying electricity to multiple operations, these energy models significantly cut carbon footprints and reduce long-term fuel costs. Farmers benefit from energy self-reliance, while environmental gains include fewer greenhouse gas emissions and reduced reliance on non-renewable resources.

Yet, the transition often requires substantial initial capital and technical know-how. Training programs and public-private partnerships can alleviate these hurdles by offering favorable financing, education, and maintenance support. When properly implemented, renewable energy infrastructures become cornerstones of energy-efficient, low-impact agricultural systems.

Eco-integrated strategies do not merely focus on energy. They also emphasize the protection of soil fertility, water quality, and biodiversity. Planting cover crops, rotating diverse species, and preserving natural habitats around farm boundaries foster ecological resilience. These methods improve pollination rates, natural pest control, and long-term soil health, enhancing yield stability and reducing the need for synthetic agrochemicals. In an eco-integrated setting, energy use complements these ecological benefits: for instance, solar-powered sensors can monitor soil moisture, enabling precise irrigation that conserves water and maintains habitat viability.

Farmers adopting eco-integration often cite economic resilience as a primary motivator. By producing on-site energy through biogas or solar systems, farms hedge against volatile fossil fuel prices. Cost savings arise from reduced expenditures on conventional inputs, while premium markets for organic or low-carbon products can boost profitability. However, these benefits are not equally accessible. Smallholder farmers may need collective ownership models or government subsidies to invest in technologies that larger operations can more readily afford. Policy interventions and robust extension services thus become critical levers to bridge these gaps.

To better visualize the complexities of eco-integrated agriculture—covering energy, ecology, and socio-economics—Table 1 synthesizes key dimensions and insights drawn from multiple case studies and research findings. This table provides a snapshot of common drivers, benefits, and barriers, enabling a structured overview of where eco-integration thrives and where further innovation or support may be required.

Table 1. Drivers, benefits, barriers of eco-integration.

Dimension	Drivers	Benefits	Barriers
Energy Efficiency	Renewable energy incentives, rising fuel costs	Reduced GHG emissions, lower operational costs, energy autonomy	High initial capital, limited technical expertise, grid constraints

Ecological Health	Soil regeneration strategies, biodiversity goals	Enhanced soil fertility, natural pest control, water conservation	Knowledge gaps, reluctance to adopt new methods, fragmented policy
Economic Viability	Market premiums, stable input costs	Profit resilience, diversified income, stronger rural economy	Unequal access to finance, uncertain ROI, lack of supportive subsidies
Social Adoption	Community-based extensions, consumer demand	Increased participation, empowered smallholders, skill building	Cultural resistance, fear of tech displacement, unequal information access

From Table 1, it is evident that eco-integrated agriculture operates at the intersection of environmental, economic, and social factors. Energy efficiency is driven by both market forces (fuel cost volatility) and policy incentives (renewable energy credits). Ecological health benefits accrue through deliberate management practices that prioritize soil quality and biodiversity. Economic viability hinges on market demand, stable input costs, and access to financial tools. Finally, social adoption necessitates education, trust, and community engagement. Overcoming barriers in any one dimension often requires a coordinated approach involving farmers, policymakers, researchers, and local communities.

Governments play a pivotal role in shaping the future of eco-integrated agriculture. Subsidies for solar panel installation, tax breaks for organic soil amendments, and grants for cooperative biogas projects can catalyze widespread transition. Furthermore, certification schemes recognizing eco-integrated practices allow producers to tap into niche markets that reward low-carbon, environmentally responsible production methods. Regulatory consistency across regions helps maintain fair competition, especially crucial when cross-border trade and carbon accounting come into play.

Nevertheless, some policies inadvertently hamper progress—for example, fossil fuel subsidies or stringent land-use regulations that deter the construction of on-farm energy facilities. As policy landscapes evolve, ensuring that incentives align with ecological objectives becomes imperative. This alignment fosters innovation, encourages private-sector investment, and raises public awareness about the merits of sustainable farming.

Innovative technologies—from greenhouse sensors and hydroponic systems to drone-based yield mapping—offer new avenues for eco-integration. However, technology transfer remains uneven. Larger enterprises can afford advanced tools and training, while smaller farms risk being left behind. Tailored extension programs that consider local knowledge, language barriers, and cultural practices are essential to achieve broad-based adoption. By integrating indigenous farming wisdom with cutting-edge science, extension services can help farmers transition smoothly without discarding traditional methods that may still hold ecological value.

Furthermore, public-private partnerships and farmer-led research networks can expedite the flow of information and reduce dependency on external consultants. Peer-to-peer learning models, where experienced practitioners share insights on eco-integration, often result in faster behavioral change than top-down directives alone.

A key advantage of eco-integration is its contribution to climate change mitigation. Reductions in fertilizer overuse lower nitrous oxide emissions, while on-site renewable energy curtails carbon dioxide release from fossil fuels. Simultaneously, agroforestry and perennial crops can sequester carbon in soils, forming part of a broader climate-smart agriculture strategy. Water management also improves, as controlled irrigation and buffer zones limit drought stress and eutrophication in surrounding waterways.

In the context of extreme weather events—hurricanes, floods, droughts—ecosystem-based farm designs can enhance resilience. Deep-rooted trees stabilize soils, wetland restoration mitigates flood impacts, and diversified crop rotations reduce vulnerability to pests or disease outbreaks. When combined with local energy generation, eco-integrated agriculture stands as a bulwark against disruptions in fuel supply and grid failures, reinforcing both food and energy security.

Scaling eco-integrated agriculture from pilot projects to mainstream adoption requires coordinated efforts. Research institutions can further refine life cycle assessments, quantifying the net energy savings and environmental gains under diverse geoclimatic conditions. International cooperation might standardize metrics, ensuring global benchmarks for eco-integration.

Simultaneously, capital markets must become more receptive to green investments in agriculture. Initiatives that bundle microloans or crowdfunding could fill financing gaps for small-scale farmers. Corporations can adopt sustainability mandates that prioritize sourcing from eco-integrated producers, thereby creating a market pull effect. As technology becomes more affordable and consumer awareness grows, the shift toward eco-integrated agriculture may accelerate, potentially redefining agricultural norms.

Ultimately, this movement embodies more than an environmental and economic approach. It reflects a philosophical commitment to harmonizing human activities with natural processes. By tapping into circular resource use and renewable energy, eco-integrated farms move closer to a regenerative blueprint that benefits not only immediate stakeholders but also global ecosystems.

4 Conclusion

Eco-integrated agriculture offers a compelling framework for modern agro-production systems, emphasizing a balanced interplay of energy efficiency, ecological health, and socio-economic benefits. By integrating renewable energy technologies, such as solar panels and biogas plants, farms can reduce their dependence on fossil fuels and mitigate greenhouse gas emissions. These energy savings dovetail with agroecological practices—from cover cropping to biodiverse field margins—that help stabilize soil fertility, conserve water, and bolster beneficial wildlife populations. The resulting synergy reinforces long-term productivity and ecological resilience, aligning agricultural outcomes with the broader imperatives of sustainability and climate adaptation.

A salient outcome of this approach is the diversification of income streams, as producers gain new revenue sources from energy sales and premium markets for environmentally certified products. However, widespread implementation hinges on supportive policies, equitable financing mechanisms, and robust extension services that can disseminate knowledge at scale. Smallholder farmers, especially, stand to benefit from collective models or cooperatives that offset high initial investments in technology and infrastructure. At the same time, consumer awareness and market demand for sustainably produced foods act as catalysts, incentivizing adoption and rewarding conscientious practices.

While technical innovations form a cornerstone of eco-integrated agriculture, human elements—education, collaboration, and policy alignment—remain equally crucial. Successful transitions require that farmers, researchers, policymakers, and local communities co-create solutions grounded in both scientific evidence and traditional knowledge. This participatory ethos can streamline the adoption of eco-integrated techniques, ensuring they are practical, culturally acceptable, and economically viable.

Looking ahead, eco-integrated agriculture has the potential to reshape not only how food is produced but also how farms engage with surrounding environments and local economies. By drawing upon renewable energy sources, employing precision agriculture tools, and championing ecological integrity, agro-production complexes can evolve into holistic systems that sustain life and livelihoods. The path toward this future necessitates continued research, innovation, and governance support, but the ultimate payoff—an agriculture sector that is productive, climate-smart, and socially inclusive—promises significant gains for current and future generations.

References

1. M. M. Akter, I. Z. Surovy, N. Sultana, "Techno-economics and environmental sustainability of agricultural biomass-based energy potential", Applied Energy **359**, 122662 (2024)
2. R. G. Araújo, R. A. Chavez-Santoscoy, R. Parra-Saldívar, "Agro-Food Systems and Environment – Sustaining the Unsustainable", Current Opinion in Environmental Science and Health **31**, 100413 (2023)
3. S. R. Gliessman, "Transforming food and agriculture systems with agroecology", Agriculture and Human Values **37(3)**, 547-548 (2020)
4. G. Brunori, G. Branca, L. Cembalo, "Agricultural and Food Economics: the challenge of sustainability", Agricultural and Food Economics **8(1)**, 1-2 (2020)
5. J. E. Bergez, A. Béthinger, C. Bockstaller, "Integrating agri-environmental indicators, ecosystem services assessment, life cycle assessment and yield gap analysis to assess the environmental sustainability of agriculture", Ecological Indicators **141**, 109107 (2022)
6. M. P. Faucon, T. Aussenac, R. Debref, "Combining agroecology and bioeconomy to meet the societal challenges of agriculture", Plant and Soil **492(1-2)**, 61-78 (2023)
7. G. Wang, R. Shi, L. Mi, J. Hu, "Agricultural Eco-Efficiency: Challenges and Progress", Sustainability **14(3)**, 1051 (2022)
8. G. Brunori, "The role of agriculture in the ecological transition", Agrochimica **67**, 35-41 (2023)
9. M. I. Khalil, B. Osborne (Eds.), "Special issue: Developing climate-resilient agri-environmental production systems", Nutrient Cycling in Agroecosystems **123(1)**, 61 (2022)
10. X. Xu, C. Qin, Y. Zhu, "Developing the agri-environment biodiversity index for the assessment of eco-friendly farming systems", Ecological Indicators **142**, 109220 (2022)
11. E. Elahi, Z. Khalid, Z. Zhang, "Understanding farmers' intention and willingness to install renewable energy technology: A solution to reduce the environmental emissions of agriculture", Applied Energy **309**, 118459 (2022)
12. O. U. Oteh, J. Mbanasor, N. M. Agwu, A. O. Oloveze, "Building sustainable agri-food market systems: Interconnectedness of social, economic and environmental factors in biofortified cassava food systems", African Journal of Science, Technology, Innovation and Development **16(2)**, 238-252 (2024)
13. M. I. Khalil, B. Osborne, "Developing climate-resilient agri-environmental production systems", Nutrient Cycling in Agroecosystems **123(9)**, 1-4 (2022)
14. L. I. Rangel Vázquez, "Bioeconomy and New Materials: Challenges and Opportunities for Sustainable Agriculture", Microscopy and Microanalysis **29(Supplement_1)**, 12 (2023)

15. C. I. Nicholls, "Agroecology: a viable path for food production in a planet in crisis", Acta Horticulturae **1355**, 5-6 (2022)

Ecotoxicological assessment of the mutagenicity level of agricultural soils based on the soil-plant system

A. L. Atoyants, *E. A.* Aghajanyan, and *R. E.* Avalyan*

Yerevan State University, Research Institute of Biology, Laboratory of General and Molecular Genetics, Yerevan, Armenia

Abstract. Under modern conditions of agricultural production, soil pollution with heavy metals (HM) and pesticides is a pressing issue, including for Armenia. Biotesting of the potential mutagenicity level of soil samples from three agricultural land plots in the coastal areas of the Lake Sevan basin was conducted in the spring-autumn period, taking into account the content of HM and pesticides in them. To determine the genetic effects in soil samples, the micronucleus test (Trad-MN) of the model test object Tradescantia (clone 02) was used. According to the main test criteria of the Trad-MN test (frequencies of tetrads with MN and frequencies of MN in tetrads), the maximum level of the studied indicators was established in soil samples of the Lichk region in all time periods of observation, where their values were 2.5-3.5 times higher, respectively. A reliable positive correlation was shown between the frequency of both test criteria and the concentration of As ($p<0.01$; $p<0.001$) in the studied soil samples.

1 Introduction

In the context of increasing anthropogenic load on agroecosystems, the state of agricultural soils, especially in terms of the content of heavy metals (HM) and pesticides, deserves more and more attention. As is known, soil is the most sensitive indicator of the ecological and geochemical state of natural areas, since it is where the migration routes of many chemical elements, including toxicants, intersect. Soils of agricultural landscapes are a source of pollution of agricultural products for living organisms, therefore, regional monitoring of soil pollution is an essential part of the ecotoxicological testing system for the state of the environment.

Despite the fact that pesticides play an important role in ensuring food security by protecting crops from pests and diseases, along with their benefits, there are many environmental and sanitary risks associated with the use of these chemicals. [1–2].

One of the most pressing issues is the accumulation of pesticides and heavy metals (HM) in the soil, where they can persist for a long time, exerting toxic effects on living organisms. [3–4]. Particularly dangerous are heavy metals such as cadmium, lead and arsenic, which are

* Corresponding author: re_avalyan@mail.ru

not subject to microbiological or chemical decomposition, making them a constant source of threat to the environment [5–7].

Pollution problems are especially relevant for resource-limited countries such as Armenia, where the agricultural sector plays a key role in the economy. There are a number of environmental challenges in the republic related to the violation of environmental safety in agriculture, which is aggravated by the impact of chemical fertilizers, pesticides and emissions from industrial facilities [8–9]. Particularly dangerous is the pollution of lands used for agricultural production, as well as the pollution of water bodies as a result of the use of chemicals [10]. These problems can lead to a deterioration in the quality of life of the population, disruption of ecosystems and loss of biodiversity.

The widespread practice in Armenia of improper use of pesticides, their excessive use to control pests or diseases, as well as mixing and disposal of pesticides without taking precautions to protect soil, water and natural vegetation may affect the ecological state of natural agroecosystems. In addition, insufficient attention is paid to the study of the environmental situation in relation to the content of HM and pesticides in agricultural soils, as well as their impact on cultivated crops, in the republic.

To solve these problems, an integrated approach is needed, which includes not only strict control over the use of pesticides, but also the use of innovative methods for monitoring environmental pollution. In this case, to determine the ecotoxicological situation in a specific natural and economic region, it is especially important not only to develop and apply express methods for assessing the degree of potential toxicity of the studied territory, but also to use reliable and sensitive test objects for the purpose of biotesting the degree of anthropogenic pollution of natural territories.

In the genetic monitoring system *Tradescantia* (clone 02) is a reliable indicator in biotesting the mutagenic effects of the natural environment. The micronucleus assay (Trad-MN) is used as the main marker bioassay for *Tradescantia*, allowing detecting genetic disorders in the process of microsporogenesis, manifested in the formation of micronuclei in microspore tetrads, under the influence of toxic substances, including pesticides and heavy metals. This bioassay is part of the International Program for Plant Tests (IPPB) under the auspices of the United Nations (UNEP) on the environment.

The aim of this work was to assess the level of mutagenicity of soil samples from agricultural lands in the coastal areas of the Lake Sevan basin using the micronucleus test (Trad-MN assay) of *Tradescantia* (clone 02) in the soil-plant system.

2 Materials and methods

The study material was soil samples from agricultural lands for growing vegetable crops in the coastal areas of the Lake Sevan basin. Soil samples were collected from three regions (Gavar – 40°35′40″N 45°11′29″E, Lichk – 40°15′62″N 45°23′06″E, Martuni – 40°13′10″N 45°30′52″E), two plots for each - the first plot of soil treated with pesticides; the second plot of untreated natural soil, located at a distance of 500 m from the first. Sampling was carried out in three seasons: a) spring (May), when fertilizers have not yet been applied to the field and sowing has not been done; b) summer (July), after fertilization; and c) autumn (October), after harvesting. All three sites were located in mountain-steppe landscapes, where mountain-chestnut soil types predominate, at an altitude of 1780–1982 meters, surrounded by untouched natural areas. Soil sampling was carried out in accordance with the methodology of agrochemical study of agricultural soils.

Soil samples were collected in dry weather conditions from a depth of 5–20 cm using the envelope method at a distance of 100–200 m from each other. In each regional soil group, at least five samples were collected from one test plot to form a combined soil sample for the purpose of a more objective assessment of the degree of accumulation of chemical components. The mass of each point sample was 200 g, the combined one – 1 kg. The soil of the ESU greenhouse (mountain chestnut soil type), located at a distance of 30-50 km from the main study areas, was used as a conditional background sample. The concentration of heavy metals and qualitative analysis of pesticides in the soil were carried out at the RA Monitoring Center using atomic absorption spectroscopy (AAS) or inductively coupled plasma mass spectrometry (ICP-MS). Pesticide analysis in the soil was carried out by gas chromatography on *PerkinElmer Clarus 680 GC* devices.

Plants of the model test object *Tradescantia* (clone 02) were used as the object of the research. The vegetation experiment in the soil-plant system was conducted in the greenhouse of YSU. Tradescantia plants were grown in pots with the studied soil samples under the same vegetation conditions: T = 20–25 °C, day/night light regime - 18 h/6 h. For each variant of the regional soil sample, three vegetation vessels with 300–400 g of soil in each were used, in which 5–7 *Tradescantia* plants at a certain stage of organogenesis were planted. The plants were watered with tap water. After 3–4 weeks, after the formation of inflorescences consisting of 16–20 flower buds, the small buds were fixed to conduct the Trad-MN assay according to the generally accepted method.

The use of the micronucleus test (Trad-MN) allows recording the occurrence of such genome disorders as chromosomal aberrations (acentric fragments or lagging chromosomes), which are recorded as micronuclei (MN) at the stage of microspore tetrads, when the process of microsporogenesis is disrupted. When testing using this method, two test criteria are taken into account: the frequency of tetrads with micronuclei and the frequency of micronuclei in tetrads. Fixation of buds was carried out using ethanol-glacial acetic acid (3:1). Temporary acetocarmine preparations were prepared. Microspore tetrads were analyzed at a magnification of 10x40 using a *Motic Images Swift M10L* microscope. The number of tetrads with micronuclei was determined and the number of micronuclei per 100 tetrads was recalculated using the standard method. For each soil sample, 3000 tetrads were examined. A correlation analysis was also conducted between the frequency of micronuclei in microspore tetrads and the concentration of chemical elements in the water samples under study. All the results obtained were statistically processed using the *Stastgraphics Centurion 16.2* computer program.

3 Results and discussion

The analysis of the results of determining the gross content of the studied chemical elements in the studied soil samples of the Gavar, Lichk, Martuni districts showed that their content was heterogeneous. Although the soil samples from the treated agricultural plots were distinguished by an increased content of TM compared to the untreated plots, for some elements in both cases there was an excess of generally accepted norms (MPC) for soil (Table 1). The table presents data only for the summer period, when the soils were exposed to the most intense impact of toxic substances. In all the studied soil samples, there was an excess of the norm of the elements - As, Co, Ni, as well as - Cr and Cu in all samples, except for the Martuni sample. The main sources of the studied elements in agrocenoses were organic fertilizers.

Table 1. Content of chemical components in soil samples (mg/kg).

Soil samples	Content of chemical components (mg/kg)								
	As	**Co**	**Cr**	**Cu**	**Fe**	**Mn**	**Ni**	**Pb**	**Zn**
Gavar	**7,2**	**155,9**	**133,5**	**73,6**	33265,3	770,3	**86,0**	17,0	89,8
Gavar/c	**4,2**	**103,2**	**109,9**	**65,3**	30907,4	704,5	**68,0**	12,0	86,4
Lichk	**10,5**	**80,9**	**140,6**	**83,6**	40194,2	834,5	**86,2**	31,5	90,5
Lichk/c	**9,0**	**74,3**	**130,0**	**78,4**	36835,8	804,8	**75,5**	27,0	87,0

Continuation of Table 1.

Soil samples	**Content of chemical components (mg/kg)**								
	As	**Co**	**Cr**	**Cu**	**Fe**	**Mn**	**Ni**	**Pb**	**Zn**
Martuni	**4,8**	**56,1**	82,4	44,2	32189,5	736,0	**60,6**	16,0	86,0
Martuni/c	**3,2**	**36,2**	60,7	31,4	22514,2	420,0	**53,4**	18,0	84,9
MAC/soil	2	20	100	55	40000	1500	50	32	100

Note: Concentrations exceeding the Maximum Allowable Concentrations (MAC) for soil are highlight in bold.

The soils of the treated plots of agricultural land differed in the content of the pesticides used (mainly Pendimethalin, Fipronil, Tefluthrin, Folpet), which have a dangerous and toxic effect on biota. Folpet is a systemic pesticide, organochlorine fungicide. It is intended for processing grapes, vegetables and fruits from pathogenic fungi. This fungicide is considered to be low-toxic to plants. Tefluthrin is an insecticide, a contact pesticide, which is used in agriculture to control harmful insects. Tefluthrin-based preparations belong to hazard classes 2 and 3 for humans and require careful handling. Fipronil is an insecticide, a contact-intestinal pesticide. It is used in the processing of seed material (seed dressing) and spraying of plants. When used moderately according to the regulations, it does not exhibit phytotoxicity. However, large-scale use of this compound is highly toxic. Pendimethalin is a soil herbicide. It has herbicidal activity only in relation to germinating seeds, therefore it is intended only for soil application. Almost completely decomposes in the soil by the end of the growing season, as a result of which it does not have a phytotoxic effect on subsequent crops. Unfortunately, this study only determined the presence of the mentioned pesticides without data on their concentrations in soil samples.

When conducting biotesting of the mutagenicity level of the studied soil samples to detect disturbances in the microsporogenesis process with the formation of micronuclei (MN) in microspore tetrads (clastogenic effect) based on the analysis of the micronucleus test of *Tradescantia*, a significant increase in the frequency of MN was shown compared to the control (conditional background) in the soil samples of all the studied areas. At the same time, the treated soils of agricultural areas showed a higher level of clastogenicity than untreated (control) areas (Table 2).

Table 2. Induction of clastogenic effects in soil samples of agricultural areas by Tradescantia (clone 02).

Soil samples	**Micronucleus frequency in sporogenous cells of Tradescantia**					
	Tetrads with MN (%±m)			**MN in tetrads (%±m)**		
	spring	summer	autumn	spring	summer	autumn
Gavar	$3,8\pm0,35^{***}$	$5,8\pm0,43^{***}$	$5,5\pm0,42^{***}$	$4,8\pm0,39^{***}$	$7,1\pm0,47^{***}$	$6,0\pm0,43^{***}$
Gavar/c	$3,2\pm0,32^{*}$	$4,7\pm0,39^{***}$	$4,5\pm0,38^{***}$	$3,8\pm0,35^{**}$	$5,7\pm0,42^{***}$	$5,1\pm0,40^{***}$
Lichk	$5,6\pm0,42^{***}$	$8,0\pm0,46^{***}$	$5,8\pm0,43^{***}$	$7,1\pm0,41^{***}$	$8,7\pm0,52^{***}$	$7,1\pm0,47^{***}$
Lichk/c	$4,9\pm0,39$	$5,1\pm0,40^{***}$	$4,7\pm0,39^{***}$	$4,1\pm0,36^{***}$	$6,0\pm0,43^{***}$	$5,7\pm0,42^{***}$
Martuni	$2,9\pm0,31$	$4,9\pm0,39^{***}$	$4,5\pm0,38^{***}$	$3,4\pm0,33^{*}$	$5,5\pm0,42^{***}$	$5,2\pm0,41^{***}$
Martuni/c	$2,7\pm0,30$	$3,5\pm0,34^{**}$	$3,5\pm0,34^{**}$	$3,1\pm0,32$	$3,96\pm0,36^{**}$	$3,9\pm0,35^{**}$
Fon	$2,2\pm0,27$			$2,5\pm0,29$		

Note: differences are significant at: **–* $p < 0.05$, ** *–* $p < 0.01$, *** *–* $p < 0.001$

In the spring, before planting vegetable crops (mainly melons), the lowest mutation rate was observed in the soil samples studied. It should be noted that the highest mutation rate was observed in the summer, when the crops were actively flowering and pesticides were applied to control pests. Despite the fact that in the autumn the soil was less polluted in terms of toxic substances, at the same time, the clastogenicity level retained high toxicity rates according to both criteria, exceeding the background level by 1.6-3.6 times. The most pronounced clastogenic effects were observed in the soil samples from the Lichk district, exceeding the background indicators by 2.1-3.6 times in the treated areas and by 2 times in the corresponding control areas. The lowest value of the soil clastogenicity level was observed in the samples taken from the Martuni district. Moreover, this trend persisted throughout the study period. Thus, based on the obtained results for both test criteria, it was shown that the soils from the treated areas had a higher level of clastogenicity than similar soils from untreated areas, and this trend was observed in all seasonal periods of monitoring.

The conducted correlation analysis between the concentration of chemical elements and the test criteria of the Trad-MN test showed the presence of a reliable positive correlation between both test criteria and the concentration of arsenic (As) in soil samples - MN in tetrads ($r=0.98$, $p<0.001$) and tetrads with MN ($r=0.84$, $p<0.05$). As is known, arsenic, along with cadmium and lead, are ecotoxicants that have a cumulative and carcinogenic effect on living organisms.

4 Conclusion

Agroecosystems are usually exposed to many types of pollutants, including biotic and abiotic byproducts of agricultural practices. Among the agricultural sources of heavy metals, pesticides are the most common. Excessive use of pesticides over a long period of time leads to the accumulation of heavy metals in agricultural soils, reduces soil fertility and, as a result, reduces plant growth and productivity. As a result of uncontrolled use, pesticide residues accumulate significantly in the soil and increase pollution, which directly or indirectly harms fauna and flora.

A comparative analysis of soil samples from agricultural lands in different regions of Armenia and a conventional background sample based on the results of the Trad-MN test of clone 02 of *Tradescantia* in the soil-plant system showed the presence of clastogenic effects in the studied soil regional groups with varying degrees of manifestation depending on the sample. This paper presents the results of the total mutagenic activity of the studied soil samples. The highest level of genetic disturbances during microsporogenesis was observed in the Lichk sample from the territory of the soil plot treated with pesticides. The absence of a correlation between clastogenic effects and the content of chemical components in the soil samples may indicate a combined multicomponent effect of ecotoxicants (additive toxic effect) in the soil-plant system and their influence on increasing the level of clastogenicity in sporogenous cells of Tradescantia. Therefore, further studies should take into account not only the content of heavy metals and pesticides, but also other environmental factors that may affect the mutagenic activity of agricultural soils.

Studies on the ecological and genetic assessment of the state of agricultural soils taking into account the content of toxic substances in them, in particular pesticides, using clone 02 of *Tradescantia* were conducted in Armenia for the first time. In the future, it is planned to continue studies aimed at identifying the long-term effects of heavy metals, heavy metals and pesticides on agroecosystems, especially taking into account seasonal changes. Based on the conducted research, the feasibility of using *Tradescantia* (clone 02) as a model test object for biotesting the degree of anthropogenic disturbance of the soil cover of agricultural agroecosystems was demonstrated.

References

1. I. Mahmood, S. R. Imadi, K. Shazadi, A. Gul, K. R. Hakeem, "Effects of Pesticides on Environment", in Plant, Soil, and Microbes **1**, 253–269 (2016), DOI: 10.1007/978-3-319-27455-3_13
2. A. K. Lysov, "Application problems of plant protection agents and ways to reduce their anthropogenic impact on the environment", AgroEcoEngineering **3(116)**, 34–51 (2023)
3. R. A. Wuana, F. E. Okieimen, "Heavy Metals in Contaminated Soils: A Review of Sources, Chemistry, Risks and Best Available Strategies for Remediation", ISRN Soil Science **2011**, 402647 (2011), DOI: 10.5402/2011/402647
4. R. Riyazuddin, N. Nisha, B. Ejaz, M. I. R. Khan, M. Kumar, P. W. Ramteke, R. Gupta, "A Comprehensive Review on the Heavy Metal Toxicity and Sequestration in Plants", Biomolecules **12(1)**, 43 (2021), DOI: 10.3390/biom12010043
5. M. M. Lasat, "Phytoextraction of Toxic Metals: A Review of Biological Mechanisms", Journal of Environmental Quality **31(1)**, 109–120 (2021), DOI: 10.2134/jeq2020.05.0193
6. N. L. Mdeni, A. O. Adeniji, A. I. Okoh, O. O. Okoh, "Analytical Evaluation of Carbamate and Organophosphate Pesticides in Human and Environmental Matrices: A Review", Molecules **27(3)**, 618 (2022), DOI: 10.3390/molecules27030618
7. S. Kroyan, "State of the main soil types in the Republic of Armenia under the conditions of climate change", Soil Science and Agrochemistry **1**, 5–18 (2019)
8. G. Tepanosyan, L. Sahakyan, D. Pipoyan, A. Saghatelyan, "Risk Assessment of Heavy Metals Pollution in Urban Environment", in Risk Assessment (IntechOpen, 2018), DOI: 10.5772/intechopen.70798
9. G. Fayvush, K. Aghababyan, A. Aleksanyan, M. Arakelyan, A. Gasparyan, M. Kalashian, L. Margaryan, S. Nanagulyan, "Biodiversity Conservation Problems", in Biodiversity of Armenia (G. Fayvush, Ed.) (Springer, Cham, 2023), DOI: 10.1007/978-3-031-34332-2_6
10. Armenian Women for Health and Healthy Environment: NGO, Review of the situation on highly hazardous pesticides (HHP) and alternatives in Armenia (Yerevan, Armenia, 2020)

Determination of the level of natural background radiation in the surface soil of Kufa farms, Iraq

Sarab J. Musa[1,*], *Noor* Ali Sami[1], *Gufran* Hade Seood[1], *Esraa* Salim Obayes[1], and *Hanan* Naji Mohsion[2]

[1]Department of Physics, University of Babylon, College of Education for Pure Sciences, Babylon, Iraq
[2]Department of Medical Physics, Faculty of Medical Applied Sciences, University of Kerbala, Karbala, Iraq

Abstract. This study measured background radiation in some sites of Kufa farms including the four directions: north, east, west, and south (October, 2023). An FS-600 dosimeter was used to measure these quantities. As shown below, the results present dose rate (in μSv/h) measured at each of the 4 locations: North: 0.1927 ± 0.005120 μSv/h West: 0.1833 ± 0.00912 μSv/h East: 0.1507 ± 0.00835 μSv/h South: 0.1587 ± 0.00794 μSv/ h These observations confirm that the radiation levels in all these areas are within allowable limits and therefore, there is no radiation hazard," the agency said. Geographic Information Systems (GIS) were used to pinpoint the precise coordinates for each measurement, offering a geographic display of the radiation levels found in proximity to these regions. the results of this study show that the levels of radiation within a certain amount of area in four directions in Kufa farms such as west, north, east, and south are low and there is no significant threat to health.

1 Introduction

Soil background radiation means naturally occurring radiation on soil and is not caused by humans. Soil is where the plants grow, the primary source of human food, therefore understanding the soil background radiation is very important. Such background radiation can impact human and plant health. So, monitoring this radiation cannot only determine soil safety but also agricultural product safety. Soil radiation is a composite of natural and artificial radiation. Natural radiation comes from naturally occurring radioactive elements in the Earth's crust, for example, radium, uranium, and thorium [1]. These elements spontaneously break down and release radiation that can fluctuate in different areas, depending on what is in the soil and the geological structure of the region. Artificial radiation is radiation that results from human activity or manmade sources including the mining of radioactive minerals as well as emissions from nuclear power plants. Soil background radiation is observed in practice by making accurate measurements with radiation detection devices, such as Geiger counters or X-ray detectors. This is because the soil composition and mineral composition density, organic content and other factors affect soil radiation intensity,

* Corresponding author: pure.sarab.hamad@uobabylon.edu.iq

at the same time, soil depth also affects soil radiation intensity, and the situation of the area (for example: industrial area, military area) also affect soil radiation intensity. For this reason, excessive levels of radiation in agricultural soil could be dangerous, when they surpass natural levels of radiation. For example, excessive radiation proved detrimental to plant growth, causing the collapse of crops or decreased agricultural production. Moreover, some of this radiation can leach into the food chain through plants or groundwater, endangering people and animals with long-term health risks. Risks of cancer, infertility, and genetic defects are some of them. However, there are studies indicating that some plants may be capable of absorbing minimal amounts of radioactive elements, which may help purify their surrounding environments to some extent of certain radioactive pollutants. Thus, research on the impacts of background radiation on agricultural soil not only concerns about possible threat but also concerns about possible opportunities by using plants for environmental pollution remediation [2]. Thus, agricultural soil background radiation measurements are an essential factor with numerous applications in sustainable agriculture, environmental science, and public health. Such measurements also help ensure that the land where crops are grown is safe and healthy for people and the environment around it [3-5].

2 Material and methods

Kufa Farms area in southwestern Iraq, is one of the important agricultural areas known for agriculture, containing fertile lands. The Kufa farms are known for the diversity of the crops included in them, including grains, vegetables, and fruits in addition to the traditional agriculture on which the people of the region rely. This region is part of the central Euphrates Plain that stretches through numerous Iraqi cities. The Euphrates River is the main source of irrigation, making the farms of Kufa one of the most important districts of Iraq in terms of agriculture. Locations were determined using GPS and drawn by GIS technical, as shown in Table 1 and Figure 1, respectively

Table 1. North Location of samples in the present region.

No.	Latitude	Longitude
1	32 02 19.79376756N	44 25 17.23349388E
2	32 02 19.48081812N	44 25 18.98948784E
3	32 02 19.11571044N	44 25 19.77186144E
4	32 02 18.75060276N	44 25 20.60639328E
5	32 02 19.63312008N	44 25 16.77450144E
6	32 02 19.57748460N	44 25 17.78706660E
7	32 02 19.58861184N	44 25 18.42131100E
8	32 02 19.34381580N	44 25 19.37824068E
9	32 02 18.89873196N	44 25 20.07924744E
10	32 02 18.77633412N	44 25 21.14744808E
11	32 02 19.67762832N	44 25 17.45325408E
12	32 02 19.75551828N	44 25 16.94140788E
13	32 02 19.61086596N	44 25 18.69948804E
14	32 02 19.63312008N	44 25 18.06524400E
15	32 02 18.79858824N	44 25 20.34629760E

https://doi.org/10.1051/e3sconf/202563302004

Table 2. West Location of samples in the present region.

No.	Latitude	Longitude
1	32 02 14.22469500N	44 25 17.10265260E
2	32 02 14.67134556N	44 25 16.74715548E
3	32 02 15.34132104N	44 25 16.43267712E
4	32 02 16.02952692N	44 25 16.09085280E
5	32 02 16.94245020N	44 25 16.08601260E
6	32 02 17.75959620N	44 25 16.05124056E
7	32 02 18.59412804N	44 25 16.05124056E
8	32 02 14.41896504N	44 25 16.90980672E
9	32 02 15.08213976N	44 25 16.54083228E
10	32 02 15.82765476N	44 25 16.18476564E
11	32 02 16.58429700N	44 25 16.04011332E
12	32 02 17.38544784N	44 25 16.10687604E
13	32 02 18.20885244N	44 25 15.96222372E
14	32 02 18.93211332N	44 25 16.31829072E
15	32 02 16.33950096N	44 25 16.05124056E

Table 3. East Location of samples in the present region.

No.	Latitude	Longitude
1	32 02 18.85491924N	44 25 21.59739984E
2	32 02 18.54196980N	44 25 22.62317880E
3	32 02 17.49880500N	44 25 22.53624816E
4	32 02 16.32036264N	44 25 21.58140468E
5	32 02 16.02126636N	44 25 21.80928756E
6	32 02 15.55837944N	44 25 21.94459320E
7	32 02 18.76520688N	44 25 22.19339496E
8	32 02 17.97518364N	44 25 22.70524116E
9	32 02 17.20741416N	44 25 22.34917416E
10	32 02 16.57317012N	44 25 21.41449824E
11	32 02 18.26448792N	44 25 22.71636804E
12	32 02 18.80971548N	44 25 21.89296344E
13	32 02 16.81885608N	44 25 21.28942992E
14	32 02 17.09764548N	44 25 22.04443092E
15	32 02 17.03961780N	44 25 21.67398228E

Table 4. South Location of samples in the present region.

No.	Latitude	Longitude
1	32 02 14.11531116N	44 25 19.63215192E
2	32 02 13.98769692N	44 25 19.38603864E
3	32 02 13.80994836N	44 25 19.05788748E
4	32 02 13.61852664N	44 25 18.53831460E
5	32 02 13.53648876N	44 25 17.83643520E
6	32 02 13.58662308N	44 25 17.49916884E
7	32 02 13.82362116N	44 25 17.19836328E
8	32 02 15.35186040N	44 25 21.53867664E
9	32 02 15.05276412N	44 25 20.96184828E
10	32 02 14.88897348N	44 25 20.49896136E
11	32 02 14.73942552N	44 25 20.13577320E
12	32 02 14.51154264N	44 25 19.67288592E
13	32 02 13.57152576N	44 25 18.18452640E
14	32 02 14.63705628N	44 25 19.89008688E
15	32 02 13.72463484N	44 25 18.75512352E

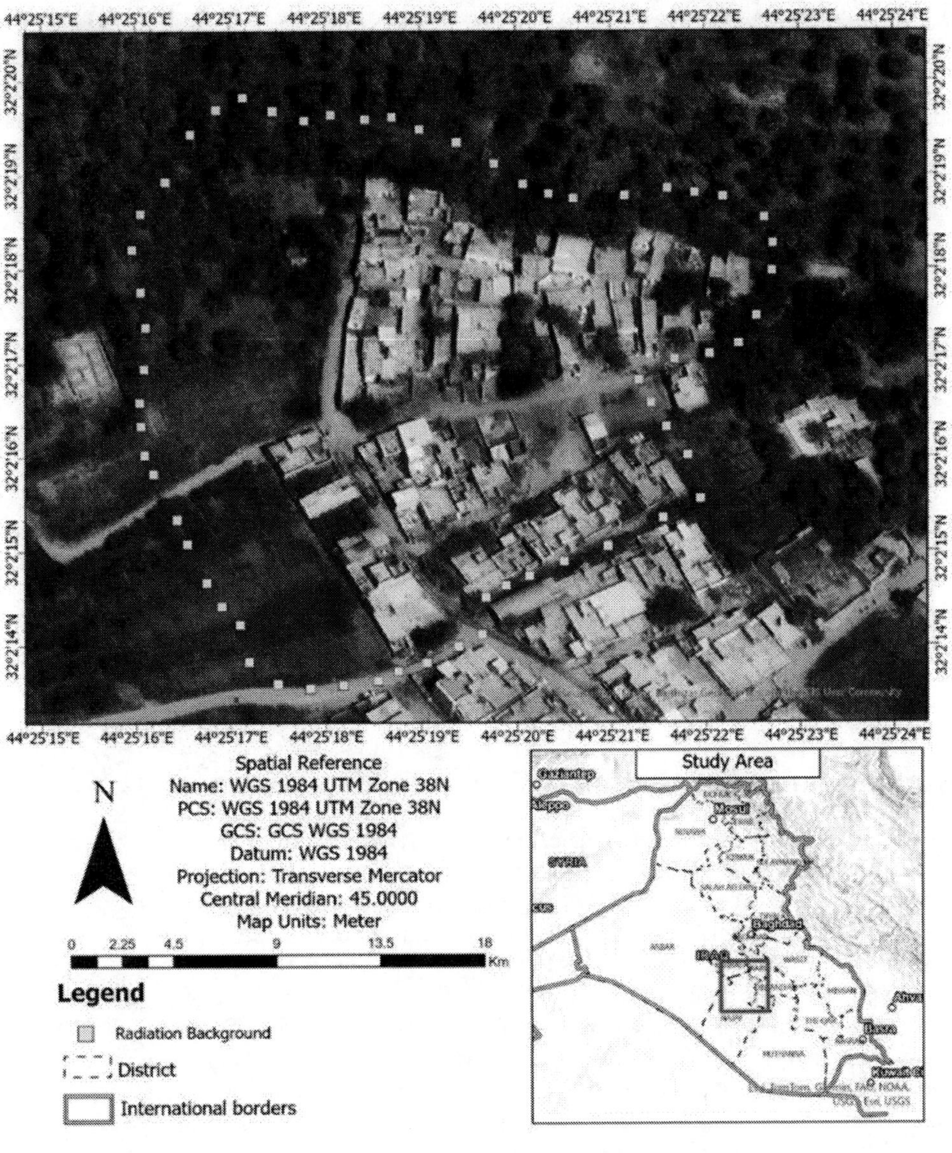

Fig. 1. Map of study area.

In this study, a total of 60 soil samples with a weight of 1 kg were collected, and all samples were collected at a depth of 0 cm, meaning they were taken from the surface layer of the soil. Samples were collected from 60 different sites distributed on farms, directed to four directions, with 15 sampling points. The geographical distribution of samples facilitated a multiple variability of measurements of a radiating background from different regions. However, the precise locations of the sample sites were determined through the Global Positioning System (GPS), which offers high accuracy in tracking the exact position of where sample sites were located. Geo-information systems (GIS) is used to plot these places on a map. The sample sites from the study can be found in tables 1, 2, 3 and 4 and figure 1. The soil samples were immediately measured for radiation level using a portable FS-600 dosimeter. This was measured with no sample preparation of the soil. For site radiation level measurements, we used the active isotopes FS-600 dosimeter, measuring 3 samples for 5-minute measurements at each site. The FS-600 dosimeter contains a Geiger-Muller tube, a type of radiation detector that senses ionizing radiation (gamma rays, X-rays, as well as beta particles). This dosimeter has an energy response range of 48 kV to 1.5 MV. Additionally, it is capable of detecting sensitivity levels ranging from 0.1 μSv/h to 10 mSv/h. Factory-calibrated to provide accurate readings of radiation levels. The study was based on measuring the levels of radiation in the sediments of different zones of the Kufa farms using accurate measuring equipment with the application of the (mapping) technique.

Fig. 2. Portable FS-600 dosimeter.

3 Results and discussion

Table No. 5 shows the results at site North and West, and Figures 3 and 4 show a comparison of the results of site North and West with the limit set by the World Health Organization, respectively. Likewise, Table No. 6 shows the results East and South Locations, and Figures 5 and 4 show a comparison of the results of East and South Locations with the limit set by the World Health Organization, respectively. The variation of the radioactive background in agricultural soil, especially in the farms of Kufa, is attributed to various sources. Here is evidence based scientific discussion:1. 1) Variations in Geological Structure The geological structure of various regions also plays an important role in determining the amount of natural radioactive elements, that is being, uranium, thorium, and potassium, present in the soil. These natural elements are present in varying concentrations in different types of soil, and they undergo radioactive decay, contributing to the total radioactive background. Soil mineral content in agricultural areas can be influenced by past geological processes (like volcanic eruption or sediments carried by neighboring rivers), which could explain the differences in radioactivity between different Kufa farms [6, 7]. Agricultural methods, including ploughing, irrigation, and fertilizer application, may also affect the distribution of radioactive elements in the soil. Adding phosphate fertilizers which may contain trace quantities of naturally occurring radionuclides, including uranium, could increase the radioactive background level in some farming areas. The disturbance of the soil may cause the movement of these minerals and thus affect the total rates of radioactivity [8]. Water Irrigation and Radionuclide Leaching The irrigation technology used in Kufa farms could contribute to the leaching of radionuclides into the soil. Groundwater and other water sources with natural radioactivity may be used for irrigation, which can lead to an increased accumulation of radioactive elements in the soil. Leaching can cause spatial variability in radioactivity, which is influenced by the distance of a given location to irrigation canals or wells [9]. Atmospheric Deposition and Precipitation Radioactive elements can be deposited onto soil through atmospheric fallout, a process that occurs when radionuclides are released into the atmosphere from natural or anthropogenic sources, such as radon gas emanating from soil or nearby industrial activity. How frequently precipitation occurs and where winds blow will also affect the allocation of these metals. High precipitation regions tend to have lower surface radioactivity as particles are leached into the soil while it may be higher in arid regions [10]. And Regional Historical and Anthropogenic Influences The proximity of Kufa to urbanized regions or historical events related to radioactive materials can probably lead to discrepancies in soil radioactivity. Historical industrial activities such as past nuclear test activities, as well as the use of radioactive materials in neighboring industries or experiments can give rise to lasting remnants in the geosphere. Similar practices, including the application of radiation in agricultural or medical applications, might increase radioactive activity in the immediate soil.

Table 5. Results of Dose rate (μSv/h) samples. North and West Locations of the present region.

No	Code	North Location Dose rate (μSv/h)	West Location Dose rate (μSv/h)
1	U1	0.20	0.19
2	U2	0.18	0.22
3	U3	0.19	0.20
4	U4	0.19	0.21
5	U5	0.20	0.23
6	U6	0.22	0.20
7	U7	0.19	0.19
8	U8	0.21	0.21
9	U9	0.20	0.22
10	U10	0.19	0.13
11	U11	0.18	0.17
12	U12	0.20	0.12
13	U13	0.20	0.19
14	U14	0.21	0.14
15	U15	0.13	0.13
Average		0.1927 ±0.005120.	0.1833 ±0.00912
Worldwide average		0.247	

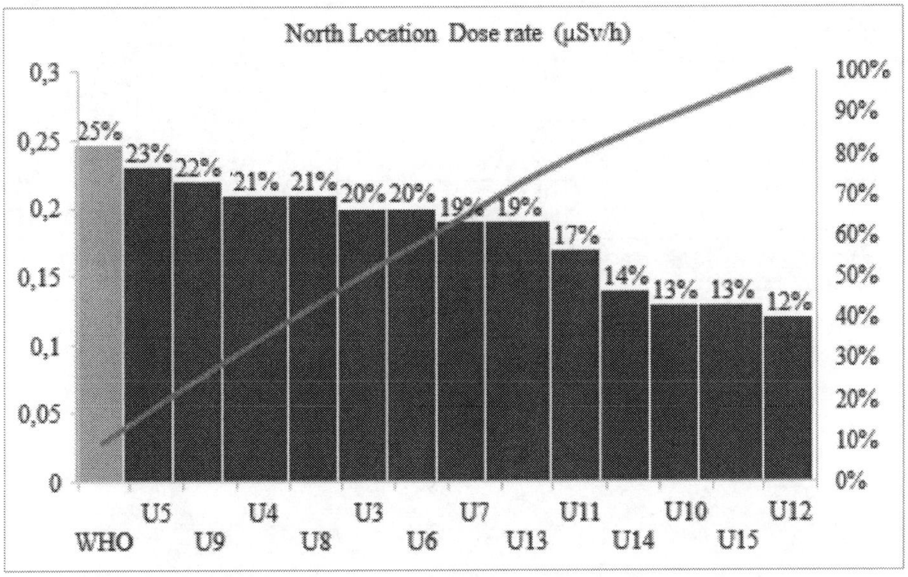

Fig. 3. Comparison of the results of radiation dose. North Locations with safe limit.

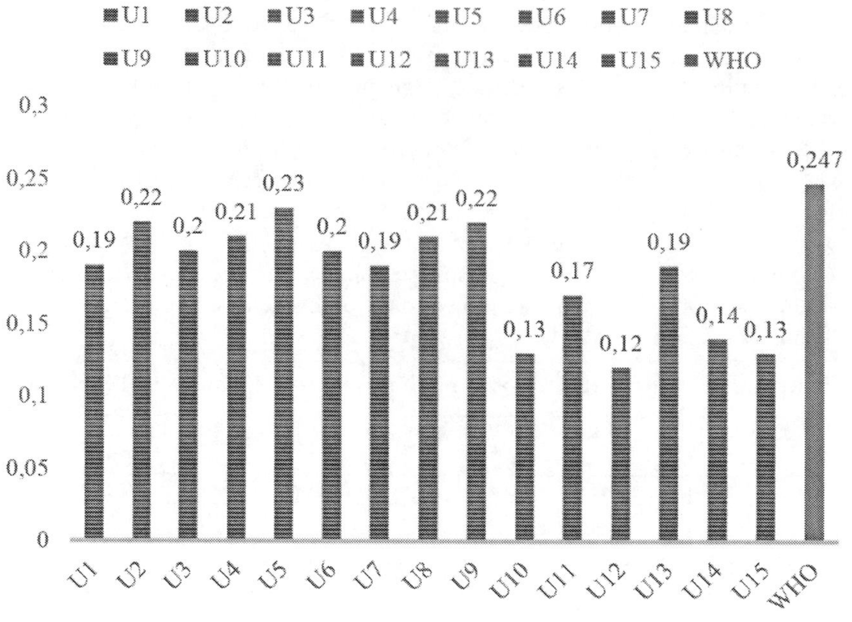

Fig. 4. Comparison of the results of radiation dose. West Locations with safe limit.

Table 6. Results of Dose rate (μSv/h) samples. East and South Locations of the present region.

No	Code	East Location Dose rate (μSv/h)	South Location Dose rate (μSv/h)
1	U1	0.12	0.14
2	U2	0.11	0.12
3	U3	0.18	0.13
4	U4	0.13	0.16
5	U5	0.14	0.21
6	U6	0.21	0.21
7	U7	0.18	0.19
8	U8	0.11	0.11
9	U9	0.19	0.17
10	U10	0.13	0.15
11	U11	0.15	0.15
12	U12	0.17	0.14
13	U13	0.19	0.13
14	U14	0.11	0.18
15	U15	0.14	0.19
Average		0.1507 ±0.00835	0.1587 ±0.00794
Worldwide average		0.247	

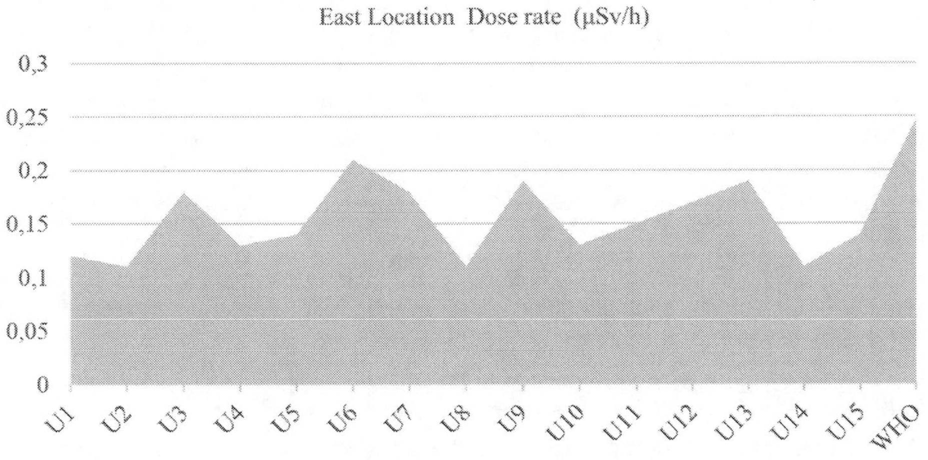

Fig. 5. Comparison of the results of radiation dose. East Locations with safe limit.

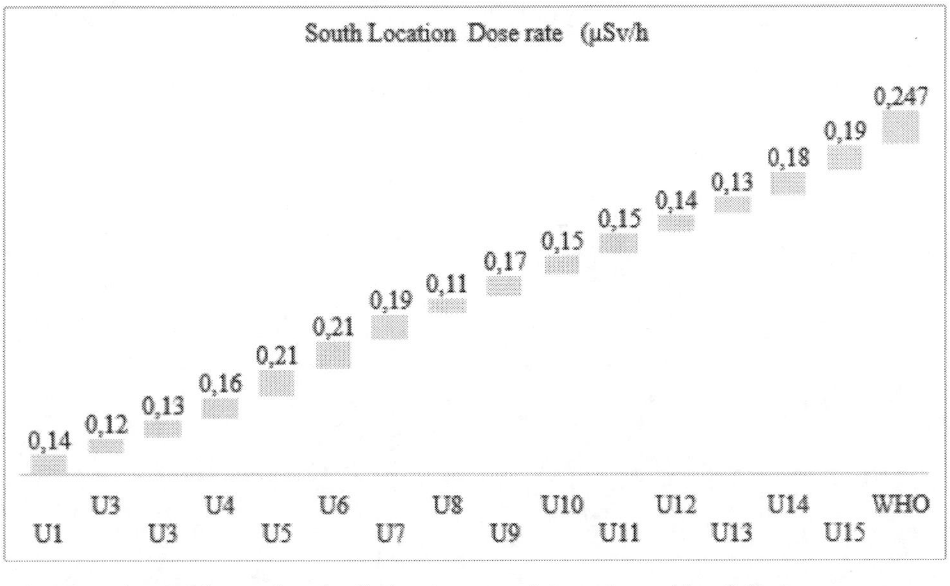

Fig. 6. Comparison of the results of radiation dose. South Locations with safe limit.

4 Conclusion

This study concludes that the radioactive background of selected Kufa agricultural areas predominantly falls below the permissible limits established by the World Health Organization. Consequently, we may conclude that the locations assessed in this study are secure from a radiation standpoint and pose no risk.

References

1. S. Machraoui, G. A. Khouqeer, F. Hosni, S. Labidi, "Radiation protection for Gabes phosphate area residents: Assessing natural radioactivity in soil and potable water samples," Journal of Radiation Research and Applied Sciences **17(3)**, 100985 (2024)

2. B. Yang, C. Pang, F. Tuo, Q. Zhou, Z. Li, "Radioactivity and elemental oxidation composition in soil from Yangjiang, a high background natural radiation area, China," Journal of Environmental Radioactivity **276**, 107447 (2024)
3. H. Alonso, J. G. Rubiano, J. G. Guerra, M. A. Arnedo, A. Tejera, P. Martel, "Assessment of radon risk areas in the Eastern Canary Islands using soil radon gas concentration and gas permeability of soils," Science of the Total Environment **664**, 449-460 (2019)
4. A. Cecconi, I. Verginelli, R. Baciocchi, "Sustainability," Sustainability **16(8)**, 3317 (2024)
5. A. A. Ibrahim, A. K. Hashim, A. A. Abojassim, "Determination of alpha activity in soil samples of agricultural college of Kerbala university, Iraq," Annals of Agri Bio Research **26(1)**, 125-131 (2021)
6. BOSEAN Company Inc., Reference Manual version 1, FS-600 Geiger Counter Nuclear Radiation Detector (BOSEAN Company Inc., 2024)
7. D. M. Dhahir et al., "Natural radioactivity levels in soil samples of some schools in Al-Shatrah city at Dhi Qar governorate, Iraq," Malaysian Journal of Science **39(3)** (2020)
8. R. Al-Jubouri, S. Al-Saadi, "Impact of Fertilizers and Farming Practices on Radioactive Elements in Agricultural Soils in Kufa," Environmental Science and Pollution Research **15(3)**, 221-227 (2018)
9. H. Ali et al., "Atmospheric Fallout and Its Contribution to Soil Radioactivity in the Kufa Region," Environmental Monitoring and Assessment **192(4)**, 215-223 (2020)
10. M. Zubair et al., "Anthropogenic Effects on Soil Radioactivity in Iraq: A Case Study of Kufa Farms," Journal of Environmental Health and Risk Assessment **11(1)**, 134-144 (2021)

Use GIS technical to draw of the background radiation in selected soil samples in the Najaf sea area

Mohammed A. Mohammed[1], *Tabarak* S. Hashesh[1], *Faisel* G. Mohammed[1], *Rand* A. Hayder[2,*], and *Maryam* H. Ali[3]

[1]Department of Remote Sensing and GIS, College of Science, University of Baghdad, Baghdad, Iraq
[2]College of Dentistry, Al-Iraqia University, Baghdad, Iraq
[3]Al Hikmah University, College Network Engineering and Cybersecurity, Ilorin, Nigeria

Abstract. The main objective of the current study is to measure and obtain the background radiation levels in soil samples taken from the Najaf Sea area where most of the soil samples were taken there. In this work we utilized an FS-600 dosimeter to measure the radiation. The radiation dose (D), the annual effective dose (AED) and lifetime cancer risk (ELCR) of each of the samples were calculated. Finally, the GIS technology was used to map the studied primary outcomes. The soil samples had an average background radiation value of 0.161 microsieverts/hour (μSv/h), 0.126 millisieverts/year (mSv/y), and 0.449 lifetime cancer risk (ELCR). The D and AED of all the samples was found to be within safe breadth from 0.247 μSv/h to 2.4 μSv/yr, value that is in line with the standards Set by the United Nations Scientific Committee on the Effects of Atomic Radiation in 2008. These results imply that background radiation in the soil in the study area does not pose a health threat.

1 Introduction

Background radiation, often referred to as natural or environmental radiation, is ionizing radiation that originates from natural sources within the Earth and the surrounding universe. It is ubiquitous and plays a critical role in the radiation exposure that individuals experience daily [1]. Among the various sources of background radiation, radiation emitted from soil is one of the most significant contributors. Soil, being a complex mixture of minerals, organic matter, water, and air, contains trace amounts of naturally occurring radioactive materials, including uranium, thorium, and radon. These radioactive elements decay over time, emitting radiation that can affect the surrounding environment and living organisms [2]. This research investigates background radiation in soil samples from the Najaf Sea region, examining its sources, variations, and potential health implications. A variety of factors, such as geographic location, soil composition, mineral content, and environmental conditions, affect radiation levels in soil [3]. As such, understanding these effects is crucial for assessing radiation exposure risks, especially in areas with higher concentrations of naturally occurring

* Corresponding author: rand.a.hayder@aliraqia.edu.iq

radioactive materials (NORM). Furthermore, the research explores the importance of monitoring background radiation in selected soils from the Najaf Sea region for public health, environmental safety and regulatory purposes [4].

This research aims to provide a comprehensive overview of current scientific knowledge regarding background radiation in soils. It highlights regional differences in soil radiation, and discusses potential risks associated with long-term exposure. Many types of research in the world focus on studying the radiation background in the environment due to its great importance [5]. In the last two decades, this research has received attention that has led to conducting many surveys on health risks to humans with the aim of understanding the distribution of radionuclides and levels of natural radioactivity in the environment. Many literary references have addressed dose measuring devices to measure the radiation background in the soil [6-8]. In this research, I will try to study the radiation background of the soil in some important sites in the Najaf Sea area, Najaf Governorate, which have direct contact with the lives of citizens to protect against radiation risks, by taking samples from different sites in the study area.

2 Materials and methods

The Najaf Sea, a wide depression located in the south-central part of Iraq, extending from the north of the city of Najaf to the southwest of the city of Al-Hirah, was submerged by the waters of the Euphrates River, and because of its large area, estimated at about 60 miles in length and 30 miles in width, it was called the Najaf Sea. Locations were determined using GPS and drawn by GIS technical, as shown in Table 1 and Figure 1, respectively.

Table 1. Location of samples in the present region.

Location code	Latitude	Longitude
1	31 57 16.99253280N	44 11 38.43967200E
2	31 56 33.56324160N	44 14 26.23062120E
3	31 51 54.65850480N	44 16 40.48760280E
4	31 56 03.76158480N	44 16 12.12814920E
5	31 54 57.18036240N	44 15 11.97264240E
6	31 53 36.78575280N	44 14 39.48729000E
7	31 51 31.27777920N	44 13 48.41416200E
8	31 50 36.06721080N	44 15 31.50986040E
9	31 53 10.39639920N	44 12 30.30311880E
10	31 54 21.24324720N	44 11 01.56563160E
11	31 55 50.74434840N	44 09 30.22018920E
12	31 56 50.97585120N	44 07 59.10016080E
13	31 58 47.72145000N	44 07 20.93822400E
14	32 00 24.85950480N	44 07 24.13261200E
15	32 00 29.27139120N	44 09 34.73291160E
16	31 58 54.21655200N	44 08 55.59938160E
17	31 59 02.58550440N	44 11 08.95930440E
18	31 50 00.51420480N	44 18 22.21549920E
19	31 55 24.15098280N	44 12 54.32184000E
20	31 55 01.53175440N	44 19 04.10206800E

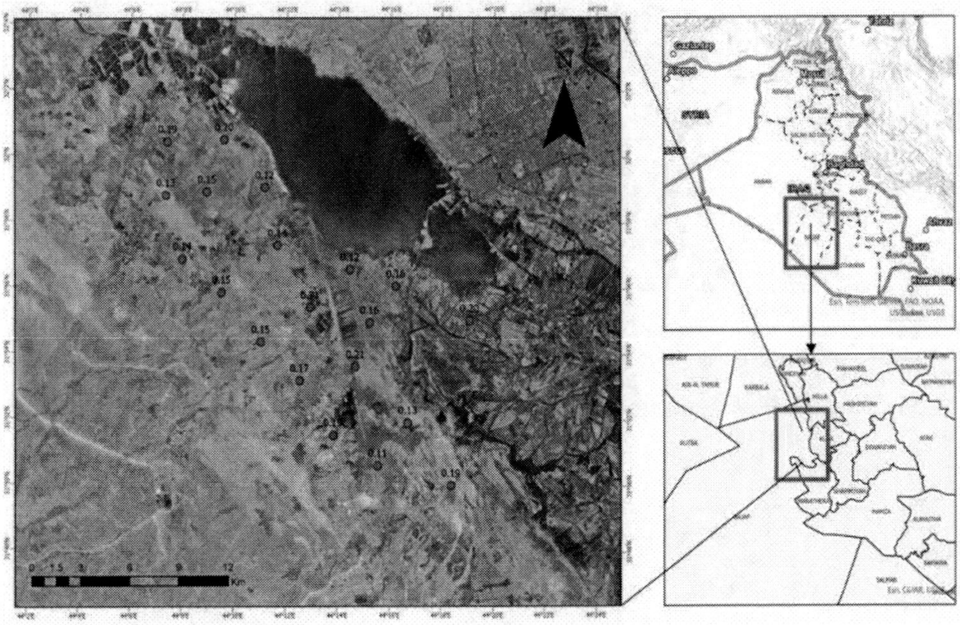

Fig. 1. Map of study area.

Twenty soil samples, each weighing 1 kilogramme and taken from a depth of 10 cm, were obtained from twenty locations in the Al Najaf Sea region, Iraq. Locations were identified using GPS and shown using GIS technology, as presented in Table 1 and Figure 1, respectively. Subsequently, soil samples were measured immediately without any preparation using a portable FS-600 dosimeter manufactured in China, with a duration of 5 minutes for three samples at each location. The FS-600 dosimeter (Figure 2) incorporates a Geiger-Muller tube that operates within an energy response range of 48 keV to 1.5 MeV and a dosage range of 0.1 μSv/h to 10 mSv/h [9]. It is utilised to detect and monitor ionising radiation, including gamma rays, X-rays, and beta particles. The manufacturer calibrates the portable FS-600.

 https://doi.org/10.1051/e3sconf/202563302005

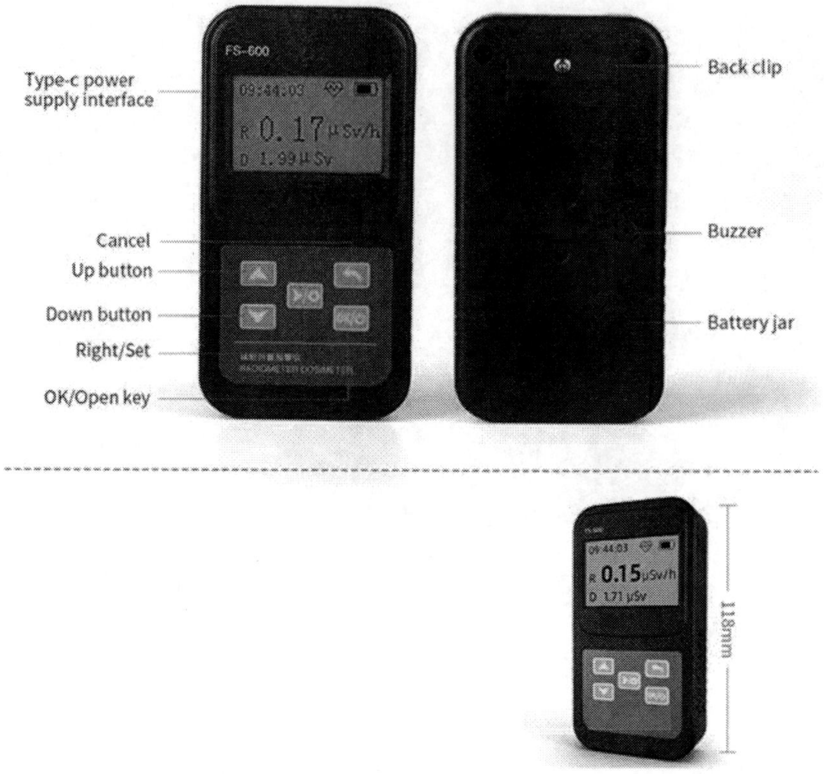

Fig. 2. Portable FS-600 dosimeter.

The absorbed dose rate in the air (D) about concentrations of terrestrial nuclei can be interpreted. straight from the dosimeter utilised in the current investigation, measured in μSv/h. The Annual Effective Dose Equivalent (AEDE) for outdoor exposure may be calculated using the below equation

$$AED_{\text{outdoor}} = D_{outdoor} \times T \times C_C \times OF_{outdoor} \quad (1)$$

Where, Doutdoor represents the absorbed dose rate outdoors (μSv/h), T denotes time in hours, specified as 8760 h, Cc is the correction coefficient, valued at 0.7, and OFoutdoor is the occupancy factor, provided as 0.20. The Excess Lifetime Cancer Risk (ELCR) factor quantifies the likelihood of developing cancer over a lifetime due to a specific amount of exposure. In this context, seventy years is regarded as the typical lifetime for a person. This risk factor is provided by

$$\text{ELCR}=\text{AED}\times\text{DL}\times\text{RF} \quad (2)$$

In this context, AEDE denotes the outdoor annual effective dose equivalent, DL represents the standard Duration of Life (70 years as previously indicated), and RF signifies the Risk Factor (1/Sv). For stochastic effects, ICRP typically employs an RF value of 0.057 for the population.

3 Results and discussion

Table No. 2 shows the results of D which ranged from 0.11 to 0.21 at a rate of 0.161 and Figure No. 3 shows a comparison of the results of D with the global limit. Also, Table No. 3

 https://doi.org/10.1051/e3sconf/202563302005

shows the results of AED, and ELCR at a rate of 0.126,0.449 respectively and Figure No. 4 shows a comparison of the results of AED with the global limit. This variation in background radiation in your soil is a combined result of many geological and environmental factors affecting where radioactive materials end up in the ground. One of the key factors is the composition of the soil itself, as soil contains various mineral components that can be rich in naturally occurring radioactive materials such as uranium, thorium and potassium-40. These naturally occurring radioactive materials are found in varying concentrations based on the type of soil and their mineral components. Also, radioactive materials are dispersed with the soil according to the geology of the area. For example, soils in regions that contain rock with high amounts of radionuclides, such as uranium or thorium, may have higher levels of radiation than soils in other regions. Some rocks are naturally rich in radioactive materials (for instance, granite and basalt rocks), so the background radiation can vary from one place to another. But sauna is not only a factor in the radiation content in the ground. These chemical processes in the soil are influenced by temperature and relative humidity, which factors into the ways radioactive materials interact with other elements found in soil. In warm and wet regions, chemical reactions might boost, making radiation degrees increase normally or reduce.

Table 2. Results of Din samples of the present region.

No.	Code	Doutdoor (µSv/h)	
		Average (µSv/h)	Error
1	R1	0.14	0.05
2	R2	0.12	0.02
3	R3	0.13	0.01
4	R4	0.16	0.08
5	R5	0.16	0.06
6	R6	0.21	0.08
7	R7	0.19	0.05
8	R8	0.11	0.01
9	R9	0.17	0.03
10	R10	0.15	0.04
11	R11	0.15	0.04
12	R12	0.14	0.07
13	R13	0.13	0.04
14	R14	0.19	0.05
15	R15	0.20	0.07
16	R16	0.15	0.05
17	R17	0.12	0.03
18	R18	0.19	0.07
19	R19	0.21	0.09
20	R20	0.20	0.08
Min		0.11	0.01
Max		0.21	0.09
Average		0.161	0.051
Worldwide average		0.247	

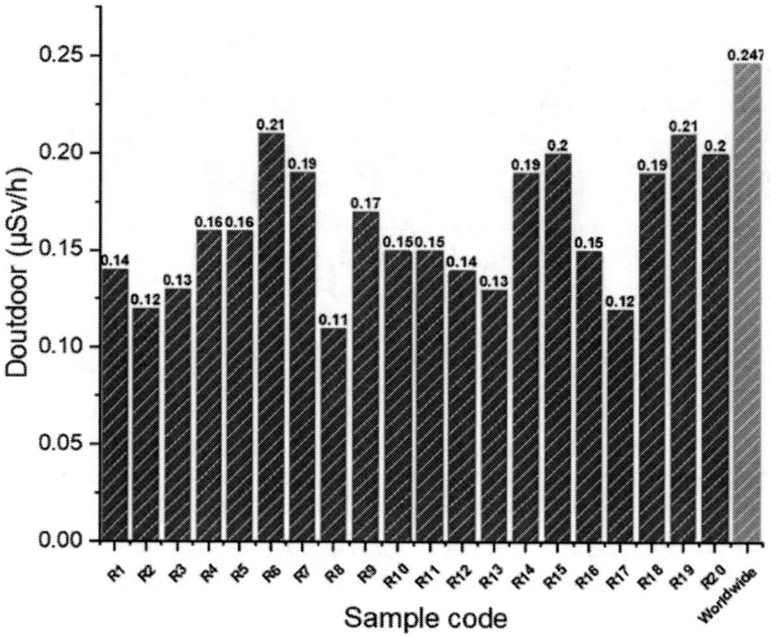

Fig. 3. Comparison of the results of radiation dose with safe limit.

Table 3. Results of AED, and ELCR in samples of the present region.

No.	Code	AED (mSv/y)	ELCR×10^{-3}
1	R1	0.112	0.391
2	R2	0.096	0.335
3	R3	0.104	0.363
4	R4	0.128	0.446
5	R5	0.128	0.446
6	R6	0.167	0.586
7	R7	0.151	0.530
8	R8	0.088	0.307
9	R9	0.136	0.474
10	R10	0.120	0.419
11	R11	0.120	0.419
12	R12	0.112	0.391
13	R13	0.104	0.363
14	R14	0.151	0.530
15	R15	0.159	0.558
16	R16	0.120	0.419
17	R17	0.096	0.335
18	R18	0.151	0.530
19	R19	0.167	0.586
20	R20	0.159	0.558
Min		0.088	0.307
Max		0.167	0.586
Average		0.126	0.449
Worldwide average		0.48	0.29

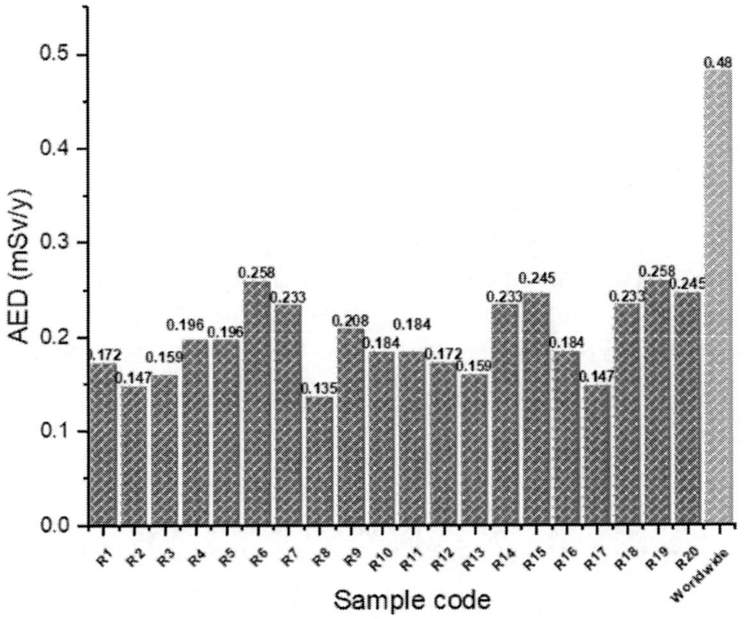

Fig. 4. Comparison of the results of AED with safe limit.

4 Conclusions

Various conclusions can be drawn with regard to the levels of natural radiations in Najaf Sea region (Najaf Governorate) and its compatibility with the safe limits set by the international standards based on the data and analyses that have been carried out in the results and discussion section.
Radioactivity Level and Radiation Safety Assessment of Soil Samples Radioactivity levels are assessed based on absorbed radiation dose and annual effective dose only, since all soil samples in the studied area were found to be compliant with international regulations, set out by the United Nations Scientific Committee on the Effects of Atomic Radiation (UNSCEAR 2008), which provide global assessments of natural radiation levels and their impact on human health.

Rules and guidelines for minimizing radiation exposure are established by the National Council on Radiation Protection (NCRP 2009).

Reassess of radiation safety at the region The soil in the Najaf Sea region is classified as radiation safe because the natural radiation background is not a significant health risk for the population and the environment. For this reason, numerous human activities including farming, building houses, and expanding cities are likely to occur here without fears of radiation.

The Efficacy of GIS technology in displaying radiation in geography of Najaf Sea The study concluded that the GIS technology was efficient and accurate in producing the three-dimensional maps of radiation for the Najaf Sea region. Using this technique enabled a distinct visualization of the radiation dose distribution within the region. Perform Geospatial analysis key event, thus increasing the reliability of the results. Providing visual tools for assessing the dispersion of radiation, thus improving the decision process for environmental and health issues.

References

1. E. Ahmed, S. S. Nemah, E. S. Obayes, G. H. Seood, A. S. Jassim, "Evaluation of radon gas air concentration in schools in Waist, Iraq," in E3S Web of Conferences **592**, 06001 (2024)
2. S. S. Nayif, E. J. Mohammed, A. S. Jassim, A. A. Abojassim, "Estimation of the radiological risk due to Radon-222 exposure in dwelling of Al-Hussainya District, Karbala Governorate," Appl. Environ. Soil Sci. **2024(1)**, 6511029 (2024)
3. B. A. Ali, A. S. Jassim, "History of the impact of radiation on the environment: Hiroshima bomb," in E3S Web of Conferences **583**, 04015 (2024)
4. F. G. Mohammed, H. J. Alatta, A. S. Mohamed, A. H. Harif, S. G. Mohammed, "Calculating land surface temperature of South Baghdad by the using Landsat 8 images," in AIP Conf. Proc. **2922(1)** (2024)
5. A. H. Ali, A. S. Jassim, A. A. Abojassim, R. J. Dosh, *Health risk assessment of radon concentrations in water samples of selected areas north of Al-Najaf Governorates* (2024)
6. BOSEAN Company Inc., Reference Manual version 1, FS-600 Geiger Counter Nuclear Radiation Detector (2024)
7. H. H. Abbas, M. J. Tarbool, H. A. Al-Noamany, A. A. Abojassim, "Assessment of radiological hazard to background radiations in area about Kufa Cement Factory," NeuroQuantology **20(4)**, 460 (2022)
8. R. J. Dosh, A. K. Hasan, A. A. Abojassim, "Background radiation in primary schools of Al-Najaf City, Iraq," Int. J. Nucl. Energy Sci. Technol. **15(3-4)**, 283–301 (2022)
9. R. R. Muneam, E. R. Wdaah, H. H. Abbas, A. A. Abojassim, "Background radiation in the Physics Department, Faculty of Science, University of Kufa," Ann. Agri-Bio Res. **28(2)**, 481–486 (2023)
10. L. A. Alasadi, A. A. Abojassim, "Natural Radioactivity in Soil Samples of Najaf City, Iraq", Arabian Journal of Geosciences **15**, 1–16 (2022), DOI: 10.1007/s12517-022-10901-0

https://doi.org/10.1051/e3sconf/202563303001

Assessment of Shatt al-Arab water quality using water indices in Basrah City of South Iraq

Ban A. Alrazaq[1], *Ebtesam F.* Khanjar[1,*], *Israa J.* Muhsin[1], and *Al-Ali* I.A.[2]

[1]Department of Remote sensing and GIS, College of Science, University of Baghdad, Baghdad, Iraq
[2]Department of Geology College of Science, University of Baghdad, Baghdad, Iraq

Abstract. This study aims to assess the state of pollution in Shatt al- Arab water by measuring the physical and chemical characteristics. The study was based on the link between field data and mathematical models derived from satellite images of Landsat and Geographic Information Systems in order to evaluate Shatt al-Arab water and to find digital maps that can be used to predict the quality of water along the river. The current study includes evaluating the water of the Shatt al-Arab, studying the concentrations of pollutants in it, and knowing the suitability of its water for different purposes. However, chemical and physical variables related to water quality were studied. These variables were compared with the World Health Organization (WHO) recommendation limits. After the processing of the satellite images, NDWI, MNDWI, NDMI, WRI and AWEI were derived using the mathematical equations for calculating each indicator. The correlation between these derived indicators and the field-measured water quality parameters was calculated and regression equations were built in order to find the best mathematical models for expressing water quality components from satellite images without needing field data. The results show that there is moderate to strong relation between the water quality parameters and derived indices.

1 Introduction

Water is a significant regular asset that is crucial for human survival and the ecosystem's health. Water is included coastal water bodies and freshwater bodies like lakes, rivers, and groundwater. Since the past few decades, the increment of anthropogenic activities specially in the modern region has had consequences for water bodies. These are the worldwide problems occurring around the world [1].

Polluted water is a sort of water containing improper physical, chemical, and biological materials which is unacceptable for drinking or home use. Water pollutants include contamination due to domestic wastes, insecticides and herbicides, food processing waste, pollutants from livestock operations, volatile organic compounds, heavy metals, chemical waste, and others [2].

Water Quality (WQ) is a term referring to the suitability of water for different uses. Any specific use will have definite requirements for the water characteristics: physical, chemical,

* Corresponding author: Ebtesam.khanjer@sc.uobaghdad.edu.iq

or biological. There is a wide variety of natural and human influences effect on water quality. The natural influence includes geological, hydrological, and climatic because they affect the quantity and the quality of water available. Their effect is generally highest when available water quantities are low and maximum use must be a limited resource; for example, high salinity is a repeated problem in arid and coastal areas [3-4].

Water indices NDWI, MNDWI, NDMI, WRI, and AWEI that are used to predict water quality parameters from satellite imagery are discussed and showed how to calculate. At last, correlation and regression are defined which are used in the study to find the mathematical model that correlates the water quality parameters measured from the field with water indices driven from satellite images.

2 Materials and methods

Conversion of raw Digital Number (DN) values to physically meaningful reflectance is a prerequisite for reliable quantitative analysis of multispectral satellite imagery. Initially, Top-of-Atmosphere (TOA) planetary reflectance for each spectral band was computed via a linear scaling of the Level-1 pixel DN (Qcal), according to:[5]

$$\rho\lambda' = M_P*Qcal + A_P \tag{1}$$

where:

$\rho\lambda'$ = Top-of-Atmosphere Planetary Reflectance, M_P = Band multiplicative scaling factor
A_P = Band additive scaling factor, Qcal = Level-1pixel value

The top of atmosphere value must be converted to true value because it does not contain a solar elevation angle correction. The conversion equation is given by:

$$P\lambda= \rho\lambda'/\sin(\theta) \tag{2}$$

where:

$P\lambda$ = True Top-of-Atmosphere Planetary Reflectance, θ = Solar Elevation Angle

Subsequent conversion of DN values to radiance addresses instrument gain and bias characteristics, thereby reducing sensor-linked noise without introducing additional error. This step employs the physical relationship [6]:

$$L\lambda= Gain\ \lambda+ Bias\ \lambda \tag{3}$$

where:

$L\lambda$ is the radiance value of the pixel, Gain: the gain value for a specific band
Bias: the bias value for a specific band

Extraction of aqueous features necessitates the computation of multiple normalized indices that exploit contrasting spectral behaviors of water, vegetation, and built surfaces. The Normalized Difference Water Index (NDWI) quantifies canopy and surface moisture by comparing green and near-infrared reflectances: [7, 8]:

$$NDWI=(Green-NIR)/(Green+NIR) \tag{4}$$

where NIR = pixels from the near-infrared region, and Green = pixels from the green channel.

The Modified NDWI (MNDWI), substituting short-wave infrared (SWIR) for NIR, enhances open-water detection and suppresses urban artefacts [9]:

$$MNDWI=(Green-SWIR)/(Green+SWIR) \quad (5)$$

where SWIR = pixels from the short-wave infrared region, and Green = pixels from the green channel.

The Normalized Difference Moisture Index (NDMI) isolates vegetation water content variations via the ratio of NIR and SWIR bands [9, 10]:

$$NDMI=(NIR-SWIR)/(NIR+SWIR) \quad (6)$$

where SWIR = pixels from the short wave infrared band, NIR = pixels from the near-infrared band.

The Water Ratio Index (WRI) leverages the reflectance dominance of water in visible bands relative to NIR and SWIR:

$$WRI=(Green+Red)/(NIR+SWIR) \quad (7)$$

where Green = pixel values from the green band, Red= pixel values from the red band, NIR = pixel values from the near-infrared band, and SWIR = pixel values from the short wave infrared band.

Finally, the Automated Water Extraction Index (AWEI) employs empirically derived coefficients to maximize separation of water from non-water, with two formulations tailored to the presence or absence of shadow.

$$AWEInsh=Green+Blue\times0.25-1.5\times(NIR+SWIR1)-0.25\times SWIR2 \quad (8)$$

$$AWEIsh=4\times(Green-SWIR1)-(0.25\times NIR+2.75\times SWIR2) \quad (9)$$

These reflectance conversions and index computations were executed in a unified processing chain, ensuring consistency across the multispectral dataset and enabling robust delineation of terrestrial and aquatic features within the study area.

3 Results and discussion

One of the common goals of water monitoring is to describe the past and present water quality conditions of a region based on monitoring data. The primary challenge is to extrapolate from a sample of water quality measurements made at a limited number of river sites (i.e. a monitoring stations) to an area containing sampling stations and a large number of un-sampled sites. Simulation refers to the use of a calibrated model to predict variables based on a set of modified

This study was carried out for assessment of Shatt al Arab water quality using the most common water quality indices, building of maps for pollutants concentrations along the river and finding mathematical relationships for water quality parameters, extracting their values from satellite images. To achieve this, several steps were taken:

1- Data collection (downloads satellite images of the study area and obtains pollutants concentrations from stations for the period (2018-2023).

2- Building of spatial-temporal database for pollutants by input data collected from stations for the study period.

3- Finding mathematical relationships between variables, several sup-steps were taken:

a-Calculation of water indices NDWI, MNDWI, NDMI, AWEI, and WRI from satellite images after images preprocessing and extracting their values at stations for every year in the study period.

b-Perform statistical analysis of the data using Matlab program (finding the correlation and regression between water quality parameters and water indices).

6- Comparing the results obtained from applying mathematical modeling with field measurement and calculating the error.

Five monitoring sites on the study area were selected for measuring the physical and chemical parameters. These stations were before Karmat Ali River (SH1), Al-Meqal (SH2), Abo Al-Khasib (SH2B), Al-Seeba (SH3), and Al- Fao (SH4). All of these stations lie in Basra city, the description of the stations with their coordinates is shown in Table (1). The measurements were taken for the parameters every month from 2018 to 2023 which were analyzed by the Iraqi Ministry of Environment for different physical and chemical parameters.

Table 1. Description of the selected stations on study area.

Station code	Location	Station coordinates	
		Longitude (E)	Latitude (N)
SH1	Before Karmat Ali River	47° 45′ 56″	30° 39′ 01″
SH2	Al-Meqal	47° 46′ 70″	30° 34′ 45″
SH2B	Abo Al-Khasib	47° 56′ 67″	30° 27′ 95″
SH3	Al-Seeba	48° 11′ 66″	30° 19′ 60″
SH4	Al-Fao	48 ° 29 ′ 04 ″	29 ° 58′ 51″

Landsat imagery with 30 m spatial resolution is acquired from the United States Geological Survey (USGS) for the study area which lies in two scenes with paths 165 and 166, row 39, for the period (2018-2023).

Converting of DN to radiance using equation (1) was done to each band in Landsat 7 imagery, and then the radiance value converted to reflectance using equation (2). Also, the converting of DN of Landsat 8 imagery bands into

reflectance was done by using equation (3). The conversion process of DN into reflectance values was done within ARCMAP GIS program spatial analysis tool map algebra raster calculator tool.

After the images were preprocessed, different kinds of the water index were calculated these are NDWI, MNDWI, NDMI, AWEI and WRI using formulas (4), (5), (6), (7) and (8) respectively. The calculation process was done using raster calculator. The reflectance index value was extracted for the five stations along the river for all study years using tool from Landsat toolbox extract multi values to points within Arc GIS.

In order to determine the nature and strength of relationship between the water quality parameters (field measurement) and the five water index (derived from satellite image) used, and thus find a mathematical models for each water quality parameter to predict their values from the satellite images without need to field measurements by using water indexes derived from satellite images, regression model was applied in which water quality parameters were considers as dependent variables and water indexes as independent variables.

Thiel's estimators are used as a non-parametric model to find out the linear regression model expressing the dependent variables in terms of independent variables. MATLAB 2018 program is used to calculate Thiel's estimators and extract the results of linear regression by applying equations (2.38), (2.39), and (2.40) respectively. The results of Thiel's estimator's model are shown in Table (2) to Table (7).

Table 2. The results of Thiel's estimator's model in 2018.

Parameters	Determination coefficient R2	Correlation coefficient R	t- test	RMSR	Model estimated, 2010
pH	0.5002	0.7072	0.2962	0.2084	pH=7.3330-0.1380 NDWI
DO	0.7664	0.8754	1.2214	0.9174	DO= 8.7857 -0.1167 WRI
PO4	0.3921	0.6262	1.1794	0.0133	PO4=0.2881+ 0.0016WRI
NO3	0.7704	0.8777	0.7716	1.4230	NO3=6.8490+2.4550MNDWI

Table 3. The results of Thiel's estimator's model in 2019.

Parameters	Determination coefficient R2	Correlation coefficient R	t- test	RMSR	Model estimated, 2011
pH	0.3395	0.5827	0.1282	0.1039	pH=7.9050 -0.0002 WRI
DO	1	1	2.6653	0.3147	DO =6.2488-0.0003 AWEI
PO4	0.6846	0.8274	0.5233	0.0192	PO4= 0.2455-0.0225 NDMI
NO3	1	1	1.2347	0.5685	NO3=7.0185+1.5695MNDWI

Table 4. The results of Thiel's estimator's model in 2020.

Parameters	Determination coefficient R2	Correlation coefficient R	t- test	RMSR	Model estimated, 2012
pH	0.8190	0.9050	1.3427	0.0779	PH =8.1000-0.1350NDWI
DO	0.4892	0.6994	1.9247	0.3381	DO=5.8300 -0.8400NDWI
PO4	0.1405	0.3749	0.2787	0.0118	PO4= 0.2950+0.3571*10-3WRI
NO3	1	1	1.0002	418.1226	NO3=10.7000+935.1500NDWI

Table 5. The results of Thiel's estimator's model in 2021.

Parameters	Determination coefficient R2	Correlation coefficient R	- test	RMSR	Model estimated, 2014
pH	1 0.2244 1	1 0.4737 1	3.9006 2.7603 4.4147	0.0800 0.0969 0.0681	8.0944-0.0976WRI 8.2419-0.2909 NDMI 8.0716-0.3416NDWI

Continuation of Table 5.

Parameters	Determination coefficient R2	Correlation coefficient R	- test	RMSR	Model estimated, 2014
DO	0.4984 0.9199 0.9511 0.5026	0.7060 0.9591 0.9753 0.7089	2.8807 2.8805 11.9327 3.4833	0.3078 0.2640 0.2292 0.2650	DO=7.6028 -1.0080 NDWI DO=7.9947-1.0726MNDWI DO=8.1941-2.9749NDMI DO=8.1072 -0.2886WRI
PO4	0.2204	0.4695	1.1710	0.0169	PO4=0.3150-0.0062WRI
NO3	1 1	1 1	3.1363 2.6669	0.7826 0.9035	NO3=8.8247+2.7910NDWI NO3=7.4175+0.7533WRI

Table 6. The results of Thiel's estimator's model in 2022.

Parameters	Determination coefficient R2	Correlation coefficient R	t- test	RMSR	Model estimated,2015
pH	0.2728	0.5223	0.4876	0.0937	pH =7.9264+0.0485NDMI
DO	1 1 0.6690 0.8877	1 1 0.8179 0.9422	5.2439 3.3021 6.5830 5.0916	0.1533 0.1784 0.2186 0.1311	DO=9.0116 -0.9005 NDWI DO=9.1500 -0.7824 MNDWI DO=9.1726-1.5299 NDMI

https://doi.org/10.1051/e3sconf/202563303001

					DO=9.3521-0.2096WRI
PO4	1	1	9.8449	0.0372	PO4=0.4603-0.3897NDMI
NO3	0.5312 0.2900 0.5497	0.7288 0.5385 0.7414	2.8457 5.4281 3.2928	0.1181 0.1162 0.0950	NO3=3.9822 -0.3766 NDWI NO3=4.0990 -0.6705 NDMI NO3=4.1320-0.0982WRI

Table 7. The results of Thiel's estimator's model in 2023.

Parameters	Determination coefficient R2	Correlation coefficient R	t- test	RMSR	Model estimated 2016
pH	0.6264 0.8281 0.9043	0.7915 0.9100 0.9510	2.9018 9.8721 4.2476	0.0977 0.0754 0.0774	pH =7.6920 +0.3287 NDWI pH =7.5721 +0.8102 NDMI pH =7.5662+ 0.0839WRI
DO	1 1 0.9958 1	1 1 0.9979 1	4.9903 3.4023 19.661 6.5886	0.5804 0.5392 0.2953 0.3774	DO=7.6800 +3.3582NDWI DO=7.3732 + 2.7474MNDWI DO=6.9003 +6.3180NDMI DO= 6.8323+0.6348WRI
PO4	0.2619	0.5118	0.5665	0.0727	PO4=0.4106+ 0.0478NDWI
NO3	0.3286	0.5732	0.4405	0.4594	NO3=4.0932+ 0.2346NDWI

The results of the regression model demonstrated that there was a significant relationship between the water quality parameters and the explanatory variables. This can be inferred from the (t) value which is greater than 0.1 and its associated (p) value which is less than 0.05.

R represent the correlation coefficient between regression model variables where values that belongs to 50% represent moderate correlation, if it is greater, the correlation is strong and if it is less, the correlation is weak.

R2 represent determination coefficient and its value determine data explanation. Its value represents the percentage of the amount that the independent variables explain from the variance in the dependent variable.

The impact coefficient (p) was recorded value less than 0.05 so that, the null hypothesis was rejected and the alternative hypothesis was accepted, which means that the regression is significant and there is an effect of the independent variables on the dependent variables, and we can predict the dependent variable through the independent variable that entered in its model.

To clarify the relationship between the field measurements and the values expected from the models, the mathematical models estimated were applied and the values expected from them were founded, then the relationships were plotted across the monitoring stations for each study year using Microsoft Excel program as shown in Figures (1) to (6).

 https://doi.org/10.1051/e3sconf/202563303001

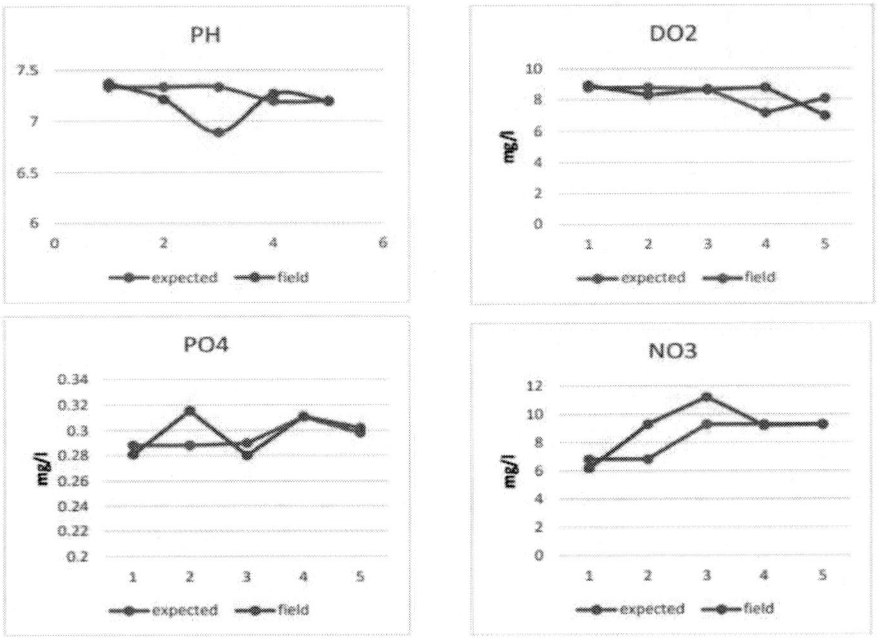

Fig. 1. The relationship between the model expected value and field for 2018.

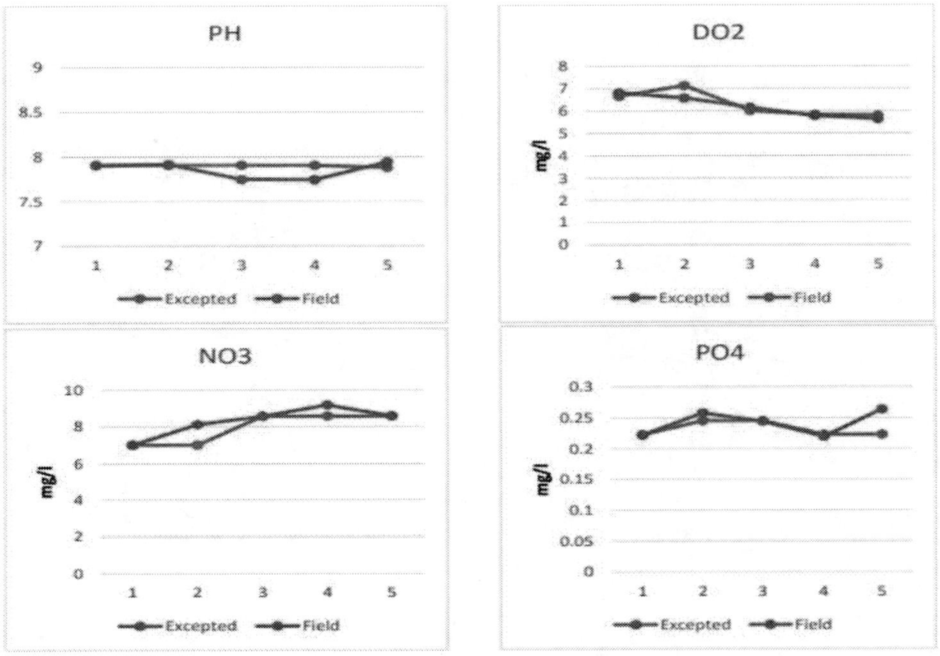

Fig. 2. The relationship between the model expected value and field for 2019.

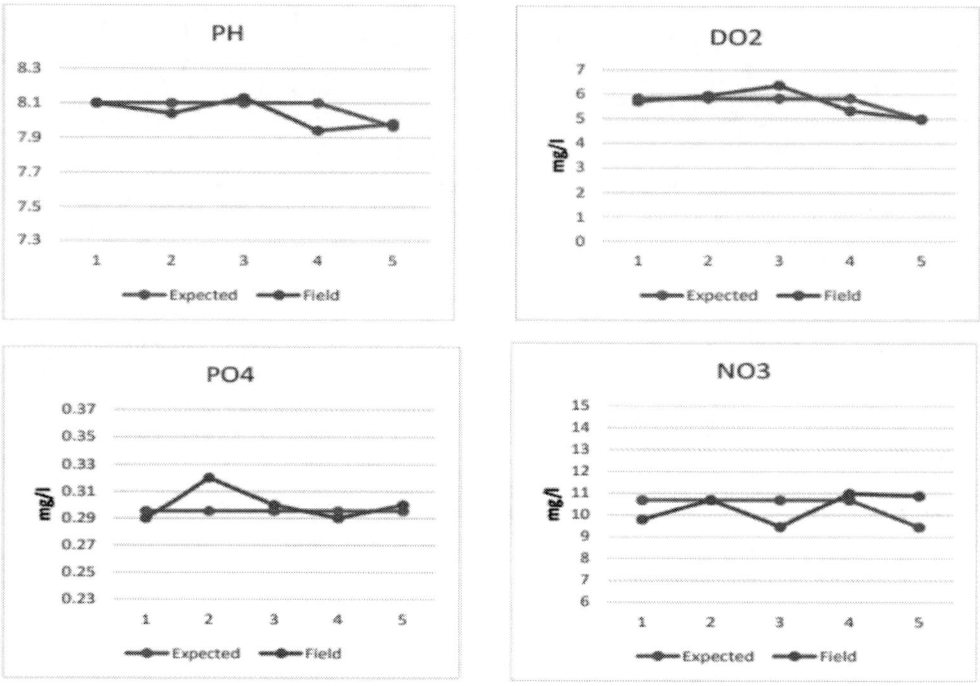

Fig. 3. The relationship between the model expected value and field for 2020.

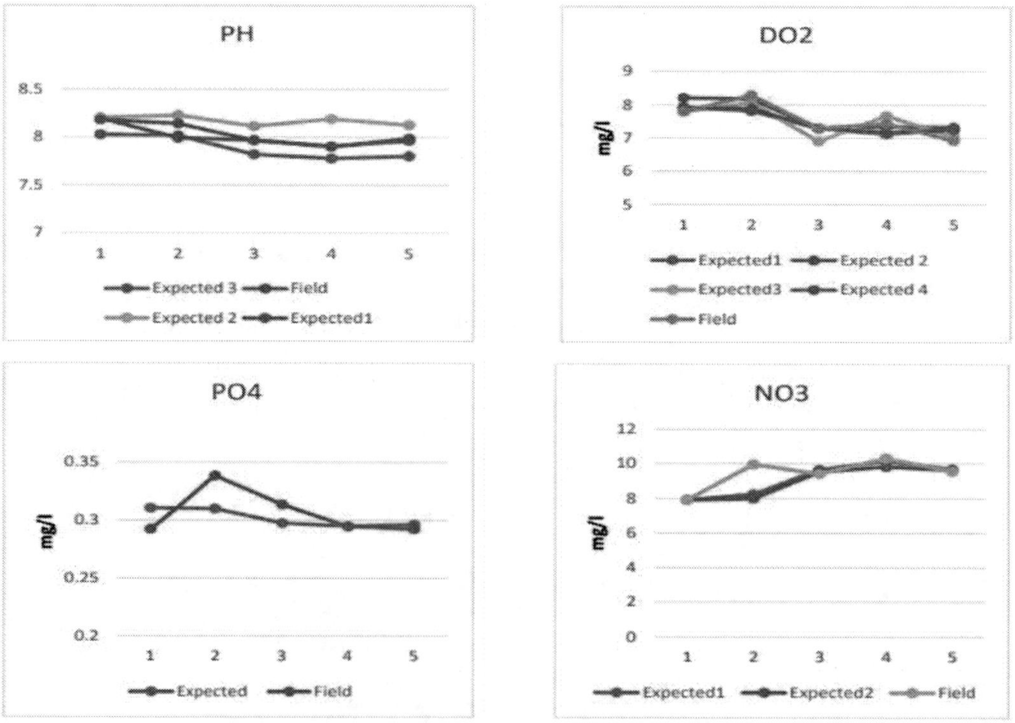

Fig. 4. The relationship between the model expected value and field for 2021.

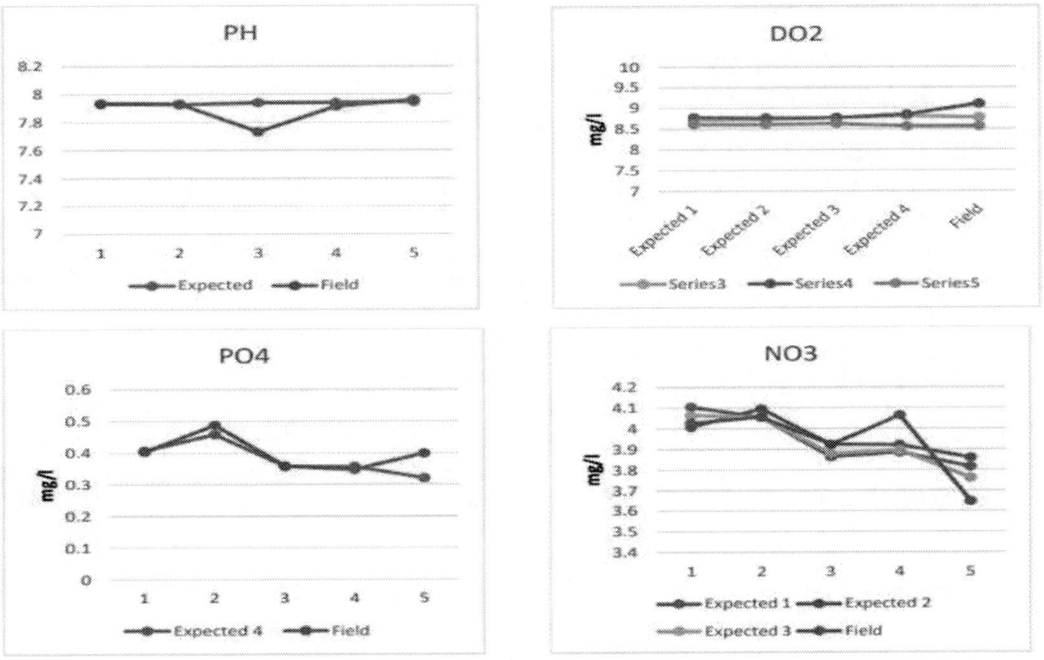

Fig. 5. The relationship between the model expected value and field for 2022.

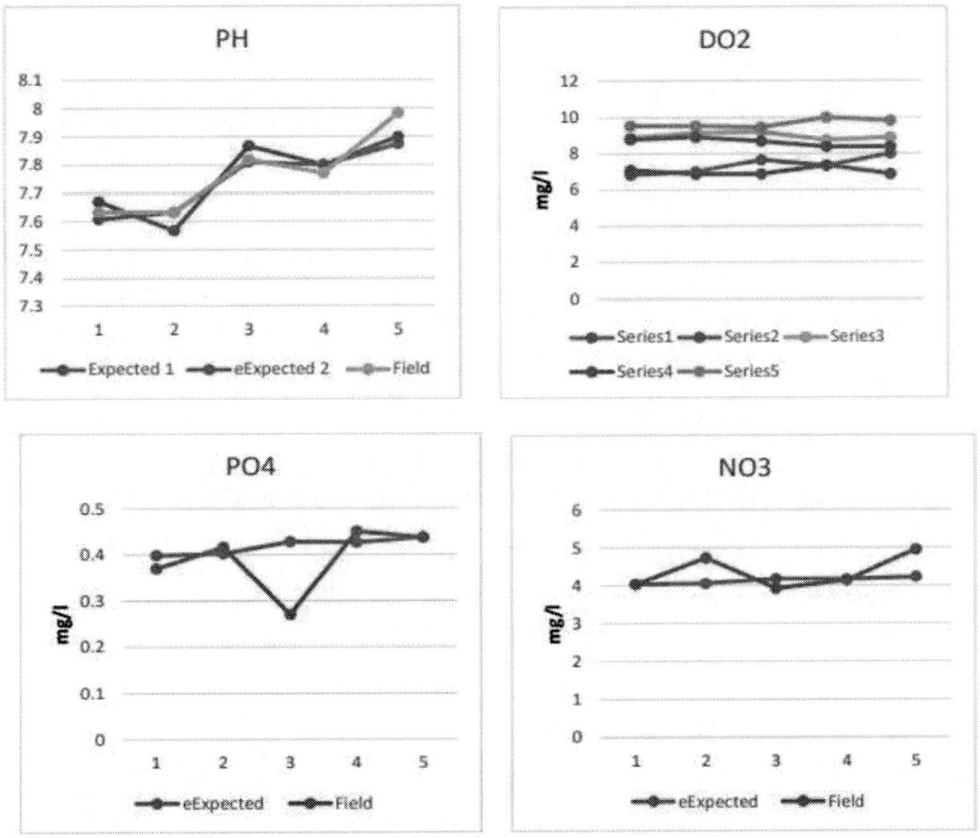

Fig. 6. The relationship between the model expected value and field for 2023.

4 Conclusions

The main goal of this study is to monitor the water quality parameters of the study area and compare them with global standards, water assessment for different uses, and build mathematical models to express the water quality parameters from remotely sensed data without need to field measurements. According to the results that have been obtained from the study, WQI of the river was observed at all stations along the river was poorly for potable water because increased in salinity and some ions, as a result of the discharge of liquid waste that is generated due to various human activities, particularly within Al-Ashar area and Abu Al-Khasib station, including untreated sewage, and the releases of sewage in large quantities from the Basra city.

The Water Quality Index (WQI) has been useful and practical in evaluating water for drinking and irrigation purposes over an extended period. The results of WQI showed that the study area suffered from water quality deterioration because a significant drop in sewage networks, manufacturers, and businesses waste going on to the water source leading to a broad collapse in the quality of water. Also, the decay in WQ of the SAR was due to a remarkable decrease in the amount of freshwater reaches from the upper tributaries of the Tigris as well as the Euphrates basin.

It can be noticed from the regression models that the most values of correlation coefficients are greater than 50% i.e. the correlation relationships between model variables are classified as moderate to strong. Also, the most independent variable used in building models is WRI which means that it is the most influential factor on the water quality parameters, while the least used is AWEI.

References

1. N. K. Tyagi, S. Tyagi, "Influences of natural and anthropogenic factors on surface and groundwater quality in rural and urban areas," Frontiers in Life Science **8(1)**, 23-39 (2015)
2. N. H. Omer, "Water Quality Parameters," in *Water Quality - Science, Assessments and Policy* (Intech Open, London, United Kingdom, 2019)
3. S. O. M. Al-Mayyahi, I. A. Al-Ali, "Assessment of Some Trace Elements in Surface Water and Soil near Al-Ahdab Oil Field, Wasit Governorate, Iraq," IOP Conference Series: Earth and Environmental Science **1300(1)**, 012024 (2024)
4. A. S. A. N. et.al., "Comparison of Land Surface Temperature Derived From Landsat 7 ETM+ and Landsat 8 OLI/TIRS for Drought Monitoring," in IOP Conference Series Earth and Environmental Science **313(1)**, 012041 (2019)
5. G. A. Al-Edhar, A. A. Al-Sadoon, E. F. Khanjer, "Monitoring the Land surface temperature LST with different seasons for Babylon City using GIS and R.S techniques," IOP Conf. Series: Earth and Environmental Science **1202**, 012018 (2023)
6. E. S. Hassan, A. F. Hasson, E. F. Khanjer, "Evaluation the spectral reflectivity of Baghdad city for different bands using R.S and GIS techniques on December 2021," IOP Conf. Series: Earth and Environmental Science **1202**, 012006 (2023)
7. M. Elhag, I. Gitas, A. Othman, J. Bahrawi, "Effect of water surface area on the remotely sensed water quality parameters of Baysh Dam Lake, Saudi Arabia," Desalination and Water Treatment **194**, 369-378 (2020)

8. Y. Hibi, "Modeling variable density flow in subsurface and surface water in the vicinity of the boundary between a surface water–atmosphere system and the subsurface," Journal of Contaminant Hydrology **234**, 103688 (2020)
9. D. Cierniak, et al., "Impact of artificial infiltration on removal of surfactants in surface water treatment process," Desalination and Water Treatment **199**, 241-251 (2020)
10. Q. Cheng, et al., "Measurement of water content at bare soil surface with infrared thermal imaging technology," Journal of Hydrology **615**, 128715 (2022)

https://doi.org/10.1051/e3sconf/202563303002

Monitoring changes in the marshes of Mesopotamia using multispectral remote sensing data

Jehan Sabeeh Mosleh *

Al-Nahrain University, Baghdad, Iraq

Abstract. The central focus addressed in this study revolves around the transformation that has occurred in wetlands. It is of utmost importance to acknowledge that the intensified desiccation of wetlands and lakes spanning over a vast expanse of 10,000 km bears immense significance for the provincial ecosystem and micro-climate, consequently impacting the well-being of individuals. These ramifications encompass alterations in precipitation patterns, consistent elevation in temperatures, particularly during prolonged and scorching summer seasons, as well as the presence of relentless dry winds unhindered by the protective presence of reed beds. The primary focus addressed within this work pertains to the transformation occurring within wetlands. In order to gain a comprehensive insight into these alterations in land cover within the designated study area, data derived from the Landsat satellite was utilized, specifically utilizing scenes characterized by path 166 and rows 38 and 39 during varying years, namely 1985, 1998, and 2016. It is important to note that Landsat imageries have the capacity to visually document marshes at a scale of 30 meters. Particularly, this investigation places particular emphasis on the dynamic changes transpiring within these marshes using remote sensing and GIS technologies as cost-effective techniques. The results show diminishing water and vegetation cover, and the land has become mostly arid. This is primarily due to human intervention including political decisions and dams on the major rivers that supply water to the region.

1 Introduction

Marshlands are some of the most vital ecosystems on the planet. This is based on the key role these ecosystems play in floods alleviation and pollutions filtration along with their role as habitats for a variety of flora and fauna. Additionally, they are economically important as it represents a local centre for recreation, fishing, grazing for cattle and reaping reeds [1-3]. These systems are considered vulnerable, as they are affected by many natural factors and anthropogenic activities such as climate change and human intervention, respectively. Over the last decades, Marshes in some regions have dramatically declined and heavily polluted, due to global warming, land-use change and dams [4, 5].

* Corresponding author: jehan.sabeeh@nahrainuniv.edu.iq

In the sake of acquiring temporal and spatial data about the Earth surface, remote sensing is the most useful tool to be utilised. In particular, in the last five decades, this technology has been widely used in many studies in this regard, for instance; land-cover/use changes or wetland areas mapping [5-7]; climate warming and the carbon cycle in environments of wetland; peatland fires and carbon releasing; and processes of hydrology in wetlands [8-10].

One of the most important wetlands on Earth is the Mesopotamian marshlands the jewel of wildlife Iraq. The Marshes of Mesopotamia lies in the south of Iraq, covers a vast area of three provinces and are locally named “Al-Ahwar”. They are formed in the basin of the bottom area of the Mesopotamia, where the Euphrates and the Tigris rivers sandwich between them broad floodplains and mainly supply the marshes along with the Karkha River from Iran. These areas have been generated as a result to accumulate the alluvial materials that come from upstream with the rivers to form a flat elevation. The three formed marshes are called "Al-Huwaiza", "Al-Amarah marsh" or "the Central" and "Al-Hammar".

The importance results in that it is the largest ecological system of its kind in the local region and Western Asia, together with a thriving home to man and nature for 5,000 years. Marshes in southern Iraq play more than one vital role. For example, but not limited to, they work as a natural treatment system for the rivers' wastewater, where fertilizers are filtered out before the water reaches the Gulf. Secondly, they have a crucial part in the conservation of regional biodiversity in the Middle East. Another important role is moderating the local climate by directly contributing to humidity and evapotranspiration rate s. During the last decades, the covered area of these marshes has changed to range between 10,500 km2 and 20,000 km2 due to different reasons.

This work aims to understand the temporal and spatial changes in the marshlands over the selected years (1985, 1998, and 2016). This is to assess the impact of these changes on the local ecosystem and environment. This study, also, utilizes remote sensing technology to provide detailed insights into land cover changes and the effectiveness of recovery efforts.

In general, the accelerated desiccation of wetlands and lakes of over 10,000 km has to have significant consequences on the provincial ecosystem and micro-climate. These ramifications will include changing rainfall patterns, invariable rise in temperatures, especially in long and hot summer seasons, and blowing dry winds without hindered of reed beds. This will affect a gigantic area of square kilometres beyond wetland's borders. Also, serious implications on human health owing to degradation of such a vast ecosystem varying from the wate r pollution and scarcity effects, to arable land desertification and to exposure to extreme thermal and toxic dust storms [1].

Concerns have become a reality, where during the late 1980s and early 1990s, the water flow of two main suppliers (Tigris and Euphrates) has been altered due to demands for fresh water and agricultural purposes. This caused a dramatic decreasing in all three marshes size by 95% and declining in the vegetation area which turned into barren land. Fortunately, what survived of this massive disaster is 30% of the north-eastern part of Al-Huwaiza marsh, which is fed by the Karkha river. This disaster is considered to be one of the greatest environmental disasters concerning wetlands.

In late 2003, politically-driven activities, which were responsible for this drainage alteration, were changed and many clay dams removed and a number of random gaps opened on the Tigris and Euphrates towards marshes. This helps the Mesopotamian marshes to be re-established and within three years 50% to 60% was recovered as well as many species of plants and animals returned [1]. To understand these changes, it is better to graphically present them on maps.

With regard to the satellite monitoring, each shuttle has spectral bands that are observing the earth surface to record multispectral imagery for the landcover. Electromagnetic waves are greatly reflected from water bodies and some of its spectra can easily penetrate vegetation. Therefore, this high variability, as ability, over marshlands relies on landcover types (vegetation, deep water, soil and so on) and their reflection coefficients. Thus, remote sensing technology presents a fabulous utility for observing wetlands [6].

The aim of this work is to re-examine the changes have been taken place in the study area, where three years have been considered to indicated these changes, using remote sensing and GIS.

Understanding climate change effects and contributing to the knowledge base in environmental monitoring.

2 Methods

There are several types of remote sensing which have applied in different study themes based on sensors and their frequency [6]. Particularly, satellite data has been employed in many of these applications for estimating the changes over a long period of time. One of the most important methods of information extraction is the classification method, which is used to classify of land-cover/use based on techniques of statistical pattern recognition that are utilised to multispectral remote sensor data [6]. The classification process of an image includes arranging all the pixels of the image in classes or categories according to shared qualities or characteristics, however, needs identifying the boundaries precisely [10].
This method has two techniques; unsupervised and supervised classification. The latter uses the algorithm of maximum likelihood and is often used for high-resolution multispectral imagery ([1, 6], which has been adapted in this report.

The key components of the methodology are representing in Figure 1. This diagram visualizes the step-by-step process involved in collecting satellite data, classifying the land cover, analysing the data, validating the results, and interpreting the changes in the marshlands.

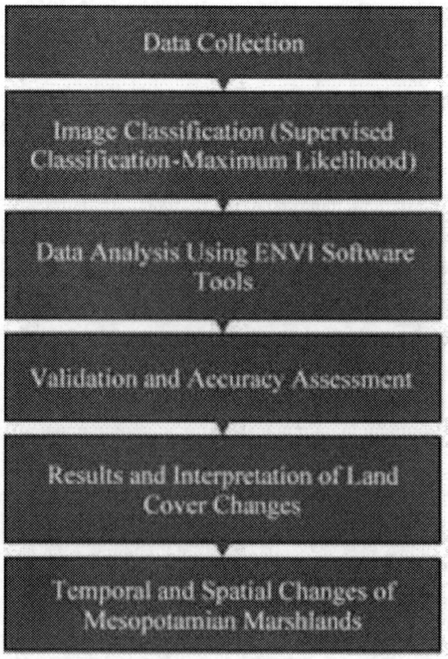

Fig. 1. Block diagram of the research method.

Data source

Table 1 exhibits related information about the acquired data from Landsat 5 and 8. In each specific date, two multispectral imageries in the same path with different raw (Path 166 and Rows 38 and 39, are coordinates of scenes) were collected in order to cover the whole study area. This required using the Mosaic technique to merge the two imageries by employing ENVI tools. This software also was used to collect the training data, which is necessary for the classification step, through the polygon method by picking manually all pixels with the same characteristic in the region of interest to merge as one class. Hence, in this study, five classes were focused on: Land Vegetation, Deep Water, Swamp Area, Bare Soil, Shallow Water. This process was applied for the three cases (1985, 1998 and 2016) to produce thematic maps in addition to record and check statistics and the accuracy level.

Table 1. Used imagery characteristics for the classification.

Data type	Path/raw	Resolution	cloud %	Bands	Date
Landsat 5 TM	166/38,39	30m	10	1-7	9/3/1985
Landsat 5 TM	166/38,39	30m	10	1-7	10/12/1998
Landsat 8 TIRS	166/38,39	30m	10	1-11	18/6/2016

Study Area

The Mesopotamian region embeds marshes and swamps in the southern part of Iraq on river systems of the Tigris and Euphrates. The study area also contains areas of very shallow water 1-2metres depth interior lakes. The lower part of the region is positioned on the 29° 55' to 32° 55' N and 45° 25' to 48° 30' E, Figure 2 shows the area, which appeared as the most widespread of its kind in the Middle East [4].

3 Results

3.1 Patterns in the Time Period 1982, 1998 and 2016

The major point dealt with in this report is the change in wetlands. To observe remotely these changes in land cover of the study area, data from Landsat satellite (scenes with path 166 and rows 38 and 39) was used. Figure 6 shows the output of the mosaic process for those two scenes in different years; 1985 and 1998. Generally, Landsat imageries record the marshes photographically at a 30 m scale, but this study is particularly concerned about changing of these marshes. Therefore, in this study, five land-cover class have been selected; Land Vegetation, Deep Water, Swamp Area, Bare Soil, Shallow Water, which are in accordance with studies related to wetlands.

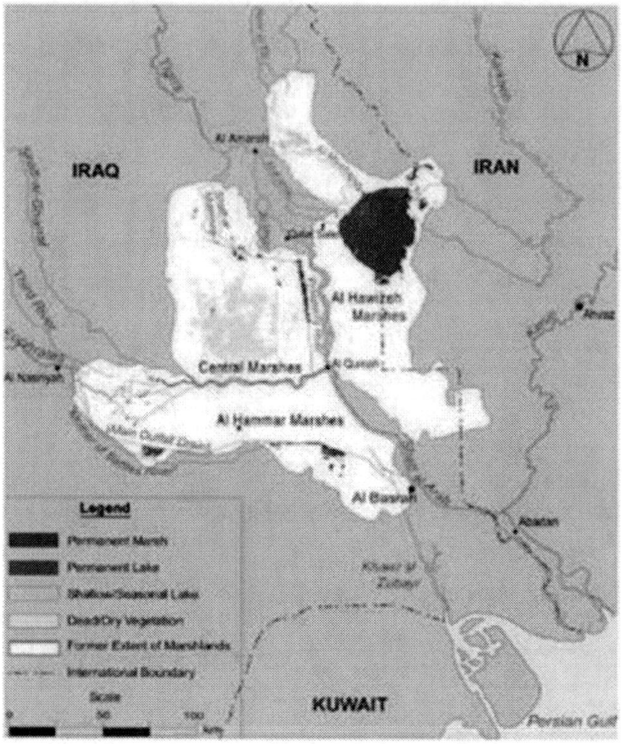

Fig. 2. Study area (Sam, 2000).

This process is too challenging, where the imagery has to be digitised initially. To identify a class, 15 to 20 distinct polygons are manually determined based on each pixel spectral characteristics and landcover in the image. Then, digitising process of data training will be input data for the next process of supervised classification. Figure 3 to 5 describe detailed land cover based on specific classes for maps of three years. These figures show clearly the changes that took place in each class of land cover.

By comparing the land cover in the years 1985 and 1998, it is obvious the effect of degradation especially on the classes of Shallow water and Swamp areas, while Deep wate r and Land vegetation classes had experienced slight changes. On the other hand, Bare Soil cover area increased by double.

 https://doi.org/10.1051/e3sconf/202563303002

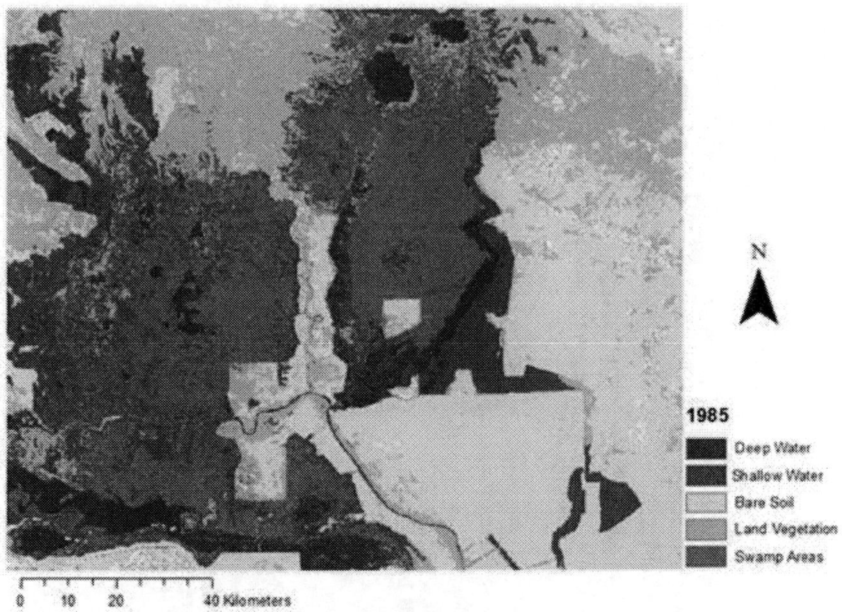

Fig. 3. Classified map of Marshlands in southern of Iraq (1985).

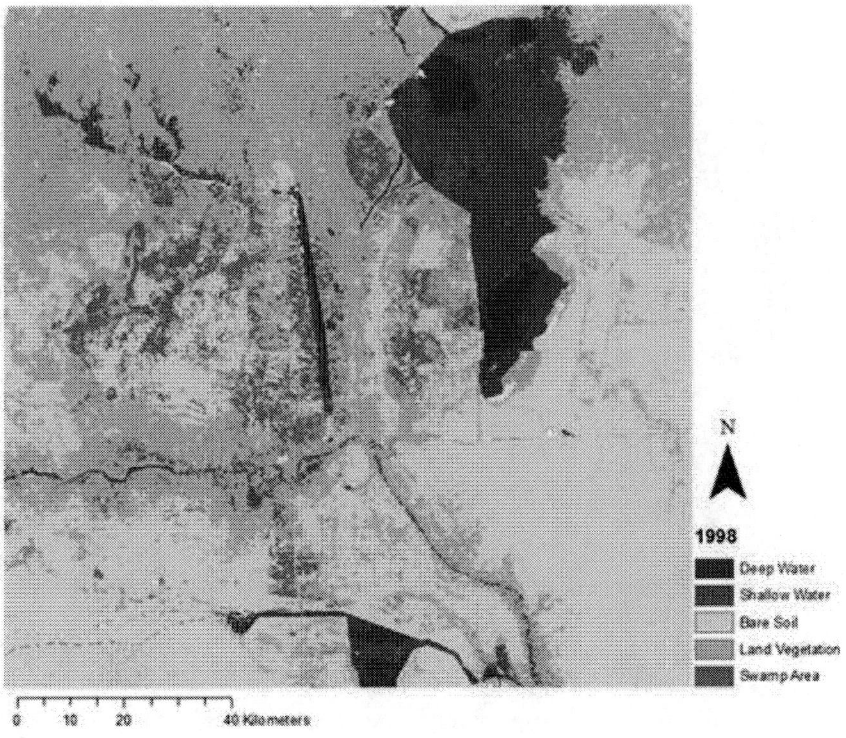

Fig. 4. Classified map of Marshlands in southern of Iraq (1998).

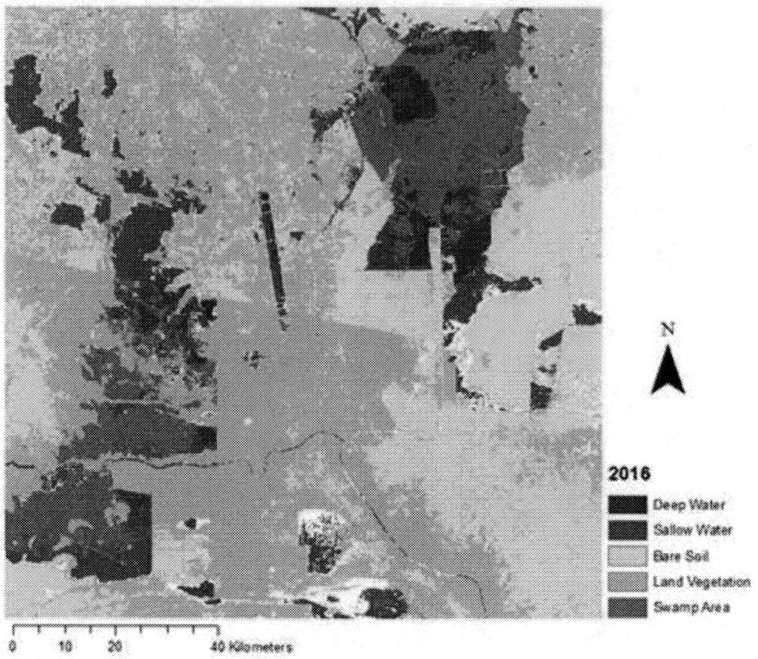

Fig. 5. Classified map of Marshlands in southern of Iraq (2016).

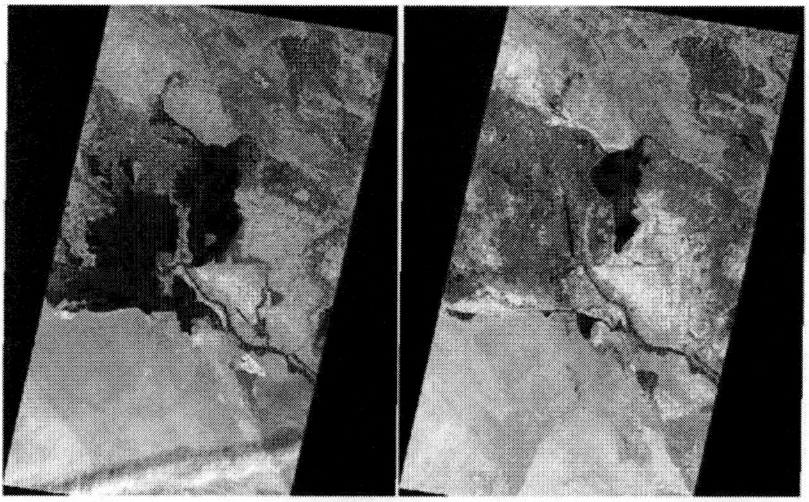

Fig. 6. Mosaicked Landsat image of the study areas in 1985 (left) and 1998 (right).
Source: Landsat 5 TM (Path 166 and Rows 38 and 39).

However, by comparing patterns of 1998 to these of 2016, patterns were slightly altered after part of the issue have been mitigated i.e. small dams. Thus, Swamp areas and Shallow waters have partially recovered, especially in Al-Hammar and Central Marshes, besides covering a wide area of previous marshes to vegetation lands.

3.2 Results accuracy assessment

The accuracy assessment of classification has been used to have an objective judgement between classified Landsat images of unknown quality and accuracy standard, as well as to carefully measure the level of detail observed within the classification (Jensen, 2015). Using such a method, it becomes possible to improve the classification accuracy by minimizing the detail amount via not to focus on a particular class but to generalizing to broad classes.

A large number of polygons has been selected to process of validation, where for each one of the selected classes, the number of polygons chosen is similar to the sample size. Table 2 displays the overall accuracy of classification processes to roughly s tart from 92% of year 1985 to reach 98% of year 2016. While Kappa Coefficients of classifications shows agreement almost perfect between the classification and data of ground truthing of each year.

Table 2. Accuracy and Kappa parameters of Confusion Matrix for all years.

Confusion Matrix Para.	Year 1985	Year 1998	Year 2016
Overall Accuracy	91.8776%	97.0310%	98.2680%
Kappa Coefficient	0.8958	0.9605	0.9726

Table 3 illustrates all omission and commission errors, i.e. error matrix, for the classification map of the study area for year 1985, that acquired from Landsat Multispectral Scanner images.

Table 3. Accuracy assessment of Confusion Matrix for 1985.

Land Cover classes	Year 1985			
	Commission (%)	Omission (%)	Producer acc. (%)	User acc. (%)
Deep Water	33.9	20.82	79.18	66.1
Shallow Water	7.51	18.14	81.86	92.49
Bare Soil	0	0.56	99.44	100
Land Vegetation	0.4	0.37	99.63	99.6
Swamp Area	6.52	0.03	99.97	93.48

3.3 Challenges and Limitation

This study concentrated on local anthropogenic circumstances that affected the marshes. However, there are other factors should be involved such as environmental conditions and regional anthropogenic, e.g. dams of the main rivers' upstream courtiers.

In term of the process of classification, it was challenging especially for wetlands which has diverse landscapes and often being complex and many overlapped land covers such as water levels and vegetation variations. To improve this process, the study area should be divided into a systematic grid or grid based on most interesting landcover which may increase the accuracy as well.

4 Discussion and conclusion

The results of this study are in agreement with literature on the wetlands in Iraq [1-3]. Based these and other researches three years 1985, 1998 and 2016 has been chosen to observe the temporal and spatial changes in the Mesopotamian marshlands. These marshlands have experienced aggressive degradation resulted in broad changes in landcover and disappearing a high percentage of marshes. Then, a recovery has been witnessed but, unfortunately, it was not satisfied, due to many reasons such as low rainfall rates and high temperatures in the area besides massive dams built on the primary water source by countries of upstream Turkey and Syria.

In order to achieve this study, remote sensing technology has been adopted as more effective and efficient method to cope with temporal and spatial changes of a broad area. The study uses a convenient and widely implemented tool that is Supervised Classification with maximum likelihood algorithm. After that, to validate the selected approach, accuracy assessment process has been performed which shows a high accuracy and almost perfect.

References

1. R. Albarakat, V. Lakshmi, C. J. Tucker, "Using Satellite Remote Sensing to Study the Impact of Climate and Anthropogenic Changes in the Mesopotamian Marshlands, Iraq," Remote Sens. **10** (2018)
2. A. K. Al-Nasrawi, I. Fuentes, D. Al-Shammari, "Changes in Mesopotamian wetlands: investigations using diverse remote sensing datasets," Wetlands **41(7)**, 94 (2021)
3. CEOBS, The past, present and future of the Mesopotamian marshes (CEOBS, 2021), https://ceobs.org/the-past-present-and-future-of-the-mesopotamian-marshes/
4. M. Guo, J. Li, C. Sheng, J. Xu, L. Wu, "A review of wetland remote sensing," Sensors **17**, 777 (2017)
5. M. K. Habeeb, M. A. Al-Shaheen, A. F. Abbas, H. A. Hamza, A. N. Okash, N. A. Hussain, P. Reiss, "The fragile ecology in Iraq's Mesopotamian marshlands endangered and restructured by a sharp increase in salinity," GSC Adv. Res. Rev. **16(1)**, 007-018 (2023)
6. A. Derdour, A. Jodar Abellan, A. Melian-Navarro, R. Bailey, "Assessment of Land Degradation and Droughts in an Arid Area Using Drought Indices, Modified Soil-Adjusted Vegetation Index and Landsat Remote Sensing Data", Cuadernos de Investigación Geográfica **49**, 65–81 (2023), DOI: 10.18172/cig.5523
7. N. Jouma, F. Dadaser-Celik, "Landscape Transformation after Irrigation Development in and around a Semi-Arid Wetland Ecosystem", Landscape and Ecological Engineering **66(1)**, 4-12 (2021), DOI: 10.1007/s11355-021-00460-z
8. N. A. Commey, Ju. Magome, H. Ishidaira, K. Souma, "Catchment-Scale Land Use and Land Cover Change Analysis in Two Coastal Ramsar Sites in Ghana, Using Remote Sensing", Water **15**, 3568 (2023), DOI: 10.3390/w15203568
9. N. Buławka, H. A. Orengo, "Application of Multi-Temporal and Multisource Satellite Imagery in the Study of Irrigated Landscapes in Arid Climates", Remote Sensing **16**, 1997 (2024), DOI: 10.3390/rs16111997
10. Nandakrishnan, P. R. Ch. Prasad, "Geoscape Characterization of Ashtamudi, Sasthamkotta, and Vembanad Ramsar Sites in Kerala, India", Water Resources **51**, 462–474 (2024), DOI: 10.1134/S009780782360273X

 https://doi.org/10.1051/e3sconf/202563303003

Optimal exploitation of water resources in Muthanna Governorate Iraq Using (RS) and (GIS) technologies

Maitham Majid Abdulameer[1,*], *Rawya* Mezaal Mahmood[2], and *Qais* Awad Omar[3]

[1]Forests and Desertification Department, Ministry of Agriculture, Baghdad, Iraq
[2]Extention and Training Office, Ministry of Agriculture, Baghdad, Iraq
[3]Expert, The deceased

Abstract. This study employs an analytical approach combined with spatial description by utilizing Geographic Information Systems (GIS) and Remote Sensing (RS) techniques. It focuses on deriving digital elevation contour lines, calculating slope direction and steepness, and determining water flow in valleys along with their expected ranks during rainfall and floods. It also suggests water harvesting measures and establishes essential criteria for implementing modern methods in project execution within the study area. Additionally, it monitors temperature variations, rainfall patterns, and demographic distribution, as well as paved roads, using LANDSAT satellite imagery and Digital Elevation Models (DEM). The results indicate a direct impact on the optimal utilization of water resources (rainfall and floods) through groundwater replenishment, erosion prevention, and the collection of water for grazing and agriculture. There is also a direct influence on surface soil layers and vegetation cover, which in turn affects the movement of herders and population distribution by selecting suitable locations based on modern techniques with reliable outcomes, unlike choices made in areas that do not utilize these technologies. Furthermore, a relationship was observed between population distribution, transportation lines, water availability, and their susceptibility to climatic conditions.

1 Introduction

The decline in environmental standards and the negative changes in climate characteristics are fundamental elements that are directly affected by human misuse, ultimately leading to the depletion of natural resources and the inability to ensure their sustainability. Al-Muthanna is the second-largest province in Iraq in terms of area, but perhaps the least populated. Its geography is predominantly desert, with the Euphrates River flowing through the center of Al-Samawah district, which is the only water source for irrigating agricultural crops that used to be the primary source of livelihood for 65% of the province's population before 2003, Now, the city of Al-Samawah and its districts aspire to economic development by supporting investments and investors, and by trying to create an environment conducive to attracting capital to generate new job opportunities. The province

* Corresponding author: maitham_1982@yahoo.com

is characterized by the availability of valuable resources for investment, especially in the industrial sector, due to the abundant and low-cost raw materials available for extraction and production, such as limestone used in cement production, as well as deposits of sodium and chlorine compounds in the form of salts used in salt production. Additionally, there are noteworthy opportunities for producing detergent. From a marketing perspective, the province is located midway to all southern Iraqi provinces, which are suffering from a shortage of construction materials, and the demand for these materials has increased significantly in recent years due to a local production halt. The province is facing a significant population crisis, making investment in the housing sector a very good opportunity for interested parties. The province also has relatively inexpensive labor available [1, 2] Al-Muthanna Province is famous for Lake Sawa, an enclosed saline lake surrounded by a natural limestone wall. Lake Sawa is fed by groundwater flowing from beneath the earth.

1.1 Research objectives

1. Identify the best areas for the presence of natural resources and their utilization for agriculture.
2. Assess the possibility of predicting climate changes accompanying the study area (floods and rainfall) and establish response measures.
3. Propose locations for the construction of dams and reservoirs for water collection.
4. Build a database and produce maps that serve as a reference for various projects in the governorate.
5. Reduce the phenomenon of sand and dust storms.
6. Mitigate environmental and living risks.
7. Develop natural pastures and livestock resources.
8. Create job opportunities and eliminate poverty among the population.

1.2 Study area problem

1. Al-Muthanna is considered the second-largest governorate in Iraq in terms of area, but it may be the least populated and economically developed.
2. Its geography predominantly features a desert character.
3. The Euphrates River runs through the center of Samawah district, which is the only water source for irrigating crops, a primary source of livelihood for 65% of the governorate's population.
4. Monitoring climatic elements (temperature and rainfall) over 30 years shows a clear increase in temperature and intensity, along with fluctuations in rainfall levels.
5. The phenomenon of sand and dust storms and the movement of sand dunes has become recurrent.
6. There has been a decline in vegetation cover, particularly the deterioration of natural pastures and livestock resources in the governorate.

1.2.1 Location and area

Al-Muthanna Governorate is located in the southwestern part of Iraq, between latitudes 29° 5' and 31° 27' North, and longitudes 43° 50' and 46° 32' East. The total area of the governorate is 5,345,450.584 hectares (Iraqi Geological Survey).

https://doi.org/10.1051/e3sconf/202563303003

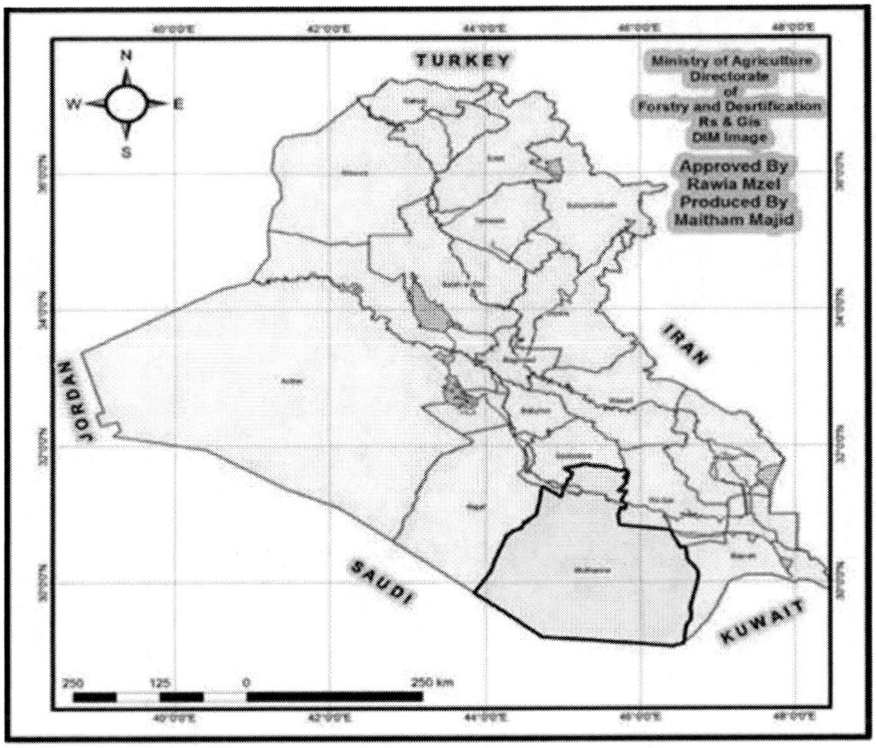

Fig. 1. Location of the studied region.

1.3 Monitoring climatic factors (Temperatures, rainfall)

Iraq's climate is characterized as semi-arid continental in some areas, particularly in the northern and northeastern parts, where winters are cold and wet, and summers are dry and warm. The remaining regions resemble a semi-tropical climate, with cold and semi-arid winters and hot, dry summers.

1.3.1 Temperature characteristics

Data from the General Authority for Meteorology and Seismic Monitoring was relied upon for a period of 30 years (1981-2010) across 30 stations in Al-Nukhayb, Al-Qaim, Haditha, Al-Anbar, Mosul, Erbil, Sulaymaniyah, Kirkuk, Sinjar, Samarra, Baghdad, Amara, Basra, Ali Al-Gharbi, Karbala, Hilla, Ramadi, Haditha, Beiji, Anah, Rutba, Tal Afar, Al-Samawah, Al-Hai, and Diwaniyah.

Thermal characteristics are among the important factors affecting overall environmental processes and economic activities. Through the data, we observe a clear increase in temperatures over the past 30 years, which necessitates addressing this serious change and proposing solutions for it.

Due to the desert location of Al-Muthanna Governorate in Iraq within the arid region, its thermal characteristics are defined by two main seasons: summer and winter. The summer lasts for eight months, from March to the end of October, while winter extends for four months, from November to the end of February. This has a significant impact on agricultural activity, as each crop has its thermal requirements needed from the germination stage to the maturity stage. Any temperature outside the minimum and maximum limits that plants can tolerate leads to damage, stunted growth, or death of the plant. [3, 4].

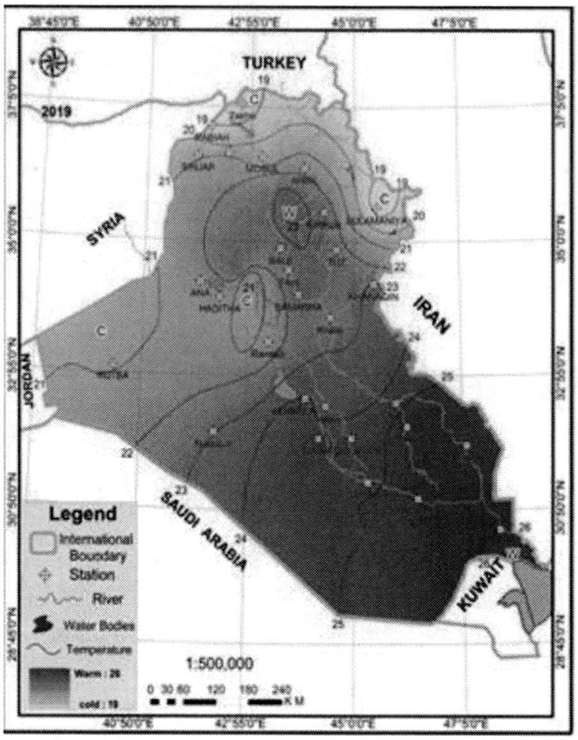

Fig. 2. Mean annual temperatures (C⁰).

1.3.2 Rainfall

The study area suffers from low and fluctuating rainfall from year to year. The average annual rainfall in the study area, based on data from the General Authority for Meteorology and Seismology over 30 years (1981-2010) and 30 stations, including Al-Nukhayb, Al-Qaim, Haditha, Anbar, Mosul, Erbil, Sulaymaniyah, Kirkuk, Sinjar, Samarra, Baghdad, Amarah, Basra, Ali Al-Gharbi, Karbala, Hilla, Ramadi, Haditha, Beiji, Anah, Rutba, Tal Afar, Al-Samawah, Al-Hai, and Diwaniya, is observed. Most of the rainfall occurs during the winter and early summer, particularly in the months of January, February, and March. Data indicate a noticeable decline in the amount of rainfall over the past 30 years, making it essential to address this serious change and devise solutions.

Winter rains are of the cyclonic type, associated with the atmospheric depressions reaching Iraq, including the study area. Their occurrence is facilitated by the rapid rise in the surface temperature of the earth due to the increased angle of sunlight and the higher number of actual sunshine hours, as well as the nature of the surface and the lack of vegetation cover.

As a result, these rains are not directly beneficial for agriculture and negatively impact the density and distribution of vegetation cover. They often fall as intense showers over short periods, similar to the rains in other arid deserts. Additionally, the evaporation rate exceeds the amount of rainfall received. Nevertheless, these rains contribute to replenishing groundwater reservoirs, which are crucial for agricultural activities in the region, as well as being important for the growth of desert plants. Given the low rainfall rates, it is necessary to drill wells and undertake water harvesting activities [3, 4].

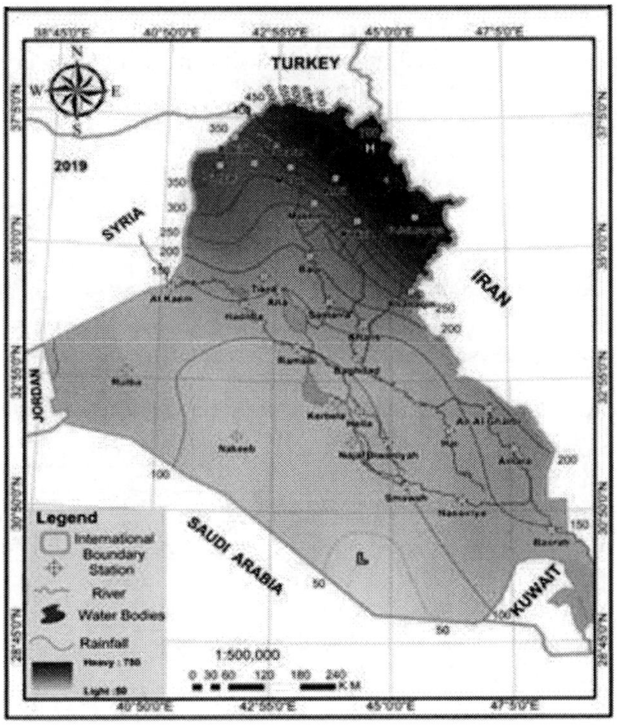

Fig. 3. Mean annual rainfall (mm).

1.4 Monitoring the deterioration of water in Lake Sawah due to climatic conditions and the extensive use of groundwater

Through satellite images, we observe a dangerous indicator of deterioration, which prompts all researchers and specialists to investigate and find solutions to this problem or to mitigate it using modern technologies.

2 Materials and methods

2.1 Monitoring process and use of geographic information systems

Land cover monitoring is part of the activities aimed at mitigating land degradation. This process begins by surveying areas susceptible to changes and assessing the state of land changes in these areas, along with the underlying causes that drive and exert pressure leading to this phenomenon. It also involves exploring ways to mitigate these impacts and the measures used to rehabilitate the degraded areas. Undoubtedly, monitoring changes requires a sound methodology and precise studies, which are of great importance. Additionally, regular and continuous monitoring and observation necessitate the use of various modern technologies and the adaptation of suitable methods to monitor and assess degradation processes. The significance of using remote sensing technologies and geographic information systems in land monitoring is evident in providing the following:

- Ensuring a comprehensive view and outlook over a wide area of land, which assists the analyst in observation and monitoring operations.
- Providing complete and continuous coverage of land cover, which helps in monitoring emergent changes occurring in the monitored areas.

https://doi.org/10.1051/e3sconf/202563303003

- The ability to study any area as needed and within the required timeframe.
- The ability to monitor remote and rugged areas that are difficult to access.
- Recording the current state of a specific area at a certain time, which serves as a reference document for that condition.
- Establishing basic criteria for project implementation.

2.2 Data used

This study was conducted to determine land cover and the most critical emergent natural factors in Al-Muthanna Governorate, southern Iraq, relying on fieldwork and Remote Sensing (RS) technology, as well as Geographic Information System (GIS) software (ARC GIS).

- **Satellite Imagery**: Satellite imagery from 2013, represented by the OLI (Operational Land Imager) sensor of LandSat 8, was used to identify the nature of the land cover in the area, as shown in Figure 1.
- **Digital Elevation Model**: A Digital Elevation Model (DEM) was utilized.
- **Image Processing Software**: The Erdas 9.1 software was used for interpreting and processing satellite images, while ARC GIS 9.2, one of the GIS software programs, was used to prepare and produce digital maps.
- **GPS Devices**: GPS devices were employed to collect field points during the field survey.
- **Office Software**: Microsoft Office software was used.
- **Field Surveys**: Field tours and surveys were conducted to obtain high-accuracy information and reduce the margin of error.

3 Results and discussion

3.1 Land cover

The method of revealing the nature of land cover and natural impacts is one of the important applications in assessing the environmental situation of any area in the world. One of the key facilitators of these applications is the use of remote sensing techniques and geographic information systems (GIS) to produce results that evaluate the environmental conditions of the study area, in addition to conducting field surveys of the grazing stations [3-5].

A Landsat 2013 satellite image was analyzed, and reference points were projected for the area to identify the nature of the land cover, which indicates that 70% of Al-Muthanna Governorate consists of barren land. Through the interpretation of the satellite image, it was noted that the desert lands contain alluvial deposits and limestone, as well as deposits of chlorides and sodium compounds used in cement manufacturing and salt production.

3.2 Soil of the study area

The soil of the study area is classified as modern desert soil, composed of various sizes and types of materials that reflect the nature of the parent rocks from which it was formed, influenced by weathering processes over varying periods. Some components reflect the effects of transport and sedimentation, either through water via floods depositing rock fragments, particularly in the bottoms of valleys and depressions, or through air, with winds playing a role in the formation of dunes and sandy soils.

Based on this, two main zones of soil can be distinguished, interspersed with secondary regions. The first zone is represented by mixed gypsum desert soil found in the northern part of the plateau within the lower valley area. Its characteristics include sandy or sandy loam soils with a course to medium texture, containing gypsum content that ranges from less than 2% to more than 50%. Its salinity ranges between 2000-6000 parts per million. In the bottoms of the valleys and some depressions, the texture is sandy loam or loamy, with calcium content approximately 19.8% [5-8].

An increase in gypsum content in the soil leads to a decrease in agricultural crop yields due to its direct impact on many soil properties. It reduces the soil's ability to retain water and nutrients, in addition to lowering the soil's clay particle content, which affects its potassium levels. Gypsum soils are characterized by the presence of hard layers that impede the movement of water, air, and roots through them. The properties of the soil in this zone, particularly in the depressions, along with the availability of groundwater and its proximity to the surface, make it susceptible to agricultural exploitation [7-10].

3.3 Digital elevation lines in Al-Muthanna governorate

Al-Muthanna Governorate was extracted from the DAM digital model of Iraq, and the digital model was projected onto the governorates of Iraq to indicate its location, as shown in Fig 4.

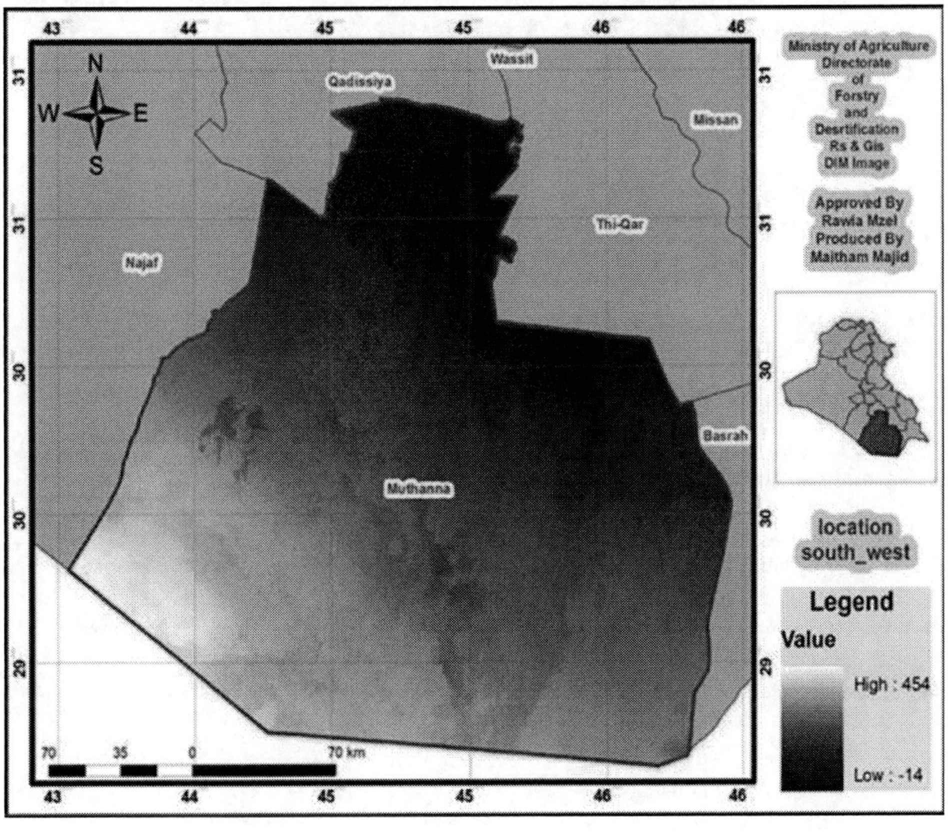

Fig. 4. Elevation lines in Al-Muthanna governorate.

https://doi.org/10.1051/e3sconf/202563303003

3.4 Contour Lines Creating applications of contour lines for the digital model

Defining these lines helps implement agricultural activities through water harvesting activities represented by crescent arches or rhomboids, and how to apply them to the project so that they are effective in exploiting water resources such as rain, as shown in Fig. 5.

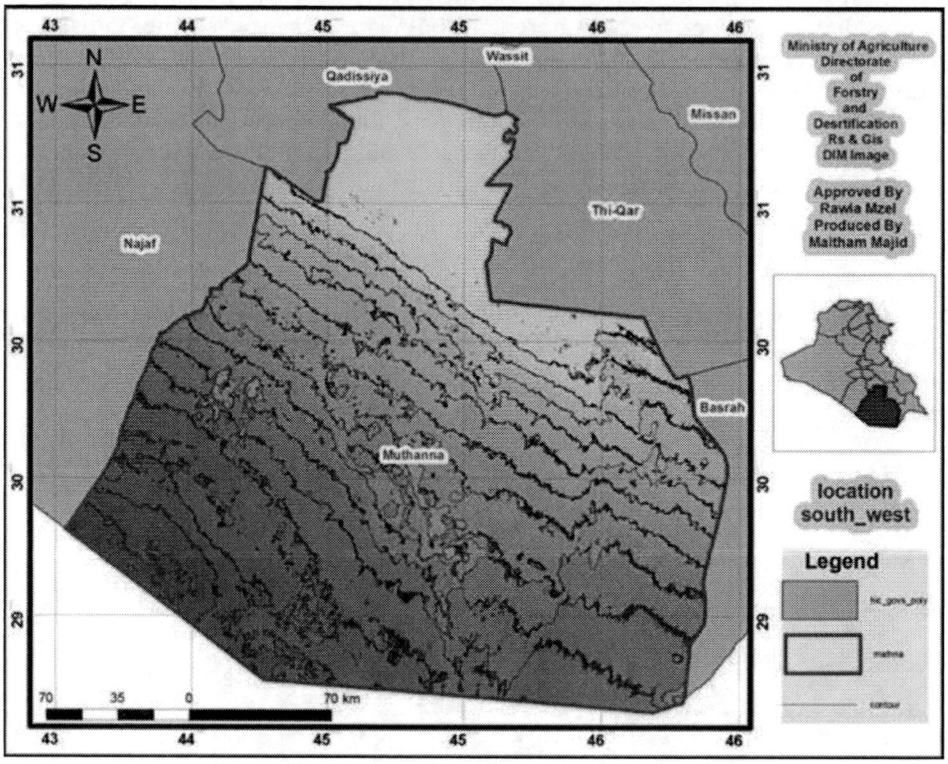

Fig. 5. Contour lines.

3.5 Water flow in valleys

Applications for the ranks of watercourses in the Digital Elevation Model (DEM) have shown that there are five ranks of watercourses in the area. In the event of rain or floods, we benefit from this in applying for flood control and understanding its course towards locations in the area, which determines the presence of residents, livestock movement, and choosing the best site for the proposed project, as illustrated in Fig. 6.

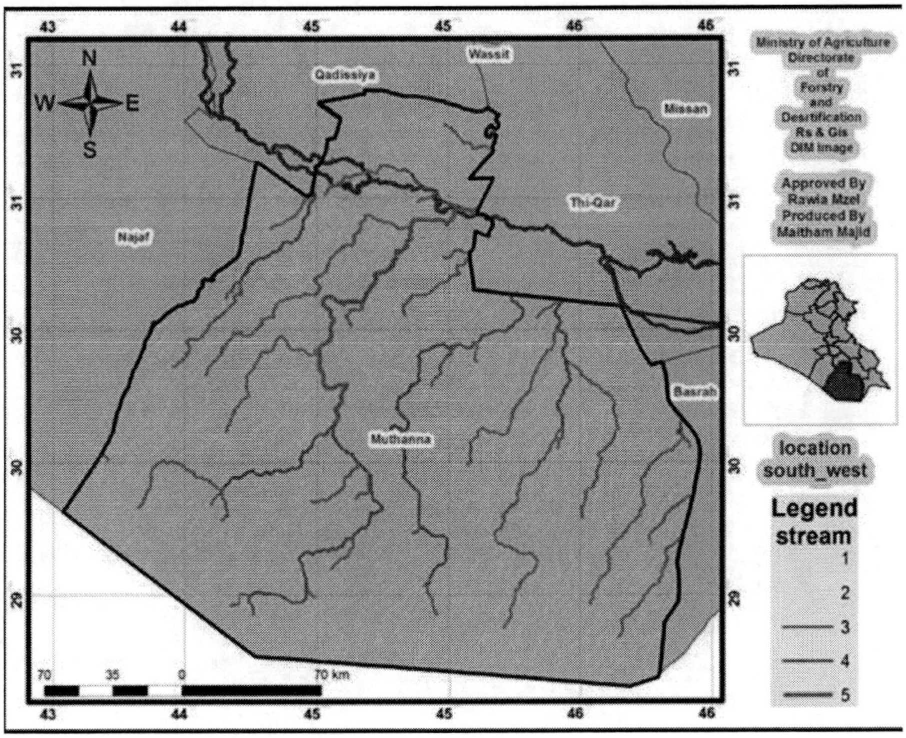

Fig. 6. Water flows.

3.6 Population distribution

Population distribution is closely related to the landforms of any region, as the surface significantly impacts economic life, especially concerning human existence. The population density decreases with elevation above sea level due to the challenges humans face in utilizing the land and the climatic difficulties associated with that elevation, as well as the difficulties in road construction. In contrast, flat areas attract populations due to the availability of factors conducive to human settlement, provided that other resources such as fertile soil and water are available. The population distribution is directly linked to areas where water sources, rivers, and watercourses are located, as illustrated in Fig. 7.

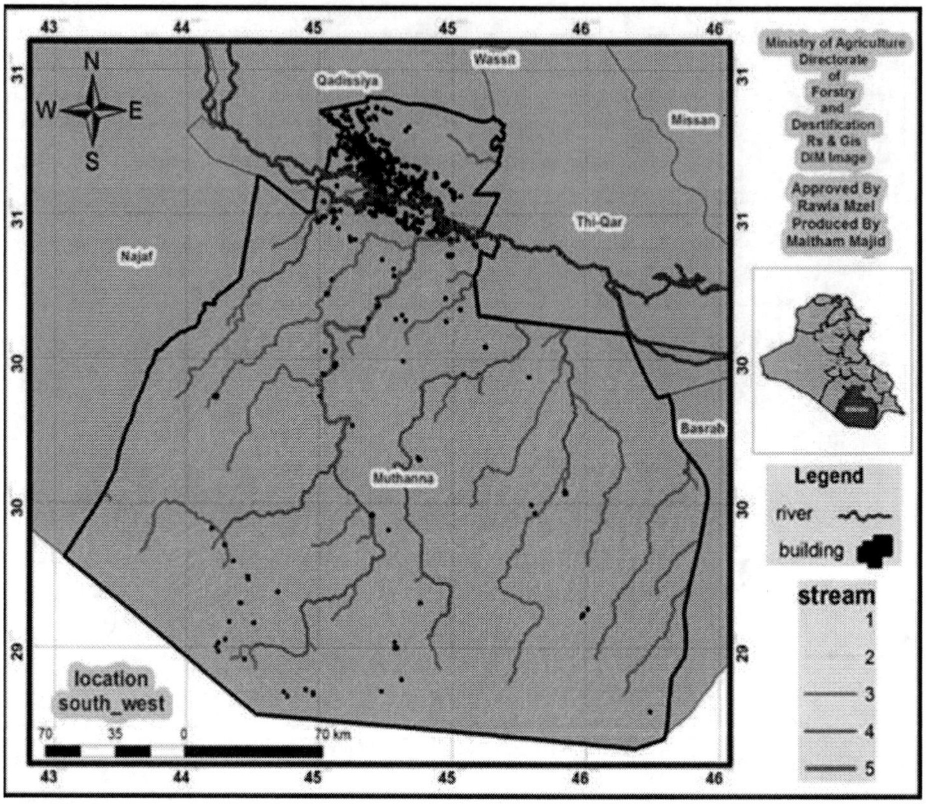

Fig. 7. Population distribution.

3.7 Transportation routes

A map illustrating paved and unpaved roads that assist in selecting a suitable location for project establishment. The importance of these routes lies in their role in accessing chosen sites and facilitating movement, as shown in Fig. 8.

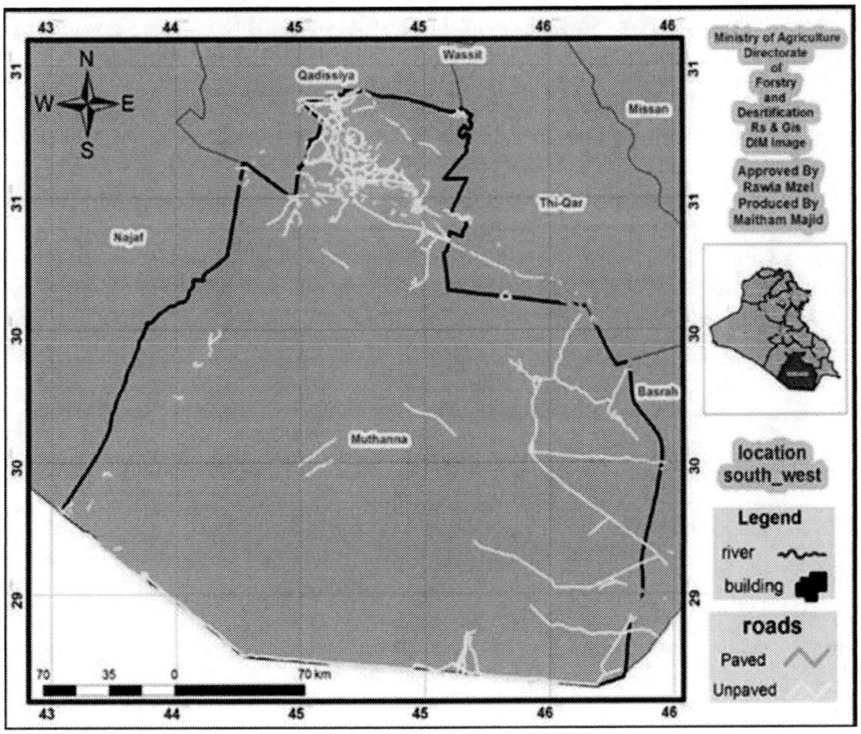

Fig. 8. Transportation routes.

4 Conclusion

1. Modern technologies yield more accurate results, reducing effort, time, and costs.
2. The rainfall rate is low, making it essential to dig wells and conduct water harvesting activities.
3. Implementing techniques such as arched or diamond-shaped structures for water harvesting due to the slight slope and flatness of the land, with results categorized into five classifications (454 - 14).
4. Establishing barriers and excavations for water harvesting based on the watercourse hierarchies, which had four ranks that could be utilized during floods and rainfall.
5. The area of desert land is estimated to be 70% of the province's total area.
6. Rising temperatures and fluctuations in rainfall rates over 30 years (1980-2010) indicating the impact of climate changes and the necessity for adaptation and mitigation of these effects.
7. Making cities and human settlements inclusive, safe, resilient, and sustainable.
8. The spread of cities and the density of population growth in the Al-Samawah district necessitate finding modern methods for resource utilization in other districts.
9. Optimal use of water resources for multiple purposes such as fish farming, watering livestock, agriculture, recharging groundwater, and preventing soil erosion.
10. Raising environmental and agricultural awareness and enhancing the skills of workers in these fields to use modern technologies.
11. Optimal use of water resources for multiple purposes such as fish farming, watering livestock, agriculture, recharging groundwater, and preventing soil erosion.
12. Building a database and producing maps that serve as references for various projects targeting the province and as a model for other provinces.

13. The degradation of Lake Sawa by 95% over the years (1985-2024) is evidence of the decline in groundwater reserves.

5 Challenges

1. The available work opportunities in desertification control programs do not match the scale of the problem.
2. There is a lack of awareness and guidance programs in both directions: effectively using modern technologies and influencing herders and farmers to successfully sustain and invest in projects correctly.
3. Limited financial support that does not correspond to the scale of projects, as they deal with harsh desert environments and remote locations.
4. Insufficient advanced equipment, programs, and ongoing training courses to keep up with modern technologies.

6 Recommendations

1. Sustainable use of land and restoration and enhancement of sustainable use of terrestrial ecosystems.
2. Sustainable water management, combating desertification, and halting land degradation.
3. Raising awareness and activating agricultural guidance in this field.
4. Utilizing modern technologies for optimal exploitation of water resources.
5. Organizing advanced and specialized courses in this field and keeping up with modern programs.
6. Conducting field visits by specialists and accurately documenting points and observations as required for a specific area.
7. Significantly employing modern technology in the development of the country.
8. Increasing the involvement of the private sector.
9. Focusing on the role of local communities, especially (youth and women).
10. Emphasizing the role of civil society organizations.

References

1. R. Maz'al Mahmoud, S. A. Hassan et al., *Atlas of Iraq's Climate* (Ministry of Transport, General Authority for Meteorology and Seismology, 2021)
2. National Action Program to Combat Desertification (Ministry of Agriculture, 2018)
3. J. M. Lyatuu, T. A. Tuni, Ch. J. Kilawe, "Factors Influencing Community's Adoption of Domestic Water Conservation Measures in Moshi Rural District, Tanzania", Sustainable Water Resources Management **9**, 184 (2023), DOI: 10.1007/s40899-023-00970-8
4. A. H. H. Ardakani, S. Shojaei, H. Siasar, M. R. Ekhtesasi, "Heuristic Evaluation of Groundwater in Arid Zones Using Remote Sensing and Geographic Information System", International Journal of Environmental Science and Technology **17**, 633–644 (2020), DOI: 10.1007/s13762-018-2104-1
5. R. Al-Siaede, "Using Landscape Analysis Techniques to Prevent Silt Accumulation in the Reservoir of the Dwerige Weir Project and Developing River Basin, Missan, South Eastern Iraq", Iraqi Journal of Science **63**, 3031–3039 (2022), DOI: 10.24996/ijs.2022.63.7.26
6. N. Gaurav, G. Singh, "Delineation of Groundwater, Drought and Flood Potential Zone Using Weighted Index Overlay Analysis and GIS for District Patna, Bihar, India",

Nature Environment and Pollution Technology **21**, 813–828 (2022), DOI: 10.46488/nept.2022.v21i02.046

7. T. Alharbi, "Mapping of Groundwater, Flood, and Drought Potential Zones in Neom, Saudi Arabia, Using GIS and Remote Sensing Techniques", Water **15**, 966 (2023), DOI: 10.3390/w15050966
8. Climatic Data (Rainfall) (1981–2010), General Authority for Meteorology and Seismology (2021)
9. P. Lemenkova, "Deep Learning Methods of Satellite Image Processing for Monitoring of Flood Dynamics in the Ganges Delta, Bangladesh", Water **16**, 1141 (2024), DOI: 10.3390/w16081141
10. USGS, Digital Elevation Model (DEM) Aster (2016)

Radiation risk assessment in water samples collected from homes in Al-Qasim and Al-Talee'a, Babylon Governorate

Mustafa M. Amin, Inaam H. Kadhim, and *Salar* H. Ibrahem*

Department of Physics, College of Education for Pure Sciences, University of Babylon, Babylon, Iraq

Abstract. Inflammable noble gas radon is colorless and odorless. Its formation is a result of uranium's breakdown processes. Radon gas quantities were measured in the drinking water of several homes in the Al-Tali'ah and Al-Qasim districts of Babylon Governorate. Our American-made alpha particle detector (RAD-7 H2O) was used for this. 20 water samples were taken; 10 came from Al-Tali'ah, and 10 from Al-Qasim. From 0.158 Bq. L-1 to 0.319 Bq. L-1, with a rate of 0.179 Bq. L-1, we measured radon concentrations; the effective annual dose ranged from 0.406 μS/y to 0.815 μSv/y, with a rate of 0.459 μSv/y. The lifetime cancer risk varied between $0.0156*10^{-4}$ and $0.0313*10^{-4}$, with a rate of $0.0175*10^{-4}$. The water samples from the locations above did not contain enough radon gas to meet the standards set by the World Health Organization in 2011 for radon concentrations and the United Nations Scientific Committee on the Effects of Atomic Radiation (UNSCEAR) for the effective annual dose. Consequently, the water supply is safe to drink, and there are no dangers to the public's health.

1 Introduction

Water is an indispensable element of life. However, its exposure to environmental pollutants leads to many health problems for living organisms [1]. Radon gas is present in the air, water, and soil; through it, people are exposed to radiation, as it is a natural source [2]. When solid, liquid, or gaseous radioactive materials merge with environmental elements, such as water, harmful materials will spread and pollute the water [3]. Among the radioactive materials is radon gas, which has many properties. It has a density (9.73 g. L^{-1}) and a boiling point (61.8 degrees Celsius), and it is chemically inert [4-5]. The percentage of humans receiving radiation is 50% due to radon gas, which has a longer half-life than other radioactive isotopes. Radon and other isotopes are created when an alpha particle decays as part of a chain reaction involving uranium decay. There is less radon in surface water compared to groundwater [6]. Radioactive elements can reach living organisms through several direct and indirect routes. The most direct route involves ingesting radioactively contaminated vegetables or water. In contrast, the low doses taken through indirect consumption lead many to believe it poses no health risks [7]. Radon gas is involved in about

* Corresponding author: ali.alhameedawi@uokufa.edu.iq

21,000 lung cancer deaths per year, 2900 of whom are non-smokers. Cancer risk increases by 16% when the radon concentration rises above zero. God mentioned it per liter. Therefore, monitoring radioactive contamination of water is of great importance. For this reason, many studies have been conducted focusing on monitoring radon concentrations in water. Also, neighboring countries and other countries have conducted studies to measure radon gas concentrations [8-15]. Our study examines the change in radon levels in drinking water in the Al-Tali'ah and Al-Qasim districts in Babil Governorate, as this region lacks such studies.

2 Materials and methods

2.1 Study area and data samples

Al-Qasim District and Al-Talee'a Subdistrict, located in Babylon Governorate in central Iraq, are part of the alluvial plain region. Al-Qasim is bordered by Al-Qadisiyah Governorate to the south, Al-Mahawil District to the north, and Al-Kifl to the west. Al-Talee'a lies adjacent to Al-Qasim and is surrounded by agricultural and residential areas characteristic of the region.

Al-Qasim is situated at approximately 32.29°N latitude and 44.45°E longitude, with an elevation of about 25 meters above sea level. Al-Talee'a, located nearby, is at approximately 32.26°N latitude and 44.43°E longitude, with an elevation of about 23 meters above sea level. These coordinates highlight their central position within the fertile plains of Iraq.

Ten samples were taken from Al-Qasim and ten from Al-Talee'a, for a total of twenty water samples that were used to measure the concentration of radon gas in these areas. The purpose of this research is to shed light on the state of the environment. The sample codes and their locations according to the GPS are presented in Table (1), while the sampling sites using GBS techniques are illustrated in Figure (1).

Table 1. Sample name, code and location within the research area.

NO.	Code Sample	precise site	
		latitude	longitude
1	T1	32.11576	44.42367
2	T2	32.11289	44.42162
3	T3	32.11159	44.42335
4	T4	32.11048	44.42560
5	T5	32.11073	44.42542
6	T6	32.11264	44.43049
7	T7	32.11181	44.43009
8	T8	32.10484	44.42569
9	T9	32.11399	44.42558
10	T10	32.17377	44.41149
11	Q11	32.29234	44.41195
12	Q12	32.18046	44.41073
13	Q13	32.25483	44.41014
14	Q14	32.18066	44.41238
15	Q15	32.18094	44.42239
16	Q16	32.18325	44.41326
17	Q17	32.18076	44.42295
18	Q18	32.19215	44.40458
19	Q19	32.12547	44.40128
20	Q20	32.12458	44.42514

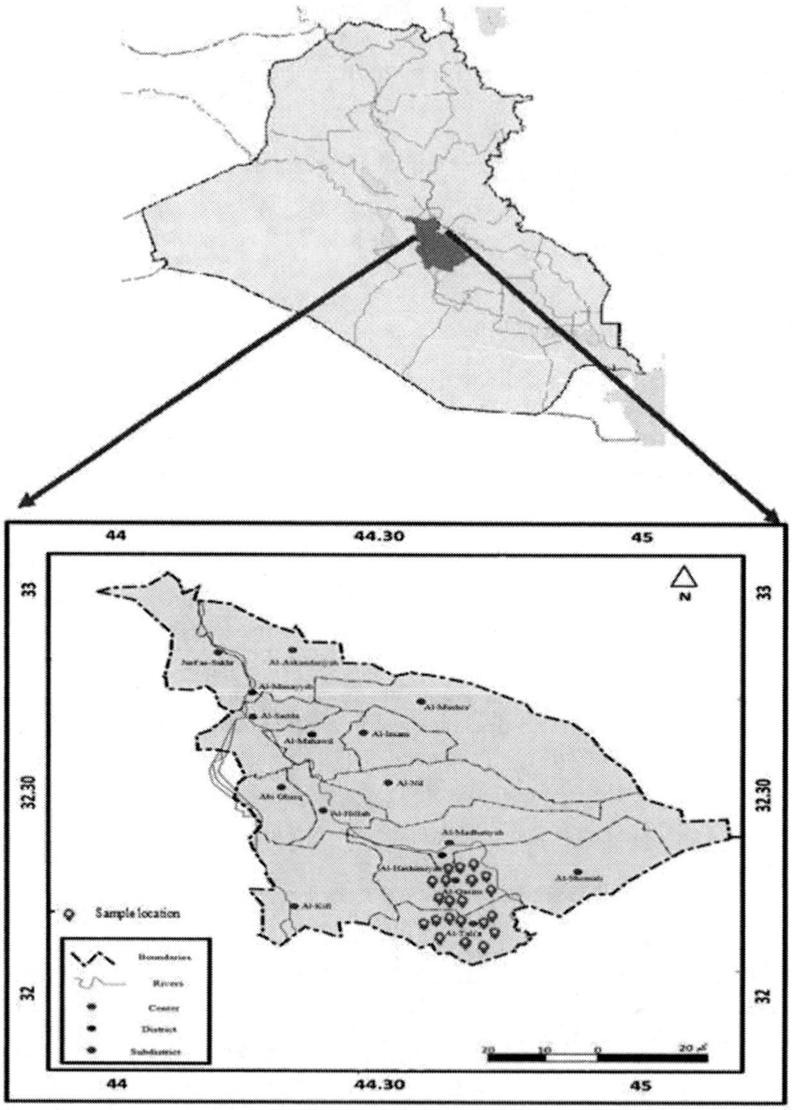

Fig. 1. Area of Study (Al-Talee'a Subdistrict and Al-Qasim District).

2.2. Measurement of radon gas

This instrument can determine how much radon gas is present in a given volume of water. Humidity below 6% is ideal to start with. The next step is to add 250 ml of the sample water. Afterward, a system Grab removes radon from the sample using a pump. The pump runs for five minutes every session. The next step is to remove the radon gas and transfer it to the RAD-7 instrument for measurement. The instrument will then pause for 5 minutes to achieve equilibrium. The procedure is repeated four times longer, totaling 30 minutes of testing for radon concentrations, standard deviation, humidity, and temperature. The operating number, cumulative spectrum, quadrupole diagram, and number of cycles are also part of it. Figure

(2)—the schematic diagram for the preparation (RAD-7)—shows the removal of 94% of radon gas from the water sample, which is an extremely high percentage.

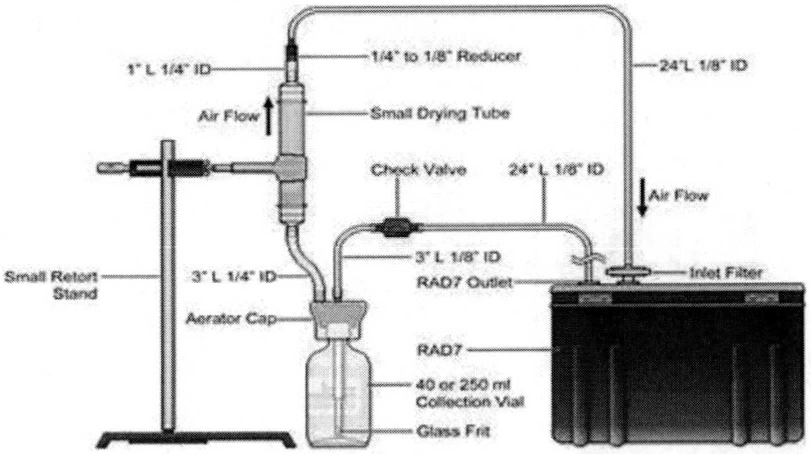

Fig. 2. Diagram illustrating the RAD-7 device for monitoring radon in air. Adapted from the reference with authorization.

2.3 Theoretical calculations

The calculation of the Effective annual dose in μSv/y was based on the consumption of {^{222}Rn} concentrations in water used for drinking by individuals in their homes, as stated in equation:

$$AED = C\left(\frac{Bq}{L}\right) * WC\left(\frac{L}{Y}\right) * DCF\left(\frac{\mu Sv}{Bq}\right) \quad (1)$$

DCF stands for the conversion factor, which is 0.0035 μSv/Bq, WC stands for the yearly how much water an individual uses, which is 2 in units of L/d, and C represents the 222Rn concentration in Bq/L.

Using equation, we can calculate cancer risk over a lifetime associated with 222Rn concentrations in residential home water used as drinking water for adults:

$$LTC = AED\left(\frac{\mu SV}{Y}\right) * DL(Y) * RF(SV^{-1}) \quad (2)$$

DL is the expected lifespan (70 years), and RF is the risk factor (0.055 Sv-1) that the ICRP has recommended.

3 Results and discussion

The measured radon levels varied from (0.158 Bq/L) in the ninth sample to (0.319 Bq/L) in the tenth sample at a rate of (0.179 Bq/L), which is a value less than the permissible limit (0.5 Bq/L) according to ICRB. The effective annual dose ranged from (0.406 μS/y) to (0.8151 μS/y) at a rate of (0.459 μS/y), which is equivalent to (0.000459 μS/y), which is lower than what is legally allowed limit set by the UN Commission on Radiological Protection and Health, which is (100 μSv/y) or (0.1 mSv/y). As for the lifetime cancer risk, it ranges from $(0.01564 * 10^{-4})$ to $(0.03137 * 10^{-4})$ at a rate of $(0.0175 * 10^{-4})$ and these values were less than the permissible limit (10^{-4}) . We note that most of the radon concentrations in the samples are close, which can be explained based on similar environmental conditions, as most of the houses where the samples of water were collected share the same unified water network. The radon concentrations may be close, or there may be other reasons, such as the accuracy of the measurement or due to the effect of ventilation and the techniques used. When the water is exposed to air during sampling, this may lead to removing a close percentage of radon from all samples. To maintain the presence of radon concentrations in the water, samples must be collected, storage containers tightly closed, and carefully stored to reduce gas loss. All of these measurements are shown in Table (2).

Table 2. Yearly effective dosage and risk of cancer over a lifetime.

No.	STUDY AREA	Sample code	Rn concentrations (Bq/L)	S. D	AED(μSv /y)	Lifetime cancer 10^{-4}
1	Taleaa	T1	0.24	0.015	0.6132	0.0236082
2	Taleaa	T2	0.16	0.013	0.4088	0.0157388
3	Taleaa	T3	0.16	0.013	0.4088	0.0157388
4	Taleaa	T4	0.159	0.013	0.406245	0.015640433
5	Taleaa	T5	0.24	0.015	0.6132	0.0236082
6	Taleaa	T6	0.159	0.013	0.406245	0.015640433
7	Taleaa	T7	0.16	0.013	0.4088	0.0157388
8	Taleaa	T8	0.16	0.013	0.4088	0.0157388
9	Taleaa	T9	0.158	0.013	0.40369	0.015542065
10	Taleaa	T10	0.319	0.018	0.815045	0.031379233
11	Qassem	Q11	0.16	0.013	0.4088	0.0157388
12	Qassem	Q12	0.239	0.015	0.610645	0.023509833
13	Qassem	Q13	0.19	0.014	0.48545	0.018689825
14	Qassem	Q14	0.16	0.013	0.4088	0.0157388
15	Qassem	Q15	0.16	0.013	0.4088	0.0157388
16	Qassem	Q16	0.159	0.013	0.406245	0.015640433
17	Qassem	Q17	0.16	0.013	0.4088	0.0157388
18	Qassem	Q18	0.16	0.013	0.4088	0.0157388
19	Qassem	Q19	0.159	0.013	0.406245	0.015640433
20	Qassem	Q20	0.16	0.013	0.4088	0.0157388
Minimum		---------	0.158	0.013	0.40369	0.015542
Maximum		---------	0.32	0.018	0.815045	0.031379
average		---------	0. 179	0.013379	0.459389	0.017514
Permissible limit		---------	0.5	---------	100	10

Table 3. Previous research on radon concentrations in Al-Kifl Subdistrict and Al-Mahawil center drinking water.

Country	**Radon Concentration (Bq/L)**
Italy	6.8
Turkey	14.13
Iran	0.93
China	0.543
Malaysia	1.95
Kuwait	2.97±1.44
City	Radon Concentration (Bq/L)
Baghdad	0.0266 ± 0.025311
Qadisiyah	0.24
Wasit	0.1888±0.0137
Hilla	0.199
Kirkuk	0.97
Thi-qar	0.175 ± 0.03
Anbar	0.139±0.03
Najaf	0.12±0.08
Current Study	0.179 ±0.013

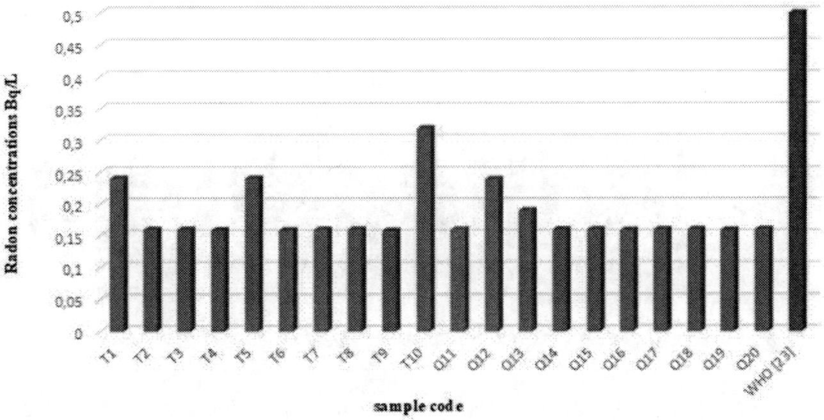

Fig. 3. ^{222}Rn concentrations in water samples from the Al-Talee'a Subdistrict and the Al-Qasim District were compared to the allowed values for drinking water.

https://doi.org/10.1051/e3sconf/202563303004

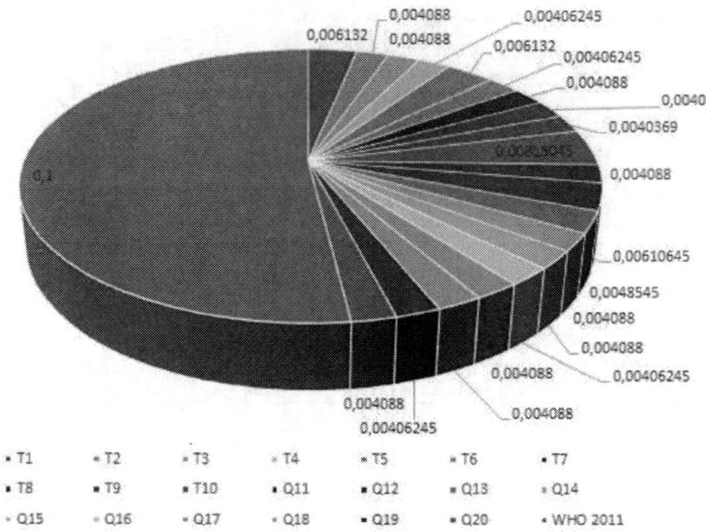

Fig. 4. Research samples from the Al-Talee'a Subdistrict and the Al-Qasim District were examined for the percentage of AED in unit (mSv/y).

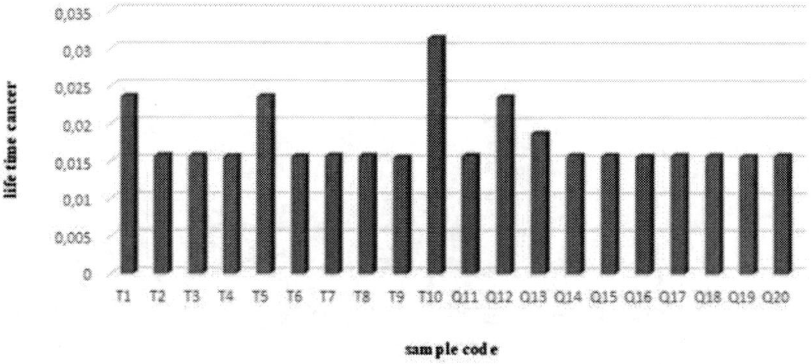

Fig. 5. ^{222}Rn concentrations in water samples were compared to the allowed limits for drinking water in the Al-Talee'a Subdistrict and Al-Qasim District to determine cancer risk factors.

4 Conclusions

It can be concluded that the RAD7 Solid State Detector is a cutting-edge and trustworthy technology due to its superior sensitivity and speed in detecting alpha particle radiation compared to other options. The results show that compared to the limits set by the EPA and WHO in 1993, the concentration of radioactive radon gas in the water samples taken from Babylon Governorate was the lowest. In addition, the Annual Effective Dose (AED) was below the allowed limit of 0.1 mSv/y or 100 μSv/y, according to UNSCEA guidelines. Furthermore, no acceptable lifetime cancer risk figure was found. This means the radioactive radon gas detected in water samples from Iraq's Babylon Governorate is entirely harmless for human consumption and other applications [16-18].

References

1. F. O. Ajibade, B. Adelodun, K. H. Lasisi, O. O. Fadare, T. F. Ajibade, N. A. Nwogwu, A. Wang, "Environmental pollution and their socioeconomic impacts," in *Microbe Mediated Remediation of Environmental Contaminants*, 321–354 (Woodhead Publishing, 2021)
2. M. Baskaran, *Radon: A Tracer for Geological, Geophysical and Geochemical Studies* (Springer, Basel, 2016)
3. C. Nuccetelli, R. Trevisi, "Radon and NORM: naturally occurring radioactive materials," in *Physical Agents in the Environment and Workplace: Noise and Vibrations, Electromagnetic Fields and Ionizing Radiation*, edited by G. Licitra, G. d'Amore, M. Magnoni (CRC Press, Boca Raton, 2018)
4. Q. B. Carroll, *Radiography in the Digital Age: Physics-Exposure-Radiation Biology* (Charles C Thomas Publisher, 2023)
5. C. R. Cothern, *Handbook for Environmental Risk Decision Making: Values, Perceptions, and Ethics* (CRC Press, Boca Raton, 2019)
6. Y. Fakhri, M. Kagoshima, G. Langarizadeh, Y. Zandsalimi, L. R. Amirhajeloo, M. Moradi, M. Bigard, M. Mirzaei, "Effective dose radon 222 of the tap water in children and adults; Minab City, Iran," Glob. J. Health Sci. **8(4)**, 234 (2016)
7. R. M. Harrison ed., *Pollution: Causes, Effects and Control* (Royal Society of Chemistry, 2023)
8. N. A. Kadhim, A. S. Mdekil, A. A. Abojassim, "Age-dependent health risk assessment for radon concentrations from drinking water available in the Iraqi markets," Egypt. J. Chem. **64(4)**, 1889–1895 (2021)
9. N. Ahmad, J. U. Rehman, M. Rafique, T. Nasir, "Age-dependent annual effective dose estimations of 226Ra, 232Th, 40K and 222Rn from drinking water in Baling, Malaysia," Water Sci. Technol. Water Supply **18(1)**, 32–39 (2018)
10. A. M. El-Taher, A. A. Abojassim, L. A. Najam, H. A. A. B. Mraity, "Assessment of annual effective dose for different age groups based on radon concentrations in the groundwater of Qassim, Saudi Arabia," Iran. J. Med. Phys. **17(1)**, 15–20 (2020)
11. M. Kaur, A. Kumar, R. Mehra, R. Mishra, "Age-dependent ingestion and inhalation doses due to intake of uranium and radon in water samples of Shiwalik Himalayas of Jammu and Kashmir, India," Environ. Monit. Assess. **191**, 1–17 (2019)
12. A. H. Alomari, M. A. Saleh, S. Hashim, A. Alsayaheen, I. Abdeldin, "Activity concentrations of 226Ra, 228Ra, 222Rn and their health impact in the groundwater of Jordan," J. Radioanal. Nucl. Chem. **322**, 305–318 (2019)
13. H. Büyükuslu, F. B. Özdemir, T. Ö. Öge, H. Gökce, "Indoor and tap water radon (222Rn) concentration measurements at Giresun University campus areas," Appl. Radiat. Isot. **139**, 285–291 (2018)
14. J. Yong, Q. Liu, B. Wu, Y. Hu, G. Feng, "Assessment of radiation dose hazards caused by radon and its progenies in tap water by the human dosimetric model," J. Water Health **19(6)**, 933–945 (2021)
15. N. F. Ismail, S. Hashim, M. S. M. Sanusi, A. T. Abdul Rahman, D. A. Bradley, "Radon levels of water sources in the southwest coastal region of Peninsular Malaysia," Appl. Sci. **11(5)**, 6842 (2021)

16. N. A. Abdulkhaleq, S. K. Dawood, K. M. Qader, S. Y. Taher, "Evaluation of radon concentrations in drinking water available in Baghdad Governorate markets, Iraq," in E3S Web of Conferences **583**, 02010 (2024)
17. M. S. Aswood, M. S. Almusawi, N. K. Mahdi, A. F. Showard, "Evaluation of committed effective dose of radon gas in drinking water in Al-Qadisiyah province, Iraq," Periódico Tchê Química **17(36)** (2020)
18. K. S. Jassim, I. H. Kadhim, N. T. Shihab, "Estimated radon concentration in drinking water samples for different regions of Hilla City, Iraq," Int. J. Nucl. Energy Sci. Technol. **12(3)**, 213–221 (2018)

https://doi.org/10.1051/e3sconf/202563303005

Utilization of remote sensing and GIS approach for surface water monitoring in the Najaf Sea, Al-Najaf providence, Iraq: an overview

Abdulkareem Husham*, *Mufid* Al-hadithi, and *Rana* Mozihim

Surveying Engineering Department, Engineering Technical College-Baghdad-Middle Technical University, Baghdad, Iraq

Abstract. This study aims to review the studies conducted on the Najaf Sea using remote sensing and geographic information systems (GIS) to develop effective water resource management plans and promote environmental sustainability. The studies focused on environmental changes, land use/land cover classification, hydrological modeling and water quality assessment of the Najaf Sea. Sentinel-2 and Landsat satellite images were the most commonly used data sources, with classification methods such as maximum likelihood algorithm and spectral indices (NDVI, NDWI, NDBI and NDSI) applied to detect land cover changes. Hydrological simulations were performed using the SWAT model, while climate classification methods analyzed regional climate characteristics. Water quality assessment was based on physical and chemical indicators, with total dissolved solids (TDS) being the most frequently evaluated parameter. GIS-based spatial analyses and statistical correlation techniques provided insights into environmental impacts, over-nutrientification and urbanization. Water quality analysis of the Najaf Sea revealed significant variations in salinity and pollution levels due to evaporation, agricultural runoff, and seasonal fluctuations.

1 Introduction

Remote sensing (RS) and geographic information systems (GIS) are essential instruments that furnish the geospatial data required to establish databases and comprehensively monitor the Earth's surface. Satellite observations help keep an eye on conditions on land, at sea, and in the air, especially during natural disasters. This improves global environmental protection, lowers the damage from disasters, and supports long-term growth. These advanced technologies systematically acquire high-resolution data, collected comprehensively over durations spanning from a few minutes to several days [1].

Remote sensing (RS) and geographic information systems (GIS) with land classification are contemporary techniques that offer novel approaches to managing intricate ecosystems. By tracking temporal changes at local, regional, and global levels, these technologies enhance our comprehension of the environment and facilitate more effective management decisions [2, 3]. The significance of the Najaf Sea area is attributed to its position as a notable

* Corresponding author: ACC0011@mtu.edu.iq

topographical landmark and a distinctive geographical feature in the Middle Euphrates region, encompassing a water swamp known as Najaf Lake at its center. The opening of the Najaf Sea to the alluvial plain on its southern side makes farming easier by making it easier to get irrigation water and giving farmers access to a lot of land around the edge [4]. The present study aims to review previous studies in which remote sensing techniques and GIS tools were used to monitor water quality, land use changes, and hydrological dynamics in the Najaf Sea. As well as studies that addressed the impact of environmental and human factors, identify research gaps, and provide recommendations for future studies to enhance water resources and management strategies.

1.1 Najaf Sea location

Najaf Sea is situated in the Najaf Governorate of central Iraq, extending longitudinally from the northern region of Najaf city to the southwestern area of Al-Hirah city, a district within Najaf. This region lies in the transitional zone between the ancient, stable Archean plateau and the contemporary alluvial plain created from the sediments of the Tigris and Euphrates rivers, situated nearby. The northern desert (Najaf Desert) flanks it to the west, while the city of Najaf, Al-Hirah, and Al-Mishkhab flank it to the east.

The area is (km^2), with a length of (60) miles, a width of (30) miles, and a depth of (10) meters above sea level. The Najaf Sea is located between latitudes 31.30° to 32.10° north and longitudes 43.30° to 44.30° east.as shown in Figure 1.

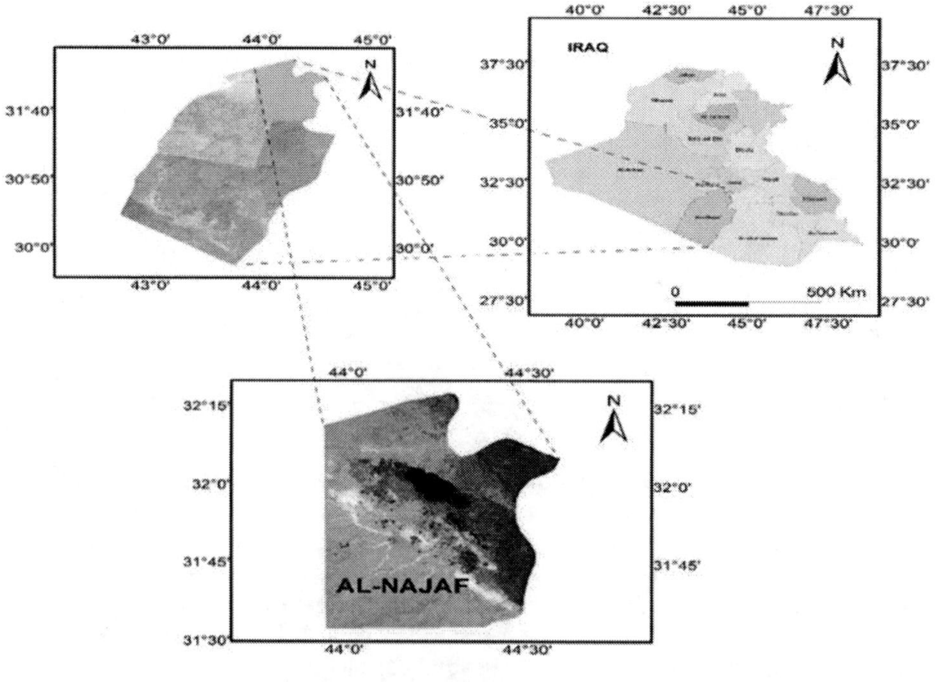

Fig. 1. Astronomical and geographical location of the Al Najaf sea boundaries.

1.2 Sources of recharging water for Najaf Sea

The water resources in the study area comprise three primary canals, along with deep and shallow groundwater wells, springs, a water reservoir, and surface runoff within the drainage basin. Table 1 shows a description and details of all the water sources that contribute to feeding the Najaf Sea.

Table 1. Overview of water resources for Najaf Sea.

Water Resource	Description
Rainfall	The region receives minimal and irregular precipitation. November recorded the highest monthly rainfall at 17.6 mm, while there was no precipitation from June to September. The cumulative rainfall from 1981 to 2013 was 100.2 mm.
Canals	The Euphrates River supplies surface water, comprising two primary tributaries: the Kufa and Abbassia rivers. The Mashkhab River divides into three primary canals: Al-Sadeer, Abojedhoe'o, and Al-Bederiya.
Wells	The region contains artesian and shallow wells, with artesian wells exhibiting a higher discharge (≥30 liters/second) compared to shallow wells, which have a lower discharge (≤10 liters/second). The total well discharge is 1.244 cubic meters per second.
Springs	There are six natural springs in the study area, yielding a total discharge of 0.05 cubic meters per second.

The canals comprise the Sudair Canal, the Abu Jedhu Canal, and the Badriyah Canal. In the area, there are 24 shallow wells, 34 artesian wells, and six springs, along with rainfall that supplements the water resources [5] (Table 2 &3).

Table 2. Wells within surrounded Najaf Sea, including their classifications and discharge rates (Q).

Well No.	Q (m^3/sec)	Well Type	Well No.	Q (m^3/sec)	Well Type	Well No.	Q (m^3/sec)	Well Type
1	0.007	Shallow	21	0.004	Shallow	41	0.033	Artesian
2	0.006	Shallow	22	0.01	Shallow	42	0.02	Artesian
3	0.007	Shallow	23	0.009	Shallow	43	0.033	Artesian
4	0.005	Shallow	24	0.008	Shallow	44	0.033	Artesian
5	0.007	Shallow	25	0.03	Artesian	45	0.03	Artesian
6	0.004	Shallow	26	0.033	Artesian	46	0.033	Artesian
7	0.007	Shallow	27	0.03	Artesian	47	0.033	Artesian
8	0.009	Shallow	28	0.03	Artesian	48	0.033	Artesian
9	0.009	Shallow	29	0.035	Artesian	49	0.035	Artesian
10	0.007	Shallow	30	0.033	Artesian	50	0.03	Artesian
11	0.01	Shallow	31	0.03	Artesian	51	0.03	Artesian
12	0.007	Shallow	32	0.033	Artesian	52	0.033	Artesian
13	0.009	Shallow	33	0.015	Artesian	53	0.035	Artesian
14	0.007	Shallow	34	0.033	Artesian	54	0.03	Artesian
15	0.008	Shallow	35	0.035	Artesian	55	0.033	Artesian
16	0.008	Shallow	36	0.033	Artesian	56	0.033	Artesian
17	0.005	Shallow	37	0.033	Artesian	57	0.033	Artesian
18	0.005	Shallow	38	0.033	Artesian	58	0.03	Artesian
19	0.007	Shallow	39	0.033	Artesian			
20	0.007	Shallow	40	0.033	Artesian			

Table 3. Springs within the study area and their respective discharges.

Spring Name	Q (lit/sec)	Spring Name	Q (lit/sec)
Al-Ghariba	6	Ghal'et Muslim	14.1
Al-Rohban	12.6	Hezbaneh	6.3
Said Sa'id	8	Al-Jarthamy	3.1

 https://doi.org/10.1051/e3sconf/202563303005

1.3 Hydrochemistry of waters

Water quality is a crucial determinant of its appropriateness for potable, domestic, agricultural, and industrial applications [5]. Water assessment aids in comprehending the reactions that govern the development of natural water chemistry [6]. Najaf sea which is being the lowest topographical features in the region, function as reservoir for the accumulation of all surface water sources According to the [7] and [8] Najaf sea Water is classified based on total hardness, physical properties and total dissolved solids (ppm) as shown in Table (4, 5 and 6).

Table 4. Physical properties of water.

Property	Range	Average	Notes
Temperature (°C)	28.7 - 34.9	31.27	The temperature of water in creeks is lower than that of marshes due to water flow.
pH	7.1 - 7.6	7.3	Moderate to alkaline, contingent upon temperature
Total Dissolved Solids (TDS)	1130 - 12690 (ppm)	4798 (ppm)	Elevated in marshes as a result of evaporation Table (3.2)
Electrical Conductivity (EC)	1525 - 17131 (μS/cm)	6478 (μS/cm)	Rises with elevated TDS values Table 3.2
Total Hardness (T.H)	650 - 7471 (ppm)	2633 (ppm)	All samples are categorized as extremely hard water. Table 6

Table 5. Classification of Water According to TDS (ppm) [7].

Water Class	Drever (1997)	Todd (2007)
Fresh Water	< 1000	10 - 1000
Slightly Saline Water	1000 – 2000	----
Slightly Brackish Water	2000 – 20,000	1000 – 10,000
Brackish Water	----	10,000 – 100,000
Saline Water	35,000	----
Brine Water	> 35,000	> 100,000

Table 6. Classification of water according to total hardness [8].

Term	Degree of Water Hardness
Soft	$0 < TH \leq 60$
Moderately Hard	$60 < TH \leq 120$
Hard	$120 < TH \leq 180$
Very Hard	$180 < TH$

2 Materials and methods

2.1 Previous studies

Enas Riyadh Wdaah et al., 2023[9] , used Landsat 5, 8, and 9 satellite images from 2005 to 2022 to look at how the land cover changed in Iraq using remote sensing and GIS methods, especially the "maximum likelihood" approach in the ERDAS IMAGINE 2015 supervised image classification software. The study revealed notable findings, with a 33.229% increase in water bodies and a 1.197% rise in arid lands, whereas agricultural lands diminished by 8.218% and clay soil decreased by 26.208%. The research validated that remote sensing methodologies were highly effective in tracking land cover alterations and facilitating sustainable resource management strategies.

Mustafa et al., 2022 utilized Sentinel-2 satellite imagery from 2016 to 2020 and employed spectral indices including NDVI, NDBI, NDWI, and NDSI, along with maximum likelihood classification for land use/land cover (LULC) mapping. The study showed important results, such as a 50% rise in areas that were flooded by high water levels and a 56% rise in plant cover, which was caused by good agricultural policies and plenty of water. Urban expansion was gradual, with developed regions rising by merely 2%. The research established that environmental alterations significantly affected agricultural, industrial, and residential sectors in the Najaf Seas.

Ataa Ali et al., 2021 , employed the SWAT model and SWAT-CUP calibration tools, utilizing digital elevation models (DEM), land use maps, soil maps, and weather data from CFSR and IMOS. The research modeled three scenarios of water level fluctuations in the Najaf Seas, contingent upon varying precipitation rates. The results showed that the SWAT model accurately predicted runoff and hydrological processes ($R^2 = 0.85$, NSE = 0.77). The three modeled scenarios comprised filling a constrained lake, filling a larger lake, and inundating the city. The findings offered significant insights for the management of water resources and the assessment of flood risk in the Najaf Seas.

Kamal et al., 2018 used weather data from the Karbala station from 1976 to 2016 and the Thornthwaite method to figure out potential evapotranspiration (PE). They also used climate classification methods from Mather (1974), Kettaneh and Gangopadhyaya (1974), and Al-Kubaisi (2004). The research indicated that the area possesses an arid climate, characterized by a total annual precipitation of 95.5 mm, compared with evaporation measuring 2828.6 mm. Groundwater flows from northwest to southeast, with a surplus occurring solely during the winter months, culminating in an annual surplus of 28.11 mm. The climate classification methods validated the region's arid characteristics.

Suhair et al., 2020 conducted a study on Lake Najaf utilizing data gathered from December 2018 to March 2019. The research assessed parameters including Chl-a, TP, TN, Pb, TSS, TDS, pH, EC, MPN, and COD. The Carlos Enrichment Status Index (TSI) and CTSI were used for classification, along with Geographic Information Systems (GIS) for spatial analysis and statistical methods like LSD and ANOVA. We assessed water quality and fertilization via field and laboratory analyses. The findings indicated that Lake Najaf was categorized as highly eutrophic (TSI > 80), exhibiting dense algal blooms and considerable degradation in water quality. We noted elevated phosphorus levels and diminished oxygen concentrations, with the most severe pollution occurring near river flows. Seasonal variations showed elevated nutrient concentrations in winter, followed by a reduction in spring due to biological activity and dilution effects.

Hussein et al. 2019[7] collected eleven water samples from the Najaf Sea region and analyzed their chemical and physical properties. The findings indicated that all sources were unsuitable for drinking, yet appropriate for irrigation, with the exception of marsh water. We recognized rainfall, groundwater, and agricultural streams as the primary sources of water, while evaporation facilitated the concentration of salts. Evaporation and regional geological features influence water quality, as the research demonstrated. The water was categorized according to hardness and sodium concentration, with the majority deemed appropriate for irrigation. The study advocated for the enhancement of water management to mitigate evaporation and salinity issues.

2.2 Remote sensing technique using for surface water monitoring

Over the past decade, it has been successfully employed remote sensing and GIS techniques instead of traditional methods to monitor the environment, particularly to monitor ecosystems such as rivers, forests, pastures, and urban areas. Numerous nations have established extensive environmental monitoring initiatives utilizing remote sensing technology, and a multitude of studies have been undertaken in Europe, the United States, and other areas. Furthermore, researchers have created numerous individual indicators for use in remote sensing for water extraction.

Previous studies have addressed a number of approaches reviewed in this study, covering a range of topics including land cover changes, hydrology, water quality, climate classification, and environmental impacts, which are summarized in the Table (7).

Table 7. Methods used in previous studies with details.

Methods used in the previous studies	
Remote Sensing & GIS Analysis	• Supervised classification using ERDAS IMAGINE 2015. • Landsat (5, 8, and 9) and multispectral satellite images for land cover classification.
Satellite Image Processing & Spectral Indices	• Sentinel-2 satellite imagery (2016–2020). • Spectral indices: NDVI, NDBI, NDWI, NDSI. • Maximum likelihood classification for land use/land cover (LULC) mapping.
Hydrological & Flood Modeling	• SWAT model with SWAT-CUP calibration. • Use of Digital Elevation Model (DEM), land use/soil maps, and climate/weather data. • Scenario-based modeling of water level changes.
Meteorological & Climate Classification Methods	• Analysis of meteorological data (1976–2016). • Thorn Thwaite method for Potential Evapotranspiration (PE). • Climate classification (Maher 1974, Kathawa, Gangopadhayaya 1974, Al-Kubaisi 2004).
Water Quality & Eutrophication Assessment	• Measurement of physical-chemical parameters (pH, TDS, TSS, TN, TP, BOD, COD). • Carlson's Trophic State Index (TSI) and GIS-based spatial analysis. • Laboratory and field testing for aquatic quality
Land Cover Change Detection	• Sentinel-2 imagery (2016–2020) analyzed using Maximum Likelihood Algorithm. • Confusion matrix for accuracy assessment. • Classification of built-up areas, barren land, clay soil, agricultural land, and water bodies.
Water Sample Collection & Salinity Analysis	• Analysis of chemical and physical properties. • Classification of water sources (rain, groundwater, and agriculture).
Environmental & Sustainability Analysis	• Use of Landsat imagery and indices (NDVI, NDWI) to examine vegetation and water resources. • Statistical correlations (Pearson) to analyze climate factors and environmental change.

3 Result and discussion

The alteration in the surface area of the Najaf Sea has been extensively utilized in numerous previous studies to evaluate the shoreline changes of the sea through satellite imagery from 2005 to 2022. The results of these studies revealed significant changes in the land cover in the Najaf Sea area between 2005 and 2022. The results showed that the area of bodies of water grew by 33.229%. This shows that the effects of more rain and the state's irrigation plans during the study period were felt. The area of agricultural land diminished by 8.218% due to agricultural neglect, whereas barren lands expanded by 1.197%, reflecting the effects of climate change in the region. Clay soils experienced a notable reduction of 26.208%, indicative of the impacts of water scarcity and adjacent environmental alterations [9]. These modifications underscore the necessity of formulating sustainable natural resource management policies to enhance land utilization and adapt to environmental alterations in the Najaf Sea and its adjacent regions.

4 Conclusion

Based on the previous reviewed studies that looked at changes in the surface of Lake Najaf, its environment and water levels using a combination of GIS and remote sensing, the following conclusions can be drawn:

- The Najaf Sea is notable for its ecological and geographical significance, serving as a crucial water source that agricultural endeavors and enhances the local climate also functioning as a primary stopover for migratory birds.
- The region encircling the Najaf Sea has experienced considerable alterations in land cover in recent years, characterized by an expanded water surface area resulting from governmental initiatives and increased precipitation, alongside a reduction in agricultural land due to neglect.
- The Najaf Sea confronts significant environmental issues, including contamination from domestic and industrial effluents, as well as pressures from unregulated water diversions.
- The Sentinel-2 satellite was the most widely used for analysis in several studies
- Total dissolved solids (TDS) was the most widely measured water quality indicator and was reported in many studies focusing on water quality assessment.

Remote Sensing & GIS Analysis was the most widely used technique, especially for land use/land cover (LULC) mapping and environmental change detection.

References

1. R. Al-Bahadeli, M. Al-Hadith, F. M. Shnewer, Y. A. Ali, "Develop a methodology for evaluation of the environmental sensitivity areas to desertification in the Maysan Province, Iraq," IOP Conf. Ser.: Mater. Sci. Eng. **745**, 012133 (2020), DOI: 10.1088/1757-899X/745/1/012133
2. M. A. Salim, S. A. Abed, *Avifauna diversity of Bahr Al-Najaf wetlands and the surrounding areas, Iraq* (2017), https://www.researchgate.net/publication/319551228
3. H. Badr Ghalib, A. B. Al-Hawash, W. S. Al-Qurnaw, B. H. Sultan, A. W. Al-Enzy, "Marshes waters sources hydrochemistry of the Bahr Al-Najaf at Najaf Province, Iraq," J. Phys.: Conf. Ser. **1279**, 012059 (2019), DOI: 10.1088/1742-6596/1279/1/012059
4. E. R. Wdaah, A. F. Nori, "Detection of land cover change in the Sea of Najaf / Iraq for the period (2005–2022) using remote sensing techniques," (2023), www.multiresearchjournal.com
5. M. Al-Helaly, I. Alwan, A. AL-Hameedawi, "Environmental investigation of Bahar Al-Najaf region using Sentinel-2 images," Eng. Technol. J. **40(5)**, 732–742 (2022), DOI: 10.30684/etj.v40i5.2245
6. A. A. Farhan, B. Sh. Abed, "Estimation of surface runoff to Bahr AL-Najaf," J. Eng. **27(9)**, 51–63 (2021), DOI: 10.31026/j.eng.2021.09.05
7. K. K. Ali, Z. A. Kadham, Evaluate the climatic conditions for the Al-Najaf-Ain-Al-Tamur area (Middle of Iraq)

8. S. Al-Sharifi, S. Al-Bakri, S. Al-Bayati, "Evaluation of eutrophication levels in Al-Najaf Al-Ashraf Lake, Iraq (biological and chemical study)," Eng. Technol. J. **38(10A)**, 1531–1538 (2020), DOI: 10.30684/etj.v38i10a.1601
9. M. H. Al-Helaly, I. A. Alwan, A. N. Al-Hameedawi, "Land covers monitoring for Bahar-Al-Najaf (Iraq) based on Sentinel-2 imagery," J. Phys.: Conf. Ser. **1973**, 012189 (2021), DOI: 10.1088/1742-6596/1973/1/012189
10. H. Sh. Al-Bahrani, A. H. Al-Rammahi, S. K. Al-Mamoori et al., "Groundwater Detection and Classification Using Remote Sensing and GIS in Najaf, Iraq", Groundwater for Sustainable Development **19**, 100838 (2022), DOI: 10.1016/j.gsd.2022.100838.

Land suitability for solar power plants in Karbala, Iraq using MCD and AHP-GIS

Mustafa S. Srayyih[1], *M.* M. Al-Gbory[3], *Ebtesam* F. Khanjar[2,*], *Noora* M. Saleem[3], and *Ban* A. Alrazaq[2]

[1]Department of Astronomy and space, Collage of Science, University of Baghdad, Baghdad, Iraq
[2]Department of Remote sensing and GIS, Collage of Science, University of Baghdad, Baghdad, Iraq
[3]Department of Physics, College of Education, Aliraqia University, Baghdad, Iraq

Abstract. The attention of developed countries is turning to utilizing renewable energy sources, especially after the fossil fuel crises and environmental pollution caused by power plants, Solar cells are considered one of the main sources of energy in smart cities, which are especially focused on the roofs of houses and the unused area that provide a continuous source of energy and reduce polluting emissions to the atmosphere, solar energy is a more today's efficient renewable resource, especially in areas with high solar energy. Solar cells turn solar energy into a source of electricity for domestic and industrial use. In this study we used using geographic information systems (GIS) and remote sensing to potentially appropriate locations for solar power plants in the holy city of Karbala, Iraq, which makes religious tourism in this city the most distinctive because it is a significant religious hub that draws many visitors from both inside and outside of Iraq. Despite other historical, cultural, heritage, and tourist attractions in the province. After determining the necessary criteria and collecting their data via the satellite images of Landsat 8 for the year 2022 and analyzing them by ArcGIS, the ideal land for the installation of solar cells was chosen using the Multi-Criteria Decision-Making Process (MCDA) and the Analytical Hierarchy Process (AHP). The results showed that the areas that are suitable and very suitable are (15.41%), equivalent to (775.76607 Km2) of the total area of Karbala governorate (5033 Km2). As a result, this model is suitable for setting the parameters chosen to choose the most promising sites in Karbala governorate that can achieve the research objectives, this study helps urban planners, designers and investors to implement and develop solar power plants within the governorate.

1 Introduction

The world population is constantly increasing and this growth requires more energy to meet daily needs [1], so the demand for fossil fuels is increasing, which is the most common type of fuel for the operation of power plants, Additionally, the main source of energy comes from the production of fossil fuels It is the main source of greenhouse gas emissions that have contributed to global climate change [2, 3], This has caused a focus on using renewable

* Corresponding author: Ebtesam.khanjer@sc.uobaghdad.edu.iq

energy sources, which are generally thought of as clean and having a far smaller impact on the environment than other energy sources. [4, 5]. A renewable and supplementary energy source that is gradually displacing non-renewable energy sources is solar energy. Due to falling solar panel prices, one of the energy sources with the quickest growth is solar power generation.[6], The average monthly solar radiation values for the region, which depends on the number of sunny days and the length of daylight hours, is one of the most important criteria for determining the installation areas of the solar system[7]. Based on previous studies in this field, the most important criteria have been identified that make the area suitable or unsuitable for the installation of solar cell stations, namely solar radiation, Elevation , slope, Aspect, land cover/ use, distance to power lines and main roads[8]. Renewable energy is the basis of smart cities to achieve a sustainable, healthy and efficient environment by reducing emissions and pollution and reducing costs by improving energy distribution and reducing waste[9]

Remote sensing is a multi-scale and multi-temporal approach that provides important data around objects and atmosphere at or close the surface of the Earth depending on radiation that is reflected or released from certain areas or objects. Satellite or airborne sensors are used in remote sensing techniques to obtain data on particular items or regions. As a result, they are dependent on geological, biological, chemical, and/or physical characteristics [10]. All forms of data are created, managed, analyzed, and mapped by the geographic information system. It is a structured collection of applications and information created to efficiently collect, store, update, process, analyze, and present all types of spatially linked data, The broad scientific issue of land-use classification of land cover has taken on a new dimension as a result of technological advancements in the fields of remote sensing (RS) and geographic information systems (GIS), which provide timely, accurate, and high quality input for decision-making while enhancing sustainable use and conservation practices of natural resources. Large geographic areas may be covered by satellite images, which also makes field operations less complicated. Analyzing the strengths and limits of the various forms of accessible sensor data allows for the selection of suitable remote sensing data for picture categorization. Satellite imagery captured by satellite sensors can be measured, analyzed, and interpreted to classify land use and land cover.

MCDM (Multiple Criteria Decision Making) technique is employed to determine which replacements given under various parameters and varied aims are the most suited, best The analytical hierarchy process (AHP) was developed in 1980 to solve unstructured problems by Saaty and is considered one of the most important MCDM methods used to determine the best choice amongst a set of alternative choices based on multiple criteria. AHP is one of the powerful and useful MCDM technologies that enables multifaceted problems for decision makers and provides high suppleness and reliability with a usable structure. The AHP technology includes important steps to enumerate Using the pairwise comparison technique to homogenize the impacts on the entire unit, the hierarchy process combines the elements of the pairwise comparison approach and the decision-making criteria technique to determine the weighted scores of the criterion. In a GIS context, suitability analysis is a geographic or GIS-AHP-based procedure used to assess if a specific location is suited for a given use. Every part of the landscape has intrinsic qualities that are, to some extent, either appropriate or unsuitable for the planned activities. This is the fundamental tenet of GIS suitability analysis. The typical map overlay approaches to the mapping and analysis of land-use suitability have significantly been advanced by the combination of MCDM techniques with GIS.A method that aggregates and converts spatial data into a final decision might be conceived of as GIS-based MCDA. This study aims to provide a GIS-based methodology for making the decision to install solar cells in Karbala governorate, and this proposed methodology was implemented using the ArcGIS program to calculate solar radiation and also analyze and classify land cover / use.

Karbala is positioned in the Mesopotamian plain 92km away from Baghdad in the southwest route and take place between Eastern Longitude (44° 25' 00- 43° 45' 00) and Northern latitude (32° 40' 00-32° 20' 00), it has an area of about 5 thousand square kilometers which characterize about 1.2% of Iraq total area. It's inhabited by more than 1.2 million residents and host millions during religious events like the Ashura and the Karbala province is considered one of the important tourist sites where thousands of visitors go throughout the year for religious tourism, as the city contains the shrine of Imam Hussein and the shrine of his brother Imam Abbas, who represent great religious importance for Shiite Muslims. Karbala governorate is also considered one of the historical cities because it contains historical monuments such as (Al-Ukhaidir Fortress) and tourist places such as (razarze Lake and Ain Al-Tamr), so it is considered a destination for visitors to archaeological tourism.

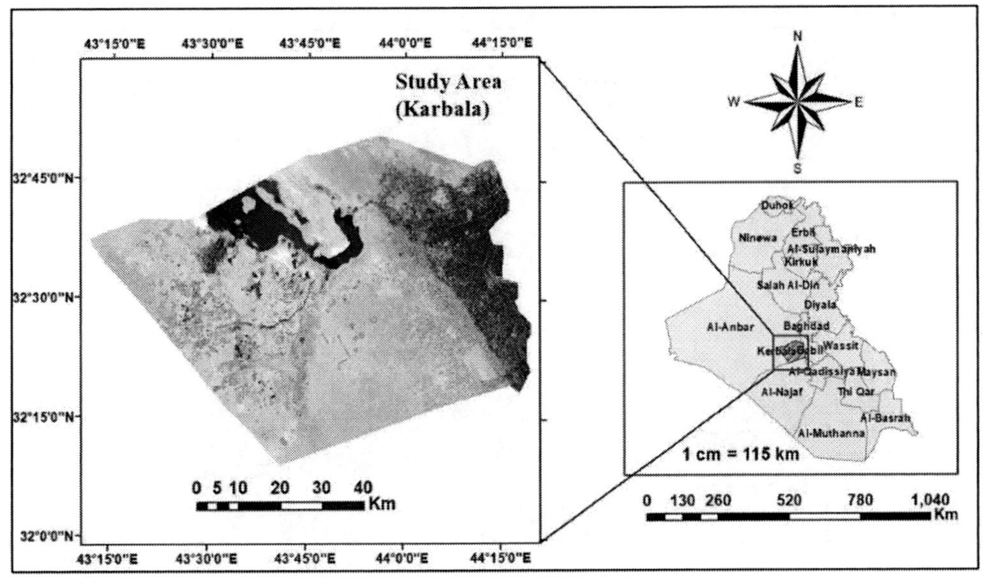

Fig. 1. Study area map.

2 Materials and Methods

2.1 Smart city and PV station

To facilitate the necessary transition to more energy-efficient cities, methodological approaches that employ alternative urban models are needed to address the expanding urbanization process and the ensuing energy demand. Maximizing the solar photovoltaic potential on appropriate rooftops and enhancing the energy balance across a smart grid are the goals of solar smart city planning. However, through integration and intensification, urban planning can play a significant role in managing energy production by utilizing solar potential on well-constructed roof and facade areas.

2.2 Data acquisition

The criteria were resulting from various data sources, including satellites, classification and websites. As shown by Table 1.

Table 1. Site criteria for installing solar power plants.

Criteria	Reference	Unit
global horizontal irradiance (GHI)	ALOS PALSAR satellite image (DEM) processed with ArcGIS	(WH/m2)
elevation	ALOS PALSAR satellite image (DEM) (ASF) processed with ArcGIS	m
Slope	ALOS PALSAR satellite image (DEM) (ASF) processed with ArcGIS	Degree
Aspect	ALOS PALSAR satellite image (DEM) (ASF) processed with ArcGIS	Degree
Land cover/use	Landsat 8 OLI/TIRS (USGS)	Unitless
Distance from power lines	Open street map	m
Distance from major roads	World Bank Data Catalog	m

In order to find the best locations for solar farms in the study area, the study used a methodology based on GIS in conjunction with the AHP method. Figure 2 shows a flowchart of the methodology.

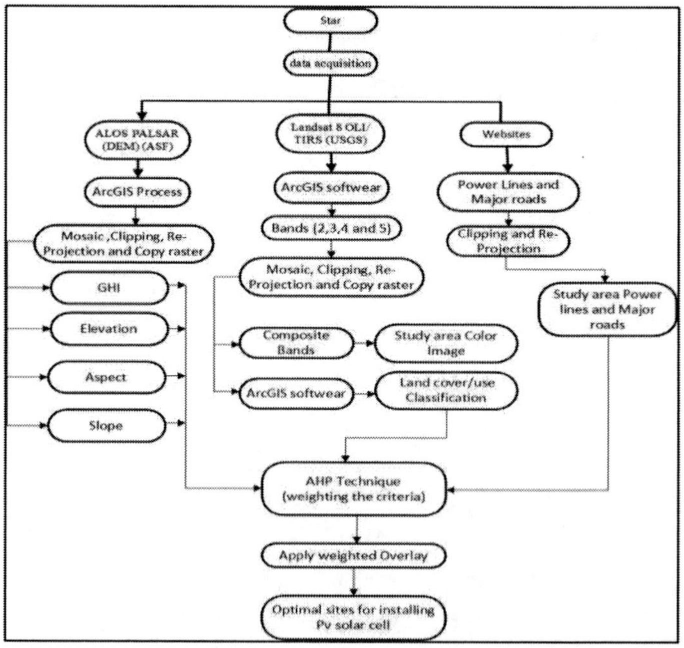

Fig. 2. Methodology flowchart for Land suitability of solar cell site selection.

2.3 Data preparation

After obtaining the data, the data is prepared using the ArcGIS program in order to obtain accurate results. Satellite images are combined using (Mosaic to new raster tool) to get a satellite image that covers the entire area of the study area. The second step is to subtract the study area from the satellite image produced by the Mosaic tool by a mask that represents the boundaries of the study area so that the results are focused only on it. The third step is to correct the spatial reference of the satellite image in this study, the spatial reference used is (UTM Zone 38n). The last step is to get rid of the NoData Value and the Background values of the satellite image. After these steps, the data is ready for analysis

2.4 Calculate criteria

• Global Horizontal irradiation (GHI): It represents the quantity of solar radiation that the Earth's surface receives. This value is of great importance in the installation of photovoltaic plants, and includes both direct radiation and reflected radiation as shown in Figure 3. This criterion calculated using (Areal solar radiation tool) contained in the ArcGIS program using ALOS PALSAR satellite images as shown in Table 2.

Table 2. ALOS PALSAR satellite images DEM data.

Scenes	Band	Path/Frame	Resolution (m)	Spatial reference
Scene 1	L-band	585/640	12.5	UTm_Zone_38n
Scene 2		585 /630		
Scene 3		586/640		
Scene 4		586/630		
Scene 5		587/640		
Scene 6		587/630		

The areas with red color represent the highest values of solar radiation during the year and are considered the best areas for installing photovoltaic plants, which recorded the highest value of 125570 (WH/m^2) and the values are graduated as shown in Figure 3 to reach the lowest value of 76619 (WH/m^2) (areas in yellow).

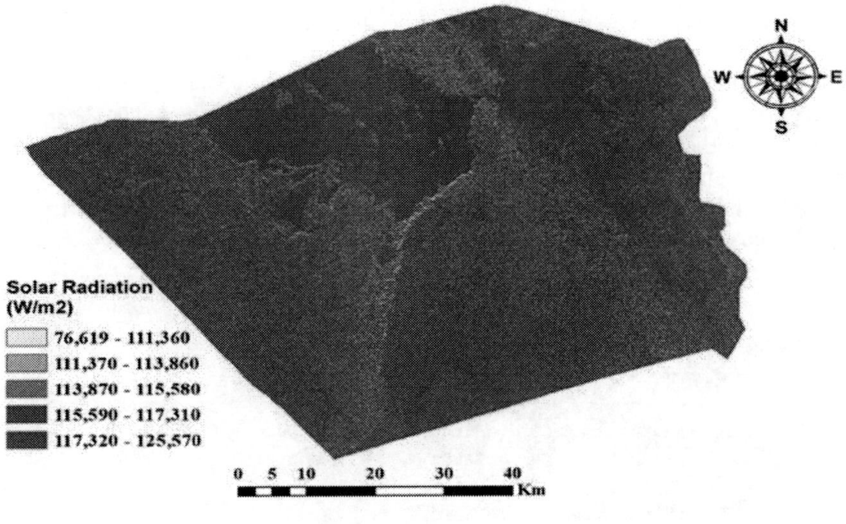

Fig. 3. Annual Solar Radiation map for Karbala 2022.

• Elevation

High-altitude areas receive more solar radiation than low-lying areas, where they receive the lowest values of solar radiation, as shown in figure 4, the highest value of altitude reaches 156 (m) and gradually decreases until it reaches the lowest area with a value of 22 (m).

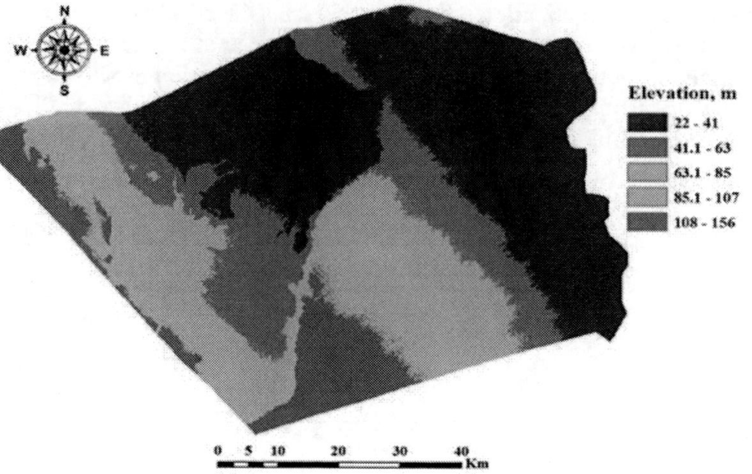

Fig. 4. Elevation values in Karbala 2022.

- Slope and Aspect

The slope is a crucial factor that needs to be properly taken care of. It illustrates how sloping or level the land is. Sitting PV solar systems are better suited to flat terrain with extremely small slopes. Because locations with extreme slopes will cast shadows on the solar panels next to them, the amount of electricity produced will be reduced. The Aspect similarly to slope indicates the steepness of the surface but with an expansion of exhibiting the direction of the slope of the surface with the accompanying angles. Although the south, southeast, and southwest directions have been selected as the most favourable for sitting PV solar power plants.

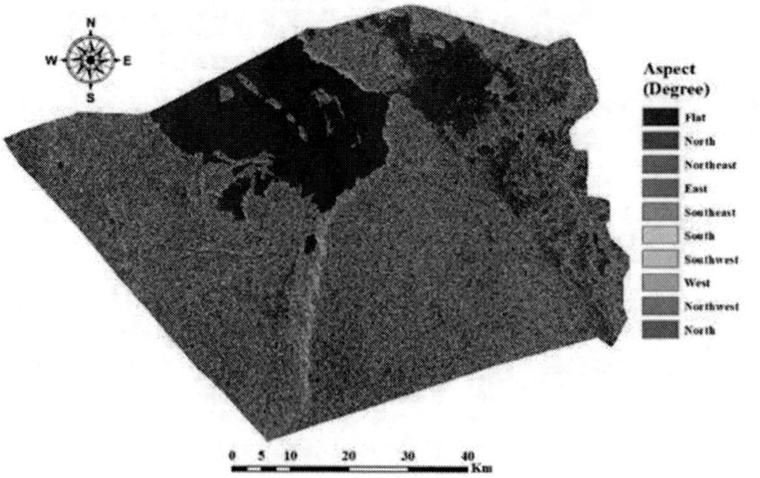

Fig. 5. Aspect map of Karbala.

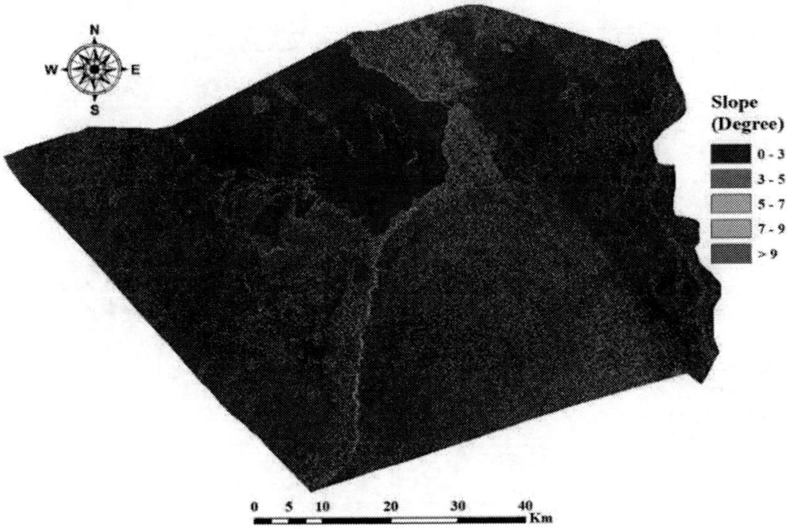

Fig. 6. Slope map of Karbala 2022.

- Land cover

Not all lands are suitable for the installation of photovoltaic panel systems, the best candidate areas for this purpose are bare lands and pastures. The ground cover of the study area resulting from the Landsat 8 satellite as shown in Table 3. The land cover is classified into 6 varieties, which are barren lands, rangelands, Crops, Built area, forest evergreen and water.

Table 3. Landsat 8 OLI Spectral Bands.

Landsat 8 Sensors	Band	Band Name	Wavelength (μm)	Resolution (m)
Operational Land Imager (OLI)	2	Blue	0.450-0.515	30
	3	Green	0.525-0.600	
	4	Red	0.630-0.680	
	5	Near Infrared (NIR)	0845.-0.885	

Table 4. Land cover classes' details.

Class	Area (Km)	Percentage %
Barren lands	3185.568	62.90
Rangelands	65.644	1.29
Crops	356.243	7.03
Built area	819.513	16.18
forest evergreen	388.712	7.67
water	248.186	4.90

 https://doi.org/10.1051/e3sconf/202563304001

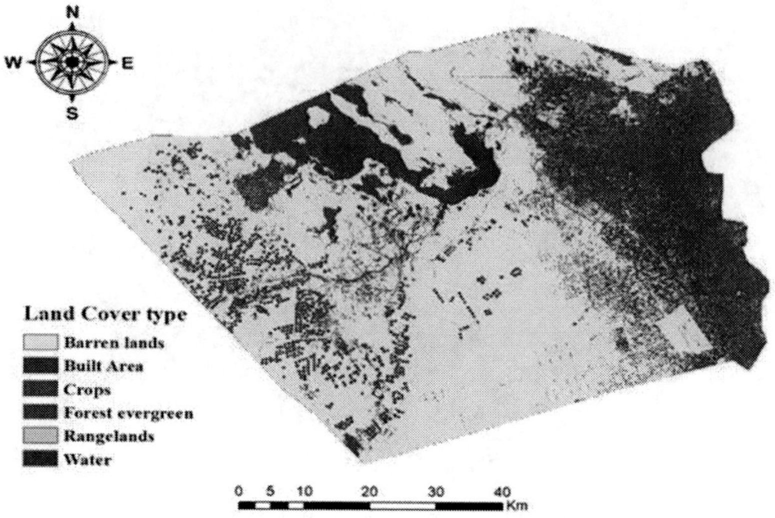

Fig. 7. Karbala Land cover Classification.

- Major roads and Power lines

Proximity to major roads and major energy transmission lines affects the cost of organization construction and the related environmental damage. Figure 8 and Figure 9 represent the map of the main roads and the map of the main power lines in order

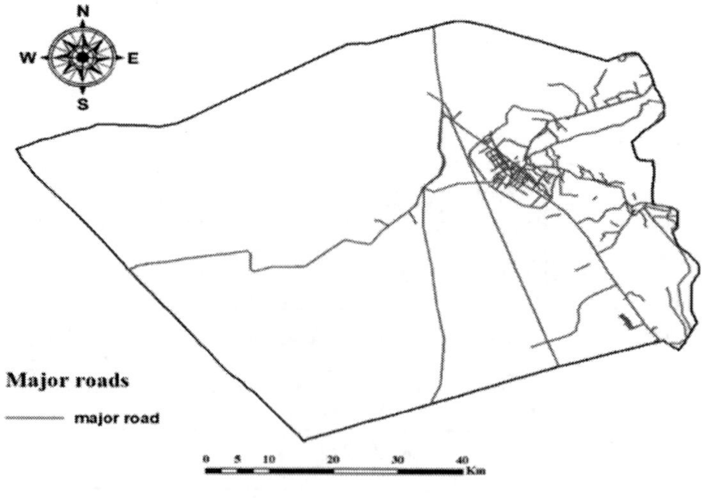

Fig. 8. Major roads Map of Karbala.

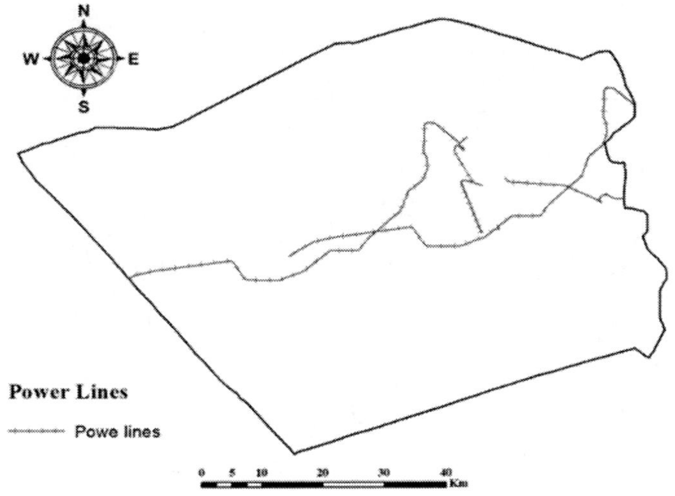

Fig. 9. Major power transmission lines.

2.5 Criteria reclassify

After calculating the values of the criteria used in this study, we classify the values of the standards based on their suitability for the installation requirements of photovoltaic regulators from 1 to 5 ratings, where Rating 1 represents the least suitability and Rating 5 the highest suitability, Photovoltaic plants require at least 4 WH/m^2 solar radiation for daily use and according to the efficiency of the system. Studies on the suitability the site for photovoltaic systems confirmed that the best slope values are less than <3%, but values exceeding them are Unsuitable, and the Aspect in the south and Southeast is very convenient. As for the land cover, barren lands and rangelands represent the best places to install photovoltaic plants, in addition to their proximity to energy transmission lines and Main Roads.

Table 5. Suitability rating criteria.

C1	C2	C3	C4	C5	C6	C7	Rating
76619-111360	>9	Water	Flat	22-41	<1	1	1
111374-113860	7-9	Built Area/ Forest Evergreen	N, NE, NW	41.1-63	1-3	3	2
113870-115580	5-7	Crops	E, W	63.1-85	3-6	6	3
115590-117310	3-5	Rangelands	SE, SW	85.1-107	6-9	9	4
117320-125570	<3	Barren Lands	S	107.1-156	>9	>9	5

Where: C1: GHI, C2: Slope, C3: Land cover/use, C4: Aspect, C5: Elevation, C6: Distance from Power lines (DFP) and C7: Distance from major roads (DFR)

2.6 Criteria rating

Pairwise comparison is essential in the use of AHP, criteria are prioritized by importance by judging them in pairs and then a pairwise comparison matrix is created. Judgments that are represented by numbers according to the Saaty scale are used for comparison.

Table 6. Pairwise Comparison scale.

Saaty scale	Definition
1	equally important
3	moderately more important
5	strongly more important
7	very strongly more important
9	Extremely more important
2,4,6,8	Intermediate values

With this approach, difficult choices are reduced to a series of side-by-side comparisons. By constructing an equal comparison matrix (A), the method also enables testing the decision's reasoning, which reduces bias in decision-making. as shown by equation No. 1. If n is the number of inputs, then the size of the array is n×n

$$A = \begin{bmatrix} 1 & a & b \\ \frac{1}{a} & 1 & c \\ \frac{1}{b} & \frac{1}{c} & 1 \end{bmatrix} \tag{1}$$

The Matrix A is combined by dividing each column's elements by the sum of the same column's elements in order to weight each condition. The new matrix's average row value establishes the relative weights that the criterion must have. The consistency ratio (CR), a metric for assessing the consistency of the weights, is a component of the AHP. The consistency index (CI) must first be calculated before the CR can be determined. (Equation (2))

$$CI = \frac{\lambda max - n}{n-1} \tag{2}$$

Where: λmax means the Subjective value of the pairwise comparison matrix, n is the number of the input criteria, The CR is ultimately determined by divided the CI by the random consistency index (RI). Saaty reported the RI values for the proper n values. To determine CR's value, Equation (3) is used.

$$CR = \frac{CI}{RI} \tag{3}$$

Table 7. The pairwise comparison matrix for the multi-criteria decision analysis (MCDA).

	GHI	Slope	Land Cover	Aspect	Elevation	DFP	DFR
GHI	1	2	3	4	5	7	9
Slope	0.5	1	1.5	2	2.5	3.5	4.5
Land Cover	0.333333	0.666667	1	1.333333	1.666667	2.333333333	3
Aspect	0.25	0.5	0.75	1	1.25	1.75	2.25
Elevation	0.2	0.4	0.6	0.8	1	1.4	1.8
DFP	0.142857	0.285714	0.428571	0.571429	0.714286	1	1.285714286
DFR	0.111111	0.222222	0.333333	0.444444	0.555556	0.777777778	1

 https://doi.org/10.1051/e3sconf/202563304001

Table 8. The standardized matrix and the weighted dissemination for the multi-criteria decision analysis (MCDA).

	GHI	Slope	Land Cover	Aspect	Elevation	DFP	DFR	Weight
GHI	0.394119	0.394119	0.394119	0.394119	0.394119	0.394119487	0.394119487	0.394119
Slope	0.19706	0.19706	0.19706	0.19706	0.19706	0.197059744	0.197059744	0.19706
Land Cover	0.131373	0.131373	0.131373	0.131373	0.131373	0.131373162	0.131373162	0.131373
Aspect	0.09853	0.09853	0.09853	0.09853	0.09853	0.098529872	0.098529872	0.09853
Elevation	0.078824	0.078824	0.078824	0.078824	0.078824	0.078823897	0.078823897	0.078824
DFP	0.056303	0.056303	0.056303	0.056303	0.056303	0.056302784	0.056302784	0.056303
DFR	0.043791	0.043791	0.043791	0.043791	0.043791	0.043791054	0.043791054	0.043791

2.7 Suitable areas

The results of suitable areas for installing solar farms had obtained by applying the Weighted Overlay tool. The results were a raster model that was classified into five categories according to the degree of suitability, as shown in Figure 10, The first category with dark green color represent the Unsuitable area (0.424565 km^2) while the last category is the red color class with a high level of suitability (3.075711 km^2). The presence of more than one Suitable site for the installation of solar photovoltaic power plants near the Holy Shrine, which can supply the electric power grid with energy, especially that it receives thousands of visitors throughout the year and millions of pilgrims on the anniversary of the Arbaeen.

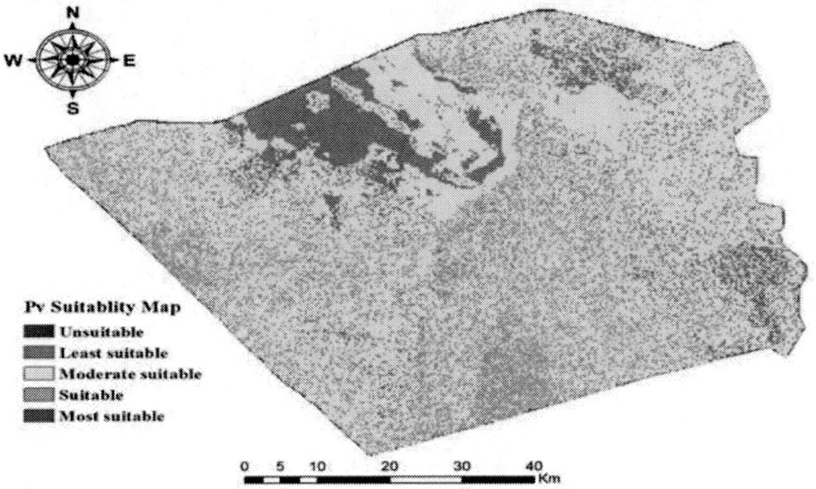

Fig. 10. Results of suitable areas for installing solar farms.

3 Results and conclusions

Based on the results obtained, Karbala governorate is exposed to large amounts of solar radiation annually, which contributes to the move towards exploiting this amount of solar radiation to generate clean electric energy and reduce the environmental impacts resulting from ordinary energy sources.

The results of using the solar energy project showed that the solar radiation standard, slope and type of land cover are among the most important criteria related to determining the location of solar farms

The results indicated that the most suitable areas have an area of (3.075711 km^2), suitable areas (772.690359 km^2) and the moderate suitable areas are (3606.027189 km^2) and the least suitable area is (650.790079 km^2) and finally, the unsuitable area is (0.424565 km^2).

References

1. P. A. Owusu, S. Asumadu-Sarkodie, "A review of renewable energy sources, sustainability issues and climate change mitigation," Cogent Eng. **3**, 1167990 (2016)
2. H. Z. Al Garni, A. Awasthi, "Solar PV power plants site selection: A review," in Advances in Renewable Energies and Power Technologies, 57–75 (2018)
3. V. Çoban, "Solar energy plant project selection with AHP decision-making method based on hesitant fuzzy linguistic evaluation," Complex Intell. Syst. **6**, 507–529 (2020)
4. N. S. Imamverdiyev, "Site selection for solar photovoltaic system installation using analytical hierarchy process model in Azerbaijan," J. Belarusian State Univ. Geogr. Geol. **2021**, 75–92 (2021)
5. Y. Choi, J. Suh, S.-M. Kim, "GIS-based solar radiation mapping, site evaluation, and potential assessment: A review," Appl. Sci. **9**, 1960 (2019)
6. S. Natya, V. Rehna, "Land cover classification schemes using remote sensing images: A recent survey," Br. J. Appl. Sci. Technol. **13**, 1–11 (2016)
7. A. Asakereh, M. Soleymani, M. J. Sheikhdavoodi, "A GIS-based Fuzzy-AHP method for the evaluation of solar farms locations: Case study in Khuzestan province, Iran," Sol. Energy **155**, 342–353 (2017)
8. J. A. Prieto-Amparán, A. Pinedo-Alvarez, C. R. Morales-Nieto, M. C. Valles-Aragón, F. Villarreal-Guerrero, "A regional GIS-assisted multi-criteria evaluation of site-suitability for the development of solar farms," Land **10**, 217 (2021)
9. Q. N. Jaleel, N. K. Ghazal, "The best school site choosing for rural areas of the Husseiniya district in Karbala province using GIS (Model Builder) Techniques," Iraqi J. Sci. **2017**, 2473–2485 (2017)
10. A. Alhammad, Q. Ch. Sun, Ya. Tao, "Optimal Solar Plant Site Identification Using GIS and Remote Sensing: Framework and Case Study", Energies **15**, 312 (2022), DOI: 10.3390/en15010312

Improvement efficiency and stability for silicon solar cells

Nagham Subhi Halboos and *Adnan* Falh Hassan*

Department of Physics, Faculty of Science, University of Kufa, Al-Najaf, Iraq

Abstract. Scientific research is advancing rapidly to enhance solar cell efficiency as an alternative to fossil fuels, which cause significant environmental damage. This study focuses on two key strategies: first, the synthesis of zinc oxide nanoparticles (ZnO NPs) using electrochemical methods due to their simplicity, speed, and cost-effectiveness; and second, utilizing ZnO as a highly efficient luminescent solar concentrator (LSC). X-ray diffraction (XRD) and transmission electron microscopy (TEM) confirmed the formation of ZnO nanorods and spherical particles. Additionally, fluorescein sodium dye was incorporated to enhance solar cell efficiency. fluorescein sodium dye and zinc oxide nanoparticles were mixed in volume ratios.The combination of 50% ZnO with fluorescein sodium dye resulted in an efficiency increase of 33.75%, whereas the standalone dye application yielded an efficiency of 25.97%. These findings represent significant progress in the pursuit of clean, efficient, and sustainable energy solutions, demonstrating the potential of nanotechnology and innovative dye applications in improving solar cell performance and mitigating environmental impact.

1 Introduction

Nanoscience has experienced remarkable advancements over the past three decades, emerging as one of the most transformative scientific disciplines due to its widespread applications in industry, agriculture, medicine, national security, and energy management [1]. Nanomaterials are defined by their structural dimensions within the nanoscale range (1–100 nm), granting them unique properties such as enhanced chemical reactivity, exceptional electrical conductivity, and superior mechanical strength. Their increased surface area, high flexibility, and enhanced optical and magnetic properties make them ideal for improving solar cell efficiency, particularly as luminescent solar concentrators (LSCs), which enhance light harvesting and minimize light loss [2, 3]. As global efforts intensify to combat environmental crises through sustainable energy solutions, photovoltaic (PV) systems have emerged as a leading renewable energy technology, offering scalable pathways to reduce carbon emissions [4]. The rapid adoption of solar power is driven by growing concerns over the ecological damage caused by fossil fuels, as well as the inherent reliability and near-limitless availability of solar energy [5, 6]. ZnO nanoparticles have gained significant attention in solar cell research as a multifunctional n-type semiconductor with a bandgap

* Corresponding author: adnanf.aljubury@uokufa.edu.iq

energy between 3.2 and 3.35 eV [7]. ZnO is widely integrated into photocatalysis, electronic devices, and renewable energy systems, in addition to commercial applications in coatings, polymer composites, sunscreens, and printing technologies [8, 9]. researchers have turned to advanced technologies such as luminescent solar concentrators (LSCs). These systems enhance photovoltaic efficiency by concentrating diffuse sunlight, thereby improving energy conversion rates and reducing material waste—an essential advancement in sustainable energy infrastructure [10, 11].

Fluorescein sodium exhibits strong absorption with a peak around 494–495 nm in water, and its molar absorptivity typically ranges between 75,000 and 95,000 M^{-1} cm^{-1}, depending on the solvent. In aqueous alkaline conditions, its fluorescence quantum yield is very high, usually between 0.92 and 0.95. However, this yield decreases in more polar solvents due to increased non-radiative relaxation rates. Fluorescence lifetimes also vary with the solvent, ranging from about 4.0 ns in water to 2.0–3.5 ns in organic solvents such as ethanol or octanol. The general trend observed is that as solvent polarity increases, the fluorescence quantum yield of fluorescein sodium decreases, primarily due to enhanced non-radiative decay processes. The exact mechanism behind this behavior is still not fully understood have also identified solvent viscosity as a contributing factor to fluorescein sodium's solvent sensitivity. In high-viscosity solvents, the motion of the flexible diethylamino groups on the xanthene ring is restricted, potentially lowering non-radiative decay rates. However, other studies have not found a direct correlation between non-radiative decay rates and solvent viscosity. Specific mode coupling between the vibrations of fluorescein sodium and those of interacting solvent molecules has also been suggested to explain the large variation in non-radiative decay rates.

Furthermore, the solvent influences the positions of the absorption and emission maxima for fluorescein sodium. For cationic fluorescein sodium in methanol, these values are 552 nm and 577 nm, respectively, whereas in water, a slight red shift occurs, shifting the respective maxima to 557 nm and 580 nm. This red shift is attributed to the greater stabilization of the highly polarizable excited state due to more favorable interactions between the dye and polar solvents, leading to a decrease in the S_0–S_1 energy level gap.

2 Materials and methods

2.1 Nanofabrication of Zinc oxide (ZnO NPs)

ZnO nanoparticles were synthesized using an electrochemical cell, a method characterized by its simplicity, high productivity, and low cost. Two zinc electrodes were immersed in a pre-prepared potassium chloride (KCl) solution, with one electrode being larger than the other. The solution was prepared by dissolving 10 g of KCl in 100 mL of deionized water, from which 5 mL were extracted. Additionally, polyvinylpyrrolidone (PVP) was added as a stabilizer to prevent nanoparticle aggregation. The stabilizer was prepared by dissolving 10 g of PVP in 100 mL of deionized water, from which 10 mL were drawn and added to the electrochemical cell. The total solution volume was adjusted to 200 mL.
The electrodes were immersed and connected to a constant voltage source ranging from 0 to 30 V, with a current range of 0 to 5 A. The applied current was set to 0.05 A, with an electrode immersion ratio of 1:2. The cell was placed on a hot plate to ensure uniform stirring and electrical conductivity. After one hour, the voltage was removed, and the solution was collected, centrifuged, and washed with deionized water followed by absolute ethanol. The nanoparticles were then dried in an oven at 70°C for one hour and subsequently calcined in a furnace at 700°C for one hour.

2.2 Characterizations of the synthesized ZnONPs

X-ray diffraction (XRD) results confirmed the formation of ZnO nanoparticles, with Miller indices (111, 100, 002, 101, 102, 110, 103, 200, 112, 201, 004, 202) corresponding to 2-theta angles ranging from 18.333° to 76.997°, forming a hexagonal crystal structure with an average crystallite size of 20 nm (Table 1, Figure 1). The XRD results matched reference card JCPDS No. 01-073-1764. Furthermore, TEM analysis revealed nanoparticle agglomerations in the form of spherical-like structures (Figure 2).

Table 1. X-Ray Diffraction data of ZnO NPs.

2-Theta	FWHM	hkl	crystal structure	D nm
18.333	0.216	111	Hexagonal	37.25488
31.699	0.397	100	Hexagonal	20.80163
34.383	0.403	2	Hexagonal	20.63479
36.196	0.42	101	Hexagonal	19.89947
47.5	0.499	102	Hexagonal	17.39344
56.586	0.443	110	Hexagonal	20.36594
62.882	0.492	103	Hexagonal	18.92563
66.315	0.381	200	Hexagonal	24.9067
67.916	0.47	112	Hexagonal	20.37832
69.089	0.537	201	Hexagonal	17.96053
72.617	0.486	4	Hexagonal	20.28475
76.997	0.507	202	Hexagonal	20.02136

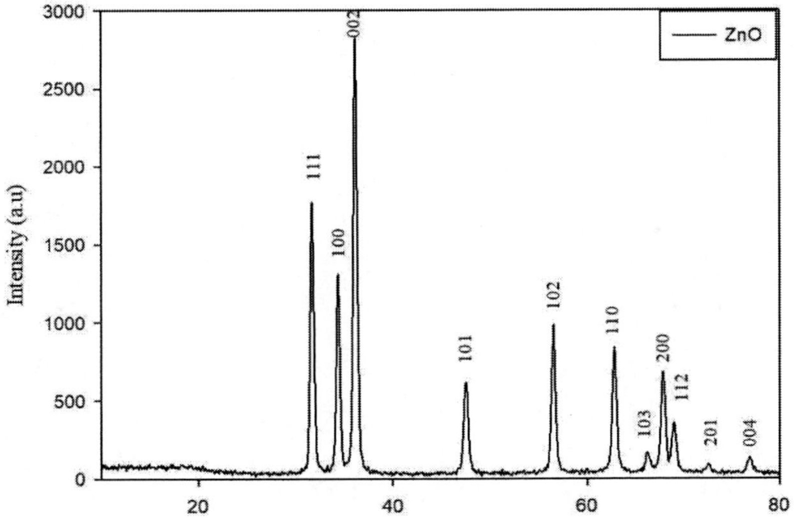

Fig. 1. XRD Pattern of ZnO nanoparticles.

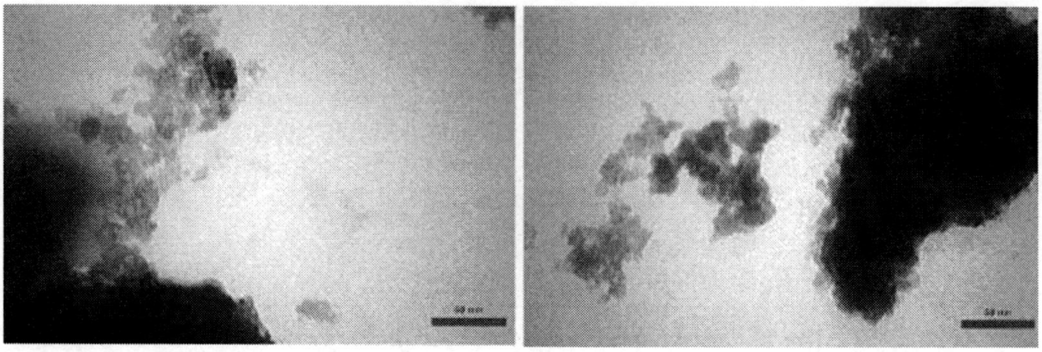

Fig. 2. TEM images of ZnO NPs.

2.3 The quantum efficiency for five concentrations of fluorescein sodium Dye

Five concentrations of fluorescein sodium dye (BDH Chemicals Ltd Poole England) were prepared a namely (1×10^{-5}M, 2×10^{-5}M, 5×10^{-5}M, 1×10^{-4}M, 3×10^{-4}M). The absorption spectra were recorded using a UV-visible spectrophotometer (SCINCO mega-2100) and the fluorescence spectra using a spectrofluorometer (F96PRO from Shanghai Kingdak). The figure (3) included the drawings from which calculated the following: maximum wavelength of absorbance (λAmax), maximum wavelength of fluorescence (λFmax), Stokes shift ($\Delta\lambda = \Delta\lambda A_{max} - \Delta\lambda F_{max}$), fluorescence lifetime (τ_f), radiative lifetime (τ_{fm}), and quantum efficiency (Q_{fm}) for all five prepared concentrations as shown in Table (2).

 https://doi.org/10.1051/e3sconf/202563304002

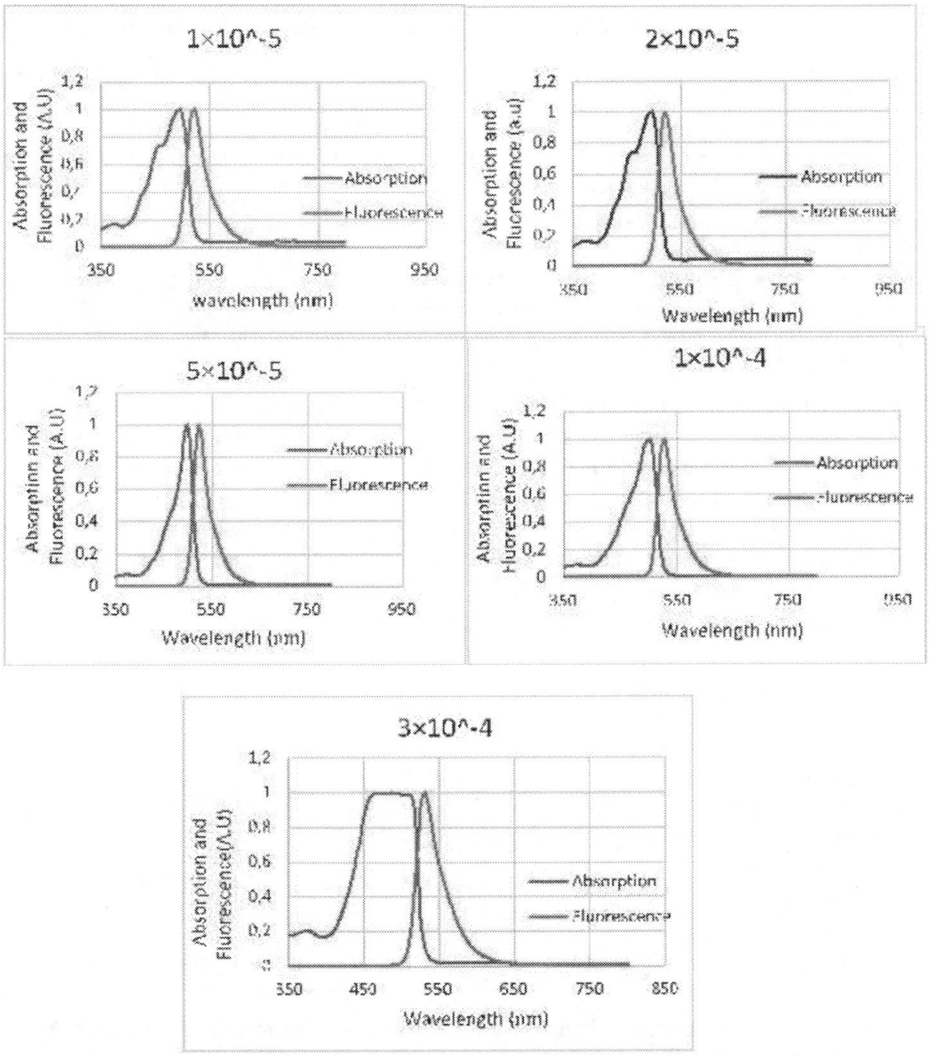

Fig. 3. Absorption and fluorescence plots for the five concentrations.

Table 2. Absorption and fluorescence parameters.

Concentration (mol/L)	λ Amax (nm)	λ Fmax (nm)	Stock Shift Δλ=λflo-λabs (nm)	The radiated Life time τfm (n sec)	The fluorescence Life time τf (n sec)	The
1×10^{-5}	485	522	37	1.13669	0.6827	60.06
2×10^{-5}	495	522	27	1.40508	0.9312	66.26
5×10^{-5}	499	524	25	2.27822	1.9483	85.52
1×10^{-4}	498	526	28	1.73994	1.2310	73.57
3×10^{-4}	486	532	46	1.11785	0.5365	48.00

3 Results and discussion

Before incorporating the prepared LSC system, the solar cell exhibited a baseline efficiency of 0.797%. Following the integration of the fluorescein sodium dye-ZnO system, the efficiency significantly improved to 1.004, driven by enhanced light harvesting and charge carrier dynamics.

The Stokes shift results for fluorescein sodium dye exhibited values ranging from 25 to 46, as listed in Table 2, with the highest quantum efficiency (Q_{fm}) recorded at 85.52%. The second improvement was achieved by adding 50% of ZnO NPs to the dye, which significantly enhanced the solar cell conversion efficiency. The highest recorded efficiency was 33.75% for the 5×10^{-5} M concentration, as shown in Table 4.

Additionally, this increase in efficiency is attributed to fluorescein sodium dye's broad spectral range (450–575 nm) in the visible spectrum. The presence of ZnO nanomaterials improved solar cell performance by scattering incident light toward the active light-absorbing center, thereby enhancing light harvesting and reducing optical losses.

From the results shown in the table4, it was found that adding concentrators to the solar cell system caused a significant change in the cell's efficiency. The highest results were recorded in the presence of nano-oxides, while the lowest were observed in their absence. This indicates that the nanoscale size of these oxides greatly enhances the efficiency of the solar cell. The reason for this is the increased surface area of the nanoparticles due to their small size compared to bulk materials such as dye in similar cases. This property led to increased light scattering toward the solar cell, Raman scattering, for example, which resulted in harvesting more of the incident light and reducing losses [6, 8].

Higher concentrations of the dye (sodium fluorescein) lead to a decrease in efficiency due to the following reasons:

Excessive light absorption and photoinduced photoexcitation: When dye concentrations increase, its molecules absorb large amounts of light, which can lead to:Impediment of electron transfer from the dye to the conduction band of ZnO (due to saturation of active sites on the nanoparticle surface).

- Increased electron-hole recombination due to the close proximity between dye molecules, which reduces charge separation efficiency.
- Self-assembly: At high concentrations, dye molecules may aggregate (such as forming dimers or agromers), reducing the effective surface area for interaction with ZnO and impairing charge transfer.

The sodium fluorescein dye with ZnO NPs interacts as follows. Photoinduced Electron Transfer:

1. The dye (sodium fluorescein) absorbs light, exciting its electrons from the ground state to the excited state. The excited electrons from the dye transfer to the conduction band in ZnO due to the compatibility of their energy levels.
2. The excited electrons transfer from the dye to the conduction band in ZnO due to the compatibility of their energy levels.
3. These electrons are then used in reduction reactions (such as the analysis of organic compounds in catalytic applications).

The role of ZnO nanoparticles: It provides a large surface area for dye absorption It has a suitable bandgap (~3.2 eV) to receive electrons from the dye.

Sodium fluorescein is exposed to degradation over time due to photochemical oxidation or chemical changes that may occur on the surface of zinc oxide nanoparticles (ZnO), leading to a decrease in its efficiency.In addition, ZnO nanoparticles tend to agglomerate over time, which limits the effective surface area available for reactions, thereby reducing their efficiency in catalytic processes.The catalytic activity of these particles may also decrease due to the accumulation of by-products resulting from chemical reactions, or due to changes in the crystalline structure of zinc oxide under the influence of operational conditions such as heat or prolonged exposure to radiation.

The instability of the substance is primarily due to:

1. Photodegradation or dissolution of the dye.

2. Aggregation or corrosion of ZnO particles.
3. Weakening of the bonds between the dye and ZnO NPs [5].

As for the fluorescent dye (sodium fluorescein), it showed the least increase in efficiency due to the large size of the dye particles, which absorbed a significant portion of the incident light, reflected another portion, and reduced scattering. Consequently, less light was harvested by the silicon cell, as shown in the Table 3.

Table 3. Solar cell parameters for fluorescein sodium only.

Concentration (mol/L)	Imax (mA)	Vmax (Volt)	FF	η%	Δη%
1×10^{-5}	347.0	0.412	0.576	0.916	14.93
2×10^{-5}	379.9	0.404	0.652	0.983	23.33
5×10^{-5}	391.6	0.400	0.624	1.004	25.97
1×10^{-4}	346.8	0.401	0.640	0.891	11.79
3×10^{-4}	334.6	0.415	0.575	0.890	11.66

Table 4. Solar cell parameters for fluorescein sodium with ZnO NPS.

Concentration (mol/L)	Imax (mA)	Vmax (Volt)	FF	η%	Δη%
1×10^{-5}	448.1	0.363	0.495	1.042	30.74
2×10^{-5}	432.0	0.377	0.528	1.044	30.99
5×10^{-5}	420.0	0.396	0.533	1.066	33.75
1×10^{-4}	402.6	0.402	0.640	1.037	30.11
3×10^{-4}	393.2	0.408	0.570	1.028	28.98

The addition of 50% zinc oxide (ZnO) nanoparticles combined with fluorescein sodium dye to a silicon-based solar cell achieves a 33.75 % efficiency enhancement through synergistic mechanisms: enhanced light absorption occurs as fluorescein sodium captures underutilized visible wavelengths, while ZnO high surface area improves dye binding and photon harvesting via 33.75%.

References

1. C. Ballif et al., "Status and perspectives of crystalline silicon photovoltaics in research and industry," Nature Reviews Materials **7(8)**, 597–616 (2022)
2. R. Grover et al., "Aligned zinc oxide nanostructures for dye sensitized solar cells: A review," in Advances in Solar Power Generation and Energy Harvesting, 9–16 (2020)
3. S. Baruah, R. A. Afre, D. Pugliese, "Effect of size and morphology of different ZnO nanostructures on the performance of dye sensitized solar cells," Energies **17(9)**, 2076 (2024)
4. M. Di Vece, "Using nanoparticles as a bottom up approach to increase solar cell efficiency," KONA Powder and Particle Journal **36**, 72–87 (2019)
5. F. Kabir et al., "Instability of dye sensitized solar cells using natural dyes and approaches to improving stability-An overview," Sustainable Energy Technologies and Assessments **52**, 102196 (2022)
6. M. Rafiee et al., "An overview of various configurations of luminescent solar concentrators for photovoltaic applications," Optical Materials **91**, 212–227 (2019)
7. A. V. Balan, P. Gopinath, V. Radhika, "Enhanced Photovoltaic Performance of Dye Sensitized Solar Cell Based on Nanocomposites of Zinc Oxide Photoanode", Journal of Ovonic Research 20, 85–92 (2024), DOI: 10.15251/jor.2024.201.85

8. S. Baruah, R. A. Afre, D. Pugliese, "Effect of Size and Morphology of Different ZnO Nanostructures on the Performance of Dye-Sensitized Solar Cells", Energies 17, 2076 (2024), DOI: 10.3390/en17092076
9. S. Noor, A. Joda, M. M. Alrufaie, "Fabrication of Solar Cells Using $ZnO/MnO_2/CdS$ Nanocomposite as a Catalyst", Journal of the Pakistan Institute of Chemical Engineers 51(1) (2023), DOI: 10.54693/piche.05112
10. F. I. Lai, Ju. F. Yang, W. Ch. Chen, "Enhancing Dye-Sensitized Solar Cell Performance with Different Sizes of ZnO Nanorods Grown Using Multi-Step Growth", Catalysts 13, 1254 (2023), DOI: 10.3390/catal13091254
11. S. A. M. Abd Noor, A. M. Ju. Al-Shamari, "Preparation and Performance of ZnO and ZnO/MnO_2 Nanostructures as Anode Electrodes in DSSCs", Indonesian Journal of Chemistry 23, 1618 (2023), DOI: 10.22146/ijc.84037

 https://doi.org/10.1051/e3sconf/202563304003

Analysis of the efficiency of enterprises in the field of energy and sustainable development: European experience

Sun Qiming[1], *Irina* Vlasova[2], *Olga* Mustafina[2], *Andrey* Rykov[2], Vera Kolchina[2], and Svetlana Moiseenko[3]

[1] Nanjing University, Nanjing, China
[2] Ural State University of Economics, Yekaterinburg, Russia
[3] Bryansk State Engineering and Technology University, Bryansk, Russia

Abstract. The energy sector plays a pivotal role in driving sustainable development across Europe. This paper analyzes the effectiveness of enterprises operating in the field of energy and sustainable development by examining European experiences. Through a bibliographic analysis of recent studies and a comparative assessment of different countries, the research identifies key factors contributing to enterprise success in this sector. The study employs a mixed-method approach, combining quantitative data analysis with qualitative case studies to provide a comprehensive understanding of the operational efficiencies and challenges faced by these enterprises. The findings highlight the importance of innovation, policy frameworks, and public engagement in enhancing enterprise effectiveness. Recommendations are provided for policymakers and industry stakeholders to foster a more sustainable and efficient energy sector.

1 Introduction

The pursuit of sustainable development has become a central objective for nations worldwide, particularly within the European context where energy enterprises are at the forefront of this transition. The energy sector is not only a critical driver of economic growth but also a significant contributor to environmental challenges such as climate change and resource depletion. Consequently, analyzing the effectiveness of enterprises within this sector is essential for understanding how to achieve both economic and environmental objectives.

European countries have been actively implementing policies and technologies aimed at promoting renewable energy and sustainable practices. Enterprises operating in this sector are influenced by a myriad of factors, including regulatory frameworks, technological advancements, market dynamics, and societal expectations. Previous studies [1,2] have emphasized the role of innovation and policy support in enhancing enterprise performance in renewable energy integration.

However, there is a need for a comprehensive analysis that encapsulates the diverse experiences of European countries in this domain. This paper aims to fill this gap by providing an in-depth examination of the effectiveness of energy enterprises in Europe, focusing on how they contribute to sustainable development goals. By leveraging bibliographic analysis and comparative assessments, the study seeks to identify best practices and challenges that can inform future strategies. The significance of this research lies in its potential to guide policymakers, industry leaders, and stakeholders in making informed decisions that promote sustainability and efficiency in the energy sector. As urbanization and energy demand continue to rise, understanding the dynamics of enterprise effectiveness becomes increasingly crucial.

2 Bibliographic analysis

The scholarly discourse on the efficiency of European energy enterprises—understood as their capacity to convert physical, financial, and knowledge inputs into outputs that advance both profitability and sustainability—has grown markedly since the adoption of the European Green Deal. The Deal itself constitutes the overarching policy frame: by codifying carbon-neutrality as a legally binding objective, it elevates energy efficiency from a desirable trait to a compliance requirement for all market actors [1]. This mandate has oriented recent research along three interlocking axes: measurement of efficiency, drivers of performance, and pathways for sustainable transition.

A first stream focuses on how best to quantify enterprise-level efficiency against multidimensional goals. Tutak, Brodny, and Siwiec employ data-envelopment analysis across 27 EU members, explicitly weighting economic output, demographic scale, and renewable penetration [3]. Their cross-sectional design reveals persistent north-south heterogeneity, motivating calls for tailored policy support. Karasek et al. extend this line by comparing 2013 and 2020 efficiency frontiers; their findings confirm a continental convergence, yet also expose new laggards among coal-dependent states [8]. To contextualise Europe internationally, Zhang et al. analyse China's power sector under explicit carbon constraints [2]. Although ex-EU, their methodology offers a transferable template for incorporating emission caps into productivity indices—a feature only gradually appearing in European studies.

Recent publications trace efficiency gains to two primary drivers: energy-system restructuring and technology diffusion. Komarova et al. map the interplay between macro-economic development and energy-intensity trends, concluding that decoupling GDP growth from consumption is attainable when renewables exceed a 25 % share [4]. Chen et al. corroborate this from a firm-level perspective: using a panel of 230 EU utilities, they show that every 1 % decrease in technical energy losses raises return on assets by 0.4 %, even after accounting for carbon-price volatility [5]. Complementing these quantitative accounts, Sozinova and colleagues explore corporate-social-responsibility (CSR) programmes in EnergyTech firms; they identify digital grid solutions and life-cycle carbon accounting as the most effective CSR investments for simultaneous efficiency and reputational benefits [6].

Efficiency improvements do not occur in a vacuum; they are steered by policy mixes that combine carrots (subsidies, tax credits) and sticks (mandatory audits, emission ceilings). Swain, Karimu, and Gråd demonstrate that EU regions with higher renewable-energy-directive (RED II) compliance also record statistically significant job creation, implying that social co-benefits can reinforce corporate buy-in [7]. Conversely, Molina Cañamero et al. catalogue the regulatory and technical barriers—grid-connection queues, fragmented permitting, and a shortage of skilled labour—that still dampen enterprise-level progress [9]. Their Delphi survey of 60 industry experts ranks harmonised cross-border grid codes as the most urgent remedy.

Finally, the comparative study by Redko et al. places EU enterprises alongside Ukrainian counterparts, illustrating how shared innovation platforms accelerate technology adoption and reduce duplication of R&D expenditure [10]. Such findings stress that efficiency is not merely a function of internal optimisation but also of inter-firm learning within regional value chains.

Taken together, the literature portrays a European sector moving from isolated best-practice cases toward systemic efficiency, propelled by binding EU goals, maturing renewable technologies, and increasingly granular performance metrics. Yet three gaps persist: (i) longitudinal micro-data that link specific managerial practices to efficiency trajectories; (ii) integrated models that couple physical network constraints with market-based price signals; and (iii) comparative assessments of policy cost-effectiveness across member states. Addressing these gaps will sharpen both academic insight and executive decision-making, ensuring that energy enterprises remain pivotal actors in Europe's broader sustainable-development agenda.

3 Materials and methods

The research adopts a mixed-method approach, combining quantitative data analysis with qualitative case studies. Quantitative data were gathered from official European energy reports, enterprise performance metrics, and sustainability indices. This data provided a foundation for assessing operational effectiveness across different countries. Qualitative data were obtained through case studies of selected European countries, including Germany, Denmark, and the Netherlands, known for their progressive energy policies and successful enterprises in the renewable sector. The case studies involved analyzing policy documents, enterprise reports, and expert interviews to gain deeper insights into contextual factors affecting enterprise performance. Bibliographic analysis was conducted using academic databases to review relevant literature published between 2020 and 2024. References [1-10] were instrumental in framing the research context and identifying key themes related to innovation, policy frameworks, and public engagement.

Data analysis involved comparing quantitative performance indicators such as energy output, efficiency ratios, and sustainability ratings. Qualitative data were analyzed using thematic coding to identify recurring patterns and insights related to enterprise effectiveness.

4 Results and discussions

The analysis of the effectiveness of energy enterprises in Europe reveals a complex interplay of technological, policy, market, and social factors that influence their performance in promoting sustainable development. This section presents the findings from quantitative data analysis and qualitative case studies of select European countries, including Germany, Denmark, Spain, and the United Kingdom, each illustrating unique approaches and challenges in the energy sector.

Technological innovation has been identified as a key driver of enterprise effectiveness in the energy sector. Enterprises that invest in research and development tend to achieve higher efficiency and competitiveness. For instance, Germany's significant investment in renewable energy technologies, particularly solar photovoltaic (PV) and wind power, has positioned its enterprises as global leaders [2]. The country's Energiewende (energy transition) initiative has spurred innovation, leading to advancements in energy storage and grid integration technologies.

Zhang et al. [3] highlight that Chinese enterprises, learning from European counterparts, have improved their operational efficiency by adopting advanced technologies and management practices. This cross-cultural technological exchange underscores the importance of innovation in enhancing enterprise effectiveness globally.

In Denmark, technological innovation in wind energy has been a cornerstone of enterprise success. Danish enterprises like Vestas have become world leaders in wind turbine manufacturing due to continuous innovation and government support [5]. The integration of smart grid technologies has further improved the efficiency and reliability of renewable energy sources.

However, enterprises face challenges in technology adoption due to high costs and rapid technological obsolescence. Smaller enterprises, in particular, struggle with the capital investment required for cutting-edge technologies [9]. Collaborative efforts between government, industry, and research institutions are necessary to mitigate these challenges.

Supportive policy frameworks significantly influence enterprise effectiveness. Stable and predictable policies reduce investment risks and encourage enterprises to invest in long-term projects. In Spain, the initial success of renewable energy enterprises was largely attributed to favorable policies such as feed-in tariffs [4]. However, sudden policy changes and reductions in incentives led to market instability and reduced enterprise effectiveness.

The United Kingdom's Contracts for Difference (CfD) scheme provides a clear and stable pricing mechanism for renewable energy, enhancing investor confidence and enterprise performance [5]. Enterprises operating under such supportive policies can plan strategically and allocate resources efficiently.

Conversely, inconsistent or volatile policies can hinder enterprise effectiveness. *Lüthi and Wüstenhagen* [1] emphasize that policy risks, such as sudden changes in subsidies or regulatory requirements, can deter investment and innovation. Enterprises must navigate these risks through diversification and adaptive strategies.

Market liberalization has introduced both opportunities and challenges for energy enterprises. Increased competition can drive efficiency and innovation but may also pressure enterprises with limited resources. In the European Union, the liberalization of the energy market aimed to create a more competitive and integrated energy system [6].

In Germany, market liberalization has allowed new entrants, including small and medium-sized enterprises (SMEs), to participate in the energy market. This inclusivity has fostered innovation but also intensified competition [9]. Enterprises must differentiate themselves through unique value propositions, such as specialized services or technological expertise.

However, liberalization without adequate regulatory oversight can lead to market concentration and reduced competition. The dominance of large enterprises can stifle innovation and limit opportunities for smaller firms [6]. Regulatory bodies must balance the promotion of competition with measures to prevent monopolistic practices.

Public engagement is crucial for the successful implementation of renewable energy projects. Enterprises that actively involve local communities and stakeholders tend to experience higher acceptance and smoother project execution. In Denmark, community ownership models and transparent communication have led to widespread public support for wind energy projects [7].

Devine-Wright [8] suggests that place attachment and identity significantly influence public attitudes toward energy projects. Enterprises that acknowledge and respect local values can mitigate opposition and foster collaborative relationships. In the UK, public resistance to onshore wind farms has prompted enterprises to engage more proactively with communities, adjusting project designs to address concerns.

Social acceptance challenges can delay projects and increase costs. Enterprises must invest in stakeholder engagement strategies, including public consultations, education campaigns, and benefit-sharing mechanisms. These efforts can enhance enterprise effectiveness by reducing conflicts and building trust.

Comparative Analysis of Selected Countries

Germany's energy enterprises have benefited from a strong policy framework, technological innovation, and public support. The Renewable Energy Sources Act (EEG) provided stable incentives that encouraged investment in renewables [2]. Enterprises have achieved high levels of effectiveness, contributing significantly to the country's renewable energy capacity.

However, challenges remain in grid integration and managing the intermittency of renewable sources. Investments in smart grid technologies and energy storage are ongoing to address these issues. Enterprises are also adapting to reductions in subsidies by exploring new business models and markets.

Denmark's focus on wind energy has resulted in enterprises that are highly effective and globally competitive. Government-industry collaboration has fostered innovation and export opportunities [5]. Public acceptance has been high due to inclusive approaches and community benefits.

Enterprises face challenges in scaling up offshore wind projects, including environmental concerns and high capital costs. Efforts to streamline regulatory processes and develop cost-effective technologies are underway.

Spain experienced rapid growth in renewable energy enterprises due to generous incentives but faced setbacks when policies changed abruptly [4]. Enterprises struggled with financial losses and market uncertainty. This case illustrates the importance of stable policy environments for enterprise effectiveness.

Recent policy reforms aim to restore confidence and support for renewables. Enterprises are cautiously re-engaging, emphasizing the need for predictable long-term strategies.

The UK's mixed approach to energy policy has created both opportunities and challenges for enterprises. The CfD scheme has supported large-scale renewable projects, but policy shifts have introduced uncertainties [10]. Enterprises are navigating a complex landscape of regulatory changes and market dynamics.

Public resistance to certain projects has prompted enterprises to enhance stakeholder engagement. The focus on offshore wind and emerging technologies like tidal energy offers new avenues for enterprise effectiveness.

Challenges and Opportunities

Challenges:

- Technological Barriers. Integrating high levels of renewables poses technical challenges for grid stability and reliability. Enterprises must invest in solutions like energy storage and demand response systems.
- Financial Constraints. Access to financing remains a barrier, especially for SMEs. High upfront costs and perceived risks deter investment.
- Policy Uncertainty. Fluctuating policies introduce risks that affect long-term planning and investment decisions.
- Market Competition. Intense competition can pressure profit margins and limit resources for innovation.

Opportunities:

- Emerging Technologies. Advancements in energy storage, smart grids, and digitalization offer opportunities for enterprises to enhance effectiveness.
- International Collaboration. Cross-border projects and knowledge sharing can accelerate innovation and market expansion.
- Sustainable Development Goals. Alignment with global sustainability objectives can open new markets and attract investment.

To enhance effectiveness, energy enterprises should.

- Invest in Innovation. Continuous R&D to improve technologies and operational processes.
- Engage Stakeholders. Proactive public engagement to build support and address concerns.
- Adapt to Policies. Monitor policy developments and adapt strategies accordingly.

- Diversify Portfolios. Explore multiple renewable sources and markets to mitigate risks.
- Collaborate. Partner with other enterprises, research institutions, and governments to leverage resources and expertise.

5 Conclusion

The exploration of the effectiveness of energy enterprises in Europe within the framework of sustainable development reveals a multifaceted interplay of factors that contribute to their success. This study emphasizes that technological innovation, supportive policy frameworks, market dynamics, and public engagement are integral to enhancing enterprise effectiveness. Technological innovation stands out as a cornerstone for enterprises aiming to excel in the renewable energy sector. The commitment to research and development enables enterprises to improve operational efficiencies, reduce costs, and stay competitive in a rapidly evolving market. The experiences of German and Danish enterprises illustrate how investment in advanced technologies, such as energy storage and smart grids, can lead to global leadership in renewable energy.

Supportive policy frameworks are critical in providing the stability and predictability necessary for enterprises to make long-term investments. Consistent policies that offer incentives and clear guidelines encourage enterprises to engage in sustainable practices and drive the adoption of renewable energy sources. The contrasting experiences of Spain and the United Kingdom highlight the impact that policy stability—or the lack thereof—can have on enterprise performance and investor confidence. Market liberalization presents a dual-edged sword for energy enterprises. While it fosters competition and can lead to innovation and efficiency gains, it also poses challenges, especially for smaller enterprises that may struggle against larger, well-established competitors. Regulatory bodies play a vital role in ensuring that market liberalization promotes fair competition without leading to market monopolization.

Public engagement and social acceptance are essential for the successful implementation of renewable energy projects. Enterprises that prioritize transparent communication and involve local communities in decision-making processes tend to gain greater public support, which can expedite project development and reduce opposition. This approach not only addresses societal concerns but also strengthens the relationship between enterprises and the communities they serve. The study identifies several challenges that enterprises must navigate to enhance their effectiveness. Technological barriers, such as grid integration of intermittent renewable sources, require ongoing investment and innovation. Financial constraints, particularly for small and medium-sized enterprises, necessitate access to funding mechanisms and risk mitigation strategies. Policy uncertainties demand that enterprises remain adaptable and vigilant to regulatory changes that could impact their operations.

Opportunities abound for enterprises willing to embrace emerging technologies and new business models. Advancements in digitalization, artificial intelligence, and energy storage technologies offer avenues for enterprises to improve efficiency and create value-added services. International collaboration can facilitate knowledge sharing and open up new markets, contributing to global efforts toward sustainable energy transitions. For policymakers, the findings suggest that creating and maintaining stable, transparent, and supportive policy environments is crucial. Policies should aim to balance the promotion of renewable energy with the realities of market dynamics and technological capabilities. Incentives should be designed to encourage innovation and investment while ensuring long-term sustainability and avoiding boom-and-bust cycles.

Enterprises are encouraged to adopt strategic approaches that include investing in innovation, engaging with stakeholders, adapting to policy environments, diversifying their energy portfolios, and collaborating with partners across the industry. By doing so, they can enhance their resilience, competitiveness, and contribution to sustainable development goals. In conclusion, the effectiveness of energy enterprises in Europe is deeply influenced by their ability to innovate, navigate policy landscapes, compete in liberalized markets, and engage with the public. The European experience offers valuable lessons on how enterprises can align their operations with sustainable development objectives while achieving economic success. By addressing challenges and leveraging opportunities, energy enterprises can play a pivotal role in the global transition to a sustainable and low-carbon energy future.

Future studies could expand on this research by exploring the long-term impacts of emerging technologies such as hydrogen energy and carbon capture on enterprise effectiveness. Additionally, investigating the role of consumer behavior and demand-side management in shaping enterprise strategies would provide a more comprehensive understanding of the energy market. Comparative analyses involving non-European countries could also offer insights into how different cultural, economic, and regulatory contexts influence enterprise effectiveness in the energy sector.

References

1. European Commission. A European Green Deal. Retrieved from https://ec.europa.eu/info/strategy/priorities-2019-2024/european-green-deal_en (2020)
2. M. Zhang, M. Zhou, Y. Huang, M. Zhou, "Efficiency and Productivity Growth of China's Power Generation Industry under Carbon Emissions Constraints", Journal of Cleaner Production **142**, 2905–2915 (2017)
3. M. Tutak, J. Brodny, D. Siwiec, "Studying the Level of Sustainable Energy Development of the European Union Countries and Their Similarity Based on the Economic and Demographic Potential", Energies **13**, 6643 (2020), DOI: 10.3390/en13246643
4. A. V. Komarova, I. V. Filimonova, A. A. Kartashevich, "Energy Consumption of the Countries in the Context of Economic Development and Energy Transition", Energy Reports **8**, 683–690 (2022), DOI: 10.1016/j.egyr.2022.07.072
5. W. Chen, M. Alharthi, J. Zhang, I. Khan, "The Need for Energy Efficiency and Economic Prosperity in a Sustainable Environment", Gondwana Research **127**, 22–35 (2024), DOI: 10.1016/j.gr.2023.03.025
6. A. A. Sozinova, T. N. Litvinova, V. N. Ostrovskaya, G. V. Vorontsova, "Perspective Directions of Corporate Social Responsibility of Energy Companies of EnergyTech for Sustainable and Environmental Development of Energy Economy", Frontiers in Energy Research **10**, 908489 (2022), DOI: 10.3389/fenrg.2022.908489
7. R. Bali Swain, A. Karimu, E. Gråd, "Sustainable Development, Renewable Energy Transformation and Employment Impact in the EU", International Journal of Sustainable Development and World Ecology **29**, 695–708 (2022), DOI: 10.1080/13504509.2022.2078902
8. A. Karasek, B. Fura, M. Zajączkowska, "Assessment of Energy Efficiency in the European Union Countries in 2013 and 2020", Sustainability **15**, 3414 (2023), DOI: 10.3390/su15043414
9. L. Molina Cañamero, E. Peñalvo-López, I. Valencia-Salazar, V. León-Martínez, "Analysis of Current Barriers and Solutions in the Implementation of Energy Efficiency Methods in the European Industry", Renewable Energy and Power Quality Journal **21**, 499–404 (2023), DOI: 10.24084/repqj21.375
10. K. Redko, O. Borychenko, A. Cherniavskyi et al., "Comparative Analysis of Innovative Development Strategies of Fuel and Energy Complex of Ukraine and the EU Countries:

International Experience", International Journal of Energy Economics and Policy **13**, 301–308 (2023), DOI: 10.32479/ijeep.14035

E3S Web of Conferences **633**, 05001 (2025) https://doi.org/10.1051/e3sconf/202563305001
GASES 2025

Spatial distribution of the ozone layer and its relationship with temperature and relative humidity over Iraq for the year 2024 using R.S and GIS techniques

Khudhur Ahmed Hussein[1,2,*], *Ebtesam* F. Khanjer[1]

[1]Department of remote sensing and GIS, College of Science, University of Baghdad, Baghdad, Iraq
[2]Department of of remote sensing and geographic information systems, College of Science, University of Baghdad, Baghdad, Iraq

Abstract. The climate of Iraq is one of the most changeable countries in the Middle East, as well as facing many environmental challenges, the most important of which is the high temperature and humidity above normal, and this may cause depletion and increase of the ozone hole, leading to major problems for humanity and other living organisms. The research aims to study the spatial distribution of the ozone layer and its relationship to temperature and relative humidity on Iraq for the year 2024 using GIS technologies, where data were obtained from satellites at 132 stations recorded by the government website (NASA Earth data). The study selected the whole of Iraq from North to South for the year (2024) over four semesters (winter, summer, autumn, spring). I use geographic information systems software to collect, process and analyze spatial data and use the SPSS (Version 26) program in statistical analysis to find out the correlation coefficient and the strength of the relationship between them. After obtaining the results, it was found that in winter the relationship between heat and humidity with ozone and the strength of the relationship is inverse and average, while in summer the ozone rate is low and the correlation relationship is also inverse and strong, and we noticed that in spring the ozone gas rate was high and the correlation relationship is directly and weak with heat and humidity, and in autumn the strength of the correlation feed was inverse and strong. The average strength of the correlation of temperature with ozone was more than humidity throughout the year, with the strength of the correlation of heat with ozone in the summer (0.57 -), in winter (0.58 -), in autumn (0.73-) and in spring (0.11).

1 Introduction

The ozone layer plays a vital role in protecting our planet, absorbing about 90% of harmful rays, as the ozone layer is concentrated in the lower part of the stratosphere [1], at an altitude of 10 to 50 kilometers, the thickness of the ozone layer varies depending on the seasons and geographical location, as it is a protective shield that protects the planet from

* Corresponding author: Khedr.Ali2309@sc.uobaghdad.edu.iq

harmful ultraviolet radiation. With the increase in climatic changes, understanding the factors affecting the thickness of this layer has become extremely important. This research aims to study the influence of basic climatic elements, such as specific heat and humidity, on the thickness of the ozone layer, using the latest available technologies in the field of remote sensing and geographic information systems [2]. This research will contribute to enhancing our understanding of the complex interactions between the climate and the ozone layer, helping to develop better strategies for Environmental Protection. Therefore, stratospheric ozone plays a pivotal role in understanding the physical and chemical processes occurring in the atmosphere, studies have well documented the depletion of stratospheric ozone since the seventies of the last century, and this was directly related to human activities. These studies were based on the analysis of aggregate ozone data and ozone characteristic data. Studies conducted in recent years have also shown a sharp decrease in the concentration of ozone over Antarctica during the spring, as a result of human activities. These results were supported by numerous studies based on ground-based measurements and Satellite Data [3]. Studies have shown that the concentration of ozone at ground level is clearly shown by the change of seasons, with the maximum concentration occurring during the spring and low levels during the winter for all the sites studied. Ozone gas also affects the dye used to pollute textiles, and destroys rubber, especially rubber used in the manufacture of automobile tires, the air quality specifications by the US Environmental Protection Agency (EPA) state.

Table 1. The effect of ozone concentration by exposure period.

Ozone concentration in the air ppm	Period of exposure	Influence
0.02	1h	Destruction of substances that include rubber in their composition
0.03	8h	Harmful effects on plants
0.05	1h	Irritation of the nose, throat, dryness of the mucous part of the respiratory tract
0.1	1h	Irritation of the bronchi of the respiratory tract
0.3	(8>)h	Skin irritation
2	2h	Severe cough and chemical changes in the proteins of the lungs
10-5	(1/4)h	Acute inflammation and pulmonary hemorrhage

2 Materials and methods

2.1 Study area

The study included 132 stations in Iraq belonging to the NASA Space Agency from the (Earth data) site [6], where Iraq is located in western Asia, and extends over large parts of the Zagros Mountain range, the Syrian desert and the Arabian Peninsula (Figure (1)). It is characterized by the diversity of its terrain between mountains, plains and deserts, which greatly affects its climate. The climate of Iraq is generally characterized as hot and dry continental, with long hot summers and short cold winters. Temperatures vary greatly between regions, with the highest temperatures recorded in desert areas and the lowest in mountainous areas, and the climate of Iraq is influenced by its diverse geographical location, as it extends over parts of the Zagros Mountain range and the surrounding deserts. While mountainous areas have a relatively mild climate, desert areas suffer from extreme heat and prolonged drought [4].

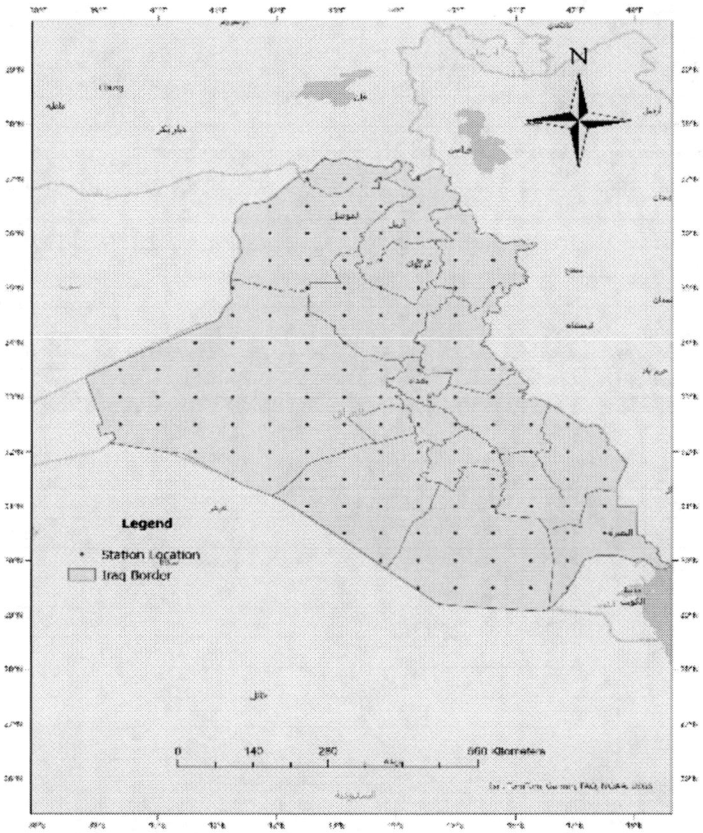

Fig. 1. 132 station sites in Iraq (2024).

2.2 Methodology and data collection

The research focuses on studying the difference in the spatial distribution of ozone gas in Iraq and its relationship with heat and Specific humidity using remote sensing techniques and geographic information systems, where the IDW algorithm was used one of the ArcGIS Pro techniques [5], through which predictive maps were created showing ozone depletion for four seasons and then a statistical analysis was conducted to understand and clarify the relationship between:

1. Acquiring ozone, temperature and Specific humidity data through satellites for the NASA Earth data website [6].

2. The stage of entering and processing ozone, temperature and Specific humidity data in ARC GIS pro by projecting data with their locations in the form of points (point) without relying on latitude and longitude data with Metric accuracy using the program GIS pro. As shown in Figure (2) and Table (1).

Table 2. Sample Virtualization station data for ozone, temperature and specific humidity.

FID	Lon	Lat	Total ozone (Dobson) 2024	Temperature (Kelvin) 2024	Humidity 2024
0	44.38	30	263.509	280.848	0.005628
1	45	30	263.97	282.204	0.005241
2	45.63	30	264.376	283.169	0.004862
3	46.25	30	264.548	282.966	0.0047

Continuation of Table 2.

FID	Lon	Lat	Total ozone (Dobson) 2024	Temperature (Kelvin) 2024	Humidity 2024
4	46.88	30	264.611	282.001	0.004637
5	43.75	30	264.759	280.282	0.005388
6	44.38	30	265.427	281.493	0.004991
7	45	30	265.622	283.239	0.004772
8	45.63	30	265.739	283.329	0.004774
9	46.25	30	265.927	282.38	0.004759
10	46.88	30	266.267	281.762	0.004837
11	43.13	31	268.521	280.118	0.0054
12	43.75	31	267.837	280.637	0.004906
13	44.38	31	267.904	282.634	0.004648
14	45	31	268.234	283.341	0.00469
15	45.63	31	268.335	282.614	0.004887
16	46.25	31	268.279	282.052	0.004936
17	46.88	31	268.249	281.466	0.005108
18	47.5	31	268.751	281.88	0.005165
19	42.5	31	272.642	279.942	0.00552
20	43.13	31	271.54	279.95	0.005358
21	43.75	31	271.267	281.016	0.004814
22	44.38	31	271.53	282.977	0.004688
23	45	31	271.579	282.755	0.004871
24	45.63	31	272.091	281.876	0.005062
25	46.25	31	271.952	281.528	0.004969

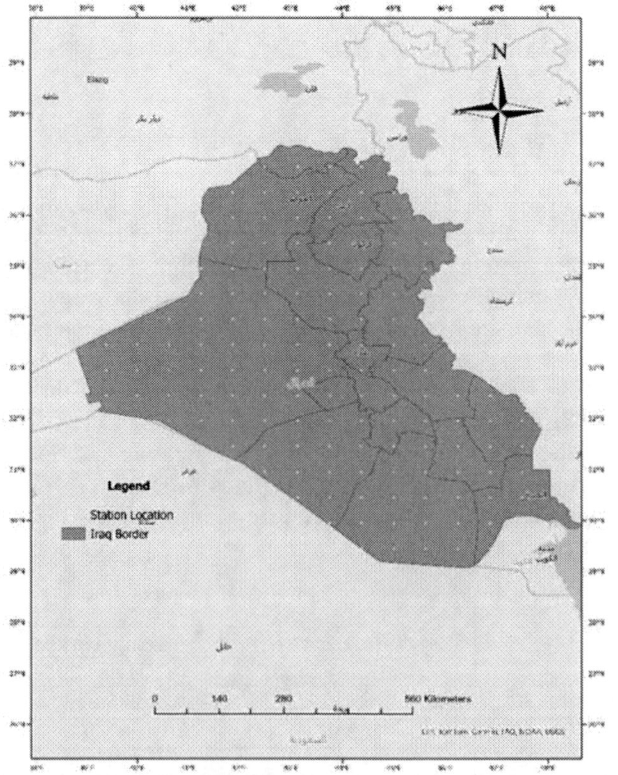

Fig. 2. Virtualization station data for ozone, temperature and specific humidity (2024).

3.The inclusion of the four-month quarterly data (January, April, July, November) representing the four quarters.

4. Using the IDW algorithm within the interpolation completion tools of the Arc GIS pro program to predict the pattern of ozone gas distribution and its relationship to the relative temperature and humidity for a year 2024.

5. Work on the analysis of layers resulting from the application of the algorithm (IDW) To estimate an unknown value at a given point based on known values of other neighboring points [7, 8].

6. Using the statistical program SPSS (Version 26) to understand the type of relationship of climatic elements with ozone gas by using the regression coefficient to clarify the relationship and the strength of the correlation (R-squared) is a statistical measure used in regression analysis to evaluate the quality of the model created. Simply put, R-squared refers to the percentage of variance in the dependent variable that can be explained by the independent variables in the model.

3 Results and discussion

The distribution of ozone gas is associated with many climatic elements, the most important of which are heat and humidity, and the effect varies from season to season, and in the recent period, sensitive devices have been used to monitor the concentration of ozone gas [9, 10]. It turned out that air pollution is not limited to environments only, but extended at regional levels, Where ArcGIS PRO v 0.3 was used in order to model all these variables to predict ozone depletion and influential climatic elements, which is considered one of the variables that cause ozone depletion, and since the aim of the research is to find out the relationship between ozone and climatic elements and their impact, and:

3.1 The behavior of ozone gas distribution at the study stations

Figure (3,4) show the behavior of ozone distribution during the year 2024 at the study stations from NASA data. Which represents the ozone change during four seasons, where it was noted that the ozone percentage is always higher in northern Iraq and begins to decline in the center and then the south, where the standard deviation rate of ozone concentration in the winter was (16.6) Dobson unit in the whole of Iraq, this means that there is a large variation in the thickness of the ozone layer may be the result of natural factors, as for the less concentrated governorates (Muthanna, Basra, Nasiriyah) and the more concentrated governorates (Duhok, Erbil , Nineveh).The standard deviation rate of ozone concentration in the autumn was in the Dobson unit in the whole of Iraq and this indicates the variation of ozone distribution is average compared to other seasons and this is due to the relative stability in weather conditions and the least concentrated governorates are (Muthanna , Basra, Nasiriyah) and the highest concentrated governorates (Nineveh , Dohuk , Erbil), and the standard deviation rate of ozone concentration in the summer (3.6) in the Dobson unit in the whole of Iraq and this indicates the variation of ozone distribution is low compared to this is due to the weather conditions , and the least concentrated governorates are (Muthanna , Najaf , Basra) and the most concentrated governorates (Nineveh , Dohuk , Erbil) , The standard deviation rate of ozone concentration in the spring was (6.81) in the Dobson unit throughout Iraq, and this indicates the variation in the distribution of ozone is average compared to other seasons, and this is due to the relative stability in weather conditions , and the least concentrated governorates are (Karbala , Anbar , Babylon) and the most concentrated governorates (Basra, Muthanna, Duhok).

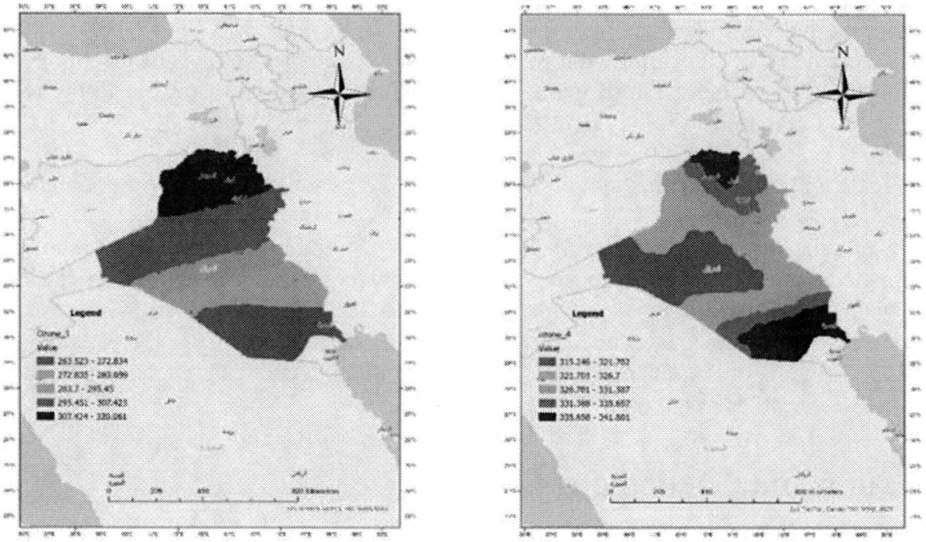

Fig. 3. Layer thickness Ozone data for January and April.

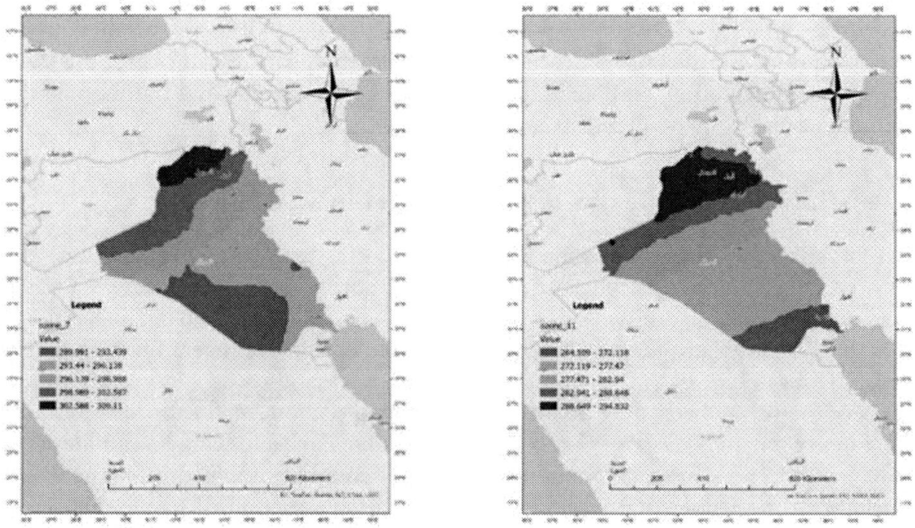

Fig. 4. Layer thickness Ozone data for July and November.

3.2 The behavior of the distribution of temperature and relative humidity at the study stations

The temperature was based on an altitude of 2 meters from the Earth's surface and the relative humidity at an altitude of 10 meters from the Earth's surface, where the highest temperature in the summer was the average rate of (302.03) Kelvin and the average autumn temperature (289.8) Kelvin and spring (287.04) Kelvin and the lowest average temperature in Iraq was in winter at (280.03) Kelvin as shown in Figure (5) and where the relative humidity in summer and spring was higher in southern and central Iraq while in winter and autumn were higher

 https://doi.org/10.1051/e3sconf/202563305001

in western Iraq as shown in Table 2,3 and Figure (6). Figures (7-12) show humidity data and temperature/humidity regression coefficient on each season.

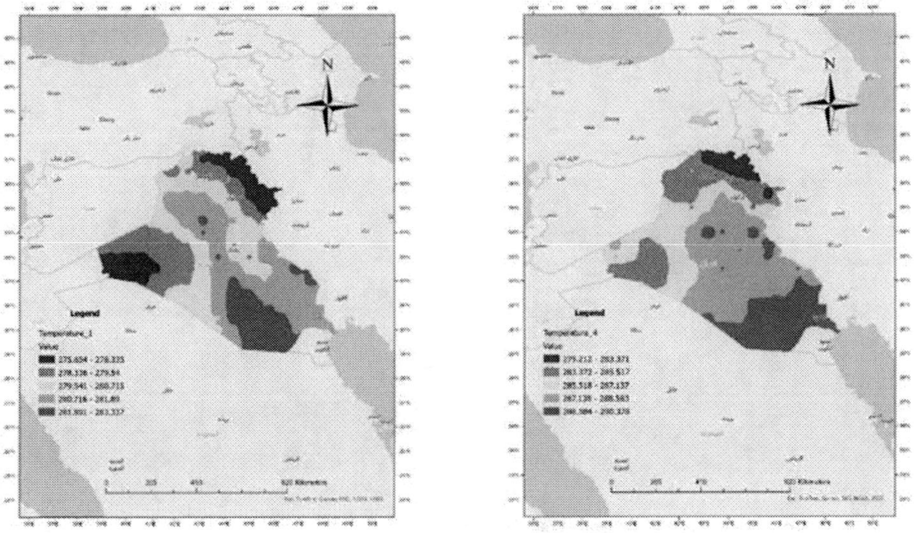

Fig. 5. Temperature data for January and April 2024.

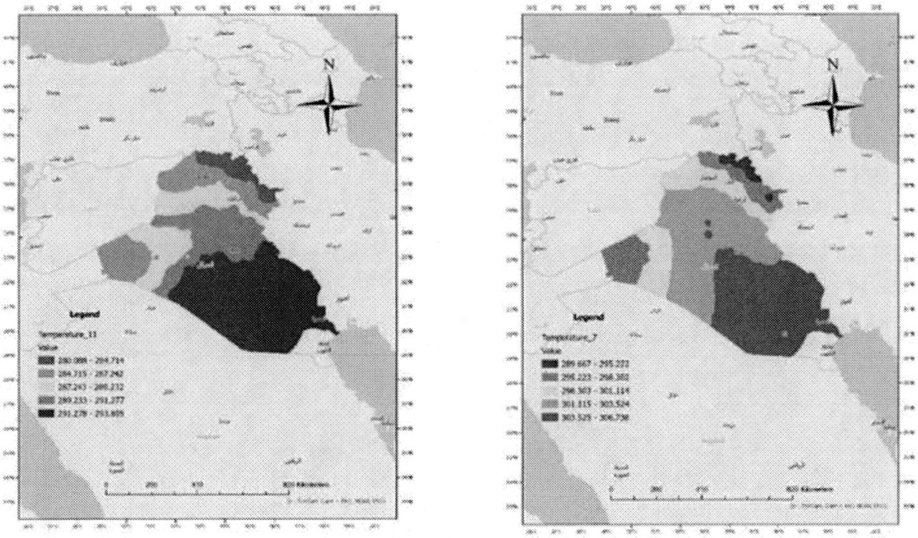

Fig. 6. Temperature data for November and July 2024.

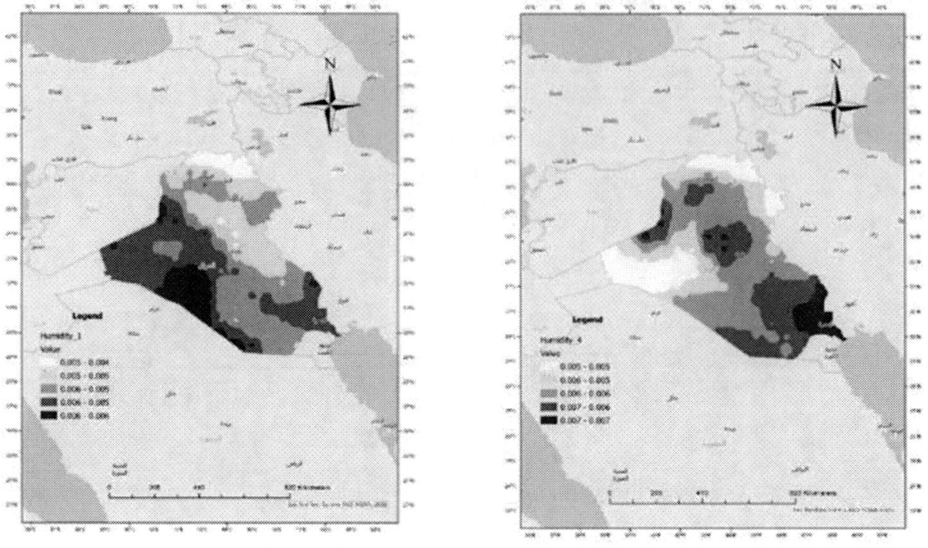

Fig. 7. Specific humidity data for January and April 2024

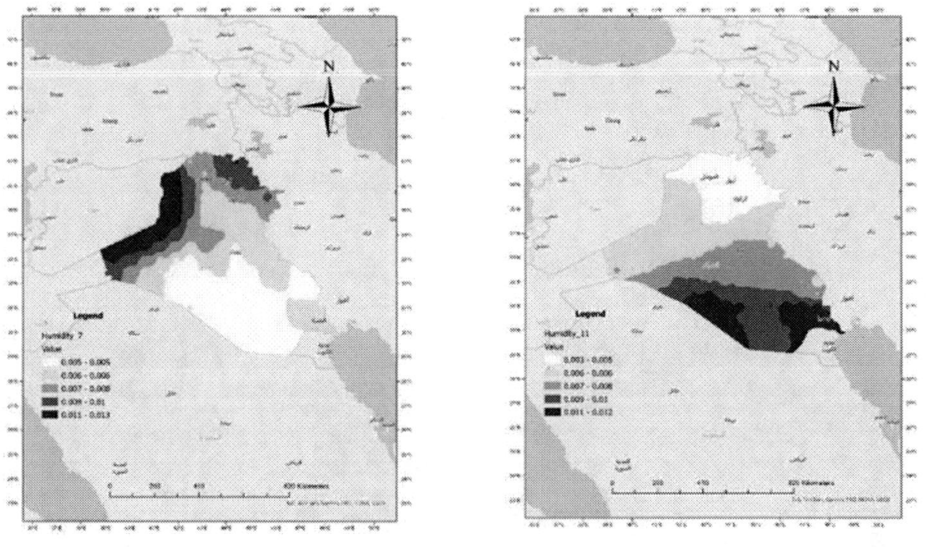

Fig. 8. Specific humidity data for November and July 2024.

3.3 Statistical analysis

Using statistical analysis to find out the strength and type of relationship of the two variables temperature and relative humidity to ozone using statistical tools in the SPSS (Version 26) program shows the following:

Winter: the relationship was the inverse of temperature, humidity and the presence of a strong correlation factor.

Summer: the relationship was inverse to temperature and humidity and the presence of a correlation factor between them.

Spring: the relationship was a function of temperature, humidity and the presence of a weak correlation factor.

Autumn term: the relationship was inverse to temperature, humidity and the presence of a strong correlation factor.

Table 3. A table showing the relationship of the climatic elements of the four seasons with ozone.

Season	Correlation Temperature (2024)	Correlation Humidity (2024)
Winter	-0.5788	-0.384
Summer	-.58.8	0.28
Spring	0.11	0.28
Autumn	-0.73	-0.79

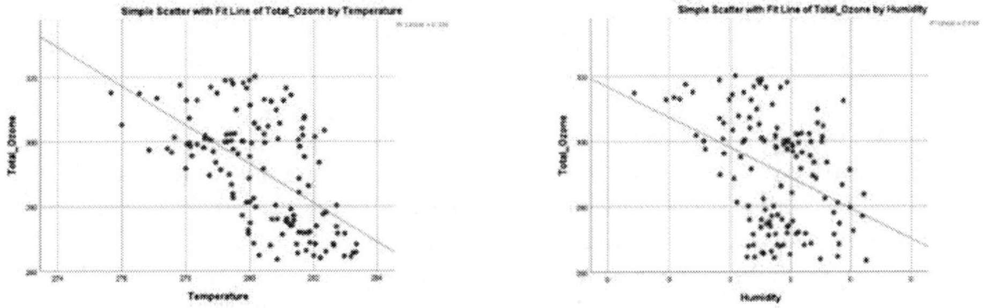

Fig. 9. Winter temperature and humidity regression coefficient (2024).

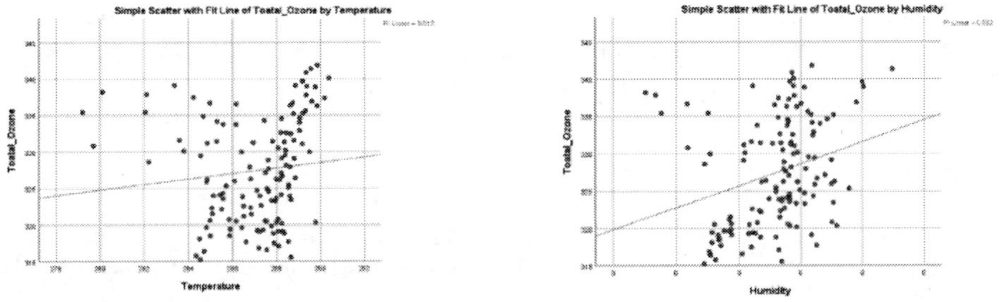

Fig. 10. Spring temperature and humidity regression coefficient (2024).

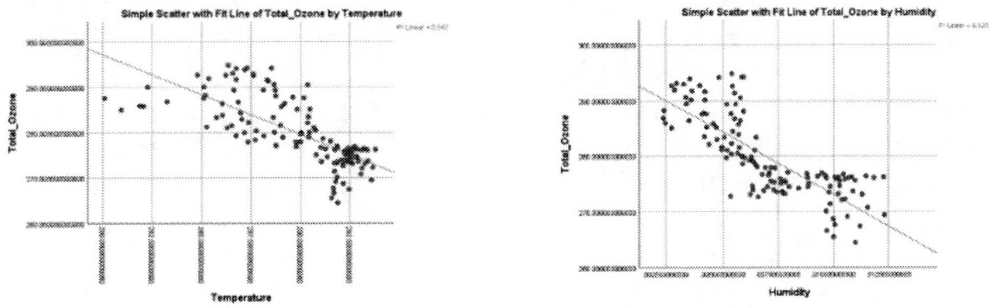

Fig. 11. Autumn temperature and humidity regression coefficient (2024).

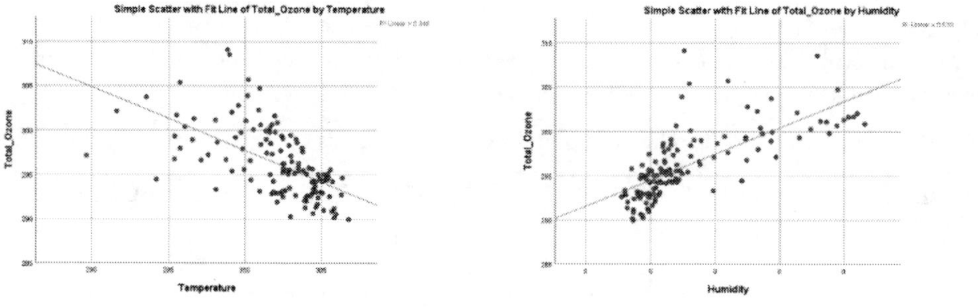

Fig. 12. Summer temperature and humidity regression coefficient (2024).

4 Conclusion

The study showed that the concentration of stratospheric ozone at the ground ozone level is clearly shown by the variability of the four seasons, with the maximum concentration occurring during the spring and a noticeable decrease in summer for all studied sites, which means that high temperature and humidity are responsible for the formation of high ozone concentrations .This study proved that stratospheric ozone values are high during the temperate seasons of ozone gas many climatic influences, and it turned out that these two variables temperature and relative humidity have an important active role and are associated with ozone throughout the seasons, where there is a relationship between the two variables humidity and temperature with ozone gas values, after conducting an analysis. Statistically, it was found that ozone is most concentrated in the spring and least concentrated in the Winter (Spring> Summer> Autumn >Winter).

References

1. H. N. Huynh, V. F. McNeill, "The potential environmental and climate impacts of stratospheric aerosol injection: a review," Environ. Sci. Atmos. **4(2)**, 114–143 (2024)
2. X. Lyu et al., "A synergistic ozone-climate control to address emerging ozone pollution challenges," One Earth **6(8)**, 964–977 (2023)
3. P. Ilić, Z. Popović, D. Nešković Markić, “Assessment of Meteorological Effects and Ozone Variation in Urban Area”, Ecological Chemistry and Engineering **S 27**, 373–385 (2020), DOI: 10.2478/eces-2020-0024
4. M. Petrus, C. Popa, A. M. Bratu, “Determination of Ozone Concentration Levels in Urban Environments Using a Laser Spectroscopy System”, Environments **11**, 9 (2024), DOI: 10.3390/environments11010009
5. K. Wanghe, et al., "FRESF model: An ArcGIS toolbox for rapid assessment of the supply, demand, and flow of flood regulation ecosystem services," Ecol. Indic. **143**, 109264 (2022).
6. Ozone and climate data, (2024), https://www.earthdata.nasa.gov
7. A. Gufron, et al., "A preliminary LSTM-IDW model for spatiotemporal hourly solar radiation estimation in tropical regions," Renew. Sustain. Energy Transit. **7**, 100105 (2025)
8. F. A. Canales, M. Payares-Fontalvo, H. Florez-Guerra, G. J. Acuña, "Geographic information systems (GIS) tools in complementarity research-estimation and

visualization," in *Complementarity of Variable Renewable Energy Sources*, edited by J. Jurasz, A. Beluco, 81–99 (Elsevier, 2022)

9. P. Newman, G. Morris, "Ch. 6.3 The Brewer-Dobson Circulation," in *Stratospheric Ozone – an Electronic Textbook*, edited by R. M. Todaro (NASA's Goddard Space Flight Center Atmospheric Chemistry and Dynamics Branch, 2019)
10. A. H. Anwer et al., "State-of-the-art advances and challenges in wearable gas sensors for emerging applications: innovations and future prospects," Chem. Eng. J. **502**, 157899 (2024)

E3S Web of Conferences **633**, 05002 (2025) https://doi.org/10.1051/e3sconf/202563305002
GASES 2025

Spatial modeling for predicting and identifying levels of climate-related disease hazards in Iraq using (GIS) Kirkuk as a model

Areej Abdlwahd Jamel Alqdori*

Al-Mustansiriya University, College of Basic Education, Department of Geography, Baghdad, Iraq

Abstract. Particulate matter (PM2.5) concentrations are a serious concern for human health. In this study, a multinomial model was proposed to predict PM2.5 concentrations, and the relationship between PM2.5, PM10, and atmospheric parameters was studied. The study was conducted in northern Iraq, including Kirkuk Governorate. Data were collected from different sources and two datasets collected in February 2020 and July 2022 were used. The model was applied to the study area within Kirkuk Governorate: Based on the July 2022 dataset, the average value of the coefficient of determination R^2 within Kirkuk Governorate was estimated to be 0.97; Based on the February 2020 dataset, the average value of R^2 within Kirkuk Governorate was estimated to be 0.98, and the prediction accuracy was 82% for July and 96% for February. Moreover, the health impacts and air quality in the area were found to range from moderate to unhealthy. The aim of this study was to develop a multinomial model to predict PM2.5 concentrations using PM10, humidity, temperature and wind speed as independent variables. The results showed a high correlation coefficient (R^2 = 0.98) between predicted and measured PM2.5 concentrations. The outputs of the (IDW) model indicate that the predicted PM2.5 concentration ranges between (35.92-47.65) µg/m3, which is an unhealthy air quality for sensitive groups in Kirkuk Governorate. The results of the study highlighted the impact of industrial areas and recommended monitoring and reducing exposure to particulate matter and pollutants emitted from factories.

1 Introduction

Atmospheric and fugitive dust—airborne particulate matter emitted during industrial processes such as mining, construction, manufacturing, and transportation—pose substantial risks to human health, environmental integrity, and infrastructure. These particles range in aerodynamic diameter from coarse (> 10 µm) to fine (< 2.5 µm), the latter capable of penetrating deeply into the pulmonary alveoli. Chronic and acute exposures to particulate matter have been epidemiologically and toxicologically linked to increased incidence of asthma, chronic bronchitis, cardiovascular disease, lung cancer, and impaired neurocognitive function. Moreover, fugitive dust often carries adsorbed toxicants—silica, heavy metals,

* Corresponding author: ghufraaneng2017@uomustansiriyah.edu.iq

asbestos fibers, persistent organic compounds—that exacerbate respiratory pathologies and systemic toxicity.

Beyond human morbidity and mortality, dust deposition degrades soil fertility, alters aquatic chemistry, and diminishes photosynthetic productivity in vegetation. Deposition on infrastructure accelerates material weathering and soiling, compromising both aesthetic and mechanical performance of buildings, solar panels, and transportation networks. Ecosystem impacts include soil acidification, metal mobilization, and habitat disruption, which can precipitate loss of biodiversity and altered trophic interactions.

Addressing these multifaceted challenges necessitates targeted research and mitigation strategies. The present study aims to (1) construct predictive models of urban $PM_{2.5}$ concentrations across a heterogeneous study region using linear regression, multinomial modeling, and geographic information system (GIS) techniques; (2) characterize large-scale spatial variability of particulate matter; and (3) elucidate links between PM exposure and human health outcomes. By integrating advanced spatial analysis with empirical monitoring data, this work seeks to inform the design of efficient dust suppression measures—ranging from water-spray dust suppression and chemical stabilizers to improved ventilation, regular equipment maintenance, and community education. Effective mitigation not only will reduce airborne burdens for workers and adjacent populations but also will preserve ecosystem function and infrastructure longevity.

Technological advances in GIS—including machine learning algorithms, cloud-based platforms, mobile data collection, and three-dimensional mapping—now permit near real-time, high-resolution analyses of pollutant dispersion patterns. Such capabilities support dynamic adjustment of control strategies and enable participatory monitoring approaches. Systematic, long-term air quality surveillance, combined with spatial modeling, underpins sustainable management plans that align with epidemiological evidence linking fine particulates to disease outbreaks and chronic health burdens.

2 Materials and methods

Spatial data acquisition and preprocessing involved the compilation of pollutant and meteorological records from five monitoring stations distributed across the Kirkuk Governorate (10139 km^2), northern Iraq. Stations provided point-form measurements of $PM_{2.5}$, PM_{10}, ambient temperature, relative humidity, and surface wind speed. Satellite imagery, aerial photographs, GPS surveys, and digital elevation models complemented in situ data to capture land cover, topography, and proximity to major dust sources (e.g., refineries, unpaved roads).

All datasets underwent quality assurance protocols—completeness checks, outlier detection, and validation against ground-truth observations. Spatial layers were harmonized to a common projection and grid resolution. Vector data (point locations, road networks, administrative boundaries) and raster data (continuous surfaces of elevation, land use, interpolated pollutant concentrations) were managed within ArcGIS 10.3.

Inverse Distance Weighting (IDW) interpolation was selected to generate continuous pollutant concentration fields from discrete station data, leveraging the principle of spatial autocorrelation whereby nearer observations exert greater influence on estimated values. IDW parameters (power, search radius) were optimized through cross-validation to minimize root-mean-square error. The resulting gridded maps facilitated the quantification of spatial gradients, identification of pollution hotspots, and assessment of exposure levels at unmonitored locations.

Quantitative modeling employed multivariate linear regression and multinomial logistic frameworks to relate $PM_{2.5}$ concentrations to predictor variables—including traffic density, industrial activity indices, land use categories, meteorological parameters, and elevation.

Model performance metrics (R^2, Akaike Information Criterion, confusion matrices for categorical predictions) guided selection of the most parsimonious models. GIS techniques enabled visualization of predicted versus observed values and supported spatial diagnostics of model residuals.

The study area's climatic context—hot semi-arid with summer maxima near 50 °C, winter minima approaching 0 °C, annual precipitation between 100–180 mm concentrated in December–April—was incorporated to account for seasonal variations in dust mobilization and pollutant dispersion. Topographic heterogeneity (elevations 115–950 m above sea level) and proximity to the Little Zab River and Lake Al-Azim were also considered, given their influences on local wind patterns and moisture regimes.

By integrating geostatistical interpolation, regression modeling, and GIS-driven spatial analyses, this methodology provides a robust framework for predicting airborne particulate concentrations, assessing human health risks, and guiding strategic placement of dust control interventions across industrial landscapes.

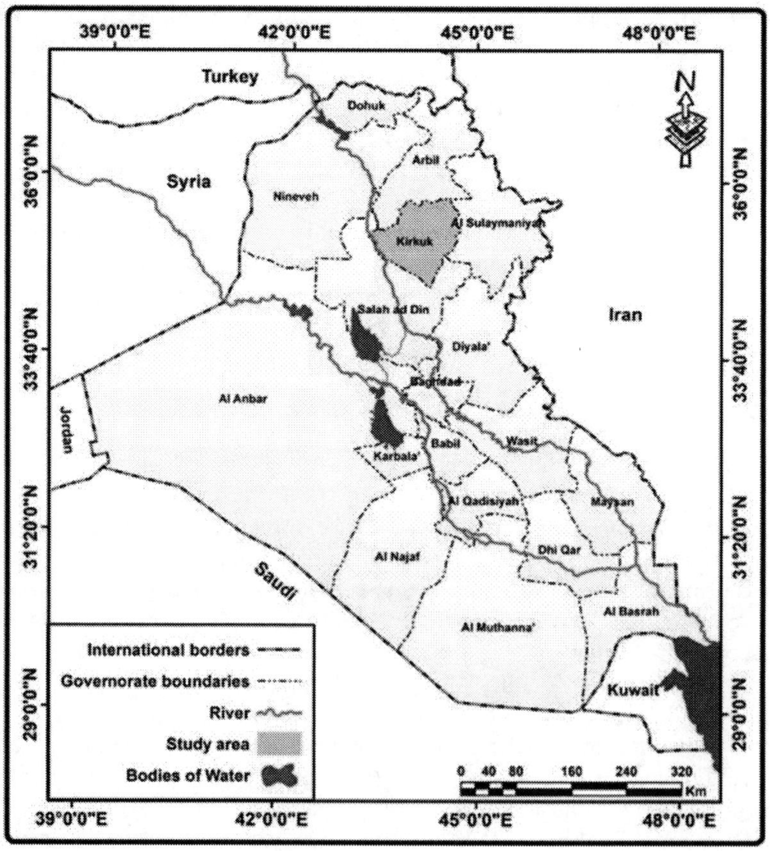

Fig. 1. Study area location.
Source: Ministry of Water Resources, General Authority for Survey 2022.

Furthermore, a remotely sensed image using the Moderate Resolution Imaging Spectrometer (MODIS) was downloaded from NASA satellite data, and additional measurements were applied using a multi-meter air quality meter inside Kirkuk (Figure 2). The information was collected using air quality monitors to collect basic information about open spaces needed to conduct pollution assessments in urban, industrial or rural areas. These monitors contain different sensor technologies to measure outdoor air pollutants such as nitrogen dioxide, ozone and particulate matter. Figure (2) shows the 10/2.5PM laser air quality monitor, which contains a pollution meter sensor, and is characterized by being (AN-5800D) rechargeable with a timer for air filtering in the home, indoor and outdoor places [1-6].

Fig. 2. Laser air quality monitoring device 2.5/10.

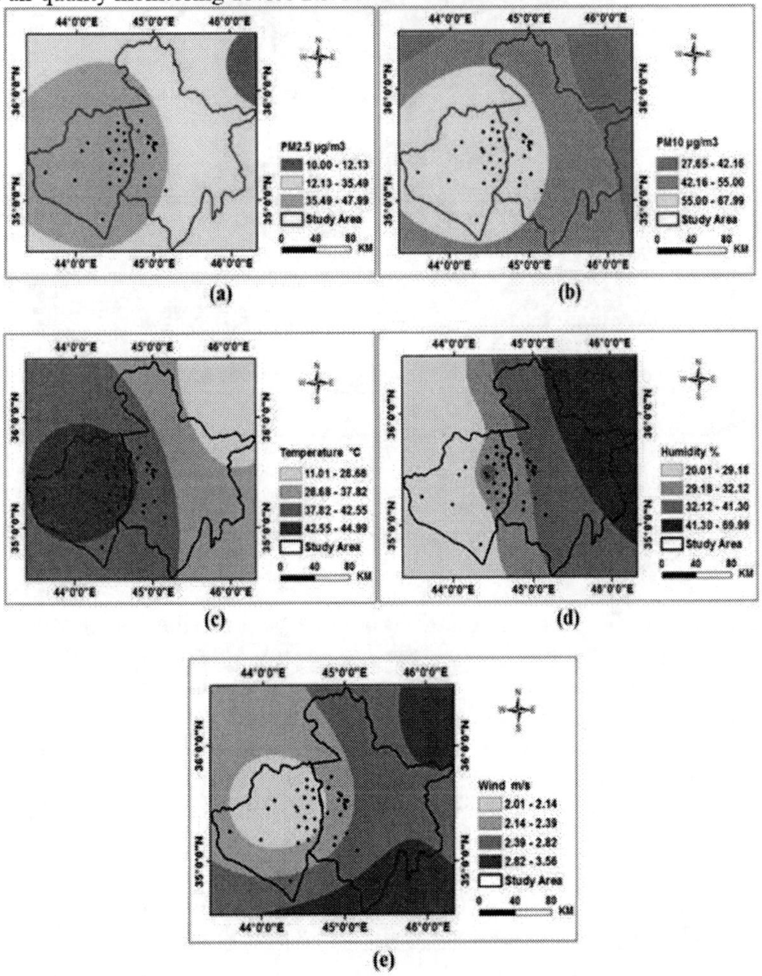

Fig. 3. Map stations of Kirkuk and its surroundings using Arc GIS 10.3 and IDW analysis on the acquired data and stations to measure air quality and weather represented by temperature, humidity, wind and dust particles 2.5 /10 PM.

The soil types in Kirkuk vary, but the most common are light alluvial soil, red-brown soil, and a little clay or sandy alluvial soil, all of which are susceptible to wind erosion [7-11]. These soils are spread around natural valleys in which water flows during the rainy season, and are suitable for rain-fed agriculture and the production of all types of crops.

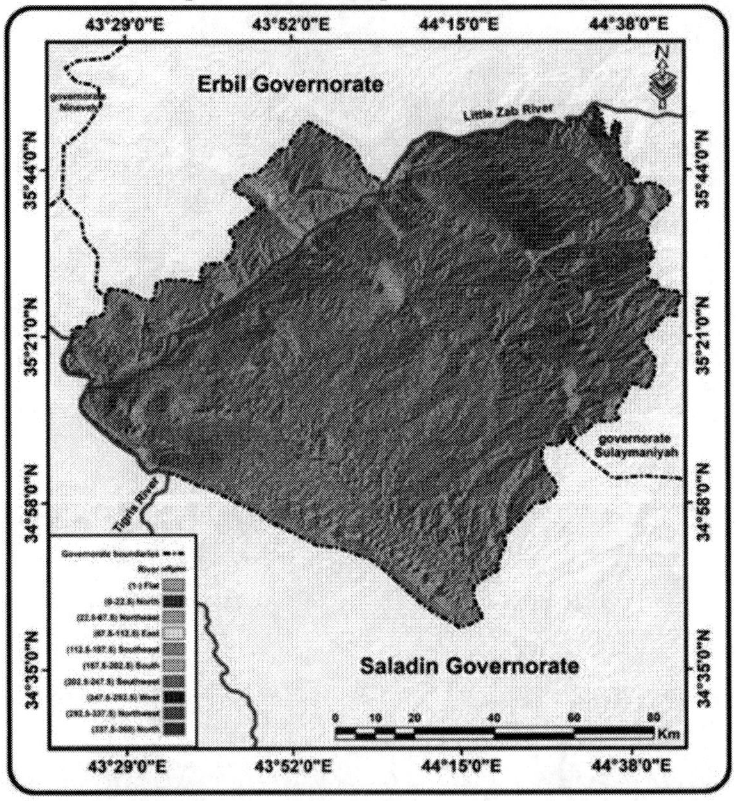

Fig. 4. The trends of surface slopes in Kirkuk.
Source: Unpublished data from the Climate Division, Iraqi Meteorological Directorate, Ministry of Transport and Communications, 2015, based on the accuracy of the Digital Elevation Model (DEM) and the outputs of the Arc GIS 10.3 program.

The performance of the proposed multiple-boundary regression model was assessed using the coefficient of determination (R^2) and the probability value (P-value). A P-value below 0.05 indicates that the association between dependent and independent variables is statistically significant. High R^2 values reflect strong model accuracy, while goodness-of-fit measures (e.g., Akaike Information Criterion) further validate model suitability.

Using July 2022 air-quality and meteorological data, Inverse Distance Weighting (IDW) interpolation in ArcGIS produced continuous surface maps for $PM_{2.5}$, PM_{10}, temperature, relative humidity, and wind speed. $PM_{2.5}$ concentrations ranged from 10.00 to 47.99 μg/m³, with 10.00–12.13 μg/m³ classified as "good" (green), 12.13–35.49 μg/m³ as "moderate" (yellow), and 35.49–47.99 μg/m³ as "unhealthy for sensitive groups" (orange). PM_{10} values spanned 27.65–67.99 μg/m³, predominantly under 55 μg/m³, indicating generally "good" air quality.

Spatial gradients of meteorological variables showed temperatures rising from 11 °C in the northeast to 45 °C in the west, humidity increasing from 20 % in the west to 70 % in the east, and wind speeds varying between 2.0 and 3.6 m/s, with higher speeds toward the northeast and southeast. T-tests confirmed that all parameters correlated significantly with $PM_{2.5}$ ($P < 0.05$).

Multiple linear regression yielded the following predictive equations:

$$\text{July 2022: } PM_{2.5} = -26.6 + 0.275\, PM_{10} + 1.404\, T + 0.124\, H - 5.57\, W \tag{1}$$

$$\text{February 2020: } PM_{2.5} = -125.1 + 1.76\, PM_{10} + 1.78\, T - 1.017\, W \tag{2}$$

Here, T is temperature (°C), H is relative humidity (%), W is wind speed (m/s). Regression outputs demonstrated strong associations between $PM_{2.5}$ and each predictor, with $R^2 = 0.82$ for July 2022 and $R^2 = 0.96$ for February 2020. These values indicate that over 80 % of $PM_{2.5}$ variability was explained by the model, confirming high predictive accuracy.

Spatial mapping of residuals and predicted $PM_{2.5}$ highlighted pollution hotspots in the southern sectors of the study area, coinciding with industrial zones and major transport corridors. February 2020 predictions also revealed widespread "unhealthy" conditions ($PM_{2.5}$ > 35 μg/m³), underscoring seasonal vulnerability. Analysis of IDW outputs for meteorological factors confirmed that temperature exerted the greatest influence on $PM_{2.5}$ levels, followed by PM_{10}, wind speed, and humidity.

Table 1. Regression outputs for July 2022 database.

NO.	Variables	Intercept coefficient	Variable coefficients	P-value
1	PM10	-26.6	0.275	0.002
2	Temperature		1.404	0.003
3	Humidity		0.124	0.012
4	Wind		5.57-	0.0008

Table 2. Regression outputs for February 2020 database.

NO.	Variables	Intercept coefficient	Variable coefficients	P-value
1	PM10	-125.1	1.76	0.002
2	Temperature		1.78	0.0087
3	Humidity		-1.017	0.00068

To mitigate fugitive dust and fine-particle emissions, recent technological innovations and control strategies include:

Dust Control Agents: Environmentally benign chemical additives bind particulate matter, increasing particle mass and reducing airborne dispersion. These agents can be applied to roads, stockpiles, and other exposed surfaces, offering long-lasting suppression under variable weather conditions.

Electrostatic Precipitators: Utilizing high-voltage electrodes to charge and capture suspended particles, electrostatic precipitators effectively remove submicron dust fractions. Widely adopted in cement manufacturing and power-generation plants, these systems achieve high collection efficiencies with low pressure drops and minimal maintenance.

Automated Dust Monitoring Systems: Real-time sensor networks measure particulate concentrations and meteorological parameters, transmitting data to centralized platforms for analysis. Automated alerts enable immediate operational adjustments—such as activating misting systems or temporarily halting dusty processes—to maintain compliance with air-quality standards.

Foam Suppression in Mining: Application of aqueous foam to haul roads and material stockpiles has demonstrated substantial reductions in dust emissions. The foam's enhanced

adhesion and moisture retention limit particle entrainment during loading, transport, and transfer operations.
Sandbag Barriers: Deployed along site perimeters or around high-traffic areas, sandbag or geotextile barriers disrupt wind flow and trap migrating dust. These low-cost, reusable barriers provide flexible containment, particularly suitable for transient construction zones and landfill boundaries.

3 Results

Polynomial modeling of ground-level $PM_{2.5}$, incorporating PM_{10} and key meteorological variables, achieved excellent performance ($R^2 = 0.97$ for July 2022; $R^2 = 0.98$ for February 2020). Using all available weather datasets significantly enhanced predictive precision, while PM_{10} emerged as a robust predictor of $PM_{2.5}$ concentrations.

Health-risk mapping classified much of the study area under “moderate” to “unhealthy” air quality categories. Sensitive populations—particularly those with pre-existing respiratory or cardiovascular conditions—face elevated risks during high-pollution episodes, most pronounced in winter months (February 2020). Spatial analyses pinpointed industrial peripheries and major transport arteries as priority control zones.

Based on these findings, we recommend a multipronged strategy:

1. Increase density of ground stations and integrate satellite-derived aerosol data to capture fine-scale pollution dynamics.
2. Enforce stricter emission limits for industrial and vehicular sources, implementing graduated penalties for noncompliance.
3. Incentivize adoption of renewable energy—solar, wind, geothermal—in place of coal and oil, reducing both greenhouse gases and co-emitted particulate precursors.
4. Develop outreach programs on pollution-avoidance behaviors and real-time air-quality advisories, empowering communities to limit exposure during peak events.
5. Plant windbreaks and vegetation buffers to intercept airborne dust and improve local microclimates.

Forecast maps for July 2022 indicate $PM_{2.5}$ levels between 35.9 and 47.7 $\mu g/m^3$, reinforcing the urgency of targeted interventions. Through the integration of advanced modeling, geospatial analysis, and innovative control technologies, stakeholders can design adaptive management plans that safeguard both human health and environmental integrity as urban and industrial activities continue to expand.).

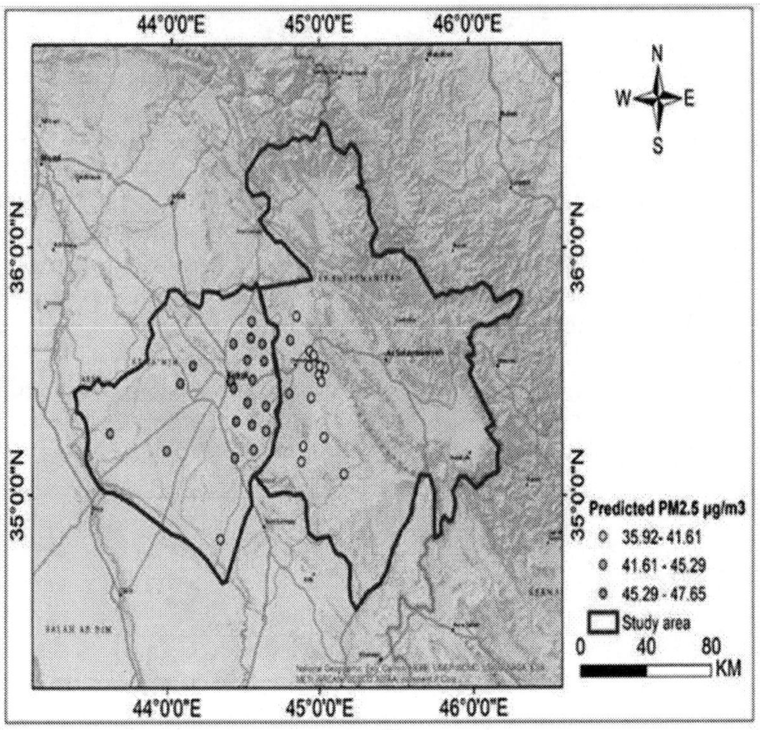

Fig. 5. Prediction map of expected PM2.5 concentrations in Kirkuk city and its surroundings (represented by part of Sulaymaniyah Governorate).

4 Conclusions

Statistical analyses confirmed the robustness of our models, with all hypothesis tests yielding p-values ≤ 0.05, thus rejecting null hypotheses of no association. Predictive mapping of $PM_{2.5}$ concentrations within Kirkuk Governorate indicated values between 35.9 and 47.7 $\mu g/m^3$—levels classified as "unhealthy for sensitive groups." This deterioration in air quality correlates strongly with rapid, unregulated industrial expansion and intensified vehicular traffic, which emerged as the primary contributor to fine-particle emissions. Long-term exposure to $PM_{2.5}$ elevates risks of cardiovascular events, chronic respiratory diseases, and lung cancer, whereas PM_{10} levels remained within "good" to "moderate" ranges, posing less immediate health concern.

Model performance was excellent: multivariate regressions achieved coefficients of determination (R^2) of 0.97 for July 2022 and 0.98 for February 2020, demonstrating high predictive accuracy. Spatial distributions of model residuals highlighted pollution hotspots near industrial zones and major roadways, underscoring the need for geographically targeted interventions. For individuals with pre-existing cardiopulmonary conditions and other sensitive populations, limiting outdoor activities during high-pollution periods is advisable.

Mitigation of fugitive dust and $PM_{2.5}$ emissions necessitates a suite of engineering and administrative controls. Industrial facilities should institute regular equipment maintenance to prevent particulate leaks; deploy dust suppression techniques—such as water sprays, chemical binding agents, and mechanical barriers—during material handling; and implement localized ventilation and filtration systems to capture airborne dust at its source. Provision of appropriate personal protective equipment (respirators, eye protection, gloves) remains essential for safeguarding worker health.

Furthermore, collaborative engagement with environmental engineers, occupational hygienists, and industry peers through conferences and professional associations facilitates the exchange of best practices and advances in dust-control technology. Among suppression technologies, water-based atomizers effectively aggregate and settle fugitive particles in mining, construction, and bulk-material handling; chemical suppressants offer durable binding where water use is constrained (e.g., coal processing); and inexpensive sandbag barriers provide flexible, reusable containment for perimeter dust control in diverse settings.

Our integration of geostatistical interpolation, regression modeling, and GIS spatial analysis proved invaluable for pinpointing emission sources, quantifying exposure gradients, and prioritizing control measures. These findings support the development of evidence-based air-quality management plans tailored to Kirkuk's climatic regime—hot, semi-arid summers and cold, wet winters—and its complex topography. Continued expansion of satellite- and ground-based monitoring networks, coupled with dynamic GIS-driven decision-support tools, will be critical for adaptive management, enabling authorities to evaluate intervention efficacy, refine emission standards, and protect public health as industrial and urban activities evolve.

References

1. M. F. O. Al-Hassani, *Dust phenomena and their impact on respiratory diseases in Al-Muthanna Governorate*, Master's thesis, Al-Muthanna University, College of Education for Humanities, 23 (2019)
2. K. W. Ali, "Spatial suitability for winter tourism in Sulaymaniyah Governorate and ways to develop it using geographic information systems GIS," University of Garmian, College of Education, Department of Geography **6(2)**, 12 (2024)
3. R. S. H. Al-Janabi, "GIS and space visualization applications to study climate change and environmental problems to achieve sustainable development in Iraq," Madad Al-Adab J. Univ. Iraq, Coll. Arts **14(SI)** (2024)
4. M. G. Al-Matar, "Using geographic information systems and spatial analysis to study leukemia in the State of Kuwait for the period 2006-2012," J. Fac. Arts Cairo Univ. **78(3)**, 29 (2018)
5. M. T. M. Al-Saadi, "Spatial modeling to analyze drought levels in southern Iraq using remote sensing and geographic information systems," Cambridge J. Sci. Res. **12**, 14 (2021)
6. H. F. Saadoun, "The impact of climate change on humans and their health from a geographical perspective," Al-Mustansiriya Univ., Coll. Educ. J. **4**, 505 (2019)
7. N. M. Hamed, "Dust phenomena and their effect on asthma in Fallujah and Rutba districts," Madad Al-Adab J. Univ. Iraq, Coll. Arts **13(33)**, 17 (2023)
8. M. H. Abdulateef, et al., "Population displacement in southern Iraq (Basra, Maysan, Dhi Qar) as a model," J. Coll. Basic Educ. **127(30)**, 76 (2024)
9. A. A. Othman, et al., "GIS-based modeling for dam site selection in the Kurdistan Region of Iraq," J. Int. Soc. Geospat. Surv. Georeconnaissance **9(4)**, 244 (2020)
10. R. L. Mustafa, S. O. Rashid, "Geographical potential of winter tourism and establishment of nature reserves in Penguin area," J. Koya Univ. Humanit. **4(1)**, 158–159 (2021)
11. M. F. Saad, "The role of the state in protecting against air pollution in Iraq," J. Environ. Stud. **16**, 94 (2017)

Physicochemical and mineralogical characterization of Sultan Uwais feldspar for sustainable glass manufacturing in earth and environmental sciences

Dinara Buranova*, *Sherzod* Matchanov, and *Elyor* Atashev

Urgench State University named after Abu Rayhan Biruni, 220100 Urgench city, Uzbekistan

Abstract. This study presents a detailed physicochemical investigation of feldspar raw materials from the Sultan Uwais deposit, which are widely applied in glass and ceramic production. Samples were characterized using scanning electron microscopy (SEM) and energy-dispersive X-ray spectroscopy (EDS). The SEM images revealed predominantly angular, medium-sized grains typical of igneous rocks, with feldspar composition determined as (in wt.%): Si – 31.98, K – 8.32, Al – 8.89, Na – 2.16. Additional X-ray phase and infrared spectroscopic analyses confirmed the mineralogical composition comprising quartz (7–10%), microcline (60–65%), albite (21–25%), and calcite (2–3%). These results demonstrate the suitability of the Sultan Uwais feldspar for high-quality glass production. The findings also provide a basis for optimizing resource utilization in the glass industry, contributing to efficient and sustainable raw material management.

1 Introduction

Today, the demand for building materials is increasing day by day. In recent years, it is an important task to provide construction processes in our Republic with building materials produced on the basis of local raw materials. The Republic of Uzbekistan has a large number of raw materials that can be used as building materials. For example, Zinelbulak talc-magnesite raw materials, Tuprokkalya barkhan sand, and raw materials from sandstone and kaolin deposits in the Khojakul raw material zone in the Republic of Karakalpakstan can be cited [1-8].

This scientific work presents the results of modern physicochemical analysis of feldspar raw materials from the Sultan Uvays deposit, which is widely used in the glass and ceramic industry. The study of its chemical and mineralogical composition and scientific substantiation for the use of this valuable raw material in the above-mentioned areas is an urgent issue.

* Corresponding author: dinara.b@urdu.uz

2 Methods

The analyses were mainly carried out using X-ray phase, infrared spectrometry and SEM methods. X-ray fluorescence analyses were carried out on a high-performance Japanese Rigaku NEX SG EDXRF spectrometer. X-ray diffraction patterns were obtained on a computer-controlled XRD-6100 (Shimadzu, Jaran) powder diffractometer under the influence of CuKα radiation. The X-ray diffraction patterns of the samples were analyzed based on the ASTM American index and the tables of X-ray indices of minerals compiled by Giller and Mikheev [9-10].

IR spectrometric analyses were performed on an infrared "IRTracer-100" Fourier-spectrometer (SHIMADZU CORP., Japan, 2017), with a spectral range of 4000÷400 cm^{-1}, a signal sensitivity ratio of 60,000:1, and a scanning speed of 20 spectra per second. The obtained spectrometric parameters were compared with tables of spectrometric parameters of minerals and analyzed in a comparative manner.

3 Results and discussion

At the initial stage of the research, the amount of elements present in the sample was determined based on the X-ray spectrum analysis of the Sultan Uwais feldspar raw sample (Fig. 1).

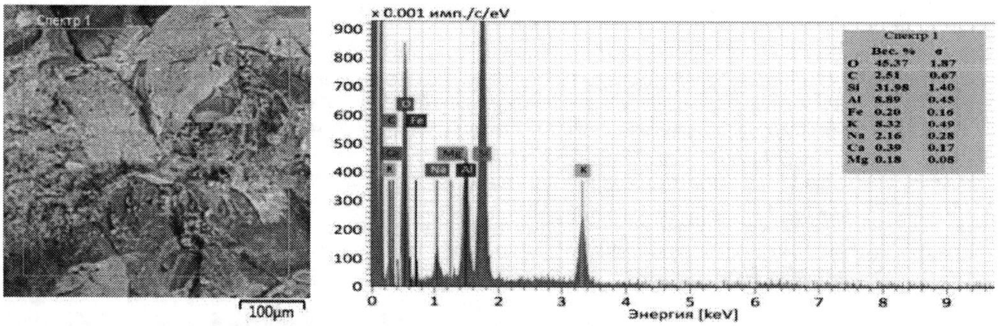

Fig. 1. Picture. SEM image and EDS spectrogram of feldspar from Sultan Uwais mine.

In the image obtained using a Scanning Electron Microscope (SEM), the grains are mainly angular and medium in size, which is a characteristic of igneous rocks. The presence of cracks and voids in the image may be the result of natural weathering processes.

The results of Energy Dispersive Spectroscopy (EDS) in the obtained analysis images show the mass fractions of the elements. The image results revealed that the feldspar composition mainly consists of wt. %, Si - 31.98 % and the elements that form the basis of feldspar are K - 8.32 %, Al - 8.89 %, Na - 2.16 %.

The detection of microcline and albite, determined based on the SEM and EDS analyses, indicated the presence of feldspar. The main components are mica, namely in the form of muscovite and biotite, and in the form of silicon oxide. Also, iron compounds in the sample are present in the form of hematite and magnetite. The composition was also studied for the presence of a small amount of calcite.

Modern physicochemical analysis experiments were conducted to study the chemical and mineralogical composition of the raw material in more detail based on physicochemical analyses and to determine the possibility of using this raw material in the glass industry.

To do this, we conducted HRD analyses of the studies (Figure 2).

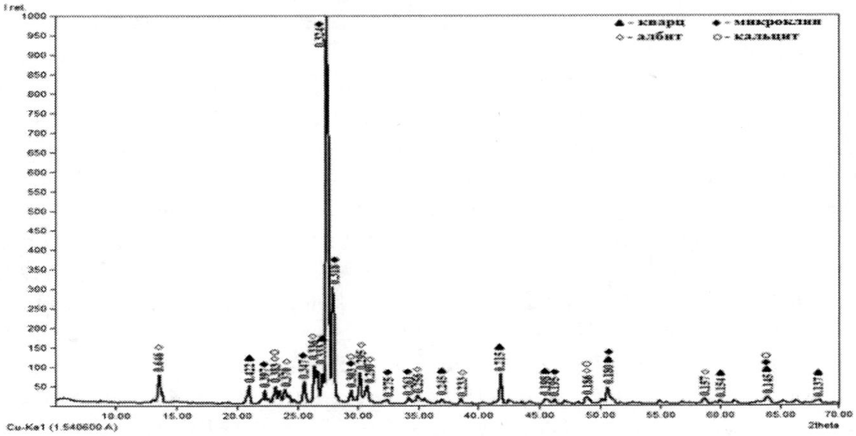

Fig. 2. Radiograph of a feldspar sample from the Sultan Uwais mine.

Analysis Results The mineralogical composition of the samples was studied using the BGMN/Profex Rietveld software package. According to the results of X-ray diffraction analysis, the mineralogical composition of the feldspar raw material was determined as follows: quartz 7.0-10.0; microcline 60.0-65.0; albite 21.0-25.0 and 2-3% calcite.

When analyzing the Sultan Uvais feldspar sample using the IR spectrum, it was found that the general appearance of the IR spectrogram contained absorption bands similar to those of natural aluminosilicate minerals. The obtained IR spectrogram of the sample is presented in Figure 3.

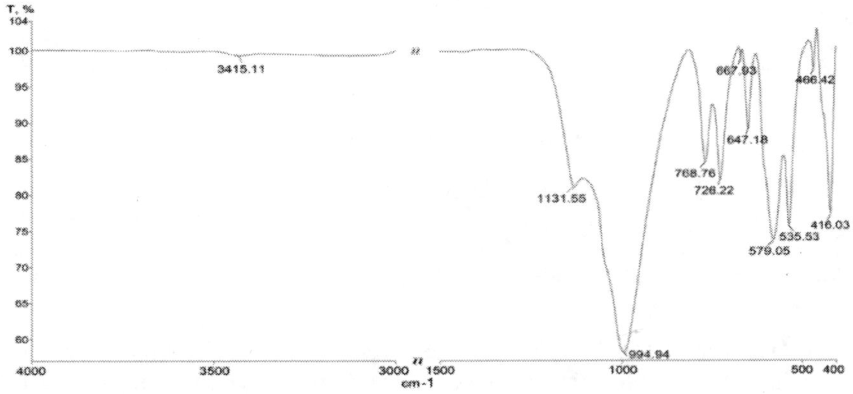

Fig. 3. IR-spectrum of sultan Uwais mine feldspar.

The results of the analysis revealed that the -OH group in the sample was manifested in valence vibrations in the region of 3415 cm^{-1}. The vibration frequencies of the absorption lines of the studied sample at 443 cm^{-1}, 717 cm^{-1}, 726 cm^{-1}, 770 cm^{-1}, 996 cm^{-1} and 1131.55 cm^{-1} belong to various bonding groups of quartz, while the vibrations in the regions corresponding to the frequencies of 416 cm^{-1}, 466 cm^{-1}, 535.53 cm^{-1}, 579.05 cm^{-1}, 647.18 cm^{-1} are characteristic of potassium-sodium feldspar.

The data obtained from the IR spectrum analysis confirm the results of the X-ray phase and SEM analyses of the sample presented above.

It should be noted that, based on the results of the analysis, it was established that the sample fully meets the requirements of GOST 22557-2019 for the mineralogical composition of feldspar. In order to determine the feasibility of using this raw material in experimental studies on obtaining enamel coatings for glass surfaces, they were subjected to thermal treatment at temperatures ranging from 900 to 1300 °C.

In order to determine the formation of new mineral crystal phases with a strong structure in the fired samples as a result of thermal treatment, X-ray and infrared spectrometric analysis methods were used. The determination of the crystal phases of the thermally treated samples was carried out by the method of comparative analysis using existing applications.

In the X-ray image of the initial sample (Fig. 2), it was found that the diffraction maximum with a high intensity at interplanar distances of 0.324 nm corresponds to the microcline mineral. Also 0.397; 0.347; 0.318; 0.303; 0.275; 0.261; 0.195; 0.180; Diffraction maxima at intervals of 0.145 nm were also found to belong to microline mineral. In the X-ray analysis of the sample, 0.422 with low intensities related to the quartz mineral; 0.333; 0.245; 0.215; 0.198; 0.180; 0.154; and diffraction maxima at 0.137 nm were observed.

Diffraction maxima belonging to albite in the composition are 0.646; 0.370; 0.336; 0.290; 0.256; 0.233; 0.186; and occurred at intervals of 0.157 nm. Diffraction maxima between 0.303 and 0.383 nm with relatively low intensity were found to correspond to calcite mineral.

As a result of the X-ray phase analysis of the sample (Fig. 4) obtained after thermal treatment at 900 °C, it was observed that the intensity of the diffraction maxima corresponding to microcline and albite mineral in the initial analysis of the raw material decreased.

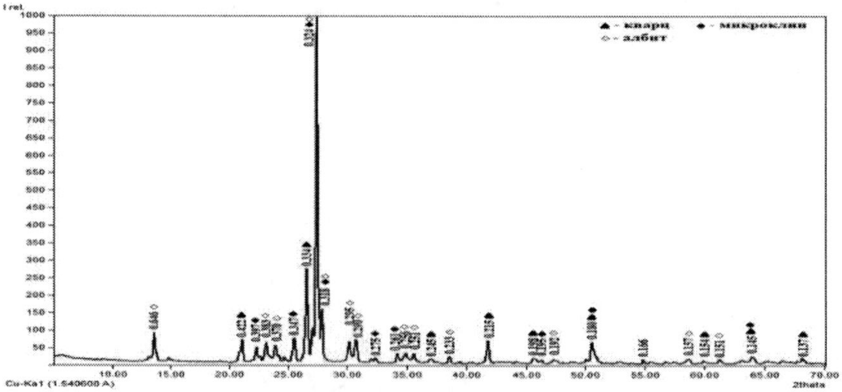

Fig. 4. X-ray image of a sample of feldspar from Sultan Uwais mine burned at a temperature of 900 °C.

It should be noted that in the X-ray diffraction pattern, the diffraction maxima belonging to the calcite mineral completely disappeared, while those belonging to the high-temperature modification of quartz slightly increased. This indicates that the albite and microcline minerals formed a partially liquid phase in the sample, making it amorphous, due to which the intensities of the diffraction maxima belonging to the quartz mineral increased.

X-ray phase analysis was performed on the samples fired at a temperature of 1300 °C at the next stage of thermal treatment, i.e., at a temperature of 1300 °C, the X-ray diffraction pattern of the fired sample is shown in Fig. 5.

 https://doi.org/10.1051/e3sconf/202563306001

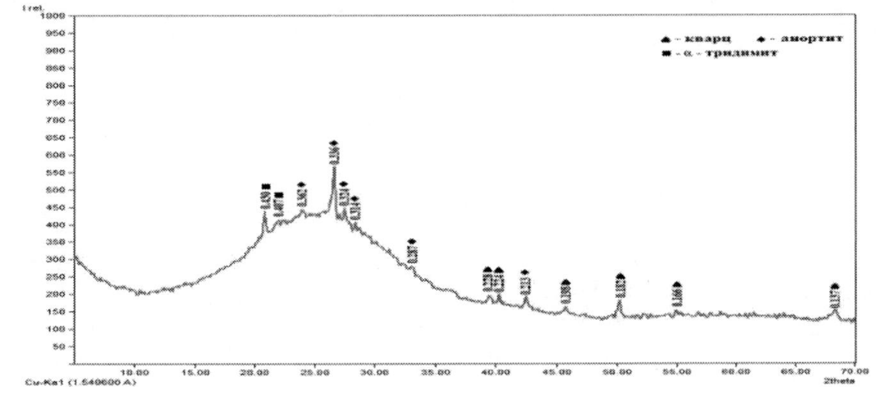

Fig. 5. Radiograph of a sample of Sultan Uwais mine feldspar fired at 1300 °C.

The X-ray diffraction pattern of the feldspar sample fired at 1300 °C showed that it became highly amorphous. The X-ray diffraction pattern with the highest intensity (0.336 nm) was found to belong to the anorthite mineral. Also, the low-intensity 0.378; The diffraction maxima in the ranges of 0.368; 0.351; 0.347; 0.324; 0.314; 0.287; 0.213 nm were also found to belong to the anorthite mineral.

The sample also contained diffraction maxima partially belonging to the high-temperature modification of α-quartz (0.228; 0.224; 0.198; 0.182; 0.166; 0.137 nm) and α-tridymite (0.430; 0.407). Thus, the X-ray diffraction pattern of the original feldspar sample revealed the presence of quartz, feldspar group minerals, albite and microcline, while the X-ray phase analysis of the sample thermally processed at 900 °C revealed a decrease in the intensity of the diffraction maxima of the feldspar group minerals. At this temperature, the low-temperature soluble microcline mineral phases partially transformed from a crystalline state to an amorphous state and the calcite mineral fragmented. As the firing temperature increased to 1300 °C, it was found that the sample mainly transformed into a glass phase.

When thermal treatment is carried out at a relatively high temperature, i.e. up to 1300 °C, it was theoretically observed that the eutectic mixtures of feldspar group and iron silicate minerals in the composition reduce the melting temperature. This indicates that the selected raw material sample can be obtained as a glassy mass at temperatures above 1100 °C.

Also, at this temperature, feldspar minerals such as microcline and albite lose their stability and decompose, forming new phases.

$$CaO + Al_2O_3 + 2SiO_2 \rightarrow CaAl_2SiO_3$$

Quartz undergoes a high-temperature polymorphism to α-tridymite at 870°C. The transition to the amorphous phase, as a result of partial or complete melting of minerals at high temperatures, is also explained by the decrease in peaks in the X-ray diffraction pattern and the flattening of the spectrum.

In order to determine the validity of the X-ray phase changes in the samples of feldspar from the Sultan Uwais deposit that were subjected to thermal treatment at 900-1300 °C, their IR spectroscopic analyses were also carried out (Figs. 6 and 7).

When comparing the IR spectrum analysis of the sample of feldspar from the Sultan Uwais deposit fired at 900 °C (Fig. 6) with the IR spectrum analysis of the initial sample of this raw material that was not subjected to thermal treatment (Fig. 3), a partial change in the intensity of the absorption bands was observed.

https://doi.org/10.1051/e3sconf/202563306001

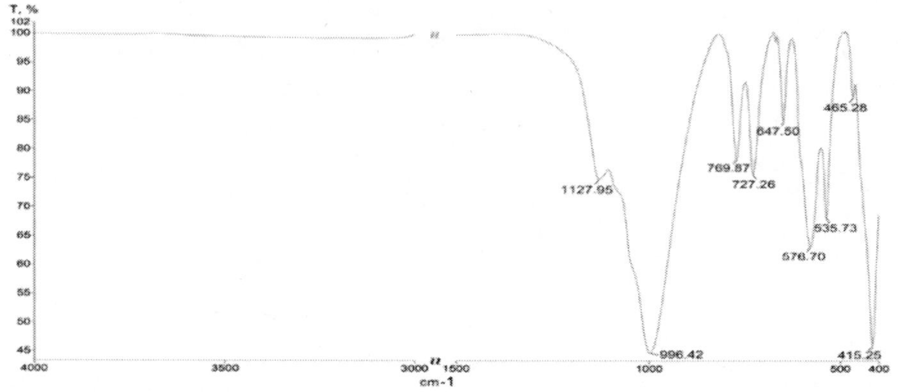

Fig. 6. IR spectrum of a sample of feldspar from Sultan Uwais mine burned at a temperature of 900 °C.

During the research, IR spectroscopic analyses were also conducted on a sample thermally processed at 1300 °C (Fig. 7).

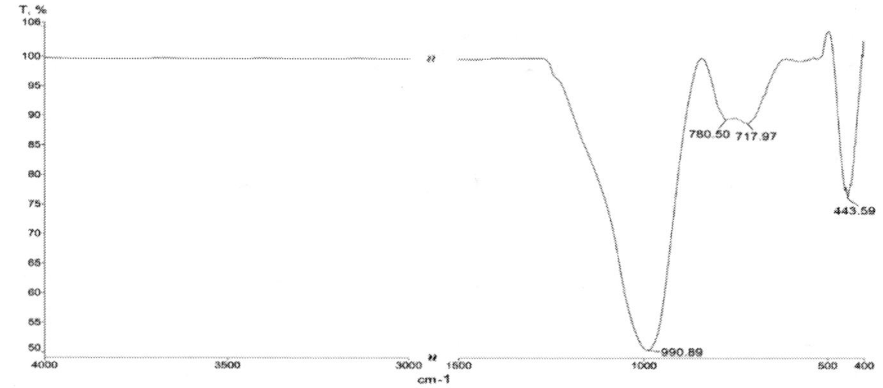

Fig. 7. IR spectrum of a sample of feldspar from Sultan Uwais mine burned at a temperature of 1300 °C.

IR spectroscopic analysis of the sample obtained at this temperature revealed that it mainly consists of vibration frequencies of absorption lines belonging to different bonding groups of quartz (443 cm^{-1}, 717 cm^{-1}, 780 cm^{-1}, 990 cm^{-1}). The decrease in the number of absorption lines is a result of the even distribution of Si and Al ions in the (Si,Al)-O-(Si,Al) bond along the bond lengths.

4 Conclusion

In order to use Sultan Uwais feldspar raw material in the glass industry, as a result of researching its chemical-mineralogical, granulometric composition and physico-chemical properties, the following conclusions were drawn:

The presence of feldspar was indicated by the detection of microcline and albite, which was determined based on the SEM and EDS analysis. It shows that the main components are mica, i.e. muscovite and biotite, as well as silicon oxide. There are also iron compounds in the sample in the form of hematite and magnetite. A small amount of calcite was also found in the composition.

According to the results of X-ray analysis, mineralogical composition of feldspar raw materials in %; quartz 7.0-10.0; microcline 60.0-65.0; albite was found to consist of 21.0-25.0 and 2-3% calcite.

XRD analysis of the raw material revealed that feldspar minerals undergo significant chemical and structural changes at high temperatures. In the sample reheated at 900 °C, calcite decomposes, releasing CO_2, and feldspar minerals partially transform into an amorphous phase. In the feldspar sample reheated at 1300 °C, the minerals decompose, forming new phases - anorthite, and quartz transforms into α-tridymite. The results of IR spectrometric analysis also confirmed that the initial sample was in a hydrated form, containing OH groups, and that at 900 °C the hydroxyl groups of the sample decomposed, and partial phase changes were observed. In the sample reheated at 1300 °C, a transition to an amorphous or glass phase was observed, which proved that this is associated with the liquefaction or recrystallization of feldspar at high temperatures.

References

1. A. M. Eminov, I. R. Boyjanov, Dj. S. Djabberganov, Al. A. Eminov, "Physical and mechanical properties of ceramic tile based on the clay – dune sand composition", Glass and Ceramics 81(3–4), 39–44 (2024)
2. Sh. K. Matchanov, A. Sh. Ruzmetova, Y. X. Yakubov, "Khodzhakul kaolins of Uzbekistan: composition, physical and chemical properties, and processing methods", Obogashchenie Rud **5**, 27–32 (2023)
3. Z. K. Babaev, K. K. Kudiyarova, Sh. K. Matchanov, Sh. Sh. Matchonov, "Feldspars of the Sultan-Uvayskoye deposit: main characteristics and processing methods", Obogasheniya Rud **3**, St. Petersburg (2022)
4. Sh. M. Masharipova, Z. R. Kadyrova, D. P. Danilovich, F. A. Ataeva, Z. A. Masharipova, "Development of ceramic brick compositions for restoration of historical heritages based on loess–defecate composition", Steklo i Keramika **97(5)**, 17–22 (2024)
5. A. Khadzhiev, F. Atabaev, A. Jumaniyozov, Y. Yakubov, "Study on pozzolanic activity of porphyrites of the Karatau deposit", in *E3S Web of Conferences* **563**, 02029 (2024)
6. H. A. Pirnapasova, E. A. Atashev, M. Zh. Zhumaniyazov, "Results of X-ray and IR spectroscopic analysis of talc-magnesite minerals of the Zinelbulok deposit", in *Proceedings of the VII International Scientific and Practical Conference: Modern Research: Theory and Practice* (ISBN 9785-00215-413-5), pp. 28 (2024)
7. A. Petrova, B. Ivanov, "Chemical Durability of Feldspar-Based Glass", Journal of Chemical Engineering **47(2)**, 99–113 (2020)
8. Y. Zhang, Y. Hu, N. Sun, R. Liu, Z. Wang, L. Wang, W. Sun, "Systematic review of feldspar beneficiation and its comprehensive application", Minerals Engineering **128(1)**, 141–152 (2018)
9. Zh. U. Myrhalykov, A. M. Yessirkepova, G. K. Issayeva, B. S. Kulbai, "To the problem of evaluation methods for synergetic effect in secondary resources management on textile industry", Izvestiya Vysshikh Uchebnykh Zavedenii.Seriya Teknologiya Tekstil'noi Promyshlennosti **355(1)**, 5–10 (2015)
10. A. Baumgartner, J. Hayden, A. Schwaighofer, B. Lendl, "In Situ IR Spectroscopy of Mesoporous Silica Films for Monitoring Adsorption Processes and Trace Analysis", ACS Applied Nano Materials **1(12)**, 7083–7091 (2018)

Determination of the composition of the Zinelbulak talc-magnesite deposit rock using modern physicochemical methods

Maxsud Jumaniyazov, *Sherzod* Kurambayev, *Elyor* Atashev*, *Arslon* Jumaniyozov, and *Mokhigul* Buranova

Urgench State University named after Abu Rayhan Biruni, 220100 Urgench city, Uzbekistan

Abstract. In this study, the elemental and oxide composition of talc-magnesite raw materials obtained from the Zinelbulak deposit, located in the Sultan Uvays mountain range of Uzbekistan, were investigated using X-ray fluorescence analysis. The results revealed the presence of the following compounds in mass percentages relative to the total mass: talc ($3MgO·4SiO_2·H_2O$) 60.97%, magnesite ($MgCO_3$) 27.74%, dolomite ($MgCO_3·CaCO_3$) 2.90%, chlorite ($5MgO·5FeO·Al_2(SiO_3)_3·H_2O$) 8.01%, and calcite ($CaCO_3$) 0.38%. The results of the IR spectral analysis of the sample confirm the findings of the X-ray fluorescence and X-ray phase analyses. To further study the composition of the raw material, differential thermal analysis and thermogravimetry methods were employed. A quantitative analysis of particle size during the grinding process of the talc-magnesite raw material from the Zinelbulak deposit was also conducted. The research results indicate the potential for utilizing these new raw material sources in the production of talc and magnesite.

1 Introduction

Uzbekistan possesses abundant raw material reserves essential for producing modern materials across various industries. Notably, the Kulatau clay deposit, the Tuproqqal'a dune sands, the Sulton Uvay feldspar deposit, the sandstone, kaolin, and phosphorite deposits in the Xo'jakul raw material zone, and the Zinelbulak talc-magnesite resources serve as primary raw material bases for diverse industrial productions. Among Central Asian countries, the Zinelbulak talc-magnesite deposit holds particular significance due to its substantial reserves, extensive distribution, feasibility of open-pit mining, and ease of processing. Developing modern technologies for processing the minerals from this deposit could simultaneously facilitate the extraction of talc and magnesite, which are critically needed in many sectors of the national economy [1-5].

Talc is the softest of minerals and is widely used in industry due to its numerous beneficial technological properties. Notably, its white color, high melting point, greasy feel, chemical inertness, ease of grinding, high absorbent properties, fluxing characteristics, fire resistance, as well as excellent heat and electrical insulation capabilities are particularly valued. Talc is

*Corresponding author: elyor.a@urdu.uz

primarily utilized in soap production, the paper industry, plastics and ceramic manufacturing, rubber technology, paint and roofing material production, cosmetics and pharmaceuticals, refractory material production, the chemical industry, metallurgy, cosmetics, baby powder production, and even in the food industry [6, 7].

Magnesite, due to its unique physical and chemical properties, is a highly sought-after material utilized across various industries. Primarily, it is used in the production of refractory materials, special cements, and fire-resistant paints. Additionally, magnesite is employed in the manufacturing of jewelry. Beyond these applications, magnesite plays a role in pulp and paper production, serves as an additive in animal feed, and is used in the production of fertilizers. In recent years, this mineral has been used in the production of new-generation construction boards. These boards offer high sound and thermal insulation, resistance to moisture and cold, and are considered environmentally friendly [8, 9].

In this scientific work, the properties of minerals from the Zinelbulak deposit have been studied using modern physicochemical analysis methods to establish a new raw material base for the critically important and unparalleled talc and magnesite needed in the aforementioned industries. It is known that China, India, Russia, and Brazil are leaders in processing talc-magnesite mineral raw materials. Scientific sources and practical data indicate that during the extraction of talc from talc-magnesite, valuable by-products such as magnesite and chlorite are also obtained [6, 8, 9]

In the central part of the Sultan Uvays mountain range, there is a talc-bearing field composed of ultrabasic, basic, acidic, and alkaline intrusive rocks, forming part of the Paleozoic basement exposed at the surface. The talc deposits of the Sultan Uvays mountains are conventionally divided into three areas: Qizilsoy, Qazg'ansoy, and Zinelbulak[1, 10].

The Zinelbulak deposit occupies the western and southwestern sections of these talc-bearing zones. n this study, the talc-magnesite from the Zinelbulak deposit, located within the Sultan Uvays mountain range, was selected as the research object. The total reserves of raw materials in this deposit are estimated at 23.47 million tons, with 7.74 million tons classified under categories A+B+C1 and 15.73 million tons under category C2 [1, 10, 11]. Additionally, some scientific sources report the reserves as 2.01 million tons (B), 4.77 million tons (C1), totaling 6.78 million tons (B+C1), and 15.73 million tons (C2) [11].

2 Methods

In the analyses, X-ray phase, X-ray fluorescence, infrared spectroscopy, and SEM methods were primarily used. X-ray fluorescence analyses were conducted using the highly efficient Rigaku NEX CG EDXRF spectrometer from Japan. X-ray diffraction (XRD) patterns were obtained using the XRD-6100 powder diffractometer (Shimadzu, Japan), with CuKα radiation, and were computer-controlled. The diffraction patterns of the samples were analyzed based on the ASTM American database and tables of X-ray diffraction indices compiled by Giller and Mikheev [12, 13].

Infrared spectrometric analyses were performed on an "IRTracer-100" Fourier spectrometer (Shimadzu Corp., Japan, 2017), with a spectral range of wavenumbers from 4000 to 400 cm^{-1}, a signal-to-noise sensitivity ratio of 60,000:1, and a scanning speed of 20 spectra per second. The obtained spectrometric indicators were compared and analyzed based on tables of mineral spectrometric indices [14]. The particle size of the crushed samples was determined by laser diffraction using the Shimadzu SALD-7500 (WingSALD II: Version 3.4.1) device [15].

3 Results and discussion

In the initial stage of the research, SEM analyses were conducted on samples obtained from the Zinelbulak talc-magnesite deposit to investigate their microstructure and surface morphological changes, and the analyses were interpreted (Figure 1).

Fig. 1. SEM images of the raw material sample at various magnifications.

In the SEM images, sample particles magnified by 50, 100, 250, 500, and 1000 times can be observed. At the 1000x magnification, large leaf-like or plate-like particles are visible in the center. Additionally, the images clearly show distinct layers. As the magnification decreases, smaller fractions ranging between 10-50 microns appear, which also exhibit layered and coarse, uneven granular or cubic shapes. This indicates the presence of silicates, such as talc and chlorite, which form "layered" structures. The coarse and irregular granular, cubic-like particles in the images are likely associated with magnesite or other hard phases. Based on the SEM images, various morphological characteristics on the sample surface were identified. In particular, signs of strong erosion caused by chemical changes over time were observed on the surfaces of magnesite grains. The presence of talc layers along the boundaries of these grains suggests that, as a result of evolutionary processes and metamorphic changes, thin talc fragments developed around the coarse magnesite grains, partially transforming the magnesite into talc.

In the subsequent stages of research, this raw material was further analyzed using an expanded X-ray fluorescence spectrometer (Figure 2).

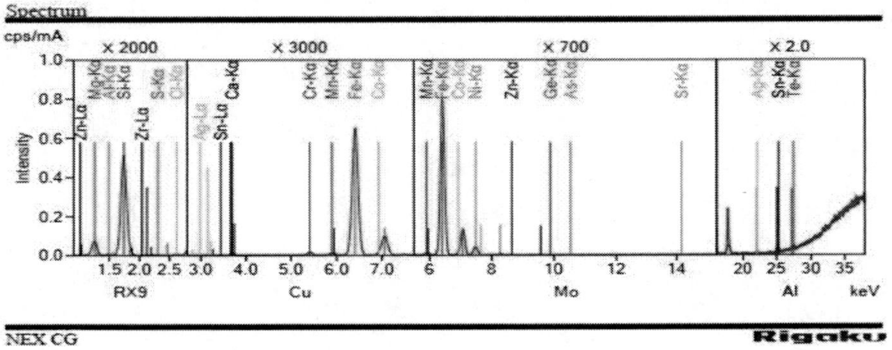

Fig. 2. Comprehensive analysis of Zinelbulak talc-magnesite using an X-ray fluorescence spectrometer.

According to the analysis results, the chemical composition of the talc-magnesite raw material, expressed in mass percentages, is as follows: SiO_2 – 37.6%; MgO – 29.10%; Fe_2O_3 – 5.90%; Al_2O_3 – 1.05%; CaO – 1.00%; SO_3 – 0.08%; MnO – 0.13%; ZrO_2 – 0.15%; NiO – 0.20%; Cr_2O_3 – 0.14%; and loss on ignition – 18.84%. The Zinelbulak talc-magnesite is distinguished from talc ores of other deposits by its higher mass fraction of MgO.

To further investigate the composition of this raw material, X-ray phase analysis was conducted. The results are presented in Figure 3.

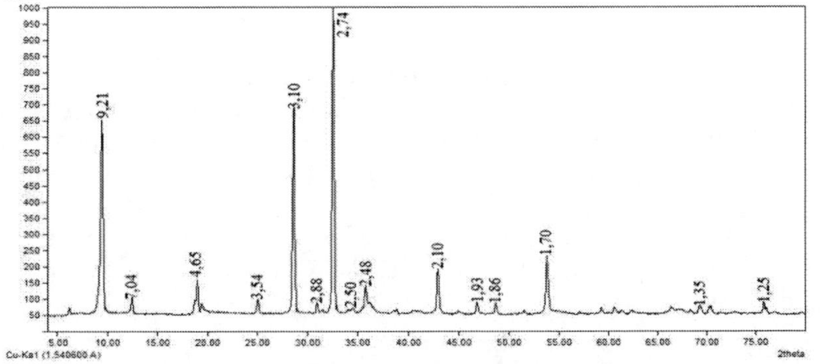

Fig. 3. X-ray diffraction pattern (XRD) of Zinelbulok talc-magnesite raw material.

According to the X-ray diffraction pattern, the peaks with the highest intensity correspond to magnesite at 2.74 d,Å and to talc at 3.10 and 9.21 d,Å. The analysis also identified talc ($3MgO{\cdot}4SiO_2{\cdot}H_2O$) with peaks at 1.65, 1.68, 1.70, 2.08, 2.10, 2.45, 2.48, 2.59, 2.60, 3.54, and 4.65 d,Å. The chlorite compound ($5MgO{\cdot}5FeO{\cdot}Al_2(SiO_3)_3{\cdot}H_2O$) was identified with peaks at 14.21, 7.04, and 4.71 d,Å, while the iron-rich talc compound (minnesotite) showed a peak at 9.59 d,Å.

The magnesite ($MgCO_3$) peaks were identified at 1.70, 1.77, 1.93, 2.10, 2.32, and 2.50 d,Å. Calcite ($CaCO_3$) was observed with peaks at 1.98, 1.86, 1.74, and 2.49 d,Å, while dolomite ($MgCO_3{\cdot}CaCO_3$) exhibited peaks at 1.25, 2.00, 2.54, 2.65, and 2.88 d,Å.

Based on the results of the X-ray phase analysis, the mineralogical quality and quantitative composition of the samples were determined using the BGMN/Profex Rietveld software package. The composition (%) of the analyzed sample is as follows: talc – 60.97%, magnesite – 27.74%, chlorite – 8.01%, dolomite – 2.90%, and calcite – 0.38%.

The IR spectrometric analysis results of the sample also confirmed the above-mentioned X-ray fluorescence and X-ray phase analysis results (Figure 4).

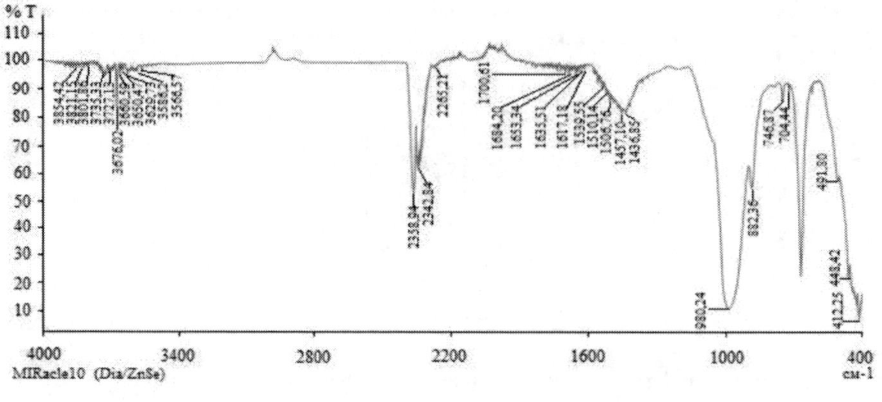

Fig. 4. Infrared spectroscopy analysis of talc-magnesite raw material.

In the IR spectral analysis, the talc crystal lattice Si-O bonds were identified in the low-intensity absorption bands at 980-920 cm^{-1}. Medium-intensity absorption bands at 1700-1600 cm^{-1} were associated with Si-O-Si bonds, while low-intensity bands at 480-450 cm^{-1} corresponded to the metal-oxygen bonds of talc and magnesite, represented by Si-O-Me and Mg-O. This confirms the layered silicate structure of talc.

The asymmetric valence vibrations of CO_3^{2-} ions associated with magnesite, dolomite, and calcium carbonate appeared in high-intensity spectral bands at 1436.85 and 1457.10 cm^{-1}, representing the main spectral region of magnesite. Deformation vibrations were observed in the low-intensity absorption frequencies in the range of 764-746 cm^{-1}, which indicate carbonate structures.

The analysis also revealed strong absorption frequencies for the hydroxyl layers of talc (OH^- ions) in the range of 3000-3700 cm^{-1}. Asymmetric valence vibrations were identified at absorption frequencies of 3676.02, 3660.59, and 3650.47 cm^{-1}, corresponding to the 3Mg, 2Mg+Fe^{2+}, and Mg+2Fe^{2+} bonds, respectively. The X-ray phase and IR spectroscopic data mutually confirmed each other.

In subsequent research, thermal (TG-DTA) analyses were conducted to further clarify the analysis results, as well as to study the thermal stability and decomposition processes of the raw material (Figure 5).

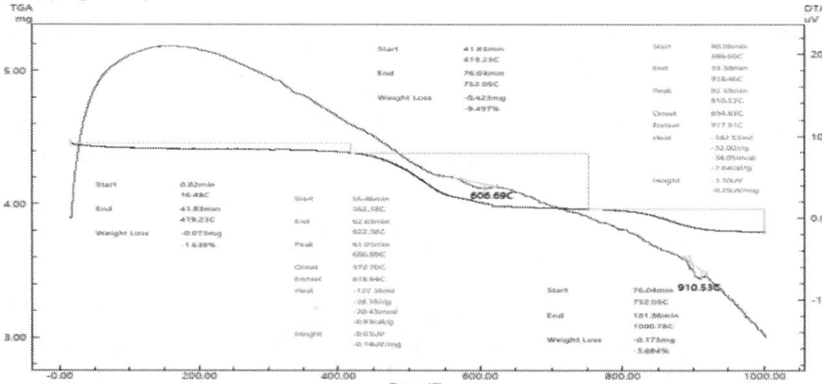

Fig. 5. TG-DTA analysis of talc-magnesite raw material.

TG-DTA analyses indicated that the changes occurred mainly in three stages. Stage 1 was observed at temperatures ranging from 16.46 to 419.32°C. This stage is associated with the release of free and bound water from the raw material sample, resulting in a weight loss of 1.639%, which corresponds to 0.0739 mg. Stage 2 occurred between 419.32 and 752.05°C.

During this phase, the sample lost 0.42 mg, or 9.497% of its weight. The first endothermic process was observed at 606.69°C, corresponding to the thermal decomposition of $MgCO_3$ and the release of CO_2. Stage 3 took place between 752.05 and 1001.78°C. The sample lost 0.1773 mg, or 3.884% of its weight. The second endothermic process was observed at temperatures between 910.53 and 917.63°C, indicating the breakdown of the talc structure and the formation of magnesium silicate (MgSiO), silicon dioxide (SiO_2), and water. This stage corresponds to the release of water and the reformation of crystal lattices.

The TGA curve in the analysis results showed a decrease consistent with the stages, indicating a total mass loss of up to 15%.

When studying the application areas of the raw material, particle size distribution is of particular importance, especially when subjected to mechanical processing and grinding. The particle size distribution of the raw material samples was determined using laser diffraction with the Shimadzu SALD-7500 (WingSALD II: Version 3.4.1) device (Figure 6).

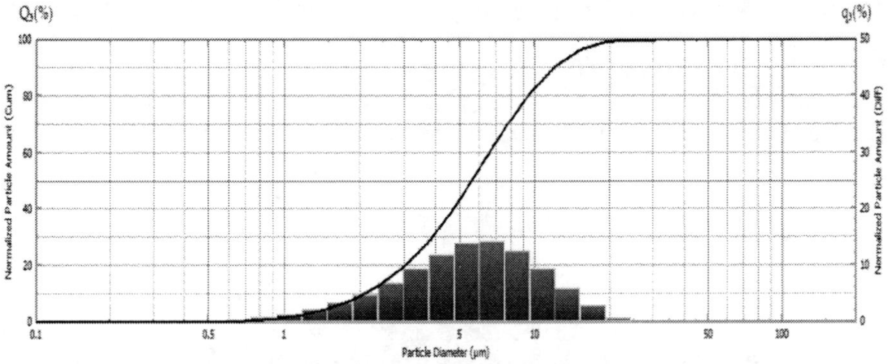

Fig. 6. Histogram of the particle size distribution analysis of ground talc-magnesite raw material.

In this histogram, the X-axis represents the particle diameter, while the Y-axis shows the cumulative distribution. The curve illustrates the percentage of particles smaller than a certain size. It is known that in ceramics and glass production, the material particle size requirement is typically 4-10 μm, while for the paint coating industry, plastics, and rubber manufacturing, the particle size requirement is 5-8 μm.

According to the histogram data, 50% of the crushed talc-magnesite raw material particles are smaller than 7-8 μm, and 90% of them are smaller than 15-20 μm. The proportion of particles smaller than 1 μm and larger than 20 μm is very low. These indicators demonstrate that the raw material meets the GOST 21235-75 standards.

4 Conclusion

The composition of minerals from the Zinelbulak talc-magnesite deposit was determined using modern physicochemical methods. The microstructure and surface morphological changes of the composition were studied using a scanning electron microscope (SEM). The obtained data indicated that the mineral mainly consists of two phases – talc and magnesite. Additionally, minor quantities of minerals such as serpentine, enstatite, quartz, hematite, and magnetite were observed. However, distinct morphological features of these minerals were not detected in SEM images, likely due to their minimal quantity or their fine dispersion across the surface.

A detailed chemical composition analysis using an X-ray fluorescence spectrometer revealed the following composition by percentage: SiO_2 – 37.6%; MgO – 29.10%; Fe_2O_3 – 5.90%; Al_2O_3 – 1.05%; CaO – 1.00%; SO_3 – 0.08%; MnO – 0.13%; ZrO_2 – 0.15%; NiO – 0.20%; Cr_2O_3 – 0.14%; and loss on ignition – 18.84%. Based on X-ray phase analysis, the mineralogical quality and quantitative composition of the raw material were determined as follows (%): talc – 60.97%, magnesite – 27.74%, chlorite – 8.01%, dolomite – 2.90%, and calcite – 0.38%.

In XRD analysis, talc exhibited strong peaks at 3.10 and 9.21 d,Å and lower intensity peaks at 1.65, 1.68, 1.70, 2.08, 2.10, 2.45, 2.48, 2.59, 2.60, 3.10, 3.54, and 4.65 d,Å. IR spectroscopic analysis confirmed the presence of hydroxyl layers (OH^- ions) with absorption frequencies in the range of 3000-3700 cm^{-1}. TG analysis indicated a 1.639% mass loss due to water release in the temperature range of 16.46–419.32°C, confirming the presence of talc.

The presence of magnesite was established through XRD analysis with a high-intensity peak at 2.74 d,Å and moderate-intensity peaks at 1.70, 1.77, 1.93, 2.10, 2.32, and 2.50 d,Å. Comparative analysis using IR spectroscopy and TG-DTA revealed that CO_3^{2-} groups were identified at 1436-1485 cm^{-1} and 2358-2524 cm^{-1} absorption frequencies. The TG-DTA analysis confirmed that magnesite underwent decarbonization in the temperature range of 600-750°C, releasing CO_2 and resulting in a 9.49% total mass loss, leading to the formation of MgO.

Due to talc's low hardness of 1 on the Mohs scale, 90% of the particles in the crushed samples were reduced to sizes smaller than 15-20 µm. This particle size distribution confirms that Zinelbulak talc-magnesite raw materials are suitable for use in ceramics, rubber, plastics, and paint materials production.

References

1. E. A. Atashev, Technology of obtaining magnesium azosuperphosphates using Central Kizilkum phosphorites and Zinelbulok magnesite raw materials, Dissertation abstract for the Doctor of Philosophy (PhD) on technical sciences (Urgench, 2022), p. 44
2. A. M. Eminov, I. R. Boyjanov, Dj. S. Djabberganov, Al. A. Eminov, "Physical and mechanical properties of ceramic tile based on the clay – dune sand composition", Glass and Ceramics **81(3–4)**, 39–44 (2024)
3. D. B. Buranova, M. Yu. Yunusov, Z. K. Babaev, "On the use of cobalt-containing metalurgical waste to obtain colored glass and glass enamel", Glass and Ceramics **79(7–8)**, 300–305 (2022)
4. Sh. K. Matchanov, A. Sh. Ruzmetova, Y. X. Yakubov, "Khodzhakul kaolins of Uzbekistan: composition, physical and chemical properties, and processing methods", Obogashchenie Rud **5**, 27–32 (2023)
5. Sh. M. Masharipova, Z. R. Kadyrova, D. P. Danilovich, F. A. Ataeva, Z. A. Masharipova, "Development of ceramic brick compositions for restoration of historical heritages based on loess–defecate composition", Steklo i keramika **97(5)**, 17–22 (2024)
6. A. Savchenok, L. Hariuzov, "Soapstone (Talc-Chlorite Schist)", in The History of Natural Stone in Saint-Petersburg (2023), pp. 177–181
7. J. Sanz, O. Tomasa, A. Jimenez-Franco, N. Sidki-Rius, "Talc", in Elements and Mineral Resources (2022), pp. 391–393
8. C. Jia, G. Allinson, S. Dai, P. Tai, "Formation process and thermodynamic mechanism of the crust in magnesite mining areas", Arabian Journal of Geosciences **15**, 580 (2022)
9. W. Abdelwahab, A. M. El-Shamy, "Exploring Talc Deposits: Physical Characteristics, Geological Factors, and Formation in Ultramafic and Mafic Rocks - A Comprehensive Review", Chemistry Africa **8**, 43–83 (2024)
10. Z. A. Babakhanova, Technological foundations of the production of refractory materials and composites based on rare earth elements, graphite and talc raw materials, Abstract of the dissertation for the Doctor of Technical Sciences (DSc) (Tashkent, 2019), pp. 61

11. D. Kh. Asabaev, F. A. Badalov, A. M. Ergashov, "Assessment of the mineral resource potential of magnesian rocks and prospects for its rational utilisation in western uzbekistan (southern tan-shan)", Oriental Renaissance: Innovative, educational, natural and social sciences **2(11)**, 259–275 (2022)
12. D. Singh, S. Singh, N. K. Rai, P. K. Singh, "Analysis of Engineering Properties of Weathered Sedimentary Rocks", in *Proceedings of the 2nd International Conference on Geotechnical Issues in Energy, Infrastructure and Disaster Management* (2024), pp. 213–227
13. J. Shaeri, M. J. Mahdavinejad, A. Zalooli, "Physico-mechanical and Chemical Properties of Coquina Stone Used as Heritage Building Stone in Bushehr, Iran", Journal Geoheritage **14**, 95 (2022)
14. S. Ghorbani, S. H. Hoseinie, E. Ghasemi, T. Sherizadeh, "Application of Leeb Hardness Test in Prediction of Dynamic Elastic Constants of Sedimentary and Igneous Rocks", Geotechnical and Geological Engineering **40**, 3125–3145 (2022)
15. M. K. Salah, H. T. Janjuhah, J. Sanjuan, "Analysis and Characterization of Pore System and Grain Sizes of Carbonate Rocks from Southern Lebanon", Journal of Earth Science **34**, 101–121 (2023)

Experiments on bleaching of kaolin from the Khodjakulskoye deposit

Sherzad Matchanov*, *Aida* Ruzmetova, and *Yusufboy* Yakubov

Chemical Technologies, Urgench State University named after Abu Rayhan Beruni, 220100 Urgench, Uzbekistan

Abstract. This paper presents the results of experiments on the bleaching of kaolin from the Khodjakul deposit. Various methods for removing coloring impurities were investigated, including acid leaching, treatment with chemical reducers, and physico-chemical methods. The effectiveness of using various reagents, such as oxalic, sulfuric, and hydrochloric acids, as well as nitric acids, was evaluated. The optimal processing conditions for achieving the maximum degree of whitening were determined. The whiteness index was established. The whiteness of the samples was analyzed using the SBDY-1P whiteness meter. The obtained results allow the conclusion that certain methods hold promise for improving the quality of kaolin from this deposit and its further industrial use. Comparative studies were conducted on the original, enriched, and acid-treated kaolins. To improve whiteness, methods regulated by GOST 21286-82 "Enriched kaolin for ceramic products. Technical requirements" were applied. The whiteness level was determined according to standard methods specified in GOST 16680-79 "Enriched kaolin. Methodology for measuring whiteness".

1 Introduction

In the Aral Sea region, there is a large reserve of natural minerals widely used in various sectors of the national economy [1, 2]. Among them, kaolin is a natural mineral applied in industries such as ceramics, paper production, cosmetics, and pharmaceuticals. Bleaching of kaolin is an important technological process that enhances its quality and broadens its areas of application. The development of new processing methods, especially environmentally friendly ones, plays a key role in improving the characteristics of kaolin and reducing its environmental impact. However, natural kaolin often contains impurities that decrease its whiteness and degrade the quality of the final product.

To increase the efficiency of using domestic kaolin mineral resources, methods of its enrichment have been studied [3, 4]. This will, in the future, improve the quality of finished products and establish the production of scarce grades of kaolin based on local raw materials, which are currently predominantly imported from neighboring countries. The reserves of natural kaolin with reduced levels of coloring oxides (Fe_2O_3 and TiO_2) are limited. The geographic distribution of kaolin raw materials in the CIS countries is shown in Figure 1.

* Corresponding author: mbsh76@mail.ru

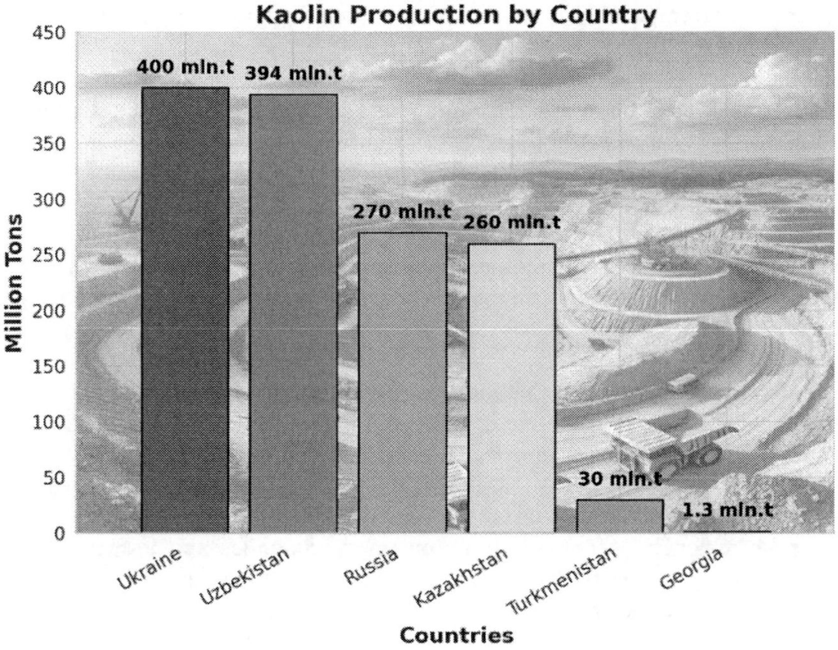

Fig. 1. Distribution of kaolin raw materials across the CIS countries.

Given the scarcity of natural kaolins with lower content of coloring oxides in the territory of the Republic of Uzbekistan, our goal is to improve the quality of kaolin from the Khodjakul deposit. To enhance the quality of kaolin, various bleaching methods are applied, including:

Chemical bleaching – this method is based on the use of reagents that oxidize or reduce the color-forming impurities, such as iron and organic substances. In oxidative bleaching, substances like hydrogen peroxide (H_2O_2), ozone (O_3), and sodium hypochlorite (NaClO) are used. They effectively break down colored organic impurities and oxidize iron-containing compounds. In reductive bleaching, chemical compounds such as sulfur dioxide (SO_2) or sodium hydrosulfite ($Na_2S_2O_4$) are applied, which reduce iron from the colored trivalent form (Fe^{3+}) to the colorless divalent form (Fe^{2+}). The presence of iron and titanium compounds in kaolins and other components of ceramic mixtures contributes to a decrease in whiteness and the emergence of undesirable hues. In kaolin and clays, iron is found in the form of free minerals (goethite, hematite, etc.), as oxide films covering the surface of silicate particles, or integrated into the crystalline structure of silicates [5].

Froth flotation enrichment allows the removal of organic impurities and iron-containing minerals by using special reagents (collectors) that selectively bind with the contaminating particles, enabling their removal in the form of foam.

In the biological method, an alternative approach is the use of microorganisms capable of synthesizing organic acids that break down the color-forming impurities in kaolin. Screening of such microorganisms and their application for kaolin treatment has shown promising results in improving its quality. The authors of the work [6] proposed a kaolin bleaching method that involves the bioreduction of Fe(III) to Fe(II) using activated microflora, followed by magnetic separation and treatment with ammonium oxalate solution. This method allows for a significant reduction in the content of color-forming impurities and an increase in the whiteness of kaolin.

Magnetic separation – this method is used to remove ferromagnetic and paramagnetic impurities. During the treatment process, kaolin passes through a magnetic field, in which iron and other metal impurities are held. This method is particularly effective when there is a high concentration of iron compounds.

One of the methods of whitening is heat treatment. Calcination in the range of 600–900°C promotes the decomposition of organic compounds and reduces the concentration of colored impurities. However, this method can lead to transformations in the mineral's crystal structure, which is not always acceptable. In industry, various technologies are used to remove coloring impurities from kaolins, including mechanical, chemical, biotechnological methods, as well as their combinations [7, 8]. Proper selection of the cleaning method ensures maximum effectiveness in removing these inclusions.

The choice of method depends on the composition of impurities, the intended purpose of the final product, and economic feasibility. In industry, a combination of several methods is often used to achieve the best effect.

Among the physico-chemical methods, kaolin enrichment in a fluidized bed stands out. Studies show that drying and selective grinding of kaolin in a fluidized bed with a heat carrier allow for effective removal of impurities and improvement of the final product quality [9].

In kaolin, the concentration of strongly magnetic or non-associated iron-containing compounds with clay minerals is usually minimal compared to the presence of iron in the form of weakly magnetic, low-crystalline, or amorphous phases. These inclusions can be located inside the aggregates of clay particles or form surface coatings, making their removal difficult using physical and chemical processing methods. For this reason, many researchers have analyzed methods of acid leaching carried out in solutions of both organic and inorganic acids. This process can be carried out either without the use of reducers or with their addition to improve the removal of contaminating components. Organic and inorganic reagents are used as chemical reducing agents for removing iron-containing impurities. During the reaction, reducers interact with (hydro)oxides of iron, converting them from an insoluble state to soluble Fe(II) compounds, which are then removed by filtering the clay minerals. The main mechanism ensuring the rapid dissolution of (hydro)oxide particles is their reduction. However, in a neutral environment, this process occurs extremely slowly, requiring the creation of optimal conditions for the effective removal of impurities. In practice, kaolin is also used as an adsorbent for bleaching vegetable oils.

2 Research methods

For the tests, a technological sample of the initial kaolin from the Khodjakul deposit was selected. The bleaching operation was carried out using methods regulated by GOST 21286-82 "Enriched kaolin for ceramic products. Technical conditions". The whiteness degree was determined according to the standard methods of GOST 16680-79 "Enriched kaolin. Method for determining whiteness." This method is based on measuring the reflectance coefficient of the surface of the analyzed sample.

For conducting scientific experiments, the selection of kaolin samples was carried out according to established standards regulating the requirements for kaolin. The average portion of kaolin was reduced to a mass of (50±5) g. All drying stages at a temperature of (105±5) °C were conducted in specialized drying equipment. After undergoing the wet enrichment stage, the material was initially air-dried, then the fragments were ground in a porcelain mortar and sieved through sieve No. I until fully passed, evenly distributed in a thin layer, and dried until a stable weight was achieved.

The whiteness of kaolin was determined using the SBDY-1P whiteness meter, which operates on the basis of the photoelectric effect and is equipped with a D65 light source (R457), intended for the analysis of the whiteness of powdered materials. For measurements, the required amount of kaolin was poured onto a sheet of paper, then lightly tamped by folding the paper in half and gently pressing it with the palm of the hand. Afterward, the sample was thoroughly mixed with a spoon, poured into a cuvette included in the leukometer set, and lightly tapped at the base to remove any voids, forming a cone-shaped pile of the material with a height of 4–6 mm. The surface of the layer was leveled with a clean glass plate without any defects, removing excess kaolin from the edges of the cuvette. Then, by pressing again, the surface was brought to the level of the cuvette edges without shifting the glass to avoid the appearance of gloss, which could distort the measurement results. The microstructure of kaolin was analyzed using the SALD-7500nano electron microscope, equipped with a semiconductor UV laser with a wavelength of 405 nm and a power of 40 W, as well as the Tescan Vega 3 Sbh scanning electron microscope, equipped with an Azteclive Lite Xplore 30 energy-dispersive microanalysis system with a tungsten cathode for thermionic emission. For sample preparation for the electron microscope examination, a one-step platinum-carbon replication method was used.

3 Results of experimental studies

The analysis of the chemical and mineral composition of the initial kaolin sample from the Khodjakul deposit showed that all the tested samples contain the same minerals, although their proportions vary, with kaolinite and quartz being the predominant minerals. The elemental composition of this kaolin sample is presented in Figure 2.

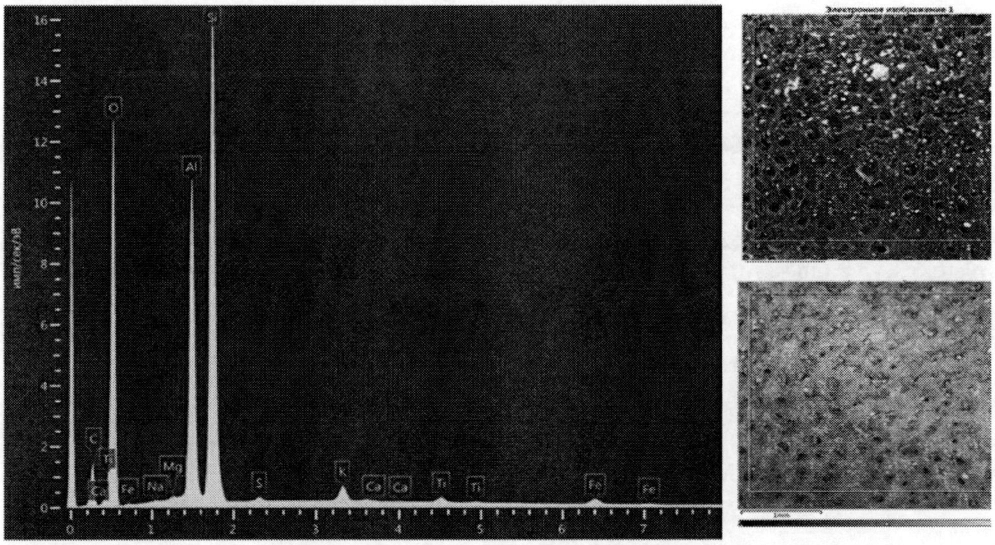

Fig. 2. Elemental composition of the original kaolin from the Khodjakul deposit.

The quality of kaolins is impaired by limiting impurities such as iron and titanium compounds. In this kaolin sample, the average content of Fe_2O_3 and TiO_2 is 1.21% and 0.71%, respectively. It should be noted that part of the iron and titanium is isomorphically incorporated into the kaolinite structure. To identify the presence of iron-containing impurities, a microstructural characterization of the initial kaolin was conducted. Figure 3 shows electron microscope images of the original and enriched kaolin from this deposit.

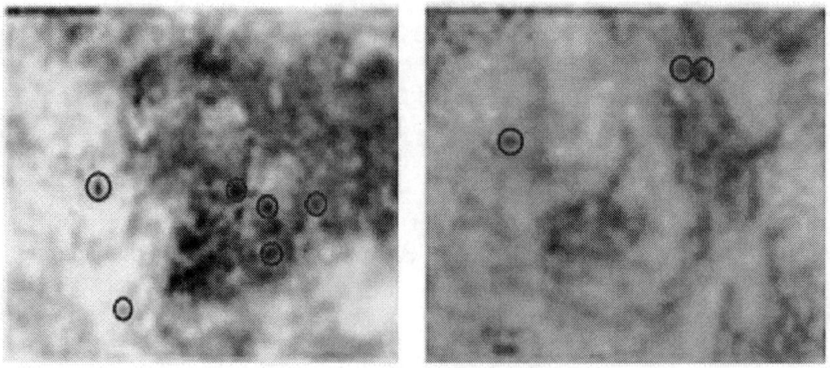

Fig. 3. Electron microscope image of the original and enriched kaolin.

The analysis of the obtained data shows that this kaolin sample belongs to the category of polymineral raw materials, and the iron-containing particles present in it have high dispersity and are represented by the mineral hematite. The enrichment of kaolin is presented in our previous works.

For this kaolin sample, the method of acid treatment was selected. In laboratory tests dedicated to the whitening of kaolin using acid treatment, H_2SO_4, HCl, HNO_3, and $C_2H_2O_4$ were used. Both the initial and the kaolin enriched by the wet method were subjected to whitening. The experiments were conducted using solutions of the specified acids with concentrations ranging from 5 to 20% at a temperature of 100°C for 2 hours, with a solid-to-liquid ratio (S:L) of 5:1, followed by multiple washings and filtration. To evaluate the whiteness level of kaolin, a blue light filter with a working wavelength of 457 nm was used. The instrument setup and measurement procedure were carried out according to the operating manual, and the obtained data were rounded to 0.1%. After completing each measurement cycle, which included 16-20 measurements, the leukometer was calibrated using a calibration plate, the whiteness of which closely matched the whiteness value of the sample under investigation.

When using H_2SO_4, HCl, and HNO_3 at the specified concentrations, the whiteness of the kaolin did not meet the requirements of GOST 16680-79 "Enriched Kaolin. Method for Determining Whiteness." The best whitening result was achieved with a 20% oxalic acid ($C_2H_2O_4$) solution. The obtained results with different acids are presented in Table 1.

Table 1. Comparative results of kaolin after acid treatment.

Samples of kaolins		**Original sample (comparison)**
Enriched kaolin		
Enriched kaolin with H_2SO_4		

Continuation of Table 1.

Enriched kaolin with HCl	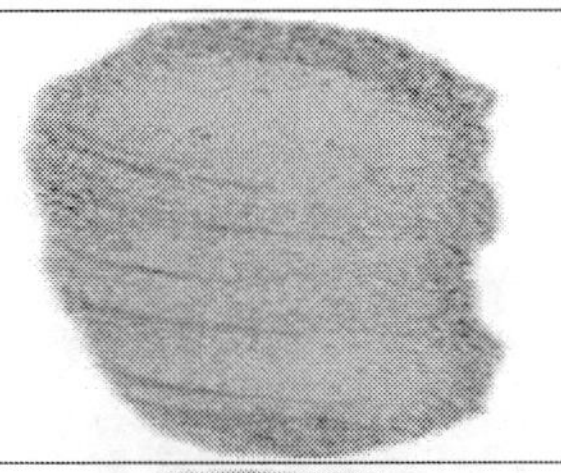	
Enriched kaolin with HNO_3	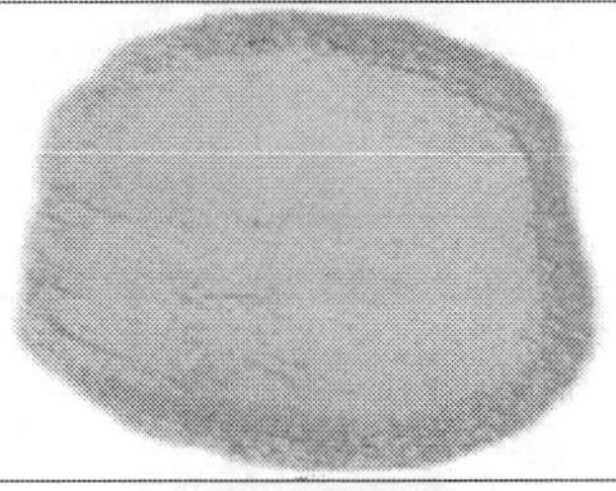	
Enriched kaolin with $C_2H_2O_4$ (Oxalic acid)	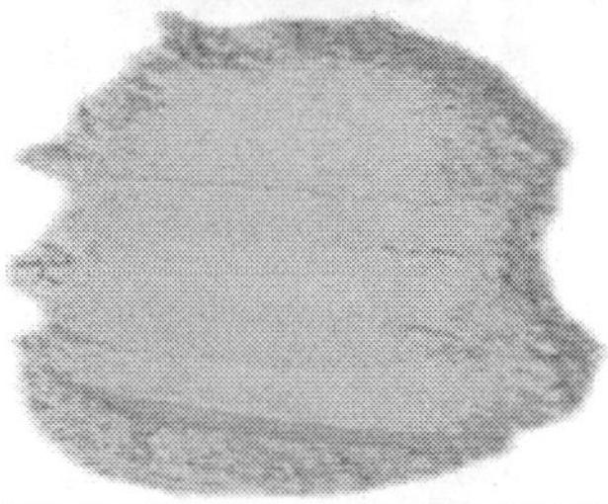	

The analysis of the obtained experimental data allows us to conclude that the reduction of unwanted impurities, including iron oxide (Fe_2O_3) and titanium oxide (TiO_2), which directly affect the whiteness of kaolin, was successfully achieved through chemical whitening. As a result of the conducted laboratory studies, it was found that the application of a 20% oxalic acid solution achieved an impurity removal efficiency of 13.20%. This conclusion is fully supported by the obtained numerical values of the whiteness index for the kaolin samples, which were determined using the high-precision whiteness meter model SBDY-1P (Table 2).

Table 2. Results of chemical whitening.

№	Raw material	Leaching conditions concentration, %		Content, %				Whiteness index, R (457)
				Fe_2O_3	TiO_2	Al_2O_3	SiO_2	
1	Non-enriched kaolin	-		1,21	0,71	31,14	64,90	68,80
2	Enriched kaolin	-		0,99	0,55	32,81	61,10	75,30
3	Enriched kaolin with $C_2H_2O_4$	$C_2H_2O_4$	5	0,97	0,50	32,93	60,99	76,66
			10	0,95	0,47	33,80	60,66	78,43
			15	0,93	0,44	35,92	60,21	80,15
			20	0,92	0,42	37,61	59,10	82,40

To clearly demonstrate the dynamics of kaolin whiteness changes at different stages of processing, specific experimental indicators should be considered. In its initial, unprocessed form, the whiteness index of kaolin was 68.80%, due to the presence of impurities that reduce its optical properties. After the wet enrichment stage, the whiteness level increased to 75.30%, indicating partial removal of contaminating compounds. The highest whiteness index was recorded after chemical treatment of kaolin: as a result of exposure to active reagents, particularly oxalic acid, a whiteness index of 82.40% was achieved, confirming the effectiveness of this purification method. Figure 4 shows a diagram of the chemical bleaching results.

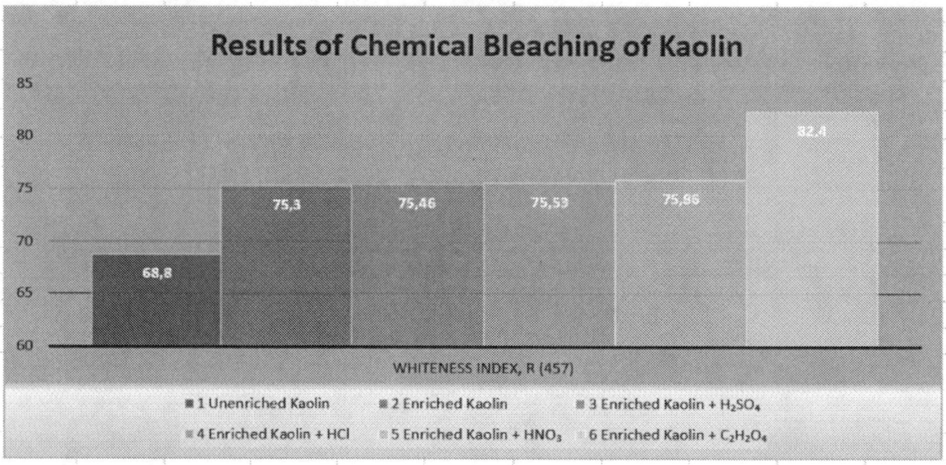

Fig. 4. Results of chemical bleaching.

Figure 5 shows the results of the elemental composition analysis of the enriched kaolin sample from the Khodjakul deposit after acid treatment, obtained using a scanning electron microscope.

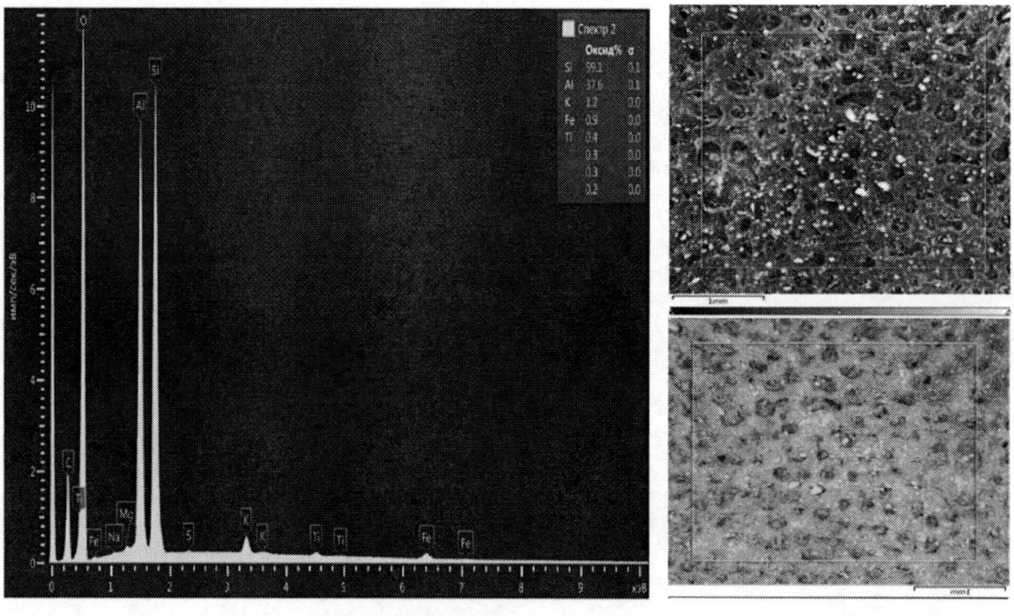

Fig. 5. SEM image of the enriched kaolin from the Khodjakul deposit after acid treatment with 20% C2H2O4.

The analysis results show that the content of Al increased from 31.14 to 37.60 compared to the non-enriched kaolin; the amount of Si decreased from 64.90 to 59.10; the amount of Fe decreased from 1.21 to 0.92; and Ti decreased from 0.71 to 0.42.

4 Conclusion

The conducted experiments on kaolin bleaching from the Khodjakul deposit involved the study of various methods for removing coloring impurities, such as iron and titanium oxides. The research showed that the most effective method was acid leaching using oxalic acid, particularly at a 20% concentration, which resulted in a significant increase in kaolin whiteness. Physicochemical methods, including magnetic separation and treatment with chemical reducers, also demonstrated certain effectiveness, but their application requires optimization of the processing conditions. The whiteness analysis of the treated samples, performed using the SBDY-1P whiteness meter, confirmed that the selected technologies could significantly improve kaolin quality and expand its potential applications. The results suggest that the developed methods can be used for industrial processing of kaolin from the Khodjakul deposit, which will increase its competitiveness and reduce dependence on imported analogs. Further research could focus on improving the technological parameters of bleaching processes and studying the environmental aspects of chemical reagent application.

References

1. S. Tadjiev, E. Atashev, "Obtaining azosuperphosphate from low-grade phosphates of the Central Kyzyl-kum", Journal of Critical Reviews **7(5)**, 472–477 (2020)
2. O. Myachina, S. Shamuratov, U. Baltaev, U. Alimov, Sh. Kurambaev, E. Atashev, T. Kuramboev, "Agrochemical efficiency of slow release phosphate fertilizers derived on the base of phosphorite activation", E3S Web of Conferences **434**, 03014 (2023)
3. E. S. Bronitskaya, S. I. Anufrieva, M. V. Kydzhy, M. A. Fokin, G. S. Grishaev, E. I. Lyubimova, V. A. Rassulov, "Enrichment of primary kaolins of the Ural-Mugodzhar Krai in order to improve the efficiency of development of the domestic mineral resource base of kaolins", Exploration and Protection of Subsoil. Scientific and Practical Journal **6**, 28–34 (2020)
4. G. I. Storozhenko, T. E. Shoeva, N. A. Dvornikov, "Technology of dry kaolin processing in fluidised bed of heat carrier", Izvestia Vuzov. Construction **7**, 64–74 (2021)
5. Y. Bagherzadeh, M. H. Golmakani, E. Z. Karimi, "Straight synthesis of alumina from kaolin by HCl acid leaching", Journal of Mining and Metallurgy Section B: Metallurgy **19**, 2250 (2023), DOI: 10.2298/JMMB230101017B
6. S. Abdelgalil, K. El-Barawy, Y. G. Xia, "The Recovery of TiO2 from Ilmenite Ore by Ammonium Sulfate Roasting–Leaching Process", Processes **11**, 2570 (2023), DOI: 10.3390/pr11092570
7. K. Pariyan, M. R. Hosseini, A. Ahmadi, A. Zahiri, "Optimization and kinetics of oxalic acid treatment of feldspar for removing the iron oxide impurities", Separation Science and Technology **55(2)**, 1–12 (2019)
8. N. Wang, H. Gu, H. Wen, "Enrichment of niobium and titanium from kaolin using an acid–alkali leaching process", Metallurgical and Materials Transactions **49**, 3552–3558 (2018)
9. S. M. Sobirov, A. M. Nishonov, A. S. Sobirov, "Use of Zarafshan and Tavaksai kaolin mineral by activation in cotton oil", Universum: Chemistry and Biology **12(78)** (2020), https://7universum.com/ru/nature/archive/item/10965
10. B. Taheri, "Iron Removal from Kaolin by Oxalic Acid Using a Novel Pre-Agitating and High-Pressure Washing Technique", Clay Minerals **58**, 224–233 (2023), DOI: 10.1180/clm.2023.11

E3S Web of Conferences **633**, 07001 (2025) https://doi.org/10.1051/e3sconf/202563307001
GASES 2025

Identifying transboundary ecotourism zones in surkhandarya region: opportunities and prospects

Shokhrukh Ravshanov[1,*], *Dunyobek* Khursanov[1], *Bobur* Malikov[1], *Elshod* Eshmamatov[1], and *Khayitboy* Abduvaliev[2]

[1]Samarkand State University, Samarkand, Uzbekistan
[2]Fergana State University, Fergana, Uzbekistan

Abstract. This article examines the issue of identifying transboundary ecotourism zones in the Surkhandarya region from both theoretical and practical perspectives. Within the framework of the study, the region's natural-geographical conditions, ecosystems, and cultural heritage are comprehensively analyzed to assess the potential for ecotourism development. Additionally, the concept of transboundary ecotourism, its regional development strategies, and integrative aspects of cooperation with neighboring countries are explored. The criteria for delineating ecotourism zones, their compliance with tourism infrastructure, and principles of sustainable development are analyzed, leading to the development of scientifically grounded recommendations. The research findings contribute to the formulation of regional ecotourism policies and provide a scientific and practical basis for the effective utilization of ecotourism resources.

1 Introduction

Transboundary research in the field of tourism and recreation has been conducted primarily in European Union countries and North America, driven by the increasing permeability of state borders. These studies are considered a crucial component in the formation of transboundary regions. Researchers such as M. Braym (2013), V. Grama (2011), A. Kovach (2011), J. V. Skott, and J. Layn (2012) have devoted significant attention to transboundary tourism and recreation within the framework of Euroregions. The works of R. Lagievskiy and D. Revelas (2004), as well as K. Vodeb (2010), directly examine the formation of transboundary zones in tourism and recreation. Meanwhile, X. Vachoviak (2006), E.K. Prokkola (2018), and D. Timoti (1995) focus on transboundary cooperation in tourism, whereas the scientific studies of N. Serich (2011) and others are dedicated to transboundary tourist routes.

The formation of transboundary territorial structures (systems, regions, clusters) in tourism and recreation, as well as the factors influencing them, has been explored in the scientific studies of A.Y. Aleksandrova (2007), A.I. Ziryanov (2012), I.I. Dragileva (2006), A.N. Dunes (2011), V.S. Kornevets (2010), D.G. Mirzexanova (2005), A.V. Mitrofanova

* Corresponding author: shokhrukh.ravshanov717@gmail.com

(2010), and D.V. Nikolayenko (2000). However, the interrelations between terms reflecting transboundary cooperation in the tourism and recreation sector have not yet been fully established. While the distinctive features of transboundary tourism zones and their role within the recreational geography system have been identified, the factors and patterns shaping transboundary formations remain partially defined and lack systematic classification. Moreover, practical research on transboundary tourism cooperation along the borders of the Republic of Uzbekistan remains limited.

However, the methodology for studying the territorial organization of transboundary tourism and recreation systems, regions, complexes, and clusters has not been sufficiently developed. Research on transboundary tourism and recreation systems in Uzbekistan has been conducted since the country's independence. During this period, international relations have expanded, and economic and cultural cooperation with neighboring countries has strengthened. These processes have created new opportunities for studying tourism and recreation systems in transboundary zones. Such research is essential for the development of this sector, as it contributes to the sustainable development of regions, the diversification of tourism and recreation services, and the rational use of natural resources [3-6].

For this reason, only in the last 10–15 years have the concepts of "(international) cross-border region" and "cross-border cluster" appeared in international practice, based on a systemic approach. These terms are often considered synonymous with the concept of "cross-border tourism and recreational zone" in recreational geography. The terms "border" and "cross-border region" are already widely used by foreign scientists. Border regions are understood as territories that demonstrate mutual connectivity and interaction between neighboring states, are directly adjacent to national borders with significant cross-border influence and have significant potential for the development of international cooperation.

Transboundary regions encompass the border areas of two or more neighboring countries, forming a distinct transboundary geographical entity. According to P. Y. Baklanov and S. S. Ganzey (2008), transboundary regions are characterized by a combination of natural resources and specific types of economic activities. Their natural foundation consists of either a single geosystem or a set of two or more regional-scale geosystems interacting within the border zone of a state. Alternatively, they can be defined as complex geographical structures that integrate specific natural resources, infrastructure, population distribution, and economic activities within the boundaries of major geosystems [1-2].

The aforementioned scholars have conducted research on transboundary regions in the context of sustainable environmental management. Therefore, their definitions emphasize that these regions are fundamentally composed of geosystems and possess a natural foundation.

2 Methodology

Transboundary regions can have complex structures and multi-level networks. Due to their location between multiple countries, they are shaped by economic, social, political, and environmental factors. D.G. Mirzexanova (2013) defines such regions as: "A territorial structure formed by two or more neighboring or closely situated countries that share a mutual interest and willingness to develop international tourism, create a unified tourism product, and establish closer interactions with tourists".

Therefore, considering a transboundary region as a "complex natural-social-territorial system with a certain degree of natural and spatial integrity," we regard the approach proposed by P.Y. Baklanov and S.S. Ganzey (2008) as well-founded.

An analysis of the most common definitions of the term "cross-border region" in various disciplines and fields of activity shows that such regions are defined primarily on the basis of economic criteria, historical, ethnic, geographical, ecological or natural unity, as well as common historical and cultural traditions. The key condition for the successful existence and development of international cross-border regions is the establishment of good-neighborly relations between states, as well as efforts to strengthen economic ties and expand social, cultural and other forms of cooperation. In addition, this process requires neighboring countries to mutually benefit from the use of their border potential.

Although the concept of a *"transboundary area"* is widely used in the context of international tourism, there is still no universally accepted definition. This term encompasses a broader meaning than classical economic or physical-geographical concepts, as international tourism relations are not limited to shared borders or common natural and cultural factors between neighboring countries. Instead, transboundary tourism develops through the complementary use of mutually beneficial resources.

Crucially, cooperation between geographically close countries with diverse natural, ethnic, and cultural characteristics is a key factor in the development of transboundary tourism. Integrated tourism products created within these areas serve as an effective tool for fostering transboundary tourism. For instance, major tourism initiatives such as the *"Eastern Russia Ring,"* the *"Great Silk Road,"* the *"Great Tea Road," "Genghis Khan's Routes,"* the *"Golden Ring of Altai,"* and the *"Baltic Amber Road"* represent successful examples of such collaboration [7-9].

A cross-border tourism area is often formed on the basis of common tourism resources – a common natural or historical and cultural heritage. Sharing these resources makes the destination more attractive to tourists and offers greater benefits to the participating regions. As a result, countries involved in cross-border tourism cooperation gain a competitive advantage over regions that do not interact with their foreign neighbors.

For instance, in the Upper Rhine region, which spans the borders of Germany, France, and Switzerland, the *Regio TriRhena* Euroregion has developed a unified tourism strategy called *"Three Countries – Four Cities – One Trip."* Another example of transboundary tourism development based on a unified natural geosystem is the international tourism route project in the Altai mountain region, which connects four countries—Kazakhstan, China, Mongolia, and Russia.

Additionally, within a transboundary tourism area, the resource potential of one country can complement the tourism opportunities of another. The concentration of historical and cultural heritage sites is often seen as a mutually enriching factor, enhancing the overall attractiveness of the region.

Thus, it is appropriate to categorize the criteria for the formation of transboundary tourism areas into four main groups: the unity, complementarity, contrast, and uniqueness of natural, historical, cultural, social, and other factors. Unlike traditional regions, a transboundary tourism area is characterized by a combination of one or more selection criteria.

Criteria for Identifying Transboundary Areas:

- **Unity of the natural geosystem and territorial integrity** (Natural and ecological resources);
- **Common historical development** (A shared historical society or similar historical development);
- **Cultural unity** (Cultural similarity or contrast in the development of ethnic groups).

These criteria represent the key factors used to define transboundary areas.

Transboundory ecotourism zones (TBEZ) are influenced by various internal and external factors that play a decisive role in strengthening interregional ties. Internal factors reflect the level of tourism development, its territorial structure, dynamics and specific management processes of TBEZ. These factors determine the internal structure and goals of the zones. External factors, on the other hand, are related to the impact of the natural and socio-economic environment on TBEZ. They can be classified according to various regional systems: natural, socio-ecological, ethno-geographical and political-geographical factors.

The creation of tourism and recreational zones in border mountainous areas presents unique challenges, primarily due to state borders. Borders serve as both physical and psychological barriers for tourists, changing the conditions for tourism development. The prospects for cross-border tourism depend on the socio-economic cooperation between neighboring countries, the presence of a common natural and cultural heritage and the level of tourist interest in neighboring states [9-10].

Surkhandarya region is currently receiving special attention regarding the sustainable utilization of its transboundary territories. This emphasis is well-founded, given the region's distinctive geographical position. As Surkhandarya shares mountainous border areas with neighboring countries, it holds significant potential for the development of transboundary tourism. The establishment of such zones is particularly important for various sectors in Uzbekistan, as it contributes to reducing the negative impacts associated with the peripheral status of many border regions.

Surkhandarya stands out for its impressive mountain ranges, including Hisar, Bobotog, Ko'hitang, and Baysun, which attract visitors with their mesmerizing landscapes, remarkable natural beauty, and curative springs. The region's towering peaks, extraordinary terrain, spectacular waterfalls, and pristine mineral springs offer an unforgettable experience for travelers.

To efficiently utilize the immense potential of the region's mountainous and foothill areas, three mountain tourism clusters are planned. The Baysun mountain cluster will include the villages of Omonxona and Darband, as well as the Zovboshi mountainous area. The Sariosiyo mountain cluster will encompass the slopes of the Xonjiza mountain range in Sariosiyo and Uzun districts, along with the Sangardak village area in Sariosiyo district. The Sherobod mountain cluster will focus on the picturesque landscapes of the Sholkon village.

3 Results

Surkhandarya region, located in the southern part of Uzbekistan, shares borders with neighboring countries where Uzbek, Tajik, Turkmen, and even Afghan national traditions have intertwined, forming unique ethnographic characteristics. To enhance cultural exchange, deepen awareness of national traditions, and stimulate ecotourism, the establishment of transboundary ecotourism zones in this region is crucial. Beyond the existing ecotourism destinations, particular attention should be given to identifying underutilized areas, developing a comprehensive resource and information database, and systematically mapping these locations. These efforts will not only contribute to the sustainable development of tourism but also help unlock the full potential of the region's natural and cultural heritage.

Surkhandarya's border areas hold great potential for the development of transboundary ecotourism zones. The region's mountain ranges—Hisar, Baysun, Bobotog, and Ko'hitang—bordering Tajikistan and Turkmenistan, are particularly rich in ecotourism, medical tourism, and speleotourism resources.

Given these factors, it can be concluded that Surkhandarya region possesses substantial ecotourism potential and has the capacity to develop multiple transboundary ecotourism zones.

Based on our research findings, we have identified three major transboundary ecotourism zones in the Surkhandarya region (Figure 1).

https://doi.org/10.1051/e3sconf/202563307001

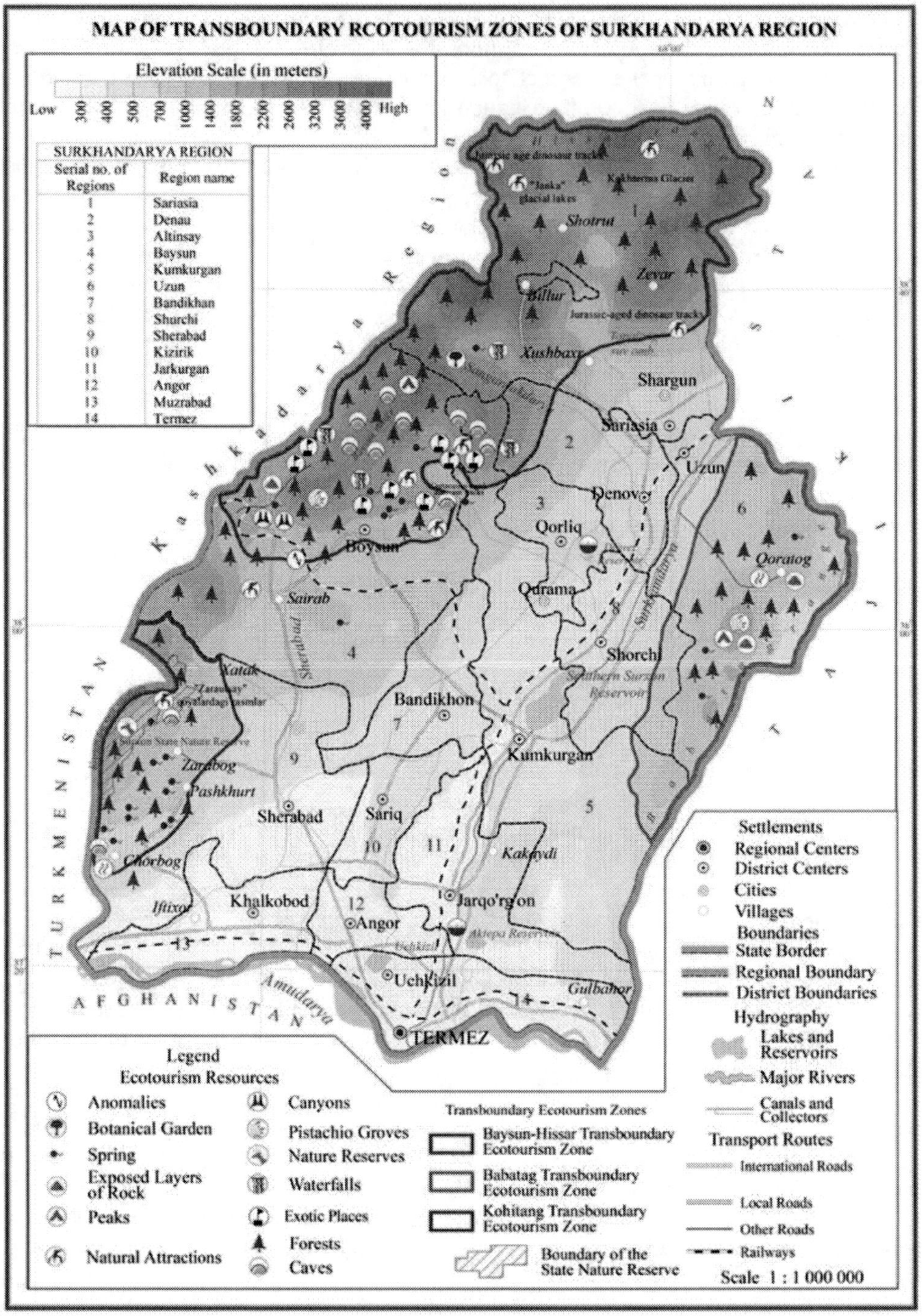

Fig. 1. Map of transboundary ecotourism zones in the Surkhandarya region.

Below, we provide a brief overview of the attractiveness of these areas.

Bobotog Transboundary Ecotourism Zone: Geographical Features and Ecotourism Potential. The Bobotog mountain range is located along the Uzbekistan-Tajikistan border, between the Surkhandarya and Kofarnihon rivers, extending approximately 125 km in length and 30-40 km in width. Geologically, it is considered a young mountain formation, composed of rock formations from the Cretaceous, Paleogene, and Neogene periods. The highest peak of Bobotog, Zarkosa, rises to 2,290 meters above sea level. The region's topography has been shaped mainly by denudation and karst processes.

Bobotog has a subtropical continental climate, characterized by relatively mild winters and hot, dry summers. The annual average precipitation ranges between 250 and 500 mm, with an uneven distribution across foothills and higher elevations. The flora of the region includes over 300 species of vascular plants, 16 of which are listed in Uzbekistan's Red Book of endangered species. The fauna is diverse, featuring rare species such as the Bukhara wild sheep, markhor, gazelle, golden eagle, and falcon.

Bobotog holds significant potential for ecotourism development. The region offers both natural and cultural tourism opportunities, including the Govurg'on Gorge, the Qorayumaloq Peak, dense juniper forests, and Chagam village, which provides insight into the local ethnic lifestyle. Future plans for ecotourism development in the Bobotog region include the establishment of camping sites, recreational areas, and eco-trails. At the same time, the conservation of natural landscapes and ensuring ecological sustainability remain key priorities.

Table 1. Ecotourism resources of the Bobotog transboundary ecotourism zone.

Ecotourism resources				
Hydrological Resources	**Geological Resources**	**Landscapes**	**Protected Natural Areas**	**Ethnographic Resources**
1. Biyasigmas Spring 2. Jiyda Spring 3. Otkamar Spring 4. Oybulok Spring 5. Shorbulok Spring 6. Sariqamish Spring 7. Tolli-Chashma Spring 8. Chagamsoy River 9. Oqsuv Spring 10. Oqsuvsoy Spring	1. Zarkosa Peak 2. Korayumalok Peak 3. Xarsang 4. Chodir Hill 5. Canyon Hill 6. Yetti Kiz (Seven Girls)	1. Scenic Tugai Forests 2. Dense Juniper Forests 3. Bobotog Pistachio Groves 4. Qizilsoy	1. Hazratbobo Pilgrimage Site 2. Govurgan and Govurgan Ruins 3. Khojaibaru 4. Gairob	1. Chagam Village

KohitangTransboundary Ecotourism Zone: Geographical Features and Ecotourism Potential. The Kohitangmountain range is located on the Uzbekistan-Turkmenistan border, in the southwestern part of the Pamir-Alay system. It is distinguished by its unique geological structure and high ecotourism potential. This range includes the southern part of the Baysuntog and extends approximately 100 km from the Amu Darya valley to the Sherobod River gorge. Its highest peak is Ayribobo, reaching 3,137.7 meters above sea level. The mountain system primarily consists of limestone and gypsum layers, with widespread karst formations. The Kohitang region has a continental climate, characterized by hot and dry summers. Its flora comprises 578 species, including Zarafshan juniper, ephedra, plane trees, pistachio, almond, and other rare species. The area also hosts diverse wildlife, with several protected and rare species. Additionally, Kohitangis rich in archaeological and cultural heritage, making it a prime location for ecotourism.

A designated ecotourism site has been established within the "Zarautsay" area of the Qizilolma section of the Surxon Nature Reserve, covering 395.6 hectares. This area is being developed with campsites, yurts, recreational zones, and eco-trails to enhance ecotourism infrastructure. Future plans for this region include not only fostering ecotourism but also ensuring the conservation of natural ecosystems.

Table 2. Ecotourism resources of the Kohitangtransboundary ecotourism zone.

Ecotourism resources				
Hydrological Resources	**Geological Resources**	**Landscapes**	**Protected Natural Areas**	**Ethnographic Resources**
1. Orinbulak Spring 2. Tahirchoq Spring 3. Sulfur Spring 4. Khalija Spring 5. Qozonbulak Spring 6. Alamli Spring 7. Salty Spring 8. Laylakon 9. Qizilolmasoy Stream 10. Golden Spring 11. Arrabulak Spring	1. Khojaikon 2. Ajinakamar Cave 3. Ayri Bobo	1. Dense Juniper Forests 2. Tangidara Landscapes	1. Surxon State Reserve 2. Zarautsoy Monuments (Petroglyphs)	1. Xatak Village

Baysun-Hisar Transboundary Ecotourism Zone: Geographic Features and Ecotourism Potential. The Baysun-Hisar mountain system is located along the Uzbekistan-Tajikistan border and is part of the western Pamir-Alay range. The Hisar Mountains stretch 200 km from east to west, with Hazrati Sultan Peak (4,643 m) being the highest point. Geologically, this mountain system consists of sandstone and crystalline rocks interspersed with granite intrusions, encompassing all the characteristic features of the Pamir-Alay ecosystems. The region contains 239 glaciers, covering a total area of 83.6 km², with the largest glacier, Chap Qaznoq, spanning 1.9 km². The climate of the region varies significantly with altitude. The foothills experience a dry and hot climate, while the higher elevations are cold and humid. Spring and summer provide ideal conditions for tourism. Ecologically, the area is significant due to its surface and groundwater resources, as well as its unique flora and fauna. The region is home to brown bears, reed wolves, mountain goats, and other species listed in the Red Data Book.

Baysuntog is one of the highest sections of the Hisar range, with some peaks reaching 3,700 meters in elevation. This range extends into the Kohitangmountains, which rise to 3,137 meters. The Baysuntog area encompasses various natural zones, with vegetation and wildlife changing as altitude increases. The region's soils include gray soils, brown and dark brown forest soils, light brown meadow-steppe soils, and grassland soils. The plant life is rich in medicinal and forage species, though anthropogenic factors have led to habitat reduction. Baysuntog is also known for its mineral water resources, including the famous

"Omonkhon Spring," renowned for its therapeutic properties. These waters offer great potential for the development of balneological tourism. The region is also home to significant cultural heritage sites, including the Teshiktosh Cave, where Neanderthals once lived, and the Amur Temur Cave, both of great archaeological value. The Khoja Gur-Gur Ota massif is a popular destination for alpinists and extreme sports enthusiasts.

With its glaciers, caves, pristine landscapes, and rich cultural heritage, the Baysun-Hisar mountain system offers immense ecotourism potential. Developing ecotourism in this region will not only provide economic benefits but also contribute to the preservation of its natural ecosystems.

Table 3. Ecotourism resources of the Baysun-Hisar transboundary ecotourism zone.

Ecotourism resources				
Hydrological Resources	**Geological Resources**	**Landscapes**	**Protected Natural Areas**	**Ethnographic Resources**
1. Qizilnur Waterfall 2. Sangardak Waterfall 3. Konibeshbuloq 4. Oqbuloq 5. G'orbuloq Waterfall 6. Xomkon 7. Panjobosoy 8. Pitti Waterfall 9. To'palang Reservoir 10. Xo'jaipok Ota 11. Qo'tir Spring 12. Dagrez Reservoir 13. Xojamayxona Spring 14. Omonxona 15. G'alchalik Spring 16. Uchko'z Spring 17. Itbuloq 18. Kanishbuloq 19. Toshbuloq 20. O'mizbuloq 21. Oynabuloq 22. Yonarbuloq 23. Tomchibuloq 24. Jo'jabuloq 25. Momobuloq	1. Boybuloq Cave 2. Teshiktosh Cave 3. The Rock Between Two Mountains 4. Xo'jaipok 5. Qorong'u Yulduz Cave (Dark Star Cave) 6. Izetskaya Cave 7. Festivalnaya Cave 8. Vyunok Cave 9. Ulugbek Cave 10. Alexander Vishnevsky Cave 11. Medveje Logovo Cave (Bear's Den Cave) 12. Hazrati Sultan Peak	1. Xo'jabuzbarak Gorge 2. Xo'ja Gur-Gur Ota Massif 3. Xonqochdi Gorge 4. Dehimalak Gorge 5. Xo'jamayxona Gorge 6. Juniper Forests of the Baysun-Hisar Mountains 7. Aliyor Arch 8. Darbant Canyon 9. Baysun Red Canyon 10. Vaxshivor Valley 11. Chulbayr 12. Oqdavon 13. Xo'jaydod 14. Ketar Gorge 15. Amir Temur Depression 16. Zavboshi 17. Baysun Magnetic Mountain	1. Xo'jaqochqorota 2. Mazor of Alisher 3. Sherakmozor 4. Issiq-Chashma (Hot Spring) 5. Jurassic-Age Dinosaur Tracks 6. Petroglyphs on the Cliffs of Laylakxona 7. Sairab Plane Trees 8. Cretaceous-Age Dinosaur Tracks	1. Baysun Bahori

4 Discussion

Research findings highlight several key aspects of establishing ecotourism zones in transboundary mountainous regions.

First and foremost, the elevation and terrain fragmentation play a decisive role in shaping ecotourism and recreational activities. The intricate geological structure and diverse natural landscapes of mountainous areas determine the extent and manner in which tourism resources can be utilized. Additionally, climatic conditions, particularly microclimatic variations, influence the seasonal dynamics of ecotourism. While high-altitude areas are characterized by cold and humid climates, the foothills experience relatively milder conditions, thereby impacting tourism activities throughout the year.

The remarkable diversity of natural landscapes further enhances the tourism potential of these regions. The geographical structure, which transitions from foothill plains to snow-capped peaks and glacial formations, allows for a broad spectrum of ecotourism activities. This diversity fosters the development of multiple tourism types, including adventure sports, scientific-ecological tourism, and nature-based recreational activities. Moreover, the region's rich biodiversity, particularly the presence of endemic plant and animal species, enhances its appeal as an ecotourism destination while supporting conservation efforts. Ecotourism programs centered on protecting natural ecosystems are vital in this regard.

A critical aspect of ecotourism development is the systematic registration and mapping of tourism resources. Establishing a comprehensive inventory of natural and cultural assets, evaluating their tourism potential, and devising scientifically grounded management strategies are fundamental steps in this process. The aesthetic and ecological value of mountain and foothill landscapes creates extensive opportunities for various tourism activities, such as trekking, mountaineering, and wildlife observation, positioning these areas as ecotourism hubs.

Above all, fostering international cooperation among neighboring countries is essential for the sustainable development of transboundary tourism regions. Joint efforts in environmental protection, tourism infrastructure development, and strategic ecotourism planning are indispensable. Such collaboration ensures the effective management of natural resources and contributes to the long-term sustainability of ecotourism in these regions.

5 Conclusion

Tourism experts increasingly recognize Surkhandarya as a region with exceptional natural richness, offering a solid foundation for ecotourism development. The region's ecotourism potential is shaped by several key factors, including the diversity of its terrain, extensive hydrological networks, distinctive karst formations, climatic variations, and diverse vegetation across its natural complexes.

The establishment of transboundary ecotourism zones is guided by four fundamental criteria: unity, complementarity, contrast, and uniqueness of natural, historical, cultural, and social elements. Unlike conventional tourism destinations, transboundary ecotourism zones are distinguished by the presence of one or more of these selection criteria, making them uniquely suited for sustainable tourism initiatives.

Surkhandarya shares borders with Tajikistan, Turkmenistan, and Afghanistan, providing vast opportunities for the cooperative use of natural and ecotourism resources. To optimize this potential, the region's mountainous and foothill landscapes have been classified into three distinct transboundary ecotourism zones: Bobotog, Ko'hitang, and Baysun-Hisar. These zones are designed to promote cross-border tourism collaboration, leveraging shared natural assets for sustainable tourism growth.

The experience of establishing transboundary ecotourism zones in Surkhandarya highlights their role in bridging economic disparities between peripheral and central regions. This makes them a vital component in shaping tourism development policies and strategic planning for the region, ultimately contributing to regional economic stability and environmental conservation.

References

1. N. M. Ardoin, M. Wheaton, A. W. Bowers, C. A. Hunt, W. H. Durham, "Nature-based tourism's impact on environmental knowledge, attitudes, and behavior: a review and analysis of the literature and potential future research", Journal of Sustainable Tourism **23(6)**, 838–858 (2015), DOI: 10.1080/09669582.2015.1024258
2. P. Y. Baklanov, "Structural features and development potential of border and transboundary regions: theoretical aspects", Regional Studies **3(61)**, 19–24 (2018), DOI: 10.17059/ekon.reg.2021-2-6
3. A. F. de Araújo, I. Andrés-Marques, L. López Moreno, "No Planet-B Attitudes: The Main Driver of Gen Z Travelers' Willingness to Pay for Sustainable Tourism Destinations", Sustainability **17**, 847 (2025), DOI: 10.3390/su17030847
4. A. N. Dunets, V. N. Ivanova, A. L. Poltarykhin, "Cross-border tourism cooperation as a basis for sustainable development: a case study", Entrepreneurship and Sustainability Issues **6(4)**, 2207–2215 (2019), DOI: 10.9770/jesi.2019.6.4(45)
5. H. Huo, J. Chen, W. Li, "Applying Ecotourism Knowledge and Destination Image in Planned Behavior Theory in Ecotourism", Polish Journal of Environmental Studies **33(1)**, 685–695 (2024), DOI: 10.15244/pjoes/172047
6. M. Á. Oviedo-García, M. Castellanos-Verdugo, M. Vega-Vázquez, F. Orgaz-Agüera, "The Mediating Roles of the Overall Perceived Value of the Ecotourism Site and Attitudes Towards Ecotourism in Sustainability Through the Key Relationship Ecotourism Knowledge-Ecotourist Satisfaction", International Journal of Tourism Research **19**, 203–213 (2017), DOI: 10.1002/jtr.2097
7. Sh. Ravshanov, D. Khursanov, "Analysis of ecotourism and recreational resources for transboundary regions in Surkhandarya, Uzbekistan", Bulletin of the National University of Uzbekistan **3(3)**, 266–269 (2024), DOI: 10.69617/nuuz.v3i3.2
8. S. V. Stepanova, "Cross-border tourist routes: the potential of Russia's North-West", Baltic Region **9(4)**, 97–112 (2017), DOI: 10.5922/2079-8555-2017-4-7
9. I. Trišić, D. Privitera, S. Štetić, M. D. Petrović, M. M. Radovanović, M. Maksin, D. Šimičević, S. Stanić Jovanović, D. Lukić, "Sustainable Tourism to the Part of Transboundary UNESCO Biosphere Reserve "Mura-Drava-Danube". A Case of Serbia, Croatia and Hungary", Sustainability **14**, 6006 (2022), DOI: 10.3390/su14106006
10. L. Trogisch, R. Fletcher, "Fortress tourism: exploring dynamics of tourism, security and peace around the Virunga transboundary conservation area", Journal of Sustainable Tourism **30(2–3)**, 352–371 (2020), DOI: 10.1080/09669582.2020.1857767

Geoecological basis of nature protection and holistic use of natural resources in the Fergana valley

Abdusamad Khamidov[1,*], *Dilfuza* Bakhtiyorova[2], *Surayyo* Saydullayeva[2], *Gulshoda* Janizoqova[2], and *Kamola* Yusupova[2]

[1]Fergana State University, Fergana, Uzbekistan
[2]Samarkand State University, Samarkand, Uzbekistan

Abstract. The article describes the period in the Fergana Valley and adjacent regions in the 1970s and early 1980s, when environmental issues and the deterioration of environmental condition were just beginning to be placed the agenda. Firstly, scientific conditions on the geographical sides of the holistic use of existed hydro resources, natural complex changes with the influence of anthropogenic factors in the effects of watering areas on the natural world was analyzed. In the late 1980s, researchers looked into and carried out a thorough analysis of the regional and regional ecological condition and natural issues that appeared in the Fergana Valley. These evaluations were the first comprehensive scientific analysis in environmental protection in the region. Many sensitive concerns were brought up at that time. In the highland regions of the Fergana Valley, various natural and natural-anthropogenic systems were formed: deserts, hills. It is an ancient irrigated oasis, a fast-growing and densely populated industrial-agrarian region with developed industry, farming, transport and other industries. It is emphasized that the growth and practical implementation in geoecological foundations of conservation and sensible application of natural resources in the Fergana Valley are relevant.

1 Introduction

Continuous development of production in the Fergana Valley, increase of anthropogenic load on the environment, constant effusion of various industry waste, transportation, construction and farming into the natural surroundings started to lead to serious deterioration in the ecological condition in the region in 1970-1980s.

In the these years, nature protection issues as well as the deteriorating ecological condition were just starting to become essential in the Fergana Valley. The first scientific ideas of Y. Sultanov and A. Maksudov on geographical aspects of logical use of available water resources, landscape changes under the man-made impact in the irrigated area and their ecological consequences were released [1].

* Corresponding author: khamidovabdusamad10@gmail.com

In the late 1980s, Y. Sultanov conducted a deep and comprehensive analysis of the valley and local ecological condition and environmental protection problems that appeared in the Valley of Fergana. This article was the initial comprehensive scientific analysis of the conservation of the valley. The writer raises many acute issues in his work. For example, all cities of conical zones have began to pollute land and surface waters with their wastes, and if that process is not prevented, the scale of water problem will develop in the future and will be a problem [2].

Taking into account, Ferghana Valley is surrounded by mountains, he noted the demand on reducing gas emissions by industrial factories to the necessary point, otherwise the environmental and socio-economic consequences will lead to disastrous events. The writer notes that because of intense land development in the region, the area of forests, thickets and pastures has sharply decreased and and this procedure is moving quickly. He also analyzes its undesirable consequences and writes that in the future there will be unpleasant geo-ecological problems, the solution of which will be difficult.

I. Abduganiev et al. (1995) showed for the first time that in the Fergana Valley, due to the prevailing altitudinal-landscape belt, the exchange of matter and vitality is provided in both directions, noting that on the mountain slopes the gravitational flow downward of everything is going nicely expressed, and on the plains - the movement of matter to the slopes under the influence of storm [3].

The intensity of of human-induced pressure and its manifestation on a regional level has emphasized the essence of reserves and small-scale micro-reserves in the Ferghana Valley for the conservation of natural complexes and the defense of their diversity and for the defense of essential inner species of geosystems.

There have always been canals and woods and ditches in the Ferghana Valley, and roads lined with thick rows of plant and non-plant trees. I. Abduganiev et al. (1995) argued that available green plants along highways absorb most of the emissions from motor vehicles and have a positive effect on the natural purification of urban air. Currently, 110-120 thousand tons are produced annually in Fergana oblast and 50-60 thousand tons in Osh oblast. Gaseous wastes rise into the air. In this condition, the function of waste assimilation on the sides of urban roads will be even more intensified. Balsamic poplar, white oak, white acacia, ailanthus, linden, common sycamore, chestnut, sycamore, sycamore, hanging birch and others are considered to be the most resistant to waste and effective in its treatment. This idea can be considered the most suitable option for air purification from waste in difficult orographic and climatic conditions of the valley, as well as in densely populated areas [4].

The nature of the Fergana Valley is under pressure from two sides: the first is the process of complex and extensive use of natural resources, the second is the pressure associated with industry and other types of social production (Sultonov, 2001). The extensive exploitation of natural resources continued until the 1990s (and in some places even today). With high population densities and shortages of space, liquid and pastures, it is evident to everyone how important it is to utilize them carefully. However, in reality, in some places we can see waste, irholisticity and qualitative degradation of materials, which results in undesirable phenomena such as poorness, synthetic scarcity, and departure from circulation. This, in turn, negatively affects sustainable farm growth, resulting in severe socioeconomic consequences. It is optimizing the impact of anthropogenic factors on the environment. The author characterizes the impact of industry, transport and construction on the natural environment as the second largest anthropogenic impact, emphasizing its importance. This is logical, and the scale, consequences and strength of this pressure are increasing. Particular attention should be paid to this in the valley context. In the future, both types of oppression should be prioritized and practical goals set accordingly. In this respect, Y. Sultanov's geo-ecological idea is characterized by depth, practicality and a wide range of impact [5].

Due to the large-scale development of farming and manufacturing in industry in Namangan province, interconnected geo-ecology problems appear in the region. A. Kazakov, K. Boymirzaev et al. (2003) reported that in recent years, due to the very low temperature of water coming from the Tokhtagul reservoir through the Naryn River, its apply to watering of agricultural crops has led to a postpone in their vegetation in the average of 20-25 days, consequently in a 25-30 % reduction in yields [6]

In this years, large areas of land in the foothills of the Kuramin and Chatkal mountains have been developed, irrigated areas have been expanded, new reservoirs and floodplain ponds (more than 20), more than 100 pumping stations, and the Big Namangan and Chust canals have been put into operation. Starting from 1994-1995 there has been an increase in groundwater level on the plains located between the hills of Yangikurgan and Uichi-Chartok districts. The authors present relevant ideas to address broad issues of great practical importance. Rising groundwater levels in the foothills and in the plains behind the mountains lead to soil salinization and waterlogging. It is said that the occurrence of this hydromorphic process in place of automorphic conditions will adversely affect the process of soil formation and development. This ecological and landscape problem should be deeply studied and solved positively. Since ancient times, people have always kept floodplains open and ready for floods. The practical appeal of the authors is that flood plains and flood plains should always be kept ready.

2 Materials and methods

Altitude-landscape zonality in the Fergana Valley is expressed quite clearly, in a classical manner. Each region has its own natural and historical features, and the diversity is vividly expressed in its environment, resources, economy and population distribution. The use of nature is also unique and regional. Regions with mountain landscapes are intimately related, and the exchange of matter and energy is of great significance in this process. From this point of view, the scientific views and ecological-geographical ideas of K. Boimirzaev and A. Nazarov (2002) are of much significance [7].

The authors identified hilly and plain landscape regions, analyzed scenery and environmental issues existing in each of them, and created alternatives for their enhancement. Modifications in the natural occurrences of the Fergana Valley as a result of anthropogenic impact make it of much practical importance to conserve some of its characteristic areas in their "natural state". According to I. Abduganiev et al. (1995), tiny reserves and micro-preserves play an important part in protecting important local features of geosystems. It is time to create such micro-reserves on the territory of each collective farm. This is a very good idea, even if it were implemented on every farm with a large territory, it would be a big undertaking.

Under conditions of land scarcity, development of foothills is one of the important opportunities for development of irrigated farming. In this regard, it is necessary to consider justifications and ideas of geographers, taking into account opinions of specialists and farmers, as this logical holistice considers natural, peasant, and economic factors in a single intricate. With regard to, A. Maksudov et al. (1998), S. Jalolov et al. (2001), A. Ibragimov et al. (2001), A. Kazakov et al. (1983, 2001), Y. Sultanov (1999), A. Nigmatov et al. (2000) [8]. The scientific views of K. Boimirzaev et al. (2002) contain quite reasonable ideas. First of all, it depends on specific natural conditions of the arid region being developed, in particular, lithologic, geomorphologic and soil conditions. During development it is important to pay attention to such factors as soil thickness, presence of salts and gypsum, and during irrigation - to such factors as lithological composition of soil, steepness of relief, efficiency of water use, level of contemporary advancements in irrigation methods. Prevention of all types of erosion, water use, prevention of erosion and improvement of fertility of the soil are the most necessary and required suggestions. They are presented in the perceptions and views of the above mentioned authors.

3 Results

Diverse natural and natural-anthropogenic geosystems were formed in the highland areas of the Fergana Valley: deserts, hills, mountains, pastures. This is an ancient irrigated oasis, fast-growing and densely populated industrial-agrarian region with developed industry, farming, transportation and other branches of production. Thus, the nature of the valley undergoes changes as a result of anthropogenic impact and unfavorable geo-ecological problems arise in it from period to period.

The main task of geographical and natural as well as geo-ecological research conducted in the region area is to assess the current state of geosystems due to develop techniques and actions that will allow holistic use of natural sources in it, maintaining an optimal balance between the natural evolution of geosystems and the fluctuation of anthropogenic impact. It is important to ascertain the function and significance of natural components in the formation of geosystems and the control of adverse environmental consequences of human economic occurrence [9].

When solving valley's geo-ecological issues it is also necessary to study the unusual characteristics of territorial dispersal of the existing environmental state. This is due to the fact that the nature and scale of anthropogenic impact on every geosystem, as well as the degree of geosystem stability according to this influence are different.

Thus, the natural condition and ways of its improvement in each geosystem are various. In particular, in hot areas geosystems of the region it is outlined by degradation of atmospheric air, water and soil cover, reduction or deterioration of land resources under the influence of water and wind erosion, salinization, poisoning, waterlogging, compaction, etc., changes in the composition of agricultural crops, vegetation and fauna in natural and natural-anthropogenic geosystems, eutrophication of water bodies, etc. [10].

4 Discussion

Water erosion, mudflows, change and pollution of soil cover, impoverishment of biogenic components, salinization and waterlogging of soils, etc. are observed in adyr geosystems.

Mountain geosystems are subject to water erosion, mudflows, deforestation, change and pollution of the natural environment during mining, landslides, rockfalls, karst phenomena, avalanches, earthquakes, changes in the construction of biogenic components and so on.

On geographical and ecological aspects of holistic use of liquid resources, landscape changes under the impact of human-caused variables in irrigated areas and their ecological consequences see Y. Sultanov (1973). Y. Sultanov (1973, 1995, 1999, 2001), A. Maksudov (1974, 1995), A. Kazakov, K. Boimirzaev (1995, 2003) and others have published their Scientific viewpoints. In reality, most of the water from the rivers flowing through the valley (Akbura, Aravon, Isfairam, Sokh, Kasonsoy, etc.) is used for irrigation before reaching the Syr Darya. Extensive irrigation and irholistic water use led to the emergence in atypical landscape characteristics in the province of Central Fergana for the valley, such as waterlogging of lands, rising groundwater table and others. As a result of irrigation of Burgundy steppes and uplands in the top part of the Sokh plateau, the water under the table has risen, which worsened the ameliorative condition. In addition, a similar condition is observed around the reservoirs located in the valley. Water resources, especially groundwater, are polluted by industrial enterprises.

In oasis geosystems, nitrogen, phosphorus, potassium and other chemical substances entering water bodies with wastewater cause deterioration of water quality and its pollution (eutrophication). All this makes it necessary to develop measures for holistic use of surface and ground waters of the valley. The emergence of hydromorphic processes related to water, replacing automorphic conditions, also leads to changes in natural landscapes. Holistic use and protection of land resources in the Fergana Valley is a task of great national importance. The lands in the valley are very limited, so it is necessary to develop and implement intensive systems of land resources utilization management. Today, land resources are decreasing from year to year due to pollution, waterlogging, salinization, soil erosion, compaction, as well as use of land for construction and other purposes. At present, the main problem is that for intensive use of land resources and soils of the Fergana Valley it is advisable to organize forest plantations in the western parts of the region to preserve the humus layer of soil and reduce wind erosion (Mirzajonov, 1973). To prevent salinization, waterlogging and other environmental problems in Central Fergana, the issues of drainage systems management and solving problems related to water erosion (gullies, karst, gullies, erosion, etc.) on mountain plains, uplands and foothill sediments are now becoming topical (Sultonov, 2001).

The western part of the valley is exposed to the "Kokand" and "Bekabad" winds. When these winds blow, they cause great damage to agricultural lands. As a result of wind erosion, cotton yields in this region have decreased from 34.5 quintals to 22.1 quintals. The main reason for this is that according to the idea of K.M.Mirzajonov, forest plantations were established in the areas of the Kokand zone since 1962 to prevent soil erosion. Later, similar incidents occurred as a result of felling of these neighborhood forests by the population. It is time to further develop this idea showing the best way to control wind erosion in valley conditions.

In addition to soil erosion, large areas of land in the southeastern, southern and southwestern parts of the valley are highly saline, which in turn reduces crop yields and leads to degradation of soil structure. Under conditions of intensive agricultural development, great attention should be paid to desalinization of these saline lands, groundwater utilization, and open and closed drainage systems [2].

Since the second half of the last century, due to the development of large territories in the Fergana Valley, the area of plants that grow naturally has greatly reduced. Development of farms by comprehensive manner concurrently resulted in reduction of pastures territory. In conditions of Fergana valley it is practical for growing feed plants for animals not only through rotation of crops, but also by organizing agrophytocenoses in farming zones. It is known that the flora of the valley is very diverse, and this diversity is especially clearly manifested in the altitudinal zonality of its landscapes. R. Shonazarov (1995) noted that there are also widespread areas of medicinal plants in the valley. Therefore, to preserve the flora and useful plants of the region, it is necessary to avoid involvement of typical landscapes in agricultural production and create agrophytocenoses [3].

Nowadays, due to the influence of human impact, more than 60% of the region territory has acquired Anthropogenic and natural-anthropogenic appearance landscapes. This is because fast mankind growth in the valley, wide development of farming, the accessibility of hydro resources, straightforward access to natural resources and others.

The idea of conservation of environmental landscapes and extensive utilization of the current anthropogenic forms of ground, and what branches of finance can be formed in each vertical valley, is carried out by A.A.Rafikov and A.A.Nazarov (2002). This idea will offer security that is both practical and scientific of the ecological and geographical basis of the plain, ridge and hill geosystems of the Fergana Valley [8].

Natural landscapes around the cities of the Fergana Valley are changing extremely with the influence of anthropogenic factors. In this regard, Kyrgyz geographers S. Ergashev and others conducted a thorough examination of the modifications in city landscapes under the impact on anthropogenic factors, showing the reduction in the region of natural landscapes in proportions [10].

The concept proposed by I. Abduganiev (1995) emphasizes that in order to improve the atmosphere of places, much focus should be on the kinds of plants to be placed. Here, the primary focus is paid to the ability of trees to absorb dust, various odors and toxic gases, and the ability of different types of trees planted side by side to rise to the sky is also taken into account.

This idea should be realized as soon as possible. As currently the level of air pollution in the cities of the Fergana Valley is very high. Implementation of this idea will simultaneously stimulate building of highways and development of hygienic protection zones around industrial enterprises.

The current ecological and geographical condition of both man-made and natural landscapes of the Fergana region and problems of their optimization are examined by local geographers, including Y. Sultanov (1965, 1974, 1995, 1999, 2000, 2001), I. Abduganiev (1995, 1999, 2000, 2001), K. Boimirzaev (1995, 1999, 2001, 2002), A. Nazarov (2001, 2002), O. Abdullaev (1995, 1999, 2001, 2002), A. Khamidov (1995, 1999, 2001, 2002), R. Kholikov (1989, 1995, 1999, 2001, 2002) and many other scientists participate. They expressed their natural-geographical and geo-ecological ideas, thoughts and views on improving the geo-ecological condition in the region. These views and system of measures were widely covered in the materials of scientific and practical republican and international conferences devoted to the ecological state, nature and holistic use of natural resources of the Fergana Valley, held in Fergana in 1995 and 1999 and in Namangan in 2001-2005.

In general, to date, natural-geographical, geo-ecological and bio-ecological studies have been conducted in various landscapes of desert, steppe, mountain and pasture areas of the Fergana Valley, as a result of which a system of directions and measures to improve the geo-ecological condition of the region and holistic use of its nature has been developed (Y. Sultanov 1995, 2001; I. Abduganiev 1995, 2001; K. Boimirzaev 1999, 2001; O. Abdullaev 1995, 2001; R. Kholikov, A. Kazakov 1995, 2001; A. Khamidov 1995, 2002; A. Nazarov 1999, 2001; S. Ergashev 1995, 2002, etc.). The time has come to develop plans and programs for 3-5-10 years at the level of regional khokimiyats in order to consistently reduce air, water and soil pollution by industrial and transport wastes, as well as the death of plants and animals due to anthropogenic factors. In this regard, the task of implementing a plan to reduce the amount of emitted waste more or less two times, and then by 75%, the long term by 90% by putting together all enterprises and organizations that pollute the environment and liquid bodies of the region with the up to date apparatus for purification of smokes, dust and wastewater or rebuilding of some of the traditional industrial enterprises according to a latest project becomes urgent [10].

Since the Fergana Valley region continues to develop economically within the path of agro-industrial development. Thus, the geoecological condition of the region should be assessed based on the above-mentioned natural-geographical and geoecological factors (Table 2), that is, up to the geomorphological and lithological and soil circumstances of flat, hilly and mountainous areas developed for farming, soil density, the presence of salts and during irrigation, more attention should be paid to the lithological makeup of the soils, the slope of the relief (especially in hilly and mountainous territories), etc. The most significant and relevant actions today are the effective utilization of territorial waters, the development of modern irrigation techniques (modified to local conditions), the maximum attention to the prevention of all types of soil erosion and at the meantime great attention to natural protection in the areas of industry, building, transport and facilities, the creation and application of holistic methods of utilizing environment and its sources, and most importantly - increasing the environmental literacy, consciousness and tradition of the population [8].

To conclude, it can be said that it is appropriate to address issues of preservation of the environment, holistic use of natural resources and development of territories not along with political and administrative boundaries, but with natural and geographical boundaries. A salient example of this is the Fergana Valley.

5 Conclusion

As written above, the results of studies conducted in the process of investigating the components of nature, including landscape, cartographic, natural and geographical and network zoning, thereby made it more significant that geoecological research in the region.

Changes in the natural environment and deterioration of the geoecological condition in the Fergana Valley depend on a number of other elements, such as its geographic position, particular geographical and natural circumstances, and current regional and local laws and rules, environment and variability of typical complexes, the population's economic activities and the distinctive characteristics of production prosperity.

The primary trends that identify the geoecological Natural-geographical research's importance in the valley are primarily the following:

- creation of different natural and natural-anthropogenic landscapes in regions with high mountains: deserts, mountains, pastures, use of terrain and mountain chains;
- with the contemporary market's utilization of surface and groundwater conditions;
- in the process of using climatic resources;
- in the use of soils in farming and the economy as a whole;
- exploitation of plant and animal resources and their disappearance or reduction of range;
- in industrial production and use of resources;
- development of transport, construction, utilities, etc.

On our part, local scientists have put forward concepts for the prevention and solution of these geoecological problems that have been formed and developed over many years. The main question now is how to implement this in real world.

In conclusion, the life itself proves that developing geoecological foundations for nature protection and holistic utilization of natural sources in the valley, assessing its geoecological importance, it is practically correct to work not on political and administrative borders, but at the degree of natural geographical boundaries.

The Republics of Uzbekistan, Kyrgyzstan and Tajikistan need to join forces to tackle geoecological issues in the Fergana Valley. It is real that each of these republics is growing towards its private independence does not let them to jointly tackle more geoecological problems, hence there are exact objective and biased explanations. It is difficult to imagine the geoecological significance of thorough investigation in the Fergana Valley without the partnership efforts of these republics. Because the people living here live on the same territory, profess the same religion, use the same water and soil, and culture and horizon are close.

The participation of natural components of the territory throughout the economic process production has established particular geoecological conditions that determine the increasing geoecological significance of comprehensive natural-geographical research. It was at this third major stage that the study of natural elements entered the branch of investigation related to the enhancement of their ecological conditions.

Specifically, if the investigation on the study of natural components conducted at the first and second major stages, based on the stage of their development in space and time, continued in complex natural-geographical directions, then in the fifth, sixth and seventh aspects of the cycle the research process began to acquire geoecological significance.

As is known, the Fergana Valley is a unique basin with an area of 78.9 thousand km2, surrounded by mountains. The territory of the republic borders on the states of Uzbekistan,

Kyrgyzstan and Tajikistan, and changes in its natural environment and the aggravation of the geoecological condition depend on a number of other factors, including the geographical location, specific natural-geographical conditions and factors, current regional and local natural patterns, stability and fluctuation of natural components or compounds, the nature of economic motion of the people, and the unusual characteristics of production development. The distinctions between these claims on problems of territorial character, utilization of their resources and protection of the environment give rise to a particular goal and subjective explanations. On the other hand, political and administrative borders cannot be an obstacle to the creation of geoecological and geological foundations for the preservation of nature and holistic use of sources; such work can yield positive results only if it is carried out within natural borders.

Currently, there are geoecological problems of the integrated use of the nature and resources of the Fergana Valley, including the development of industrial production, transport, construction and public utilities. From the point of view of the geoecological significance of solving these problems, local scientists have developed common perceptions, but administrative boundaries hinder their implementation in work.

It requires the cooperation of this region's countriesto bring intricate natural-geographical studies to the degree of geoecological investigation in solving geoecological issues. Only under the situation of all-encompassing holistic use of the nature and resources of the Fergana Valley, its protection, will the geoecological significance of complex natural-geographical research increase. In order to identify and demonstrate the geoecological significance of network and natural-geographical research in the Fergana Valley, as well as an accurate assessment of the current geoecological condition, we carried out the this work:

- data that has been completely and partially examined via maps of field studies conducted in the region, reporting information and archival data of organizations and institutions conducting research in the Fergana Valley of the Republics of Uzbekistan, Tajikistan and Kyrgyzstan, as well as observational resources collected in the field;

- The results of geological and geomorphological studies, average long-term meteorological data, data on improving the classification of soil types, stages of development of natural-geographical studies, periodization of formed geoecological problems, concepts for solving geoecological problems, dynamics of studies on nature conservation and holistic use of natural resources have been developed and presented in the form of tables, - Maps have been compiled for the network of meteorological stations operating in the region, the dynamics of the reduction of the area of natural vegetation, the dynamics of the development of hydrological studies, some comprehensive studies conducted, improvement of natural-geographical studies, leveling of the Fergana Valley and the surrounding mountains and the Southern Fergana's scenery (Oston-Karachatyr), improvement of natural-geographical zoning, the current ecological state of nature conservation.

References

1. O. I. Abduganov, B. Azamova, A. Mirzaliev, "Problems of ecological tourism development in protected natural areas", Economy and Society **12-1(91)**, 708–712 (2021)
2. Y.-J. Ahn, Z. Juraev, "Green Spaces in Uzbekistan: Historical Heritage and Challenges for Urban Environment", Nature-Based Solutions **100077** (2023), https://doi.org/10.1016/j.nbsj.2023.100077
3. F. Yang, H. Yuan, N. Yi, "Natural Resources, Environment and the Sustainable Development", Urban Climate **42**, 101111 (2022), DOI: 10.1016/j.uclim.2022.101111
4. N. Dudley, H. L. Timmins, S. Stolton, J. E. M. Watson, "Effectively Incorporating Small

Reserves into National Systems of Protected and Conserved Areas", Diversity **16(4)**, 216 (2024), https://doi.org/10.3390/d16040216

5. T. T. Gatiso et al., "Sustainable protected areas: Synergies between biodiversity conservation and socioeconomic development", People and Nature **4(4)**, 893–903 (2022), https://doi.org/10.1002/pan3.10326
6. R. E. Grumbine, J. Xu, "Five steps to inject transformative change into the post-2020 global biodiversity framework", BioScience **71(7)**, 637–646 (2021), https://doi.org/10.1093/biosci/biab013
7. A. M. Mahmud, M. G. Nosirov, T. Mukimov, D. S. Normuradov, Kh. Khodjibabayev, B. Ismailkhujaev, S. Urokov, "Multi-faceted analysis of land use impact on rangeland health: Insights from normalized difference vegetation index assessment in stream, road, and mining areas", Journal of Ecological Engineering **26(1)**, 196–203 (2025), doi:10.12911/22998993/195472
8. R. Pirnazarov, S. Usmonova, "Damaged lakes in Uzbekistan and adjacent mountainous areas, their research and the aggravation of problems associated with them", FarSU Scientific Reports **3**, 282–289 (2024)
9. S. N. Mikhailovna, S. L. Nailevich, K. R. Ilsurovich, "Ecological Problems and Financing Environmental Protection Actions", Caspian Journal of Environmental Sciences **19**, 739–744 (2021)
10. Statistical Newsletter of Fergana Region, Statistics Department of Fergana Region (Fergana, Uzbekistan, 2022), https://farstat.uz/uz/

E3S Web of Conferences **633**, 07003 (2025) https://doi.org/10.1051/e3sconf/202563307003
GASES 2025

Geoecological foundations strengthening the stability of the national system of protected areas of the republic of Uzbekistan

Yusupjon Akhmadaliev*, *Olimjon* Abduganiev, *Elyorbek* Makhkamov, *Dilshodbek* Kosimov, and *Nargiza* Komilova

Fergana State University, Fergana, Uzbekistan

Abstract. This study analyzes the current state, spatial structure, and development prospects of the system of protected natural areas (PNAs) in the Republic of Uzbekistan, based on the objectives set by the Kunming-Montreal Global Biodiversity Framework (GBF). Structure, coverage and integrity of the existing network of PNAs in Uzbekistan determined significant disparities among them. The analysis of them confirmed that in the regions of Uzbekistan, the coverage of PNAs by the main IUCN categories (I, II, III, IV and V) is 1.5% in Namangan region, 0.2% in Fergana region, and 0% in Andijan region were analyzed. The formation of a system of protected natural areas in these regions is guided by scientific principles aimed at preserving biological and landscape diversity, improving the ecological situation, and ensuring the rational use of natural resources. To enhance the functionality, completeness, and integrity of the protected natural areas system in the Fergana Valley, proposals were developed for establishing PNAs types.

1 Introduction

Today, the "Great Acceleration" offers countless opportunities and benefits to human society, but is also leading to the disappearance of nature and its biodiversity at an alarming rate. The world prioritizes sustainable development by conserving biological and landscape diversity, preventing ecological and social problems and improving the environment through the conformation of international ecological networks. The concept of sustainable development points out that providing the necessary environmental sustainability, as well as, opportunities for ecotouristic development are defined by the creation of a integrated system of different levels in the regions of Uzbekistan [1, 2].

Andrea Santangeli et al. (2023) suggest that while the existing protected area network can play a role in slowing the decline of species occupancy rates, it is not sufficient on its own to stop the biodiversity crisis. To enhance the effectiveness of protected areas and reverse biodiversity loss, efforts must focus on expanding coverage, improving connectivity, and strengthening management practices

.

* Corresponding author: yu.ahmadaliyev@gmail.com

Equally important is the need to rectify historical socio-economic disparities between the Global North and the Global South, ensuring equitable financial and institutional support for biodiversity initiatives in developing regions.

Empirical evidence suggests that biodiversity conservation efforts should extend beyond merely expanding the coverage of global protected areas. A more comprehensive approach is required, one that integrates biodiversity considerations into all sectors of society, ensuring its mainstreaming across economic and policy frameworks. Furthermore, conservation strategies must be adapted to working landscapes, fostering sustainable land-use practices that support both ecological integrity and human livelihoods [7].

2 Materials and methods

The World Database of Information played an essential role in the development of the international MET system. Since 1950, MNEs number in the world has increased 20-fold and their sizes have increased 13-fold. In 2025 February, WDPAs issue contained 305,196 items of information, of which 303,313 were from multinational corporations. Of these, 293,260 relate to MNEs with area facilities and 11,936 to point facilities. The data cover 244 countries and territories. The coverage of terrestrial and inland water bodies as MEAs in the Protected Planet (WDPA) report database for February 2025 is 16.44%. 8.34% of ocean and sea areas have been designated as MEAs (Figure 1). The data shows that the area of protected areas in the Republic of Uzbekistan is 42,788 km2 and the coverage rate is 9.53% [1].

Despite a number of economic-social commitments in these directions at the international level, the qualitative reduction of BM continues. The realization of Goals 14 and 15 set out in the Sustainable Development Resolution 2030 of UN is inextricably connected to the Aichi Strategic Plan's implementation. In 2022, at COP-15 adopted the Kunming-Montreal, that set the goal of expanding of Specially Protected Areas system in the world [3, 8]. Thus, looking beyond 2020 it will be essential to ensure that future targets are not only ambitious but also measurable across all aspects of what makes protected areas effective. Associated with this will be a need for target setting to prescribe and support the collection of data to assess and evaluate future targets [6].

Fig. 1. Protected areas and OECMs of the world.

According to the Ministry, Uzbekistan has 7 nature reserves (222,397 ha), 1 complex (landscape) reserve (628,300 ha), 2 biosphere reserves (111,670.61 ha), 12 natural parks (4,257,661 ha), 11 natural monuments (3,311.7 ha), 12 wildlife sanctuaries (1,081,396 ha) and 1 "Jayron"preservation (16,522,000 ha) with a total area of 6,321,258.21 ha. There are also 66 forestry farms, 5 forestry hunting farms, 4 research and experimental stations and 9 specialized forestry farms with a total area of 11,738.1 thousand ha (Table 1) [2].

The main criteria and principles for the organization of the IET system were developed in the Convention on Biological Diversity (1992) and the Common European Strategy (1995). To date, international organizations and researchers have developed many criteria and methods for comprehensive assessment of the IET system. For example, international organizations such as WWF, IUCN, WCPA, TNC and others have assessed IBM as a reliable tool for protecting nature, ensuring sustainable use, maintaining ecological stability and etc. WWF has implemented a methodology to assess the effectiveness of nature conservation in the MEOPs and their regional system. In recent years, there has been a shift from the approach of creating protected areas in individual regions to the concept of nature conservation as a whole. These approaches are aimed at assessing the key functions of the SMEs and determining the effectiveness of the system. The analysis of the system of specially protected natural areas is carried out according to the following indicators: the coverage level and land categories.

According to Melissa Penagos Gaviria et al. (2022), an analysis of Asian regions revealed that Western Asia had the lowest values for both the Percentage of Landscape and the Proximity Index. Their study found that only 15% of Cost-Effective Zones'(CEZs) in Asia were designated as protected areas. While a larger number of CEZs were situated in interior regions, the highest-priority zones (9%) were concentrated along national borders. The authors emphasize the importance of expanding the coverage of protected areas, with a particular focus on enhancing connectivity—such as through transboundary conservation initiatives—to ensure the preservation of their ecological functions [10].

3 Results and discussion

The Aichi Strategic Plan 2010-2020's goals and targets (covering 17% of terrestrial ecosystems, 10% of coastal and marine areas) have not been equally realized in all countries. In addition, discussions are underway regarding the new Global Biodiversity Framework, which seeks to ensure the protection of at least 30% of the planet's terrestrial and marine areas by 2030 [2].

Table 1. Protected Natural Areas (PNAs).

I – category PNAs		III – category PNAs		IV – category PNAs		V – category PNAs	
State reserves (ha)		**National natural parks (ha)**		**Natural monuments (ha)**		**Wildlife sanctuaries (ha)**	
1. Zaaminsky	26,840	1. Zaamin	24,110	1. Vardanzi	124	1. Arnasay	63,300
		2. Ugam-Chatkal	506,941	2. Mingbulak	1,000	2. Dengizkul	50,000
2. Nuratis	17,752			3. Chust	96	3. Karakir	30,000
		3. Zarafshan	2,426.5	4. Yazyavan steppes	1,471.5	4. Sudochye-Akpelki	280,507
3.Gissarsky	(80,986)	4. Khorezm	21,687.5	5. Akbarobod	39.5	5. Mubarek	264,469
	78,986	5. Kitab geological	3,938	6. Zilkha	22.2	6. Aktau	15,420
4. Kyzylkum	10,311			7. Bustonbuva	8.5	7. Karnabchul	25,000
		6. "Southern Ustyurt"	1,447,143	8. Yangibazaar	470	8. Koshrabat	16,500
5. Surkhanskiy	23,802			9. Paykent	30	9. Nurabad	40,000
		7. "Central Kyzylkum"	1,200,000	10. Varakhsha	7	10. Kumsulton	4,900
6. Chatkal biosphere	24,706			11. Urungach	43	11. Khadicha	11,300
		8. "Aralkum"	1,000,000	**Total**	**3,311.7**	12. Barsakelmes (2022)	280,000
7. Aktau-Tamdinssky	40,000	9. "Pap"	10,000	**State biosphere reserves (ha)**		**Total**	**1,081,396**
Total	**(224,397) 222,397**	10. "Upper Tupalang"	27,851	1. Lower-Amudarya	68,717.8	12.BSP "Djeyran"	16,522
II – category PNAs (ha)						**Total for category**	**1,097,918**
Complex (landscape) reserves		11.Balbatag	12,064	2. Ugam-Chatkal	42,952.8	**National parks (ha)**	
"Saiga" KLZ	628,300	12. "Omankutan"	1,500			Durmen	32,4
Total	**628,300**	**Total**	**4,257,661**	**Total**	**111,670.6**	**Total**	**32,4**
Today in the republic there are 7 state reserves, 1 complex (landscape) reserve, 12 natural parks, 1 national park, 11 natural monumants, 2 state biosphere reserves, 1 specialized nursery "Jeyran". Their total area is 6,321,258.21 hectares. According to the classification of the International Union for Conservation of Nature (IUCN), the area of protected areas of categories I – IV and biosphere reserves is 6.321 million hectares or 14.08% of the total area of the country.						**VII – category PNAs**	
						Territoriesnfor management of selected natural resources	11,738,100
						Total	**11,738,100**

**PNAs in Uzbekistan by categories*

According to IUCN requirements, the main focus is to take into account categories: I, II, III, IV. When analyzing the composition of the existing IUCN network in Uzbekistan by natural-geographical regions, it was determined that there are significant differences between them. For example, a higher percentage of MES (I, II, III and IV) than the national average is characteristic of the Tashkent region (37,5 %) and Karakalpakstan (22,2 %). The share of specially protected natural areas, which constitute the core of ecological networks of our republic, is 11,41%, including biosphere reserves – 0,24%, state reserves – 0,49%, landscape reserves – 1,40%, natural parks – 9,48%. In this respect, natural parks are characterized by optimal sizes. The activity of natural parks is mainly organized in Tashkent, Samarkand, Namangan, Khorezm, Jizzak, Kashkadarya, Navoi, Surkhandarya, Karakalpakstan (Figure 2, 3).

Natural monuments of republican significance (0,007 %) are located mainly in Fergana, Namangan, Khorezm, Bukhara and Tashkent oblasts. The area of protected areas is 2,4 %; they have not been established in Andijan, Fergana, Syrdarya, Khorezm, Surkhandarya and Tashkent oblasts. Forestry enterprises (20,7 %) have been established practically in all regions of our republic. This will ensure the integrity of ecological corridors on the scale of Uzbekistan (Table 2). According to the above analysis, the lowest indicators by areas and category composition of TPL in our republic are found in Syrdarya, Andijan, Fergana and Namangan. Representativeness of NWOs in these regions indicates that they do not meet the requirements of the Biodiversity Conservation Strategy [1].

In the Fergana Valley, it was limited to the creation of one national natural park and a total of 7 natural monuments of national monuments (Mingbulak, Chust, Central Fergana, Yazyavan steppes, Akbarabad, Zilkha, Bustonbuva). This situation limits the possibility of organizing large TPLs. The indicator of coverage of the main categories of TPL is 1,5 % in Namangan, 0,2% in Fergana and 0% in Andijan. Analysis of the composition of TPLs in these regions by category, area and landscape shows that they do not fully meet the requirements of ecological networking, and TPLs are very fragmented. In 2022 year, the Pop Nature Park (IUCN, II; 10,000 ha) was established in Namangan, while in Fergana and Andijan a protected regime for wildlife conservation and large-scale protected areas have not been established. Therefore, it is advisable to initially conduct comprehensive studies on increasing the area of protected areas up to 17% by administrative units and natural-geographical regions. The above analysis shows that improvement of the system and structure of SMEs in Namangan, Fergana and Andijan regions is one of the urgent tasks.

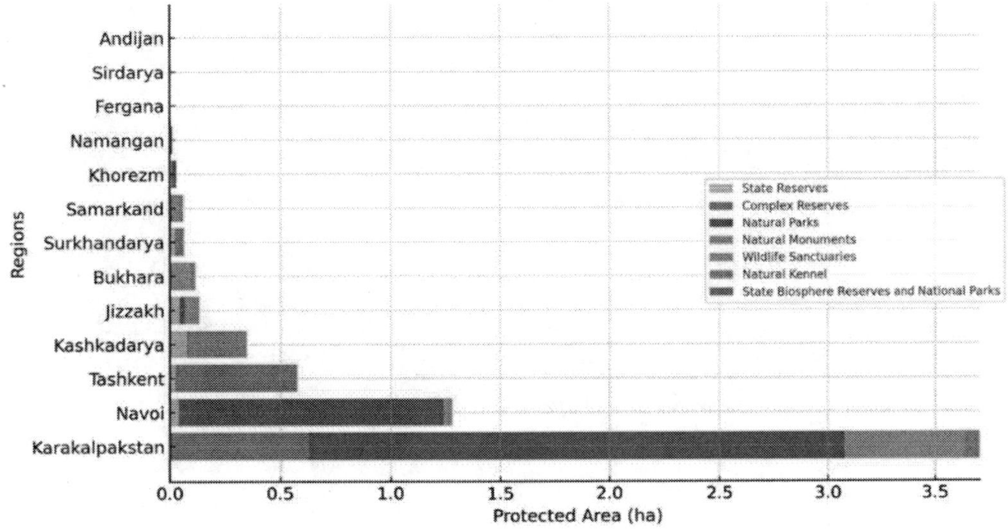

Fig. 2. Protected Natural Areas in Republic of Uzbekistan (1=1000000 ha).

According to Nigel Dudley et al. (2024), conventional wisdom suggests that large protected areas are more effective than smaller reserves. This perspective is based on island biogeography theory, economies of scale, and the necessity of maintaining viable populations of large predators and wide-ranging species. However, this viewpoint tends to overlook evidence showing that strategically located smaller reserves play a crucial role for many species, particularly in landscapes that have been significantly altered by human activity. Additionally, many countries are either unwilling or lack the capacity to establish new large protected areas [4].

According to Gatiso et al. (2022), balancing conservation efforts with socioeconomic development objectives is crucial for sustainability; however, it continues to be a topic of intense debate. Protected areas (PAs) are often regarded as vital in addressing these potentially conflicting goals. Nevertheless, there is insufficient evidence to confirm whether PAs are successfully achieving both objectives at the same time [5,9].

In recent years, researchers have shown that specially protected natural areas, organized as separate territories or objects, cannot fully ensure the protection and rational use of nature [2]. METs are an important component of global environmental policy, and their effective management serves to ensure ecological balance and sustainable development. Therefore, at the international level, MET system development is coordinated by organizations such as UNEP, WCMC, WCPA, IUCN and WWF.

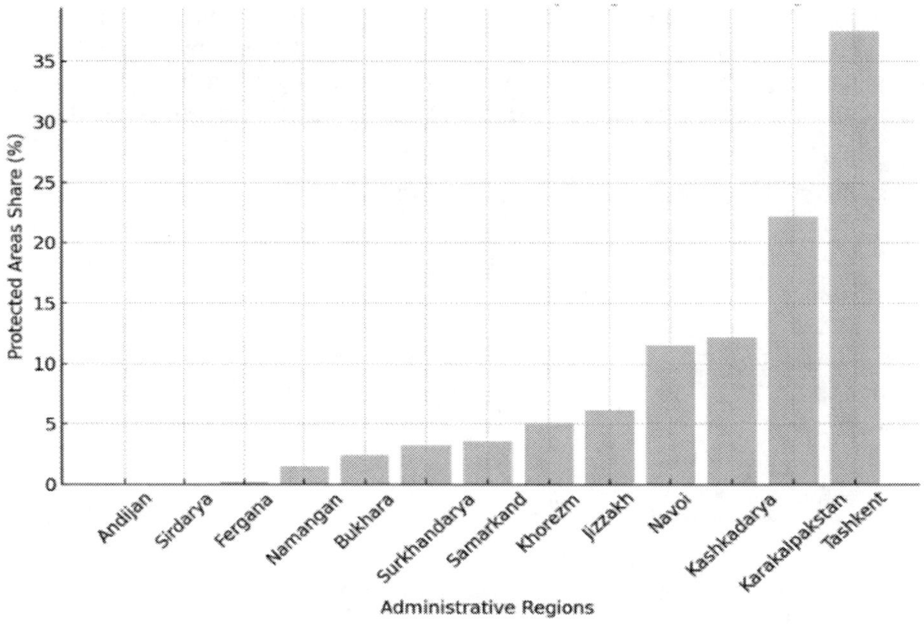

Fig. 3. Protected Natural Areas share of the Republic of Uzbekistan (%).

The system of specially protected natural areas should be able to solve such tasks as preservation of unique natural objects and complexes, genetic fund of plants and animals, optimization of the negative impact of human activity on nature, study of natural processes and environmental monitoring, development of ecological tourism, increase of the level of environmental awareness and education.

Problems of nature protection can be solved only if they are integrated into programs and plans of sustainable development. Expansion of the areas of protected natural territories is defined as one of the priority directions in the strategic plans of Uzbekistan on sustainable development and biodiversity conservation, as well as in other normative and legal documents. Uzbekistan has undertaken to set its own national targets for the implementation of the Kunming-Montreal Global FPB. Nowadays, 14,08% of the territory of Uzbekistan is classified as IUCN categories: I, II, III, IV (Table 2). This indicator characterizes the fulfillment of tasks set by the State Program for the development and expansion of the MOST system for 2021-2028 for the period up to 2025. However, this indicator is below 30% (target 11 Aichi CBD) recommended by the Global Biodiversity Framework of Kunming-Montreal. At the same time, within the administrative-territorial units of Uzbekistan, there are various limitations regarding the size, overall structure, organization and principles of financing the system of ecological educational institutions.

Today, one of the urgent tasks is to achieve the establishment of all categories in all regions of Uzbekistan (especially in Syrdarya, Andijan, Fergana and Namangan) and to improve them at the level of ecological networks. Therefore, attention was paid to analyzing the composition of IUCN and the territory they occupy based on the requirements set by IUCN and the Convention on Biological Diversity. To increase the coverage of IET in the Fergana Valley up to 17%, a comprehensive assessment of the ranges of species included in the "Red Book" of Uzbekistan and landscapes with the highest "biodiversity" is necessary.

By giving such "reserves" the status of protected natural areas, protection of rare and endangered species can be ensured.The continuity of the ecological corridor in the Fergana Valley can be ensured by using river beds and bends, river valleys of different sizes, watersheds and mountain ranges, protective forest strips, water protection zones, water bodies, green zones along highways and settlements, as well as groves and hedgerows around agricultural crops. In addition, ecological corridors may consist of several small isolated areas.

4 Conclusions

The indicator of ensuring sustainable development of economic, social and political spheres, is determined by the level of solving environmental problems. Therefore, when solving these issues, special attention is paid to the creation of an ecological network structure that will serve as the basis for preserving biodiversity and ensuring environmental sustainability.

1. Analysis of the composition, occupied area and landscapes of specially protected natural areas of the Fergana Valley (Namangan, Fergana and Andijan regions) by category shows that they do not meet the requirements set out in the Uzbekistan's Strategy for the Conservation of Biological Diversity for the period 2019-2028. The lowest level of coverage of SMEs in our republic (17%) was observed in the Namangan, Fergana and Andijan regions. This indicator for the regions of the Fergana Valley is the lowest in comparison with other regions of Uzbekistan, and from the SMEs that perform the function of the central core, only one national natural park and seven natural monuments have been created. In the Fergana Valley, the potential for expanding the ENP turned out to be quite high in the Pap, Sokh, Uzbek, Chust, Yangikurgan and Besharik districts. In areas where the possibility of creating ENPs is extremely small and insignificant, it is advisable to create micro-reserves, natural monuments, ecological zones with a strict regime of nature management, and expand ecological corridors.
2. The highest level of "focus on biodiversity" falls on the Kuramin and Chatkal mountains of the Fergana Valley, Sokh and Shakhimardan, Central Fergana regions, the upper reaches of the tributaries of the Chodak river, the Syr Darya taiga, the Kampirkak, Kirtashtov, Teshiklitash, Chilustun and Karatov mountains. These territories are located in the Pap, Sokh, Uzbekistan, Besharik, Chust and Yangikurgan districts and provide an opportunity to organize the main categories of ENPs (I, II, III, IV).
3. Creation of new EIAs, implementation of projects based on the analysis of environmental experts (local scientists) and GIS, as well as the connection of the distribution areas of species included in the "Red Book" and landscapes with the highest "biodiversity orientation" through ecological corridors are urgent tasks. Considering the socio-economic capabilities of the Fergana, Namangan and Andijan regions, as well as their geoecological and natural-geographical conditions, this is possible to improve the system and form ecological networks by creating state reserves, natural parks, sanctuaries and micro-reserves.

References

1. Y.-J. Ahn, Z. Juraev, "Green Spaces in Uzbekistan: Historical Heritage and Challenges for Urban Environment", Nature-Based Solutions **100077** (2023), https://doi.org/10.1016/j.nbsj.2023.100077
2. N. Dudley, H. L. Timmins, S. Stolton, J. E. M. Watson, "Effectively Incorporating Small Reserves into National Systems of Protected and Conserved Areas", Diversity **16(4)**, 216 (2024), https://doi.org/10.3390/d16040216
3. T. T. Gatiso, L. Kulik, M. Bachmann, A. Bonn, L. Bösch, A. Freytag, M. Heurich, K. Wesche, M. Winter, I. Ordaz-Németh, T. Sop, H. S. Kühl, "Sustainable protected areas: Synergies between biodiversity conservation and socioeconomic development", People and Nature **4(4)**, 893–903 (2022), https://doi.org/10.1002/pan3.10326

4. J. Geldmann, A. Manica, N. D. Burgess, L. Coad, A. Balmford, "A global-level assessment of the effectiveness of protected areas at resisting anthropogenic pressures", Proceedings of the National Academy of Sciences **116(46)**, 23209–23215 (2019), https://doi.org/10.1073/pnas.1908221116
5. R. E. Grumbine, J. Xu, "Five steps to inject transformative change into the post-2020 global biodiversity framework", BioScience **71(7)**, 637–646 (2021), https://doi.org/10.1093/biosci/biab013
6. A. C. Hughes, R. E. Grumbine, "The Kunming-Montreal Global Biodiversity Framework: what it does and does not do, and how to improve it", Frontiers in Environmental Science **11**, 1281536 (2023), https://doi.org/10.3389/fenvs.2023.1281536
7. M. Penagos Gaviria, Ż. Kaszta, M. S. Farhadinia, "Structural Connectivity of Asia's Protected Areas Network: Identifying the Potential of Transboundary Conservation and Cost-Effective Zones", ISPRS International Journal of Geo-Information **11(7)**, 408 (2022), https://doi.org/10.3390/ijgi11070408
8. C. F. Rammelt et al., "Impacts of meeting minimum access on critical earth systems amidst the Great Inequality", Nature Sustainability **6(2)**, 212–221 (2023), https://doi.org/10.1038/s41893-022-00995-5
9. A. Santangeli et al., "Mixed effects of a national protected area network on terrestrial and freshwater biodiversity", Nature Communications **14(1)**, 5426 (2023), https://doi.org/10.1038/s41467-023-41073-4
10. A. A. Tishkov, "'Characteristic Space' and 'Characteristic Time' as a Key Categories of Biogeography", Izvestiya Rossiiskoi Akademii Nauk. Seriya Geograficheskaya **4**, 20–33 (2016), https://doi.org/10.15356/0373-2444-2016-4-20-33

Geoecological principles and criteria for evaluating the efficiency of regional touristic-recreational systems of the Fergana valley

Yusupjon Akhmadaliev*, *Olimjon* Abduganiev, *Elyorbek* Makhkamov, *Khayitboy* Abduvaliev, and *Zafar* Madaminov

Fergana State University, Fergana, Uzbekistan

Abstract. In the article analyzes comprehensive assessments of the geoecological, natural, historical, economic and ecological potential of the Fergana region in the tourism development. Touristic-recreational potential of the nature of Fergana region were studied and recommendations for its use, methods of assessing the touristic-recreational potential of the Fergana region have been improved based on GIS, taking into account regional characteristics. Touristic-recreational potential of Fergana region was evaluated based on natural factors, historical factors, economic factors, touristic factors, ecological situation and negative factors. The role and importance of each administrative-territorial unit in the touristic-recreational potential of Fergana region was determined. On the basis of the thematic and comprehensive evaluation of the touristic-recreational potential of Fergana region, the private (natural, historical, economic) and general typology (highest, high, medium, low) of the administrative units of the region was distinguished. With the help of this evaluation method, the results of the touristic-recreational potential determined by the specific potential of each typology can be used in the development of promising programs aimed at the development of the sector.

1 Introduction

Tourism in the world has become a global activity that knows no political, ideological, geographical and cultural boundaries. Nowadays, special attention is paid to scientific research on the use of recreational and tourist resources and the recreational and tourist potential of the territory, as well as the creation of promising programs for the development of recreation and tourism based on territorial organization and comprehensive assessment. Tourism and recreation occupy a worthy place in the state economy along with agriculture, industry, transport and other macroeconomic sectors. In the world, touristic-recreational activity is one of the leading and rapidly developing sectors in terms of economic development and creation of new jobs. According to the World Tourism Organization (UNWTO) and the World Council on Travel and Tourism (WCTT), the number of jobs in the tourism sector will reach 350 million by 2025 [4, 8].

* Corresponding author: yu.ahmadaliyev@gmail.com

Recent studies indicate that the global ecological tourism market is projected to grow by 13.5 %, reaching $428.97 billion by 2028 (WTTC, 2023). Ecotourism's dual role in supporting environmental conservation and alleviating poverty has been widely acknowledged [1, 4]. At the United Nations Sustainable Development Summit held in 2015, all member states of the UN unanimously adopted the "Sustainable Development Goals", which include 17 goals and 169 targets. Based on the background of the coexistence of crises and opportunities, the Sustainable Development Goals were aimed to create a comprehensive framework for the long-term prosperity and development of mankind. The purpose was to fully mobilize the power of various countries to carry out actions from the three dimensions of the society, economy, and environment and to provide continuous benefits for the present and future [9].

It is important to raise the economic and social potential of the regions through the development of tourism. For this, research is needed to develop the geoecological basis of using the natural, historical, economic opportunities of the regions for the purpose of tourism and recreation. In order to effectively organize tourism and recreation, it is necessary to evaluate the territory separately in terms of complex and components. Factors can be divided into two types according to the nature of Fergana valley defining and creating touristic-recreational opportunities of the region.

Touristic-recreational potential (TRP) is a combination of tourism and recreation resources, their territorial structures, and conditions that help meet the tourism and recreation needs of tourists. The assessment of touristic-recreational potential is the initial stage of the research and shows the main features of the area. This creates an opportunity to realistically assess tourism and recreation in the Fergana valley and develop geoecological principles and criteria for evaluating the efficiency of regional touristic-recreational systems.

2 Materials and methods

It is possible to ensure the effective geoecological development of touristic-recreational activities by planning and designing tourism and recreation in the touristic-recreational area. Scientists have different approaches to the analysis of the definitions of the touristic-recreational object and the area. Nowadays, concepts such as touristic-recreational resource, touristic-recreational potential, regional touristic-recreational systems are widely used in geoecological research of tourism and recreation [2, 5, 10].

Evaluation of TRP is carried out mainly in 4 directions: medical-biological, psychological, technological and economical. Based on the geographical nature of the research, attention was paid to natural factors. However, other indicators were also studied to expand the possibilities of using the evaluation results of efficiency of regional touristic-recreational systems in practical works. One of the most effective methods in touristic-recreational evaluation of regional touristic-recreational systems is the integrated complex evaluation method [3].

A resource-based approach to the study of the geoecological essence of tourism and recreation has been put forward by many scientists and their foundations have been developed. M.A.Sarancha was said that the touristic-recreational potential of the area is a set of geoecological opportunities for the organization of touristic-recreational activities in regional touristic-recreational systems. D.Getz noted 150 touristic-recreational models and divided those models into three groups: theoretical, planning and forecasting models of touristic-recreational development in regional touristic-recreational systems [5].

M.V.Gudkovskikh noted levels of regional touristic-recreational systems: touristic-recreational area, touristic-recreational district, touristic-recreational system, touristic-recreational cluster and touristic-recreational route [10]. In the assessment of tourism and recreation potential were developed by many researchers and we used the improved balance method. The balance method is carried out in the following order: 1. Selection of the territorial unit; 2. Defining the main criteria; 3. Adjustment of rating scale; 4. Evaluation of general and individual results; 5. Analysis of results; 6. Final comprehensive assessment; 7. Comparison of results and zoning of the territory according to the level of touristic-recreational potential; 8. Development of the touristic-recreational potential map of the region. The role of touristic guides in sustainable development of tourism can be understood through Social Exchange Theory (SET), which explains how reciprocity, trust, and mutual benefit shape social relationships.

The aim of this scientific research is to study and explore the geoecological and touristic-recreational potential of Fergana valley and to develop proposals and recommendations for evaluating the efficiency of regional touristic-recreational systems of the Fergana valley. The object of this scientific research is natural, natural-historical, socio-economic conditions of Fergana valley. In determining the touristic-recreational potential of the Fergana valley, along with its natural conditions, a special place is occupied by natural-historical, religious and cultural monuments, craft centers, ethnographic and ethnoecological traditions, which are a unique heritage of cultural development. Such objects serve to organize such types of touristic-recreational activities as cultural-informative, historical-informative, religious, ethnographic, and also perform an educational function. This helps to optimize touristic routes and increase the level of interest.

In the balance method, the following blocks of equal importance are distinguished in TRP assessment: natural, historical, economic, touristic, geoecological and all negative factors of them. Each block consists of their components, and components include a group of evaluation parameters. The sum of the points for each group of evaluation parameters constitutes the total value of the component points. Therefore, the final result of the assessment indicates the share of each region in the total regional potential. In the end, the results of each block are summarized. The final result for each block does not represent the sum of points, but the share of each regional touristic-recreational systems in the general potential.

Tourism technology innovation acts as an intermediary in promoting sustainable development in the tourism industry through green credit. The impact of green credit on the sustainable development of the tourism industry in different regions of the world [6, 7].

3 Results and discussion

Fergana region is a unique region with its unique natural conditions, diverse landscapes, clean air, healing waters, and medicinal plants. Its fauna and flora are unique and stand out from other regions of our republic and Central Asia. Ensuring environmental safety is of great importance in the process of using nature. In this regard, it is necessary to determine how much of the natural landscape can be used in various sectors of the economy. The main attention is paid to assessing or taking into account the ability of landscapes to perform socio-economic functions determined by society. In determining the touristic-recreational potential of the Fergana region, along with its natural conditions, a special place is occupied by natural, historical, religious and cultural monuments, craft centers, ethnographic and ethnoecological traditions, which are a unique heritage of its cultural development. Such objects serve to organize such types of touristic-recreational activities as cultural-introductory, historical-introductory, religious, ethnographic, and also perform an educational function.

The desert and steppe areas of the region also have their own unique flora and fauna and are protected as natural monuments. They are home to various types of reptiles, including venomous snakes and desert tortoises, birds such as the hawk, the hawk, and mammals such as the yellow-tailed deer and the lynx. These areas require the creation of new tourist routes for ecotourism enthusiasts interested in flora and fauna. In order to effectively organize recreational and tourist activities in this area, the following should be implemented: A system of trails for ecotourism routes will be developed in the protected desert area of the Central Fergana region. Signs will be installed at appropriate points along the trails to indicate the location of shelters, drinking water, and other necessary items for recreation and information about the area. Rafting on the Shakhimardan River (together with the Aksuv and Koksuv streams), motorized and rowing boats on Kurbankul and Yashilkul, speleotourism in Teshik Tash, Altynbeshik Cave and Konigut Cave, trekking, mountaineering, ski jumping, freestyle, snowboarding, skiing, and more can be organized on the mountains and slopes near the Yordan and Shakhimardan valleys.

Agrotourism is also considered one of the new sectors for Fergana region, and approximately 6-7% of tourists visiting rural areas account for the share of tourists. The region has well-developed agricultural sectors such as cotton growing, grain growing, vegetable growing, gardening, livestock breeding, sheep breeding, poultry farming, cocoon farming, and beekeeping. Grain, cotton, vegetables, and melon crops are planted in the fields. Apricots, pomegranates, figs, apples, dates, and high-quality grapes are grown in the gardens. It is necessary to develop agrotourism routes that will promote "brand" plants such as Rishton apricot, Chungara rice, Kokand cherry, Altiarik cucumber, Kuva pomegranate, Yozyavo melon, and Akkurgan fig and ensure direct tourist participation in their cultivation.

The main mountain tourist objects of the Fergana Valley are located in the border areas of the Republics of Uzbekistan, Kyrgyzstan and Tajikistan and are of transboundary importance. This has an impact on the decrease in tourist flows. In the transboundary tourist zones of the Fergana Valley, extreme types of tourism can be established in the form of mountain tourism, speleotourism, ecotourism, rafting and other mountain sports. The creation of transboundary recreational and tourist zones in cooperation with neighboring republics will allow not only to develop the regional economy, but also to preserve the natural conditions of the region.

If we pay attention to the territorial location of the steppes in the Fergana Valley, the main part of them corresponds to the river basins in the mountain and hilly regions of the valley. In such areas, cutting down forests and polluting water bodies in turn accelerates the occurrence of drinking water shortages, floods, landslides. Due to the ethno-ecological culture of the indigenous people, various dangers of nature were prevented by establishing sacred shrines in these areas. Examples of such regions are Shahimardan and Yordon regions in the Fergana Valley.

By geoecological principles and criteria regional touristic-recreational systems of the Fergana valley made it possible to combine the administrative units of the region into 4 groups: **the highest**, ***high***, *medium* and low regional touristic-recreational systems (Table 1).

Table 1. Touristic-recreational potential of regional touristic-recreational systems (%).

№	***Administrative-territorial units***	Na	HC	SE	T	***Input***	E	Ne	***Output***	Total	***Share***
1	**Fergana d.**	8,01	6,42	5,5	7,82	***27,73***	1,85	5,38	***7,23***	20,5	***10,3***
2	***Margilan city***	3,71	9,70	6,7	5,25	***25,38***	5,56	4,23	***9,79***	15,59	***7,8***
3	***Kokand city***	3,77	10,00	7,3	5,72	***26,83***	7,41	4,81	***12,22***	14,61	***7,3***
4	***Sokh d.***	7,00	3,63	4,1	4,55	***19,27***	1.85	4,42	***6,27***	13	***6,5***
5	***Quva d.***	6,04	5,31	5,3	7,00	***23,68***	5,56	5,38	***10,94***	12,74	***6,4***
6	*Fergana city*	3,71	6,70	7,7	5,83	***23,93***	7,41	4,05	***11,46***	12,47	***6,2***
7	*Beshariq d.*	6,63	5,73	4,5	5,48	***22,37***	5,56	5,77	***11,33***	11,04	***5,5***
8	*Buvayda d.*	4,60	5,45	5,0	5,25	***20,27***	3,70	5,58	***9,28***	10,99	***5,5***
9	*Rishtan d.*	5,02	5,59	5,1	5,60	***21,31***	5,54	5,19	***10,73***	10,58	***5,3***
10	*Qoshtepa d.*	5,14	4,47	5,1	4,67	***19,38***	3,70	5,77	***9,47***	9,91	***5,0***
11	*Yozyovon d.*	5,90	3,63	4,3	5,13	***18,93***	3,70	5,58	***9,28***	9,65	***4,8***
12	*Quvasoy city*	5,80	3,77	5,2	6,42	***21,23***	7,41	4,81	***12,22***	9,01	***4,5***
13	*Altiariq d.*	5,31	4,89	5,2	5,13	***20,57***	7,41	5,19	***12,6***	7,97	***4,0***
14	Uchkuprik d.	4,90	4,19	5,0	4,90	***18,96***	5,56	5,38	***10,94***	8,02	***4,0***
15	Bagdad d.	4,54	3,91	4,5	4,10	***17,08***	3,70	5,58	***9,28***	7,8	***3,9***
16	Dangara d.	5,70	3,63	4,8	4,90	***19,01***	5,56	5,77	***11,33***	7,68	***3,8***
17	Toshloq d.	4,12	6,28	5,3	4,43	***20,16***	7,41	5,38	***12,79***	7,37	***3,7***
18	Uzbekistan d.	5,97	3,63	5,1	4,32	***19,02***	7,41	5,38	***12,79***	6,23	***3,1***
19	Furqat d.	4,13	3,07	4,2	3,50	***14,89***	3,70	6,35	***10,05***	4,84	***2,4***
	Total	100	100	100	100	***400***	100	100	***200***	200	***100***

** - the table is calculated by the author*

Another important aspect of tourism development in Fergana region is the presence of large historical centers of national crafts. There have been settlements in the Fergana valley since ancient times. Even today, several types of crafts have been preserved in the region. In particular, ancient types of crafts such as pottery, knife making, embroidery, wood carving, and carpet weaving are practiced by the only skilled craftsmen in the republic. For example, Rishtan pottery in Fergana valley.

In gastronomic tourism, a unique brand has been formed in Rishton district, Buvayda district, Yozyavon district and Fergana district. Rishton district is famous not only for its "rishtan pilaf", Buvayda district for its "wild bird meat", Yozyavon district for its "balyk", and Fergana district for its "sopaki (diet) pilaf". Rishton chefs also cook pilaf in almost all teahouses in Fergana city. "Pilaf centers" have been established in the cities of Fergana, Margilan, and Kokand. The unique dietary pilaf of Fergana district is also quite famous.

According to the natural touristic-recreational potential of Fergana region, Sokh, Beshariq have the highest index, Uzbekistan has a high index, while in general touristic-recreational potential, Sokh is high, Beshariq is medium, Uzbekistan is low index. Beshariq, Rishtan, Buvayda, Toshloq, which have a high index in the historical and cultural touristic-recreational potential of Fergana valley, have a medium and low index in the general potential. This can be explained by the low economic and social indicators, the environmental situation and the high indicators of negative factors (Figure. 1).

The indicators of Fergana district, which has the highest recreational and touristic potential in Fergana region, were calculated using the above formula (in %):

$$10{,}3 = \frac{(8{,}01 + 6{,}42 + 5{,}5 + 7{,}82) - (1{,}85 + 5{,}38)}{2}$$

In order to raise the potential and brand of Fergana region in the field of tourism and recreation, as well as its position at the national and international levels, the touristic-recreational systems with the lowest share in the touristic-recreational potential of the region (Altiariq, Dangara, Tashlak, Uchkuprik, Bagdad, Uzbekistan, Furkat), it is necessary to develop recommendations for the development of such types as agrotourism, gastronomic tourism, sports tourism, shop tourism, ethnographic tourism.

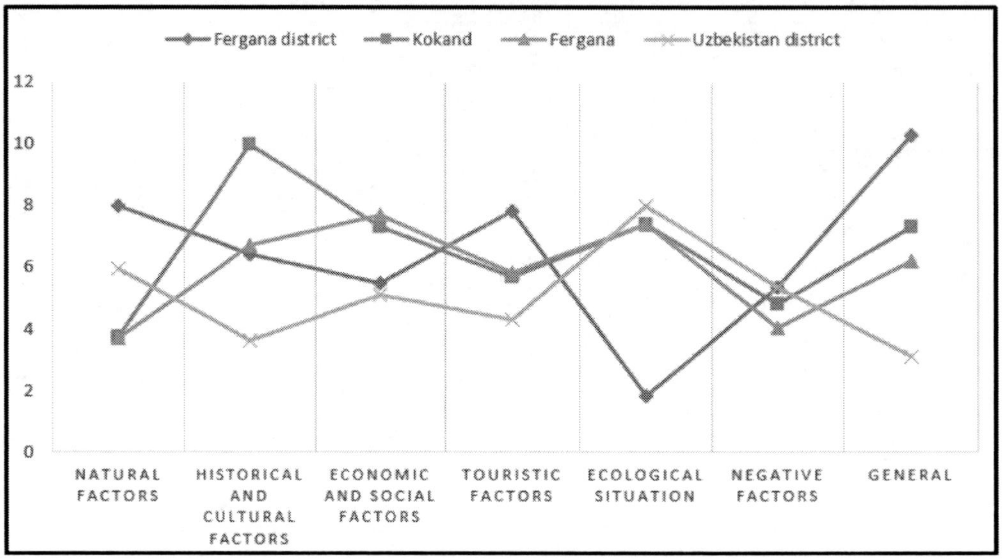

Fig. 1. Touristic-recreational potential analysis of touristic-recreational systems.

The natural conditions of the Fergana region, the meeting of almost all natural phenomena in the region of Central Asia in the region create the diversity of its ecotourism potential and the possibility of permanent organization. Such routes can be organized in Fergana district itself in 3 directions: Fergana-Vadil-Shahimardon-Yordon, Fergana-Avval-Ettibuloq, Fergana-Satkak-Chimyon (Figure 2).

https://doi.org/10.1051/e3sconf/202563307004

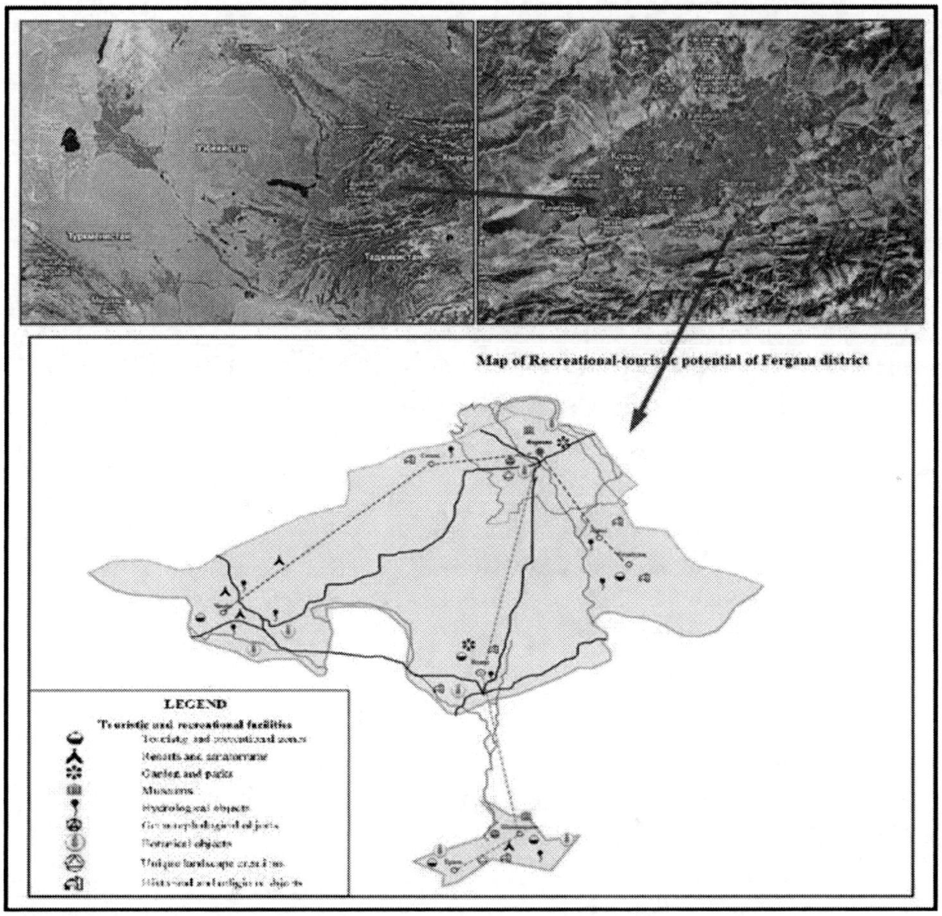

Fig. 2. Map of Touristic-recreational potential of regional touristic-recreational systems (Fergana district, Uzbekistan (1:750 000)).

1. Fergana-Vadil-Shahimardon-Yordon (3-5 days). In the Vadil region, the main tourist and recreational facilities are the riverside recreation areas, botanical natural monuments, and the unique natural conditions of Shahimardon. Watching the sunrise behind the mountains in the early morning in the Shahimardon-Yordon region. Trekking (hiking in the mountains). Visiting Qurban Lake, Yashil Lake and boating in the lake water. Descending from the village of Yordon to the Kuksu River Valley. Visiting and studying the Chakka-Tomar hydrological monument, the Ancient Moraine geological monument, the Kuktash Rock and the picturesque Yordon River Valley. Relaxing on the banks of the Shahimardon River.

2. Fergana-Avval-Yettibulak (1-3 days). Guest houses organized in about 800 households located around streams and springs in the Avval-Yettibulak region serve mainly local tourists during the holiday season. The majority of the population of Fergana region satisfies their recreational and tourist needs in this area in the summer.

3. Fergana-Satkak-Chimyan (1-3 days). The Chimyan and Satkak hills, springs and streams are the main tourist and recreational objects.

4 Conclusion

1. The balance method of evaluating the touristic-recreational potential was improved taking into account the natural, historical, economic, geoecological and negative indicators of the

regional touristic-recreational systems. This guarantees the improvement of the thematic touristic-recreational directions and the mechanisms of their organization based on each regional touristic-recreational systems.

2. On the basis of the thematic and comprehensive evaluation of the touristic-recreation potential of touristic-recreational systems of Fergana valley, the private and general typology of the regional touristic-recreational systems of the Fergana valley was distinguished. With the help of this evaluation method, the results of the touristic-recreational potential determined by the specific potential of each typology can be used in the development of promising programs aimed at the development of the sector.

3. In Fergana valley touristic-recreational potential for regional touristic-recreational systems with high landscape-ecological indicators and high natural-historical touristic-recreational potential touristic-recreational routes were established and the possibilities of expanding the use of the existing ones were identified with geoecological principles.

References

1. A. Abdurakhmanova, F. Ahrorov, "The economic and social impacts of ecotourism on local employment and income: A case study of rural Samarkand, Uzbekistan", Regional Science Policy & Practice **17(3)**, 100180 (2025), https://doi.org/10.1016/j.rspp.2025.100180
2. S. Gaonkar, S. V. Sukthankar, "Measuring and evaluating the influence of cultural sustainability indicators on sustainable cultural tourism development: Scale development and validation", Heliyon **11(4)**, e42514 (2025), https://doi.org/10.1016/j.heliyon.2025.e42514
3. Indicators of Sustainable Development for Tourism Destinations A Guidebook (UNWTO), http://www.e-unwto.org
4. Statistical newsletter of Fergana region, Statistics Department of Fergana region (Fergana, Uzbekistan, 2022), https://farstat.uz/uz/
5. J. Sun, D. Chen, "Green credit, tourism technology innovation, and sustainable development of tourism", International Review of Economics & Finance **99**, 104004 (2025), https://doi.org/10.1016/j.iref.2025.104004
6. Z. Xu, J. Peng, "Recognizing ecosystem service's contribution to SDGs: Ecological foundation of sustainable development", Geography and Sustainability **5(4)**, 511-525 (2024), https://doi.org/10.1016/j.geosus.2024.05.001
7. M. V. Gudkovskikh, "Tourism in the Tyumen Region: Potential and Territorial Organization", (Abstract of the Dissertation, Candidate of Geographical Sciences, Perm, 2018)
8. K. Yegemberdiyeva, Y. Yushina, A. Khen et al., "Assessment of the natural-recreational resources of the akmola region (based on the example of the shchuchinsk-borovoye resort area) for the purpose of sustainable development of tourism", Geojournal of Tourism and Geosites **30(2)**, 818-826 (2020), DOI 10.30892/gtg.302spl06-510
9. A. Makhmudova, "Organizational and economic reasons preventing the development of ecological tourism in Uzbekistan", Journal of Advanced Research in Dynamical and Control Systems **12(6)**, 1217-1220 (2020), DOI 10.5373/JARDCS/V12I6/S20201162
10. N. Dunets, A. K. Volkova, N. Ye. Ramazanova, E. V. Rygalov, "The ecological and tourist frameworks of the low mountain zone as the basis for the formation of the territorial structure in the southern part of Altai Krai", Geojournal of Tourism and Geosites 54(2 supplement), 842-848 (2024), DOI 10.30892/gtg.542spl08-1259

https://doi.org/10.1051/e3sconf/202563307005

Analysis of traffic patterns between Karkh and Rusafa

Ishraq Mahdi Lateef[1], *Alaa* Hussein Abed[1], and *Noor* Moutaz Asmael [2*]

[1]Civil Engineering Department, College of Engineering, Nahrain University, Baghdad, Iraq
[2]Highway and Transportation Engineering Department, College of Engineering, Mustansiriyah University, Baghdad, Iraq

Abstract. Rapid and radical changes in urban activities lead to changes in spatial travel patterns. Recently, users of the road network in Baghdad have faced difficulty in mobility between Karkh and Rusafa due to the congestion of bridges connecting the east of the city to the west across the Tigris River during peak hours. Accordingly, this study aims to evaluate the operational conditions of bridges during the morning and afternoon peak hours by calculating V/C to measure the level of service with the possibility of analyzing the improvement of the impact on accessibility between Karkh and Rusafa areas by adding two new bridges on the Tigris River using GIS software to find the best path with the shortest travel time between the start and end points of the O-D trip through the software algorithm. A traffic survey of Baghdad bridges was conducted over two periods to analyze the flow patterns. Then, the Baghdad road network with its bridges was drawn using ArcGIS software, and their properties were entered. The analysis results showed that most of the bridges operate at the service level F, which indicates that the value of (v/c) is greater than one. The highest average travel time savings upon completion of the two bridges will be for the areas south of Baghdad (5-9) minutes.

1 Introduction

Bridges serve multiple functions but primarily transport people and goods between two sides by crossing rough terrain. Bridges are often seen as an indicator of progress in technical ability and engineering skills as well as a symbol of the economic potential of a particular city, region, or country. This is often why many urban bridges have become the most important infrastructure in some cities. [1]. Several studies have been conducted to improve the road network in different locations in Baghdad. Some of these studies used the Highway Capacity Manual and other programs in the evaluation and improvement process. The Baghdad Municipality [2] has conducted several studies, including the Master Plan for the City of Baghdad (MPBC) prepared by Polservice in 1967, the Baghdad Comprehensive Development Plan 2000 (CDP) prepared by Polservice in 1973, the Baghdad Comprehensive Transport Study (BCTS), prepared by Scott Wilson Kirkpatrick & Associates in 1980, the

* Corresponding author: noor_moutaz@uomustansiriyah.edu.iq

Baghdad Integrated Capital Development Plan (IICDP) prepared by JCCF in 1987, and the Baghdad Comprehensive Development Plan 2030, prepared by Khatib and Eilmi in 2007.

These previous studies have proposed a hierarchy and pattern of roads with capacity depending on the functional and design classification for road networks, railways, and public transport schemes that set the framework for development and transportation in Baghdad. The increased number of trips generated and the number of trips attracted by the city of Baghdad have led to the development of road network problems. Hence, the trend of traffic simulation and modeling emerged to evaluate and find solutions to transportation problems (road networks), so many Iraqi researchers studied the evaluation and development of road networks in some areas of Baghdad city in different ways by collecting data, including the method of collecting data from the questionnaire form to predict the production of trips to evaluate the damages resulting from traffic congestion and find appropriate solutions to reduce the extent of damages using multiple programs. To assess the congestion conditions in several Baghdad streets through traffic congestion assessment criteria, including the speed performance index or the reliability measures index (travel time index, buffer time index), and data collection using GPS devices [3, 4]. To develop road networks by predicting the demand for transportation, whether using traffic volumes or the traditional method and software packages that simulate these systems and the ability to model their behavior, including geographic information systems (GIS) and TransCAD [5, 6]. Travel time and ease of access were predicted using GIS [3, 4].

Accessibility can be measured through the accessibility index from the gravity model in the GIS program [5]. The GIS program was used to model transportation networks to measure accessibility and evaluate a part of the public transportation network in Baghdad before and after the improvement was developed by adding a train line. The urban expansion has led to increased traffic on the network, which has led to a deterioration in the network condition, especially during peak hours. This study aims to evaluate the levels of traffic congestion of the existing bridges connecting the two sides of Baghdad city over the Tigris River. This includes time loss, increased traffic accident rates, and increased fuel consumption [6], as population growth outpaces infrastructure, leading to a decline in the quality of service.

Therefore, this study to analyze and evaluate the bridges and find solutions for them can help the transportation authorities and traffic management in leading the city of Baghdad towards a more efficient road network using the established LOS evaluation criteria and calculating the number of traffic lanes to discharge these traffic volumes in corresponding with the Scott Wilson's comprehensive transportation study for the city of Baghdad. In addition, the impact of implementing the southern and central Baghdad new bridges over the Tigris River will be evaluated through travel time measures using ArcGIS 10.8. which reduces travel time on the bridges (Al-Jumhuriyah, Al-Jadriya, 14 July, Two Stories, and Al-Dawra).

2 Materials and methods

A well-planned research methodology is crucial for ensuring that the research findings would meaningfully contribute to reducing the negative effects of congestion of Baghdad bridges. The current research utilized the study approach presented in Figure 1.

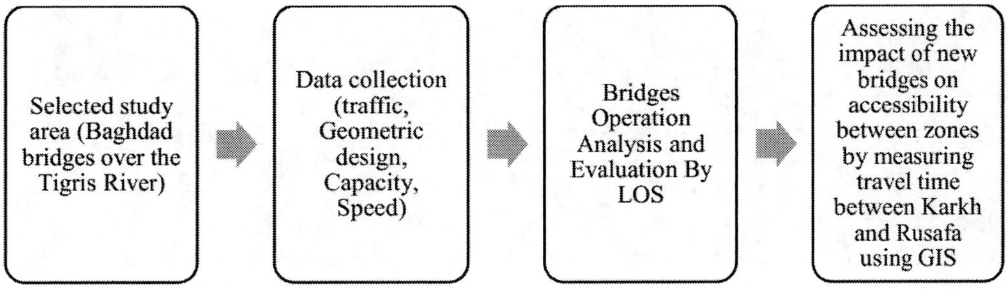

Fig. 1. Steps of research.

Baghdad city was chosen as a study area for several reasons: it is the capital of Iraq. It has the largest urban population in Iraq (population 9,418,800 in 2023), reaching 11,347,000 people in 2030 according to the Comprehensive Development Plan 2030 addressed by Khatib and Eilmi. The city of Baghdad has shown rapid urbanization and economic growth. The area of the city of Baghdad is approximately 920 km^2, and it is located between longitude (44° 34'13" East) (44° 11'12" East) and latitude (33° 19'30" North) (33°). 11'02 North (Amant Baghdad), Baghdad city consists of 15 municipalities distributed between Al-Karkh and Al-Rusafa parts, and there is a river that divides the city into two parts: Al-Karkh and Al-Rusafa, which are linked to each other by 13 bridges as shown in figure2. According to Paul Service's study, the network consists of five levels of the road (TSF, TSE, TSO1, TSO2, and TSO3) with different functional and geometric characteristics.

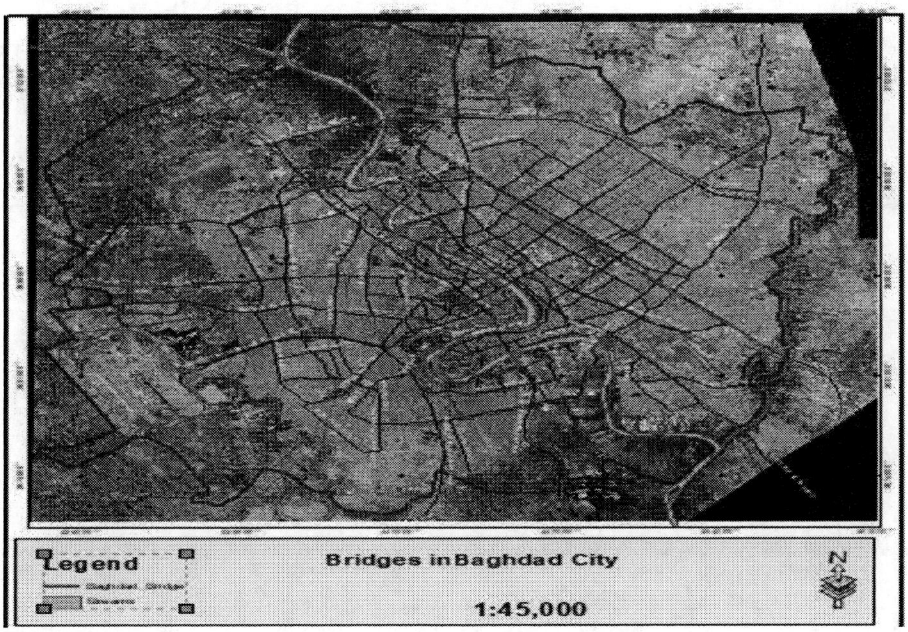

Fig. 2. Existing Bridges of Baghdad.

The Rusafa part is located east of the Tigris River and occupies 43% of the area of the city of Baghdad. It includes 8 municipalities, and the population of the Rusafa part is 57% of the population of the city of Baghdad. The Karkh part is located west of the Tigris River and occupies 57% of the area of the city of Baghdad. It includes municipalities 7, and the population of the Karkh part is 43% of the population of the city of Baghdad. 13 Bridges connect the Karkh and Rusafa.

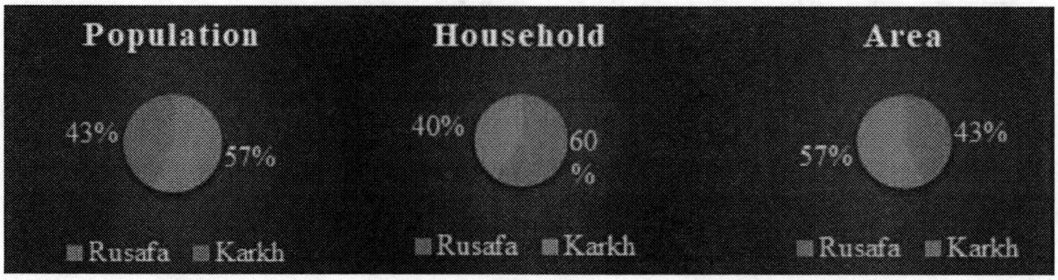

Fig. 3. Rusafa and Karkh properties.

The Tigris River is one of the natural landmarks in Baghdad. It divides Baghdad into two parts: Al-Rusafa and Al-Karkh. There are 13 bridges linking the east and west of the city as shown below in Table 1, respectively, from the north of the city, the Al-Muthanna Bridge, or the North Bridge, to the Dora Bridge in the south of the city. Most of these bridges are two-way, except for the one-way Shuhada Bridge, which transports traffic from the Rusafa side to Al-Karkh within the Center Business District CBD. Likewise, the Al-Ahrar Bridge operates in one direction, transporting traffic from Al-Karkh towards Al-Rusafa.

Table 1. Location of Bridges according to a comprehensive transportation study.

ID	Street	Bridge location
1	Al-Muthanna Bridge	Part of the Fourth Ring Road transfers traffic volumes to and from the Army Canal freeway and Ibn Manzur Street.
2	Imams Bridge	Transferring traffic volumes to and from Imam Al-Aazam Square and Imam Square
3	14 Ramadan Bridge	Part of the third ring road transports traffic volumes to and from Al-Dallal Square and 14 Ramadan Street through the 14 July freeway.
4	Sarrafiya Bridge	Part of the first ring road Transports traffic volumes to and from Safi Al-Din Al-Hilli Street and Buratha Street.
5	July 17 Bridge	Transports traffic volumes to and from the Iraqi Fighter Tunnel and Al-Talae'a Square
6	Shuhada Bridge	Transfers traffic volumes to and from Marouf Al-Rusafi Square and Al-Shuhada Square
7	Al-Ahrar Bridge	Transfers traffic volumes to and from Hafez Al-Qadi Intersection and King Faisal I Square
8	Al-Rashid Bridge	Transfers traffic volumes to and from Al-Senak Square and Al-Salihiya Intersection
9	Ajumhuria-Bridge	Transfers traffic volumes to and from Tahrir Square and Al-Shawaf Intersection
10	July 14 Bridge	Transports traffic volume to and from the Suspension Bridge Square and 14th of July Square.
11	Jadriya Bridge	Transports traffic volumes to and from Al Jadriya intersection, Al Nakheel Street.
12	Two stories Bridge	Transports traffic volumes to and from Ahmed Orabi Square and Qatar Dew Street
13	Dora Bridge	The second part of the ring road transports traffic volumes to and from Agadir Square and Atba Bin Ghazwan Road.

2.1 Data collection sources

The main objective of data collection is to collect all the information required to show the traffic flow pattern on the bridges in Baghdad city. The main types of this data are from different sources:

1- Characteristics for all sectors, such as the number of households, population, and growth factor for each year obtained from the Amant Baghdad prepared by the Baghdad Comprehensive Development Plan 2030 prepared by Al-Khatib and Alami 2007.
2- Maps of Baghdad city (road, intersections, bridges, and boundary Baghdad city) provided by the Amant Baghdad
3- Geometric features (e.g., lane width per direction, number of lanes per direction, etc.).
4- Traffic flow data, which includes vehicle traffic volume data based on traffic data recorded by applying surveillance camera footage obtained from the Iraqi Intelligence Agency's Electronic Control and Monitoring Center during peak hours (7:30 to 8:30 am) and (1:30 to 2:30 pm) for workdays (Sunday, Tuesday, and Wednesday) and classified by type. (Passenger car, bus, truck). Passenger car: Any vehicle with only four tires. Truck and bus: Any vehicle with more than four tires. This type of vehicle is converted to a passenger car using a factor (PCU) equal to (1.5).

2.2 Traffic flow pattern

The current study mainly focused on evaluating and improving the existing infrastructure (bridges) to reduce the negative effects of congestion and improve the level of service for transportation on these bridges, so the peak hour volume, which represents the design hour volume (DHV), will be analyzed. The field traffic counting started on working days (Sunday, Monday, and Tuesday) from 7-9 am and 11 - 3 pm and the peak hours (design hour volume) were from 7:30 am to 8:30 am and from 1:30 to 2:30 pm for both directions from Rusafa to Karkh and vice versa. The data was organized according to the relationship between the vehicle class and the driving direction. The traffic was calculated, and three types of vehicles were classified: passenger cars, buses, and trucks. Bridges classification, base capacity, and F.F.S, data were collected according to the comprehensive transportation study of Baghdad, can local policies be based on the urban street class if it is not possible to measure the free flow speed F.F.S [7].

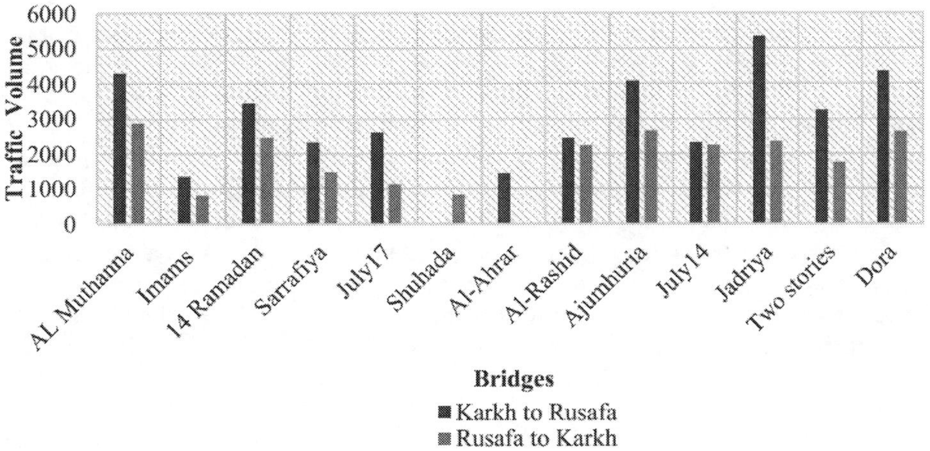

Fig. 4. Traffic Volume of Tigris River Bridges at morning peak hours.

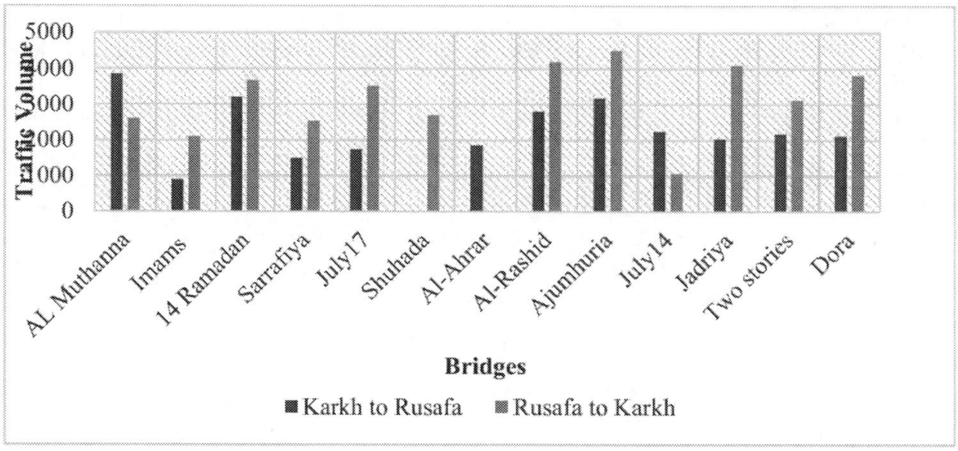

Fig. 5. Traffic Volume of Tigris River Bridges at Afternoon Peak Hours.

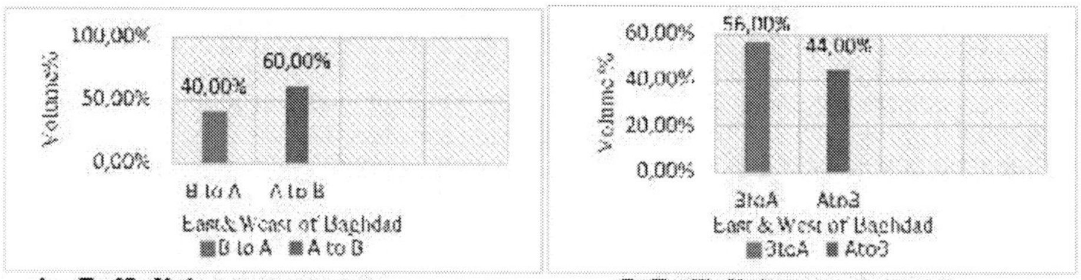

A. **Traffic Volume percentage at morning peak hours**
B. **Traffic Volume percentage at afternoon peak hours**

Fig. 6. Percentage of Traffic Volume crossing Bridges between each side of Baghdad.

3 Results

This section describes the data analysis to evaluate congestion metrics using peak-hour traffic databases. Figures 4 and 5, respectively, above show the flow on bridges in Baghdad during the periods from 7:30 am to 8:30 am and from 1:30 pm to 2:30 pm during the morning and afternoon peak hours for the two directions from Rusafa to Karkh and vice versa. The maximum flow of Al-Jadriya Bridge is during the morning peak hour for the direction of traffic from Karkh to Rusafa due to the diversity of the number of trips coming from Karkh towards Rusafa from the origin point to the destination point, including business trips, and education is large in Rusafa due to the diversity of land uses for educational, governmental, and commercial purposes in addition to residential and recreational use. The educational centers are represented by large universities such as the University of Baghdad, Al-Nahrain, Al-Mustansiriya, the University of Technology, and others, as well as state institutions and the largest commercial center, Shorja. It also includes the largest medical city in Iraq compared to the land uses of the Karkh part, most of which are residential and recreational. During the evening rush hour, the Republic Bridge located in the central business district (CBD) suffers from traffic congestion from Rusafa to Karkh. In addition to the large number of trips to the bridges located in southern Baghdad for the direction of traffic coming from Karkh to Rusafa. Figure 6A shows the percentage of vehicle volume as the number of vehicles reached (35673) pc/h at a rate of (60%) of the total flow during the morning peak hours (7:30 AM - 8:30 AM). Therefore, the diversity of the number of trips coming from Karkh towards Rusafa, from the origin point to the destination point, includes business trips and educational trips. Figure 6B shows that the percentage of traffic volume is 56 % during the afternoon peak hour (1:30-2:30) in the direction coming from Rusafa to Karkh as a result of trips from the destination workplace to the original home.

3.1 Evaluation of Tigris River Bridges Service

Congestion is a major problem for land transport and a major challenge for transport policy at all levels. Several congestion measures have been reviewed to determine the congestion level, considering different performance criteria. Based on these criteria, congestion measures can be classified into five categories: (a) speed, (b) travel time, (c) delay, (d) level of services, and (e) congestion indicators [8, 9], as shown in Figure 7. Moreover, the Baghdad Roads Department does not use specific measures to determine the level of congestion in Baghdad City, so statistics on the transport network are not widely available

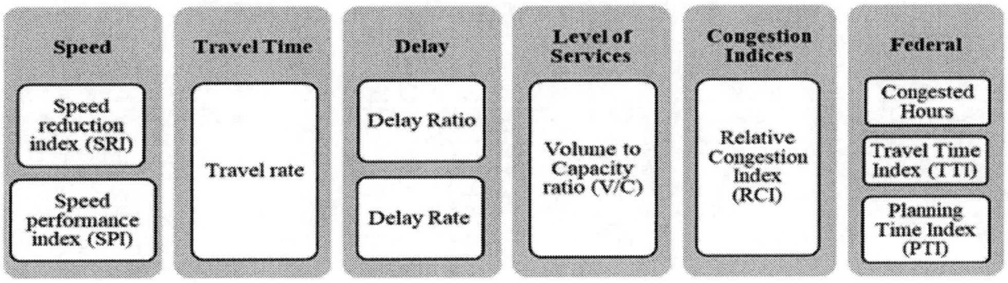

Fig. 7. Congestion measures in different categories.

In this paper, the level of service is a criterion for the evolution of the bridges in Baghdad city is shown. This indicator is used to communicate decision-makers to the bridge operation of the facility's performance. Various traffic criteria, such as density, average Travel speed, and volume-to-capacity ratio, can be used to determine the level of services. The LOS of bridges can be determined by the scale intervals of the volume-to-capacity ratio (V/C) and the threshold LOS with the V/C ratio shown in the following values in Table 4 [10].

Table 2. Level of service (LOS) based on the corresponding V/C ratio.

Los Class	Traffic State and Condition	V/C Ratio
A	Free flow	0-0.60
B	Stable flow with unaffected speed	0.61–0.70
C	Stable flow but speed is affected	0.71–0.80
D	High-density but stable flow	0.81–0.90
E	Traffic volume near or at capacity level with low speed	0.91–1.00
F	Breakdown flow	>1.00

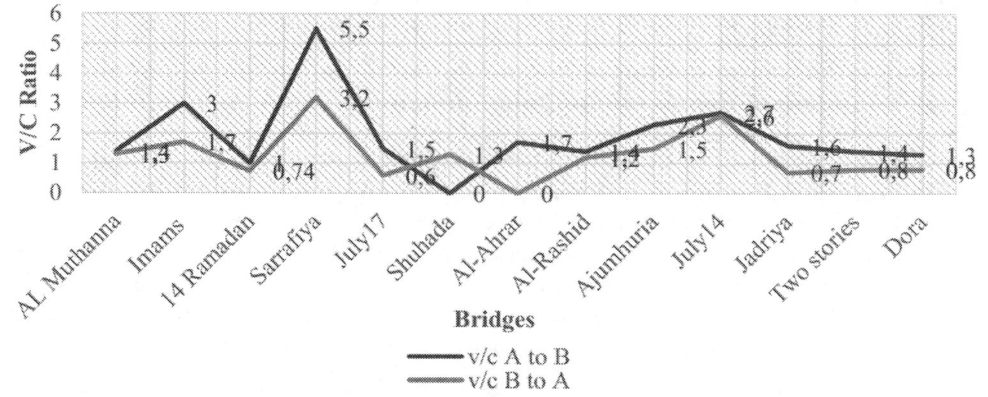

(A) V/C Ratio at morning peak hours

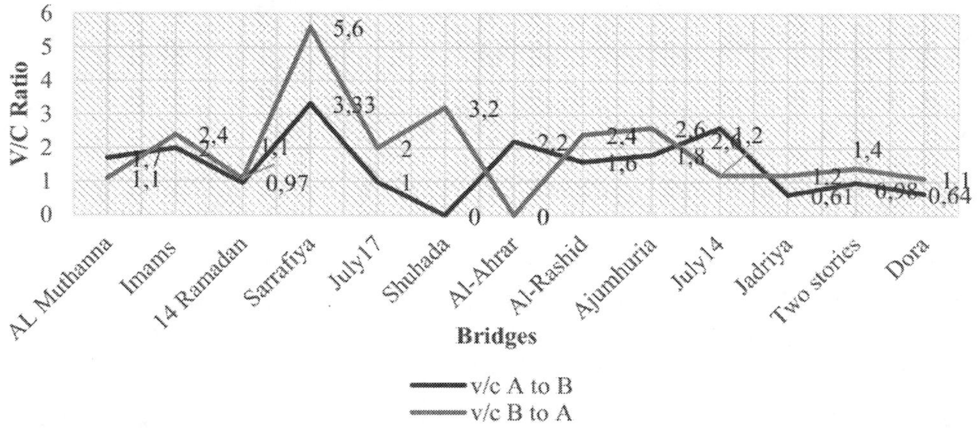

(B) V/C Ratio at afternoon peak hours

Fig. 8. Ratio V/C on Bridges.

Table 3. Level of service (LOS) for Tigris River Bridges based on the corresponding V/C ratio.

	Bridge	Rusafa to Karkh		LOS Karkh to Rusafa	
		Duration	LOS	Duration	LOS
1	Al-Muthanna Bridge	7:30-8:30 1:30-230	F F	7:30-8:30 1:30-2:30	F F
2	Imams Bridge	7:30-8:30 1:30-2:30	F F	7:30-8:30 1:30-2:30	F F
3	14 Ramadan Bridge	7:30-8:30 1:30-2:30	C F	7:30-8:30 1:30-2:30	F E
4	Sarrafiya Bridge	7:30-8:30 1:30-2:30	F F	7:30-8:30 1:30-2:30	F F
5	July 17 Bridge	7:30-8:30 1:30-2:30	A F	7:30-8:30 1:30-2:30	F F
6	Shuhada Bridge	7:30-8:30 1:30-2:30	F F	7:30-8:30 1:30-2:30	One Direction
7	Al-Ahrar Bridge	7:30-8:30 1:30-2:30	One Direction	7:30-8:30 1:30-2:30	F F
8	Al-Rashid Bridge	7:30-8:30 1:30-2:30	F F	7:30-8:30 1:30-2:30	F F
9	Ajumhuria-Bridge	7:30-8:30 1:30-2:30	F F	7:30-8:30 1:30-2:30	F F

Continuation of Table 3.

10	July 14 Bridge	7:30-8:30 1:30-2:30	F F	7:30-8:30 1:30-2:30	F F
11	Jadriya Bridge	7:30-8:30 1:30-2:30	B F	7:30-8:30 1:30-2:30	F B
12	Two stories Bridge	7:30-8:30 1:30-2:30	C F	7:30-8:30 1:30-2:30	F E
13	Dora Bridge	7:30-8:30 1:30-2:30	C F	7:30-8:30 1:30-2:30	F B

The service level on the different bridges was observed during the morning and evening peak periods, with traffic congestion varying from time to time, as shown in Table 5, where the majority of the bridges had a level of service F = V / C > 1, which means severe to extreme congestion conditions in the direction of movement from Karkh to Rusafa and vice versa. The current poor operating traffic conditions result in slow speeds, long journey times, and increased vehicle queues when demand is greater than supply. The current number of lanes is 27 in one direction for all 13 Baghdad bridges, and the number of vehicles transported from both sides of Karkh to Rusafa is 37,111 pcu/hour. This is the largest volume calculated during the afternoon peak hours; each lane transports approximately 1,374 pcu/hour/lane, which is greater than the maximum discharge of some bridges depending on the bridge classification. To improve traffic on Baghdad bridges by increasing the number of traffic lanes or adding additional new bridges and underwater tunnels between the two sides (Karkh and Rusafa).

3.2 Analysis of the impact of new bridges on travel time between(zones) Karkh and Rusafa using GIS

Analysis of the impact of new bridges for accessibility between (zones) Karkh and Rusafa was done using ArcGIS. Access mobility by the road network for the optimal journey where analyzed by travel time measure. In addition to improving the efficiency of the road network by decreasing congestion in the current road network and increasing the capacity of the city road networks, Baghdad has recently witnessed major developments in infrastructure related to projects implemented by the Roads and Bridges Authority, which include adding new two bridges linking Karkh to Rusafa south and central of Baghdad on the Tigris River as follows:
1- A bridge with two traffic lanes in both directions connecting Abu Nawas Street in Rusafa within the Karrada Zone TAZ is 11 with South Street at the intersection of the square Al-Nusour within the Mansour Zone TAZ is 8 currently under construction.
2- Gaza Bridge with three traffic lanes in both directions connecting the Al-Zafaraniya area in Rusafa within the Karrada zone TAZ is 11 and transferring traffic to Al-Dora Al-Yusufiya Road within Al-Dura Zone TAZ is 10 currently under construction. The location of the bridges is shown in Figure 9.

This section of the study analyzes the impact of two new bridges on the accessibility between zones using travel time in ArcGIS by analyzing road Baghdad networks and finding travel time based on distances and the operating speed of the trip path between the start and end points using the GIS. For this purpose, Baghdad city was divided into 15 traffic analysis zones (TAZ), 8 areas east of the Tigris and 7 regions west of the Tigris, connected by 13 bridges before improvement and becoming 15 bridges after improvement as shown in Figure 9 and Table 6 with the following values. The center of each traffic analysis zone (TAZ) was determined to calculate the travel time from the start center of the trip (TAZ) (origin) to the end center of the trip (TAZ) (destination).

An analysis of the road network and its associated bridges, numbering 13 bridges before implementation conditions of the new bridges, from the TAZ area (origin) to the destination and vice versa was conducted to determine the travel time during morning and evening peak hours using GIS, as appears in the road network analysis in Figure 10. The results of the network analysis showed that travel time varied depending on the trip activities and land use, with travel time for the morning peak ranging from A to B (15-60) min and from B to A (13-48) min and travel time in the afternoon peak hour from A to B (18-53) min, from B to A (18-60) min.

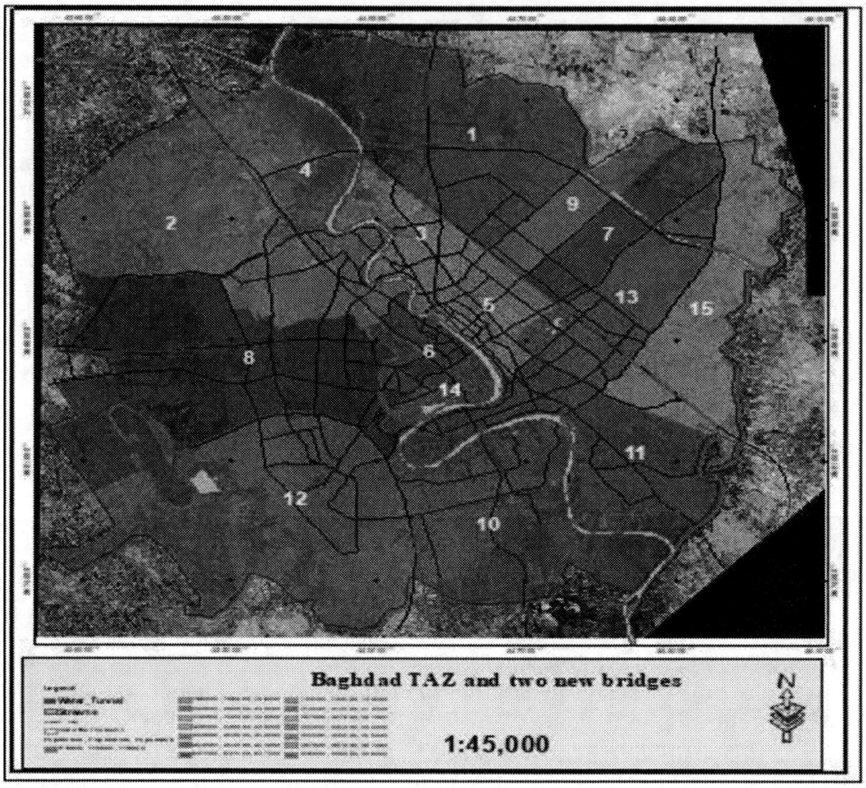

Fig. 9. Baghdad TAZ and two new bridges.

Table 4. TAZ Baghdad.

Rasafa (east of Tigris River)	**TAZ NO.(B)**	**Karkh (west of Tigris River)**	**TAZ NO.(A)**
Sha'ab	1	Al Shuala	2
Adamiyah	3	Kadhimiya	4
Al Rasafa	5	Karkh	6
Sadr 1	9	Mansour	8
Sadr 2	7	AL Dora	10
Karada	11	Al Rasheed	12
BaghdadALjadeedah	13	G.Z&AirportRoad	14
Alghadeer	15		

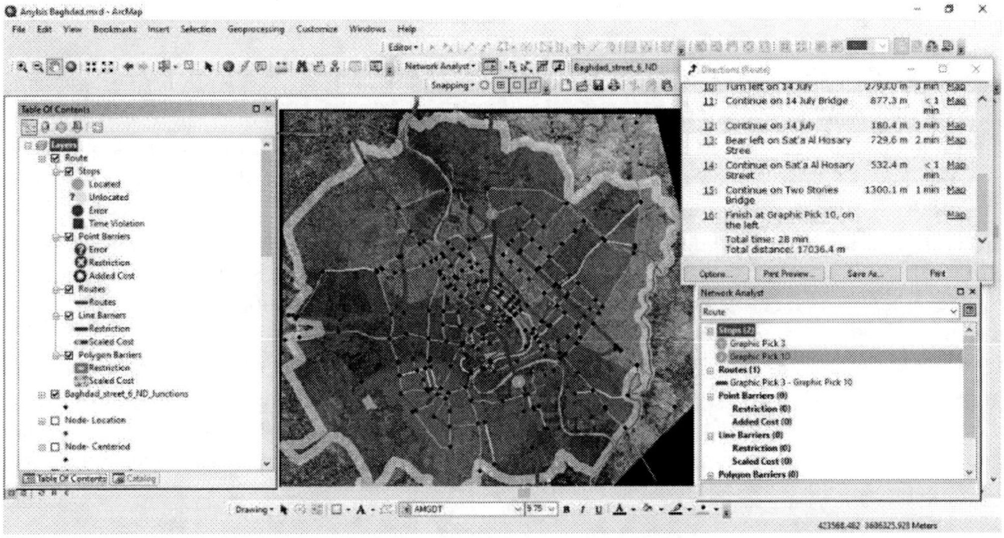

Fig. 10. Analyzing Road networks between two zones to find travel time and distances by GIS.

Figure 11 shows a comparison between before and after-implementation conditions of the new bridges, by average travel time between Karkh and Rusafa, and vice versa during morning and evening peak hours. According to Figures 11(A, B), the highest average travel time savings upon completion of the two bridges will be for the areas south of Baghdad (on both sides of Karkh and Rusafa, where travel time is reduced by 5 minutes during the morning peak hour for areas (10, 11, 15, 14, and 9 minutes during the afternoon peak hour for areas (15, 13, 10 and 8). This is a valuable result, as this area receives significant daily traffic from many other parts of Baghdad. The presence of these two bridges reduces congestion on the bridges (Al-Jumhuriyah, Al-Jadriya, 14 July, Two Stories, and Al-Dawra).

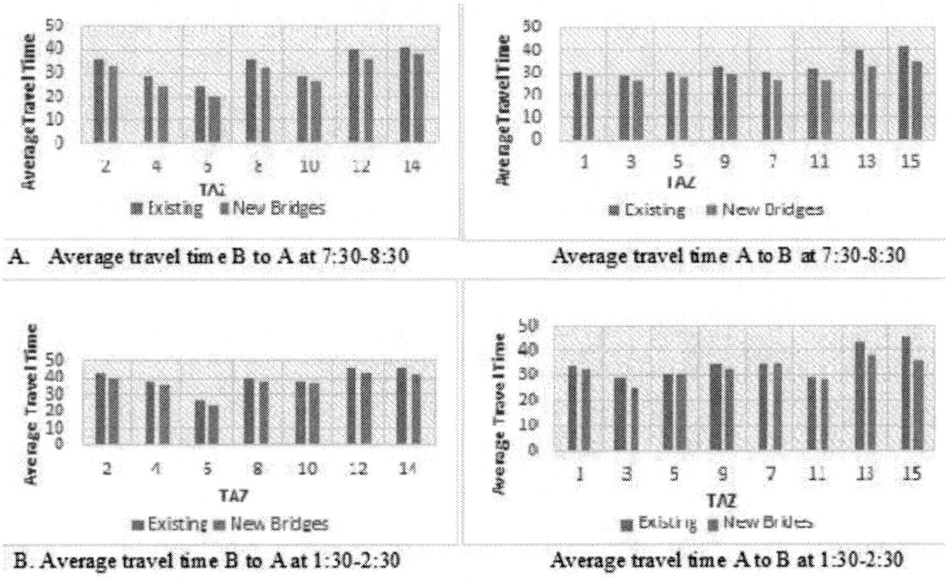

A. Average travel time B to A at 7:30-8:30 Average travel time A to B at 7:30-8:30

B. Average travel time B to A at 1:30-2:30 Average travel time A to B at 1:30-2:30

Fig. 11. Comparison between Average Travel Time (minutes) for Zones 15 for the Existing Condition and under the improvement bridges System.

4 Conclusions

Large trips move from West Baghdad to workplaces in East Baghdad and vice versa during the morning rush hour, then reverse the movement again during the afternoon rush hour. At the same time, the change in land use affects the transportation that attracts trips from inside and outside Baghdad. Therefore, it is noted that Baghdad bridges suffer from congestion during the morning peak hours, with a traffic volume of 35,673 pcu/h in the direction from Karkh to Rusafa from 7:30 pm to 8:30 pm, and the afternoon peak from 1:30 pm to 2:30 pm with a traffic volume of 37,111 pcu/h in the direction from Rusafa to Karkh. Most of Baghdad bridges operate at service level F because demand is higher than supply V/C ≥1. To reduce travel time when moving between Rusafa and Karkh by increasing the number of lanes between the two sides (Karkh and Rusafa) via bridges or underwater tunnels. We noticed an improvement in the travel time between Rusafa and Karkh after adding two bridges, bringing the number of bridges to 15 instead of 13 for the bridges located south of Baghdad, where the reduction in travel time ranged between (3-9) minutes after the completion of the construction of the Gaza Bridge and the Abu Nawas Bridge.

References

1. M. Salamak, K. Fross, "Bridges in urban planning and architectural culture," Procedia Engineering **161**, 207–212 (2016)
2. M. G. Abbas, R. J. Al-Shammari, M. A. Mahdi Al Fartusi, "The Role of River Transport in Reducing Congestion Urban Transport in the City of Baghdad as a Case Study", Periodicals of Engineering and Natural Sciences **8**, 2263–2273 (2020)
3. N. M. Abid, S. S. Hussain, "Transportation network planning using simulation: Case study Al Mansour city," in *2017 2nd IEEE International Conference on Intelligent Transportation Engineering (ICITE)*, 272–279 (2017)
4. M. Z. Mohamedmeki, M. Al-Mumaiz, "Improving the transportation system in Baghdad city," IOP Conference Series: Materials Science and Engineering **1067(1)**, 012087 (2021)
5. K. T. Falih, A. J. Mohammed, A. S. Hasan, "A GIS-based Network Analysis for Truck Vehicles in Baghdad's City Road Network," in *2022 International Congress on Human-Computer Interaction, Optimization and Robotic Applications (HORA)*, 1–5 (2022)
6. I. Talib, Z. Nassrullah, L. Abduljaleel, "A Case Study on Reducing Traffic Congestion–Proposals to Improve Current Conditions", Civil Engineering Journal **9**, 2456–2466 (2023), DOI: 10.28991/cej-2023-09-10-07
7. U. C. Odiakose, S. D. Iyeke, "Traffic Congestion Analysis of Asaba Road Using Volume to Capacity Ratio and Speed Performance Index," Journal of Applied Sciences and Environmental Management **28(4)**, 1315–1326 (2024)
8. M. Alhamdany, A. H. K. Albayati, "Statistical Modeling of Time Headway on Urban Roads: A Case Study in Baghdad", Engineering, Technology and Applied Science Research 12, 8584–8591 (2022), DOI: 10.48084/etasr.4878
9. M. Arif, K. Gupta, "Application of Graph-Based Model for the Quantification of Transport Network in Peri-Urban Interface of Burdwan City, India", Spatial Information Research 28, 447–457 (2020), DOI: 10.1007/s41324-019-00305-w
10. S. K. Singh, A. Saraswat, "Design Service Volume, Capacity, Level of Service Calculation and Forecasting for a Semi-urban City," Rev. d'Intelligence Artif. **33(2)**, 139–143 (2019)

Assessment of uninterrupted pedestrian facility, case study of Al-Mutanabbi Street

Farah Ahmed Subhi[1], and *Abeer* Khudhur Jameel[2,*]

[1]Civil Engineering Department, Al-Shaab University, Baghdad, Iraq
[2]Highway and Transportation Department, College of Engineering, Mustansiriyah University, Baghdad, Iraq

Abstract. Walking is a sustainable transportation mode as it is the most affordable way to reduce the environmental and social effects of the transportation system. Walking can be used for transportation and recreational purposes. To encourage walking, it is necessary to improve walkability and walking behavior. The main aim of this paper is to assess the walkability of Al-Mutanabbi Street in Baghdad City, which is located in a cultural and historical area. The measure of effectiveness is the pedestrian level of service PLOS. In this paper, new indicators reflecting the perception of pedestrians have been included in the assessment method to overcome the limitation of the traditional method of PLOS determination. A questionnaire form was used to collect the data. The questions related to walkability are 21 questions grouped into seven groups. The responses of pedestrians were converted to scores to find the overall PLOS, which was C. However, the scores of some variables were low, reflecting the weakness in walking performance, such as the small spaces, low speed, and frequent conflicts with other pedestrians. The safety and security in the study area have the highest scores because no vehicles are allowed to pass the street. This PLOS method is more comprehensive than the traditional methods because it accounts for variables that other methodologies neglect.

1 Introduction

Walking is the most affordable way to reduce the negative consequences of using motorized vehicles, such as air pollution, traffic congestion, and health issues. It is a sustainable transportation mode with many individual and social benefit. While walking is a kind of human transportation, walkability measures how pedestrian-friendly a location is. Urban designers and planners are using this concept to create a sustainable pedestrian-based environment for communication, recreation, and shopping. Walkability is the foundation of sustainable cities. It is a "green" mode of transportation that reduces crowding, conserves energy, and eliminates air and noise pollution [1, 2]. In addition, it promotes equity by using renewable resources and noise and balancing regional growth by safely and healthily addressing society's access and development needs. These have been recognized by the WHO and UN, which promote sustainable transportation, liveable cities, and physical activity. The

* Corresponding author: abeer_khudhur@uomustansiriyah.edu.iq

Sustainable Development Goals (SDG) of Goal 3 "Ensuring healthy lives and promoting well-being for all at all ages", Goal 9 "Building resilient infrastructure", and Goal 11 "Making cities and human settlements inclusive, safe, resilient and sustainable" require walking promotion. Therefore, attention has been paid to walking, especially after transit-oriented development (TOD) was introduced [2-4].

The pedestrian facilities are the space for walking in the form of a sidewalk, a raised space on the side of the road to provide optimal service to pedestrians [4] or a walkway, an off-street separated exclusively pedestrian facility [6,7]. However, pedestrian interactions are challenging to understand because they are highly related to human behavior and the surrounding environment [7,8]. Researchers and planners consider two primary purposes for walking: utilitarian and recreational. Utilitarian walking trips have specific destinations, such as work or school. Recreational and personal trips do not have particular destinations, including walking for leisure and cultural and social activities [4]. Providing pedestrian-oriented infrastructures and facilities is considered a contributing factor because it reflects the quality of the walking path. It includes sidewalk length, pedestrian fencing, walkway width, paving condition, landscaping, signing, and lighting. Route aesthetics and cultural and commercial buildings increase pedestrian satisfaction, especially for recreational purposes. Safety and weather conditions also increase the recreational walking frequency [4, 9-11].

1.1 Walkability and pedestrian level of service

To measure the extent to which the quality of road infrastructures is perceived for walking, pedestrian level of service (PLOS) has been used as the most common measure of effectiveness in assessing the quality of pedestrian facilities' operations [5]. It is highly used to support better decision-making and to highlight limitations and motivators to walking. PLOS is more specific measure than walkability. The variables considered in PLOS considered by Turoń et al. as a main scale of walkability and could be say it is more specific than walkability. Scholars considered the PLOS is an interpreting of the quality of the pedestrian environment while Meeder defined the PLOS as a link between walkability which is a scale of social-economic, land use and accessibility characteristics to pedestrian-friendly places. Guzman et al. expressed the relation between the walkability and built environment characteristics as complex to interpretation. The PLOS might be the best interpretation of the effect of the geometric design characteristics and pedestrians flow characteristics on the walkability. Dovey and Pafka supported the definition of Henson who showed that the PLOS concept prioritized the metrics of individual pedestrian space and safety at pedestrian facilities such as sidewalks, walkways and pedestrians, crossing; while walkability encompasses the concepts of pedestrian access, street interconnectivity, and the density and diversity of land use which is broader than PLOS concept. Urban planners focus more on the walkability studies while transportation engineers pay attention to PLOS.

Among the developed methodologies to measure the PLOS for walkway, the Highway Capacity Manual (HCM) methodology is the oldest and most valid one . The HCM methodology is based on the quantitative measures of the effectiveness of MOEs, which are different according to the type of pedestrian facility. However, the limitations of this methodology have been highlighted in the studies that involved factors that HCM is not considered, such as land use, accessibility, quality of pedestrian facilities, comfort, and safety [1]. In addition, walking is highly linked to people's cultural context and behavioral intention. The influence of perceptions of walking facilities and other measures related to walkability, such as safety, security, and attractiveness, have been taken into account recently by researchers and decision-makers. Questionnaires were designed to collect data from pedestrians. Specific methodologies have been adopted to analyze these data and measure perceptions. The findings of most of these studies demonstrated that both the physical and perceptual attributes must be considered and employed by transport planners and decision-makers.

Talavera-Garcia and Soria-Lara proposed the "Quality of Pedestrian Level of Service," which aimed to integrate a broader range of factors than the traditional methods. These factors were grouped into accessibility, safety, comfort needs, and attractiveness. This study concluded that this method has been more effective in comprehending most of the factors that transport planners in Granada (Spain) considered in transportation and walking promotion strategies. Zannat et al. developed a wide-ranging framework to measure pedestrian facilities' qualitative and quantitative impact in the Chittagong Metropolitan Area in Bangladesh on perceived PLOS. The principal factors investigated were accessibility, safety, comfort, and attractiveness. It was found that accessibility and attractiveness significantly affect the footpath, carriageway, and transit, whereas safety has an insignificant effect on the footpath. Ujjwal and Bandyopadhyaya developed satisfaction rating models for walkability conditions using an ordered probit model. Data were collected through a survey of 550 pedestrians in two cities in India. The most significant factors were safety, pedestrian infrastructure, convenience, security, encroachment, and walking comfort. Asadi-Shekari et al. investigated a group of pedestrian indicators derived from the various walking guidelines, specifically for campus areas in the Universiti Teknologi Malaysia (UTM). The investigated factors were related to the available infrastructures, such as lane width, presence of barriers, crossing area, landscapes, supporting service, furniture, surface conditions, and marking. The results were different from one location to another. However, it produced a valuable model to identify current service issues and propose improvements for a higher level of pedestrian service.

Wibowo and Nurhalima examined the applicability of four methods of PLOS assessment on the sidewalks and walkways in the Bandung Institute of Technology campus: HCM, Trip Quality, Gainesville, and Australian PLOS. The HCM methodology considers only the capacity of pedestrian facilities. At the same time, the Trip Quality method focused on the qualitative PLOS measures such as pleasures, safety, comfort, the complexity of network and space, shade trees, transparency, and physical characteristics. Ratings of 1 to 5 are applied, reflecting poor to excellent condition. The total rating scores are averaged for PLOS grades from A to F. The third method, Gainesville PLOS, was explicitly developed for Gainesville (Florida). This method grouped indicators into six categories: facility type, vehicle-pedestrian conflict, supporting facilities, vehicular level of service (VLOS), maintenance, and multimodal transport. The total scores ranged between 1 to 21 and were divided into six levels, A to F. The fourth method is the Australian PLOS method, which considers three factors: physical characteristics, location factors, and user factors. By comparing the four methods, Wibowo and Nurhalima found that the four methods did not consider user characteristics such as age, gender, income, and perception. In addition, the four methods considered the infrastructure characteristics, but the HCM lacks the inclusion of land use, safety, and comfort indicators. They concluded that no single method could produce a comprehensive assessment. Abdulhussein and Jameel investigated the impact of the role of infrastructure as a Latent Variable (LV) in the transport mode choice. They found that the provision of cycling fences and pedestrian bridges are the most significant variables. Consequently, it is recommended that these results be considered in the provision and improvement of cycling and walking infrastructures to promote active transportation.

1.2 Walkability assessment

Walkability was measured in previous research objectively, subjectively, or both. Subjective measurements employ interviews and surveys to assess user impression, while objective measurements use physical attributes. This classification has been utilized in several studies. The most accurate walkability measures are NEWS and GWI.

Saelens et al. designed the NEWS Worldwide, which is the most extensively used subjective walkability neighborhood environment measure. A special inventory form containing 83 questions has been designed for NEWS survey; they cover residential density, land use mix–diversity, retail and leisure access and closure, street connectivity, pedestrian infrastructure (sidewalks and walkways), aesthetics, traffic safety, and criminal safety. Further research adopted NEWS with modifications to meet the objectives of their studies; such as the Australian NEWS survey, Ghent NEWS and the abbreviated (NESW-A) which excluded some groups of questions that are not required in their studies.

The World Bank interns created the Global Walkability Index (GWI) to assess international walkability. Key indicators include 14 parameters reduced to nine parameters in three groups: safety and security, convenience and attractiveness, and policy support. The safety and security indicator measures the number of pedestrian fatalities from traffic accidents and crimes. Maintenance, cleanliness, facilities, amenities, walking path difficulties, and crossings rated convenience and attractiveness.

1.3 The research question and aim

The research question is: How can the perception of using walking be incorporated into the methodology of PLOS determination to extend its concept to the walkability concept? And what indicators can be used to reflect the perception of walkers for recreational walking trips? This research aims to qualify road users' perception of walking in a selected area in Baghdad City, Iraq, in terms of Pedestrian Level of Service (PLOS) to assess the walkability and identify issues from a pedestrian point of view.

This paper's novelty is that it includes indicators that reflect the perception of pedestrians in the methodology of PLOS determination. This is essential in studying and promoting walking for recreation in areas with unique characteristics. The outcome of the research will establish methods and metrics to study the relationship between walking behavior, physical environment, and city sustainability to improve urban design and "walkability" knowledge and practice.

2 Materials and methods

The main steps followed to achieve the study's aim started with selecting the study area and then defining the general situation through monitoring and observation. After that, data were collected through a designed questionnaire form and field observation. The details of the assessment stages are shown in the subsequent subsections. Figure 1 shows the methodology flowchart.

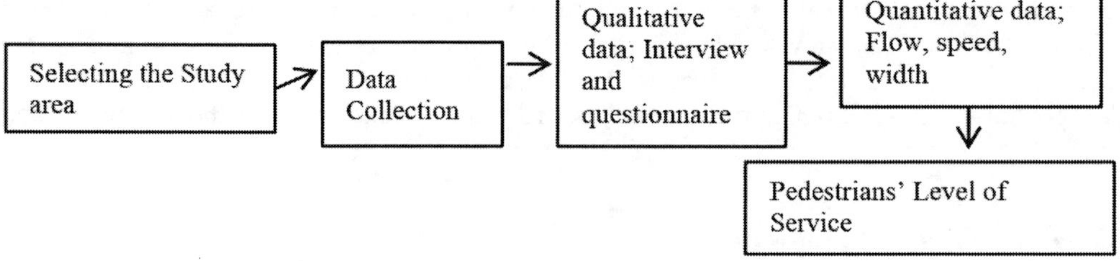

Fig. 1. The Methodology of the research.

2.1 Study area

The study area is Al-Mutanabbi Street, a 300 m street and one of the most ancient streets in Baghdad City, Rusafa Side. It was built and established during the Abbasi Era to be a meeting place for poets and writers, and then it was expanded to be a center of bikeshares. It is now considered a center of recreational, cultural, and historical activities. Visitors come to this street to buy books and visit the folklore restaurants, cafes, historical places, and the Tigris River bank, which is located at the end of this street. Al-Mutanabbi Street is exclusive for walking; no vehicles are moving on it. It is usually crowded with pedestrians during the day and evening, especially on Thursday, Friday, and Saturday. The sidewalks on both sides of the road are only adequate during the day, but in the evening, they are closed by shopkeepers. Figure 2 shows a plan view of the study area.

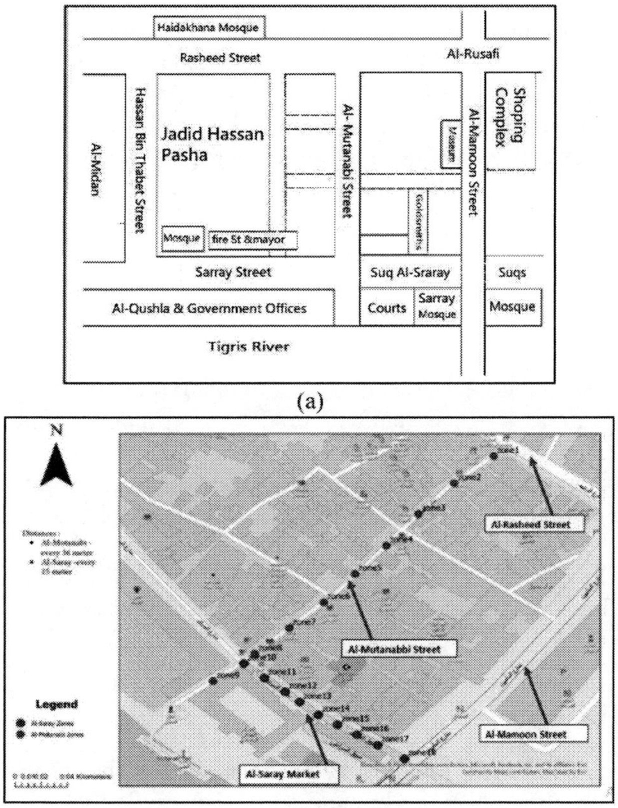

Fig. 2. The study area, Al-Mutanabbi Street (created by the author).

2.2 Data collection

Roadside interviews with pedestrians visiting the streets were carried out to ask them about the general situation of Al-Mutanabbi Street, the main factors motivating them to walk on the street, the drawbacks, and their recommendations to improve the walkability of the street.

A roadside questionnaire was designed to consist of two parts. The first part contains the respondents' demographics, including age, sex, income, occupation, and whether the respondents own a car. The second part includes psychological factors and awareness of physical factors (Factors affecting people who choose to walk in the city center. The questions of the questionnaire were selected based on previously designed survey forms such as The Neighborhood Environment Walkability Scale (NEWS) and the Global Walkability Index (GWI). The selection of questions is based meeting the aim of this research which focuses on cultural and historical areas and encompasses the metrics of PLOS and walkability concepts. The selected questions were 21 questions grouped into seven groups, shown in Table 1. The answers are designed on a six-Likert scale, from 1 representing "Excellent" to 6 representing "very bad".

The preliminary form of the questionnaire was subjected to a pilot test by being distributed to experts in transportation and urban planning. It was also distributed to about 30 visitors to give feedback and identify unintelligible, repeated, non-understandable and out of scope questions. Based on the feedback and comments of the experts, the questions have been modified to the final forms.

The questionnaire was designed in electronic form using KOBO software, which is used to create questionnaires and collect online and offline responses in the field using Mobile phones. The authors asked randomly selected visitors about the questionnaire questions and entered their responses through KOBO software. The respondents were first asked if they agreed to conduct the survey and to record their reactions for academic research purposes. They were also free to answer the questions about their income and educational level as they were optional. They also asked if they had any limitations regarding responding to the survey before recording their answers.

The size of the respondents was 322 people; it is checked according to Equation (1) to meet the sample size with a confidence interval = 1.96

$$N_{min} = (SK/E)^2 \tag{1}$$

The results were analyzed using the IBM SPSS Microsoft program.

Where: N_{min}=Sample size, S=Standard deviation =10, K=Confidence interval=1.96, E=Coefficient = 2.

The responses were then subjected to a Reliability test to measure the stability and consistency of the answers. This was done using Cronbach's Alpha via the SPSS IBM Statics 25 program; the stability coefficient should be close to one. The validity test was also measured by multiplying the reliability coefficient by itself. The result of this test showed that Cronbach's Alpha was 0.809, reflecting the significant internal consistency between the questions.

2.3 Analysis of the data

The responses obtained from the previous step were analyzed; The participants demographics were analyzed firstly to identify the limitations of the results that will be obtained in the analyzing of the second part of the questionnaire.

2.3.1 Analsys of the respondents' demographics

In this section, the results of the respondents' demographics; including age, gender, income, academic level, and having own car will be presented.

According to the age of the participants, the highest participation was from people with age of 20 to 45 years which their percentage was about 79%. This has been expected because older people may could not use walking for not providing elderly facilities in the area. Younger people were expected to be more, probably the attention to visit cultural areas and the parents' restrictions for young people are reasons for low percentages of young people visiting Al-Mutanabbi Street.

Regarding the gender of walkers, the percentage of female was about 54%, which is slightly greater than the percentage of. This means the gender restrictions was not a main issue in walking at the study area.

The monthly income range of about 57% of respondents was 500,000 to 1,000,000 Iraqi dinars; which is higher than the average monthly income in Iraq. This reflects the positive attitude of people with this range of monthly income to walk. On the other hand, people with monthly income less than the average were only 19% of the visitors. This may reflect the attitude pf people within this range of monthly income to make recreational trips.

According to the participants' academic profession levels, it has been reported that about 70% of the walkers had bachelor's degrees. This is expected because the nature of the case study area is more attractive for people who have high profession levels.

About 59% of participants has no private cars while car-owners were about 41%. The latter percentage is optimistic as it reflects the positive perception of people to walk even when thy have a car.

2.3.2 Analsys of the walakbility indices

To analyze the second part of the questionnaire responses, a score ranging from 0 to 5 has been allocated to each answer based on a method developed by Gallin (2001); 0 is allocated for the worst scale, while six is allocated for the best scale. Table 1 shows the six-scale of each question adopted from Gallin.

This method requires determining each question's weight (W), reflecting to what extent each variable affects the assessment results. The expert's opinions were used to identify the weights. Another questionnaire was designed to ask 22 experts about the subjective weights out of 5 of the 21 variables in the first questionnaire, subjectively based on their experience in urban transportation planning.

3 Results

The experts' answers for each question were averaged, and the results are shown in Table 1.

To determine the overall level of service, PLOS bands have been determined by assuming the worst condition has a zero score for all elements, and the best condition has a full score of 5 for all elements with a total of 350.5 points. The total score is the maximum score, and it is divided by six to find the interval length. The determined interval length is 58, and the PLOS bands are shown in Table (2). The sum of the multiplication of the weight of each variable by its score was divided by six and entered into Table 2 to determine the PLOS, which is C, as shown in Table (3) and Figure 3.

Table 1 The categories and variables of walkability with their scores and weights.

Categories	Variables	W	Scores					
			0	1	2	3	4	5
Physical Characteristics	Width (at day)	3.9	No path	0–<1 m	1.0–<1.5 m	1.5–<2.0 m	2.0-<3.5 m	≥ 3.5
	Width (at night)	3.5	No path	0–<1 m	1.0–<1.5 m	1.5–<2.0 m	2.0-<3.5 m	≥ 3.5
	Quality of surface	3.1	Extremely poor	Poor	Moderate	acceptable	Good	Excellent
	Obstacles per km	3.8	>10	8-10	5-8	2-5	<2	0

Continuation of Table 1.

Categories	Variables	W	Scores					
			0	1	2	3	4	5
Safety and Security	Pedestrians crossing	3.4	Not provided	Few / poorly located	Some / reasonably well located	Adequate /reasonably well-located	Supplied at an adequate frequency	Extremely adequate
	Pedestrian' crashes	3.8	Very High	High	Moderate	Few	Very few	No
	Crossing control system	3.6	No	Poor	Ineffective signals	effective at some crossing	Effective at all	Extremely Effective
	Feeling of secure	3.4	Insecure	Poor security	Reasonable	Natural	Good	Very good
	Visibility	3.6	No lighting	Poor	Moderate	Good	Very good	Excellent
Elderly and Disabilities	Facilities for Elderly and Disabilities	2.7	Very poor	Poor	Moderate	Acceptable	Adequate	Extremely Adequate
Pedestrians' movement	Speed of walking	2.4	Very slow	Slow	Moderate	Reasonable	High	Very high
	Number of pedestrians	3.4	> 350	226 -350	151-225	100-150	50-99	<50
	Pedestrians' conflicts	3.1	Very high	High	Moderate	Few	Very few	No
	Space between pedestrian	3.2	Very poor	Poor	Moderate	Acceptable	Adequate	Extremely Adequate
	Delay	3.4	Very high	High	Moderate	Limited	Very short	No
Presence of Supporting facilities and amenities	Recreational activities	3.2	Very poor	Poor	Moderate	Acceptable	Adequate	Extremely Adequate
	Historical and cultural activities	3.6	Very poor	Poor	Moderate	Acceptable	Adequate	Extremely Adequate
	Walking furniture	2.9	Very poor	Poor	Moderate	Acceptable	Adequate	Extremely Adequate
	Supported facilities	2.9	Very poor	Poor	Moderate	Acceptable	Adequate	Extremely Adequate
Weather condition	Protecting from bad weather	3.8	Very poor	Poor	Moderate	Acceptable	Adequate	Extremely Adequate
Accessibility	Accessible to the assessed area	3.4	Very poor	Poor	Moderate	Acceptable	Adequate	Extremely Adequate

Table 2. The PLOS bands.

PLOS	Total Scores
A	≥290
B	232 - <290
C	174 - <232
D	116 - <174
E	58 - <116
F	0 - <58

Table 3 The results of the data analysis, the score of indicators, and the overall PLOS.

Factor	Weight	Score	Scores×weight
Width (at day)	3.9	3	11.7
Width (at night)	3.5	5	17.5
Quality of surface	3.1	4	12.4
Presence of Obstacles	3.8	0	0
The presence of pedestrians crossing	3.4	5	17
Pedestrian' crashes	3.8	5	19
Crossing control system	3.6	5	18
Feeling of secure	3.4	4	13.6
Visibility	3.6	4	14.4
Facilities for Elderly and Disabilities	2.7	3	8.1
Speed of walking	2.4	2	4.8
Number of pedestrians	3.4	0	0
Pedestrians' conflicts	3.1	2	6.2
Space between pedestrian	3.2	2	6.4
Delay	3.4	3	10.2
Presence of Recreational activities	3.2	4	12.8
Historical and cultural activities	3.6	3	10.8
Walking furniture	2.9	3	8.7
Supported facilities	2.9	3	8.7
Protecting from bad weather	3.8	3	11.4
Accessible to the assessed area	3.4	3	10.2
Total			221.9
PLOS			C

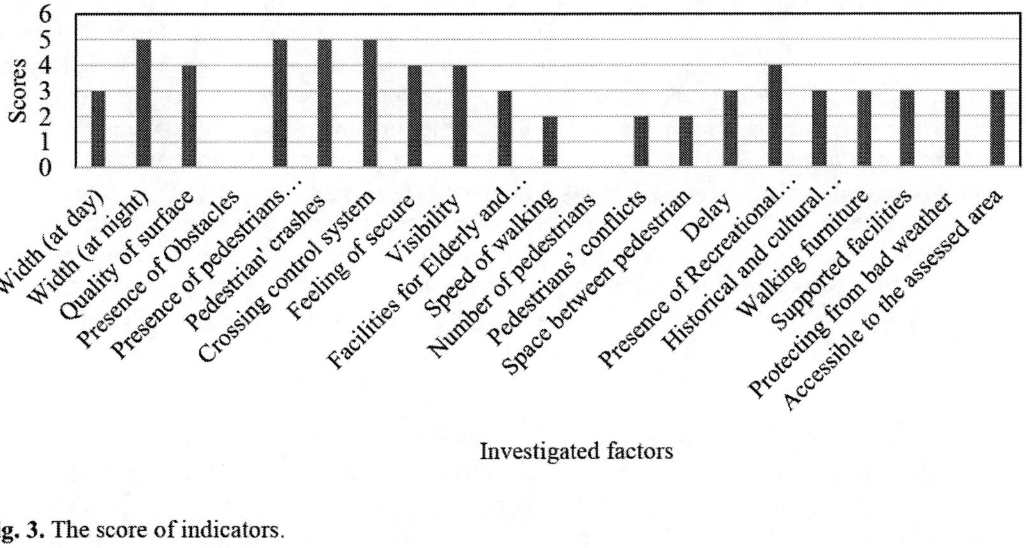

Fig. 3. The score of indicators.

It can be observed that the highest scores were allocated to the perception of the width of the street in the evening because the space of the main walkway is allocated for pedestrians only, while in the day, the bookseller's occupied spaces for their packages; therefore, the scores are less. Safety and security variables also have the highest score because the street is exclusively for walking, and there are no cars crossing the street; therefore, the feeling that pedestrians are absolutely safe from motorized vehicles. The second highest score, 4, is allocated to variables of the quality of the surface. The questionnaire was distributed in 2023 when the street was rehabilitated, and the surface quality was adequate without observed failing. The folklore visibility of the street and the presence of recreational activities also score four, reflecting the attractiveness of the walkers to this street for these reasons. The lowest scores, 2, were allocated to the speed of pedestrians, the space per pedestrian, and the conflict between pedestrians. This may result from the high walker volume in the study area, resulting in lower speed and small space for each pedestrian. A slower walking speed for recreational is more acceptable than a slow speed for transportation. Therefore, the visitors were not rated with 0 or 1. It can be seen that a score of 3 has been allocated to the supportive facilities such as seats and weather-protective shells. Another issue is highlighted, which is the accessibility to the study area. It has been observed that the streets approaching Al-Mutanabbi Street have issues of traffic congestion and unsuitability for walking. In addition, the nearest public transportation stop is far away. According to these results, improvements should be proposed to increase the PLOS and the scores for the highlighted issues. Regarding the width of the walking street, it is suggested that the booksellers have space for their packages in the surrounding areas to make the spaces available for walking only. It is also suggested that seats and other supported facilities be provided. It is also recommended to improve the PLOS in Al-Rasheed Street, Al-Midan Street, and Al-Mamoon Street to improve the accessibility to the study area.

4 Conclusion

The perception of walking is a significant variable that should be considered when assessing pedestrian safety. The designed questionnaire includes the most relevant questions used in the previously designed questionnaire to the characteristics of the study area. The results of the respondent's analysis highlight the strengths that attract pedestrians to walk in the street, such as safety, amenities, and recreational activities. The study area is more attractive in the evening because there is more available space for pedestrians than on days. It also highlights the drawbacks that need improvements, such as space and the supplied facilities. The surface quality is also reported as a drawback, so rehabilitation is required. The weather is recorded as an issue at a lower rate, and the walkers suggested providing shelters to protect them from hot and sunny weather in summer and rainy weather in winter. The analysis results could be presented to planners and decision-makers so that they can take them into account in further programs and strategies for walkability improvements on this street and other street.

The main limitation of this research is that its scope is limited to a specific study area with historical and cultural characteristics. In addition, the methodology could be applied to walking for recreational purposes only; different factors should be considered for walking for transportation. Furthermore, and according to the results of demographic characteristics of the participant, the results could be limited to people with age 20 to 45 years, having Bachelor degree, with monthly income slightly above the average monthly income in IRAQ (I QD500, 000 to IQD1000,000).

The perception and attitude could be considered in further research to study the behavior of trips with various modes of transportation. In addition, these factors could be incorporated into the mode choice models to get more comprehensive and accurate results. Other methodologies could be used to determine the level of service and compare the results with the commonly mused methodologies to investigate the most thorough method that produces results that meet the guidelines of various policies.

References

1. R. Rafiemanzelat, M. I. Emadi, A. J. Kamali, "City sustainability: the influence of walkability on built environments," Transportation research procedia **24**, 97-104 (2017)
2. S. Ridha, S. Ginestet, S. Lorente, "Adopting A Sustainable Urban Design to Improve Thermal Comfort in An Arid Climate," Journal of Engineering and Sustainable Development **27(2)**, 171–179 (2023), https://doi.org/10.31272/jeasd.27.2.2
3. W. D. Lee, W. Ectors, T. Bellemans, B. Kochan, D. Janssens, G. Wets, K. Choi, Chang-Hyeon Joh, "Investigating Pedestrian Walkability using a Multitude of Seoul Data Sources," Transportmetrica B-Transport Dynamics **66(1)**, 1-20 (2018)
4. M. Z. Muttaqin, A. Munawar, M. Zudhyirawan, "Assessment of HCM 2010 and HBS 2002 for Pedestrian Lane Quality in Indonesia (Case Study: Sardjito Hospital and Sudirman Office, Yogyakarta)," Turkish Journal of Computer and Mathematics Education **12(14)**, 4646-4652 (2021)
5. A. Banerjee, A. K. Maurya, G. Lämmel, "A Review of Pedestrian Flow Characteristics and Level of Service over Different Pedestrian Facilities," Collective Dynamics **3**, 1–52 (2018), https://doi.org/10.17815/CD.2018.17
6. D. Pemberton, *Are our suburbs walkable for the elderly? Focus on walking for transport* (Master's thesis, Itä-Suomen yliopisto, 2022)
7. Z. Wang, D. Ettema, M. Helbich, "Objective environmental exposures correlate differently with recreational and transportation walking: A cross-sectional national study in the Netherlands," Environmental research **194**, 110591 (2021)
8. A. Colom, S. Mavoa, M. Ruiz, J. Wärnberg, J. Muncunill, J. Konieczna, D. Romaguera, "Neighbourhood walkability and physical activity: moderating role of a physical activity intervention in overweight and obese older adults with metabolic syndrome," Age and ageing **50(3)**, 963-968 (2021)
9. G. Ferrari, A. O. Werneck, D. R. Silva, I. Kovalskys, G. Gómez, A. Rigotti, M. Fisberg, "Perceived urban environment attributes and device-measured physical activity in Latin America: an 8-nation study," American Journal of Preventive Medicine **62(4)**, 635-645 (2022)
10. F. A. Subhi, A. K. Jameel, "Investigating the Key Components of Walking Behaviour in Baghdad City," in IOP conference series. Earth and environmental science **1232(1)**, 012052 (2023), https://doi.org/10.1088/1755-1315/1232/1/012052
11. K. Turoń, P. Czech, M. Juzek, "The concept of a walkable city as an alternative form of urban mobility," in Scientific Journal of Silesian University of Technology. Series Transport **95**, 223-230 (2017), https://doi.org/10.20858/sjsutst.2017.95.20

Determination the changes in water and vegetation covers for Ain Al Tomar district using remote sensing techniques

Rusul Ahmed Efan[1, *], and *Sundus A. Abdullah* Albakry[2]

[1]Remote Sensing Unit, College of Science, University of Baghdad, Baghdad, Iraq

[2]Research and Technology Center of Environment, Water and Renewable Energy, Scientific Research Commission, Ministry of Higher Education and Scientific Research, Baghdad, Iraq

Abstract. Ain Al Tamor district is one of administration units of Karbala province in middle Iraq. The aim of the study is analyze the changes in vegetation area using the Normalized Difference Vegetation Index (NDVI) and the Normalized Difference Water Index (NDWI) for 1990, 2000, 2013 and 2024. Landsat 5 TM and Landsat 8 OLI used to extract the vegetation and water covers for study period. The results of the study showed that there is a clearly increase in the NDVI for Ain Al Tamor in 1990, 2000, 2013, 2024 ranged from 45.2 km^2, 67.2 km^2, 146.3 km^2 and 266.3 km^2, respectively. The study area depends on ground water for irrigation of the crops, especially the palm orchards. On the other hands, NDWI values recorded decrease values for the same years about 515.5 km^2 in 1990, 414.3 km^2 in 2000, 161.5 km^2 in 2013 and 183.1 km^2 in 2024. These fluctuations of water area due to decline of precipitation values in the study area and shrink of surface water. The study of vegetation and water covers is important, especially in desert and semi desert regions for stability of local populations, rural development, sustainable development and increasing agricultural cover.

1 Introduction

Effective monitoring and land cover change management are essential for environmental sustainability. The earth's land cover has changed quickly, unpredictably, and uncontrollably, due to population growth, urbanization, agricultural decline, climate unpredictability, and other causes. These shifts have caused natural disasters, agricultural mismanagement, biodiversity loss, deforestation, and degradation [1, 2]. Around the world, urbanized zones are experiencing fast expansion. The rate of urbanization is accelerating in developing countries like Iraq. The land use/cover (LULC) has undergone significant alterations due to human activities and growth urban populations, both spatially and temporally. Conversely, land use describes how people make use of the land and its resources, including urbanization, mining, logging, agriculture, and pastoralism [3].

* Corresponding author: rusul.ifan2309@sc.uobaghdad.edu.iq

The transformation of various land cover types is termed LULC and is the outcome of intricate interactions between the natural world and people [4]. Planning and land administration operations require the LULC [5]. Understanding how the nature of phenomena changes over time requires the ability to detect LULC alteration happened due of human activities and natural occurrences, the LULC change occurs frequently [6].

The majority of Iraq experiences hot, dry, and subtropical climate [7]. Water scarcity, desertification, heat waves, sand storms, droughts, soil and water salinization, have all occurred in Iraq in recent years [8, 9]. Iraq is one of the country's most vulnerable to the economic and physical consequences of climate change [10]. According to the United Nations Environment Program (UNEP), Iraq is the fourth most vulnerable country in the world to the effects of climate change. The Geographic information systems (GIS) and Remote Sensing (RS) are significant technologies that use image analysis to offer environmental data for a range of urban applications [11, 12]. The importance and novelty of the study lies in its use of a number of spectral indicators such as the Normalized Difference Water Index (NDWI) and the Normalized Difference Vegetation Index (NDVI). The importance and potential of GIS and RS as strong and efficient instruments for tracking land changes and evaluating deterioration have shown by the advancement of geographic systems and modern technologies. Planners and land managers can now use a set of sensors that operate at different image scales thanks to substantial advancements in RS technology.

Due of its medium geographic resolution and extensive time series, Landsat data employed to monitor changes in land cover. To ascertain the specifics of the objectives, including water, vegetation cover, and soil types, spectral precision is necessary. Because of the variations in spectrum reflectance, these materials distinguished from one another. Because of the absorption process by chlorophyll, Plants exhibit a low spectral reflectance for the visible spectrum's red and blue wavelengths, whereas their reflectivity is higher for the near-infrared green wavelength. When the water contains sediments, plankton, or large levels of chlorophyll, its reflectivity rises within the visible region. Pure water typically has a low reflectivity. The natural environment and human economic activity greatly impacted by LULC in Iraq and its modifications. One crucial metric for tracking changes around the world is the surface vegetation index. As a result, the distribution and diversity of plants and trees, as well as the type of canopy, have drawn the attention of several researchers. Simultaneously, the vegetation index is employed to evaluate environmental concerns, including land-use surveys, crop products, and land-cover assessments.

Satellite images used to monitor the relationship between natural events and human activities and to detect changes in LULC. The RS optical data, which includes aerial photography and satellite sensor images, used to determine LULC dynamics. The outcomes of LULC mapping used for global, regional, and local mapping, change detection, and leading landscape scales. One-way to conceptualize the classification of remote sensing images is as a joint effort of image processing and classification technologies. Multispectral image categorization is a crucial technique in the quantitative study of remotely sensed images. The characteristics of a pixel (image element) from a multispectral image usually recorded on many spectral channels. Numerous studies address the application of NDVI and NDWI in various geographical areas. Hashim et al., 2029 used Landsat 2 MSS and Landsat 8 OLI to study NDVI and LULC in wetlands in southern Iraq between 1977 and 2017. Taloor et al., 2021 computed the NDVI and NDWI over the Ravi basin, India, and parts of Pakistan, using Landsat 8 OLI. The NDVI and NDWI of Al Hillah city were computed using Landsat 5 and 8 images for 1990, 2000, 2013, and 2022. The objectives of the current study aims to use Landsat 5 T and Landsat 8OLI images to calculate the NDVI and NDWI values of Ain Al Tamor district, and estimate the area of these indices using ArcGIS 10.8.

2 Materials and methods

2.1 Study area

Ain Al Tamor is a district located in the western part of Karbala province, in southern Iraq. It is located about 50 kilometers northwest of Karbala city, As shown in Figure (1), making it an important agricultural and rural area within the governorates. This district is notable for its historical significance, natural landscapes, and its role in local agriculture. The study area focuses on this district to analyze environmental changes, including vegetation, water resources, and land use dynamics, which are essential to comprehending the sustainability and ecological health of the area. Ain Al Tamor is located at approximately 32°.45 N latitude and 43°.36 E longitude.

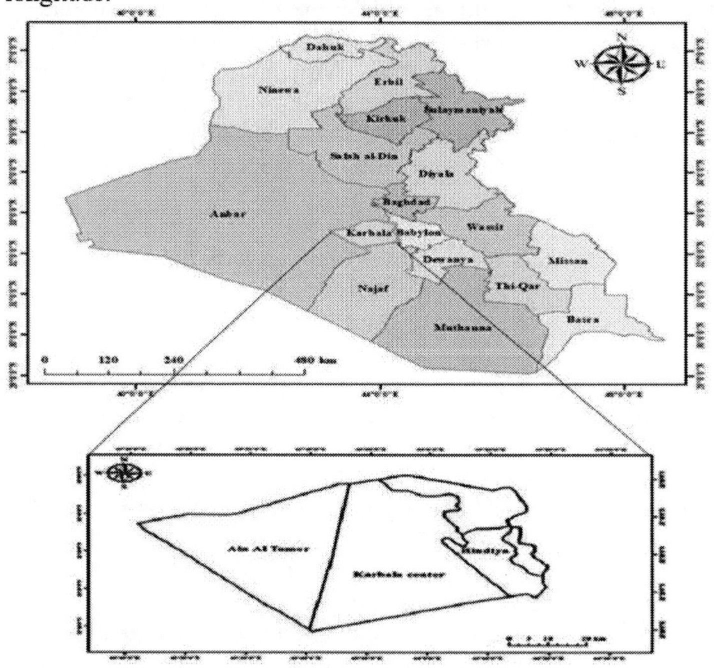

Fig. 1. Map of Karbala province in middle of Iraq.

2.2 Data accessibility

The study employed Landsat 5 TM and Landsat 8 OLI from the United States Geological Survey (USGS) website (https://earthexplorer.usgs.gov/). It provided the satellite image widely used in environmental investigation. The current study used four images from Landsat 5 TM and 8 OLI satellite, which covered the study area, as shown in Table 1. The images acquired on March 29, 1990, and March 29, 2000. While, the Landsat 8 images acquired on March 22, 2013, and April 2, 2024. All images taken when the weather was clear and almost cloud-free. The four Landsat images have orbit data with a path/row system of 166/38 and 168/37, respectively, for Ain Al Tamor district.

Table 1. Information of Landsat images acquired in current study.

No.	Satellite Image	Path/Row	Acquisition Data	Source
1	Landsat 5 TM	168/37	March 29, 1990	https://earthexplorer.usgs.gov/
			March 7, 2000	
2	Landsat 8 OLI	168/37	March 22, 2013	
			April 2, 2024	

2.3 Normalized Difference Vegetation Index (NDVI)

It is the most accurate way to process satellite images of vegetation displays for diagnosing, detecting, and removing the condition of overlap between the vegetation class and other kinds of land cover, which often overlap. Its foundation is the plants' ability to reflect light. While Near Infrared wavelengths have low reflectivity, Red wavelengths have high reflectivity.
The NDVI value falls between -1 and 1. Greater NIR levels, which indicate dense vegetation, represented by higher NDVI values. Equation (1) illustrates the percentage of the overall difference between spectral reflections at Red (R) and Near-Infrared (NIR) wavelengths.

$$NDVI = \frac{(NIR-R)}{(NIR+R)} \quad (1)$$

2.4 Normalized Difference water Index (NDWI)

In order to identify surface waters in wetland areas and to make it possible to determine the amount of surface waters, McFeeters initially proposed the NDWI in 1996. The range of the NDWI value is - 1 to 1. The NDWI value of water bodies is more than 0.5. Equation (2) used the difference between green and red spectral bands to determine the NDWI:

$$NDWI = \frac{(Green-R)}{(Green+R)} \quad (2)$$

Figure 2 shows the flowchart of methodology the current study.

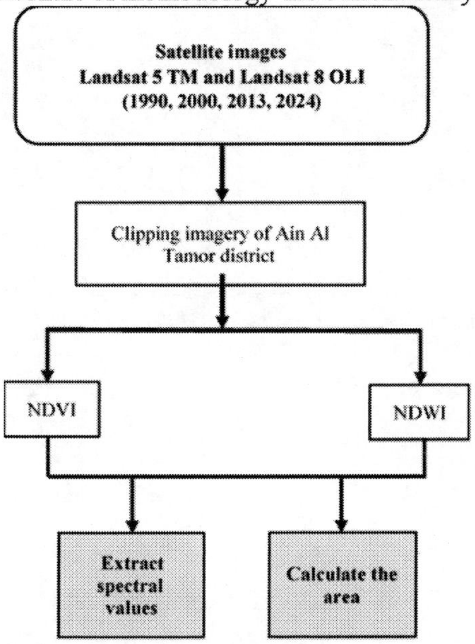

Fig. 2. The flowchart of methodology the current study.

 https://doi.org/10.1051/e3sconf/202563307007

3 Results and discussion

3.1 NDVI

Figure 3 shows the NDVI values of Ain Al Tamor district for 1990, 2000, 2013 and 2023 using images of Landsat 5 TM, and Landsat 8 OLI.

- 1990 - 2000: NDVI values showed a steady increase during this period, indicating a consistent improvement in vegetation cover across Ain al Tamor. This increase attributed to favorable climatic conditions Landsat 5 images in 1990 between 0.30 and -0.20, including adequate rainfall, and the availability of water for irrigation. The agricultural areas near the Euphrates River likely benefitted from these conditions, supporting vegetation growth. The Landsat 5 image's NDVI scale in 2000 ranged from 0.30 to 0.24.
- 2000 - 2013: The trend of continued increase in NDVI values persisted, although at a slower rate compared to the previous decade. This suggests that while vegetation health remained stable or improved slightly, the growth rate slowed down. Factors such as improved farming practices, better irrigation techniques, and local continued positive trend in vegetation. In 2013, The Landsat 5 image's NDVI ranged from 0.32 to 0.18.
- 2013 - 2024: NDVI values showed a significant increase during this period, indicating a marked improvement in vegetation health and density. This increase attributed to:
- Better irrigation system: The implementation of contemporary irrigation methods, including drip irrigation, enhanced the amount of water available for agriculture and lessened the effects of water scarcity. Landsat 8 OLI indicates that in 2024, it was between 0.63 and 0.30.

-Climate variability: Increased rainfall in some years may have boosted vegetation growth, contributing to higher NDVI values.

-**Land restoration efforts**: Increased efforts to restore degraded lands and implement sustainable agricultural practices likely played a role in enhancing vegetation cover.

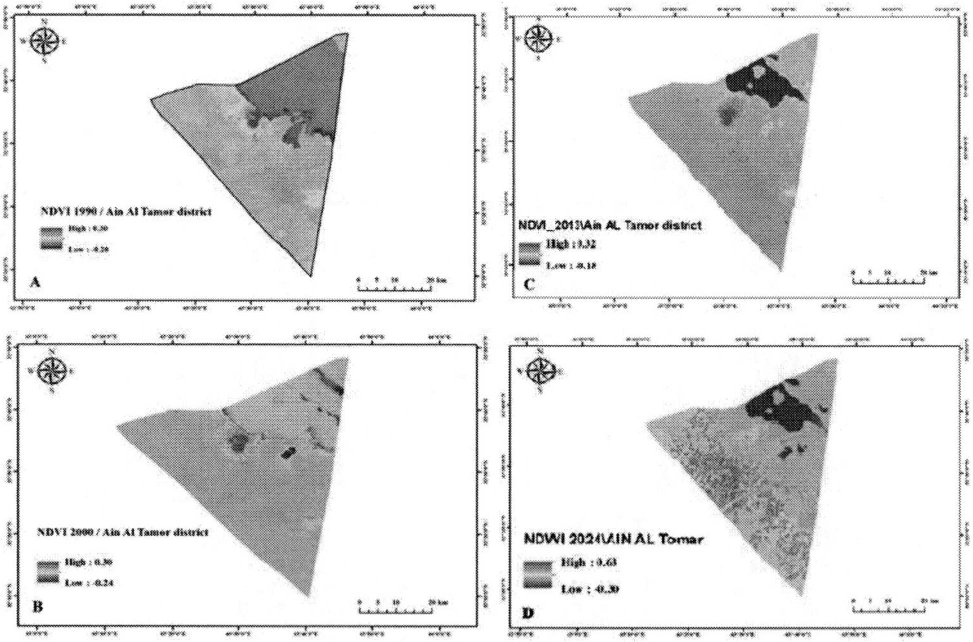

Fig. 3. NDVI values of Ain Al Tamor district in (A) 1990, (B) 2000, (C) 2013, and (D) 2024.

 https://doi.org/10.1051/e3sconf/202563307007

3.2 NDWI

Figure 4 illustrates the NDWI values of Ain Al Tamor district for 1990, 2000, 2013 and 2023 using images of Landsat 5 TM, and Landsat 8 OLI.

- 1990: In 1990, NDWI values were high 1990 Landsat 5 TM pictures showed that the NDWI values for this time frame were (0.24 to 0.31), indicating that water availability was at a relatively good level. The Euphrates River and local water bodies provided sufficient water resources for both agricultural and domestic use, resulting in positive NDWI values. This period reflects a time of abundant water resources supporting vegetation growth and agricultural productivity.
- 2000-2013: NDWI values showed a gradual decline during this period, indicating a reduction in water availability. The Landsat 5 image's NDWI scale in 2000 ranged from 0.24 to 0.32. The NDWI values for the year 2013 were (0.20 to 0.32), according to Landsat 5 TM photos. Several factors contributed to this decline:

-Decreased rainfall: A period of reduced rainfall, particularly in the early 2000s, led to lower water levels in rivers and other water sources.

-Over-extraction of groundwater: Increased demand for irrigation and domestic use likely led to over –extraction of groundwater, contributing to a decline in surface water availability.

-Upstream water diversion: water diversion for large-scale agricultural projects and damming of rivers upstream may have reduced the flow of the Euphrates River, affecting water availability in Ain Al-Tamor.

- 2024**:** NDWI values showed a marked recovery in 2024, indication an increase in water availability. NDWI ratio as determined by the Landsat 8 OLI satellite's aerial picture categorization (0.34-0.55). This recovery attributed to several factors:

-Improved water management practices: The introduction of more efficient irrigation systems and better water distribution techniques helped optimize water use and reduce wastage.

-Increased rainfall: Some years saw higher-than-usual rainfall, which contributed to the replenishment of water resources.

Government initiatives: Local and national efforts to improve water conservation, such as the construction of water reservoirs and better management of water resources, likely contributed to increased water availability.

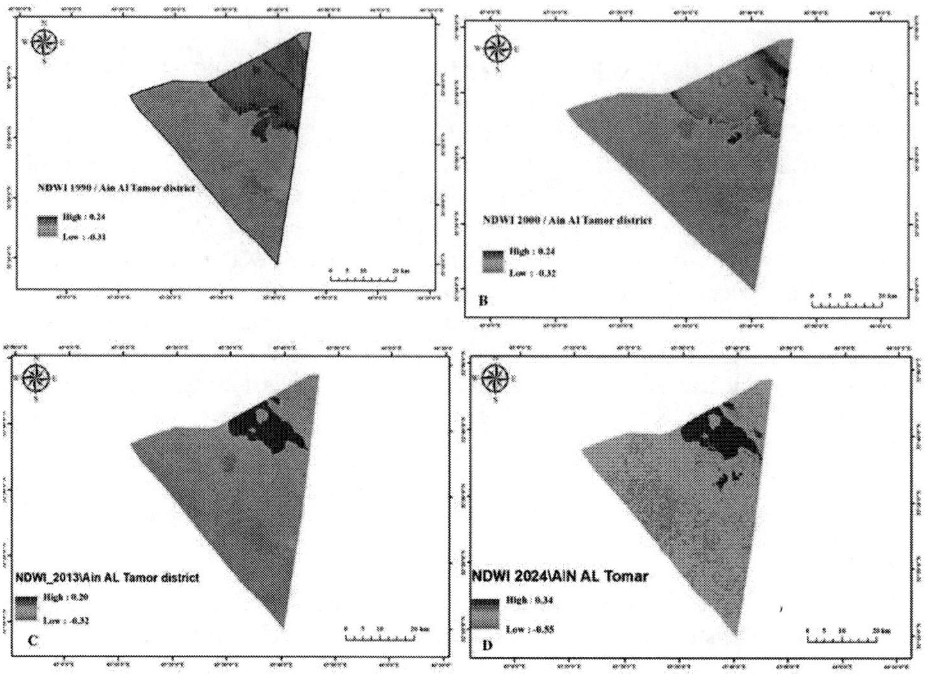

Fig. 4. NDWI values of Ain Al Tamor district in (A) 1990, (B) 2000, (C) 2013 and (D) 2024.

Table 2 shows the area and vegetation and water covers, which calculated based on spectral analysis of NDVI and NDWI. The results illustrated that area of NDVI increased from 1990 to 2024 form 32.6 km2 to 266.3 km2, respectively. On the other hand, area of water decreased dramatically from 512.7 km2 in 1990 to 183.1 km2 in 2024. The reason of this decline of water due to shrink of Razzaza lake, which lies northeastern Ain Al Tamor district.

Table 2. The area and vegetation and water covers during study periods.

Spectral index	**Area (km^2)**			
	1990	2000	2013	2024
NDVI	32.6	67.2	146.3	266.3
NDWI	512.7	414.3	161.5	183.1

The main conclusions and their ramifications are as follows:

NDVI (Vegetation Health): The consistent rise in NDVI values between 1990 and 2024 indicates that Ain Al Tamor's vegetation health has gradually improved. Numerous causes contributed to the growth of vegetation:

-Water availability: Because of advancements in irrigation and water management, agricultural water supplies are now more reliable, promoting the growth of plants.

-Favorable climate: Although there was a brief dry spell in the area, recent rainfall trends have generally benefited the plants.

-Sustainable agricultural practices: The observed good NDVI trends probably influenced by initiatives to repair degraded land, employ effective irrigation, and adopt sustainable agricultural methods.

Water Availability (NDWI): A period of water scarcity in Ain Al Tomar reflected in the reduction in NDWI values between 2000 and 2013. Numerous causes contributed to this decline:

-Water shortage and decreased rainfall: Surface water availability decreased because of decreased rainfall and increased demand for waterresources in agriculture.
-Excessive groundwater extraction: The reduction in water availability might cause by excessive groundwater withdrawal for agriculture.
- The 2024 rebound in NDWI values points to the restoration of water resources as a result of recent advancements in water management and more rainfall.

4 Conclusion

The temporal change of vegetation and water in Ain Al Tamor district was clear in the current study. The results showed that consistent rise in vegetation health (NDVI) demonstrated by the analysis of NDVI and NDWI for Ain Al Tamor from 1990 to 2024.Although there were issues with water scarcity in the area from 2000 to 2013, both vegetation and water availability have improved in recent years. Despite the clear decline of Lake Razzaza in recent years, groundwater has become the dominant source of irrigation for agricultural crops. Sustainable farming methods, more rainfall, and improved water management are all credited with these advancements. According to the findings, preserving ecological health and guaranteeing the resilience of agricultural systems in Ain Al Tamor requires effective water management, climate adaptation plans, and sustainable land use. Future maintenance of these encouraging trends will depend on ongoing initiatives to enhance land restoration, water conservation, and agricultural methods.

References

1. S. M. Zaidi, A. Akbari, A. A. Samah, N. S. Kong, J. I. A. Gisen, "Landsat-5 time series analysis for land use/land cover change detection using NDVI and semi-supervised classification techniques," Pol. J. Environ. Stud. **26(6)**, 2833-2840 (2017)
2. B. Ahmed, M. F. Allawai, "Using remote sensing and GIS in measuring vegetation cover change from satellite imagery in Mosul City, North of Iraq," in IOP Conference Series: Materials Science and Engineering **757(1)**, 012062 (2020)
3. S. S. Saud, S. A. Abdullah, B. M. Hashim, "Assessment of sustainable urban expansion with land use and land cover changes for Al-Hillah City using remote sensing and GIS techniques," Iraqi Journal of Physics **21(4)**, 66-76 (2023)
4. E. A. Mohammed, Z. Y. Hani, G. Q. Kadhim, "Assessing land cover/use changes in Karbala city (Iraq) using GIS techniques and remote sensing data," J. Phys. Conf. Ser. **1032(1)**, 012047 (2018), DOI: 10.1088/1742-6596/1032/1/012047
5. A. Rasul, H. Balzter, C. Smith, "Spatial variation of the daytime surface urban cool island during the dry season in Erbil, Iraqi Kurdistan, from Landsat 8," Urban Clim. **14**, 176-186 (2015)
6. B. M. Hashim, A. Al Maliki, M. A. Sultan et al., "Effect of land use land cover changes on land surface temperature during 1984–2020: a case study of Baghdad city using Landsat image," Nat. Hazards (2022), DOI: 10.1007/s11069-022-05224-
7. H. Tao, B. M. Hashim, S. Heddam et al., "Megacities' environmental assessment for Iraq region using satellite image and geospatial tools," Environ. Sci. Pollut. Res. (2022), DOI: 10.1007/s11356-022-24153-8
8. World Bank Group, Iraq: Country climate and development report (World Bank Group, 2022), https://hdl.handle.net/10986/38250

9. UNEP, GEO-6: Global environment outlook: regional assessment for Asia and the Pacific (UNEP, 2016)
10. Z. K. I. A. B. Fouad K. Mashee Al Ramahi, "Estimation of Suaeda aegyptiaca plant distribution regions at Iraq using RS & GIS applications," Iraqi J. Sci. **58(2a)**, 767-777 (2017)
11. H. M. Abduljabbar, "Satellite images fusion using modified PCA substitution method," Ibn Al-Haitham J. Pure Appl. Sci. **30(1)**, 29-37 (2017)
12. A. J. Abdalkadhum, M. M. Salih, O. Z. Jasim, "The correlation among land cover spectral indices and surface temperature using remote sensing techniques," in IOP Conference Series: Materials Science and Engineering **1090(1)**, 012024 (2021), DOI: 10.1088/1757-899X/1090/1/012024

 https://doi.org/10.1051/e3sconf/202563307008

Evaluating of spatial interpolation techniques for accurate air quality prediction: an overview

Sarah K. Abdulkareem*, *Mufid* Alhadithi, and *Wadhah* Amer

Middle Technical University, Engineering Technical College- Baghdad, Department of Surveying, Baghdad, Iraq

Abstract. Air pollution is a primary environmental concern in modern society. It threatens human health, ecological balance, and climate stability. The leading causes of air pollution include urban growth, industrial activities since the Industrial Revolution, fossil fuel emissions, motor vehicle traffic, inferior fuels, and power plants. This study reviews existing research aimed at assessing air quality by using Geographic Information Systems (GIS) methodologies to monitor air quality, which has helped to address air pollution problems. The study found that spatial interpolation techniques in Geographic Information Systems (GIS), specifically the Inverse Distance Weighting (IDW) and Kriging methods, effectively assess air quality by predicting concentrations of various air pollutants, such as SO_2, NO_2, CO, and $PM_{2.5}$, in unmonitored regions.

1 Introduction

Air pollution is a worldwide difficulty at all levels. The global epidemiological study on air pollution's impact demonstrated that gaseous pollutants and particulate matter possess significant potential to induce severe health effects, including respiratory and cardiovascular disorders and cardiopulmonary mortality.

Air pollutants can be categorized as primary as well as secondary. The primary air pollutants are toxic compounds that enter the atmosphere directly due to natural phenomena or human activity. A secondary air pollutant is a product of a chemical reaction between two or more compounds in the atmosphere. A secondary pollutant is formed from the reaction of some elements of the atmosphere with the primary pollutant [1].

Air quality has changed dramatically and has become excessively polluted in modern times. This phenomenon can be attributed to the burning of fossil fuels, rapid industrialization, and the increasing use of private transportation [2].

The atmosphere consists of many compounds, namely carbon monoxide (CO), fine particulate matter (PM), sulfur dioxide (SO_2), ozone (O_3), and nitrogen dioxide (NO_2), which have a direct impact on air pollution levels [3].

Geographic Information Systems (GIS) techniques store, manage, and analyze various spatial data, providing robust analyses of spatial problems [4-6].

* Corresponding author: ACC4007@mtu.edu.iq

Various interpolation techniques are used to determine levels of air quality parameters at unmonitored sites from monitored data from known sites [7].

To create a high-resolution map, spatial interpolation is performed. This geostatistical method provides a means of representing spatially continuous phenomena. Many studies have used spatial interpolation to predict pollutant concentration values for areas without data. Several studies have found that inverse distance weighting (IDW) and kriging methods are the most common methods for mapping air quality data [8].

2 Literature review

Kanakiya et al. (2015) evaluated the distribution of air pollutants such as (PM_{10}, $PM_{2.5}$, O_3, and Nox) using spatial interpolation methods, including (IDW), kriging, and Spline .Due to the limited monitoring stations, the researchers used (IDW) technique. The results revealed high concentrations of fine particles that exceeded the permissible limits. Pollutant levels were higher on working days due to heavy traffic activity, while pollutant concentrations decreased on weekends [9].

Owkash et al. (2016) estimated temporal and spatial variation of air quality using spatial interpolation techniques, particularly (IDW, Kriging, and Spline). The researchers accrued data on (SPM, NO_2, SO_2) concentrations from air quality monitoring stations dispensed in Mumbai city. The interpolation outcomes were compared with the observed data to consider the overall performance of spatial interpolation models. The existing study discovered that the consequences of the IDW and Kriging approach were very close to the observed data. Also, the concentrations of (PM_{10} and NO_2) have been excessive compared with (SO_2) concentration [10].

Oludari et al. (2016) utilized repeated transportable gas detectors to decide the spatial and temporal characteristics of air quality related to traffic activity, particularly concentrations of (CO, NO_2, NOx, and SO_2) in Nigeria. Air pollutant maps have been generated using the Kriging interpolation model. These maps confirmed that the pollution concentrations had been above the specified standards. The outcomes of the air quality index verified that the air was unhealthy, particularly at avenue intersections, industrial areas, and areas with excessive site visitor density [11].

Mohammed et al. (2016) utilized Historical data and GIS to improve the air quality monitoring network in Riyadh City. Existing measuring stations were excluded, missing data gaps were filled using different interpolation methods, and pollutant distribution maps for (NO2, SO2, NOx, and CO) concentrations were generated. The results showed that IDW and Kriging methods were the most reliable in estimating air pollution compared to other methods[12] .

Xuanyi et al. (2018) used spatial interpolation (IDW), spline, and Kriging techniques to analyze daily air quality indices and PM_{10} and $PM_{2.5}$ concentrations collected from monitoring stations in Shanghai. The results showed that pollutant levels were high during the rainy season and low during the summer. In general, the IDW method was the best at predicting $PM_{2.5}$ concentrations, while Kriging was effective in evaluating PM10 levels, and the Spline method was the worst at interpolating pollutants [13].

Zainab et al. (2018) used Geospatial survey techniques, which include the integration of mapping and GPS using ArcGIS software to assess the distribution of pollution. Measurements were made from November 2015 to March 2016 using specialized mobile tools. Distribution maps generated by ArcGIS 10.2 show that the concentrations of (TSP) and (PM) exceeded national and international requirements. (CO) and (O_3) levels remained within the permissible range. Other air pollutants such as (SO_2) and (NO_2) were slightly higher than the World Health Organization (WHO) standards but within national standards. The study demonstrated the impact of gases and industrial waste on air quality [14].

Akbar & Seyed. (2018) compared several deterministic and geostatistical interpolation methods to assess the spatial distribution of air quality represented by concentrations of (O_3, PM_{10}, $PM_{2.5}$, NO_2, SO_2, AQI, and CO). The results of (ME) and (RMSE) were used to determine the most effective interpolation method. In general, geostatistical methods were the best for mapping the distribution of air pollutants. SO2 concentrations were lower than other pollutant levels, and AQI was high during the rainy season [15].

3 Methods

Spatial interpolation is a widely used technique for estimating values at unobserved locations within a study area, using data taken from observed points. The following are several different interpolation methods used in previous studies

3.1 Inverse Distance Weighted (IDW) interpolation

It is a statistical approach that predicts cell values using a mathematical calculation based on the mean value of sample data points within the adjacent range of the necessary estimated cell. IDW is expressed using the Eq (1) below:

$$Zj = \frac{\sum_{i=1}^{n} w_i z_i}{\sum_{i-1}^{n} w_i} \text{ and } w_i = \frac{1}{d_{ji}^{p}} \quad (1)$$

where n is the total number of points in the neighborhood with known values; Z_j is the concentration value at the prediction jth point; W_i is the weight of the observed ith point; d_{ji} is the distance between the ith and jth points; and p is the power.

3.2 Kriging interpolation

This technique considers both geographic autocorrelation and data variation. It predicts unknown values of adjacent data points, with weights determined by a variogram model. Kriging is often divided into two types: ordinary and universal. Ordinary kriging assumes a fixed unknown average in the local neighborhood. However, universal kriging assumes a generic polynomial direction model. Ordinary kriging is the most popular and extensively used approach. Universal Kriging is often utilized when a trend is certain, and the trend can be stated. Kriging method formulated with Eq (2):

$$z(u) = \sum_{i=1}^{n(u)} \lambda_i z(u_i) \quad (2)$$

where Z(u) is the value of the unknown sample point at the u location vector; u and u_i are known and unknown sample point location vectors indexed by i, respectively; n(u) is the total number of sample points of surroundings to estimate the value of Z(u); $Z(u_i)$ is the values of known sample points; $\lambda_i(u)$ is kriging weight assigned for datum $Z(u_i)$ to minimize the variance for prediction.

3.3 Spline interpolation

This method minimizes the curvature of the overall surface to estimate values, resulting in a smooth surface that reliably passes through the input locations. The Spline interpolation method is ideal for predicting low and high values not included in the sample data due to the spline dilation effect and thus estimating values above, above, and below the lowest and lowest values recorded in the data set. Spline interpolation technique using Eq (3) below:

https://doi.org/10.1051/e3sconf/202563307008

S (x, y) = T (x,y) ljR rj N j=I (3)

Description: S: Interpolation point;
j: 1, 2, and (n)
N: number of samples
lj: Coefficient found in the linear equation system
rj: the distance between (x,y) and another JT point
T(x,y) and R (r) are defined differently depending on the choice.

3.4 Nearest Neighbor

This method uses the nearest coordinates to determine the effect of a scattered point on the producing cells. Also known as point sampling or proximity interpolation, it is used to perform multivariate interpolation in one or more aspects or dimensions. It considers only the nearest point and leaves out all surrounding points since this interpolation technique accurately estimates values for an unsampled point from its sampled neighbors.

3.5 Validation of the interpolation method

Cross-validation was used to select the best method by comparing spatial interpolation methods. This method temporarily removes one observation from the data set, and then any interpolation method is chosen to estimate the value of the deleted data. After that, the original value is returned to its location, and the same method is used for all data.

3.6 Comparison of spatial interpolation methods

Spatial interpolation is the process of estimating unknown values at unknown points based on known values at other nearby points. This process is widely used in geographic information systems (GIS) to analyze data and generate continuous maps from point data. Spatial interpolation methods vary in mathematical principles, assumptions, and accuracy as shown in the table 1

Table 1. Comparison of the most prominent spatial interpolation methods.

(IDW)	Kriging	Spline	Nearest Neighbour
A straightforward technique that yields reasonable estimates of pollution levels by utilizing adjacent monitoring stations and identify pollution hotspots and generate predictive maps indicating areas of pollution concentration.	This method considers the spatial correlation of variables, resulting in improved interpolation accuracy. It is commonly employed in soil research, environmental science, and ecology.	The most effective method for predicting smoothly fluctuating surfaces of phenomena such as temperature.	Allocates pollution observations from the closest tracking station to the unmeasured location, regardless of distance. The fundamental disadvantage of this technique is the disregard for measurements from surrounding locations.

Continuation of Table 1

(IDW)	Kriging	Spline	Nearest Neighbour
Uses distance to the predicted site to estimate the unmeasured position.	Uses both the measured points' relative positions to one another and their distance from the forecast location to determine weights. Since a fitted model is essential for the measured locations.	It maintains characteristics and allows for straightforward calculations with limited sample sizes.	

4 Conclusions

The study showed that many global and local studies have succeeded in addressing the problem of air pollution by using spatial interpolation techniques to predict levels of air pollutants in unmonitored locations, which can be summarized as follows:

1. Primarily, (IDW) and kriging interpolation are among the most widely used methods in GIS for mapping the spatial distribution of pollutants, including concentrations of (NO_2, SO_2, CO, NO, PM_{10}, $PM_{2.5}$).
2. The nearest neighbor interpolation method is no longer commonly used for estimating pollutant concentrations, especially in areas with uneven distribution of monitoring stations.
3. It has been found that using spatial interpolation techniques, pollutant levels are estimated, pollution hotspots are identified, and health risks are assessed, thus solving the issue of air pollution and providing healthy environments in many parts of the world.
4. The accuracy and reliability of air quality assessments are enhanced when monitoring data is combined with spatial interpolation techniques.

References

1. E. F. Khanjer, L. Awad, W. A. Hassan, B. S. Ismael, "Prepare maps for greenhouse gases with some weather elements for Baghdad city using data observation and Arc-GIS techniques," in IOP Conference Series: Materials Science and Engineering **757**, 012031 (2020), DOI: 10.1088/1757-899X/757/1/012031
2. Z. F. Ali, D. Salam, G. Pirisi, K. Kiss, "Assessment of air quality and consequent in Erbil, Iraqi Kurdistan region based GEE, GIS, and remote sensing techniques," in E3S Web of Conferences **436**, 10007 (2023), DOI: 10.1051/e3sconf/202343610007
3. M. Al Mentashri, W. El Zerey, "Spatial interpolation of toxic air pollutants in Jeddah City, Saudi Arabia," J. Nat. Prod. Res. Appl. **1(2)**, 63-72 (2021), DOI: 10.46325/jnpra.v1i02.16
4. L. Balaji, M. Muthukannan, R. K. Devi, "A GIS-based study of air and water quality trends in Madurai City, India," Nat. Environ. Pollut. Technol. **21(1)**, 21-32 (2022), DOI: 10.46488/NEPT.2022.v21i01.003
5. A. M. Rashed, F. M. Alattar, "Applying geographic information systems (GIS) techniques to estimate surface area of Al-Razzaza Lake, Karbala province, Iraq," in IOP Conference Series: Earth and Environmental Science **1300**, 012008 (2024), DOI: 10.1088/1755-1315/1300/1/012008

6. R. Al-Bahadeli, F. M. Shnewer, Y. A. Ali, "Develop a methodology for evaluation of the environmental sensitivity areas to desertification in the Maysan Province, Iraq," in IOP Conference Series: Materials Science and Engineering **745**, 012133 (2020), DOI: 10.1088/1757-899X/745/1/012133
7. A. Kumar, S. Dhakhwa, A. K. Dikshit, "Comparative evaluation of fitness of interpolation techniques of ArcGIS using leave-one-out scheme for air quality mapping," J. Geovisualization Spat. Anal. **6(1)**, 9 (2022), DOI: 10.1007/s41651-022-00102-4
8. P. Yudison, "Development of indoor air pollution concentration prediction by geospatial analysis," J. Eng. Technol. Sci. **47(3)**, 306-319 (2015), DOI: 10.5614/j.eng.technol.sci.2015.47.3.6
9. R. S. Kanakiya, S. K. Singh, U. Shah, "GIS application for spatial and temporal analysis of the air pollutants in urban area," Int. J. Adv. Remote Sens. GIS **4(1)**, 1120-1129 (2015), DOI: 10.23953/cloud.ijarsg.102
10. A. Kumar, I. Gupta, J. Brandt, R. Kumar, A. K. Dikshit, R. S. Patil, "Air quality mapping using GIS and economic evaluation of health impact for Mumbai City, India," J. Air Waste Manage. Assoc. **66(5)**, 470-481 (2016), DOI: 10.1080/10962247.2016.1143887
11. O. H. Adedeji, O. Oluwafunmilayo, T.-A. O. Oluwaseun, "Mapping of traffic-related air pollution using GIS techniques in Ijebu-Ode, Nigeria," Indones. J. Geogr. **48(1)**, 73 (2016), DOI: 10.22146/ijg.12488
12. S. Mohammed Mujtaba, T. Husain, B. Alharbi, "Optimization of air quality monitoring network using GIS based interpolation techniques," J. Environ. Prot. (Irvine, Calif.) **7**, 895-911 (2016), DOI: 10.4236/jep.2016.76080
13. X. Zhang, R. Shi, M. Chen, "Comparative study of the spatial interpolation methods for the Shanghai regional air quality evaluation," in Remote Sensing and Modeling of Ecosystems for Sustainability XV, SPIE **10773**, 188-196 (2018), DOI: 10.1117/12.2320402
14. Z. Mohammed, A. Ziboon, A. Kamal, M. Alfaraj, "Urban air quality evaluation over Kut city using field survey and geomatic techniques," in MATEC Web of Conferences **162**, 05023 (2018), DOI: 10.1051/matecconf/201816205023
15. A. Eslami, S. M. Ghasemi, "Determination of the best interpolation method in estimating the concentration of environmental air pollutants in Tehran city in 2015," J. Air Pollut. Health **3**, (2018), DOI: 10.18502/japh.v3i4.402

Geographical dimensions of the concentrations of gases polluting the air of the city of Basrah

Suha Waleed Mustafa *

University of Basrah, College of Education for women, Department of Geography, Basrah, Iraq

Abstract. Air pollution is the most dangerous type of environmental pollution due to its easy spread over a wide area and because of its danger to population communities, the study aims to discuss the temporal and spatial dimensions of the concentrations of some gases polluting the air of the city of Basra (carbon monoxide CO, nitrogen dioxide NO_x, sulfur dioxide So_2, lead compounds Pb and Voc_s volatile organic compounds) to identify the causes of air pollution, as 360 readings were taken distributed over (6) sites covering the study area comprehensively, and each site has 12 readings with 4 readings during one hour at a height ranging from (1.5-2) meters, taking into account the distance from direct sources of pollution during Recording the readings for the period from September (2021 to August 2022), the study reached a number of conclusions, including the high concentrations of polluting gases above the permissible environmental limits, the highest concentrations were recorded at the site (3), followed by the Site (6) and (1) due to the proximity to the sources of pollution represented by electric power plants, and the winter recorded the highest values of polluting gases.

1 Introduction

Pollution is defined as the introduction of substances that cause a negative change to the gaseous, biosphere, rocky, or water that make up the environment through human interaction with it directly or indirectly may result in a destructive effect on the environment, and this waste product or excess energy are classified as environmental pollutants [1].The environment is defined as the set of factors affecting the organism that determines the life system and form an interconnected ecological unit [2].The pollution has been divided according to its severity into several levels [3] :

1. Acceptable pollution: a degree of pollution that is not affected by the ecosystem and does not usually have any danger.
2. Hazardous pollution: the one that exceeds the permissible limit.
3. Destructive pollution: the degree to which the environmental balance collapses and pollution takes many forms, including water, noise, and air, which is what the study means.

The air is considered polluted if there is any change in its composition or if some impurities or other gases are mixed with it in an amount that may harm human life [4].Air

* Corresponding author: ali.alhameedawi@uokufa.edu.iq

pollution is defined as the presence of organic and inorganic substances in the air that change its properties, and the problem of air pollution is considered old, but its impact and seriousness increased with the progress of man over time and the problem worsened with the advent of the Industrial Revolution and the use of technology [5]. It is also defined as (a change in the ratio of natural air components of solid and gaseous elements, whether increasing or decreasing and causing a harmful effect on the components of the ecosystem) [6].

2 Materials and methods

In general, the rates of air pollution increase in crowded places with constant traffic, where the rates of carbon monoxide, lead, and other pollutants resulting from the burning of fossil fuels increase as well as the main source represented by the oil industries [7].

First: - the problem of studying:

The problem of the study lies in answering the following questions:

1. Is there an increase in air pollution in the city of Basra?
2. Is there a temporal and spatial discrepancy in the city of Basra?
3. Does air pollution have an impact on the ecological balance and exceed the permissible environmental limits?

Second: - the hypothesis of the study:

The air of the city of Basra is polluted with high concentrations of the gases under study, and their concentrations witness a temporal and spatial variation exceeding the permissible environmental limits.

Third: - objectives of the study:

The importance of the study stems from the identification of the temporal and spatial distribution of the concentrations of gases polluting the air of the city of Basra and the objectives can be summarized in the following points:

1. Identify the causes of air pollution in the city of Basra
2. Identify the times when pollutant concentrations increase in summer and winter
3. Follow-up of the evolution of the proportions of pollutants by tracking the concentrations of polluting gases during the period of sample collection
4. Focus on the need to monitor air quality on an ongoing basis to avoid risks affecting the city and the health of the population.

Fourth: - location of the study area:

The city of Basra is located between two latitude circles (30°.34 30,25,30°30°) North and in arcs of length (47,52,30_47,43,30°) East map (1) bordered on the north by Al-hartha district and the Karma River on and bordered on the South by Abu al-khasib district and Zubair district and bordered on the East by Shatt al-Arab district and Shatt Al-Basra from the West, the area of the city of Basra formed (5116,8) hectares [7, 8]. with a population (1,218,251 inhabitants, according to estimates in 2016, in (54) residential neighborhoods [4].

Fifth: - materials and methods of work:

1-The area was divided into six sectors for the study and the samples used in the analysis were extracted. The coordinates of the studied sectors (map 2) were taken by a GPS device, as shown in Table (1). The concentrations of polluting gases (Co carbon monoxide, NO_x nitrogen oxides, So_2 sulfur dioxide, PB lead, and Voc_s volatile organic compounds were measured on site for one hour for each pollutant every 15 minutes one reading was recorded by 4 readings per hour after taking the final rates of each pollutant and thus the number of readings studied is 360 readings with 12 samples (12 months) for each pollutant (5 pollutants) in (6) locations distributed to cover the study area with due regard The time period for recording concentrations was from September 2021 to August 2022, at an altitude ranging (1.5-2 meters). gas concentrations were recorded and measurements were taken by a set of

https://doi.org/10.1051/e3sconf/202563307009

gas detection devices. Gas detector device type: Drager CMS. It was used by using special slides for checking gases (carbon monoxide, nitrogen oxides, lead, volatile organic compounds).

2-Tetra Crowcon Gas Detector. It was used to measure gas (sulfur dioxide), a device that contains an acoustic alarm when the concentration of gas in the air reaches a dangerous rate [9, 10].

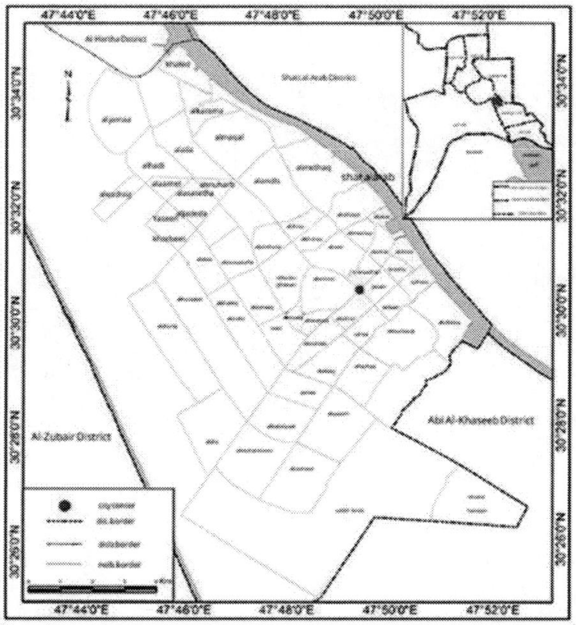

Fig. 1. Location of the study area.

Table 1. Coordinates of sample locations.

1	The first location (stronghold)	33°30′N	46°47′E
2	The second site (Andalusia):	32° 30′N	47°47′E
3	The third location (al-Hussein):	29°30′N	46°47′E
4	Fourth location (Republic):	N 31°30′	47°47′E
5	The fifth location (publican):	31°30′ N	50°47′E
6	The sixth position (the kiss):	27°30′N	48°47′E

Source: Field study

Sixth: - results and discussion:

The air quality in the city of Basra was studied and the concentration of each of the gases mentioned was determined to find out if it exceeded the permissible environmental limits and when it may pose a clear danger, Table (2) is seen as follows:

Table 2. Environmental Determinants of permissible pollutant concentrations in the air.

The polluter	Environmental limiter/ppm
Carbon monoxide CO	9
Carbon dioxide SO_2	0.03
Lead Pb	1.5
VOC_s volatile organic compounds	0,25
Nitrogen oxides NOx	0,01

Source (1): Iraqi Ministry of Environment, Air Pollution Department, Iraqi Environment Reality Report, 2018.

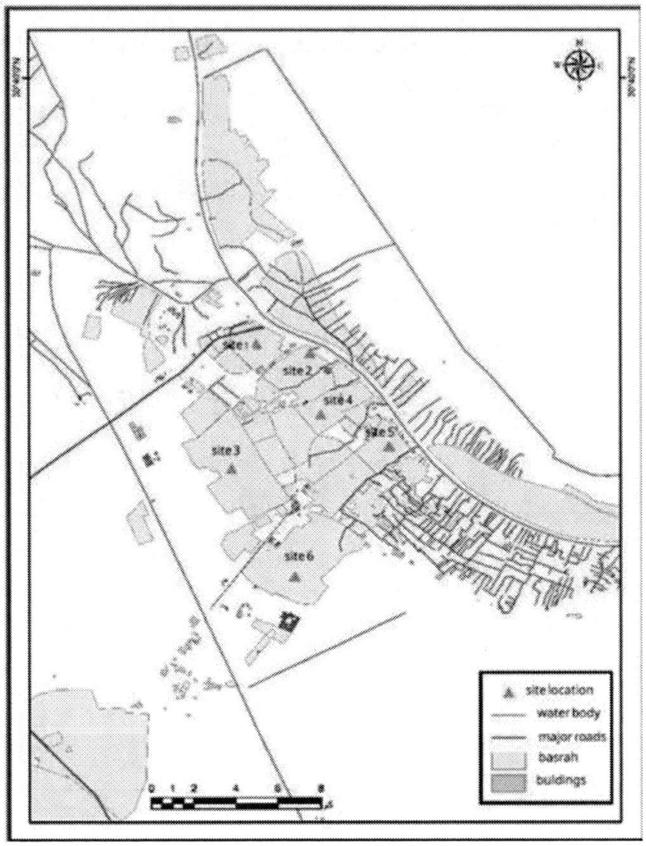

Fig. 2. Gas pollutant sampling sites.

3 Results

3.1. Carbon monoxide CO

A colorless and odorless gas that dissolves in blood plasma as much as it dissolves in water and is produced from incomplete combustion of fuel containing organic substances. gasoline fuel is the main source of carbon monoxide pollution by up to 97.9%, and this gas is released directly into the air as a result of numerous industries and the use of fuel. It is clear from Table(3) and Figure(1) the variation of gas concentrations spatially across the sites designated for the study, where the percentage of CO gas at Site 3 is higher than the rest of the sample collection areas, followed by Site (6) and comes in last place1, the temporal variation in the concentrations of CO gas is also observed over the months of sampling, so the pollutant effect at Site 1 increases in January, where its value reached (7.9) ppm and the lowest value reached in July and August, where the value was (3) ppm, and at Site (2) the temporal variation of Gas CO is also observed, increasing in the month of January with a value of 9.9 ppm and in the month of June the value reached(5.5) ppm, which is the lowest value recorded in the study area, and the variation is also being made on the rest Sample collection areas at Site (3) recorded the highest value of the pollutant in January also with a value of (20.2) ppm, the lowest in November with a value of (15.6) ppm due to overcrowding of the area and significant gas retention in it, and at Site (4) the highest value was recorded in April with a value of (7.2) ppm and the lowest percentage was in November and December with a value of (5.1) ppm. The time variation continued At Site (5), where the highest value of gas was recorded in January at (7.2) ppm, the lowest value was recorded in November at (4.9) ppm, and at Site 6 the highest values of CO-gas were recorded in February and were (11.6) ppm and the lowest was in September at a value of (5.5) ppm.

It is clear from the readings in Table (3) that the upper limits of gas concentrations exceeded the permissible limit in January in general.

Table 3. Temporal and spatial variation of (CO) ppm concentrations in Basra city (2021-2022).

Months/sample locations	Location 1	Location 2	Location 3	Location 4	Location 5	Location 6
January:	7.9	9.9	20.2	6.6	7.2	11.5
February	7.2	9.01	19.2	5.6	7	11.6
March	6.1	7.02	19.1	6.1	6	11.1
April	5.3	8.04	16.3	7.2	5.9	11.2
May	5.1	8.02	16.4	7.6	5.1	11.1
June	4.2	5.5	16.7	7.3	5.8	9.2
July	3	6.6	17.7	5.4	6.7	8.5
August	3	6.1	19.5	5.6	5.4	9.3
September	3.7	5.9	16.5	3.6	6.6	5.5
October	3.9	6.5	16.1	6.5	5.5	7.5
November	4.5	7.5	15.6	7.5	4.9	9.1
December	6.9	9.1	16.8	9.1	6.1	9.1
The rate	5.1	7.4	17.5	7.4	6.0	9.5

Source: field study 2021, 2022

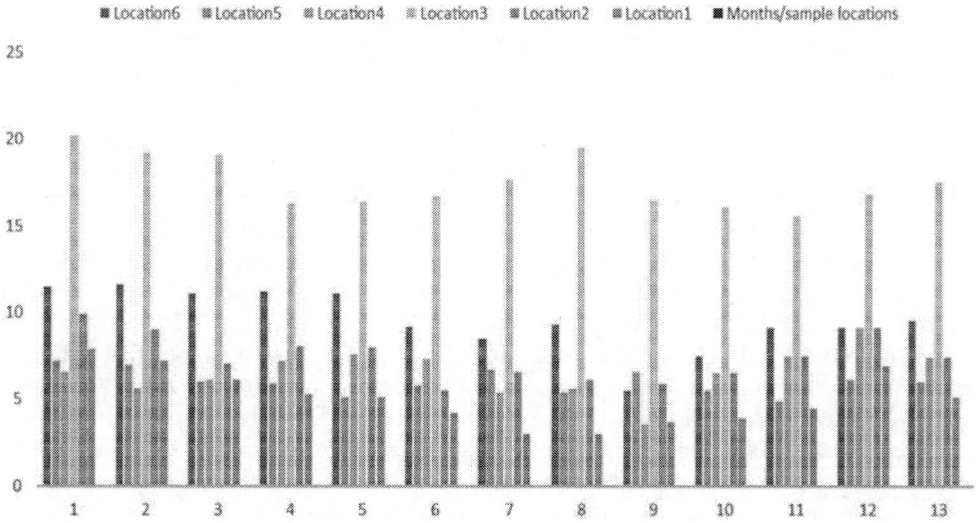

Fig. 3. Temporal and spatial variation of (CO) ppm gas concentrations in Basra City 2021, 2022.

It is reported that the impact of any pollutant on human health depends on four factors:

* General health of the person
* Length of exposure
* The degree of concentration of the pollutant
* The nature of the pollutant

This leads to the fact that the degree of health impact of the pollutant varies depending on the level of its concentration in the air and the amount inhaled by the person, and in the case of inhalation of CO gas above the permissible rate, which is (9 ppm for 8 hours), (35 ppm for one hour) this leads to mixing with the blood and adheres to hemoglobin and then it prevents the blood from transporting oxygen, in this case, carbon-oxygen (cohb) * carboxyhemoglobin, a stable compound formed in red blood cells when inhaling carbon monoxide suffocation occurs in red blood cells and due to constant exposure to gas emissions, it leads to many diseases, including Hypoxia, neuropathy, cardiac diseases. Applying the health standard to the results of collecting samples, it is concluded from the previous table that the recorded levels of carbon monoxide in the air of Basra can pose a threat to the health of individuals, whether those with allergies and shortness of breath, children and the elderly and the reason for the increase in the concentration of CO gas and exceeding the permissible limit is the presence of various industries and power plants whose smoke exit laden with CO gas aggravates its concentration in the air, as well as the frequent traffic, which is one of the main causes of the spread of polluting gases.

3.2 Nitrogen dioxide gas (NO_x)

They are inorganic gaseous compounds that are colorless and odorless, but when they merge and mix with suspended particles, they become reddish and are sharp, foul-smelling, and very toxic gases, and the most important causes of their emission are the exhausts of transport Media, oil industries and incomplete combustion processes from industrial areas. From Table(4) and Figure (2), it is clear the concentrations of gases and spatial variation at the sites from which samples were collected, where Site (4) recorded the highest percentages of pollution from gas, followed by Site (3) and comes in last place, Site (2), it is noted Also, the temporal variation in pollutant concentrations over the months of sampling increases the percentage of gas pollution at Site (1) in January and December by a value of (0.5)ppm, the lowest value was recorded in June, July, August and September, and their gas concentration reached (0.2)ppm, at Sit(2) the maximum limit of the pollutant reached in December with a concentration of (0.1)ppm and the minimum limit of the pollutant in December by a value of (0.01) ppm, and at Site(3) the highest concentration of (0.7) ppm the concentration decreased in the first October and reached a value of (0.3) ppm, and in position 4 the highest concentration was detected in January and reached (0.8) ppm, the lowest concentration was recorded in September with a value of (0.06) ppm ،In position 5, the highest concentration was recorded in the months of January, December, March and October, with a value of (0.2)ppm, the lowest readings were recorded in June with a value of (0.01) ppm, and in position 6, the highest value was recorded in January, reaching(0.5) ppm, the lowest values were recorded in the months of November and October with a concentration of (0.1) ppm.

In terms of the health impact on humans, scientific research has confirmed that the human lung is not affected if it is exposed to gas at a concentration of 100 PPT or less for periods between 20 minutes to two hours if it is exposed to a concentration of 1.5 ppm for three hours and a concentration of 7.5 ppm for two hours, the symptoms of the disease are also observed when people are exposed to a gas concentration of 500-250 PPT for more than half an hour. The release of nitrogen oxides also leads to infection with influenza infection and can reduce the ability of the blood to retain oxygen. From the study of the health standard of (0.53) ppm and comparing it with the results of the following table and figure, in some sampling areas, it is slightly higher than the standard.

The reason for the increase in gas concentration in the city air is due to an increase in the number of vehicles that release their exhaust, so the gas rises to the air layers and its value exceeds the permissible limit, as well as the presence of gas-electric stations, which are the main reason for the release and increase of NO_x in the air of the study area.

Table 4. Temporal and spatial variability of NO_x ppm concentrations in the city of Basra (2022 - 2021).

Months	Location 1	Location 2	Location 3	Location 4	Location 5	Location 6
January:	0.5	0.1	0.7	0.8	0.2	0.5
February	0.3	0.06	0.7	0.5	0.1	0.3
March	0.3	0.09	0.6	0.6	0.2	0.4
April	0.3	0.07	0.6	0.5	0.09	0.3
May	0.3	0.06	0.5	0.4	0.08	0.3
June	0.2	0.07	0.5	0.4	0.01	0.3
July	0.2	0.08	0.6	0.3	0.07	0.3
August	0.2	0.08	0.5	0.1	0.03	0.3
September	0.2	0.03	0.4	0.06	0.1	0.2
October	0.5	0.03	0.3	0.07	0.2	0.1
November	0.4	0.03	0.5	0.1	0.3	0.1
December	0.5	0.01	0.6	0.4	0.2	0.3
Rate	0.3	0.06	0.5	0.4	0.1	0. 3

Source: field study 2021, 2022

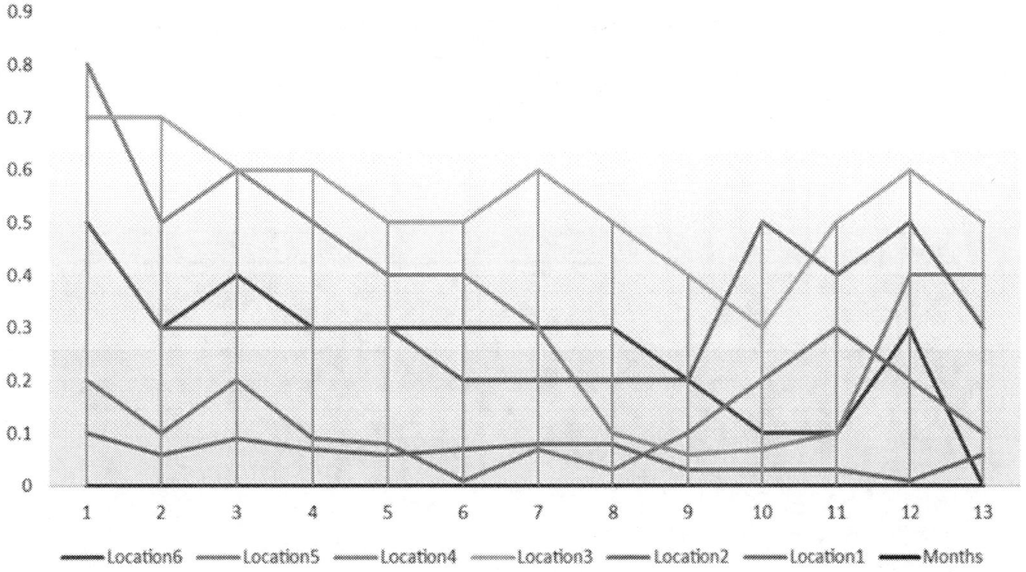

Fig. 4. Temporal and spatial variation of NO_x ppm gas concentrations in Basra city 2021:2022.

3.3 Sulfur dioxide gas SO_2

SO_2 is one of the most dangerous air pollutants, especially when the relative humidity is high, it is transformed by photochemical reactions into sulfur trioxide SO_3. and the excess oxidation of sulfur dioxide, which occurs in the presence of a reaction catalyst such as nitrogen oxides, forms sulfuric acid, and then acid rain, poses a danger to the environment. sulfur dioxide also affects the mucous membranes causing inflammation of the organ If its concentration reaches 5 ppm, this indicates serious contamination. Table (5) and Figure(3) show the spatial variation of gas concentrations in different sample collection areas, where Site(3) recorded the highest concentration of gas, followed by Site (1) and in last place Site (4), then the effect of (temporal variation) is manifested by the different timing of collecting samples from the air at Site (1) the highest concentration was in January and March with a value of (0.7) ppm and the lowest concentration was in August with a value of ppm(0.1), and at Site(2) the highest value in January and February with a value of 0.6) ppm and the lowest concentration the following months (August, September and October) amounted to (0.1) ppm, and in position 3 the largest value of gas concentration was reached in three months, namely January, February and April with a value of (0.8) ppm and reached less The value for the months of September and October at a concentration of (0.3)ppm, and in position 4 the highest value of the gas concentration was for three months, January, February and March, it was (0.5) ppm, the lowest concentration was in the months of August, September, October and December at a value of (0.3) ppm and in position 5 the highest value of gas was in the months of January and February at a concentration of (0.6) ppm and the lowest concentration was for the month of October at a value of (0.1)ppm and finally Position 6 and the highest concentration and February with a concentration of (0.7) ppm and the lowest concentration in September and October with a value of (0.2) ppm, which is the chronology of concentrations .

Table 5. Temporal and spatial variation of (SO_2) ppm concentrations in the city of Basra (2021:2022).

Month/locations	**Locations 1**	**Locations 2**	**Locations 3**	**Locations 4**	**Locations 5**	**Locations 6**
January:	0.7	0.6	0.8	0.5	0.6	0.7
February	0.6	0.6	0.8	0.5	0.6	0.7
March	0.7	0.5	0.7	0.5	0.5	0.6
April	0.4	0.5	0.8	0.4	0.5	0.6
May	0.4	0.3	0.6	0.4	0.3	0.6
June	0.3	0.3	0.5	0.4	0.3	0.6
July	0.2	0.3	0.4	0.4	0.4	0.5
August	0.1	0.1	0.4	0.3	0.3	0.3
September	0.2	0.1	0.3	0.3	0.2	0.2
October	0.2	0.1	0.3	0.3	0.1	0.2
November	0.3	0.4	0.5	0.4	0.3	0.4
December	0.5	0.5	0.7	0.3	0.5	0.5
The rate	0.4	0.4	0.6	0.4	0.4	0.5

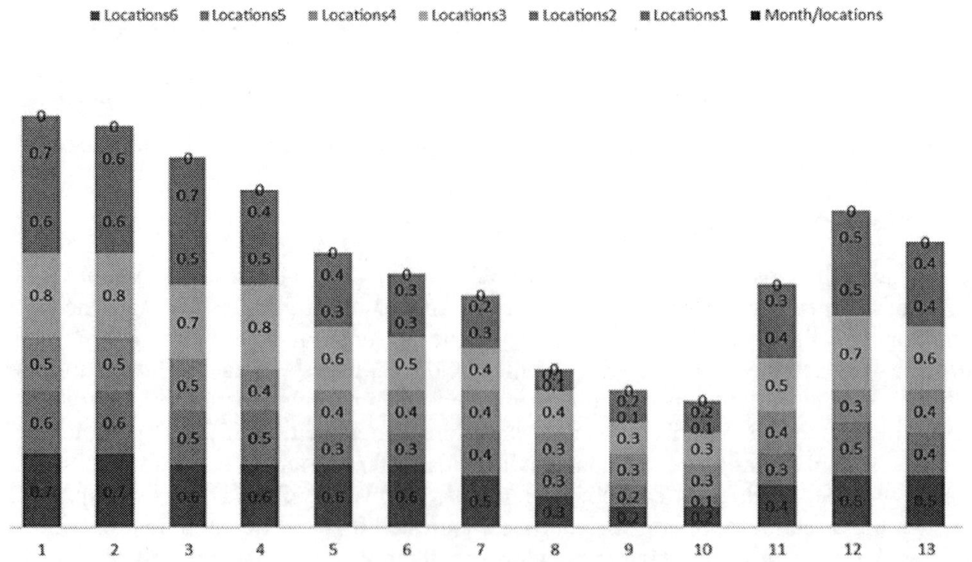

Fig. 5. Temporal and spatial variation of SO_2 ppm gas concentrations in Basra City 2021, 2022.

In terms of health impact, sulfur dioxide gas has a toxic effect at high concentrations, causing serious respiratory infections, as well as being caused by acid rain as a result of the emission of sulfur oxides into the atmosphere and the formation of sulfuric smog, the main cause of cosmic disasters.Comparing the sanitary standard of gas, the concentrations of sulfur dioxide gas pose a clear danger if it exceeds the health rate, the annual rate is (0.03) ppm, the 24-hour standard is (0.14) ppm, and the three-hour standard is (0.5)ppm, and comparing it with the results of the study with the previous table, we find that in some areas the percentage of gas rises above the permissible limit, which poses a clear danger starting from Site (2)through Site(3), Site(4), Site (5), and site (6). The increase in sulfur dioxide values in the air of the city of Basra is explained by an increase in fuel combustion from car exhausts, an increase in the number of factories, and the presence of gas and thermal power plants, which leads to high concentrations of pollutants in the atmospheric air.

3.4 Pb lead minutes

Industrial cities contain high percentages of lead minutes, which is the emissions of transport and transportation exhaust pollutants by more than 60% and is the main reason why cities are polluted with lead minutes when it is emitted from the exhaust in the form of lead bromide and remains stuck in the air for a long time in the form of a fog cloud over parts of the city. As for the lead present in the air, most of it comes from the exhausts of gasoline-powered cars containing its compounds, which are tetraethyl and tetraethyl lead. Lead, together with the exhaust gas, escapes into the air in the form of solid objects with a diameter of less than (1) micrometer.

It is clear from Table (6) and Figure (6) the spatial variation of gas concentrations at different sample collection sites, the highest concentrations were recorded for Site (3), followed by Site (6), and the last site, Site (4), then the temporal variation is visible on the different months of sample collection at Site 1, the highest concentration was recorded in January with a value of (19.9ppm)and the lowest in October with a value of 9.6) ppm, site(2) recorded its highest concentration of gas in January with a value of (12.1) ppm, while the lowest percentage in July was (5.5) ppm, and on site(3) the highest gas concentration was recorded in January at (35.1) ppm and the lowest percentage was recorded in June at (17.2) ppm ،On site (4), the highest percentage was recorded in January with a value of (7.3) ppm, and the lowest percentage was in August and September with a value of (1.4) ppm, due to the increased concentration of lead minutes.

Table 6. Temporal and spatial variation of (Pb) ppm concentrations in Basra city (2021-2022).

Month/Location	Location 1	Location 2	Location 3	Location 4	Location 5	Location 6
January	19.9	12.1	35.1	7.3	11.5	18.1
February	17.1	11.2	32.3	7.1	11.7	18.2
March	17.9	11.4	31.2	6.2	10.5	18.7
April	15.8	10.7	29.1	6.7	9.3	19.9
May	16.1	10.1	27.1	5.7	9.1	17.9
June	16.7	5.8	17.2	5.6	9.04	15.9
July	13.4	5.5	18.1	2.5	7.9	15.7
August	13.3	6.1	17.9	1.4	8.8	16.5
September	12.2	6.2	19.8	1.4	9.7	17.4
October	9.6	7.1	20.7	2.3	9.2	18.3
November	12.9	8.4	25.2	4.5	8.1	18.2
December	15.1	10.1	29.1	6.6	10.1	20.1
The rate	15.0	8.7	25.2	4.8	9.6	17.9

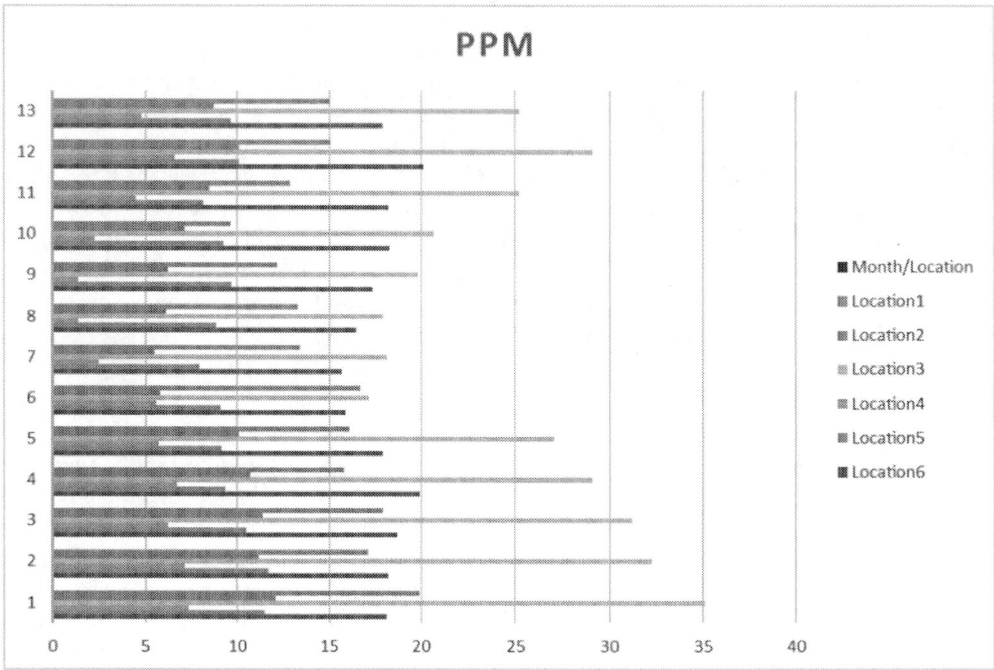

Fig. 6. Temporal and spatial variation of gas concentration (pb) ppm in the city of Basra (2021-2022).

Pb has experienced an increase in huge numbers and emits many smoke and car exhausts that pollute the atmospheric air are also major industries scattered around the city that release huge amounts of pollutants and gases that affect the quality and quality of air. On Site (5), the highest concentration was recorded in February with a value of 11.7 ppm, and the lowest concentration was in July with a value of (7.9) ppm, while Site 6 recorded its highest value in December with a value of (20.1) ppm and the lowest was in January with a value of (18.1) ppm.

From the health side, the impact of lead in the city of Basra can be doubled by the amount of solar radiation that the city receives daily, and lead is a heavy metal, the inhalation of which continuously leads to the accumulation of lead atoms in the body and then may lead to poisoning, it also has negative effects on the nervous system and liver and kidney functions, and excessive exposure to it may lead to blindness or death.

Lead is one of the heavy elements and pollution due to it is caused by the inability of its nucleus to decompose, unlike other chemical pollutants, as it enters into an acceptable cycle and moves between parts of the living environment, and the risk of lead exposure can be through inhalation or ingestion mixed with food or water

By applying the health standard to the results of the collection of available samples, in some locations, it has exceeded the permissible limit, which is the quarterly rate of (1.5) ppm and the annual rate of (0.5) ppm, which may lead to serious consequences for Human public health.

3.5 VOC_s volatile organic compounds

It is one of the most dangerous pollutants found in the open air, which is a group of chemicals, the most important sources of which are gasoline and ethylene, and also found in some detergents and fresheners, where emissions of particulate matter and organic solvents pose a great danger to those around them, which negatively affect air quality and human health if

 https://doi.org/10.1051/e3sconf/202563307009

they exceed their limit. Table (7) and Figure (7) show the effect of spatial variation of gas concentrations at different sites for sample collection, the highest concentration of the pollutant was recorded at Site 3, followed by Site (1), and in last place, Site 4, and also shows the effect of temporal variation on the difference in the timing and months of sample collection, as shown at Site(1), the highest concentration value was recorded in January with a value of (12.3) ppm, the lowest percentage was recorded in April with a value of ppm(5.6), and at Site 2, the highest concentration of the VOCs compound was recorded in January with a value of (12.3) ppm and the lowest percentage was recorded in April with a value of ppm (5.6), and at Site 2, the highest concentration of the VOCs compound was recorded in December second and February.

Table 7. Temporal and spatial variation of VOCs ppm concentrations in Basra city (2021-2022).

Months/locatio ns	Location 1	Location 2	Location 3	Location 4	Location 5	Location 6
January	12.3	4.3	19.3	2.1	3.3	6.2
February	12.1	4.3	17.9	2.3	4.1	6.3
March	9.5	3.3	17.6	2.1	4.2	5.3
April	5.6	2.5	17.7	1.9	3.9	5.1
May	5.8	1.2	17.4	1.8	3.2	5.6
June	5.7	0.9	18.1	1.7	3.1	5.9
July	6.2	0.6	12.3	0.9	2.9	3.8
August	6.1	0.7	14.5	0.3	2.1	3.4
September	5.9	1.1	16.6	0.9	1.9	2.2
October	10.5	3.2	16.6	1.1	2.1	1.5
November	10.6	4.1	14.7	1.2	2.8	2.8
December	11.7	4.1	17.1	1.9	3.1	4.9
The rate	8.5	1.5	16.7	1.5	3.1	4.4

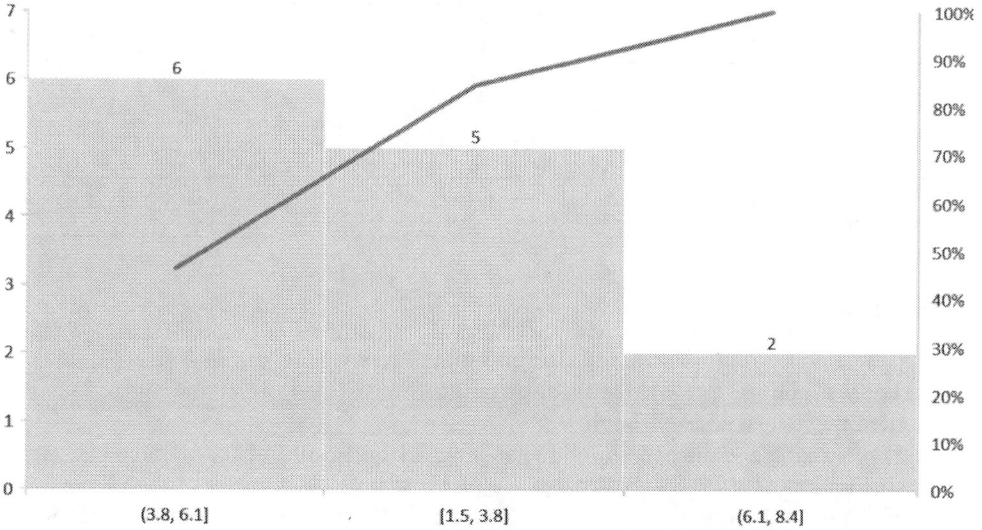

Fig. 7. Temporal and spatial variation of VOCs ppm gas concentrations in Basra city (2021-2022).

The lowest value was recorded in July with a value of (0.6) ppm, and in site(3) the highest concentration value of the Voc$_s$ compound was recorded in January with a value of (19.3)ppm and the lowest value was in July with a value of (12.3) ppm, in site(4) the highest concentration was from the share of February with a value of (2.3) ppm, and the lowest share was from the share of August with a value of (0.3) ppm, and in Site(5)the highest concentration was recorded from the share of March with a value of (4.2) ppm the lowest concentration for September was (1.9)ppm, site 6 recorded the highest concentration in February with a value of (6.3) ppm, and the lowest concentrations were recorded in October with a value of (1.5) ppm.it is dangerous to be in the air, as it greatly affects the air quality (iaq) and is very harmful to human health if it exceeds the environmentally permissible limits, it affects the respiratory system and causes lung damage, eye and nose irritation, it also affects the nervous system and liver damage, and by applying the health standard we find that it has exceeded the permissible limit in the air, which may cause many health and environmental hazards.

7-Seventh: - assessment of air quality and degree of pollution in the city of Basra.

It is clear from Table (8) that the levels of gas concentrations in the city of Basra after applying a formula to make a classification of air quality in the study area, and the sites from which samples were taken were arranged according to their degree of pollution, it turns out that the third site is the most in the intensity of pollution and increase the concentrations of Significantly, the temporal variation according to the different months of sample collection is also evident, so the most polluted month is January.

Table 8. Classification of the degree of pollution based on the calculated average of pollutant concentrations in the air of the city of Basra.

Gases	Location 1	Location 2	Location 3	Location 4	Location 5	Location 6
CO	5.1	7.4	17.5	5.8	6.0	9.5
Degree of contamination	Acceptable	Acceptable	Very high	Acceptable	Acceptable	High
NO_X	0.3	0.06	0.5	0.4	0.1	0. 3
Degree of contamination	High	Acceptable	High	High	Acceptable	High
SO_2	0.4	0.4	0.6	0.4	0.4	0.5
Degree of contamination	High	High	Very high	High	High	Very high
Pb	15.0	8.7	25.2	4.8	9.6	17.9
Degree of contamination	Very high	High	Danger	High	Very high	Very high
VOCs	8.5	2.5	16.7	1.5	3.1	4.4
Degree of contamination	High	Acceptable	High	Acceptable	Acceptable	Acceptable

4 Conclusions

1. The highest concentration of the rate of polluting gases was recorded at Site (3) due to its proximity to the gas power plant, followed by Site (6) and (1), while sites (2) and (4) recorded the lowest concentration.
2. Carbon monoxide and nitrogen dioxide gas have a significant impact on the air of the city of Basra and most sites.
3. It turned out that carbon monoxide, sulfur dioxide, and nitrogen oxides exceeded the permissible limit in most locations.
4. The concentration of lead minutes is significantly higher than the environmentally permissible limit.
5. All mentioned pollutants are dangerous for air quality and quality.
6. Prove how likely the concentration of pollutants can become dangerous to human health.
7. Pollutant concentrations varied chronologically depending on the timing of sample collection.

5 Recommendations

1. The need to constantly monitor the quality and quality of air.
2. The need for the concerned authorities to focus on monitoring the amount of air pollutants and to know their concentration in the air and how dangerous they are.
3. Trying to control the causes of air pollutants by obliging factories to install filters that purify the resulting air.
4. Converting work to renewable energies instead of fuels that cause serious air pollution.
5. Try to reduce the number of vehicles in the city and replace them with environmentally friendly means of Transportation.
6. Increasing the amount of afforestation for the city as the plant acts as an acceptable filter for some pollutants.
7. Establish strict laws to limit pollutants and impose penalties on those who trespass on them.
8. The use of unleaded fuel for vehicles.
9. Raising the awareness of the public about the seriousness of air pollution.

References

1. M. C. P. Byblos, "A geographical analysis of the reality of industrial activities polluting the air in the city of Hilla and their environmental impacts," J. Humanit. **1(24)**, 24 (2017)
2. Et et al., "Implementation of environmental management systems within government bodies and institutions in the state of Kuwait," J. Environ. Stud. Res. **9(1)**, 66–73 (2019)
3. P. Z. T. Badri, "Download a report to study the concentrations of air pollutants in the city of Nasiriyah and their impact on the respiratory system," Arabian Gulf **48(3)** (2020)
4. K. M. Jihad, X. Othman, "The impact of human activities and their role in air pollution of the urban environment in Dhi Qar governorate," J. Univ. Babylon Humanit. **29(10)**, 136–152 (2021)
5. Doski, "The impact of meteorological standards on air quality and other pollutants in the city of Duhok, Iraq," Iraqi J. Agric. Sci. **51(4)**, 1160–1172 (2020)
6. O. Z. Jassem, M. Saleh, A. A. For Aibi, "The preparation of World Vision 2 images for map modernization projects – Dujail/Al-obaiji (Iraq/north of Baghdad) as a case study," J. Eng. Technol. **39(12)**, 1828–1836 (2021)
7. "The polarization of urbanization and its impact on the urban structure of the city of Basra," Reggio **11(9)** (2021)
8. S. Mustafa, "The environmental impact of supplying the air of the city of Basra with carbon monoxide gas," Arabian Gulf **48(3)**, 3 (2020).
9. Mohammed, *Air quality survey and health risk assessment around an oil field east of Baghdad, Iraq* (2021)
10. A. A. Atiyah, B. C. Jones, "Assessment of the metallurgical and chemical properties of dust flying in the air in Iraq," Appl. Sci. **2(9)**, 1614 (2020)

 https://doi.org/10.1051/e3sconf/202563307010

Geographical distribution of tourist sites in Najaf Governorate and study of available potentials for their development

Ibtihal Abdullal Azeez, *Bneen* Nassr Ahmed and *Amtihan* Azeez Bjwo *

University of Kufa, Geomatics Technology Center, Al-Najaf, Iraq

Abstract. Tourism in Iraq in general and Najaf Governorate in particular has occupied a distinguished position among other activities, in addition to its civilizational and cultural effects due to the increase in demand and the increase in tourists, and thus increased interest in the tourism movement that provides a resource with hard currencies and reduces the unemployment rate and develops and develops vast regions.Therefore, the research aimed to reveal the tourist sites in the governorate and how to develop and develop them. The research structure dealt with three topics, the first of which came with the theoretical framework of the research from the problem, importance and goal, then stating the concept of tourism and its importance. The second topic focused on highlighting the most important tourist sites in the governorate that witnessed wide demand, especially at the present time, and were identified using the (GIs) program. Then the third topic dealt with the potential, whether natural or human, affecting tourism in Najaf Governorate, after identifying the strengths and weaknesses using SWOT analysis. Then the research came out with a set of conclusions and recommendations that can be applied by decision-makers to develop tourism more in the governorate.

1 Introduction

Geographical factors are essential and effective factors in attracting tourists to any region, including natural topographic and climatic features that help people to be comfortable, in addition to water such as rivers, springs and eyes, as well as the presence of religious shrines and monuments that contribute to attracting tourists. Najaf Governorate is one of the important centers in Iraq and has proven this through what it achieves economically and socially in terms of tourism revenues that contribute to supporting the national economy, as it receives millions of local and foreign visitors, which supports the national economy that helps in achieving sustainable development goals.

Research problem

The research problem revolves around tourism in Najaf Governorate and the possibility of developing it, and many questions arise from us

* Corresponding author: amtihana.alabrahimi@uokufa.edu.iq

A - What are the tourist sites that can be invested and developed in the governorate and how to distribute them?

B - What is the extent of the availability of natural and human capabilities in Najaf Governorate that are capable of sustaining tourism activity in it?

Research hypothesis

In light of the problem, the study assumes that Najaf Governorate has many tourist sites and many natural and human capabilities that can be invested to develop the tourism aspect in it.

Research objectives

1- Explaining the importance of the existence of a diversity of tourist sites in Najaf, including religious, archaeological and natural sites that can be developed.

2- The research aims to highlight the factors of tourist attraction and the available capabilities to develop this aspect in the governorate.

- Identifying the strengths and weaknesses in the governorate and the challenges facing tourism in it 3

2 Materials and methods

2.1 The boundaries of the study area

Najaf Governorate is located as an astronomical location between longitudes (50, 42 - 44, 44) east, and between latitudes (50, 29 - 21, 32) north, while its relative location is between two natural regions, which are the alluvial plain, which includes the eastern parts, and the western plateau, which includes the western parts, and it is bordered to the north by Karbala and Babylon, to the east by Muthanna and Qadisiyah governorates, and to the west by Anbar, while to the south and southwest it is bordered by Saudi Arabia. The area of the study area is about (28824 km2), equivalent to (6.6%) of the area of Iraq, which is (434128 km2). The governorate has three districts and seven sub-districts, which are Najaf District, followed by two sub-districts, which are Al-Haidariya and Al-Shabaka, then Kufa District, which also has two sub-districts, which are Al-Hurriyah and Al-Abbasiya, and the last is Al-Mundhirah District, and its sub-districts are Al-Mashkhab and Al-Qadisiyah See map No. (1). This central location of the region has a significant impact on the concentration and activity of tourism in it, along with the presence of other foundations [1].

Fig. 1. Boundaries of the study area.

Economic effects

These effects contribute to increasing the visibility of tourism importance as it contributes as a source of income in various tourism activities and the income of individuals working in it [2-4], in addition to providing many job opportunities for the unemployed and raising their standards of living, as well as the occupancy rates of all accommodation centers throughout the year. The tourism product contributes to non-material wealth such as the tourist uses of archaeological landmarks and climate, so tourism is an effective and influential tool in establishing and developing industries [5].

Social, cultural and political impacts

These impacts are represented in increasing the importance of tourism, as it leads to developing the cultural and tourist awareness of citizens and developing their sense of belonging to the homeland and preserving the heritage and antiquities of their country. The low level of tourism awareness among the public is one of the dangerous factors that hinder tourism from performing its role in developing society. Therefore, achieving quality is very important and leads to behavioral advancement among tourists and hosts. This tourism expansion must be accompanied by a process of development and definition of the cultural values of the countries receiving tourists and the customs, traditions, crafts, industries and folk arts that exist in them. This plays an effective role in protecting cultural values, and tourism activity also enhances opportunities for understanding, human communication and exchange of ideas between peoples.

Environmental impacts

Tourism activity leads to preserving the environment as a resource for tourism, and deepens human awareness in preserving it against visual, water and air pollution, as well as preserving traditions and heritage because this adherence increases the influx of tourists and thus increases distinction from other countries [6].

2.2 Geographical distribution of tourist sites in Najaf Governorate

The shrine of Imam Ali (peace be upon him)

The shrine is the first and most prominent religious shrine established around it, and the nucleus of the city of Najaf, the old city, was established. It is a focus around which various human activities are concentrated, where the rituals of visitation and rituals of worship are practiced inside the shrine to draw closer to God (Almighty) by the residents of the city and the thousands of residents who come to it, whether from inside or outside Iraq, during the various religious occasions distributed throughout the months of the year. On the last occasion, the number of visitors coming to the holy shrine reached (18) million visitors. There are several stages of construction and expansion that the shrine has gone through since its establishment until today. As for the number of workers in the shrine, it is constantly increasing, reaching (15 thousand) people who work to serve the Al-Haidariyah Shrine with all forms of services [7].

Kufa Mosque

Kufa Mosque is considered one of the historical and archaeological mosques in Iraq, as it is one of the oldest mosques in the Islamic world. This mosque was built by Saad bin Abi Waqqas in 19 AH / 739 AD, and was the first building in the city of Kufa, which became the largest Islamic capital and the seat of the Caliphate of Imam Ali (peace be upon him). This mosque has a great status among Muslims, especially Shiites. The mosque is connected from the eastern and western sides to two courtyards surrounded by arcades and rooms, as the northeastern side is connected to a building with a spacious courtyard, and it has two shrines. The first contains the shrine of Muslim bin Aqeel (peace be upon him), the ambassador of Imam Hussein (peace be upon him). He enjoys a distinguished and prestigious status among Shiites, so much so that they hold a funeral for him every year. The other contains the shrine of the martyr Hani bin Urwa, one of the supporters of Imam Ali (peace be upon him), who was buried next to the Emirate House in Kufa. The believers built a shrine for him located behind the grave of Muslim bin Aqeel on its northern side, and today it is considered one of the well-known shrines. And there is a lot of intention, and next to this shrine is another shrine, which is the shrine of Al-Mukhtar Abu Ubaidah Al-Thaqafi, who was one of the notables of the Shiites of Iraq and took revenge for Imam Al-Hussein (peace be upon him), so he has a special status and he ruled Kufa for a period and then was killed in the year 67 AH at the hands of Musab's army and was buried next to the mosque. The courtyard is surrounded on all sides by a row of arches, and behind each arch is a small iwan leading to a room. It was built to accommodate visitors and those in seclusion in the mosque. Behind it is a long corridor within which are located the shrine of Imam Ali, the shrine of Prophet Noah, the shrine of Imam al-Sadiq, the shrine of Imam Zayn al-Abidin, the shrine of al-Khidr, and the shrine of Prophet Ibrahim (peace be upon them). There are shrines in the middle of the mosque, such as the shrine of Prophet Muhammad (peace be upon him), the shrine of Prophet Adam, and the shrine of Gabriel (peace be upon them). In addition to these shrines, there are two low buildings: the ship or oven, which is in the middle of the mosque's floor, and it is the place from which Noah's ship set sail. Visitors descend to it by a ladder that leads to an open place and leads the visitor to a small room with a prayer niche. As for the other building, it is the house of the basin, which is a crypt extending under the floor of the mosque, ending with two stairs for entering and exiting. This place has an ancient dignity. Then, outside the mosque and next to it, there is a modest building, which is the house of Imam Ali (peace be upon him), and he used to live in it during his life until his death, where visitors enter with apparent sadness and tranquility, and it contains a small courtyard and two arched rooms written on them (the room of Hassan and Hussein, peace be upon them), then the house ends with a room in which there is a well from which visitors seek blessings, after which they move to visit the shrine of Maytham, who is one of the loyal companions of Imam Ali (peace be upon him) who was killed and crucified on the trunk of a palm tree and his location is close to the house of the Imam, then they take a considerate look at the ruins of the Emirate Palace, which was transformed into a private house for the princes and caliphs and then demolished out of pessimism due to the large number of people who killed each other in it, and the Iraqi Antiquities Directorate recently revealed its remains, which are adjacent to the southern side of the Kufa Mosque, and its side is (177 m) long, and its average thickness is (6 m3) [8].

Al-Sahla Mosque

It is one of the largest mosques built in the city of Kufa in the first century AH, and its traces and memory are still alive today. History indicates that the Banu Al-Muzaffar were the real builders of the mosque, and they are a group of Ansar who settled in Kufa, so the mosque was initially called the Banu Dhafar Mosque, then it was called the Al-Sahla Mosque because of the flatness of the mosque's land and the lands surrounding it. The mosque is visited by many visitors from all over the world, and statistics indicate that they are constantly increasing, on the one hand, due to its sanctity, and on the other hand, because it contains many shrines that have a place in the hearts of its visitors, including the shrine of Imam Mahdi (may God hasten his reappearance), the shrine of Jaafar Al-Sadiq, Al-Khidr, the Prophet Idris, and the Prophet Ibrahim (peace be upon them) [9].

Al-Hanana Mosque

It is one of the mosques of Najaf Al-Ashraf that is visited by neighbors and visitors. It is one of the three places where Imam Al-Sadiq (peace be upon him) prayed, and the location of the head of Imam Al-Hussein (peace be upon him). It is about (2) km and its area is (7400 m2), and this mosque was a church that Christians used to go to for worship and to perform their rituals after these events that were mentioned, it became a mosque for worship and prayer and believers from different places go to it. [10]

The shrine of Kamil bin Ziyad al-Najfi al-Kufi

He is one of the companions of the Commander of the Faithful (peace be upon him), and the owner of a famous supplication named after him which he learned from the Imam, and his shrine is located in the (al-Lathwiyah) area and this area was a prison for al-Nu'man before Islam, and after Islam it became a cemetery for the notables of Kufa from the followers, scholars, companions and governors including Khabab, Adi and Sa'sa'a al-Abdi, and its location is in the middle of the al-Hanana neighborhood to the east and is about (2) km away from the shrine of the Commander of the Faithful (peace be upon him), and its area is (5400) m2 and it has a high dome and a wide and spacious courtyard and the reconstruction operations are continuing in it to change its features and expand it.

Wadi Al-Salam

It is the largest cemetery in the world and one of the important Muslim cemeteries located in the southwest of Najaf Al-Ashraf, and it contains more than six million graves, which is the highest estimate in the world, so it was included in the UNESCO World Heritage List. The cemetery was known as (Wadi Al-Salam), which means that it is safe from all the pests and insects of the earth. The cemetery occupied an area estimated at (91608) hectares, or (13%) of the city's area of (7700) hectares and more. Its importance lies in addition to being a place of visitation for many residents of the city and outside it, as it created several economic activities and a large group of people and occupied them in work and jobs, as it included a number of washrooms and burial offices and many people work in it, in addition to containing shops, stalls, food and fruit vendors, children's toys, and others.

Churches in Najaf

Najaf was previously the city of churches and monasteries, some of which were later transformed into mosques. It is the city of peaceful coexistence and included three Thirty churches are spread across its land, and their ruins are still visible to this day. These churches and monasteries were known for their beautiful flower gardens and vast orchards, extending from the Najaf airport area to the city of Al-Hirah. These churches include the largest Christian cemetery in the country, with an estimated area of 1,416 dunams in the Umm Khashm area.

Al-Hirah City

This city is known to be an extinct Arab kingdom, and it is 15 km from the city of Kufa. It is one of the most important archaeological sites as it includes many palaces and ancient monuments that indicate the existence of this ancient kingdom extending from Anbar and Hit in the north to the Arabian Gulf in the south, and east of the Euphrates River to the edges of the desert in the west. Among these palaces in the city are (the White Palace and the Khawarnaq Palace, which were built by King Numan bin Al-Mundhir, the Al-Uthaib Palace, the Sandan Palace, the Al-Sudairiyya Palace, and due to the abundance of its trees and blackness, it was named by this name, the Al-Zawraa Palace, the Al-Adassin Palace, and others, in addition to the presence of more than 30 monasteries that Christians visit for worship.)

Khan Al-Shilan and Khan Al-Rahba

Najaf is famous for the presence of inns, which date back to the Ottoman era, where travelers to the holy shrines in Najaf and Karbala rest after coming from different regions of the Islamic world. Some of these inns are outside the city, such as the (Al-Rahba) inn, which is located on the Hajj road. Al-Bari and the inn were restored and built with stone and plaster, while the (Sheelan) inn is located inside the city, and in addition to housing inside the inn, there were important institutions for commercial exchange and monetary transactions. Today, the inn has become a heritage museum with an estimated area of (2000 m2) and a height of (12 m).

Najaf Sea

The sea, with its natural and human characteristics, represents the most suitable areas for tourism in Najaf. The location of the area is to the south and southwest of Najaf and to the right of the road linking Najaf and Al-Hirah, in the form of a natural extension of the alluvial plain and in the form of a tongue across the plateau south to the north of Al-Mundhirah. This area was chosen for tourism for several aesthetic and environmental reasons, including that the spread of tourist services from hotels and restaurants within the narrow alleys of the city makes it unsuitable for housing. Therefore, it is necessary to choose another suitable location, which is the Najaf Sea area, which includes a different environmental diversity of animals and plants and is qualified to be a natural reserve area. In recent years, the area has witnessed a wide turnout from the residents of the governorate and other governorates for tourism and summer vacations. In light of this, places have been prepared to help increase enjoyment and spend time relaxing in it. The picture shows Najaf Sea between the past and the present. A plan has been developed for several proposals, including the establishment of special complexes for games to entertain children, estimated at (1123) hectares. And boats in number (11), and residential projects were proposed with an area of (459.3) hectares, in addition to commercial investment from malls with an area of (987.7) hectares, and this revives the economy of the region and the most important thing taken into consideration is the maintenance of the main and internal transportation roads surrounding the region to ensure ease of access and it is under implementation.

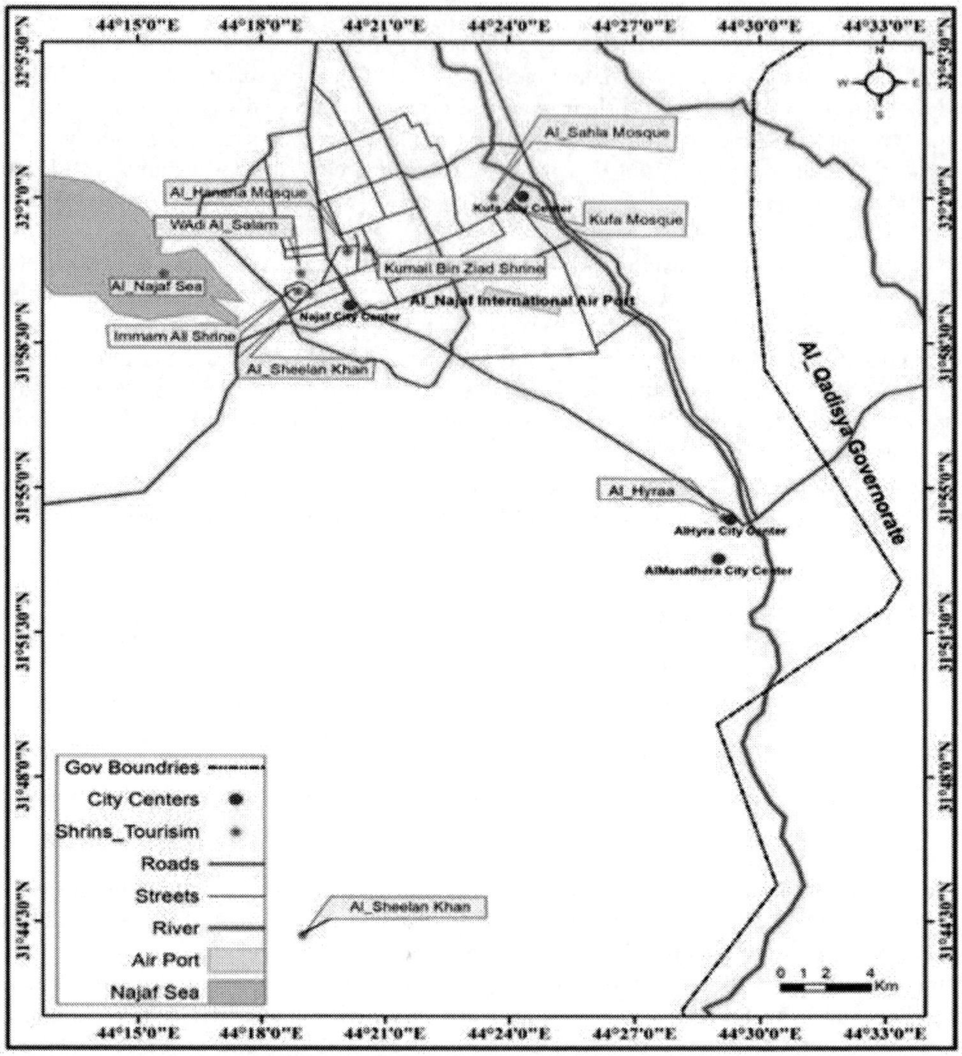

Fig. 2. Tourist sites in Najaf Governorate.

2.3 Natural and human potential of Najaf Governorate

Location

The geographical location of the country has multiple effects on tourism. Its location in terms of latitude determines the type of climate and the type of plants and animals, whether wild or non-wild, as well as the length and shortness of the day. The climatic character of the place has an impact on the type and nature of tourism. The location of Najaf Governorate is in the southwestern part of Iraq, and its geographical extension is north-east-southwest, as its eastern edges are located within the alluvial plain, while its western edges are within the eastern edges of the plateau. It takes a shape similar to a rectangle with a short side bordering Saudi Arabia. The area of the governorate is (28,824) km2. It is one of the governorates of the Middle Euphrates and shares its characteristics with it, so the residents of the neighboring governorates prefer it for tourism. It is surrounded by Karbala and Babylon from the north, Saudi Arabia from the south, Muthanna and Qadisiyah from the east, and Anbar from the west. Therefore, it is also one of the governorates of the southern desert.

Climatic characteristics

Climatic characteristics are one of the most important factors affecting the emergence of tourism, as their importance is highlighted in determining how to invest in activities, whether they depend on natural resources or man-made, in the case of suitable climates, they can be invested in practicing tourism activities such as springs, eyes, plants, wild animals and archaeological sites, while other climates of bright sun, suitable temperatures and fresh air are the most suitable climates in the governorate and qualified for the development of desert tourism. The International Travel Federation confirmed in (1972) that the climate comes first in attractive natural scenery, as areas with a suitable climate are fixed tourist and natural areas, with the duty of providing ease of access using efficient transportation, and that the study area falls within the desert climate with winter rains (Bwhs), based on the (Köppen) classification, and the prevailing is the presence of two main seasons, winter and summer, as shown in Table (1), and the characteristics of each season can be invested according to its suitability for a type of tourism.

Table 1. Average climatic characteristics in Najaf Governorate for the period (1993 – 2023).

The most famous	Actual radiation	Minimum temperature (°C)	Maximum temperature (°C)	Wind speed m/s	Relative humidity	Rainfall amount	Inspection (mm)
January	6.4	12	17.9	1.1	67.6	15.9	81.6
February	7.2	14.9	21	1.6	58.6	15.7	112.5
March	7.9	19.6	26.2	2	49.2	12.4	188.6
April	8.4	25.8	32.4	2.9	42.6	14.4	261.4
Mays	9.5	23.2	39.3	2	32	4.1	368.6
June	11.1	36.5	43.9	3.5	25.6	0	464.1
July	11.3	39	46.3	3.5	23.8	0	512
August	10.9	38.4	45.9	2.9	24.8	0	477.3
September	9.9	33.9	42.3	1.5	29.8	0	350.6
October	8.1	27.9	35.7	1.3	41	6.8	242.9
November	6.9	19	25.6	1.1	57.1	19	122.8
December	5.9	13.8	19.7	1.1	66.4	15.1	80.4
Average	8.6	25.3	33	2	43.1	8.6	2719.8

Source: Researchers based on the Ministry of Transport, the General Authority of Meteorology, the Climate Department, unpublished data 2024

Surface characteristics

The surface and its characteristics are among the factors affecting human activities. Defining the topography of the region is important to know the surface ruggedness, the nature of the rocks, the quality of the soil, and how suitable it is for establishing various service or tourism projects, in addition to the impact of these characteristics in attracting tourists. The study area is divided between the alluvial plain and the western plateau, as the latter represents the largest part, as the alluvial plain occupies an area estimated at (3424 km2) of the governorate's area, i.e. by (5%), and extends in the northeastern part and is characterized by flatness and gradual slope from north to south.

As for the plateau area, it occupies an area of (35922 km2), and we must shed light on the Badia region in the governorate, where the tourist attractions are located, as its surface shapes are characterized by their diversity and the uniqueness of their formations in addition to the splendor of its natural scenery. This contributes to enriching tourism in the governorate, as Najaf Governorate includes a vast desert that could be a destination for tourists if it is optimally exploited.

In addition to the aforementioned advantages, there are also the customs, traditions, and cultural heritage of the region's inhabitants and handicrafts, the most important of which is embroidery. These potentials could be a destination for tourists from different regions, but they must be invested in well and complemented by hotels, restaurants, and means of transportation, in addition to security. It is necessary to emphasize desert tourism activities and how they contribute to the development of sustainable development.

Population

Changes in the size and growth of the population lead to multiple economic, social and demographic changes. Therefore, population characteristics have a clear impact on the development of tourism activity from two aspects: the first is the provision of tourism services and the second is the impact on tourism movement. The most important age groups that affect tourism movement are between (15-64) years due to their ability to move, work and travel. It is noted that at the level of administrative units in Najaf, their number varied from one unit to another and as a result of many factors, including religious, economic and natural factors. Given the religious and historical status of the city center of Najaf, it always comes in first place, as it reached about (390,525) people during the year (1997), while it rose to (874,790) people, while the Kufa district came in second place, as its population reached (131,882) people in (1997) and rose to (387,089) people in the year (2020). Its religious status and the presence of shrines in it are the reason for this increase, and this in the districts affects the growth of tourism activity and movement.

Water resources

Water resources of all kinds constitute the main tributary of tourist attractions, and the amount of water present and available in it affects the creation of more opportunities and possibilities for tourists to practice water activities such as swimming, fishing, and the use of sailing and steam boats.

One of the most important tourist attractions in Najaf Governorate is the water represented by the Euphrates River, which passes from the northeastern side. The river's course extends in the administrative area about (70) km. When it enters the governorate, it branches to the south of the city of Al-Kifl at a distance of (2) km into two branches. The first is the western part of the river, which is the Kufa River, which passes through the cities (Kufa, Abu Sakhir, Al-Mashkhab, Al-Qadisiyah) in succession. Several streams branch off from it, including (Hajjat River), which is (30) km long. Then the Najaf Sea streams branch off from the river, which are (Al-Hashemi, Al-Badri, Abu Jadhu', Al-Ghaziyah). The second is the eastern part, which is known as (Al-Abbasiya or Al-Shamiya River), which runs within the Al-Abbasiya and Al-Hirah regions and branches into several streams, but they are fewer in comparison to the Kufa River streams and their distance is shorter. The rivers of the study area are characterized by several features for their development as a tourist destination, the most important of which is that the river passes near the urban centers of the governorate and sometimes penetrates into them, such as the city of Al-Mashkhab, in addition to the Kufa River passing near the religious shrines in the city and is only a short distance away. The river is also characterized by its width and the abundance of its meanders in the governorate, which gives it a beautiful view, as well as the picturesque views resulting from the river meeting the surrounding green banks, as there are many picturesque aspects of the Kufa and Abbasiya rivers when they pass through the Kufa and Al-Mundhir districts, where the blue of the water blends with the banks, as well as the spread of orchards on the banks of the river.

Transportation

The relationship between tourism and transportation is a close relationship, as without transportation, the tourist cannot reach different areas and the tourism industry cannot flourish without it, as the function of transportation is to link markets to areas receiving tourists, and transportation services are of two types: either external transportation from the tourist's home country to the receiving country or internal transportation, i.e. within the host country from the airport to the hotel and to tourist and archaeological sites.

The governorate has a network of internal roads, tunnels and bridges, in addition to external land transportation routes linked to five governorates as in Table (2). Then, Najaf Airport has a qualitative leap that has contributed to increasing the number of visitors to the governorate from different countries, and despite its modernity, it comes in second place in the number of flights after Baghdad International Airport, which shows the

importance of tourism, whether religious or cultural, which the region enjoys, which leads to saying that transportation is a pivotal factor in the development and prosperity of the tourism movement.

Table 2. Main and secondary transportation routes in Najaf Governorate.

s	Road tracks	Level	Length (km)
1	Kufa / Kafl / Babylon	major	36
2	Najaf / Karbala	major	60
3	Najaf / Abu Sakhir / Ghamas	major	44
4	Quarry Road	secondary	22
5	Kufa / Abu Sakhir / (East of the Euphrates)	secondary	27
6	Kufa / Cement Factory / Abu Sakhir	secondary	13
7	Kufa / Abbasiya	secondary	5
8	Abbassia / Freedom	secondary	10
9	Abu Sakhir / Mishkhab	secondary	12
10	Meshkhab / dip	secondary	8
11	Mishkhab / Qadisiyah	secondary	16
12	Qadisiyah / Shanafiyah	secondary	12

Recreational components

These components are represented by the presence of green areas in the province, as they provide psychological comfort for the tourist and reduce the severity of nervous tension, including parks and public gardens that give a special beauty to the province and their area in Najaf province is about (2493050) m, including Al-Ghari Park on the Najaf-Karbala road and the Ishtar Entertainment Complex, the Sultana Complex and others. As for the other recreational aspect for the tourist, it is his place of residence, as tourist accommodation services are among the basic and important services that return hard currency to the province and also enhance economic relations. Therefore, the city of Najaf has many urban stages, including the number of hotels. In 2003, it was 93 hotels, but at the present time, according to statistics, their number has reached 385 in 2018, and the most important thing that distinguishes a group of them is that they are tourist hotels, meaning they have recreational facilities and gardens that provide the tourist with comfort and recreation in his place of residence.

3 Conclusions

1 - Through the map of the location of Najaf Governorate, it is clear that it is a link between the central and southern governorates, and this represents a tourist center of gravity and also proximity to places of external demand.

2 -The study showed the presence of many tourist sites in Najaf Governorate, which vary between religious, natural and human site

3- The transportation study showed that the region has an internal and external road network that connects its parts to other governorates. Also, the presence of Najaf International Airport increased the number of tourists in it and facilitated access to and from the governorate

4 - There is a major role for water tourism in Najaf Governorate, represented by the presence of the Kufa River and the Najaf Sea, which are considered among the most important tourist sites in Iraq and are sites of attraction and attraction for tourists from inside and outside the governorate

5 -The population distribution and density of Najaf Governorate is linked to areas containing religious tourist sites, including Najaf and Kufa, more than other areas.

References

1. L. Fertas, S. S. Lahlah, M. N. Briki, "Test of the New Approach of Taxonomic of Tourist Resources for Development, Case of the Province of Setif, Algeria", Geojournal of Tourism and Geosites **43**, 878–886 (2022), DOI: 10.30892/gtg.43306-900
2. N. Świdyńska, M. Witkowska-Dąbrowska, "Indicators of the Tourist Attractiveness of Urban–Rural Communes and Sustainability of Peripheral Areas", Sustainability **13 (12)** (2021), DOI: 10.3390/su13126968
3. Z. T. Imrani, A. I. Huseynzade, B. A. Bilalov, "Priority Development Areas of Nature Tourism Resources in Shaki-Zagatala Economic and Geographic Region", Geojournal of Tourism and Geosites **54 (2)**, 921–926 (2024), DOI: 10.30892/gtg.542spl16-1267
4. R. Abilash, T. Milton, "Study on Tourism Promotion and Infrastructure Availability in Tamil Nadu with Special Reference to Tanjore District", International Journal of Health Sciences, **6**, 9401–9417 (2022), DOI: 10.53730/ijhs.v6ns1.7139
5. P. Jeelani, H. Rashid, Sh. A. Shah, "Sustainability in Tourism Demands a Mode Shift: Appraisal of Heritage Tourism; a Safe Bet for Developing Regions", Natural Resources Forum (2024), DOI: 10.1111/1477-8947.12409H.
6. Abdul Hamid, M. S. Al-Khafaji, *Iraq Tourist Guide* (Ministry of Culture and Tourism, Republic of Iraq, 2022)
7. K. A. Al-Fatlawi et al., *Najaf Al-Ashraf 2050: A Future Vision* (1st ed., Beirut, 2023)
8. M. N. Hussein, "Geographical factors affecting tourism movement in Najaf," J. Arts Al-Muthanna **136**, 436 (2021)
9. A. Al-Moussawi, H. Jaaz, "Natural and human characteristics of the western plateau in Najaf and its relationship to the exploitation of available resources," J. Res. Univ. Kufa Coll. Educ. Girls **5**, 267 (2021)
10. H. Al-Shadidy, M. Al-Hadi, *The Tourist Spatial Environment of Najaf Governorate* (Center for Urban and Regional Planning, University of Baghdad, 2020)

Happy City strategy, satisfying individuals' needs by sustainable urban design (Zurich and Dubai as models)

Tiba M Makki [*], and *Zainab* Khalid

Department of Architecture, College of Engineering, University of Baghdad, Baghdad, Iraq

Abstract. Well-being and prosperous living are among the most significant goals of sustainable development, and development strategies should be directed toward satisfying individuals' needs and living in happy cities. The research explores the potential for establishing a comprehensive strategy by examining different approaches to urban planning and their impact on individuals' happiness. The research problem is the absence of a clear, standardized vision for developing cities that satisfy individuals' needs and promote happiness. The hypothesis suggests exploring urban development strategies for the happiest cities in the world to satisfy individuals' needs and promote happiness. Relying on the concept of the happy city, sustainability, and individuals' needs and examining the success of Zurich and Dubai cities, the research aims to explore urban elements within cities that satisfy individuals' needs and promote happiness and seeks to develop a comprehensive framework for a happy city strategy that can be applied to promote happiness for Iraqi cities using measurable indicators to assess their achievement levels and improve urban well-being. Using a descriptive-analytical and comparative approach, the findings show that sustainable urban design can positively impact individuals' happiness by satisfying their needs and presents a comprehensive, actionable strategy to develop cities towards happiness relying on built and natural environmental elements that promote sustainable developments and happiness, which are crucial for the practical application of these findings.

1 Introduction

The pursuit of a sustainable future necessitates a paradigm shift that prioritizes not only environmental protection but also the well-being and happiness of human populations. Many studies suggest that urban environments should be a place to meet not only basic human needs but also psychological needs, create a sense of satisfaction, promote social interaction, create a stable economy, and promote environmental stewardship [1-5], achieving these objectives with a clear, standardized framework remains undetermined. "The Happy City" is defined as the way cities affect individuals' moods by boosting the good stuff and avoiding the bad. A city is a combined system that should be designed to support human well-being by recognizing the seen and unseen factors that affect individuals' physical and subjective

[*] Corresponding author: ali.alhameedawi@uokufa.edu.iq

well-being and control their behavior. It is a state of satisfaction with social, economic, environmental, and urban governance factors [6-7]; this concept is gaining attraction, which highlights the need to re-evaluate sustainable strategies beyond the rational use of natural resources, products, and energy in a way that protects the environment from being harmed [8], to include the equity of well-being for all generations [3]. The strong association between well-designed urban spaces and happiness emphasized this crucial tie [9,10].

Despite that, a clear, standardized framework for developing cities that prioritize happiness is currently unavailable, which represents the research problem; the research addresses this gap by investigating how sustainable urban strategy can satisfy individuals' needs and promote happiness. Using descriptive-analytical and comparative approach for Zurich city from Switzerland (ranked 2nd in the golden group) of the happiest cities in the world, and Dubai city (ranked 90th among the silver group), which represents the first regional city that manages to be in the happy city index [9]. The research aims to: 1- explore urban elements within cities that satisfy individuals' needs and promote happiness. 2- develop a comprehensive framework for a happy city strategy that can be applied to promote happiness in Iraqi cities using measurable indicators to assess their achievement levels and improve urban well-being. To achieve these objectives, the research will explore Human needs and their relation to the happy city, sustainable development regarding the happy city, as well as studying city experiences towards promoting happiness for individuals. Figure 1 shows a flow chart for the research plan.

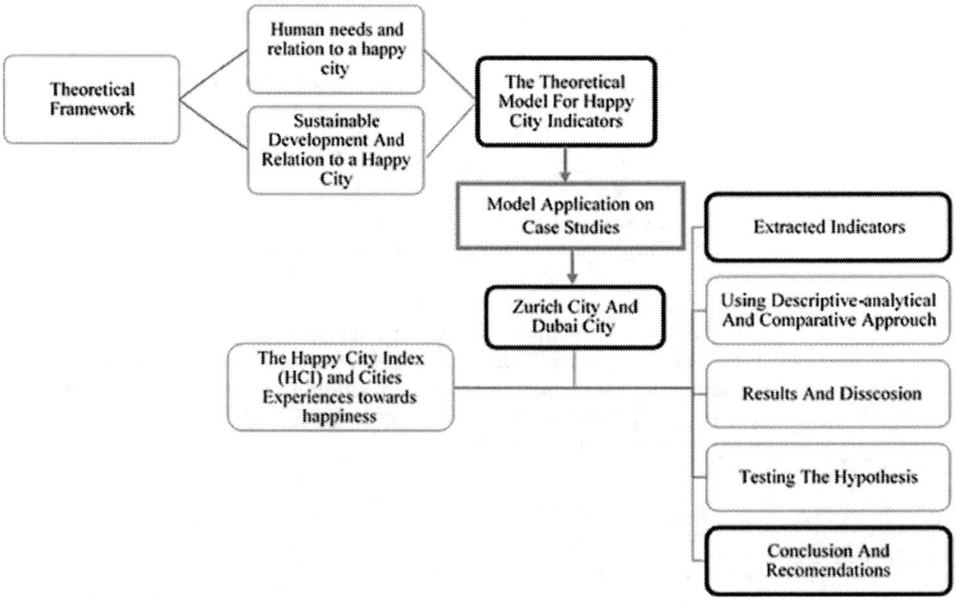

Fig. 1. Flow chart for the research plan.

2 Materials and methods

2.1 Human needs and relation to a happy city

Individuals' levels of happiness are directly affected by city functions when they provide the basic amenities that improve quality of life, support connections to places, interact with others, and support health and wealth. Moreover, a good transportation system, access to recreational places, and child-friendly areas are significant in forming a happy city [10], that human needs can be satisfied by the urban landscape, whether it was basic needs or psychological needs, and create a sense of satisfaction. Being satisfied with a place is based on daily activities and leisure, and the socially effective bonds in that place create place attachment; both will achieve a liveable place and social cohesion and improve the overall quality of life [1]. Human needs are classified according to many psychological theories, starting with Abraham Maslow's theory of human needs. He argued that human needs are arranged in a hierarchical order into five categories: basic or physiological daily needs like food, water, and shelter; the security needs by feeling secure and protected including health and financial, social and belongingness needs including family and friends bonds, self-esteem and recognition needs, and self-actualization needs when achieving personal goals when the lower needs are met, individuals will seek to satisfy the more advanced psychological needs and so on [11]. Clayton Alderfer altered Maslow's theory of human needs into a theory of Existence, which includes physical and material desires; Relatedness, which focuses on social bonds (family and friends as a desire; and Growth needs, which is about living a meaningful life, his theory gave a priority on esteem and performance, these needs vary from one individual to another. Quality of life is achieved after the juxtaposition of human basic and psychological needs, as well as the satisfaction that is enhanced by cities, to create a clear model through the physical and social factors that shape the quality of life in the city. Individuals need living spaces, recreation, and restoration that can be met through the walk, contact, and access to natural landscape elements, such as green spaces, parks, street plants, grassed areas, forests, and riversides. Individuals also have social needs like culture, place identity, community identity, and public space, theoretically, which can promote social solidarity by supporting diversity and inclusion to shape happiness in the urban landscape. The relationship between human needs and the urban landscape was studied in 2007, 90 studies were conducted in 23 different countries, resulting in classifying human needs into nature needs and interaction needs, and these needs can be satisfied through the interaction between out-door environmental factors and individuals, by determining its physical and natural elements and their direct impact on fulfilling their nature needs, or by indirect impact, both can affect individuals well-being and behavior and respond to through a group of urban settings.

These factors can boost Happiness according to individual nature's needs (physiological and psychological needs) through tangible factors presented through contact with natural green and blue environments and contact with the physical built environment. Whereas intangible factors meet individuals' interaction needs (social needs) through urban qualities [4, 5]. These urban qualities promote vibrant and diverse social interactions, participation, and a sense of community identity, urban resilience, and foster social-cultural events. The tangible and intangible urban factors were identified by Kevin Lynch in 1969 in his famous book "The Image of the City" as components of the urban landscape, that the built environment has physical components with physical qualities that give the environment what he called "the imageability," this quality (shape, color, or arrangement) affect individuals to create a strong mental image that can facilitate making a well-defined, strongly structured and very useful mental image that perceived by individuals. It can also be called legibility. These components are the structure of cities, which are represented by the shape, color, and physical arrangements of their facilities. Also, it can be a spatial or pattern relationship between the observer and his environment. The identity of place has features that identify an object to be recognized as a separate entity, different from other objects, and has the meaning of individuality. The meaning: that an object must have practical or emotional meaning to the observer that creates a suitable relation between individuals and their environment [6,7].

These elements are not isolated from one another; Districts are regulated with nodes, framed by edges, pierced by paths, and decorated with landmarks that work together as raw materials for environmental images to create a satisfying form. These elements are considered as urban cues that are responsible for creating a relationship between the urban landscape and individuals and establishing their environmental image and their own environmental experience in their urban landscape. Some urban cues have a psychological impact on individuals, satisfying individuals' need for recreation and restoration like parks and green spaces that can act as therapeutic elements within an urban landscape, with easy access and use by individuals, offering many physical and psychological benefits as well as social and economic benefits for individuals in their communities which led to promoting the quality of the urban landscape and overall quality of life. Urban settings with aesthetic qualities can affect an individual's mood by triggering one or more of the individual's five senses directly by creating liveable spaces that contribute to individuals' happiness and improve their mental health. Furthermore, safety and security are essential human needs for both physical and subjective well-being. Feeling safe is a sense of security and freedom from fear or anxiety, which can support an individual's flourishing and promote optimal functioning in their environment. Feeling safe can affect an individual's ability to interact, pursue goals, respond to stress, and achieve overall well-being and happiness by promoting traffic safety, crime prevention policies, and safe neighborhoods.

Accordingly, Human needs are basically drawn from Maslow's Hierarchy of needs, that have five categories: basic or physiological daily needs, security needs, social and belongingness needs, self-esteem and recognition needs, and self-actualization needs when achieving personal goals, and the happy city is a place that can satisfy human needs by: promoting efficiency in basic daily needs as well as walkability, accessibility, and connectivity) that represent the physiological daily needs, satisfying the need for aesthetic preferences, recreation, restoration, safety, and security that describe the psychological needs, and by satisfying the social needs like social solidarity are achieved by promoting diversity, inclusion, place identity, and urban resilience. the tangible and intangible urban factors play crucial role in satisfying human needs. Figure 2 shows Human needs that are satisfied by the happy city.

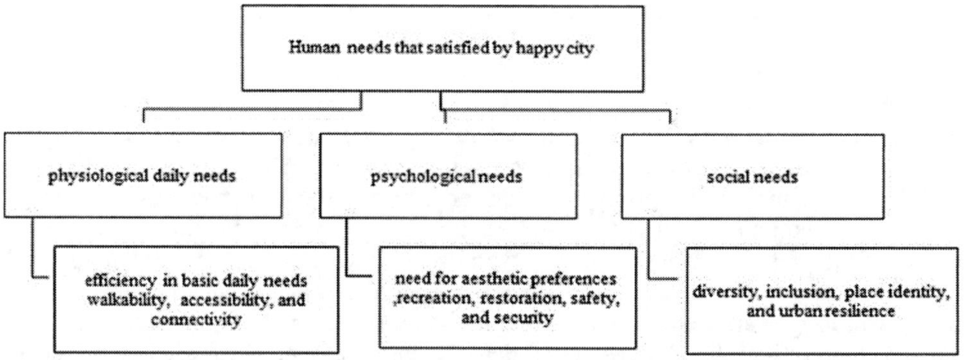

Fig. 2. Human needs that are satisfied by a happy city.

2.2 Sustainable development and relation to a happy city

The happy city concept and sustainable development are naturally interconnected; sustainable developments aim to promote individuals' quality of life by preserving natural environments and resources for the next generations. Building a happy city requires a comprehensive perspective, using all potential, collective efforts, and resources to create environments that respond to individuals' various social, psychological, and physiological needs and that affect individuals' happiness. The concept is based on strategies in the principles of identity by cultivating a sense of belonging and shared purpose, efficiency by ensuring accessibility to essential services that promote a sense of satisfaction, collectively by encouraging individual community engagement that enhances a sense of unity, and self-control which authorized individuals and communities to form their surroundings, promoting agency and accountability, which positivity improve the quality of life for all [9]. On the other hand, sustainable development also emphasizes meeting the needs of the present generation without compromising the ability of future generations to meet their own needs, focusing on social equity, economic stability, ecological balance, as well as respond and reflect the cultural values and community traditions and governance factors that promote sustainable policy for decision-making with a longer-term focus. The integration of sustainable design in happy cities is achieved in strategies such as green architecture, walkability, and the use of renewable energy systems, which simultaneously improve individuals' quality of life and mitigate environmental impact by fostering urban biodiversity and reducing urban heat island effects, which in turn improves thermal comfort for residents.

Moreover, creating pedestrian-friendly routes and public transportation nodes enhances social interaction, reduces CO2 emissions, and supports sustainable mobility [10, 11]. Furthermore, a Happy city and sustainable urban development share many objectives, especially for Safety, Mobility, Culture and physical activity, and Public Services; promoting safe urban environments, comfort, and security in a place is associated with more social interaction, which is a significant factor that affects individuals happiness, moreover, sustainable urban development that focuses on preserving urban parks and green spaces, reduce air pollution may obtain a good quality as well as promoting health and well-being of individuals. Using green elements like trees, landscaping, and flowers has a significant environmental and health role and invites residents to spend more time outdoors, healthy activities and have more social interaction; it also has an aesthetic and planning role that these elements separate areas and identify accessibility and connectivity in urban landscapes. Moreover, green urban areas provide urban agriculture with various crops that ensure food safety and promote overall health. Smart city strategies have a significant role in promoting efficiency for individuals that satisfy basic daily needs by using accessible, secured, efficient, smart physical infrastructures; at the same time, happy city objectives focus on environmental liveability and life-ability environmental liveability, including good air quality, pollution and waste treatment, beautiful public green space, social equity, sense of safety, supportive relationships, and stable economy, affordable housing, employment, cultivating culture and leisure for diverse groups of individuals, while, Life-ability includes individuals physical, mental well-being, equal opportunities for education and health services which align with smart city characteristic. Improving urban walkability does not only mitigate environmental impacts of transportation but also can positively affect individuals' health and well-being. Another significant urban sustainable development strategy that aims to create an urban environment that supports well-being, ensuring efficiency and social interaction is the (15 or 20 - minutes city concept), where the city's aspects are organized within a 15 or 20 -minutes radius, that residents would have sufficient time for themself, and for cultivating stronger social relationships, and more liveable cities. Besides that, it offers opportunities for car emission reduction; many economic benefits can be achieved by creating job opportunities, reducing traffic, and addressing climate-associated risks to generate more wealth and improve urban resilience. Moreover, the long-term planning adoption for the concept may result in a higher quality of life, promoting sustainable mobility, encouraging walking and cycling, and higher cultural results that, in turn, will promote the creation of urban parks and public spaces like squares within the city that also enhance urban aesthetics for the city and promote more happiness for its residents.

As shown before, happy city and sustainable urban development share similarities in objectives that can be translated into action plans that focus on satisfying individuals' needs to promote happiness within cities, using urban natural and built environments as indicators

 https://doi.org/10.1051/e3sconf/202563307011

that contribute to the forming of the happy city model. These indicators are shown in table (1) below.

Table 1. The theoretical model for Happy City indicators.

Individuals' needs	**The theoretical model for Happy city indicators**	
Physiological needs	**Main indicator**	**Secondary indicator**
Efficiency in Basic daily needs	Efficient urban design	15 or 20 - minutes city model & densification
		Offering variable housing options
	Efficient energy consumption and climate protection	Mitigate Climate change
		Use renewable energy
Walkability, accessibility, and connectivity	Walkability	Friendly -walkable environments
	Accessibility and connectivity	Access to health care services
		Access to educational services and libraries
		Access to Digital city & infrastructure
		Access to public transport
Psychological needs	**Main indicator**	**Secondary indicator**
Aesthetic preferences	Urban aesthetics	Urban aesthetic qualities
Need for recreation and restoration.	Built Recreational places	Playgrounds & parks
	Natural Recreational places	Proximity to green environment for recreation and restoration purposes
		Proximity to the blue environment for recreation and restoration purposes
	Environmental protection	Conservation of natural area
Safety and security needs	Safety by design	Traffic safety
		Crime prevention policy
		Safe neighborhood with a distinctive identity
	Safe and stable Economy	Attractive economic environment
Social needs	**Main indicator**	**Secondary indicator**
Social cohesion	Diversity and inclusion	Expanded cultural life
	Urban resilient	Resilient cultural operation
	Social-cultural events	Public spaces for various cultural events

2.3 The Happy City Index (HCI) and Cities Experiences towards happiness

It has been designed to assess cities based on their commitment to cultivating and supporting happiness through education, policies, economy, mobility, environmental protection, access to green areas, and innovation, which are significant components for official policies, stands as a model for applying solutions, have a effects individuals' quality of life, and to avoid establishing the "single ideal model" for this index, by focusing on the various conditions that are needed by individuals, communities, and cities as a whole to flourish [9]. The index defined the most significant urban activities that have a direct impact on happiness into five theme categories: citizens, Governance, environment, economy, and mobility. The health area will be added to the 2025 index to measure cities' happiness.

As there is no clear, defined strategy for promoting a happy city, the research aims to explore urban development strategies that were followed by two of the most recognized cities in the world that achieved the top in the Happy City Index 2024 (Zurich city from Switzerland as an international model and Dubai city from UAE, which was chosen for being the first Arabic city that manages to be in the happy city index. Zurich City focuses on fostering the well-being of its residents, while Dubai City promotes luxury, modernity, and a high-energy lifestyle. Many differences are associated with socio-cultural lifestyle and weather conditions between these cities. However, both Zurich and Dubai show excellent models for sustainable development that may contribute to their high ranking in the Happy City Index [9]. The research aims to explore the urban indicators that contribute to the foundation of a comprehensive model for the happy city strategy that can be applied to promote happiness in Iraqi cities using measurable indicators to assess their achievement levels and improve urban well-being, using a descriptive-analytical and comparative approach on two cities that were selected for this study, Zurich in Europe and Dubai in the middle east. As shown below.

2.3.1 Zurich City, Switzerland

The city covers 92 km2 (35 sq. miles) as a total area, which is divided into 12 municipal districts with a population of (434‘000); Zurich has the highest number of public fountains in the world (1,224). The Happy City Index 2024 ranked the city of Zurich as the 2nd happiest city in the world. The Strategies of Zurich **2020-2040** are aimed at maintaining a high quality of life and sustainable growth for the economy and productivity in four primary areas of action by promoting stable livelihoods, Attractive urban space, good coexistence, and an efficient city. The city of Zurich adopts a people-centered approach to promote quality of life for its residents by creating an open and vibrant society that supports cultural activity and encourages residents to participate in a wide range of different high-quality cultural activities, providing affordable housing options through many non-profit organizations, the city is also committed to social diversity in its various neighborhoods, to foster coexistence and social stability through a wide range of cultural, playful, sports, leisure and recreational activities that contribute to social cohesion and encourage social inclusion. The city is committed to ensuring accessibility to educational and health services for diverse groups of the population; the city is also committed to maintaining attractiveness for business and jobs in many sectors by transparent, efficient planning and approval services and supporting high-tech, innovation, and research job opportunities. In the case of environments, biodiversity, and climate challenges, Zurich city has “The Urban Green Program” to ensures to mitigate climate change effects and increase biodiversity by 2035 by integrating urban areas with a variety of natural green and blue elements to enhance invisible services offered by nature like Recreation and restoration in urban parks and forests, purified cool air by city trees, maintain food and biodiversity protection that have a direct and indirect contribution to individuals well-being.

Also, Zurich aims to reduce greenhouse gas emissions and to be climate-neutral by 2040, using negative emission technologies potential, refusing non-renewable energy sources, and using locally available energy, reducing energy consumption per person in the city to 2000 watts as a global climate protection policy. Moreover, Quality of life in the city is supported through high-quality public urban space that can adapt to social change with efficient amenities and the creation of new green and open spaces that ensure safety and security for all residents through digital crime prevention technologies and efficient emergency services, also promote efficient, sustainable urban mobility and enhance satisfaction with public transport and pedestrian and bicycle traffic by encourage walk, cycle, the use of climate-neutral public transportation options and reducing car dependence, ensuring a safe effective and high-capacity traffic system by far-sighted planning and an innovative traffic management system. It is also committed to sustainable development policies using efficient and modern application procedures in maintaining its roads, railways, public spaces, building stock, infrastructure, energy and water supply, waste management, water treatment to the highest standards, as well as using smart digital infrastructure to provide safe and efficient services, furthermore, preserve and promote the natural foundations of life and nurture the physical, mental and social well-being of the entire population .moreover, provide high quality public urban space and create new green open spaces to support physical, subjective and social well-being. Figure (3) below shows the green spaces in Zurich city that make it more environmentally friendly, in addition to the road network that facilitates transportation to and from critical areas and services in the city. The following table will explore Extracted indicators for Zurich's development strategies 2020-2040 according to areas determined by HCI 2024-2025.

Table 2. Extracted indicators for Zurich development strategies 2020-2040 according to areas determined by HCI 2024-2025 (by the researcher).

Main area	Extracted indicators for Zurich development strategies
mobility	promote efficient, sustainable urban mobility, and enhance satisfaction with public transport
	Encourage walking, cycling, the use of climate-neutral public transportation options, and reducing car dependence.
	ensuring a safe, effective, and high-capacity traffic system
citizen	adopting a people-centered approach to promote quality of life for its residents by supporting social life.
	promoting social diversity, participation, and cultural opportunities to foster coexistence
	Wide range of cultural, playful, sports, leisure, and recreational activities that contribute to social cohesion and encourage social inclusion
	Ensure accessibility to educational and health services for diverse groups of the population.
environment	reduce greenhouse gas emissions
	using negative emission technologies potentials to be climate-neutral by 2040
	refusing non-renewable energy sources and using locally available energy
	Enhance the environmental quality of air and water with a sustainable, energy-efficient water and waste management system.
	reducing energy consumption per person in the city to 2000 watts as a global climate protection policy
	integrating urban areas with a variety of natural green and blue elements to enhance invisible services offered by nature like Recreation and restoration in urban parks and forests, purified cool air by city trees, food, and biodiversity protection
health	preserve and promote the natural foundations of life and nurture the physical, mental, and social well-being of the entire population
	Providing high-quality public urban space by creating new green open spaces to support physical, subjective, and social well-being
economy	maintain attractiveness for business and jobs in many sectors by transparent, efficient planning and approval services and supporting innovation and research job opportunities
	High-tech innovation to support industrial, production, and processing sectors.
Governance	providing affordable housing options through many non-profit organizations
	Upgrading the city's amenities, services, and infrastructure with innovative technologies.
	Ensure safety and security for all residents through digital crime prevention technologies and efficient emergency services.

2.3.2 Application of theoretical framework on Zurich city

In order to establish The Happy City Strategy Model, extracted indicators from Zurich city will be analyzed according to the theoretical model for the Happy City indicators that the research suggests its contribution to promoting happiness to determine the active and inactive indicators that may contribute to the establishing of the strategy: using an objective scale where N= no evidence, M= moderate evidence, I=ideal evidence as shown in table (3) below.

Table 3. Application of the urban happiness theoretical model in Zurich city.

Urban happiness theoretical model			**Evaluation**			**Explanation**
Physiological needs	**Main indicator**	**Secondary indicator**	**N**	**M**	**I**	
Efficiency in Basic daily needs	Efficient urban design	15 or 20-minute city model and densification	*			There is no evidence for adopting a 15 or 20-minute city model.
		Offering sustainable housing options			*	Priority to providing affordable housing options through many non-profit organizations
	Efficient energy consumption and climate protection	Mitigate Climate change			*	Reduce greenhouse gas emissions
		Use renewable energy			*	reducing energy consumption per person
Walkability, accessibility, and connectivity	Walkability	Friendly - walkable environments			*	encourage walking, cycling, the use of public transportation options
	Accessibility and connectivity	Access to health care services			*	ensure accessibility to health services for diverse groups of the population
		Access to educational services and libraries			*	ensure accessibility to educational services for diverse groups of the population
		Access to Digital city & infrastructure			*	upgrading the city's amenities, services, and infrastructure with smart technologies.
		Access to sustainable public transport			*	Promote efficient, sustainable urban mobility and enhance satisfaction with public transport.

 https://doi.org/10.1051/e3sconf/202563307011

Continuation of Table 3.

Psychological needs	Main indicator	Secondary indicator	N	M	I	Explanation
Aesthetic preferences	Urban aesthetics	Urban qualities		*		moderate evidence for upgrading urban aesthetic qualities
Need for recreation and restoration.	Built Recreational places	Playgrounds & parks			*	Providing high-quality public urban space by creating new green open spaces to support physical, subjective, and social well-being
	Natural Recreational places	Proximity to green environment for recreation and restoration purposes			*	integrating urban areas with a variety of natural green elements
		Proximity to the blue environment for recreation and restoration purposes			*	integrating urban areas with a variety of natural blue elements
	Environmental protection	Conservation For Natural Areas			*	Enhance the environmental quality of air and water with a sustainable, energy-efficient management system.
		Protecting Biodiversity			*	maintain protection for biodiversity
Safety and security needs	Safety by design	Traffic safety			*	ensuring a safe, effective, and high-capacity traffic system
		Crime prevention policy			*	ensure safety and security through digital crime prevention technologies and efficient emergency service
	Safe and stable Economy	Attractive economic environment			*	maintain attractiveness for business and jobs

Continuation of Table 3.

Social needs	Main indicator	Secondary indicator	N	M	I	Explanation
Social solidarity	Diversity and inclusion	Expanded cultural life			*	promoting social diversity, participation, and cultural opportunities to foster coexistence
	Urban resilient	Resilient cultural operation			*	adopting a people-centered approach to promote quality of life by supporting social life.
	Social-cultural events	Public spaces for various cultural events			*	Wide range of cultural, playful, sports, leisure, and recreational activities that contribute to social cohesion and encourage social inclusion

2.4 Dubai City, UAE

One of the richest of the seven emirates of the United Arab Emirates and it is the capital of Dubai emirate, located between (UL: 54°59'18" E, 25°22'00 N" and LR: 55°33'38" E, 24°51'06" N), the city was established in 1971 with a total area of 80km2 and 80,000 population, which expanded to 1,335km2 with a 1,905,000 population. Many nationalities live in the city, and it is an excellent place for tourism, jobs, and recreation. Dubai city ranked 90th in the Happiest Cities Index 2024 in the silver group and represented the first Arabic city in the Happiest City Index [9]. Dubai Emirates has submitted a spatial roadmap for the next twenty years with many objectives, the Dubai 2020- 2040 Urban Master Plan, to advance flourishing, sustainability, and cohesiveness through a people-centric approach. The Plan has identified many key spatial intervention opportunities to achieve these objectives to enhance the sense of place by leveraging the diversity of population and economy, enhancing environmental sustainability by increasing the number of natural green areas, developing ecological corridors that create natural connections between urban and ruler area and expand natural conservation areas, enhance the environmental quality of air and water with sustainable, energy-efficient water and waste management system, reducing noise and UHI, floodplain management, promote urban agriculture to Increase food security, as well as Preserve and enhance waterfronts, waterways, and inland lakes, promoting efficiency in the existent and future infrastructure and energy conception and use renewal energy sources, offering a sustainable, cost-effective cooling strategy, maximize public transit investments, and provide affordable, sustainable housing complexes.

The strategy also has an ambitious initiative to transform the city to be pedestrian-friendly by 2040, to enhance pedestrian mobility in the city, and promote sustainable urban development through "Dubai Walk Master Plan" by constructing 6,500 kilometers of walkways, with 3,300 kilometers as new paths and 2,300 kilometers of existing renovating paths, with additional 900 kilometers planned for development after-2040. To improve walkability, connectivity, and accessibility by 110 pedestrian bridges and tunnels across the emirate. In the case of well-being, the Dubai government has announced the Quality-of-Life Strategy 2033', which aligned with the Dubai 2020- 2040 Urban Master Plan, prioritizes the well-being of individuals (nationals, residents, and visitors), society, and the city, which is the ultimate objective of the development plans, to support comprehensive wellbeing, to enhance Dubai status as one of the world's best cities in terms of living destination and wellbeing that offers its resident with a healthy, active, and enjoyable lifestyle, build a society that promotes identity and cultural diversity, and provide high-quality opportunities for entertainment.

The plan included well-being districts, creating a new generation of parks, enhanced walkability, development of outlying areas and beaches, landscape improvements, and promoting accessibility for health, sports, and entertainment facilities, safety, security, and emergency services for all. Moreover, it enhances the infrastructure for existing neighborhoods, upgrading the city's amenities and services, promoting sustainable mobility options for a healthier lifestyle, and enhancing social interaction places such as shops, mosques, and parks, with More than 115 km of pedestrian and cycling paths, implanting many trees and plants, and many economic opportunities that offer investment for locals with high-tech, innovation to support industrial, production and processing sectors. The plan also upgrades streets and ensures traffic safety with various types of urban furniture like shaded canopies and public seats, restrooms, and high-quality recreational spaces for families. Furthermore, it promotes the '20-minute city' concept, which is designed to ensure access to shops, transportation services, and recreational spaces via soft or sustainable mobility options within a 20-minute journey for more efficiency as well as featuring innovative design elements that satisfy various needs of the city and distinguish the identity of each district, including sustainable activities to foster culture and social well-being as well as arts and cultural spaces.

Furthermore, submitting a Strategic plan for justice-2026 to achieve an ambitious future by focusing on high-tech innovation, happiness, and digitization that aims to ensure effective law enforcement and enhance community trust, efficiency, and transparency of the specialized legal and penal system. Table (4) shows the extracted indicators for Dubai's development strategies according to the main six areas defined by the HCI 2024-2025

Table 4. The extracted indicators for Dubai development strategies according to the main six areas defined by the HCI 2024-2025.

Main area	Extracted indicators for Dubai development strategies
mobility	transform the city to be pedestrian-friendly by 2040, to enhance pedestrian mobility
	promoting sustainable mobility options for a healthier lifestyle
	constructing and renovating many pathways to improve walkability, connectivity, and accessibility across the Emirate
citizen	promoting accessibility to health, sports, and entertainment facilities, as well as safety, security, and emergency services for all.
	Enhance social interaction in places such as shops, mosques, and parks,
	leverage from the diversity of population and economy
	Featuring innovative design elements that satisfy various needs of the city and distinguish the identity of each district
	sustainable activities to foster culture and social well-being, as well as arts and cultural spaces
	Build a society that promotes identity and cultural diversity,

Continuation of Table 4.

Main area	Extracted indicators for Dubai development strategies
environment	increasing the number of natural green areas,
	Develop ecological corridors that create natural connections between urban and rural areas.
	Expand natural conservation areas,
	Enhance the environmental quality of air and water with a sustainable, energy-efficient water and waste management system.
	reducing noise and UHI, floodplain management
	Promote urban agriculture to Increase food security.
	Preserve and enhance waterfronts, waterways, and inland lakes,
	promoting efficiency in energy conception and use of renewal energy sources,
	offering a sustainable, cost-effective cooling strategy,
	developing outlying areas and beaches,
	landscape improvements, implanting many trees and plants
health	prioritize the well-being of individuals (nationals, residents, and visitors), society, and the city
	offers its residents with a healthy, active, and enjoyable lifestyle
	Offer high-quality opportunities for entertainment by creating a new generation of parks,
economy	many economic opportunities that offer investment for locals
	High-tech innovation to support the industrial, production, and processing sectors.
governance	provide affordable, sustainable housing complexes
	Upgrading the city's amenities, services, and infrastructure with innovative technologies.
	Promote the '20-minute city' concept.

In order to establish The Happy City Strategy Model, extracted indicators from Dubai City will be analyzed according to the theoretical model for the Happy City indicators that the research suggests contribute to promoting happiness to determine the active and inactive indicators that may contribute to the establishment of the strategy: using an objective scale where N= no evidence, M= moderate evidence, I=ideal evidence as shown in table (5) below.

Table 5. Application of the urban happiness theoretical model in Dubai City.

Urban happiness theoretical model			Evaluation			Explanation
Physiological needs	Main indicator	Secondary indicator	N	M	I	
Efficiency in Basic daily needs	Efficient urban design	15 or 20-minute city model and densification			*	promoting the '20-minute city' concept
		Offering sustainable housing options			*	Priority to provide affordable, sustainable housing complexes
	Efficient energy consumption and climate protection	Mitigate Climate change			*	offering a sustainable, cost-effective cooling strategy,
		Use renewable energy			*	promoting efficiency in energy conception and use of renewal energy sources,
Walkability, accessibility, and connectivity	Walkability	Friendly - walkable environments			*	constructing and renovating many pathways to improve walkability, connectivity, and

 https://doi.org/10.1051/e3sconf/202563307011

						accessibility across the Emirate
	Accessibility and connectivity	Access to health care services			*	promoting accessibility to health, sports, and entertainment facilities, as well as safety, security, and emergency services for all.
		Access to educational services and libraries			*	
		Access to Digital city & infrastructure			*	upgrading the city's amenities, services, and infrastructure with smart technologies.
		Access to sustainable public transport			*	promoting sustainable mobility options for a healthier lifestyle
Psychological needs	**Main indicator**	**Secondary indicator**	**N**	**M**	**I**	**Explanation**
Aesthetic preferences	Urban aesthetics	Urban qualities			*	much evidence for upgrading urban qualities
Need for recreation and restoration.	Built Recreational places	Playgrounds & parks			*	Creating a new generation of parks, developing ecological corridors that create natural connections between urban and ruler area
	Natural Recreational places	Proximity to green environment for recreation and restoration purposes			*	improving landscape by implanting many trees and plants, expanding natural conservation areas, increasing the number of natural green areas,
		Proximity to the blue environment for recreation and restoration purposes			*	Developing outlying areas and beaches, Preserve and enhancing waterfronts, waterways, and inland lakes,
	Environmental protection	Conservation For Natural Areas			*	Enhance the environmental quality of air and water with sustainable, energy-efficient management system,
		Protecting Biodiversity			*	maintain protection for biodiversity
Safety and security needs	Safety by design	Traffic safety			*	ensuring a safe, effective, and high-capacity traffic system
		Crime prevention policy			*	Ensure safety and security through digital crime prevention technologies and

						efficient emergency service.
	Safe and stable Economy	Attractive economic environment			*	Maintain attractiveness for business and jobs.
Social needs	**Main indicator**	**Secondary indicator**	**N**	**M**	**I**	**Explanation**
Social solidarity	Diversity and inclusion	Expanded cultural life			*	Build a society that promotes identity and cultural diversity,
	Urban resilient	Resilient cultural operation			*	sustainable activities to foster culture and social well-being, as well as arts and cultural spaces
	Social-cultural events	Public spaces for various cultural events			*	Enhance social interaction in places such as shops, mosques, and parks,

3 Results and discussion

The final step in creating the happy city strategy model is to compare the results from Zurich and Dubai city to determine both cities' percentage models that may contribute to the creation of the model, as shown below:

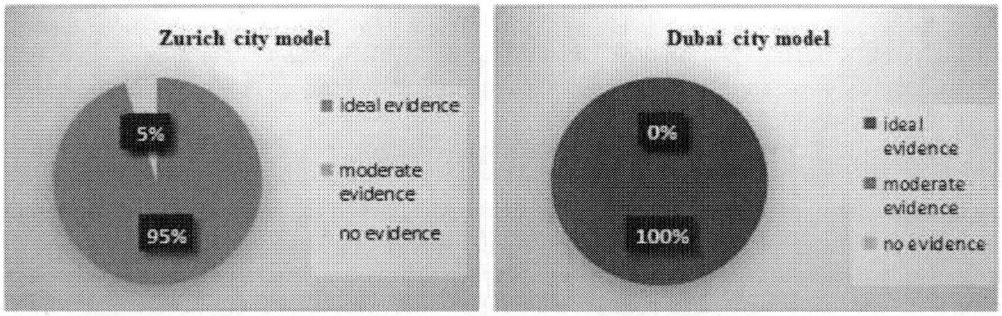

Fig. 3. Zurich and Dubai percentage models.

Accordingly, the action plans of Zurich City Urban Development Strategy 2020-2040 are basically focused on sustainable developments, creating a digital city with digital infrastructure, enhancing social diversity and inclusion, social interaction by participation in various cultural activities, and promoting individuals' well-being, it aims to maintain the high quality of life, and sustainable growth for economic and productivity by adopting a people-centered approach to promote quality of life. The model achieved 95% of the theoretical indicators of happy city strategy.

Dubai action plans are based on many urban strategies that also focus on sustainable urban development, like the Dubai 2040 Urban Master Plan, Dubai Walk Master Plan, Quality-of-Life Strategy 2033, and Strategic Plan for justice-2026; these strategies are focused on promoting efficiency by urban design by transforming towards 20 minutes-city concept to satisfy individuals basic daily needs, focusing on offering green spaces, commercial centers and recreational facilities that enhance human well-being with 20-minutes' walk or cycle and develop integrated sustainable housing complexes. The model achieved 95% of the theoretical indicators of happy city strategy.

Both cities achieved the highest ranks in the happy index and high percentage depending on the theoretical model; therefore, these results highlight the significance of sustainable development strategies in satisfying individuals' needs by:

- providing sustainable and flexible means of mobility to improve walkability, connectivity, and accessibility, which not only mitigates environmental impacts of transportation but also can positively affect individuals' health and well-being.
- prioritize the well-being of individuals, society, and the city to develop vibrant, healthy environments and inclusive communities by upgrading the city's amenities, services, and infrastructure with innovative technologies and double green and leisure areas. Studies show that mental health and well-being are supported by any leisure- and transport-related physical activity for all ages by proposes a direct effect on brain health, cognitive function, including memory and attention, positive effects on mood and stress resilience by releasing endorphins which are responsible for feeling happy and improves sleep durations, as well as social interaction and strengthening community ties.
- Enhance environmental qualities such as air and water with sustainable, energy-efficient management systems and sustainable energy consumption, as well as preserve green natural public and private spaces like gardens, forests, grasslands, rocks, waterfronts, and water resources. These environments have a significant impact on individuals' recreation and restoration needs as well as entertainment related to different age groups
- promote a safe environment using efficient high-tech innovations that support individuals' happiness and enable the satisfaction of the most essential individual needs; promote sustainable activities to foster culture and social well-being; and create arts and cultural spaces that satisfy individuals' social needs.
- enhance efficiency by urban design by reducing the time needed for daily demands and reduce car dependence by promoting more efficient public transportation options that can be reached within 20 minutes of walking or cycling, which can improve overall walkability, accessibility, connectivity, social interaction, mitigate environmental impact and sustainability

The results drawn from Zurich and Dubai's experiences by exploring their urban development strategies show that urban sustainable development strategies that are followed by those two cities match the theoretical framework model of the happy city that focuses on sustainable urban developments, people-centered design approach and promotes efficiency in daily demands, that have a significant impact on overall happiness. The results have supported the hypothesis and present a comprehensive model of the substantial role of sustainable urban development strategies in satisfying individuals' physiological, psychological, and social needs and promoting happiness. This model can be applied to promote happiness in Iraqi cities.

4 Conclusions

This research seeks to explore urban elements within cities that satisfy individuals' needs and promote happiness to develop a comprehensive framework for the happy city strategy that can be applied to promote happiness in Iraqi cities as well. Individual needs can be classified into physiological needs (efficiency in basic daily needs as well as walkability, accessibility, and connectivity), psychological needs (the need for recreation, restoration, safety, and security), and social needs like social solidarity that achieved by promoting diversity, inclusion, and place identity and urban resilience; both can be satisfied with tangible and intangible urban factors, this study explores the urban development's strategies followed by Zurich and Dubai city, which is ranked among the happiest cities in the world 2024. Various action plans that focus on sustainable urban design approaches to increase urban livability and enhance individual satisfaction by promoting efficiency, walkability, accessibility and connectivity, efficient energy consumption, and climate protection to satisfy individuals' physiological needs, as well as satisfying individual psychological needs for restoration, recreation, the aesthetic needs, and safety and security needs, the research finds significant role of sustainable urban development strategies in satisfying individuals' needs and promote happiness.

This study recommends that future interventions should focus on promoting happiness and well-being for individuals by Implementing the 20-minute city concept that can lead to a happier urban community through more efficient, socially connected, and sustainable strategies. Moreover, satisfying individual social needs by promoting diversity and inclusion, urban resilience, and encouraging more social-cultural events. The research also suggests improving happiness and well-being by ensuring access to green public spaces and recreational safe areas within a short distance that contributes to individuals' overall quality of life. Future research should focus on studying the outcome drawn from the urban strategies that are followed by the happiest cities in the world and their impact on individuals' well-being and happiness. Ultimately, creating a happy city and livable urban environment demands prioritizing the satisfaction of individuals' various needs through the natural and built environments through well-defined, sustainable–happy urban strategies.

References

1. L. A. Shaheen, "Smart happy sustainable cities definition and challenges," Int. J. Eng. Sci. **11(7)**, 1–16 (2022), DOI: 10.9790/1813-1107010116
2. M. M. Hassan, P. A. Tedong, A. M. Khir, Z. Shari, Z. Ponrahono, M. P. Sharifudin, "Reimagining cities: paving the path to happy city," Int. J. Acad. Res. Bus. Soc. Sci. **13(10)**, 418–420 (2023), DOI: 10.6007/IJARBSS/v13-i10/18818
3. Swiderski, The happy city movement," Impact Entrepreneur (2021), https://impactentrepreneur.com/the-happy-city-movement/
4. H. D. Musa, M. R. Yacob, A. M. Abdullah, M. Y. Ishak, "Enhancing subjective well-being through strategic urban planning: development and application of community happiness index," Sustain. Cities Soc. **38**, 184–194 (2018), DOI: 10.1016/j.scs.2017.12.030
5. Oxford Learner Dictionaries, "Sustainable," (2024), https://www.oxfordlearnersdictionaries.com/us/definition/english/sustainable
6. S. Mirzaei, A. Zangiabadi, "Studying and complying dimensions, indicators, and variables related to a happy city," Int. Rev. Spat. Plan. Sustain. Dev. **9(2)**, 94–111 (2021), DOI: 10.14246/irspsda.9.2_94
7. S. Wren-Lewis, S. Abdallah, C. M. O'Keeffe, Z. Travers, "Happy City Index 2016 Report," Centre for Thriving Places (2016), https://www.centreforthrivingplaces.org/wp-content/uploads/2016/10/Happy-City-Index-2016-Report-FINAL.pdf
8. L. Joshy, "Motivation: concept and theories," in *BPAE-142 Organisational Behaviour* (Indira Gandhi National Open University, New Delhi, 2022)
9. World Of Work, Alderfer's ERG theory of motivation (2019), https://worldofwork.io/2019/02/alderfers-erg-theory-of-motivation/
10. J. Szołtysek, "Meeting places in the urban strategy to build a happy city: a mixed research approach," Adv. Intell. Syst. Comput. **20(6)**, 40 (2018)
11. S. Samavati, P. Desmet, *Happy Public Space: A Guide with 20 Ingredients to Design for Urban Happiness* (Delft University of Technology, Delft, 2022)

Monitoring the distribution of parks and land cover changes along the Tigris River using spatial analysis techniques

Sundus A. Abdullah Albakri[1,*], *Zina* F. Kadhim[2]; *Sura* I. Gburi[2], *Muhend* K.Samer[2], and *Faisel* G. Mohammed[1]

[1]Dept. of Remote Sensing, University of Baghdad, College of Science, Baghdad, Iraq
[2]Dept. of Astronomy and space, University of Baghdad, College of Science, Baghdad, Iraq

Abstract. This study examines the use of GIS and Remote Sensing in monitoring vegetation growth of Baghdad in Al- Risafa limited in the region Abu Nawas that alignment Tigris River. The various land features of land for the adopted area were be analyzed by spatial analysis techniques, Classification the images of land sat 8 will be considered. The techniques of classification that adopted in this work; including supervised with unsupervised techniques that are commonly used to produce land cover features. The results produced from the NDVI images gave a good indication of vegetation degradation through the period (2019 - 2023). For year 2019 the area of Rare plant 269100 m^2, Low plant 603000 m^2, and the medium plant was 344700 m^2 total. Found that plant less area, and low plant area increases through the period (2019–2023) years. That is indicates decrease in vegetation area in the region of Abu Nawas with the decreasing rate of medium plant (12.6-8.3).These results can be due several reasons, including climate change, which causes a rise in temperatures and pollution in the waters of the Tigris River due to water scarcity and lack of interest in plowing the river on a regular basis, as well as due to urban growth represented by the large number of vertical buildings that impede air access to plants and humans alike. The parks proposed in this study, based on spatial analysis techniques depending on the classification methods, were distributed according to the selected sites, and their area ranged along the Tigris River between (172893 m^2 -340006 m^2) for the period.

1 Introduction

Sustainable environment is considered the most important goals that must be achieved for human living and the formation of human settlements. Among the most important environmental components are the presence of clean water and rivers, as the availability of green cover, gardens and parks in all cities are considered suitable conditions for human existence as well as the presence of an environment for animals to live in. Many researchers have now used remote sensing techniques to search for suitable places for the presence of water and the distribution of green cover and to conduct spatial analysis to determine the

* Corresponding author: Sundusalbakri70@gmail.com

concentration of water. In Iraq, the Tigris and Euphrates rivers are considered among the most important vital rivers for living.

The Tigris River is Western Asia's second-longest river. The Tigris River, is 1850 km long and rises in eastern Turkey. It flows through densely populated areas, particularly Baghdad, which is home to almost 8 million people. The river enters the city from the north and flows southward as part of the alluvial plain, with an average flow rate of 540 m3/s from 2005 to 2020 [1]. The amount of water in our nation's rivers has drastically decreased in recent years due to the extensive human and economic activities and the great and growing need for water as one of the components of life on water. Increased sewage and the absence of a treatment plant are two major causes of the low river discharge, as is the pollution of the Tigris River water by the transfer of pollutants that are harmful to aquatic and human life and their impact on the vegetation cover as a result of human activities and dead animals in the water, such as: household sewage, hospitals and industrial factories that empty their polluted water directly into the river without any treatment, so it passes through the city of Baghdad and threatens the ecosystem of plants and living organisms and contributes to killing aquatic life and poisoning fish and others [2].

Compared to other Iraqi cities, Baghdad is thought to be the most densely populated city. It is situated on a plain between 31 and 39 meters above sea level [1]. So its aerial extension is unrestricted by any natural boundaries. The Tigris River splits the city in half, with the eastern half being Rusafa and the western half being Karkh. The eastern boundary of the area is formed by the Diyala River, which joins the Tigris River southeast of Baghdad [3].

Study Area

Tigris River located in the northern part of the city serves as the recharge region for the 24 km long Army Canal, which is flows into the southern part of the Diyala River. Figure (1) shows the Tigris river in the geological map of Iraq.

Abu Nawas street area was chosen in Baghdad Governorate for this study. The capital, Baghdad, is considered the first city and the center of Iraq which is located in central of Iraq between (longitude 44°2157.96E; latitude 33°1854.72N) as shown in Fig (2) [4]. Geographic information system (GIS) methods are a useful and effective tool for planning and analyzing spatial data. In a digital mapping context, GIS stores geographic information (data) [5]. To visualize spatial information and relationships, data or other layers of information can be superimposed on top of the digital base map. Better viewing and comprehension of physical features and the linkages influencing a specific crucial environmental condition are made possible by GIS. To determine different environmental parameters and effect analysis, factors including vegetation, aspects, and slope steepness can be observed and superimposed. Abu Nawas area is a recent wide, shallow sloped levee extending south to the bend bend eastward to flow around in the form of a large oxbow as shown in Fig. (3) [4].

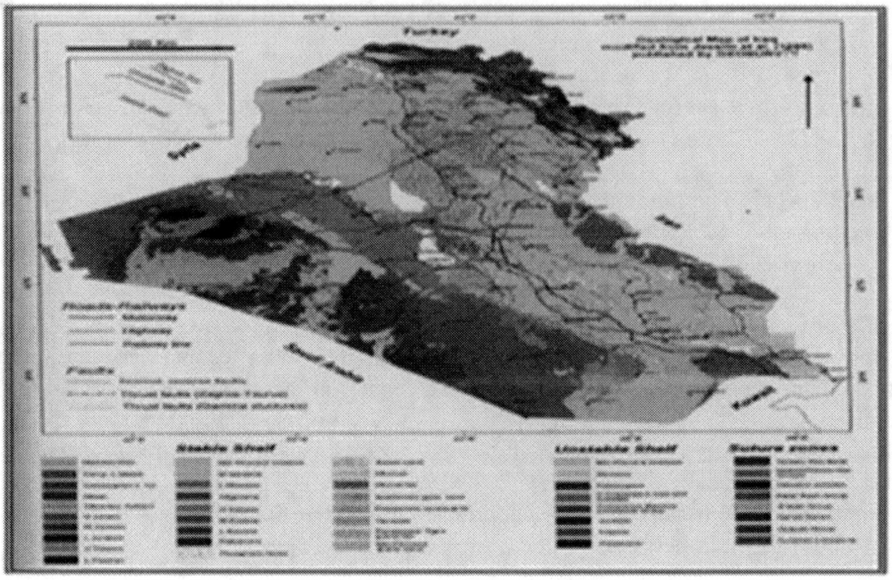

Fig. 1. Represents Geological map of Iraq contains Tigris River [2,3].

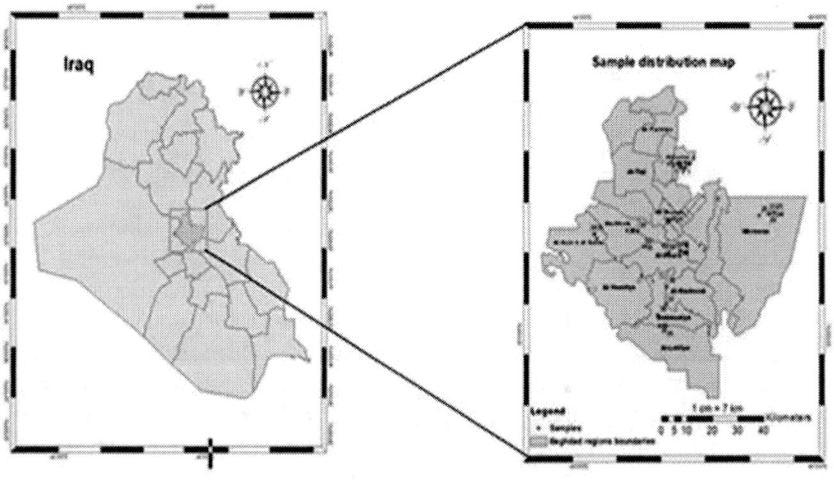

Fig. 2. Shows the location of Study area on the map of Iraq.

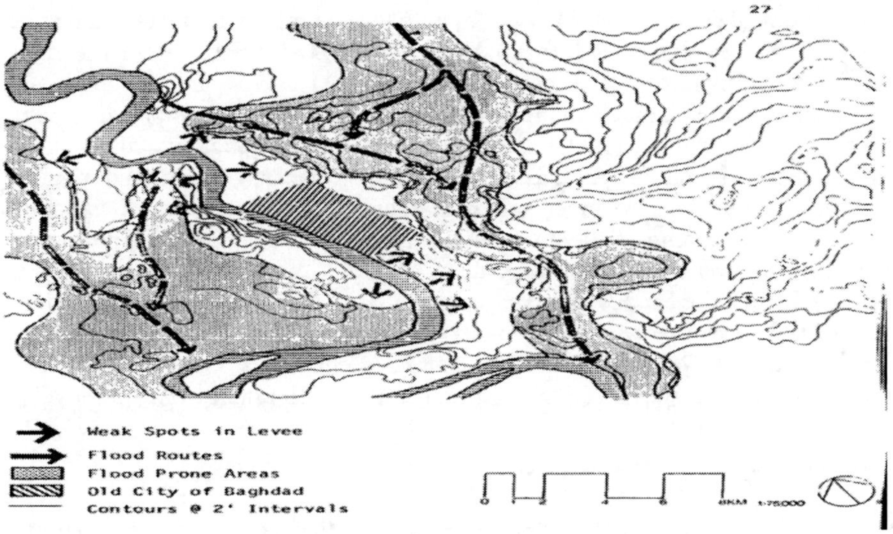

Fig. 3. The area behind the levee, and flood routes [5].

Two or three spots in the Abu Nawas natural levee are weak structurally and show extensive overflows, where the river has deposited large alluvial fans [4].

2 Materials and methods

-In this work, Certain GIS methods can be used based on their significance and the nature of the business. It is possible to estimate, generate, and assess the spatial analysis of land cover distribution using the categorization approaches. According to the facts and information available for the study area, each spectral class is distinct due to the classification being influenced by the urban planning characteristics of the examined area (rural or urban).
- Arc Map 10.8 (GIS) was used to retrieve the authorized maps. Through the intensity of radiation, reflection, and diffusion of the feature chosen for the investigation, the physical qualities influence the characteristics of the phenomenon under study.
-Normalized Difference Vegetation Index (NDVI) is studied widely to describe changes in vegetation index in landscapes around the world [3], [6].
-The source is a satellite video with a resolution of 30 square meters for the year (2019-2020-2021-2022-2023) and processed using the ARC MAP 10.8 (G.I.S) software.

2.1 Remote sensing

In order to identify and classify distant objects or materials, the concepts of remote sensing (RS) entail the detection and measurement of radiations of various wavelengths that are reflected or emitted from their surface [7]. RS is the process of gathering pictures or other information about an item using measurements taken from a distance. Examples of this include satellite imaging and aerial photos [5].
However, only optical pictures obtained by airborne or space-borne sensors are taken into consideration in this context. The scientific community has been using remote sensing technologies more and more in the past few decades to characterize and track a range of systems at the local or global level. Multi-spectral imagery, such as Thematic Mapper, has replaced pure visual imagery, such as aerial images, in this technology.

The spatial properties of the data across different digital layers can be analyzed using GIS [7]. GIS is a rapidly developing information technology that allows the creation of databases with geographical data that may be used for both the natural environment (such as population distribution and environmental factors) and human settlements (such as demographic databases). Most significantly, sustainable management can be ensured by combining the two types of databases. In addition to the analytical description of geographical issues, GIS will continue to advance as a crucial acquisition and analysis tool. It will also be used in environmental planning, impact assessment, disaster management, and remote sensing monitoring. Through a number of case studies, Environment, Development, and Sustainability has examined the application of remote sensing and GIS in the sustainable management of tropical coastal ecosystems [8]. Multispectral image classification is the primary technique used in the graphical and spatial analysis of remotely sensed pictures. A multispectral image's pixel (image element) attributes are usually recorded on several spectral channels. Following processing and correction, satellite data analysis entails registration, classification, and change detection utilizing the compared satellite data [3].

2.2 Supervision classification

Supervised classification is a type of machine learning that employs a classification algorithm and predefined training data. In supervised remote sensing, the classification process is frequently employed as the classification algorithm [6]. The most convincing aspect of the Landsat image classification method is the supervised component, which serves as the training sample. To evaluate thick vegetation normalize difference vegetation index (NDVI). The Normalized Difference Vegetation Index (NDVI) is a measure of how much vegetation exists in (NDVI). Low NDVI values (0.1 and below) denote barren rock, sand, environments. Moderate amounts indicate bushes and grassland (0.2 to 0.3), whereas high levels indicate tropical rain-forests and dense forests for a range of values between 0.6 and 0.8 [7].

Plants, that are solid and have a long near IR (NIR) reflectivity, around 0.7-1.3 μm. Chlorophyll is a substance found in greenery that powerfully reflects green (0.5-0.6 μm), red (0.4-0.5 μm), and blue (0.4-0.5 μm) . High Ruddy retention and long NIR reflectance are the two criteria used to determine the NDVI.

NDVI can be calculated by using the formula in equation (1) [10]:

$$NDVI = \frac{(NIR-R)}{(NIR+R)} \tag{1}$$

Where NIR: near infrared, R: red color spectrum,the NDVI value is in the range of -1 to 1. Higher NDVI readings represent greater NIR levels, which suggest thick vegetation [3]. NDVI classification was used to extract the total area of vegetation in the Abu Nawas area adjacent to the study area located in Baghdad. It was revealed through this technique that there was a difference in the distribution of vegetation cover through the period (2019– 2023) years in terms of Plant less, Rare plant, Low plant, Medium plant, and the total plants as illustrated in the following: tables (1), (2), (3), (4), and (5).

Table 1. The area of vegetation types for the year 2019.

Class	Area (m^2)	(%)
Plant less	1814400	59.86
Rare plant	269100	8.88
Low plant	603000	19.89
Medium plant	344700	11.37
The total	3031200	100

Table 2. The area of vegetation types for the year 2020.

Class	Area (m^2)	(%)
Plant less	1812600	59.80
Rare plant	291600	9.62
Low plant	567000	18.71
Medium plant	360000	11.88
The total	3031200	100

Table 3. The area of vegetation types for the year 2021.

Class	Area (m^2)	(%)
Plant less	1766700	58.28
Rare plant	373500	12.32
Low plant	548100	18.08
Medium plant	342900	11.31
The total	3031200	100

Table 4. The area of vegetation types for the year 2022.

Class	Area (m^2)	(%)
Plant less	1595700	52.64
Rare plant	379800	12.53
Low plant	673200	22.21
Medium plant	382500	12.62
The total	3031200	100

Table 5. The area of vegetation types for the year 2023.

Class	Area (m^2)	(%)
Plant less	1566000	51.66
Rare plant	522900	17.25
Low plant	689400	22.74
Medium plant	252900	8.34
The total	3031200	100

3 Results and discussion

Through satellite images of the Abu Nawas area that alignment Tigris River in the city of Baghdad and using the NDVI program, the following results were obtained as shown in figures below. Figure (4) shows the rates of vegetation areas with rate of plants for year 2019. The areas of Plant less was1814400 m^2, Rare plant 269100 m^2, Low plant 603000 m^2, and the medium plant was 344700 m^2. The data analysis from GIS system helps planner in Planning and Managing the environmental hazards and risks. In order to plan and monitor the environmental problems, the assessment of hazards and risks becomes the foundation for planning decisions and for mitigation activities [6]. Horologic Properties of the Tigris River shows that the annual flow of the River for the period (2005–2020) ranged between (15.96 and 98.02) (m^3/s) with a mean annual flow (46.71) (m^3/s) for the Tharthar -Tigris Canal [6]. While for the Baghdad site, the range was between (399.84 and 711) m^3/s with a mean annual flow (525.50) m^3/s.

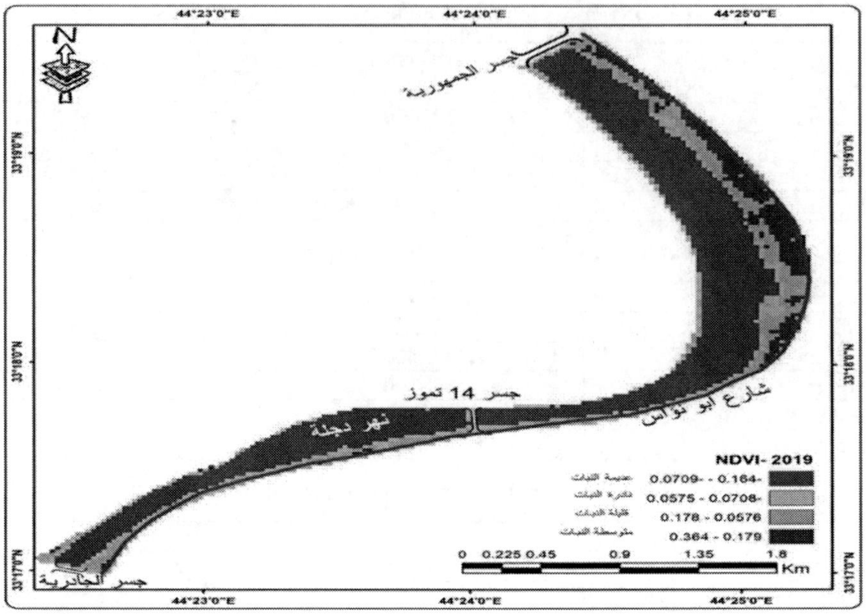

Fig. 4. The rates of vegetation areas with rate of plants that found in the study area that adopted (2019).

The results showed that the rates of vegetation areas with rate of plants for year 2020 as shown in Figure (5), the areas of Plant less was 181440 m^2, Rare plant 269100 m^2, Low plant 548100 m^2, and the medium plant was 360000 m^2.

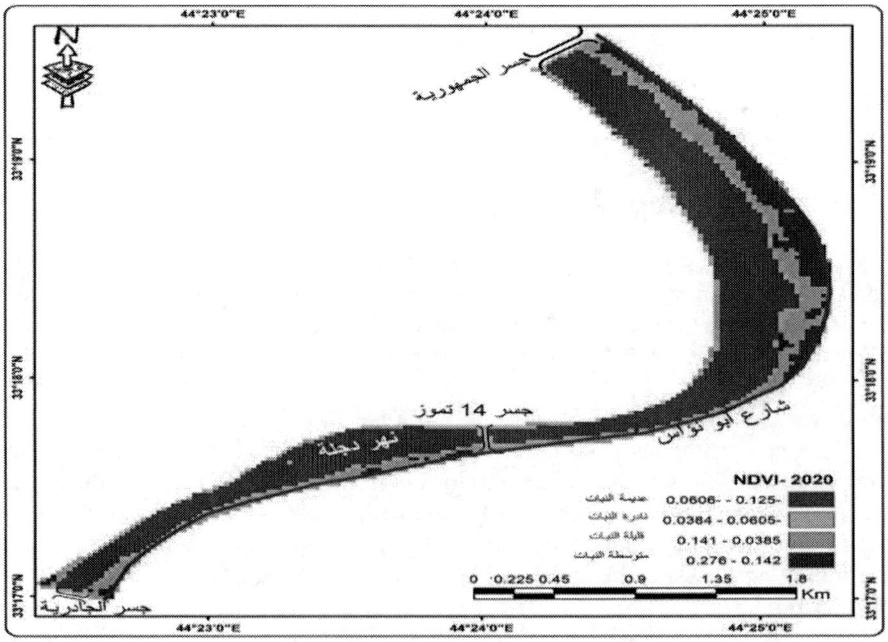

Fig. 5. Illustrated the rates of vegetation areas with rate of plants that found in the study area that adopted (2020).

 https://doi.org/10.1051/e3sconf/202563307012

Figure (6) shows the rates of vegetation areas with rate of plants for year 2021. The areas of Plant less was1814400 m^2, Rare plant 269100 m^2, Low plant 548100 m2, and the medium plant 342900 m^2 .

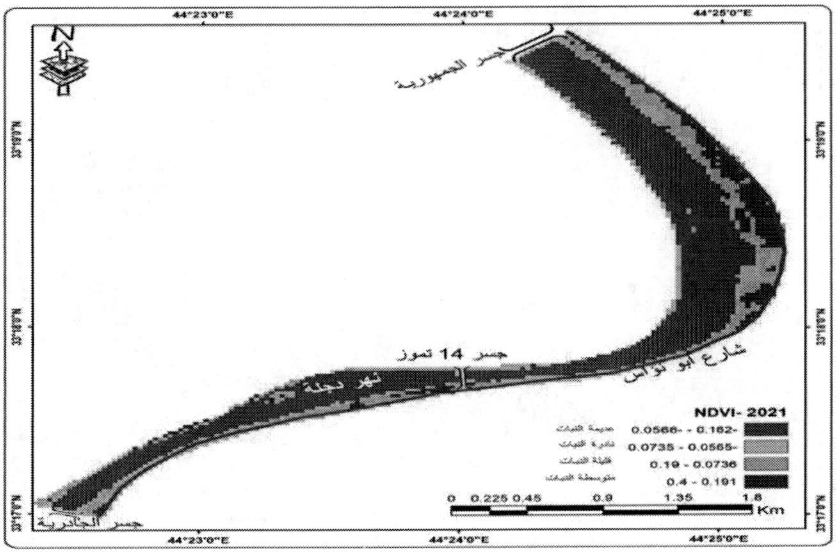

Fig. 6. Illustrated the rates of vegetation areas with rate of plants that found in the study area that adopted (2021).

The rates of vegetation areas with rate of plants for year 2022 discussed in Figure (7), the areas of Plant less was1814400 m^2, Rare plant 269100 m^2 (12.5%), Low plant 548100 m^2 (22.21), and the medium plant was 382500 m^2 (12.62).

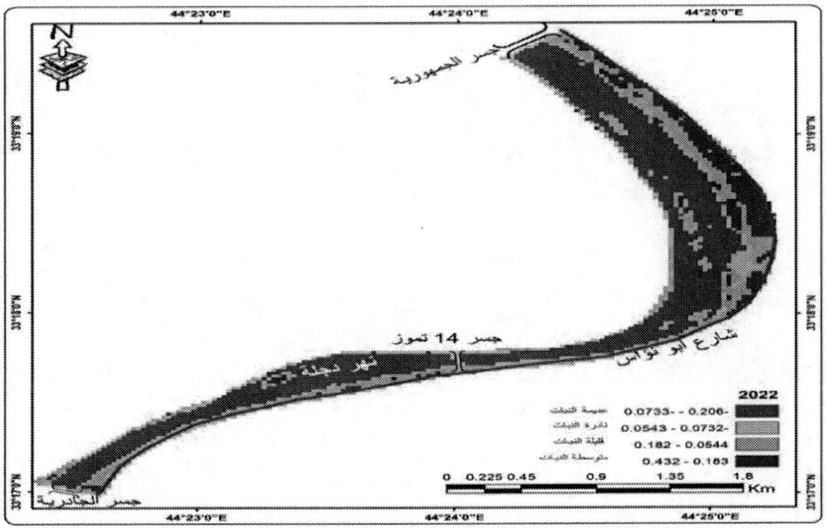

Fig. 7. Illustrated the rates of vegetation areas with rate of plants that found in the study area that adopted (2022).

Finally, the rates of vegetation areas with rate of plants for year 2023 was illustrated in figure (8). The rate of Rare plant 269100 m^2 (17.25 %) with NDVI value (0.0543-0.0732), Low plant 548100 m^2 (22.74%) With NDVI value (0.182-0.0544), and the medium plant was

 https://doi.org/10.1051/e3sconf/202563307012

252900 m^2 (8.34 %) with NDVI value (0.432-0.183). Also the results estimated that plant less area, and low plant area increases through the period (2019 – 2023) years as shown in Figure (9) and (10).

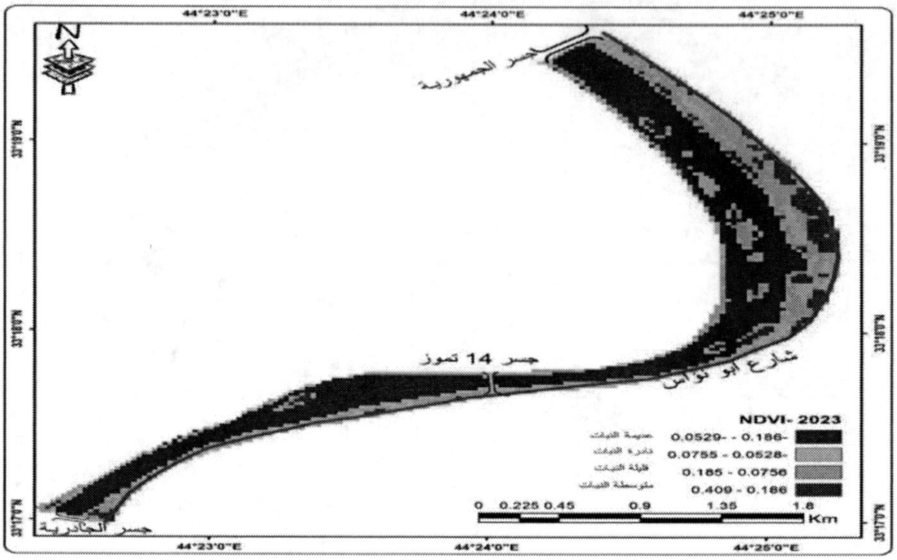

Fig. 8. Illustrated the rates of vegetation areas with rate of plants that found in the study area that adopted (2023).

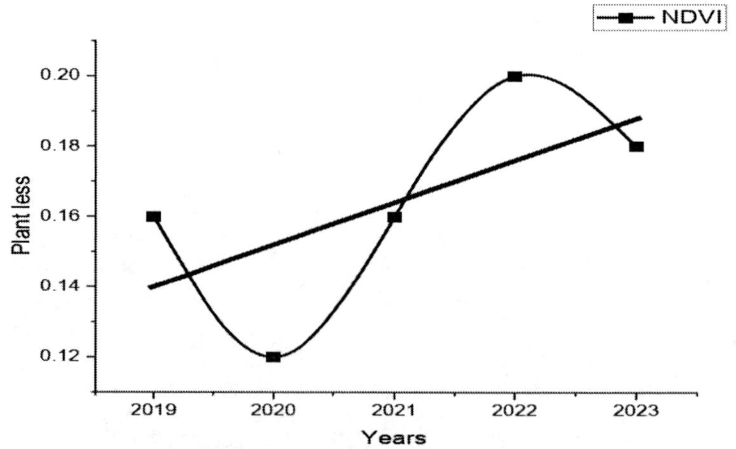

Fig. 9. Illustrated the plant less rates of vegetation areas with variation through the period (2019– 2023) years, found in the study area that adopted. Abu Nwas that alignment Tigris river.

Based on the spatial analysis and classification of NDVI of vegetated areas using supervised classification techniques, a proposed map was prepared for the distribution of parks according to the available areas, data as the results showed that proposed parks can be distributed between area (1728931 m^2) (340006m^2) for the periods (2019-2023) as shown the Figures (10) and (11).

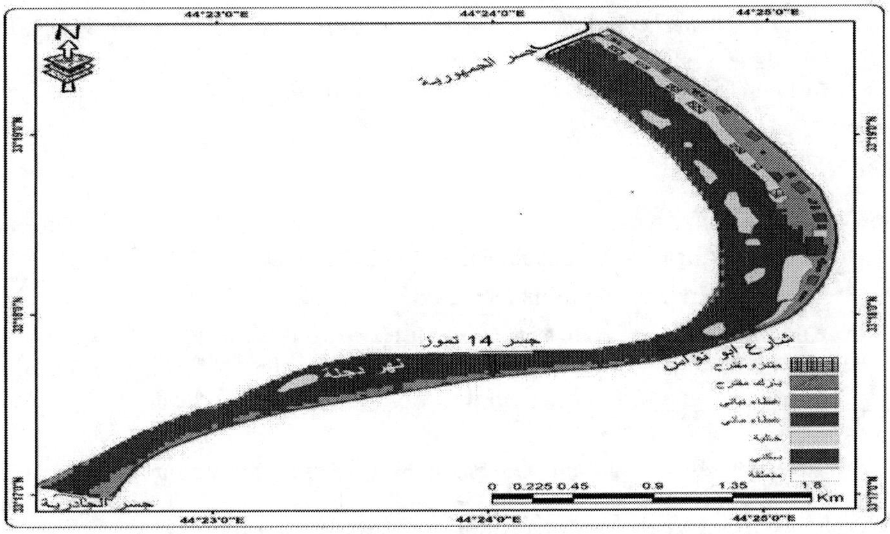

Fig. 10. Illustrated Locations of parks and parks proposed for the best year, which is 2023.Park area = 22,609 square meters, garden area = 29,405 square meters.

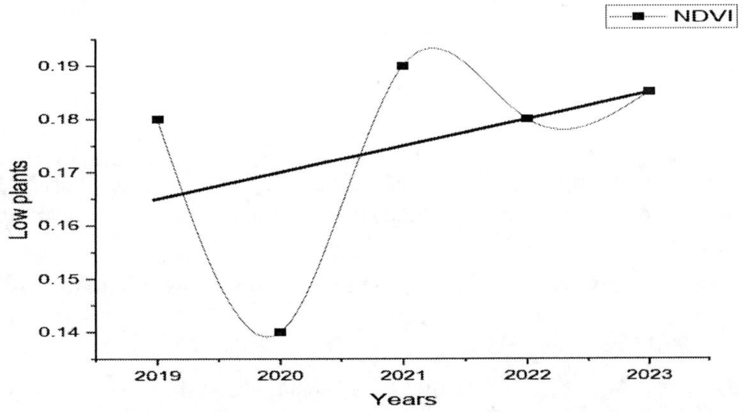

Fig. 11. Illustrated the low plant rates of vegetation areas variation through the period (2019 – 2023) years found in the study area that adopted. Abu Nawas that alignment Tigris river.

4 Conclusions

In this study, the focus was on using the spatial analysis and classification technique for the available data in order to create a proposed map for the future and current distribution of parks along the study area of the Tigris River (Abu Nawas is the case study).

The following conclusions were reached:

1.From this study by reviewing the pictures of NDVI images, conclude that the total area of vegetation plant, through the period (2019 – 2023) years was 3031200m^2.

2. The plant less, and the low plant in increasly, it indicates decrease in vegetation area in the region of Abu Nawas. Perhaps for several reasons, including climate change, which causes a rise in temperatures and pollution in the waters of the Tigris River due to water scarcity and lack of interest in plowing the river on a regular basis, as well as due to urban growth represented by the large number of vertical buildings that impede air access to plants and humans alike.

Using remote sensing techniques and spatial analysis, some of the proposed parks were distributed along the Tigris River from the Abu Nawas side, at a distance ranging between (1728931 m^2) and (340006m^2) for the adopted periods as shown in the area images ..

References

1. S. Oleiwi, M. Al-Dabbas, "Assessment of contamination along the Tigris River from Tharthar-Tigris Canal to Azizziyah, Middle of Iraq," Water **14**, 1194 (2022)
2. GIS, Al-Mostansiriyah J. Arab Int. Stud. **68**, 77 (2024)
3. S. A. Abdullah, K. H. A. Abbas, A. Alaa H. Harif, A. Alaa R. F., H. S. S., "Sustainable urban distribution of educational institutions and population density in Baghdad City using remote sensing techniques," in IOP Conf. Ser.: Earth Environ. Sci. **1202**, 012015 (2023)
4. E. A. Mohammed, Z. Y. Hani, G. Q. Kadhim, "Assessing land cover/use changes in Karbala city (Iraq) using GIS techniques and remote sensing data," J. Phys. Conf. Ser. **1032(1)**, 012047 (2018), DOI: 10.1088/1742-6596/1032/1/012047
5. Al-Hussein Nabeel Hussein, "Monitoring vegetation degradation in Baghdad," Al-Kut Univ. Coll. J. **7(2)**, 182–184 (2022)
6. S. A. Abdullah, T. R. Aziz, "Detect environment pollution by GIS for some regions in Baghdad City," J. Phys. Conf. Ser. **2114**, 012024 (2021)
7. F. K. Mashee Al Ramahi, Z. K. I. A. B., "Estimation of Suaeda aegyptiaca plant distribution regions at Iraq using RS & GIS applications," Iraqi J. Sci. **58(2a)**, 767–777 (2017)
8. T. A. Dhamin, E. F. Khanjer, F. K. Mashee, "The effect of temporal resolution of climatic factors on agriculture degradation in southern Baghdad by applying remote sensing data," Iraqi J. Sci. **64(2)**, 994–1006 (2023)
9. E. S. Hassan, A. F. Hasson, E. F. Khanjer, "Monitoring changes in the spectral reflectivity of Baghdad City and the impact of the atmospheric elements on it using R.S and GIS," in IOP Conf. Ser.: Earth Environ. Sci. **1223**, 012023 (2023)
10. G. A. Al-Edhar, A. Ab. Al-Sadoon, E. F. Khanjer, "Monitoring the land surface temperature LST with different seasons for Babylon City using GIS and R.S techniques," in IOP Conf. Ser.: Earth Environ. Sci. **1202**, 012018 (2023)

Residential mobility in Basrah city: the spatial relationship between its drivers and variables by using statistical analysis

Abbas Abdul-Hussein Kadhim Alaidani*, and *Makram* Jamal Abbas Al-Badran

University of Basrah, College of Education for Social Sciences, Department of Geography, Basrah, Iraq

Abstract. Residential mobility plays a vital role in raising urban growth and expansion. Basrah city witnessed a residential movement of many families, whether from the same neighborhood or others. This movement is motivated by several factors effectively. Several families move from one place to another, influenced by a set of motives that encouraged the movement. However, research on the motivations and drivers behind residential movement is limited. Thus, the study aims at determining drivers of residential mobility in Basrah city in 2023. It applies statistical analysis (SPSS.V.28) and methods of geographical information systems (GIS) in order to measure the drivers behind the residential mobility. The results show that several households move within the same neighborhood or between other ones due to various reasons such as an increase in family size or income, a desire to move to large house, high-cost rent, a change of marital status (marriage), a desire to move closer to family and relatives or to schools of sons or the workplace, a suffer from car parking that is located near sources of pollution and neighborhood dissatisfaction.

1 Introduction

Residential Mobility is a social urban phenomenon that has a geographical dimension. It is influenced by households' social and economic features. This is the case when families move seeking proper house and what privileges that house may offer. On the other hand, some households move in order to be close to their workplace. However, some movements are driven by the systemic decisions of agents who work in housing or real estate such as state of real estate market and its capacity in receiving segments of population, housing policies, construction techniques and ways to develop them and new payment methods that are compatible for demand. Thus, the study aims at examining motivations and reasons that urge households to move from one place to another as factors of residential mobility are regarded as determinants to choose housing and ways of living.

The Problem of the Study

* Corresponding author: abbas.Kadhim@uobasrah.edu.iq

What are the motivations that urge people to move from their residences to another ones? Is it possible to rely on variables of residential mobility to predict the future residential mobility in the city by using statistical methods?

The Research Hypothesis

There are structural, environmental, social and economic factors that motivate the residential mobility in the city. The study applies a statistical approach to predict the increase in population, area and families to know the future movements of the population.

The Importance of the Study

The study aims at examining reasons of residential mobility, determining its directions and future implications in order to organize residential areas and distribute population systematically and properly.

The procedure of the Study

The study follows a descriptive approach to characterize the reasons and drivers that lead to the residential mobility. In addition, quantitative and statistical methods are used to analyze data and measure population density such as statistical analysis (SPSS) and geographical information systems (GIS). The SPSS is applied to uncover latent root, regression equation and correlation coefficients in order to measure the size of variance in the residential mobility and examine the relationship among variables in the city. The chosen sample size is 5%, which is 4986 of the city's population.

Limits of the Study

Basrah is located between latitudes (30.0° - 36.0°) north, and longitudes (48.00° - 48.00°) east, as indicated by Map (1). It is bordered to the north by Karmat Ali and Al-Hartha district, the east by Shatt Al-Arab district, the south by Abu Al-Khaseeb district, and the west by Shatt Al-Basra District and Al-Zubair District. Map (1).

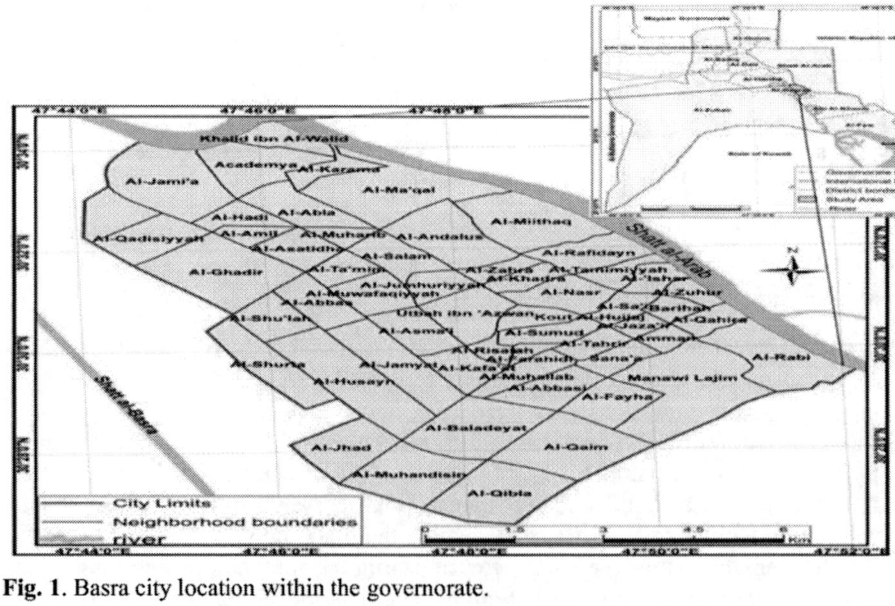

Fig. 1. Basra city location within the governorate.
Source: Based on (1) Basra Municipality Directorate, Cities Organization Division, and GIS Division, 2022. (2) Spatial analysis of GIS Data, Arc GIS.10.4.

2 Materials and methods

2.1 Motivations of residential mobility

The study classifies drivers of residential mobility in Basrah into voluntary and involuntary ones in which the voluntary one is divided into stimulant and adaptation movement as indicated below:

2.1.1 Voluntary motivations

Generally speaking, voluntary mobility is motivated by reasons such as family, neighborhood or employment. People usually move or relocate from one housing to another voluntarily seeking more calm and active social life by selling their existing house and buying an alternative piece of land in modern neighborhood where they can build another one. In addition, individuals tend to move to larger home, (the design, quality and materials of building, number of rooms), or dissatisfaction with the neighborhood or a desire to move to housing closer to shopping centers, family and relatives. Therefore, voluntary residential mobility reflects people's attempts to obtain a house leaving their first residential dwelling. Thus, whenever they obtain a new job that adjust their income and help them rent or buy another housing, a lot of people consider to move to a housing of high social rank [1]. Therefore, the motivations of voluntary residential mobility are divided into:

a-Stimulating Movements (Residential)

Stimulating motivations are residential moves that are not directly motivated by issues related to residential feature but to family-related factors i.e. transformations that households go through which resulted from nonresidential decisions [2]. Therefore they are called resulted moves adjusting for education, household income, marital status, employment status and number of children in the household. Thus in Basra, stimulating moves have the highest percentage (54.8%) in 2024 as illustrated in table (1) and figure (1). Many studies show that most moves motivated by a change in the number of family, as in Meddina, in which it is (16.3%) of the total movements in the city.

b-Adaptation Movements

Adaptation movements are resulted from a set of housing services and motivations. These include dissatisfaction with housing conditions, the residential neighborhood due to its social and urban disorder and level of quality of services in the neighborhood. In addition people move due to the difficulty they face in their previous house when they go to their work, shopping centers, family and relatives. According to table (1) and figure (1), adaptation movements come second with a percentage (41.4%) from the total number of voluntary movements in Basrah city.

2.1.2 Involuntary movements (coercive)

Involuntary movements are relocations resulted by forced and exceptional circumstances that not taken into account by the family which urge it to change the previous dwelling.

Table 1. Motivations of Residential Mobility in Basrah in (2023).

N	Motivations of Residential Mobility	Type of Movement	Number	%
1	Increase in number of children	Stimulant movements (Residential)	2731	54.8
2	High rental Prices			
3	Improvement in family income			
4	Change in marital status (Marriage)			
5	Oldness of Previous house			
6	Obtaining better house	Adaptation Movements(Residential)	2065	41.4
7	Move close to schools of sons			
8	Move close to the new workplace			
9	Move close to family and relative			
10	Obtaining public land or house			
11	Neighbors dissatisfaction	Involuntary Movements	190	3.8
12	Suffer from car parking			
Total			4986	100

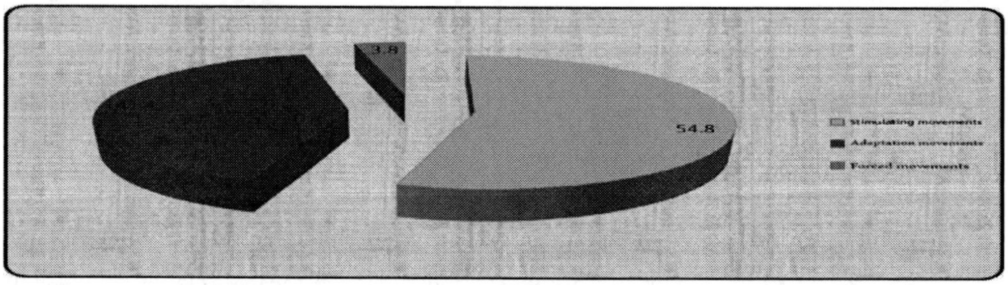

Fig. 1. The Relative distribution of motivations of residential mobility in Basra city for the year (2023) according to the sample.

The reasons behind movements include fire, the demolition of the house as a result of natural disasters such as earthquakes and volcanoes, or as a result of establishing urban construction projects and paving roads. These are all motives that lead to Involuntary movements beyond control. These movements constitute the lowest percentage in Basrah city (3.8%) of the total motives in the study area. In light of the above, the motives for residential movement in the city of Basra will be addressed and discussed in detail through their spatial relationship between the variables of residential movement using statistical analysis criteria.

2.2 Motivations of residential mobility and its spatial relationship among variables of residential mobility using methods of statistical analysis in Basrah city

Residential Mobility is motivated by several motivations which include an increase in the number of family members, high rental prices, an improvement in the family income, the change in social status (marriage), the previous residence conditions, obtaining a better residence or a public plot of land or house, moving close to schools of children, the new workplace or to family or relatives, neighbors' dissatisfaction and suffering from parking spaces that are close to sources of pollution. On the other hand, people move due to the improvement in the relative standard of living, which contributed effectively to the increase in residential mobility. Thus, according to the data obtained from the questionnaire, it appears that there is a clear disparity in the justifications and motives for residential mobility, as shown in Table (2) and Figure (2). It is noticed that the increase in the number of family members come first, obtaining a better housing second, a change in social status (marriage) third, while last rank was for suffering from parking spaces. The motivations will be discussed in detail as follows:

2.2.1 Increase in family size

The change in the number of children in family is one of the general reasons that urges people to move to another house within the city. It is also one of the factors that leads to housing crisis given that the change in the number of children influences the size of housing unit. The more the number of children in the family increases the more responsibilities that family has. Therefore, it is expected this factor urges some sons, who marry and become fathers, to move to another house away from their families. Central Agency for Public Mobilization and Statistics indicates that 40% of families in half of the Iraqi governorates suffer from people crowding in the same room and increased number of families in the same housing unit [3].

By analyzing Table (2) and Figure (2), it is found that the increasing in the number of family members ranked first among the rest of the motivations with a percentage of (16.3%). However, when it is classified according to the residential regions in the city, it is noticed that the weak region records the highest percentage among the other regions i.e. (21.9%), followed by the medium transitional region in second place i.e. (14.7), the ambitious region came in third place i.e. (13.1%), while the good transitional region received a percentage of (12.4%).

By analyzing the correlation matrix of the residential mobility variables in Basra city, as shown in Table (3), it becomes clear that the increase in the number of family members amounted to (1) and is related to the variable of the high rental cost of the previous residence with a positive relationship that is not strong (0.608), a positive relationship with the improvement in the family income (0.886), a strong positive relationship (0.965) with the change in social status (marriage), a positive relationship that is not strong (0.706) with the variable of the residence conditions, a strong positive relationship with each of the variable of obtaining a better house (0.880) and the variable of move close to the schools of the children (0.910), the new workplace (0.871) and to family and relatives (0.797), and a moderate positive relationship with suffering from parking spaces and its closeness to pollution sources for the previous residence (0.660), whereas the increase in the number of family members is associated with a weak relationship with obtaining a public plot of land or housing which amount to (0.410), and the variable neighbors dissatisfaction (0.515).

Table 2. Motivations for residential mobility according to residential regions in the city of Basra for the year 2023 (according to the sample).

Motivation	**Regions**										
	Ambition region	**%**	**The good transitional region**	**%**	**Average transitional region**	**%**	**The Weakness region**	**%**	**Total**	**%**	**Rank**
Increase in family size	142	13.1	116	12.4	190	14.7	366	21.9	814	16.3	1
High Rental Prices	67	6.2	93	9.9	112	8.7	63	3.8	335	6.7	8
Improvement in family income	121	11.1	68	7.3	132	10.2	129	7.7	450	9	6
Marriage	142	13.1	103	11	153	11.9	280	16.7	678	13.6	3

https://doi.org/10.1051/e3sconf/202563307013

Housing Conditions	96	8.8	72	7.7	106	8.2	179	10.7	453	9.1	5
Obtaining better house	171	15.7	154	16.5	183	14.2	195	11.7	703	14.1	2
Moving close to schools of children	53	4.9	44	4.7	48	3.7	96	5.7	241	4.8	10
Moving close to new workplace	102	9.4	57	6.1	64	5	113	6.8	336	6.7	7
Moving close to family and relatives	40	3.7	52	5.6	86	6.7	142	8.5	320	6.4	9
Obtaining public house	114	10.5	136	14.5	167	12.9	48	2.9	465	9.3	4
Neighbors Dissatisfaction	23	2.1	21	2.3	33	2.6	41	2.5	118	2.4	11
Suffer with parking	16	1.5	19	2	17	1.3	20	2	72	1.4	12
Total	1087	100	935	100	1291		1672	100	4986	100	-

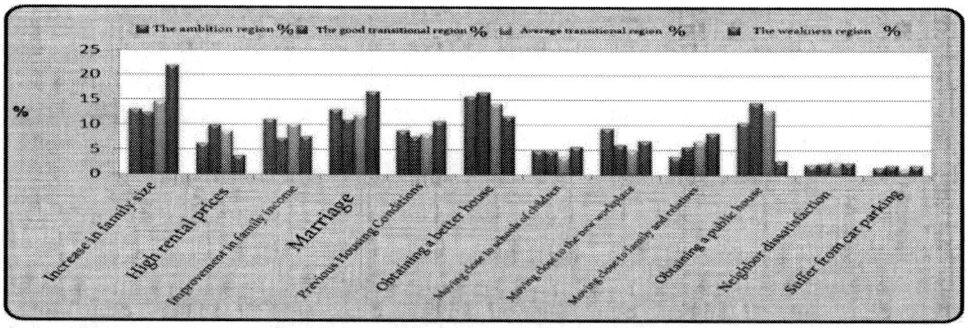

Fig. 2. Motivations of residential mobility in Basra city for the year 2023.

2.2.2 High Rental Prices of Housing

The rental prices of housing have increased recently for several reasons such as the rise in land prices within Basra city due to the large population growth, the lack of residential plots of land and a low income level. Furthermore the high cost of building materials and the absence of direct and indirect government support in building houses by granting loans to citizens, the homeowner's search for the highest profit, such as the population density in the city, and the location in relation to the desired areas that the population tends to move towards. Thus, the factor of the rise in housing rents has encouraged the process of housing mobility, especially neighborhoods characterized by high housing rents or moving towards neighborhoods whose rents are average or lower than those previous housing in proportion to the level of income earned by the head of the house, so the process of movement continued for these reasons and some individuals who formed a small family, i.e., newly married couples.

By examining Table (2) and Figure (2), this motivation ranks eighth with a percentage of (6.7%) of the total motivations for residential mobility in the study area. The good transitional region ranks highest with a percentage of (9.9%), the average transitional region comes second with a percentage of (8.7%), then the ambition region comes third with a percentage of (6.7%), and finally the weakness region with a percentage of (3.8%) of the total moving families.

By analyzing the data in Table (3), which shows the matrix of motivations for mobility, it is noted that the variable i.e. high rental prices is associated with the variable of increase in the number of family size with a positive relationship that is not strong (0.608), a positive relationship with it (0.699) with the improvement in family income, a weak positive relationship (0.579) with the marriage, a negative relationship (0.416) with the housing conditions , a medium positive relationship (0.612)with obtaining a better dwelling, a weak positive relationship with the high rental prices and moving of close to the schools of children (0.533), and a weak positive relationship (0.598) with moving close to the new workplace. Moving close to family and relatives has a weak positive relationship (0.512), a positive relationship(0.511) with obtaining a public plot of land or house, a negative relationship with lack of harmony with neighbors with a relationship (0.417), a medium positive relationship (0.653) with suffering with parking spaces.

2.2.3 Improvement in family income

The improvement in family income affects the process of residential mobility, because changing the residence requires an increase in the income, especially from the lower neighborhood to a higher. Families with a medium or low income are looking for a residence that matches the income they acquire monthly and meets the requirements of that family. As for families with a high income, they are looking for a house that may be larger in terms of area and better than the previous residence, or they are looking to leave the neighborhood in which the family lives to a more upscale neighborhood and distinguished locations in the city, such as the movement of residents from neighborhoods of the weak region and the middle transitional region to neighborhoods of the good transitional region and the region of ambition. In Table (2) and Figure (2), it is noticed that the improvement in family income ranks sixth with a percentage of (9%) of the total motives for residential mobility. The ambition region has the highest percentage of (11.1%), the intermediate transitional region (10.2%), the weakness region (7.7%), and the good transitional region with a percentage of (7.3%). By analyzing the correlation matrix for the variables of residential mobility in the city of Basra shown in Table (3), it becomes clear that the improvement in family income, as its value is (1), is linked to the variable of the increase in the number of family members with a strong positive relationship (0.886), a medium positive relationship (0.699) with the variable of the high rental prices of the previous residence, an a strong positive relationship (0.899) with (marriage), a weak positive relationship (0.572)with the housing conditions, a strong positive relationship with both obtaining a better housing (0.891) and moving close to the schools of children(0.980), a positive relationship with move close to the new work place (0.831), a negative relationship with the move close to family and relatives (0.675), a weak negative relationship with obtaining a public plot of land or housing (0.345), a weak positive relationship with neighbors dissatisfaction (0.501), and a medium positive relationship (0.717) with suffering with parking spaces.

2.2.4 Change of marital status (marriage)

Married of the head of household generally affects the family stability and increases residential mobility. Unlike if the head of the household is a widower, divorced or unmarried, in which case the process of residential mobility decreases [4]. Thus, marriage is one of the most important motives affecting residential mobility the reasons that push residents to move searching for an independent home of its own, tranquility, comfort and reassurance and peace.

By analyzing Table (2) and Figure (2) It is clear that this motive status occupies the third place with a percentage of (13.6%) of the total motives for residential mobility. As for the residential regions in the city, the weak region obtains the highest percentage among the other regions with a percentage of (16.7%), the ambition region the second place with a percentage of (13.1%), the medium transitional regions came in third with a percentage of (11.9%), and the lowest percentage was for the good transitional region with a percentage of (11%).
By analyzing the correlation matrix of the residential mobility variables in Basra city in Table (3), it becomes clear that the variable marriage is (1) and is associated with the variable of increasing in the number of family with a strong positive relationship (0.965), a weak positive relationship (0.579) with the variable of the high rental prices of the previous residence, a strong positive relationship (0.899) with the improvement in the family income, a medium positive relationship (0.688) with the housing conditions, a strong positive relationship with obtaining a better residence (0.911), and with a strong positive relationship between changing marital status through marriage and the variable of proximity to the place of education of children (0.936), with a positive relationship with moving close to the new workplace (0.887), a positive relationship (0.795) with the variable of moving close to family and relatives, a very weak relationship with obtaining a public plot of land or housing (0.315), a weak positive relationship (0.481) with neighbors dissatisfaction, and a medium positive relationship (0.644)with suffering with parking.

2.2.5 The previous house conditions

The area of the housing unit greatly affects the phenomenon of residential mobility, because there is a general desire among most movers to live in large-area homes, consistent with the size of the family and their desire to appear with a distinguished social status, and to compensate for the state of crowding that they suffered from in their small homes, especially the old residential neighborhoods, as most of those housing units have become uninhabitable due to their old age and that their rehabilitation or reconstruction is an economically unfeasible process, so families began to search for better housing in order to move to it and live in it, so they worked Some families sell their homes at attractive prices because their locations are distinguished, and many of them have been converted to other uses such as commercial use, but we find that some other residents of the old residential neighborhoods in the city prefer to stay in the same home after its reconstruction because of its proximity to the central commercial area, in addition to the nature of the good relations with the neighbors that extend for many years, especially in some neighborhoods of the old city, the homes located in Al-Abbasi, Al-Ashar, and others[5].

By examining Table (2) and Figure (2), it is found that the motive for residential mobility, obtaining better housing, ranked fifth with a percentage of (9.1%), and the weak region came first with a percentage of (10.7%), and the ambition region came second with a percentage of (8.8%), then the medium transitional region with a percentage of (8.2%), and finally the good transitional region with a percentage of (7.7%). By analyzing the correlation matrix of the residential mobility variables in Basra city in Table (3), it becomes clear that the variable of old and narrow housing reached a value of (1) and is associated with a medium positive relationship with the variable of the increase in the number of family members (0.706), and the variable of the high rental cost of the previous housing is associated with a negative relationship (0.416), and is associated with the improvement in the income level with a weak positive relationship (0.572), and is associated with the change in social status (marriage) with a positive relationship (0.688), and a not-so-strong positive relationship with the variable of obtaining a better housing (0.515), and is associated with a weak positive relationship between old and narrow housing and the variable of proximity to the place of children's education (0.560), and is associated with a medium positive relationship The proximity of the residence to the new workplace (0.704), and is positively related to the variable of the residence's proximity to family and relatives (0.787), and is negatively related to obtaining a plot of land or government housing (0.299), as well as the lack of harmony with neighbors with a negative relationship (0.371), and is negatively related to the suffering with parking lots and their proximity to sources of pollution for the previous residence (0.356).

2.2.6 Obtaining better housing

Obtaining better housing is related to the increase in the level of income. The higher the level of income, the more families look for a better housing that is compatible with the changes that accompanied the economic conditions of the family; because the increase in individual income means an increase in the standard of living that motivates individuals to achieve or satisfy their personal needs and desires, and the goal of owning suitable healthy housing comes at the forefront of these needs, we find that most residents of developing countries are unable to own suitable housing for them because their income is less than (100) dollars annually [6]. It is noted from Table (2) and Figure (2) that the motive for residential mobility, obtaining better housing, ranked second with a percentage of (14.1%) of the total motives for mobility in the study area. The good transitional region obtained the highest percentage of (16.5%), the ambition region came second with a percentage of (15.7%), then the average transitional region came third with a percentage of (14.2%), and finally the weakness region with a percentage of (11.7%).

By analyzing the correlation matrix of the residential mobility variables in Basra city in Table (3), it becomes clear that obtaining a better housing has a value of variable (1) and is associated with a positive relationship with the variable of the increase in the number of family members (0.880), with the variable of the high rental cost of the previous housing is associated with a medium positive relationship (0.612), a positive relationship (0.891)with the improvement in the income, with the change in social status (by marriage) with a strong positive relationship (0.911), a not so strong positive relationship with the variable of the age and narrowness of the housing area (0.515), with a strong positive relationship with the variable of proximity to the place of education of the children (0.920), a positive relationship with the moving close to the workplace (0.833), with a moderate positive relationship with the proximity of the residence to family and relatives (0.658), with a negative relationship with obtaining a public plot of land or housing (0.321), with a negative relationship (0.333) with neighbors dissatisfaction , with the variable of suffering with parking lots with a moderate positive relationship (0.635).

2.2.7 Moving close to the schools of children

Moving close to the place of education of children is one of the reasons for its superiority and the preparation of the necessary environment for their success. has become one of the motives for the residential movement that affected Basra city, especially families that were outside the city limits or in the districts and sub-districts, as they moved temporarily to one of the neighborhoods and then moved for the second or third time near educational institutions [7]. This is clear from the analysis of Table (2) and Figure (2) We find that the motive of proximity to the place of children's education occupied the tenth place with a percentage of (4.8%) of the total motives for residential mobility. As for the residential regions in the city, the region of weakness occupied the highest percentage among the other regions with a percentage of (5.7%), while the region of ambition obtained second place with a percentage of (4.9%), while the good transitional region came in third with a percentage of (4.7%), and finally came the average transitional region with a percentage of (3.7%).

By analyzing the correlation matrix of the residential mobility variables in Basra city, it becomes clear that the variable of proximity to the place of children's education is associated with the variable of the increase in the number of family members with a strong positive relationship (0.910), a positive relationship that is not strong (0.533 with the variable of the high rental cost of the previous residence, with the variable of improvement in the income level with a positive relationship (0.890), a strong positive relationship (0.936) with the marriage, a weak positive relationship with the Previous housing conditions (0.560), a strong positive relationship with obtaining a better residence (0.920), with a positive relationship with moving close to the work place, new (0.846), and with a medium positive relationship with the variable of moving close to family and relatives (0.685), a negative relationship (-0.215)with obtaining a public plot of land or housing, a negative relationship (-0.326) with neighbors dissatisfaction, and a medium positive relationship (0.660) with suffering from parking lots.

2.2.8 Moving close to the workplace

The proximity of the residence to the workplace is one of the most important motives for residential mobility in Basra city, as the distance affects the employee due to the fatigue they suffer from because of the road. in addition to bearing the financial consequences, as well as the problem of traffic congestion in which a lot of time is wasted in order to reach the workplace, as long distances and severe congestion affect the employee's health through morning tension for fear. These problems are reduced if the workplace is close to the residence, and the distance of arrival is short, which makes the employee more productive.

Thus, he also becomes able to deal with work problems calmly as the distance of his arrival to the residence is short. This is what urge many families to change their place of residence and move to another residence. It is clear from the analysis of Table (2) and Figure (2) that this motive occupies the seventh place with a percentage of (6.7%) of the rest of the motives for residential mobility. In terms of the residential regions in the city, the region of ambition occupies the highest percentage among the other regions with a percentage of (9.4%), while the region of weakness obtains the second place with a percentage of (6.8%), while the good transitional region came in third with a percentage of (6.1%), and the medium transitional region with a percentage of (5%).

It is clear from the analysis of the correlation matrix of the residential mobility variables in the city of Basra that moving close to the new workplace has been positively correlated with the variable of the increase in the number of family members (0.871), a weak positive relationship (0.598) with the high rental prices of the previous residence, a positive relationship (0.831) with the improvement in the family income, a strong positive relationship (0.887)with changing the social status (by marriage), and a medium positive relationship with the housing conditions (0.704), a strong positive relationship with each of obtaining a better residence (0.833), moving close to the new workplace and to the place schools of children(0.846), Moreover it is positively related to the variable of moving close to family and relatives (0.722) and negatively related to obtaining a public plot of land or housing (0.322). It is also negatively related to the variable of neighbors dissatisfaction (0.340), and weakly related to the variable of suffering with car parking (0.563).

2.2.9 Moving close to family and relatives

When most people think about moving and choosing another residence, they tend to move closer to their family or relatives in order to meet some of the social needs of the individual, including social communication or assistance. When the spatial distance between families is short, it indicates the depth of the high relationships and kinship between them. Likewise, the role of proximity has an impact in revitalizing social relations and their development and what results from them in terms of similarity in social and cultural values. This confirms, on the other hand, that rural social values, customs and traditions are still

It is clear in the community of Basra city that residents who belong to the same social class prefer to coexist with each other, because there are common denominators between them, as residents feel safe when living with a group of residents similar to them in social class, and this is what can be observed in the neighborhoods of some Iraqi cities, as some residents prefer to live near family, relatives and friends even if they have the financial capabilities that enable them to buy a better home in neighborhoods desirable to residents, but they want to move towards the neighborhoods where their relatives or some friends are present, and this phenomenon lies more in moving within the neighborhood itself [8].

It is clear from the analysis of Table (2) and Figure (2) that the motive of moving close to family and relatives occupies the ninth place with a percentage of (6.4%) of the total motives for residential mobility. In terms of residential regions in the city, the region of weakness occupies the highest percentage among the other regions with a percentage of (8.5%), the average transitional region obtains the second place with a percentage of (6.7%), the good transitional region in third place with a percentage of (5.6%), and finally the region of ambition with a percentage of (3.7%). By analyzing the correlation matrix of the residential mobility variables in Basra city, it becomes clear that the moving close to family and friends is associated with the variable of the increase in the number of family size with a positive relationship (0.797), a weak positive relationship (0.512) with the high rental cost of the previous home, a medium positive relationship with it (0.675) with the improvement in the family income, a strong positive relationship (0.833) with the change in social status (marriage), a positive relationship with the housing conditions (0.787), a medium positive relationship with obtaining a better home (0.658), moving close to family and friends and to the schools of children (0.685). It is also positively related to the variable of moving close to the new workplace (0.722), negatively related to obtaining a public plot of land or housing (0.219) and suffering with car parking (0.475), and a weak positive relationship (0.573) with neighbors dissatisfaction.

2.2.10 Obtaining a public plot of land or housing

The local government contributes to solving the housing problem not only through distributing residential land but also by providing the infrastructure for this residential land. Plots of land have been distributed in successive years and to different categories of the population. When preparing the basic plans for cities, the areas allocated for residential use are usually determined based on the current actual need and the expected future housing demand for the target year. These residential plots are distributed and various services are provided for them within a specific time schedule. Any delay in this will lead to confusion in this sector and the exacerbation of the problem over time [9].

It is clear from the analysis of Table (2) and Figure (2) that the motive for obtaining a plot of land or government housing occupied the fourth place with a percentage of (9.3%) of the remaining motives for residential mobility. As for the residential regions in the city, the good transitional region came in first place with the highest percentage among the other regions with a percentage of (14.5%), the average transitional region obtained the second place with a percentage of (12.9%), the ambition region came in third place with a percentage of (10.5%), and finally the weakness region came in second place with a percentage of (2.9%).

By analyzing the correlation matrix of the residential mobility variables in Basra city, it becomes clear that obtaining a public plot of land or housing is negatively related to the variable of the increase in the family size (401), a weak positive relationship (0.511) with the high rental cost of the previous housing, negatively related to the improvement in the family income (0.345), the housing conditions (0.299), a positive negative relationship (0.315) with the marriage and moving to the schools of Children (0.215). Moreover there is a negative relationship with obtaining a better housing (0.321), moving close to the new workplace (0.322), to family and relatives (0.219), neighbors dissatisfaction (0.433), and with the variable of suffering with parking (0.294).

2.2.11 Neighborhood dissatisfaction

Families have an ambition to establish good relations with neighbors; since that has a psychological impact on the individual's life. It also has a social significance when choosing a residence, it is the product of the authentic social environment from which those neighboring families originate, as well as the cultural level to which they belong, in addition to the economic situation and its direct role in giving the living characteristic to the residents of that residential neighborhood, and then that relationship will differ according to these factors, as we often notice in popular residential neighborhoods that the relationship between neighbors is strong and based on cooperation and social communication, unlike upscale residential neighborhoods where the state of cooperation and communication is limited or little, depending on the circumstances of the residents of the residential neighborhood in terms of their view of the relationship with the neighbor [10] . so the state of disharmony may be It is counterproductive and forces some families to move from their current residence to another residence, and perhaps to another neighborhood, due to being far from their neighbors. It is clear from the analysis of Table (2) and Figure (2) that we find that the motive of disharmony with neighbors occupied the eleventh place with a percentage of (2.4%) of the rest of the motives for residential mobility, and the medium transitional region acquired the highest percentage among the other regions with a percentage of (2.6%) of the rest of the motives for residential mobility in the city of Basra, while the weak region obtained the second place with a percentage of (2.5%), while the good transitional region came in third with a percentage of (2.3%), and the lowest percentage was for the ambitious region with a percentage of (2.1%).

By analyzing the correlation matrix of the residential mobility variables in Basra city, it is noted that neighbors dissatisfaction is associated with the variable of the increase in the number of family members with a weak positive relationship (0.515),
a negative relationship (0.417) with the variable of the high rental cost of the previous residence, a weak positive relationship (0.501)with the improvement in the family income, the change in social status (marriage) with a negative relationship (0.481), a negative relationship with the housing conditions (0.317), obtaining a better residence (0.333), moving to the close to schools of children (0.326), to the new workplace (0.340), obtaining a public plot of land or housing (0.433) and suffering with car parking (0.411). It is also related with the variable of moving close to family and relatives with a weak positive relationship (0.573).

2.2.12 Suffering from car parking spaces and their close to pollution sources

Transportation is the movement of goods and people from one place to another. Interest in transportation has increased in recent years in interaction with other cities. Transportation is a standard for measuring similarities and differences between cities, and movement is an indicator for measuring the connection between cities and neighborhoods. Therefore, transportation plays an important role in the movement of rural settlements from one place to another, in addition to its effect on the expansion of one use at the expense of another use, or changing the pattern of that use.

Basra is one of the cities that witness multiple traffic jams, as a result of the increase in the number of means of transportation that contributed to the increase in the amount of gas emissions. The growth in the number of cars has become one of the most important urban problems in the city, especially after the change in the economic situation and the entry of large numbers of cars, as the number of cars in the city of Basra reached (99165) cars in the year (2023) [11].

It is clear from the analysis of Table (2) and Figure (2) that the motive of suffering from parking lots and their proximity to pollution sources occupied the last place with a percentage of (1.4%) of the remaining motives for residential mobility. As for the residential regions in the city, the weak region and the good transitional region occupied the highest percentage among the other regions with a percentage of (2%) for each of them. As for the ambition region, it obtained the second place with a percentage of (1.5%), while the medium transitional region came last with a percentage of (1.3%).

By analyzing the correlation matrix of the residential mobility variables in Basra city, it becomes clear that the suffering with car parking has been associated with the variable of the increase in the number of family members with a medium positive relationship (0.660), the high rental cost of the previous residence (0.653). However it has a positive relationship (0.717) with the improvement in the family income, the change in social status (marriage) (0.644) and obtaining a better housing (0.635). on the other hand it is associated with a negative relationship with the housing conditions (0.356), obtaining a public plot of land or housing (0.294), moving close to family and relatives (0.475) and neighbors dissatisfaction (0.411). The variable of moving to the close to schools of children (0.660) is linked with a weak positive relationship to the moving close to the new workplace (0.563).

2.3 The impact of future trends in population and area growth on residential mobility in Basrah city

Population is one of the most important main indicators in anticipating the future of cities and estimating their need for space and various services, because the development of population size and growth rates contribute directly and fundamentally to estimating those needs based on the individual share of them according to the local planning standards approved by the Iraqi Ministry of Planning. Thus, estimating the size of the city population and basic needs contributes to creating an advanced urban environment with its distinctive characteristics [12]. Therefore, identifying the number of residents and the number of families will lead us to estimate the rates of residential mobility, which is directly proportional to the availability of the number of housing units allocated to accommodate the increasing population size with the increase in areas that have a greater possibility for the family to obtain a suitable plot of land to live in, which encourages an increase in residential mobility in the future. To estimate the future population size, the researcher set the target year, which is the year (2037). The population is calculated and then the expected number of the city of Basra can be found according to the population projection equation. Therefore, it is necessary to know the future population numbers through the population estimates process and the area of the city to develop strategic plans for housing and future movements for them in a manner that is consistent with achieving a spatial balance between the existing services and the number of residents in the city, which is as follows:

2.3.1 Estimating the population of Basra city for the period (2023-2037)

Population dynamics expressed by population growth is one of the most important demographic phenomena that various studies, especially urban geography, seek to identify its components, calculate its rates and the possibility of predicting them. the extent of its impact on urban space, as it has a large and effective role in drawing a future picture of what the city's content and general functions will be like, and then forming an initial conception about the size of the investments that will be employed in the city and the approximate amount of space needed for the growth of land uses in it, by identifying the possibilities of population growth in it as an economic, administrative and cultural center. Therefore, planning for sound growth achieves harmony between the population and the urban environment in which it lives until the distant future, and this is done by planning the future growth of the population and its relationship to the urban situation and the current and future service aspect [13].

According to table (3) the population of Basra city for the base year (2023) reached (1,355,639) people. According to the high growth rate (2.6%), it will rise in the target year to (1,889,485) people for the year (2037). With the prediction of population growth in the city and its spatial expansion, it is required that it has a sufficient amount of space and urban land uses such as residential use, represented by the establishment of new neighborhoods that alleviate population pressure in the city center and contribute to controlling the process of residential movement by developing appropriate solutions to reduce pressure on community services in the future and in a manner consistent with its future needs.

2.3.2 Estimating the future area of Basra city

Estimating the area of urban use and the area it will occupy in the future on the city's land and its geographical area represents an important planning step to identify the spatial requirements for urban and service land uses in particular, even if in an estimated form. It will provide appropriate data for the expansion of service centers in the direction occupied by the population of the city's land, in addition to identifying

Table 3. Expected population and area in Basra city for the period (2023-2037).

Year	Expected area in hectares	Projected population	Expected number of families	Expected number of housing units
2023	8858.3	1355639	225939.8	149171
2024	9073.0	1388174	231362.4	384502.3
2025	9290.8	1421491	236915.1	438007.8
2026	9513.8	1455606	242601	491423.3
2027	9742.1	1490541	248423.5	544838.8
2028	9975.9	1526314	254385.6	598254.3
2029	10215.3	1562945	260490.9	651669.8
2030	10460.5	1600456	266742.7	705085.3
2031	10711.5	1638867	273144.5	758500.8
2032	10968.6	1678200	279700	811916.3
2033	11231.9	1718477	286412.8	865331.8
2034	11501.4	1759720	293286.7	918747.3
2035	11777.5	1801953	300325.6	972162.8
2036	12060.1	1845200	307533.4	1025578
2037	12349.6	1889485	314914.2	1078994

The spatial direction that the city will take during its future expansion operations, on the other hand[i], so the annual growth rate and the share of one person were relied upon according to the foundations and standards of urban planning so that the required areas can be determined until the target year. Through the data in Table (3), it becomes clear that the city area in (2023) has reached (8858.3) hectares, and will increase in the future as a result of the increase in the number of residents. Moreover, it is expected that the area of the city will reach (12349.6) hectares in (2037) and an increase in the area of up to (3469) hectares over the base year (2023). Thus, the increase in the number of residents, especially in cities, will generate great pressure on their lands, therefore it is necessary to work on developing future plans to increase the number of residential neighborhoods. This is done by expanding the city horizontally and vertically, which will be reflected in an increase in the number of units. The residential area will become an attraction factor for many families that will encourage the process of residential movement in the city as a result of the changes occurring in the population numbers and economic and social developments. Hence, it needs continuous policies to regulate this growth, otherwise its growth will be random without meeting the different needs of the population.

2.3.3 Estimating the Future Number of Housing Units in Basra city until the year (2036)

Residential mobility has become one of the modern problems that the study area suffers from. Therefore, future plans must be developed for a future vision in light of which the number of housing units that the city needs is determined. This is one of the basic steps of great importance in confronting the housing crisis that the city of Basra suffers from in light of the continuous increase in the number of residents compared to the number of housing units, especially since there is a close relationship between the size of the population and the size of housing units, as a result of an increase in the number of expected families and the number of housing units in the future. Therefore, these expectations aim to identify the future need for housing units that will contribute to the process of residential mobility in a greater way than it is at the present time. Hence, predicting the number of Housing units are consistent with the number of families in the future, which has an important role in creating a kind of balance between them, and thus encourages residential mobility according to a future vision. According Table (3), the number of housing units reaches (149,171) housing units for the year (2023), and it will rise to reach (1,078,994) housing units for the year (2037). Therefore, by knowing the number of housing units, it is possible to identify the current housing deficit and estimate the future need for the number of required housing units.

According to the data, the average family is (6.2) individuals which reflects the links Social in urban areas with limited income, as a result of the change in the relative economic situation and social customs and traditions represented in moving to a new residence immediately after marriage. These reasons and others will increase residential mobility in the future.

3 Results

1. The study shows that the residential mobility is motivated by a group of drivers in which the stimulating movements amounts for 54.8%, adaptation movements 41.4%, whereas involuntary movements 3.8% from the total number of the motivations of residential mobility in Basrah in 2023.
2. The motivations play a significant role in urging people to move within the residential districts in the city in which an increase in family size amounts for 16.3%, obtain a better home 14.1% whereas a change in martial state (marriage) 13.6%, a suffer from car parking 1.4% from the total number of the movers in Basrah.

3. By using methods of statistical analysis (SPSS.28), it is evident that there is a between the motivations of residential mobility and a set of independent variables (an increase in family size, obtain a better housing, land or general housing)
4. The study utilizes modern methods to determine implications of residential mobility such as methods of geographical information systems (GIS) and statistical analysis (SPSS.V.28). The results show that there will be an increase in population, households, area, number of housing units and residential neighborhoods.

4 Conclusions

1. It is obligatory for the government to provide population with proper housing especially for those with lower income and solve population pressure in some areas. This is done by determining the actual demand for housing market to create programs, plans and studies that contribute in changing the residential reality by opening new horizons for urban growth trends in Basrah city.
2. Working out an organizational planning for the housing realty in the city that cope with the population growth that is resulted from the weak structural features that will be responsible for solving housing reality.
3. The government and other institutions should attempt to develop housing sector, set future plans to expand the city and determine the residential mobility, its features and directions as it is one of the important pillars in the process of urban planning of the city.
4. It is necessary to pay a special attention for city's urban planning, restructure its layout basics that corresponds to the current population growth.
5. Studying reasons that lie behind the residential mobility in Basrah neighborhoods and its implications for housing i.e. effects of residential mobility on rise of prices of residential land and rented homes in order to find suitable solutions.

Creating database that include information and data regarding residential mobility in Basrah and applying scientific methods in order to plan to develop Basrah city that help in predicting size of population for providing services according to followed planning criteria and taking into account the housing planning of all neighborhoods to provide basic services that are necessary population and land uses.

References

1. D. Tampubolon, M. Irsan, Harlen et al., "Preference of Spatial Mobility and Reside in the Suburbs Indonesia", International Journal of Sustainable Development and Planning 17, 1479–1487 (2022), DOI: 10.18280/ijsdp.170512
2. M. Jiménez-Espada, Ju. M. V. Naranjo, F. M. M. García, "Identification of Mobility Patterns in Rural Areas of Low Demographic Density through Stated Preference Surveys", Applied Sciences (Switzerland) 12, 10034 (2022), DOI: 10.3390/app121910034
3. A. M. Abdulwahab, G. M. Aboud, I. Obaid et al., "Regional Transportation's Planning Implications for Large Cities", IOP Conference Series: Earth and Environmental Science 1232, 012054 (2023), DOI: 10.1088/1755-1315/1232/1/012054
4. A. Cortez-Ordoñez, A. B. Tulcanaza-Prieto, "Are We Back to Normal? A Bike Sharing Systems Mobility Analysis in the Post-COVID-19 Era", Sustainability 16, 6209 (2024), DOI: 10.3390/su16146209

https://doi.org/10.1051/e3sconf/202563307013

5. L. Lin, T. Chen, "Activity Space and the 15-Minute Neighborhood: An Empirical Study Using Big Data in Qingdao, China", Journal of Transport and Land Use 16, 1–17 (2023), DOI: 10.5198/jtlu.2023.2159
6. G. Lesteven, D. Cissokho, P. Pochet et al., "Daily Mobility in Urban Peripheries: The Role of Clandestine Taxis in Dakar, Senegal", Sustainability 14, 6769 (2022), DOI: 10.3390/su14116769
7. T. Echiburú, R. Hurtubia, Ju. C. Muñoz, "Role of Perceived Satisfaction and the Built Environment on the Frequency of Cycle-Commuting", Journal of Transport and Land Use 14 (1) (2021), DOI: 10.5198/jtlu.2021.1826
8. R. Wolniak, "European Union Smart Mobility–Aspects Connected with Bike Road System's Extension and Dissemination", Smart Cities 6, 1009–1042 (2023), DOI: 10.3390/smartcities6020049
9. R. E. Saber, D. M. Dief-Allah, N. O. Ahmed et al., "The Correlates of Smartphone Use for Transportation Purposes in Egypt and Lebanon", Frontiers in Sustainable Cities 6 (2024), DOI: 10.3389/frsc.2024.1340372
10. M. M. Monteiro, J. De Abreu E Silva, N. Afonso et al., "Residential Location Choice and Its Effects on Travel Satisfaction in a Context of Short-Term Transnational Relocation", Journal of Transport and Land Use 14 (1) (2021), DOI: 10.5198/jtlu.2021.1952
11. A. Alonso, A. Monzón, Ia. Aguiar, A. Ramírez-Saiz, "Explanatory Factors of Daily Mobility Patterns in Suburban Areas: Applications and Taxonomy of Two Metropolitan Corridors in Madrid Region", ISPRS International Journal of Geo-Information 12, 16 (2023), DOI: 10.3390/ijgi12010016
12. S. Yasir, T. Ahmad, A. Enam, "An Analysis of the Harassments and Challenges Faced by the Public Transport Users in a Developing Country of South Asia", Frontiers in Built Environment 8 (2022), DOI: 10.3389/fbuil.2022.1049121
13. H. Masoumi, "Residential Location Choice in Istanbul, Tehran, and Cairo: The Importance of Commuting to Work", Sustainability 13, 5757 (2021), DOI: 10.3390/su13105757

E3S Web of Conferences **633**, 07014 (2025) https://doi.org/10.1051/e3sconf/202563307014
GASES 2025

Spatial analysis of air pollution concentration (radon, humidity, VOCS) in the homes of the Mitham Al-Tammar neighborhood in Kufa

Salam Ali Hassan[1,*], *Raghad* Mazhar Abdul Hamid[1], *Saja* Nafea Yousif[2], and *Ali* Saeed jassim[3]

[1]University of Baghdad, College of Education Ibn Rushd for Human Sciences, Baghdad, Iraq
[2]National center for population and demographic studies, University of Baghdad, Baghdad, Iraq
[3]University of Kufa, Al-Najaf, Iraq

Abstract. The study aims to determine the concentrations of radon, humidity, and CO2 in residences within the Mitham Al-Tammar neighbourhood of Kufa, utilising an Airthings 2960 Wave Plus device in the living rooms of each household. The concentration of radon varied, with a minimum value of 13 ± 3.9 Bq/m³ and a maximum value of 90 ± 5.8 Bq/m³, yielding an average rate of 35.8 ± 4.889 Bq/m³. The concentrations of total VOCs in residences ranged from 69.26 ppb to 810.14 ppb, with a mean of 321.4855 ± 40.4. The humidity percentages exhibited a minimum concentration of 51.68% and a maximum of 122.97%, averaging 76.594 ± 4.298. The concentrations of volatile organic compounds and humidity in the majority of residences exceeded the thresholds established by the World Health Organization and the US Environmental Protection Agency. This serves as proof that these dwellings lack adequate indoor air quality.

1 Introduction

Air contamination is developing into a significant issue which is damaging atmospheric condition of particulars and physical condition of them. Air pollution varies neighborhood by neighborhood through the city. Now the industrial development and absence of waste treatment causes contamination of the environment; causing to the dissemination of the maladies. Our residents' people are dealing with various air pollution and the buildings are polluted due to urban pollution [1]. Air quality is one of the most important factors in human health and environmental impact. Air Pollution is one of the essential environmental problems around the world, especially in most Iraqi cities. Carbon dioxide, humidity, and radon, among other air pollutants requiring careful and continuous study [2] can have a major impact on the health of residents inside homes. Our research is being performed to spatially analyze the distribution of these psychosocial agents in the houses of the Maytham Al-Tammar area of Kufa, as one of the urban and living areas that have a high degree of change and pollution in the air through expansion and activity witnessed in the region, and which we will explain as in the following sections. Apart from these, pollutants such as tin-lead,

* Corresponding author: Salam.A@ircoedu.uobaghdad.edu.iq

lead oxide, unburned hydrocarbons, nitrogen dioxide and sulfur dioxide pollutants significantly affect environmental aspects in both urban and rural areas, due to the increase of serious environmental problems that threaten life in general and human health in particular.

In this study [3] we are identifying the effect of spatial air circulation in the sampling point on the concentration of air pollution in indoor ai, therefore spatial analysis techniques such as point pattern analysis, Nearest neighbor distance, join count [4]. Two of the most common pollutants in the home include carbon dioxide, which is produced when the fuel burned in many domestic and industrial processes, and radon, a radioactive natural gas that is one of the invisible home polluters that may build up in our homes over the years and threaten the health of those who live there. Humidity as an environmental factor interacts with air pollutants and increases the susceptibility to respiratory diseases.

Maytham Al-Tammar neighborhood is one of the residential neighborhoods located in the city of Kufa, which is affiliated with the city of Najaf. It is considered one of the most important and prosperous neighborhoods in the city. The neighborhood is located west of the center of Najaf city. It is one of the urban areas that witnessed recent development in Najaf. The neighborhood maintains many services such as schools, health centers and public facilities, which are some of the characteristics that make it a comfortable place to live. The northern side of the neighborhood contains several other residential neighborhoods such as the Al-Amir neighborhood and the Al-Jabal area. It can also be easily reached from the east and the center of Najaf city. The area is also famous for its proximity to the main roads that connect the city of Najaf to different areas within the Najaf Governorate.

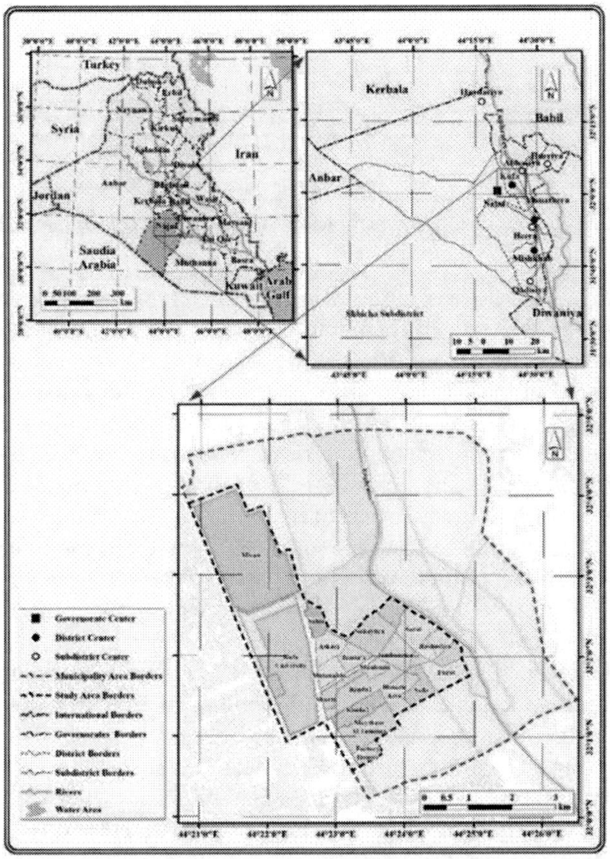

Fig. 1. The location of the city of Kufa in the province of Najaf, Iraq.

 https://doi.org/10.1051/e3sconf/202563307014

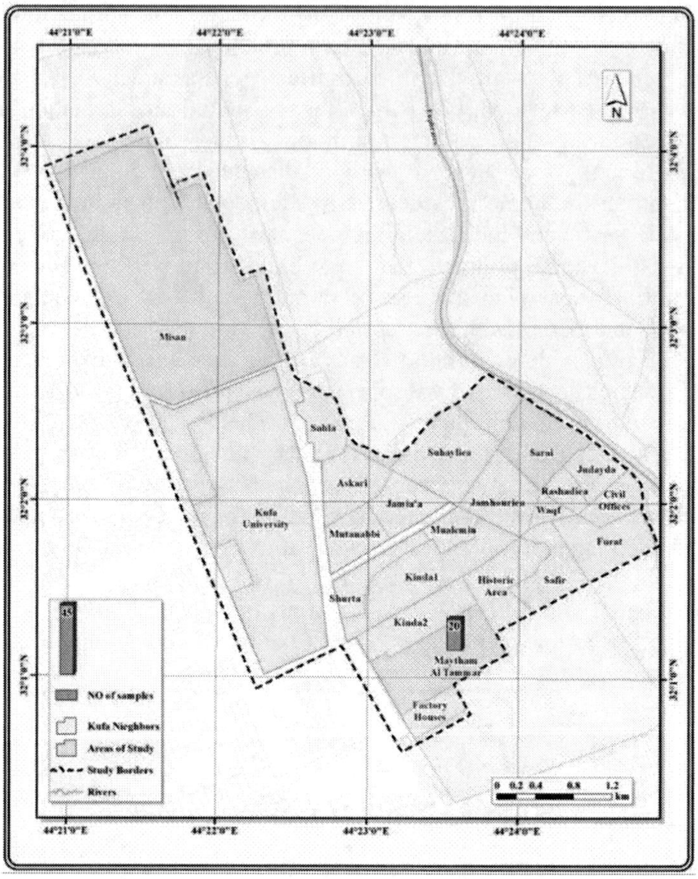

Fig. 2. Spatial distribution of air pollution samples in the Mitham Al-Tammar neighborhood in the city of Kufa.

A sample of 1% of the houses in the Mitham Al-Tammar neighborhood in Kufa, Najaf Governorate, Iraq, was selected. In each house, (radon, carbon dioxide, and humidity) were measured using a device Air things 2960 Wave Plus in the living room of each house.The apparatus was situated in the living room of each residence for a duration of 24 hours.The Airthings 2960 Wave Plus is a device developed to assess indoor air quality, encompassing radon concentrations. This is the method for measuring radon levels.The Wave Plus employs a sensor that identifies radon by ionisation. Radon gas experiences radioactive decay, releasing alpha particles. The device's sensor quantifies these particles to assess radon levels in the atmosphere.Ongoing Surveillance It consistently assesses radon concentrations in the atmosphere and delivers temporal data on radon levels. This assists users in comprehending the variations in radon levels in various locations within the home or structure.Radon Measurement Unit The gadget generally presents radon concentration in picocuries per litre (pCi/L) or Becquerels per cubic meter (Bq/m^3), which are conventional measures for quantifying radioactivity levels in the atmosphere.Battery-Operated and Wireless This is a battery-powered gadget that connects to the Airthings app via Bluetooth, facilitating convenient monitoring and access to past data.The Airthings 2960 Wave Plus facilitates ongoing monitoring and instantaneous alarms, enabling customers to safeguard their indoor air against elevated radon levels.The device can measureHumidityThe device includes humidity sensors that are usually capacitive or resistive sensors The sensors respond to variations in properties of the sensing material due to the environmental humidity[5].

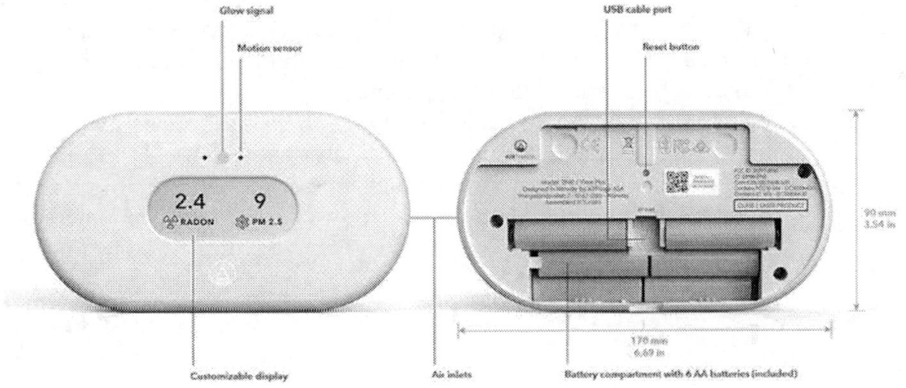

Fig. 3. Air things 2960 wave plus.

2 Results and discussion

Table 1. Results of concentrations of Concentration Rn in homes (Bq/m³) Mitham Al-Tammar neighborhood in Kufa.

House No	House Code	Concentration Rn in homes (Bq/m³)
1	M	90 ± 5.8
2	M	18 ± 10.8
3	M	36 ± 8
4	M	14 ± 5.6
5	M	75 ± 14
6	M	13 ± 3.9
7	M	47 ± 8.5
8	M	21 ± 16.8
9	M	32 ± 9.6
10	M	19 ± 13.3
11	M	51 ± 7.5
12	M	62 ± 5.6
13	M	22 ± 10.8
14	M	31 ± 4.8
15	M	20 ± 14
16	M	56 ± 3.9
17	M	17 ± 8.5
18	M	22 ± 19.8
19	M	55 ± 9.6
20	M	15 ± 7.5
Average		35.8 ±4.889

Table No. (1) shows the results of radon gas in the houses of the Mitham Al-Tammar neighborhood in Najaf Governorate, where the lowest concentration value was 13 ± 3. Bq/m³ and the highest value was 90 ± 5.8 Bq/m³ at a rate of 35.8 ±4.889 .The results showed that some houses were below the limit set by the WHO [6] and most of them were below the limit set by the WHO [6]. Figure 4 shows the differences and compares the results with the WHO. The reason for the difference in radon gas concentrations in homes is due to Geology and soil composition. Some areas have more uranium or thorium in rocks and soil, which release more radon into the air.

Home ventilation: Homes with effective ventilation systems (like open windows or forced ventilation) have the potential to lower radon concentrations, whereas poorly ventilated homes will be at risk for retaining higher amounts of radon. Construction and structures: Houses built on basement or underground foundations have more exposure to radon gas from the soil, resulting in radon concentration in more homes.

Seasonality: Seasonal changes in outside air and radon concentrations may occur since radon concentrations can increase due to closed windows and the use of heating systems during cold weather and Local aspects. And certain areas are simply more prone to high radon levels based on weather or wind patterns that allow radon to build up in homes [7].

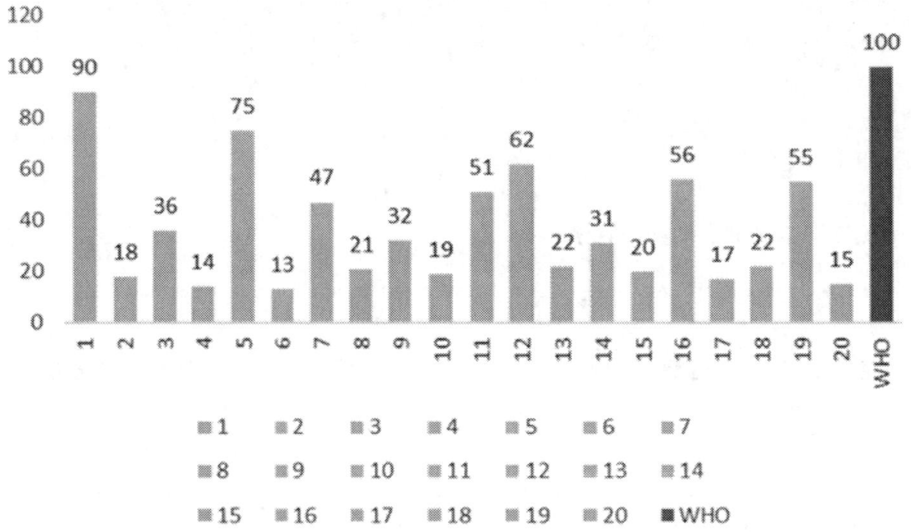

Fig. 4. Comparison of the results of Concentration Rn (Bq/m³) in homes Mitham Al-Tammar neighborhood in Kufa with the global limit of the WHO.

Table 2. Results of concentrations of Concentratio Vocs (ppb) Mitham Al-Tammar neighborhood in Kufa.

House No	House Code	Concentration Vocs (ppb)
1	M	810.14
2	M	766.12
3	M	200.29
4	M	450.10
5	M	140.10
6	M	170.22
7	M	326.77
8	M	540.87
9	M	460.60
10	M	668.53
11	M	69.26
12	M	286.33
13	M	524.57
14	M	185.50
15	M	141.22
16	M	172.53
17	M	106.80
18	M	386.14
19	M	342.37
20	M	191.25
Average		321.4855 ±40.4

As we can see in Table 2, the concentrations of total VOCs in homes were in the range of 69.26 ppb to 810.14 ppb with an average of 321.4855 ± 40.4. Most of the results were found to be higher than the WHO limit. Figure 5: Comparison of VOC results in homes in the Maytham Al-Tammar neighborhood in Kufa with the WHO global limit. Emissions from building materials, household products [8] combustion sources (such as tobacco smoke and unventilated heaters), and office equipment are responsible for very high levels of VOCs in homes [9].

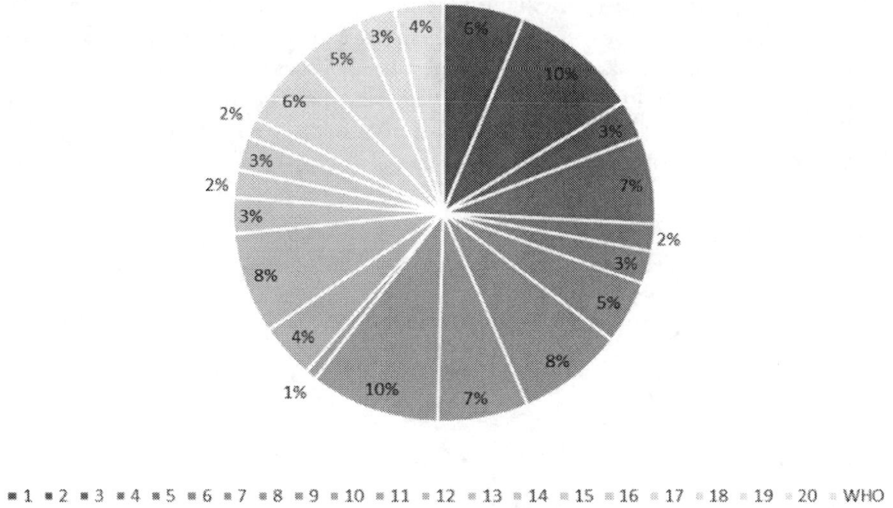

Fig. 5. Comparison of the results of VOCS in homes Mitham Al-Tammar neighborhood in Kufa with the global limit of the WHO.

Table 3. Results of concentrations of Concentration Humidity % Mitham Al-Tammar neighborhood in Kufa.

House No	House Code	Humidity %
1	M	51.68
2	M	75.12
3	M	68.48
4	M	71.42
5	M	72.59
6	M	96.36
7	M	65.57
8	M	82.95
9	M	52.65
10	M	29.92
11	M	83.87
12	M	92.61
13	M	122.97
14	M	63.93
15	M	97.90
16	M	71.87
17	M	79.24
18	M	84.22
19	M	90.30
20	M	78.23
Average		76.594 ±4.298

Table No. (3) shows the results of Humidity % in the houses of the Mitham Al-Tammar neighborhood in Najaf Governorate, where the lowest concentration value was 51.68 % and the highest value was 122.97 % at a rate of 76.594 ±4.298 .The results showed that some homes were above the limit set by the World Health Organization.The humidity concentration in homes differs due to a combination of factors that impact the indoor humidity distribution. First and foremost [10,11] we focus on ventilation: adequate ventilation will distribute the humidity evenly throughout the room, while the humidity will only accumulate in certain locations (for example, in the kitchen and bathroom) during a lack of ventilation. "It is also very region specific; homes that are in very humid areas (like coastal areas) will have greater variances between levels of humidity.Temperature also plays a role in how much humid air can hold moisture, with warm air having more capacity to retain moisture than cold air; this can also contribute to different humidity levels depending on the seasons or the different temperatures inside the home.

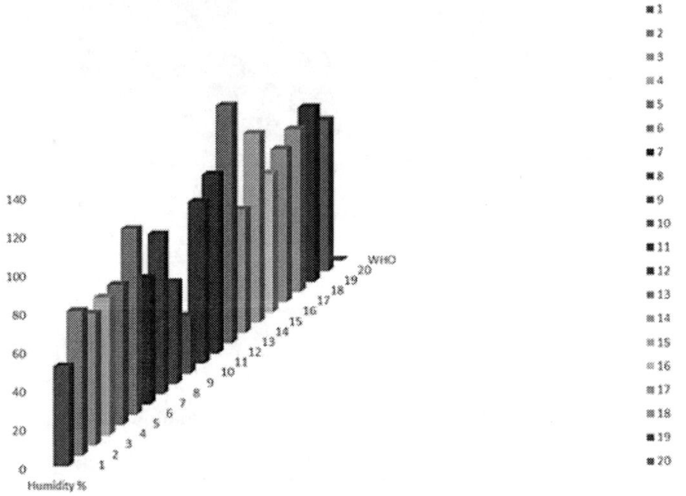

Fig. 6. Comparison of the results of Humidity % in homes Mitham Al-Tammar neighborhood in Kufa with the global limit of the WHO.

3 Conclusion

The levels of radon gas, volatile organic compounds and humidity were measured in a number of houses in the Maytham Al-Tammar neighborhood in Najaf. The results of radon gas were within the permissible limit, and the results of volatile organic compounds and humidity for most houses were higher than the limit approved by the World Health Organization and the US Environmental Protection Agency. This is considered evidence that these houses do not enjoy good indoor air quality.

References

1. S. Ozcan, N., K. M. Cubukcu, *The relationship between urban air pollution and urban planning decisions* (2018)
2. S. S. Nayif, E. J. Mohammed, A. S. Jassim, A. A. Abojassim, "Estimation of the radiological risk due to Radon-222 exposure in dwelling of Al-Hussainya District, Karbala Governorate," Appl. Environ. Soil Sci. **2024(1)**, 6511029 (2024)

3. S. Vardoulakis, E. Giagloglou, S. Steinle, A. Davis, A. Sleeuwenhoek, S. Galea, K. Dixon, J. O. Crawford, *Indoor exposure to selected air pollutants in the home environment: a systematic review* (2020)
4. H. T. Abed, S. A. Rahi, R. A. Hayder, A. S. Jassim, A. A. Abojassim, "Evaluation of concentration (Rn222, VOCs, CO2) for dust storm samples in Najaf Governorate, Iraq," in EPJ Web Conf. **318**, 06007 (2025)
5. World Health Organization, WHO ambient air quality database, 2022 update: status report (World Health Organization, 2023)
6. A. A. H. Mohsen, A. A. M. Kadhim, H. S. M. Hassan, J. M. H. Al Mosawy, A. J. M. Nasrawi, "The neurological manifestations in children with COVID-19 infection," Childs Health **19(5)**, 298–303 (2024)
7. A. A. H. Mohsen, A. J. Manji, J. M. Hashim, R. M. Obeid, "The significance of Helicobacter pylori infection as a cause of recurrent abdominal pain in children," Curr. Pediatr. Res. **22(1)**, 82–87 (2018).
8. J. Sexton et al., "Thermal and humidity effects in buildings during different seasonal cycles," J. Build. Perform. **9** (2018)
9. C. Johansson et al., "Moisture control strategies and the role of insulation in reducing indoor humidity levels," Build. Res. Inf. **44** (2016)
10. N. G. Yousif, M. G. Yousif, A. A. U. Mohsen, H. S. El-Bakaa, M. H. Younise, A. N. Altimimi, A. M. Hassan, "Prospective single center analysis of outcome stem cells transplants in patients with cerebral palsy," Pol Merkur Lekarski **51**, 339–345 (2023)
11. R. Al-Tameemi, Societies Woven in Reeds: Reconstructing the Cultural Landscape of Nippur and the Iraqi Marshlands Through the Lens of John H. Haynes's Photographic Catalog, PhD thesis (University of Cincinnati, 2019)

 https://doi.org/10.1051/e3sconf/202563307015

Spatial analysis of indoor air pollution in the housing of Al-Abbasiya district in Najaf governorate

Muhammad Dawoud Salman *

Ibn Rushd College of Education, University of Baghdad, Baghdad, Iraq

Abstract. The research aims to know the concentration of carbon dioxide and volatile organic compounds in the indoor air in the homes of Al-Abbasiya district in Najaf, concentration of carbon dioxide ranged where the lowest concentration value was 466.81 ppm and the highest value was 1204.88 ppm at a rate of 875. ±82.88 the concentrations of the total volatile organic compounds in the houses were in the range of 59.26 ppb to 477.82 ppb with an average of 273. 40±29. Most of the results obtained in this study were higher than the standard set by the World Health Organization for air quality and the US Environmental Protection Agency.

1 Introduction

Indoor air quality is one of the leading public health issues. Related factors that influence indoor air quality from indoor operation and outdoor circumstances Air pollution in urban areas is one of the biggest challenges restricting the sustainability of urban living, which makes research in indoor air pollution necessary [1]. A city dwelling almost certainly registered high indoor air pollution levels for CO2 and Total VOCs, Months of research showed the air we breathed was killing millions. In the Twenty-first century more focus was placed on protecting the environment in regard to control of indoor and outdoor pollutants and gas emissions. In various closed environments, such as houses, schools, hospitals, and other buildings, IAP has been establishing its research as a significant concern for environmental and public health [2, 3]. Indoor pollutant sources vary purposefully, and they include outdoor, human, biological, furnishings and materials used for building construction [4]. The abstract of the new study states: The present study aims at conducting statistical spatial assessment, related to the air pollution within the housing in the district of Al-Abbasiya, in the Najaf city of Iraq. The study aims to clarify the indoor air pollution problem in some residential buildings in the Al-Abbasiya junction in order to contribute to the general effort to identify the indoor air pollution state in different city buildings and then conduct questionnaire for the construction materials used in the construction of these houses [5]. Thus, this paper attempts to (a) investigate CO2 and volatile organic compounds (VOCs) levels in the dwelling of the subjects; it also attempts to assess indoor air pollution by

* Corresponding author: mohammed.d@ircoedu.uobaghdad.edu.iq

comparing CO2 and VOC concentrations found in the dwelling fieldwork with established standard thresholds [6].

2 Materials and methods

2.1 Study area

Al-Abbasiya is an Iraqi city, administratively part of the Kufa district, located inside Najaf Governorate, with an area of 2.39 square kilometers. Al-Abbasiya city is located 14.8 kilometers east of Najaf, the capital of Najaf Governorate. It is situated between latitudes 32.04° and 32.03° and longitudes 44.26° and 44.29° [7, 8].

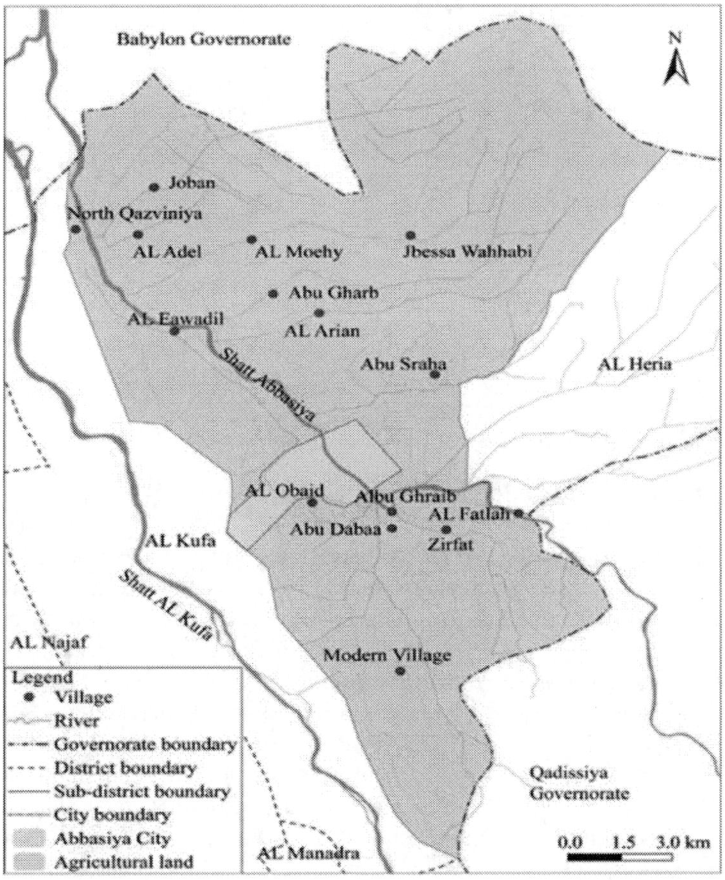

Fig. 1. Map of study area.

2.2 The practical part

20 houses were selected in the Al-Abbasiya district in Najaf Governorate, and in each house (CO2, VOCs) were measured using a device we AIR THINGS technique. The equipment was positioned in the living room of each residence for a whole day. VOC Measurement: Airthings devices quantify VOC levels with a sensor equipped with a reactive layer that alters its electrical resistance upon exposure to diverse airborne substances [9]. This sensor detects variations in VOC concentrations through resistance changes, yielding an estimated VOC

level. CO2 Measurement: Airthings devices utilize NDIR (non-dispersive infrared) technology for CO2 detection. The NDIR sensor measures CO2 levels by quantifying the infrared light absorbed by CO2 molecules in the atmosphere, yielding an accurate CO2 measurement.

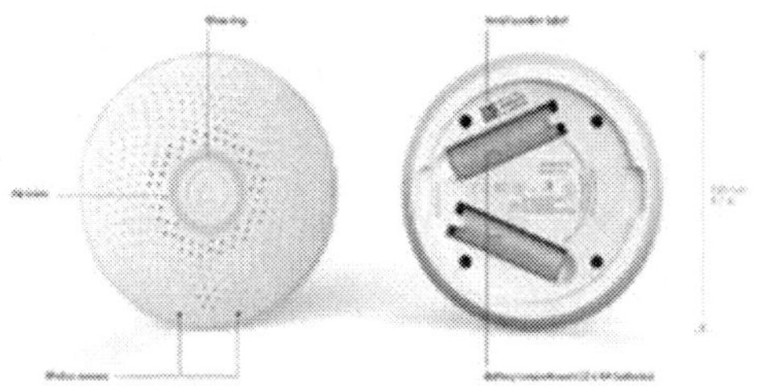

Fig. 2. Airthings.

3 Results and discussion

Table 1 shows the carbon dioxide results in the housing of Al-Abbasiya district in Najaf Governorate, where the lowest concentration value was 466.81 ppm and the highest value was 1204.88 ppm at a rate of 875. ±82.88. The results showed that some homes were below the limit set by the US Environmental Protection [10]. Agency and most were above the limit set by the US Environmental Protection [10]. Figure 3 shows the differences and compares the results with EPA. The reason for the difference and increase in carbon dioxide concentration in homes is due to:Human Respiration Humans are the largest source of CO_2 found within buildings due to respiration. Being the prone heater, oxygen gets inhaled and CO_2 exhaled, so the accumulation of this gas can be considerable in closed-off, less-aired areas. High levels of CO_2 often signal inadequate ventilation per person in a space. The contribution of various household activities and appliances to elevated concentrations of CO_2 can be realized considering the combustion activities. Gas stoves, fireplaces and other fuel-burning devices release CO_2 into the indoor environment. Even homes with attached garages may have higher CO_2 levels from the ingress of vehicle exhaust, Poor ventilation: Adequate ventilation is critical to dilute indoor pollutants, including CO_2. If ventilation systems are insufficient or malfunction, CO_2 emitted from occupants and combustion processes can cumulate and degrade indoor air quality. Impact of Elevated Indoor CO_2A significant body of research has associated high indoor CO_2 concentrations with impaired cognitive function and a variety of health symptoms, highlighting the importance of achieving adequate ventilation. Building Design and Energy Efficiency Measures. Many modern buildings prioritize energy efficiency in their design, resulting in tightly sealed structures that limit the exchange of air with the outside environment. And while those designs save energy, they can also lead to increased indoor CO_2 levels if there isn't sufficient ventilation incorporated.

Managing the delicate balance between energy efficiency and indoor air quality is critical, as too low ventilation would result in the build-up of CO_2 and other indoor pollutants. External environmental conditions could also affect indoor CO_2 concentrations. For example, in urban areas the infiltration of ambient air with elevated outdoors CO_2 in it can enhance indoor air concentrations. On the other hand, where outdoor CO_2 levels are already low, good ventilation can help lower indoor levels. For this reason, some understanding of the local environmental context is necessary for evaluating levels of indoor CO_2.

Table 1. Results of concentrations of co2 in the housing of Al-Abbasiya District in Najaf governorate.

Home number	Home code	Co2 (ppm)
1	r1	1204.88
2	r2	521.51
3	r3	810.27
4	r4	1104.48
5	r5	792.36
6	r6	914.58
7	r7	693.75
8	r8	939.17
9	r9	1146.36
10	r10	1188.34
11	r11	1029.17
12	r12	502.88
13	r13	606.80
14	r14	842.55
15	r15	757.103
16	r16	558.71
17	r17	952.02
18	r18	833.26
19	r19	466.81
20	r20	637.00
Average		875. ±82.88

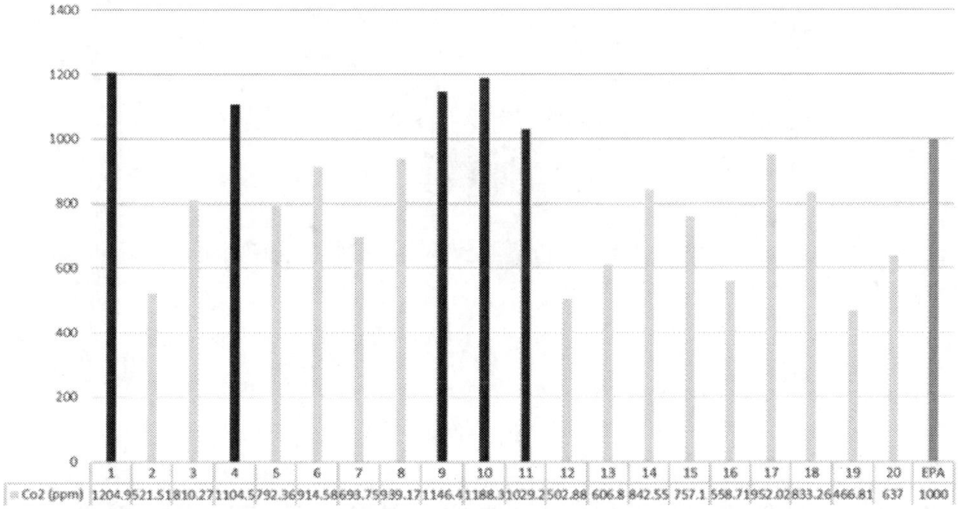

Fig. 3. Comparison of the results of CO2 in the Housing of Al-Abbasiya District in Najaf Governorate with the global limit of the EPA.

As we can see in Table No. 2, the concentrations of the total volatile organic compounds in the houses were in the range of 59.26 ppb to 477.82 ppb with an average of 273. 40±29.The World Health Organization has set that limit. The emissions of Building Materials, Household Products, combustion sources (e.g., tobacco smoke and unvented heaters), and office equipment are responsible for very high levels of VOCs in homes. Under such conditions, emissions from coatings worsen including, but not limited to, VOCs and nitrates from multiple end-of-life processes.

Table 2. Results of concentrations of VOCs (ppb) in the housing of Al-Abbasiya District in Najaf governorate.

Home number	Home code	Vocs (ppb)
1	r1	218.54
2	r2	477.82
3	r3	379.39
4	r4	319.39
5	r5	120.20
6	r6	120.19
7	r7	426.13
8	r8	439.77
9	r9	320.50
10	r10	368.63
11	r11	59.26
12	r12	486.45
13	r13	424.59
14	r14	145.55
15	r15	131.82
16	r16	132.53
17	r17	186.90
18	r18	286.14
19	r19	244.37
20	r20	181.05
Average		273. 40±29

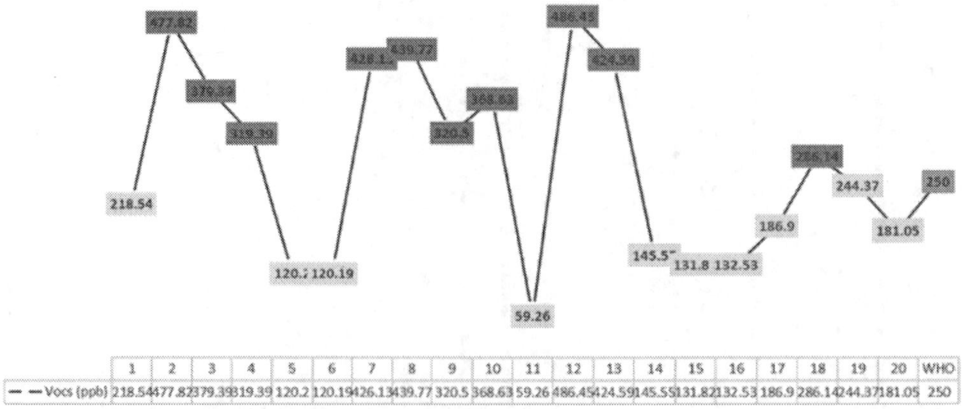

Fig. 4. Comparison of the results of VOCS in the Housing of Al-Abbasiya District in Najaf Governorate with the global limit of the WHO.

4 Conclusion

Carbon dioxide and volatile organic compound concentration levels were measured in a number of houses in the Al-Abbasiya district housing in Najaf, the results of which were higher than the limit approved by the World Health Organization and the US Environmental Protection Agency, this is considered evidence that these residences are not experiencing good indoor air quality.

References

1. C. F. Campos, et al., "Assessment of genotoxicity of air pollution in urban areas using an integrated model of passive biomonitoring," Environ. Pollut. **355**, 124219 (2024)
2. S. S. Nayif, E. J. Mohammed, A. S. Jassim, A. A. Abojassim, "Estimation of the radiological risk due to Radon-222 exposure in dwelling of Al-Hussainya District, Karbala Governorate," Appl. Environ. Soil Sci. **2024(1)**, 6511029 (2024)
3. S. Vardoulakis, E. Giagloglou, S. Steinle, A. Davis, A. Sleeuwenhoek, S. Galea, K. Dixon, J. O. Crawford, *Indoor exposure to selected air pollutants in the home environment: a systematic review* (2020)
4. N. S. Ozcan, K. M. Cubukcu, "The relationship between urban air pollution and urban planning decisions," Asian J. Qual. Life **3**, 181–192 (2018), DOI: 10.21834/ajqol.v3i11.134
5. J. Javed, *Portable solution of air quality in indoor environments: process life cycle of indoor air quality*, Master's thesis (Tampere University, Finland, 2019)
6. L. R. López, et al., "CO2 in indoor environments: from environmental and health risk to potential renewable carbon source," Sci. Total Environ. **856(2)**, 159088 (2023), DOI: 10.1016/j.scitotenv.2022.159088
7. Y. Ma, S. Wang, X. Duan, "Recent advances in direct gas–solid-phase photocatalytic conversion of CO2 for porous photocatalysts under different CO2 atmospheres," Chem. Eng. J. **455**, 140654 (2023), DOI: 10.1016/j.cej.2022.140654
8. M. Hosoda, E. D. Nugraha, N. Akata, R. Yamada, Y. Tamakuma, M. Sasaki, S. Tokonami, "A unique high natural background radiation area–dose assessment and perspectives," Sci. Total Environ. **750**, 142346 (2021)
9. R. Al-Tameemi, Societies Woven in Reeds: Reconstructing the Cultural Landscape of Nippur and the Iraqi Marshlands Through the Lens of John H. Haynes's Photographic Catalog, PhD thesis (University of Cincinnati, 2019)
10. S. A. Gignoux-Wolfsohn, et al., "New framework reveals gaps in US ocean biodiversity protection," One Earth **7(1)**, 31–43 (2024), DOI: 10.1016/j.oneear.2023.12.014

Neighborhood segregation and urban resource distribution: a path toward sustainability in Ana

Mohammed Hammad Abdulateef*

University of Baghdad, College of Education for Human Sciences Ibn Rushd, Baghdad, Iraq

Abstract. Spatial disparities in resource distribution across urban neighborhoods are a critical factor influencing sustainable urban development. This study investigates the spatial variation of neighborhood characteristics in the city of Ana, focusing on the distribution of social and infrastructural resources. Utilizing geospatial techniques and tools, including Anaconda and Matplotlib, we analyze data from 2024 to assess the accessibility of key urban resources in neighborhoods of varying sizes and population densities. The findings reveal that larger neighborhoods with better infrastructure and social networks exhibit higher resilience and greater access to sustainable resources. In contrast, smaller neighborhoods face compounded challenges, including resource scarcity and spatial segregation, which hinder their contribution to broader urban sustainability goals. These results underscore the need for policymakers to address spatial inequalities and promote balanced resource distribution to support the long-term sustainability and resilience of urban communities.

1 Introduction

In developing countries, where planning is limited, the equitable distribution of educational services is becoming increasingly important. Most educational services are unevenly distributed based on current land use, leading to large educational disparities: some areas are underserved, while others are overserved. Neighborhood planning models are crucial to ensuring a balanced distribution of services. However, this distribution model is highly dependent on exploration, especially when education professionals face the challenges of dealing with complex contexts.

The neighborhood concept has been widely used in planning literature and practice, referring to a spatially divided area in urban and rural contexts.

Many of the principles of the spatial variation model of education in the neighborhoods of Ana City stem from the new urban expansion, and it is argued that this model prioritizes possibilities such as classical means of distribution and transportation, in contrast to previous neighborhood planning models. The spatial variation of educational characteristics is logically justified to deal with the population numbers and the urban growth associated with it in today's urban expansion. The denser the neighborhood, the greater the volume and types of services it requires; therefore, the greater the urban intervention affected by it. This research is based on the nature of the study area (Ana) on the principles of proximity, density,

* Corresponding author: Mohammed.hammed@ircoedu.uobaghdad.edu.iq

diversity, and digitalization complementing the value of the city, all of which are necessary to measure the spatial variation of educational characteristics of Ana city. A study was conducted in nine selected neighborhoods of Ana City to address these principles using physical, functional, and socio-economic aspects. When talking about schools, see Figure 1.

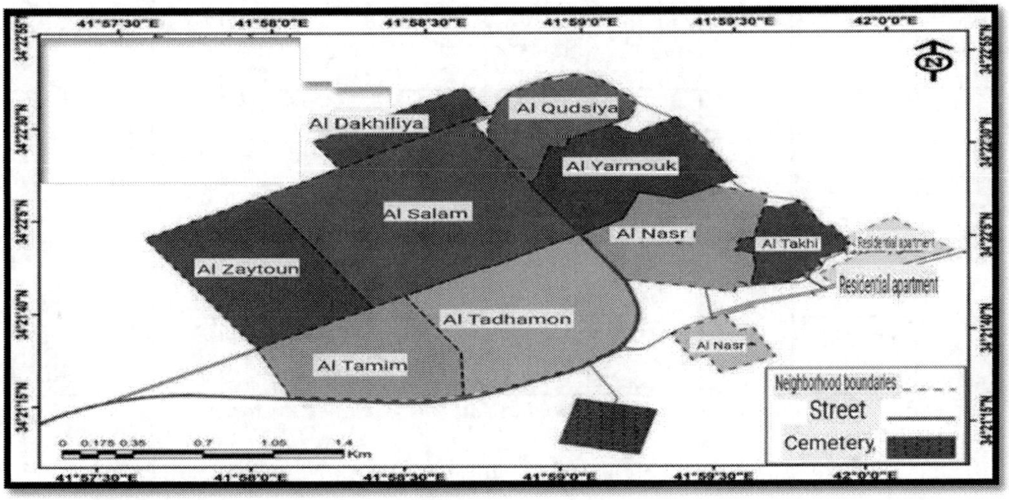

Fig. 1. The neighborhoods of the city of Ana.

2 Materials and methods

Technical development has made analysis using modern programs and techniques an important matter. Here, libraries of programs were used [1], namely Anaconda and Matplotlib. These libraries are based on the python program as an important basis for reaching new results [2,3].

Table 1. Population characteristics distribution table (educational level) for the city of Ana, age of 12 and older.

Neighborhood	Population %	He doesn't read or write %	R^2	% Reads and writes	R^2	% Maj Elementary	R^2	% Medium	R^2	% Preparatory	R^2	% Diploma	R^2	% Bachelor's	R^2	Postgraduate studies	R^2
Brotherhood (Tamimi)	9.6	10	1 +	13.3	1 +	8.1	1 -	8.6	1 -	5.9	1 -	4.8	1 - -	4.1	1 -	5.7	1 +
Victory	22.5	40	1 +	6.6	1 -	15.7	1 -	16.1	1 -	27.1	1 +	29.8	1 +	25.7	1 +	24.8	- 1 +
Yarmouk	16.2	10	1 -	13.3	1 +	9.3	1 -	12.4	1 -	9.9	1 -	10.5	1 -	9.2	1 - -	11.0	1 -

https://doi.org/10.1051/e3sconf/202563307016

Al-Qadisiyah	18.2	0	1 -	20	1 +	18.1	1 -	15.5	1 -	13.2	1 -	16.3	1 -	21.6	1 +	15.5	1 -
Peace neighborhood	19.2	0	1 -	3.3	1 -	15.7	1 ع	17.3	1 -	19.2	-	11.5	1 -	13.9	1 -	15.9	- 1 -
Solidarity	8.6	0	1 -	16.6	1 +	10.5	1 +	8.0	1 -	8.6	-	7.6	1 -	14.4	1 +	8.6	- -
Tamim (formerly crossing)	1.9	10	1 +	6.6	1 +	7.0	1 +	4.3	1 + -	3.3	1 +	9.6	1 + -	5.1	1 +	6.1	1 +
Interior	0.17	0	1 -	6.6	1 +	3.5	1 +	4.3	1 +	2.6	1 +	2.8	1 + -	1.0	1 +	3.0	1 +
Al-Zaytoun	3.2	30	1 +	13.3	1 +	17.5	1 +	13.0	1 -	9.9	1 +	6.7	1 - -	4.6	1 +	8.9	1 +

Source: Researcher's work based on data from the Ana Municipality Directorate.

Figure 2 shows a 3D visualization of demographic data, showing the relationship between population percentage, percentage of bachelor's degree holders, and percentage of diploma across neighborhood (Fig.2)]3[

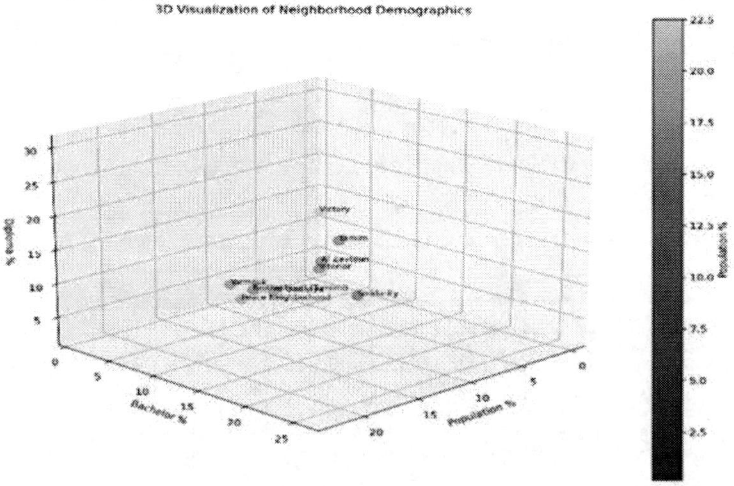

Fig. 2. A different 3D visualization with a more dynamic perspective and additional features.

Successfully enhanced 3D visualizations, showing educational metrics across neighborhoods using a surface graph and a scatterplot with solid lines [4].

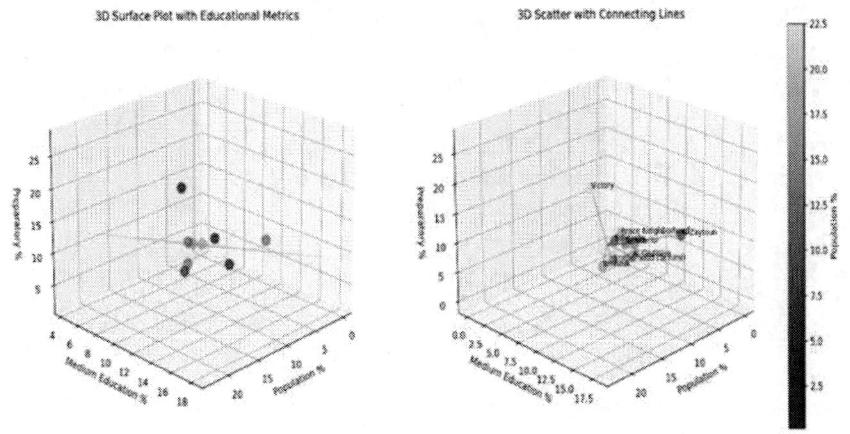

Fig. 3. 3D visualizations, showing educational metrics across neighborhoods.

2.1 Library anaconda

The table provides data on various neighborhoods, focusing on population percentages and educational attainment levels. Below is a breakdown of the columns and their meanings:

Columns in the Table:

Neighborhood: Names of neighborhoods (e.g., Brotherhood, Victory, Yarmouk, etc.).

Population %: The percentage of the population in each neighborhood.

He doesn't read or write %: The percentage of individuals who are illiterate.

Reads and writes %: The percentage of individuals who can read and write but have no formal education.

Major Elementary %: The percentage of individuals who have completed elementary education.

Medium %: The percentage of individuals who have completed middle school education.

Preparatory %: The percentage of individuals who have completed preparatory school education.

Diploma %: The percentage of individuals who have obtained a diploma.

Bachelor's %: The percentage of individuals who have obtained a bachelor's degree.

Observations

Victory Neighborhood has the highest population percentage (22.5%) and the highest percentage of illiteracy (40%).

Interior Neighborhood has the smallest population percentage (0.17%) and relatively low educational attainment across all levels.

Al-Qadisiya has the highest percentage of individuals with a bachelor's degree (21.6%).

Solidarity Neighborhood has a relatively high percentage of individuals who can read and write (16.6%) but lower percentages in higher education levels.

Based on the analysis of the educational data across different neighborhoods, I've outlined both the table contents and created a flowchart for analyzing the data using Anaconda:

- Create visualizations of the educational distribution
- Perform statistical analysis of the data
- Focus on specific neighborhoods or education levels
- Generate Python code to implement the analysis

Observations

Victory Neighborhood has the highest population percentage (22.5%) and the highest illiteracy rate (40%).

Al-Qadisiya has the highest percentage of individuals with a bachelor's degree (21.6%).

 https://doi.org/10.1051/e3sconf/202563307016

Interior has the smallest population percentage (0.17%) and the lowest educational attainment across all levels.

Solidarity has a relatively high percentage of individuals who can read and write (16.6%) but lower percentages in higher education levels [5-7].

To analyze the table and create a 3D flowchart for Anaconda, I will first structure the data into a DataFrame, analyze it for insights, and then use a 3D visualization library to create a 3D flowchart representing the analysis process [8-10].

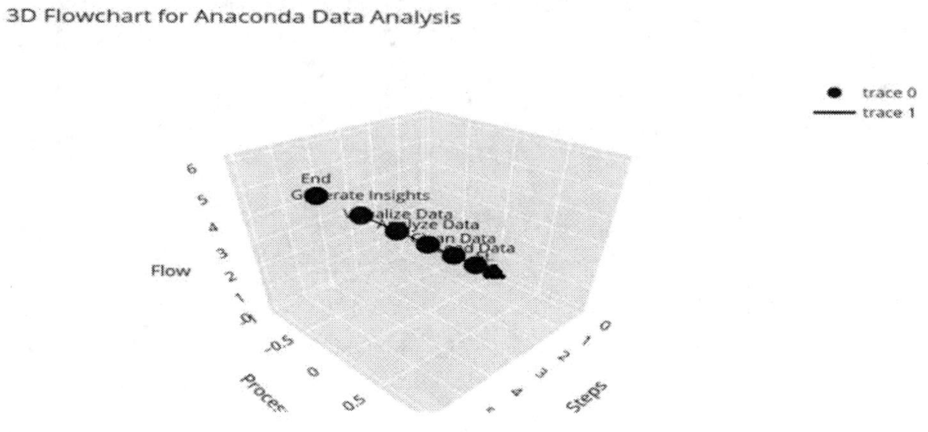

Fig. 4. 3D flowchart for Anaconda data analysis.

2.2 Library matplotlib

Key observations from the data:

Victory has the highest population percentage at 22.5%

Al-Qadisiyah and Peace neighborhood show 0% for "He doesn't read or write"

Victory shows high percentages in higher education categories (Diploma: 29.8%, Bachelor's: 25.7%).

The R^2 columns contain "1+" or "1-" indicators, suggesting some kind of statistical significance or correlation measure.

The smallest population percentage is for the Interior at 0.17%.

The table appears to be analyzing educational attainment levels across different neighborhoods, with various metrics for literacy and educational achievement. Would you like me to focus on any particular aspect of this data or perform a specific analysis?

The table shows educational statistics across neighborhoods, with Victory having the highest population (22.5%) and education levels, while Interior has the lowest population (0.17%).

I will extract the table data from the image, analyze it, and then use Matplotlib to visualize the data.

The stacked bar chart will visualize the distribution of education levels across neighborhoods, providing a clear comparison of literacy and educational attainment. Let us display the chart:

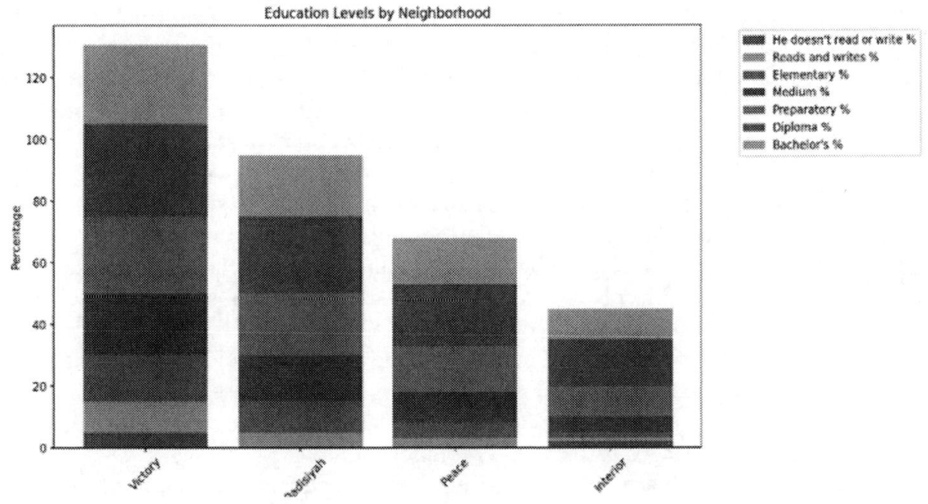

Fig. 5. Education levels by Neighborhood.

A 3D surface plot showing the relationship between neighborhoods, education levels, and percentages:

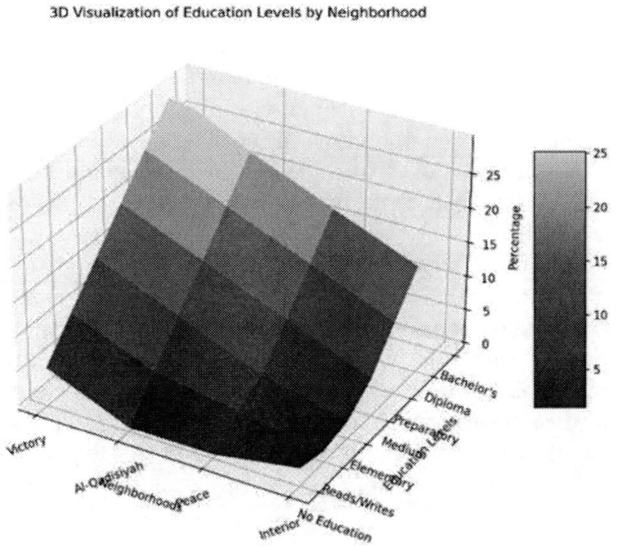

Fig. 6. 3D visualization of educational data.

This 3D visualization allows us to see the educational distribution across neighborhoods with:

X-axis: Different neighborhoods

Y-axis: Education levels

Z-axis: Percentage values

Color gradient: Indicates the percentage intensity

3 Results

It is clear from Table 1, which is related to (the spatial distribution of educational characteristics of the neighborhoods of the city of Ana), that the relationship between the area and the educational distribution characteristics in the neighborhood (Al-Takhi (Al-Tamim), for those who do not read and write, is a strong relationship in a direct direction, i.e. the more the area of the land increases, the more the number of those who do not know how to read and write increases and vice versa, while the relationship with those who know how to read and write for the same aspect was also a direct relationship, as for the relationship with the primary stage, the relationship was completely in an inverse direction, i.e. the more the area of the land increases, the percentage of the primary stage decreases and vice versa, while the intermediate stage came with a complete relationship in an inverse direction as well, as for the preparatory and diploma stages, both came with a strong relationship in an inverse direction, while the relationship between the area of the land and postgraduate studies came with a complete relationship in a direct direction. As for the Al-Nasr neighborhood, the relationship between the area and those who do not know how to read and write was complete in a direct direction, while the relationship between the area and those who know how to read and write and primary, intermediate, and higher was an inverse relationship, while the preparatory and diploma stages were a direct relationship.

The relationship in the Yarmouk neighborhood for the educational stages of illiterate and writing, primary, intermediate, preparatory, Diploma, and higher education was an inverse relationship, while the stage of those who are good at reading and writing was a direct relationship.

As for the Qadisiyah neighborhood for the educational stages of illiterate and writing, primary, intermediate, preparatory, diploma, and higher education was an inverse relationship, while the stage of those who are good at reading and writing was a direct relationship. In the Al-Salam neighborhood, the strength of the correlation for all stages was represented by a complete relationship in an inverse direction.

As for the Al-Tadamon neighborhood, all the correlations were complete but in different directions, as they were in the stages of illiterate, intermediate, and diploma in an inverse direction, the stages of read and write, primary and bachelor's degrees were in a direct direction, while the rest of the stages were equal. In Al-Tamim neighborhood (formerly Al-Obour), the relationship for all stages appears to be complete in a direct direction.

The strength of the relationship also appears in the interior district, where the relationship is completely in a direct direction for all stages except for the stage that concerns those who cannot read and write, where it is completely in the opposite direction. Finally, the relationship appears in the Zeitoun district, where all stages are completely in a direct direction except for two stages, which are the intermediate and diploma stages, where the relationship appears completely in the opposite direction. As for the industrial district, no representation of it appeared.

4 Conclusions

The research data reflects that with the difference in the percentage of educational characteristics of the city of Ana and its spatial distribution in its neighborhoods, which reflects the sustainable human development of those neighborhoods and the cultural level and type of profession and its reflection on the uses of agricultural land, the neighborhoods that are characterized by large areas showed the lowest educational characteristics, and the lowest percentage of the number of learners, and this is clear from Table 1, while small areas recorded the highest percentages of learners in the study area.

Here we can say that the area has an important and major role in determining the level of educational characteristics in the Ana area, as we can, through this study, meet the educational needs of each neighborhood as a neighborhood for the research area.

References

1. A. Abojassim, R. R. Munim, Chapter-4 Hazards of Heavy Metal on Human Health, Multidisciplinary **51** (2020)
2. V. Masindi, P. Mkhonza, M. Tekere, Remediation of Heavy Metals, 419-454 (2021)
3. A. A. Alkufi, M. H. Oleiwi, A. A. Abojassim, Current Analytical Chemistry **20(3)**, 175-182 (2024)
4. T. S. Skogheim, K. V. F. Weyde, S. M. Engel, H. Aase, P. Surén, M. G. Øie, G. D. Villanger, Environment International **152**, 106468 (2021)
5. Y. L. Yu, W. Y. Yang, A. Hara, K. Asayama, H. A. Roels, T. S. Nawrot, J. A. Staessen, Hypertension Research **46(2)**, 395-407 (2023)
6. J. C. Petit, N. Mattielli, J. De Jong, E. Bouhoulle, W. Debouge, P. Maggi, R. Suzanne, Science of the Total Environment 174763 (2024)
7. B. Pumallanqui-Ramirez, J. Li, L. Rueda-Torres, J. Rosales-Rimache, Advances in Public Health **2024(1)**, 5517405 (2024)
8. L. Heireman, P. Van Geel, L. Musger, E. Heylen, W. Uyttenbroeck, B. Mahieu, Clinical Biochemistry **50(18)**, 1317-1322 (2017)
9. M. Del Rio, C. Rodriguez, E. Alvarado Navarro, C. Wekumbura, M. B. Galkaduwa, G. M. Hettiarachchi, C. Sobin, Plos One **18(6)**, e0287406 (2023)
10. S. Shekhawat, V. C. Janu, P. Singh, P. Sharma, K. Singh, Journal of Trace Elements in Medicine and Biology **68**, 126853 (2021)

E3S Web of Conferences 633, 07017 (2025) https://doi.org/10.1051/e3sconf/202563307017
GASES 2025

The impact of air pollution on human health in the city of Baghdad: towards achieving sustainable development goals through adopting smart city solutions

Muna Abdulnabi Abbas*

University of Baghdad, National Center for Population and Demographic Studies, Baghdad, Iraq

Abstract. Air pollution refers to the contamination of indoor or outdoor environments by any chemical, physical, or biological agent that alters the natural characteristics of the atmosphere. This research aims to study the relationship between air pollution in Baghdad city, environmental sustainability and quality of life, as the Iraqi capital suffers from high levels of air pollution due to industrial factors, transportation, and emissions resulting from daily activities. This problem greatly affects the general health of the population, contributing to the spread of respiratory and heart diseases. The research addressed the various dimensions of the impact of air pollution, including its impact on the urban environment and the loss of biodiversity. It also identified how the deterioration of air quality hinders sustainable development efforts and negatively affects the quality of life in the city. In addition, the research reviews innovative strategies to mitigate air pollution, such as improving public transportation, promoting green spaces, and using clean energy technologies. The research concludes that there is an urgent need to understand the complex relationship between air pollution in Baghdad, environmental sustainability and quality of life, which will enable decision-makers and society to take effective steps to improve environmental and health conditions in the city.

1 Introduction

Highlighting the problem of air is one of the most serious small challenges facing many cities around the world, including the city of Baghdad, where Iraqi trends witness high levels of other additions as a result of industrial factors, transportation, which can be about the activities of this pollution does not only affect public health except, but extends to include environmental sustainability and quality of life in general. Many distinctive observations have allowed air participation, but studies that activate the air in the city of Baghdad were limited. We refer to studies close to this research. It is indicated in the privacy of the source of the distribution of carbon aerosol PM2.5 in Baghdad, Iraq, that starting from the partial content in the highlands is the main in calculating air quality in the city of Baghdad [1]. A specific period for taking micrometer diameter fine diameter (PM2.5) from September 2012

* Corresponding author: muna.abbas@ircoedu.uobaghdad.edu.iq

to September 2013 and was analyzed for 24 hours (to know more details about the orphans and the method of searching for results, read the intended research) and the results showed that the main correctness of the aerosol in Baghdad is gasoline and separation engines, and this is attributed to the use of resistant carbon For generators that operate on the components of it, in addition to contributing to burning wood, the study concluded that colored, sulfates, gasoline and diesel are the ones that dominate the atmosphere of the city of Baghdad by 81%, as well as aromatic hydrocarbons, multiple rings to weave are the highest in the city of Baghdad from urban areas in the world.

In another study conducted by [2], the researcher addressed to the Assessment of Particulates and Heavy Metals Concentration Distribution in Ambient Air around Al-Dora Refinery in Baghdad City, Iraq.

In this study, the Dora oil refinery in Baghdad and its surrounding neighborhoods were selected to assess the level of air pollution and its potential health risks. It is believed that the refinery has contributed significantly to the increase in chronic diseases and has a harmful effect on soil and plant life in the surrounding areas. The study aimed to determine the concentrations of heavy metals (chromium, cadmium, lead, and vanadium) and total suspended particles (TSP) in the refinery and its surrounding areas.

The results of the study showed that the amounts of TSP in the air surrounding the Dora refinery were high and exceeded the permissible national and international limits. By comparing the current results with those obtained from previous years, there was an increase in its TSP concentrations for this year. Moreover, the levels of heavy metals were much higher than the local standards and exceeded the international permissible limits. Heavy metals showed an upward trend in the following order: V>Pb>Cd>Cr. Such high concentrations of heavy metals can greatly affect human life and lead to many chronic diseases. The research results also indicated that the AQI score was high and ranged from "very unhealthy" to "very critical", indicating very low air quality. The TSP score ranged from "hazardous" to "very unhealthy", indicating that the air is congested with particulate matter and heavy metals that are hazardous to human consumption.

In the same context, a recent study conducted by [3] dealt with (Study of the Vertical Dispersion Coefficient of Air Pollution Under Different Atmospheric Stability Conditions over Baghdad). This study aims to build the six specialized curves to calculate the vertical dispersion coefficients at different stability and distances from the measurement point, which can be represented by the urban atmosphere of Baghdad. The study concluded that the vertical dispersion coefficient plays an important role in measuring the dispersion of pollutants that cause instability in the atmosphere. The dispersion coefficient is mainly produced by atmospheric disturbances that have great importance in mixing pollutants and spreading them horizontally or vertically. Despite the importance of this topic, studies that have addressed the relationship between air pollution in the city of Baghdad, environmental sustainability and quality of life are still limited. Therefore, the research aims to explore this relationship more deeply and shed light on the different dimensions of the impact of air pollution on the lives of residents. Through this, strategic recommendations can be provided to improve air quality and enhance the sustainability of the city, which contributes to improving the quality of life for all residents.

2 Materials and methods

Air pollution arises primarily from anthropogenic sources, including industrial emissions, vehicular exhaust, inefficient fuel combustion, and unregulated waste processing. These activities release a complex mixture of contaminants—namely fine and coarse particulate matter ($PM_{2.5}$ and PM_{10}), nitrogen oxides (NO_x), sulfur dioxide (SO_2), carbon monoxide (CO), ozone precursors, and volatile organic compounds (VOCs)—into the atmosphere, resulting in the deterioration of ambient air quality. The inhalation of these pollutants has been conclusively linked to increased incidence of respiratory illnesses, such as chronic obstructive pulmonary disease and asthma, as well as cardiovascular pathologies, including ischemic heart disease and stroke. Consequently, the burden of disease attributable to outdoor air pollution places considerable strain on healthcare systems and incurs significant socioeconomic costs.

Beyond human health, atmospheric contamination disrupts ecosystem function by impairing photosynthesis, stunting plant growth, and diminishing biodiversity. Acid deposition resulting from SO_2 and NO_x transforms forest soils, alters nutrient cycling, and accelerates corrosion of built infrastructure. Additionally, secondary aerosol formation contributes to regional haze, reducing solar radiation reaching the earth's surface and perturbing climatic cycles. The multifaceted impacts of air pollution thus underscore the imperative for comprehensive mitigation strategies to sustain both environmental integrity and public welfare.

In this study, we conducted a systematic evaluation of existing and emergent policies and technological interventions aimed at curbing air pollutant emissions. Our methodological approach integrated quantitative modeling of pollutant dispersion with scenario-based analyses of policy implementation in an urban megacity context, exemplified by Baghdad. Data were sourced from the World Health Organization, national environmental agencies, peer-reviewed literature, and satellite-based remote sensing platforms, ensuring robust representation of emission inventories and health outcome statistics.

We assessed interventions across three domains: transportation, energy production and consumption, and waste management. For the transport sector, we modeled the adoption rates of electric and hybrid vehicles, expanded public transit networks, and implementation of low-emission zones. In the energy domain, we examined the transition from coal and oil to renewable sources—namely solar and wind—coupled with building-level energy efficiency measures, including retrofitting and smart metering. Waste management strategies included improvements in municipal solid waste collection, incineration with flue-gas cleaning systems, and regulated disposal of agricultural residues.

Our findings suggest that an integrated policy portfolio can achieve a reduction of up to 40–60% in annual $PM_{2.5}$ concentrations under a high-adoption scenario. In Baghdad, where baseline $PM_{2.5}$ levels frequently exceed WHO guideline values by two to threefold, such reductions are projected to translate into a 20–30% decrease in pollution-related morbidity and mortality over a decade. Moreover, the co-benefits of greenhouse gas mitigation and energy security reinforce the value of investing in clean technologies.

Economic analysis indicates that the cost of implementation, when amortized over the lifecycle of interventions, is offset by savings in healthcare expenditure and productivity gains from improved population health. For instance, retrofitting residential and commercial buildings for energy efficiency exhibits a benefit–cost ratio of 2.5, driven by lower energy bills and reduced hospital admissions. Similarly, electrification of public transport fleets yields long-term net benefits, particularly when supported by renewable energy integration.

3 Results and Discussions

Despite these positive outcomes, several barriers to implementation persist. Institutional fragmentation, limited regulatory enforcement, and insufficient financial incentives hinder the scale-up of clean solutions. Additionally, public awareness and stakeholder engagement remain critical; community-level participation in monitoring air quality and adopting behavioral changes—such as reduced private vehicle use—can amplify policy impacts.

To address these challenges, we recommend a multi-tiered governance framework encompassing national legislation, municipal ordinances, and local community action. Key measures include establishing stringent emission standards, incentivizing private sector investment through tax credits and green bonds, and deploying real-time air quality monitoring networks enabled by Internet of Things (IoT) sensors. The integration of smart-city platforms can facilitate data-driven decision-making, enabling authorities to identify pollution hotspots and allocate resources efficiently.

The table represents the recommended air quality guideline levels and interim targets for various pollutants:

Table 1. Air quality guidelines.

Pollutant	Average Time	Interim Target 1	Interim Target 2	Interim Target 3	Interim Target 4	Air Quality Guideline Level
Fine Particulate Matter (PM2.5)	Annual	35 µg/m³	25 µg/m³	15 µg/m³	10 µg/m³	5 µg/m³
Fine Particulate Matter (PM2.5)	24 hours (a)	75 µg/m³	50 µg/m³	37.5 µg/m³	25 µg/m³	15 µg/m³
Coarse Particulate Matter (PM10)	Annual	70 µg/m³	50 µg/m³	30 µg/m³	20 µg/m³	15 µg/m³
Coarse Particulate Matter (PM10)	24 hours (a)	150 µg/m³	100 µg/m³	75 µg/m³	50 µg/m³	45 µg/m³
Ozone (O3)	Peak Season (b)	100 µg/m³	70 µg/m³	–	–	60 µg/m³
Ozone (O3)	8 hours (a)	160 µg/m³	120 µg/m³	–	–	100 µg/m³
Nitrogen Dioxide (NO2)	Annual	40 µg/m³	30 µg/m³	20 µg/m³	–	10 µg/m³
Nitrogen Dioxide (NO2)	24 hours (a)	120 µg/m³	50 µg/m³	–	–	25 µg/m³
Sulfur Dioxide (SO2)	24 hours (a)	125 µg/m³	50 µg/m³	–	–	40 µg/m³
Carbon Monoxide (CO)	24 hours (a)	7 mg/m³	–	–	–	4 mg/m³

(a) Represents the 99th percentile (i.e., 3-4 exceedances per year).
(b) Peak season is determined by the maximum daily average for an 8-hour period over six consecutive months with the highest ozone concentration during the six months, https://www.who.int/news-room/fact-sheets/detail/ambient-(outdoor)-air-quality-and-health

Through the photos taken of the Dora refinery, as well as the South Baghdad Power Plant, and to know the most important gases that result from that factor, we met with Assistant Professor Dr. Haider Mohammed Abdul Hamid from the University of Baghdad / College of Engineering, Department of Environment, confirmed to us that these gases contain harmful carbonates that can appear in the form of fog or clouds and work to increase the concentrations of gases such as O, carbon dioxide, methane, hydrocarbons, sulfur oxides, nitrogen oxides, suspended particles, and heavy metals attached to suspended particles, as well as soot - a black dust formed from the incomplete combustion of wood and fossil fuels and known as soot. These gases greatly affect air pollution and thus the health of the city's residents and the pollutants they are exposed to, represented by the burning of waste in Rashid Camp and the outskirts of Baghdad and its suburbs. There are many landfills and unsanitary landfills in Nahrawan, Nabaa, and Taji, in addition to the random burning they are exposed to, especially at night, which contributes to harmful gas emissions, taking into account the weather phenomena represented by changing temperatures, changing wind directions, and increasing humidity rates, all of which contribute to increasing the percentage of air pollution in the city of Baghdad (Meeting with the Professor of Environment at the College of Engineering, University of Baghdad (Assistant Professor Dr. Haider Muhammad Abdul Hamid) on 11/17/2024.).

From the photos taken at the Dora refinery, as well as the South Baghdad power station, and to identify the main gases emitted from these facilities, we met with Dr. Haider, an environmental professor at the College of Engineering, University of Baghdad, who confirmed that these gases include (hydrocarbons, sulfur oxides, nitrogen oxides, carbon monoxide, carbon dioxide, suspended particles, heavy metals adhered to suspended particles, and soot - a black dust formed from the incomplete combustion of wood and fossil fuels known as soot). These gases significantly affect air pollution and, consequently, the health of the city's residents, who are exposed to pollutants from burning waste in the Rashid camp and the outskirts of Baghdad, as well as numerous landfills in those areas.

The accelerating burden of environmental pollution and anthropogenic climate change poses profound threats to human health and ecosystem integrity. Ambient air contamination now exceeds the World Health Organization's guideline limits for more than 99 % of the global population, contributing to an estimated 6.67 million premature deaths in 2019 alone. Both chronic and acute exposures to fine particulate matter, nitrogen oxides, sulfur dioxide and ozone have been unequivocally linked to increased incidence of respiratory pathologies (such as asthma, chronic obstructive pulmonary disease and lung cancer), cardiovascular morbidity, and neurological impairments. These airborne toxicants not only compromise pulmonary and cardiac function but also facilitate systemic inflammation, oxidative stress and endothelial dysfunction, thereby heightening vulnerability to strokes, myocardial infarctions and neurodegenerative disorders.

Concurrently, progressive global warming engendered by elevated concentrations of greenhouse gases exerts cascading effects that compound air-pollution-related health risks. Rising temperatures disrupt atmospheric chemistry, alter pollutant dispersion patterns and intensify the formation of photochemical smog. Heatwaves magnify ozone generation, exacerbate heat-stress–induced cardiovascular strain and increase mortality. Moreover, alterations in precipitation regimes and sea-level rise undermine agricultural productivity, threaten freshwater availability and expand the geographic range of vector-borne diseases, thereby amplifying the burden of climate-sensitive illnesses and malnutrition.

Beyond human health, air pollution precipitates widespread ecological damage. Deposition of acidifying compounds—primarily nitrogen and sulfur oxides—leads to soil acidification, declines in forest vitality and freshwater acidification that jeopardizes aquatic fauna. Persistent organic pollutants and heavy metals bioaccumulate, impairing reproductive success and driving biodiversity loss. Ozone depletion, primarily resulting from chlorofluorocarbon emissions, diminishes stratospheric UV-B shielding, increasing risks of skin cancers and impairing terrestrial and marine photosynthetic processes. Wildlife are further displaced by habitat degradation and forced migration, contributing to population declines and local extinctions.

Mitigation of these interlinked crises demands multi-faceted interventions. At the individual level, reducing reliance on private motor vehicles—particularly for short-distance travel—and embracing mass transit can substantially lower combustion-derived emissions. Energy conservation through efficient appliances and mindful electricity usage reduces the demand for fossil-fuel power generation. Transitioning to renewable energy sources—solar, wind and geothermal—offers scalable reductions in greenhouse gas and co-emitted pollutant outputs. At the policy level, stringent emissions standards, urban planning that prioritizes green spaces and investment in low-carbon infrastructure are critical.

Given the essential role of clean air in sustaining life, coordinated global action is imperative. By integrating emission reduction targets, strengthening air quality monitoring and fostering sustainable energy transitions, societies can alleviate the health and environmental burdens of pollution and climate change, thereby safeguarding human well-being and ecological resilience for future generations.

Suggestions to implement smart cities solutions to achieve sustainable development goals in the city of Baghdad:

1. Using monitoring technologies and smart sensors to monitor air pollution levels, where this data can help make rapid and effective decisions to reduce emissions and to improve air quality in the city, it is important to monitor the city's environmental pollution constantly to make it healthier and suitable for living with the smart sensing system is the solution Which can be used to monitor the city for different purposes, one of which is pollution monitoring in major cities, the Internet -based system has to monitor air pollution with great capabilities to ensure less energy consumption during sensing and sending with minimal delay and better reliability in actual time [4-7].

2. Encouraging the use of sustainable transportation by using smart transportation systems to implement smart traffic management systems to improve traffic and reduce congestion, which leads to reducing vehicle emissions and public transportation plays an important role in achieving global goals for sustainable development as effective public transport systems contribute to models Sustainable urban development by providing accessible mobility options at reasonable prices, which are necessary to enhance economic growth [8-10].

3. The use of smart irrigation techniques to develop green spaces in the city, which contributes to improving air quality and reducing temperatures.

4 Conclusion

Air pollution represents a significant environmental challenge confronting the city of Baghdad, with profound implications for the health of its inhabitants and the broader ecosystem. Contaminated air is a primary contributor to a myriad of health issues, including respiratory diseases, allergies, and cardiovascular disorders.

The air pollution in Baghdad can be attributed to various sources, such as vehicular emissions, coal-powered electricity generation stations, and industrial pollutants. These factors collectively contribute to the elevated concentration of fine particulate matter in the atmosphere, which poses severe health risks.

The impact of air pollution on the health of Baghdad's residents is evident in the rising incidence of asthma, respiratory infections, and increased premature mortality rates. Furthermore, air pollution adversely affects the surrounding environment, including wildlife and plant life.

To address this critical issue, it is imperative that the local government and civil society take substantial measures to mitigate airborne pollutant emissions. Strategies should include enhancing public transportation systems, adopting clean energy sources, and promoting the use of advanced environmental technologies in industrial practices.

In conclusion, improving air quality in Baghdad is not just an environmental goal, but rather a necessity to ensure the health and safety of society, and to achieve sustainable development that benefits current and future generations through joint work, the city can overcome the challenges of air pollution and achieve a healthy and sustainable environment.

References

1. A. Saleem, E. J. Abdullah, "Assessment of particulates and heavy metals concentration distribution in ambient air around Al-Dora Refinery in Baghdad City, Iraq," Iraqi Geol. J., 149–160 (2024)
2. R. A. Al-Naser, M. H. Al-Jiboori, "Study of the vertical dispersion coefficient of air pollution under different atmospheric stability conditions over Baghdad," Asian J. Water Environ. Pollut. **21(6)**, 173–180 (2024)
3. A. Ali, "A framework for air pollution monitoring in smart cities by using IoT and smart sensors," Informatica **46(5)** (2022)
4. Kumar et al., "Implementation of smart LED lighting and efficient data management system for buildings," Energy Procedia **143**, 173–178 (2017)
5. T. Ahuja, V. Jain, S. Gupta, "Smart pollution monitoring for instituting aware travelling," Int. J. Comput. Appl. **145(9)**, 4–11 (2016)
6. J. Liu, H. Yan, M. White, X. Huang, *A comparative analysis of perceptions of insecurity in Milan and Beijing metro stations*, Front. Archit. Res. (2025)
7. E. L. Birch, "The new urban agenda's road map for planning urban spatial development: tangible, manageable and measurable," U. N. Chron. **53**, 8–11 (2016)
8. P. Lu, J. Miao, L. Yang, S. Dou, L. Yang, C. Wang, Y. Guo, "Cohort profile: China undergraduate cohort for environmental health study," BMC Public Health **24(1)**, 828 (2024)
9. X. Xiao, M. Gao, "Overview of climate change, air pollution, and human health," in *Air Pollution, Climate, and Health*, 3–12 (Elsevier, 2021)
10. World Health Organization, "Ambient (outdoor) air quality and health," (2024), https://www.who.int/ar/news-room/fact-sheets/detail/ambient-(outdoor)-air-quality-and-health

The impact of environmental pollution on mental health: a comparative study between urban and rural students

Bilal Tariq Hussein Allwan*, *Elaf* Hameed Mousa, *Zahraa* Tariq Battal, *Tayseer* Hindy, and *Rami* Jundi

University of Anbar, College of Education for Humanities, Dept. of Educational and Psychological Sciences, Baghdad, Iraq

Abstract. The environment and its issues have become of great importance, with environmental concerns imposing themselves on various levels, whether local, regional, or international. Environmental pollution and its risks are not confined to the borders of the country responsible for it but extend to affect other nations as well. This has encouraged countries to hold conferences and seminars to establish laws aimed at reducing this growing threat, which endangers all living beings, including humans, who are the primary cause of these problems. Environmental pollution is considered one of the most pressing issues of the modern era, threatening all nations. Iraq, in particular, has not been isolated from this pollution. Moreover, the wars, economic sanctions, and the complete collapse of infrastructure have worsened the situation, leaving negative impacts on the Iraqi environment. This research aims to examine the impact of environmental pollution on mental health through a comparative study between students in urban and rural areas.

1 Introduction

Air pollution is one of the most dangerous types of pollution, with its hazards crossing administrative and political borders. Its major sources include smoke and gases emitted by power generators, factories, and vehicle exhausts. Additionally, gases released from waste and garbage dumped in certain areas, as well as dust storms, are significant contributors to air pollution [1].

The dangers of air pollution extend to all living organisms, infrastructure, industrial products, climate, and the ozone layer. Humans are directly affected by air pollution, leading to numerous diseases that have become prevalent in the modern era, such as respiratory diseases (asthma and allergies), chronic cough, colds, neurological disorders, heart diseases, eye infections, and lung cancer. All these health issues impact an individual's mental health and psychological well-being [2].

The increasing population, unsustainable development in various sectors, and the lack of modern technological solutions for pollution control have all left negative effects on the

* Corresponding author: ali.alhameedawi@uokufa.edu.iq

environment. Moreover, Iraq's environment has been severely affected by recurring crises, leading to pollution of essential environmental elements such as air, water, and soil [3].

As a result, pollution has led to numerous health and safety problems, increasing the prevalence of diseases categorized as "environmental pollution-related illnesses." Pollution, in all its forms, primarily harms human health, yet paradoxically, humans themselves are its primary cause. This is largely due to the pursuit of higher financial profits without considering the resulting environmental consequences, coupled with a lack of environmental awareness. To combat pollution, two essential steps must be taken: first, enacting or enforcing existing environmental protection laws and ensuring proper monitoring and enforcement mechanisms; second, fostering environmental awareness among individuals through educational institutions. Just as a farmer plants a seed and nurtures it into a strong tree over time, environmental awareness must be cultivated within individuals so that it grows and becomes deeply rooted.

Mental health is just as important as physical health. In fact, physical well-being alone cannot guarantee a person's happiness without good mental health. Furthermore, prioritizing mental health is an integral part of promoting overall public health. While general health focuses on physiological and organic factors, mental health primarily concerns psychological well-being, a field closely related to psychiatry [4-6].

University students represent a crucial demographic for any nation, as they play a key role in societal progress and development. They are an essential segment of society, responsible for carrying the burden of national development and facing various life challenges, especially after completing their studies and entering the job market. A university degree has become a necessity for securing employment in an increasingly competitive labor market and is now considered a fundamental criterion for leadership in almost every aspect of life.

Therefore, it is crucial to engage with university students in a way that acknowledges their psychological development and encourages them to maximize their academic and professional potential. For this demographic to actively contribute to societal transformation, they must enjoy sound mental health, which enables them to understand their purpose, unify their personality, and function efficiently on physical, intellectual, and social levels. A mentally healthy individual approaches life with confidence and security.

Studies have shown that students with poor mental health experience diminished motivation for learning and struggle academically. Since these students will occupy important societal positions in the future, it is essential that they possess psychological balance and adaptability.

Research Problem

The current research problem can be defined through the following questions:

- Does residence (rural or urban areas) affect the mental health of university students?
- Does gender (male or female) have an impact on students' mental health?

Research Significance

The significance of this research lies in addressing a highly crucial issue affecting people's lives and health—environmental pollution—which has become a major threat to humanity. Ironically, humans themselves are the primary cause of this issue due to a lack of awareness, insufficient understanding, and, in some cases, deliberate neglect of the dangers posed by pollutants.

The increasing levels of environmental pollution in Iraq, in all its forms, emphasize the urgent need not only for legislation and the formation of environmental organizations but also for comprehensive governmental support for institutions responsible for social upbringing. These institutions play a crucial role in educating and raising awareness among people, ultimately fostering sustainable development and creating a broad environmental consciousness within society, especially among children. Among these institutions, the educational system stands out for its fundamental role in preparing future generations, instilling environmental responsibility, and ensuring sustainable environmental preservation [7].

Moreover, mental health is an equally important field that has gained significant attention from scientists and researchers. It is as vital as physical health, as it contributes to shaping well-balanced individuals who can effectively cope with life's pressures, overcome challenges, and achieve harmony both within themselves and in their interactions with society [6, 8, 9].

Research Objectives

The current research aims to explore: the impact of residence (rural vs. urban areas) and gender (male vs. female) on the mental health of university students.

This will be examined through testing the following hypotheses:

1. There is no statistically significant effect of residence (urban or rural areas) on the mental health of university students.
2. There is no statistically significant effect of gender (male or female) on the mental health of university students.
3. There is no statistically significant interaction between residence (urban or rural areas) and gender (male or female) in influencing the mental health of university students.

2 Materials and methods

Research Scope

The current research is limited to students of Al-Anbar University during the 2024-2025 academic year, based on the variables of gender (male, female) and residence (urban, rural areas).

Definition of Terms

Mental Health:

- Kafafi (1995): Defined mental health as "a state of balance and integration between an individual's psychological functions, leading them to behave in a way that allows self-acceptance and societal acceptance, resulting in a reasonable degree of satisfaction and adequacy".
- Al-Farakh & Taym (1999): Defined mental health as "a relatively stable state in which an individual is personally, emotionally, and socially well-adjusted to themselves and others. In this state, the individual can achieve self-fulfillment, utilize their abilities and potential to the fullest, cope with life's demands, maintain a well-integrated and stable personality, and exhibit normal behavior, allowing them to live in peace and harmony".

Research Population:

The research population consists of students of Al-Anbar University, totaling 21,945 students, distributed according to the variables of gender (male, female) and residence (urban, rural).

Below is Table **(1)**, which illustrates the distribution of the study population based on gender and residence.

Table 1. The distribution of the study population based on gender and residence.

Residence	**Gender**		**Total**
	Male	**Female**	
Urban Areas	5274	8990	14264
Rural Areas	2840	4841	7681
Total	8114	13831	21945

Research Sample

The sample of the current research consists of 400 students, selected randomly with an unequal distribution. See Table (2) for the distribution details.

Table 2. Distribution of the research sample based on gender and residence.

Residence	Gender		Total
		Female	
Urban Areas	96	164	260
Rural Areas	52	88	140
Total	148	252	400

Research Tool

Mental Health Scale:

In order to achieve the objectives of the current research, and after reviewing a number of relevant literatures and previous studies presented in Chapter Two, the researchers adopted the Mental Health Scale developed by Yasser Nizam Majid Said (2003). Below is a description of the scale.

Description of the Mental Health Scale

The scale consists of 119 items, distributed across 20 indicators of mental health, as shown in Table (3).

(You can now include the specific details of each of the 20 indicators or any further elaboration in the following table.)

Table 3. Behavioral Components of Mental Health in Its Final Form and the Number of Items Designed to Measure Them in Both the Initial and Final Versions.

I	Behavioral Components of Mental Health	Number of paragraphs
1	The individual's sense of tranquility and safety	7
2	Self-awareness	7
3	Self-acceptance and extension of this feeling	7
4	Insight into matters	5
5	Satisfying desires in an acceptable, non-exaggerated way	6
6	Having realistic life goals	6
7	Ability to adjust goals and pursue alternative objectives	5
8	Learning from past experiences in new situations	6
9	Rational perception of reality	5
10	Mental flexibility	6
11	Ability to form close relationships with others	7
12	Feeling responsibility towards others	5
13	Enjoying a degree of independence from others	5
14	Adopting a unified moral framework (a unified philosophy)	7
15	Feeling happiness and a desire for life	5
16	Self-control and managing intense emotions	7
17	Taking responsibility for one's actions and deeds	6
18	Feeling vitality and energy	6
19	Enjoying psychological and physical health	7
20	Self-reliance in problem-solving	5
Total		**119**

The researchers set five graded alternatives for each paragraph, which are (always, often, sometimes, rarely, never), and when corrected, they give the grades (4, 3, 2, 1, 0) respectively for the positive paragraphs, and vice versa for the negative paragraphs.

Verification of the Psychometric Properties of the Mental Health Scale

Validity of the Scale

The validity of the scale was calculated using the following methods:

- **Judgmental Validity (Content Validity):**The initial version of the scale was presented to a group of 16 psychology professors. After considering the feedback and observations of the judges, the scale was finalized in its current version, consisting of (119) items.

This process ensures that the items on the scale are representative of the domain they aim to measure, providing an accurate reflection of mental health.

Construct Validity

- **Discriminative Power of the Items:**The researchers calculated the discriminative power by ranking the sample members' total scores from highest to lowest. The two extreme groups were identified based on their total scores, with (27%) of the sample in each group, totaling (214) individuals (108 in the high group and 108 in the low group). The t-test for independent samples was then used to calculate the significance of the differences between the means of the two groups for each item of the scale. The t-values calculated ranged from (3.596 to 15.493), all of which were greater than the critical t-value of 1.96 with (214) degrees of freedom at a significance level of (0.05), indicating that all items were statistically significant and effectively discriminated between the two groups.
- **Internal Consistency (Item Validity):** The researchers used Pearson's correlation coefficient to calculate the validity of each item, by correlating each item's score with the total score. Since the item scores were continuous and graded, the validity sample consisted of 400 students. The results showed that all correlation coefficients were statistically significant at the (0.05) significance level, with values ranging from (0.376 to 0.638), which were greater than the critical value of (0.098) with (398) degrees of freedom at the (0.05) significance level, confirming the internal consistency of the scale.

Reliability of the Scale

The researchers calculated the reliability using both the Test-Retest Method and Cronbach's Alpha, as follows:

Test-Retest Method:

To calculate reliability using this method, the scale was re-administered to a reliability sample of (30) individuals, with a time interval of (14) days between the first and second administration. Pearson's correlation coefficient was then calculated between the scores from the first and second applications. The resulting correlation coefficient was (**0.93**), which is considered a strong indicator of the stability of individuals' responses on the scale over time.

Alpha-Cronbach Method:

The reliability using this method was calculated based on the responses from the main sample of (400) individuals. By applying the Cronbach's Alpha formula, the resulting Alpha coefficient was (0.91), which indicates a very good level of reliability and can be relied upon for conducting research.

Statistical Methods:

The results of the research were analyzed using the Statistical Package for Social Sciences (SPSS).

3 Results and Discussion

Classification of Sample Members on the Mental Health Scale

In order to classify the sample members according to mental health levels, the researchers used a criterion-referenced assessment, which means evaluating an individual's performance in comparison with their peers within the sample. Consequently, the mean and standard deviation were calculated, and the results are presented in the following table:

Table 4. Mean and standard deviation of the mental health scale

Health Status	Number	Mean	Standard Deviation
	400	255.165	28.27

Thus, Three Levels were Formed as Follows:

1. **Low Level**: Individuals whose scores are less than 227 (Mean - 1 Standard Deviation).
2. **Medium Level**: Individuals whose scores range from 227 (Mean - 1 Standard Deviation) to 283 (Mean + 1 Standard Deviation).
3. **High Level**: Individuals whose scores are 284 or above (Mean + 1 Standard Deviation).

Table 5. Distribution of sample members according to mental health levels.

	Percentage	Number	Level
Health Status	Low	105	26%
	Medium	163	41%
	High	132	33%
	Total	400	

Here's the translation of the provided text:

The table shows the distribution of sample members according to their levels of harm avoidance. It was observed that (163) individuals from the sample have an average level of mental health, representing (41%), and (132) individuals (33%) had a high level of mental health. Only (105) individuals (26%) had a low level of mental health.

Secondly: Identifying the differences in mental health according to the variables of gender (male – female) and residence (urban – rural).

The researchers relied on two-way analysis of variance for unequal samples. Below is a presentation of the research hypotheses that were tested and the results that were reached, as shown in Table (6).

Table 6. The means and standard deviations of the sample participants' scores on the mental health scale according to the variables (gender, residence).

Residence	gender	Mean (Average)	Standard Deviation	Number
Urban areas	Males	221.326	29.584	96
	Females	222.547	31.483	164
	Total	221.93	30.54	260
Rural areas	Males	288.943	26.438	52
	Females	287.578	25.048	88
	Total	288.31	25.389	140
Total	Males	255.13	28.01	148
	Females	255.184	28.26	252
	Total	255.2	28.27	400

The researchers used a two-way analysis of variance to identify the differences in mental health according to the variables (gender, housing). The results were as follows, as shown in Table (7).

Table 7. Two-way Analysis of Variance for the Independent Variables of the Study and Their Interactions.

Source of Variation	Sum of Squares	Degrees of Freedom	Mean Squares	F-Ratio
Gender	5.028	1	5.028	0.832
Housing (Urban/Rural)	439.677	1	439.677	72.79
Gender x Housing interaction	9.714	1	9.714	1.61
Error	2392.3	396	6.04	
Error	6681.14	399		

**The tabular F-value is (3.93) at a significance level of (0.05) and degrees of freedom (1, 396).*

First Hypothesis / Gender (Male – Female) does not affect mental health: The results indicated that there was no effect of gender on mental health, as the computed F-value was not statistically significant at the (0.05) level. Therefore, the alternative hypothesis was rejected, and the null hypothesis was accepted, indicating that gender does not affect mental health.

Second Hypothesis / Housing (Urban Areas – Rural Areas) does not affect mental health: The results indicated that the effect of housing on mental health was statistically significant at the (0.05) level. Therefore, the null hypothesis was rejected, and the alternative hypothesis was accepted, suggesting that housing does affect mental health, with rural area students showing better mental health compared to urban area students. The mean score for individuals living in rural areas for mental health was (288.31), while the mean score for individuals living in urban areas for mental health was (221.93). The researchers explained this result by noting that the natural environment in rural areas, free from pollution, dust, and dirt, has a positive impact on mental health compared to urban areas, where the nature of the environment, including the presence of factories and industrial areas, negatively affects mental health.

Third Hypothesis: The interaction of housing (urban, rural) with gender (male, female) does not affect mental health. The results indicated that there was no statistically significant effect at the (0.05) level. Therefore, the interaction of housing with gender does not affect mental health, and the null hypothesis was accepted while the alternative hypothesis was rejected based on these results.

4 Conclusions

Based on the results of the research, the researchers recommend the following:

1. Prioritizing environmental protection within the core focus of the state and all governmental institutions, not just relying on the Ministry of Environment and Health. There is a need to support development, improve living standards, and reduce poverty, which causes environmental pollution and the spread of mental illnesses.
2. The government should activate environmental laws and increase monitoring of the industrial, economic, and private sectors, particularly in wastewater discharge and other related practices.
3. Focusing on waste recycling in all its forms, which helps prolong the life of the environment and natural resources and improves human mental health.
4. Utilizing clean energy to protect the environment and reduce reliance on fossil fuels.
5. Establishing bilateral and multilateral agreements with neighboring countries to preserve the environment in all its forms and improve the health sector, ensuring environmental protection.

6. Media outlets and related institutions should educate the public by disseminating environmental culture principles and raising awareness about its preservation. Health institutions should also carry out necessary tests for residents of polluted areas.
7. Emphasizing the importance of mental health and its impact on students' behavior, organizing seminars and workshops to guide students on how to deal with different environmental events, ensuring their adaptation to the university environment.
8. Providing students with sufficient attention and support so they can overcome their mental health challenges and cope with life pressures in their personal and academic lives.

References

1. A. Y. Mustafa et al., *Mirrors of Sociology of Youth and the Environment. Empowerment and Empowerment Debate* (The Elites and National Competencies for Reform Foundation, Foundation Book No. 1, 2016)
2. K. O. Ahmed, Sh. Q. Aziz, P. H. Karim, "Contemporary Trends toward Environmental Issues in the Kurdistan Region of Iraq", Prace i Studia Geograficzne **67**, 109–130 (2022), DOI: 10.48128/pisg/2022-67.1-07
3. A. Ventriglio, A. Bellomo, I. Di Gioia, "Environmental Pollution and Mental Health: A Narrative Review of Literature", CNS Spectrums (2020), DOI: 10.1017/S1092852920001303
4. M. A. Abdel Rahman, N. N. Ibrahim, "Environmental pollution and the necessity of environmental awareness," J. Arts **135**, 411–428 (2020)
5. M. S. Attia, K. Y. Tayel, Z. N. Shata, S. S. Othman, "Psychosocial profile of institutionalised street children in Alexandria, Egypt: a comparative study with school children," J. Child Adolesc. Ment. Health **29(2)**, 103–116 (2017)
6. N. A. Mohidem, Z. Hashim, "Integrating Environment with Health: An Islamic Perspective", Social Sciences **12**, 321 (2023), DOI: 10.3390/socsci12060321
7. A. Y. Mustafa, et al., *Mirrors of Sociology of Youth and the Environment. Empowerment and Empowerment Debate* (The Elites and National Competencies for Reform Foundation, Foundation Book No. 1, 2016)
8. O. K. Bishoge, A. K. Aremu, D. D. Ajayi, S. Mfinanga, "The Efforts Made to Solve Environmental Health Problems in Developing Countries: A Case from Mtwara Town in Tanzania", Journal of Public Health and Development **20**, 253–266 (2022), DOI: 10.55131/jphd/2022/200219
9. J. X. Wang, X. Q. Liu, "Climate Change, Ambient Air Pollution, and Students' Mental Health", World Journal of Psychiatry 14, 204–209 (2024), DOI: 10.5498/wjp.v14.i2.204
10. A. Núñez-Delgado, W. Ahmed, E. Bontempi, J. L. Domingo, "The Environment, Epidemics, and Human Health", Environmental Research **214 (Part 4)**, 113931 (2022), DOI: 10.1016/j.envres.2022.113931

Thoron gas clusters in the indoor air of primary schools of Al Najaf city: a geographic analysis

Anwar Qasim Ahmed[1], *Ali* Kadhim Hussein[2], *Ali* Abid Abojassim[3,*], and *Ali* Saeed jassim[4]

[1]Department of Ecology, Faculty of Science, University of Kufa, Al-Najaf, Iraq
[2]Department of basic Science, College of Dentistry, University of Kufa, Al-Najaf, Iraq
[3]Department of Physics, Faculty of Science, University of Kufa, Al-Najaf, Iraq
[4]Department of Geology, Faculty of Science, University of Kufa, Al-Najaf, Iraq

Abstract. In this study, thoron gas concentrations were measured in a primary school in the city of Najaf using a CR-39 detector. This work can be divided into four sections: the first segment involved using the Global Positioning System to find latitude and longitude for all points. In part two, determination was made on 220Rn gas concentrations in samples from a primary school. These values are varied between 0.22 Bq/m3 and 9.00 Bq/m3 with an average value of 2.99 ± 1.80 Bq/m3 lower than the recommended value (below UNSCEAR 2000 safety standard of 40 Bq/m3). The third part of which involves using ArcMap GIS 10.8, interpolation to identify varying of thoron and the kernel density estimation (KDE) for mapping thoron concentrations along province, the result of density is lowest through some districts in Al Meelad (Sch52,67,97), Al Zahraa (Sch41), and Al Hasan (Sch56,66). While the density increases rapidly to reaches a peak in Al Jdeidat (Sch 1,2,3,5,6,11,17,23,29,33,36,42), Al Mualimeen (Sch12), Al Muthanna (Sch35), AL Rahmaa (Sch 86,87,91), Al Jamiya (Sch69,70,72,73), and Al Hindiyeh (Sch46,48). Finally using SaTScan v10.1.3 February 2024 of the spatial scan statistic was identified in 3 locations were exposure to thoron concentration in Al Najaf city.

1 Introduction

Radon, the silent killer [1-4]. This radioactive gas is a natural occurrence, born from the decay of uranium and thorium minerals nestled within earth's rocks and soil— stealthy in its way of creeping into human environments and becoming one of the primary sources for natural radiation exposure. Radon causes second only to smoking in terms of lung cancer causation yet standing strong as an entity itself with distinct identity markers including 222Rn (radon gas) and 220Rn (thoron gas): two common isotopes where radon gas emerges from uranium's radioactive decay while thoron results due to thorium's decay — each with their own half-lives that paint a different picture altogether when combined tell us something profound about radiation dose composition: 85% coming from natural sources (including radon) which implies 15% must be man-made. The emission of alpha radiation gas by radon 222 and its progenies 218Po and 214Po, along with their chain lengths: this demonstrated

* Corresponding author: ali.alhameedawi@uokufa.edu.iq

how the damage to DNA takes place within lung cells and how ultimately lung cancer manifests in human beings [5,6]. It was on record that there was a significant correlation exhibiting lung cancer cases and the prevalence of radon gas in indoor air: this occasion confirmed by the World Health Organization. The statistics provided by the United States Environmental Protection Agency (EPA) emphasised more alarming, approximately 21,000 deaths occur due to radon gas exposure indoors. This quantity of deaths is ten times higher than deaths caused by air pollution. The recommendations by EPA and WHO agencies for standard radon concentrations are different but equally essential— 148 Bq/m3 and 100 Bq/m3. Amongst various studies that were achieved on building materials, granite was found to be a major emitter of radon gas compared to the rest of studies although it contains radioactivity like all building materials, albeit insignificantly [7]. In houses, emissions play a pivotal role towards radon concentrations apart from the surrounding soil: the emissions play constitute major contributors thus suggesting where mitigation efforts should be focused and reduced of the emissions. Building material emissions account for most indoor radon levels as highlighted earlier (i.e., granites have higher level of emissions). This research focuses on exploring and identifying 220Rn concentration hotspots within the indoor air of primary school buildings in Najaf, Iraq— detected through CR-39 detectors. Alongside these schools' geographic clustering, we also assessed the thoron concentration interpolation using GIS, SaTScan software, and Kernel Density Estimation (KDE).

2 Materials and methods

2.1 Data collection

The geographical administrative regions of Al-Najaf are defined by the Najaf Urban Planning Department. Population statistics for each area are decided by the Ministry of Planning and Al-Najaf Governorate Bureau of Statistics. We achieved the survey within 100 schools located in central Al-Najaf to detect thoron using CR-39 detectors, which are manufactured by Track Analysis Systems Ltd., UK. Hence, our project is divided into four sections: The first part is about determining the latitude and longitude of each measurement point through GPS. The second part concerning the employing interpolation inverse distance weighting (IDW) techniques to estimate unsurvey areas no measured of thoron data from those with known locations.

KDE is widely used in geography, and it examines the concentration of point pattern data, thus forming the third part where thoron density range is determined. At the last part, Poisson model-based spatial scanning statistics (another approach of spatial analysis) helps in detecting geographical clusters of thoron. This makes up a comprehensive survey on geographical clustering of thoron: ending with the implementation of two different approaches within this study.

2.2 Interpolation with inverse distance weighting (IDW)

A spatial analysis was conducted to understand the thoron distribution in the area. In this concern, we formed maps of distribution using a specific interpolation technique known as Inverse Distance Weighted (IDW) implemented in ArcMap GIS 10.8. This method operates under the assumption that proximal values hold closer relations than those located farther apart; thus, the value at an unsampled point (as a reference point) is determined through a weighted average calculation based on surrounding known values. These weights are inversely proportional to the distances between the sample and prediction locations. The IDW technique is governed by a particular mathematical formula as follows:

$$Z = \frac{\sum_{i=1}^{n} \frac{1}{(d_i)^p} Z_i}{\sum_{i=1}^{n} \frac{1}{(d_i)^p}} \tag{1}$$

Were;

Z is the estimated value of the prediction point;

Zi is the measured value of the sample point;

di is the distance between the sample and the prediction point;

p is the performance parameter;

n is the number of sample points.

When the sample points are regularly distributed dispersed, the IDW interpolation technique produces better results compared with than when the sample points are grouped where point closed to other [8]. Therefore, this technique was used in this study because the sample points are well distributed.

2.3 Accuracy evaluation

In order to verify the applicability of the IDW method, RMSE and MRE items are further calculated to evaluate the accuracy of the interpolation of the above inverse distance weighted (IDW) method. The root mean square error (RMSE) of the thoron concentration estimate is calculated using formula (2):

$$RMSE = \sqrt{\frac{\sum_{i=1}^{n} (th_{esti} - th_{meas})^2}{N}} \tag{2}$$

where the thmeas is the thoron concentration measured by CR-39 at different locations in Al-Najaf, the estimated represents the thoron concentration that estimated by the IDW method, and N represents the number of samples in the study area. The accuracy of the previous method was also evaluated by calculating the mean relative error (MRE) using the following formula (3):

$$MRE = \frac{1}{N} \sum_{i=1}^{N} \left[\frac{th_{esti} - th_{meas}}{th_{meas}} \right] * 100\ \% \tag{3}$$

The coefficient of determination (R2), RMSE, and MRE of interpolation using the inverse distance weighted (IDW) method were determined.

2.4 Kernel density estimation (KDE)

Kernel Density Estimation (KDE) is a nonparametric method for estimating the probability density of a random variable in its domain from a sample of the random variable. It is often used as an alternative to the histogram. Definition of the estimator($\widehat{f_h}$) for every point x on the variable's domain is given by the following equation (4).

$$\widehat{f_h} \frac{1}{Nh} \sum_{i=1}^{N} k\left(\frac{x - x_i}{h}\right) \tag{4}$$

where k is a kernel function which must respect the three conditions below, and x and h are scalars:

$$k(x) > 0 \; if \; x < h$$

$$k(x) = 0 \; if \; x \geq h$$

$$\int_{-\infty}^{+\infty} k(x) = 1$$

The number of kernel functions is vast; among the most notable are the Epanechnikov, Quartic, Triangle, Tricube, Triweight, Cosine, and Uniform functions. Although the Gaussian function stands as the most widely used one, it is defined in the domain (−∞; +∞) which violates certain conditions (specific case of non-compact kernels). The bandwidth chosen for KDE represents another parameter: it determines the level of smoothness of the function. A high bandwidth value implies a wider spread of each data sample mass over a broader interval. However, we aim to select an appropriate bandwidth that captures well the variation in density of random variable avoiding both underfitting and overfitting problems since bandwidth plays major role toward estimated density by this method [9].

2.5 A Spatial Scan Statistic

The spatial scan statistic that stated by [10], which computed by the SaTScan software, seeks out census block groups where cases deviate from what would be expected under the null hypothesis number of cases is simply proportional to population density. The statistic operates by taking a small number of counties from different locations and of varying sizes (encompassing 1–50% of the population at risk) and then counting the number of cases inside and outside a specific county. By Monte Carlo hypothesis testing, it pinpoints locations with significantly more or fewer events than anticipated, making it well adapted for disease surveillance and radiation exposure without needing any prior assumptions about the locations to be identified. Moreover, it considers the spatial point at which populations at risk are unevenly distributed, evaluates multiple tests that seek clusters in different areas, examines high or low incidence areas based on various considerations of confounding factors. As there is no need to prespecify the size and location of the window for scanning, SaTScan's spatial risk scan statistic is widely used for disease surveillance at local levels using LLR-based assessment of high-risk areas; therefore, the LLR is given by:

$$LLR = log\left[\frac{\left(\frac{c}{n}\right)^{c}\left(\frac{C-c}{N-n}\right)^{C-c}}{\left(\frac{C}{N}\right)^{C}}\right] \tag{5}$$

where LLR is the log-likelihood ratio, N and n are the total population in the region and window, respectively, and C and c represent the number of cases or events in the region and window, respectively.

3 Results and discussions

3.1 Interpolation and Kernel Density Estimation (KDE) analysis

In this study, thoron concentrations in 100 primary school buildings at different locations in Al- Najaf were measured using CR-39. The latest results in the study area showed that thoron concentrations in building samples were below the recommended level (below the UNSCEAR 2000 safety standard of 40 Bq/m3). All measurement periods were taken through the months of (March, April, May 2022). Using the ArcMap 10.8 program, a sample map of school locations was created, as shown in Figure 2 (top panel). Whereas the data is collected at discrete locations, spatial interpolation is often used to create continuous data. According to equation (1), the inverse weighting method (IDW) was used to create an interpolated thoron map, as shown in Figure 2 (bottom panel).

According to equation (2), the kernel density estimate (KDE) represents the density of thoron concentrations in the province, as shown in Figure 3 (top panel). It can be found that the lowest density is obtained in some areas such as Al Meelad (Sch52,67,97), Al Zahraa (Sch41) and Al Hasan (Sch56,66). The density increases sharply and reaches the peak in Al Jdeidat (Sch 1,2,3,5,6,11,17,23,29,33,36,42), Al Mualimeen (Sch12), Al Muthanna (Sch35), AL Rahmaa (Sh 86,87,91), Al Jamiya (Sch69,70,72,73) and Al Hindiyeh (Sch46,48). Table (1) shows the results of the measurement of thoron concentrations in 100 primary schools in Al-Najaf city. Its values ranging between 0.22 Bq/m^3 and 9.00 Bq/m^3, with an average value of 2.99 ±1.80 Bq/m^3. The estimated Thoron IDW ranged from 0.39 Bq/m^3 to 8.99 Bq/m^3 with a mean of 3.03 ±1.82 Bq/m^3, while the estimated Thoron KDE ranged from 0.44 Bq/m^3 to 18.67 Bq/m^3 with a mean of 8.27±4.28 Bq/m^3.

The results were validated using the IDW method with 100 samples, with an RMSE of 0.7185 Bq/m^3 and a MRE of 209.2549%. The high correlation of Thoron estimation with the measured independent variable Thoron using the linear regression algorithm is shown in Figure 1.

Table 1. Results of ^{220}Rn (Thoron) measured and estimated in primary school in Al- Najaf city.

Sample code	School name	Thoron-measured in Bq/m3	Thoron-IDW in Bq/m3	Thoron-KDE in Bq/m3	Longitude in meter	Latitude in meter
Sch 1	Alghaffari	0.90	0.91	10.38	436150.90	3539909.00
Sch 2	Altahdhib	1.35	1.54	14.88	435944.20	3539177.00
Sch 3	Malik Aliashtir	3.15	3.16	16.16	436064.30	3539492.00
Sch 4	Halif Alquran	5.40	5.39	12.87	438415.80	3540403.00
Sch 5	Aishab Alkasa	2.70	2.69	7.70	436456.20	3539630.00
Sch 6	Alhaidariya	1.80	1.80	7.53	436900.30	3540156.00
Sch 7	Altaysir	3.15	3.10	9.10	438040.70	3542154.00
Sch 8	Aleasifa	4.50	2.70	16.00	438584.10	3541338.00
Sch 9	Dabel Al Khuzaie	1.80	1.84	8.66	437753.40	3539855.00
Sch 10	Damascus Al'asasia	2.70	1.87	8.66	436954.30	3541674.00
Sch 11	Alsaadiq	2.02	2.04	13.40	437108.50	3539453.00
Sch 12	Baghdad Al'asasia	4.50	4.47	16.75	437675.50	3539487.00
Sch 13	Alghari	1.80	1.81	9.45	438410.00	3538993.00
Sch 14	Alamam Alhadi Al'asasia	3.60	3.58	9.64	437237.60	3541669.00
Sch 15	Eidun	4.95	4.89	9.26	438586.60	3539133.00
Sch 16	Mustafa Jawad	1.80	1.81	6.65	438162.70	3542366.00
Sch 17	Alwathbi	1.35	1.72	10.87	437318.70	3539070.00
Sch 18	Sharif Al-Razi	4.95	4.96	15.26	438536.50	3540399.00
Sch 19	Alshaeb	2.47	2.47	7.80	437260.10	3545330.00
Sch 20	Al-Safi Najafi	2.70	2.71	12.65	437418.10	3540301.00
Sch 21	Badr Al-Kubra	0.67	0.67	4.55	436555.70	3542982.00
Sch 22	Umm Qasr	0.45	0.45	6.69	437645.00	3545648.00
Sch 23	Alzainabiya	4.50	4.36	16.31	436100.70	3539435.00
Sch 24	Sikina	3.15	3.14	4.19	435403.20	3540360.00
Sch 25	Ramallah	1.80	1.80	4.94	435289.50	3539809.00
Sch 26	Alfadila	2.25	2.25	6.08	437282.40	3540847.00
Sch 27	Alsanabil	2.70	2.67	8.98	437555.30	3542139.00
Sch 28	Fath	1.35	1.42	9.69	437122.10	3541664.00
Sch 29	Aleafa	0.22	0.39	15.52	436205.90	3539477.00

Continuation of Table 1.

Sch 30	Bilqis Al'asasia	3.15	2.71	12.35	437418.10	3540301.00
Sch 31	Al-Faraged	2.70	2.70	9.10	438584.10	3541338.00
Sch 32	Almaqasid	3.19	3.19	18.67	437533.90	3539752.00
Sch 33	Dijula	4.05	3.93	15.86	436003.50	3539253.00
Sch 34	Mustafa Gamal El Din	1.35	1.36	7.01	438195.40	3542141.00
Sch 35	Alanisaf	5.40	5.40	13.94	438062.60	3540571.00
Sch 36	Birdaa	1.80	1.81	12.08	435724.30	3539683.00
Sch 37	Al-Bariq	3.60	3.60	16.14	437773.10	3539563.00
Sch 38	Alamiani	3.15	3.15	10.45	438284.50	3539077.00
Sch 39	Nahj Al-Balaghah	3.15	3.15	3.50	439542.80	3541498.00
Sch 40	Alshiyma	1.80	1.80	5.57	436966.20	3542333.00
Sch 41	Alrisala	1.35	1.35	2.07	440067.10	3540535.00
Sch 42	Alturath Alearabiu	4.72	4.65	15.64	437178.20	3539517.00
Sch 43	Tabuk	0.90	0.90	4.88	436376.50	3542842.00
Sch 44	Almakasib	4.05	4.05	4.32	439420.70	3539935.00
Sch 45	Safin	4.05	4.05	4.56	439918.00	3542398.00
Sch 46	Altadhia	7.65	7.63	10.98	438173.50	3546738.00
Sch 47	Habib Bin Mazahir Al Asadi	3.15	3.15	8.70	437051.90	3548134.00
Sch 48	Fataa Alaslam	5.40	5.40	9.41	437569.00	3546526.00
Sch 49	Almasoudi	2.70	2.70	3.92	434719.80	3544694.00
Sch 50	Haifa	3.60	3.59	4.76	436583.30	3546154.00
Sch 51	Alhuru Alriyahiu	6.30	6.29	8.25	436527.90	3547370.00
Sch 52	Alyaequbi	4.05	4.05	4.56	435123.50	3546295.00
Sch 53	Alainsar	3.15	3.15	3.71	439450.00	3539128.00
Sch 54	Almujd	4.50	4.50	12.16	438880.20	3540828.00
Sch 55	Alsahl Aliakhdaru	1.80	1.80	3.87	434740.70	3545084.00
Sch 56	Altabirsiu	0.90	0.91	3.27	440709.10	3540836.00
Sch 57	Alabirar	1.35	2.47	5.31	436215.20	3545075.00
Sch 58	Almuhajirin	1.80	1.80	2.81	440263.90	3541439.00
Sch 59	Alrahma	1.57	1.57	5.86	436926.10	3546038.00
Sch 60	6 Kanun	5.40	5.40	8.48	437684.30	3547671.00
Sch 61	Albayeat Alkubraa	2.70	2.71	7.42	436707.10	3547077.00
Sch 62	Sawr	2.25	2.25	2.21	439985.00	3538170.00
Sch 63	Albaraq	1.80	1.81	6.79	437014.80	3548494.00
Sch 64	Alraazi	4.05	2.47	5.31	436215.20	3545078.00
Sch 65	Aliraq Alhuru	2.02	2.02	9.29	437727.20	3547168.00
Sch 66	Saeed Bin Jubair	1.80	1.71	3.14	440751.00	3540817.00
Sch 67	Aldhaariat	0.90	0.91	3.20	434860.70	3545810.00
Sch 68	Alrabab	0.90	1.87	8.66	436951.70	3541683.00
Sch 69	Altawhid	4.50	4.50	10.21	438828.50	3544025.00
Sch 70	Ibrahim Al-Khalil	1.35	1.38	7.23	439024.40	3544753.00
Sch 71	Aalhaqu Almubin	4.95	2.47	7.80	437260.10	3545330.00

Continuation of Table 1.

Sch 72	Alamam Alrida	9.00	8.99	12.40	439249.00	3544157.00
Sch 73	Alrafah	2.74	2.73	4.50	439075.40	3544962.00
Sch 74	Ali Al-Akbar	2.70	2.71	8.74	439427.90	3542919.00
Sch 75	Aliaskandaria	4.95	4.94	6.97	436211.00	3538606.00
Sch 76	Aljamie	4.95	4.95	9.84	437436.70	3548248.00
Sch 77	Alshahid Mahdi Alhakim	1.80	1.80	4.42	437404.10	3544440.00
Sch 78	Albalad Al'amin	2.25	2.47	9.71	438923.40	3542786.00
Sch 79	Khayr Albariya	3.15	2.07	3.98	437920.60	3543540.00
Sch 80	Alduea' Almustajab	0.90	2.07	3.98	437920.60	3543539.00
Sch 81	Eabuwd Ghafla	6.30	6.25	12.38	435028.70	3543777.00
Sch 82	Eadnan Zuin	0.40	0.41	1.92	436911.50	3538568.00
Sch 83	Aliaetimad	6.30	6.27	12.38	435028.70	3543777.00
Sch 84	Altaqwaa	1.35	0.90	4.88	436376.50	3542842.00
Sch 85	Albayinat	1.80	0.90	4.88	436376.50	3542842.00
Sch 86	Almawlaa Almuqadas	2.25	5.85	13.30	435429.10	3542789.00
Sch 87	Alsaafaat	5.85	5.85	13.30	435429.10	3542789.00
Sch 88	Abi Talib	6.30	6.10	10.06	439004.80	3542798.00
Sch 89	Mohammed Jawad Mughniyeh	3.15	3.15	6.35	437972.20	3545979.00
Sch 90	Sayf Alhaqi	2.70	5.85	13.30	435429.10	3542789.00
Sch 91	Alrusul	2.70	5.85	13.30	435429.10	3542789.00
Sch 92	Altasnim	0.45	0.45	0.44	434270.20	3548592.00
Sch 93	Alsalam	3.60	3.55	5.35	436949.20	3544255.00
Sch 94	Almathir	8.55	8.54	8.41	435106.20	3549344.00
Sch 95	Jarf Alnasr	0.90	0.92	5.11	436969.20	3544094.00
Sch 96	Alamam Zayn Aleabidin	1.80	1.80	1.77	435412.30	3550484.00
Sch 97	Alnasamat	2.25	2.25	2.53	435005.50	3547121.00
Sch 98	Alshahid Karim Alkhaqani	3.60	3.60	5.40	438734.50	3546279.00
Sch 99	Aleawali	0.45	0.46	1.09	437154.10	3549829.00
Sch 100	Eata' Alnajaf	2.25	2.25	2.34	436722.20	3550365.00

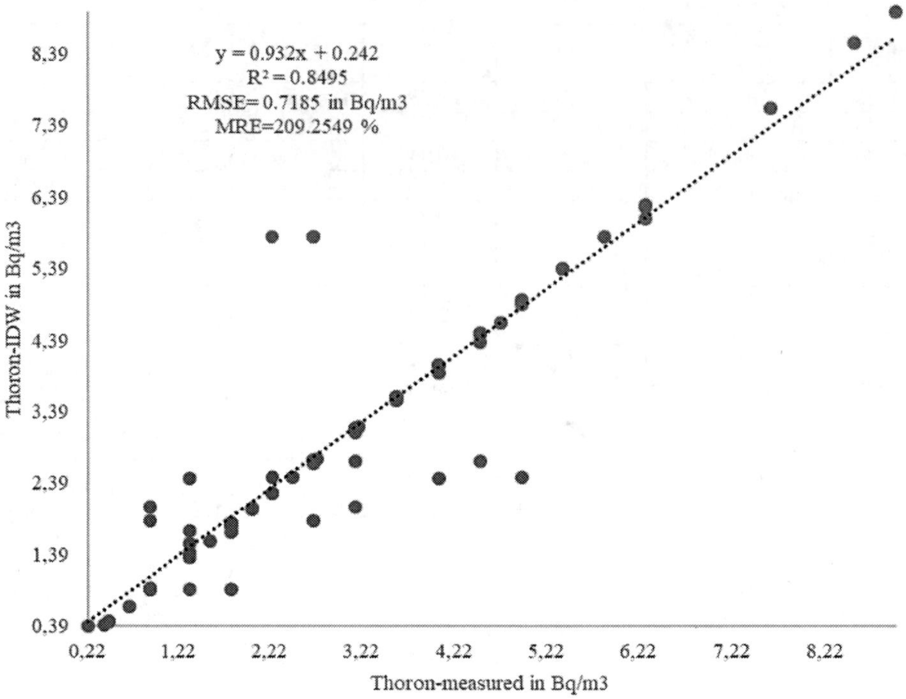

Fig. 1. Linier regression algorithm for thoron-measured with predict thoron using the inverse weighting method (IDW).

3.2 Count the population that exposed to thoron concentration

According to equation (5) the log-likelihood ratio of the spatial scan statistic identified 3 locations were exposure to thoron concentration in Al-Najaf city. We detected one statistically significant geographic cluster with risks of thoron 50 percent at risk (**Table 2**, **Fig. 3** bottom panel). The primary cluster was found in Al Karama (Sch7), Al Hannana (Sch26), Al Saha (Sch7), Al Ghadeer (Sch16,34), Al Furat (Sch78,88), Al Aadala (Sch45,74), Abu Khaled (Sch20,30), Al Muthanna (Sch35), Al Hussein (Sch10,48,68), Al Mualimeen (Sch12), Alskaan (Sch8,31), Al ashtraky (Sch54),Al hawraa (Sch44), Adan (Sch4,18), Al Saad (Sch26), Al Ameer (Sch39,58) and AL-murahlen (Sch40), where 992 cases were observed, whereas 959 were expected with a relative risk of 1.06 ($p < 0.041$). Secondary insignificant clusters (25 percent) were located in Al Muthanna (Sch35), Al Mualimeen (Sch12), Alskaan (Sch8,31), Al ashtraky (Sch54), Al hawraa (Sch44), Adan (Sch4,18) and Al Saad (Sch26), where 548 cases were observed, whereas 528 were expected with relative risk of 1.04 ($p < 0.628$). Other locations with insignificant clusters (10 percent) were found in the areas of Alamara (Sch24), Almushraq (Sch24), Alhuwesh (Sch25), and ALburaq (Sch25), a total of 231 cases were observed, which was slightly higher than the expected number of 220 cases. The relative risk for these areas was calculated to be 1.05, indicating a minimal increase in risk ($p < 0.999$). When assessing the risk of thoron radiation, it is important to compare it to radon radiation. Thoron progeny, due to their contribution to inhalation radiation doses, pose a threat that is approximately four times greater than that of radon progeny at equivalent dosages. Furthermore, thoron daughters have a longer lifespan in the air compared to radon daughters, which results in higher concentrations and prolonged exposure. This increased exposure to thoron can have negative effects on human health. [11].

The daughter of thoron, 212Pb, has a relatively long half-life of 10 hours, allowing it to be absorbed by the lungs and expose other organs to radiation. Therefore, it is more likely to cause harm. While it is generally important to consider thoron gas and its progeny, the dosage of thoron gas is often significant smaller than that of decay products and can be disregarded unless the activity concentration of thoron gas is exceptionally high compared to decay products. Various-sized aerosols can attract different nuclides that lead to decay products.

This implies that an aerial event consists of a small portion that is not attached and a larger fraction that is attached. The radioactive offspring of radon isotopes in the air pose a significant danger compared to the inert gases themselves. Even non-volatile metal isotopes can be found in the air, and some are inhaled and deposited in the lungs. The high-density ionizing alpha-particle radiation from these decay products, lodged in the bronchial epithelium of the lungs, presents a serious risk of lung cancer. Numerous techniques have been employed by researchers to investigate indoor thoron concentration [12-15]. Changes in ventilation rate, temperature, and pressure are often the main factors influencing variations in indoor radon concentrations. The ventilation rate affects the rate of air exchange between indoor and outdoor environments. When radon becomes trapped indoors, particularly during temperature inversions or when there is inadequate natural or artificial ventilation in homes or schools, it can reach hazardous levels of concentration. For the majority of individuals who spend a significant amount of time indoors, these indoor spaces serve as the primary contributors to their exposure to ionizing radiation within their homes, workplaces, and schools. Among these locations, schools rank as the second most influential source of radiation exposure for students and staff members.

Recent studies have indicated that children are more susceptible to the harmful effects of radiation compared to adults, particularly at low doses. This is since children spend more time in school and are more vulnerable to radiation exposure than adults, increasing their risk of developing lung cancer. To ensure the safety of students in schools and other facilities, it is crucial to conduct assessments of radon and thoron levels in as many school and home premises as possible. In Al-Najaf city, many school buildings are old and have been neglected, making them more susceptible to environmental factors. The focus of this study is on children aged at 6 to 12 who attend school for five hours a day, five days a week in Iraq. By establishing thoron concentrations in primary school buildings, particularly in Al Najaf Province, Iraq, this study aims to gather valuable information about population exposure and establish a database to achieve studies in future.

Table 2. Thoron concentration clusters analysis for population exposure per 100,000 in Al Najaf city.

Location	Percent of the Population at risk	Observed exposure	Expected exposure	Relative risk	p-value
Al Karama (Sch7), Al Hannana (Sch26), Al Saha (Sch7), Al Ghadeer (Sch16,34), Al Furat (Sch78,88), Al Aadala (Sch45,74), Abu Khaled (Sch20,30), Al Muthanna (Sch35), Al Hussein (Sch10,48,68), Al Mualimeen (Sch12), Alskaan (Sch8,31), Al ashtraky (Sch54), Al hawraa (Sch44), Adan (Sch4,18), Al Saad (Sch26), Al Ameer (Sch39,58) and AL-murahlen (Sch40).	50 %	992	959	1.06	0.041
Al Muthanna (Sch35), Al Mualimeen (Sch12), Alskaan	25 %	548	528	1.04	0.628

(Sch8,31), Al ashtraky (Sch54), Al hawraa (Sch44), Adan (Sch4,18) and Al Saad (Sch26).					
Alamara (Sch24), Almushraq (Sch24), Alhuwesh (Sch25), and ALburaq (Sch25).	10 %	231	220	1.05	0.999

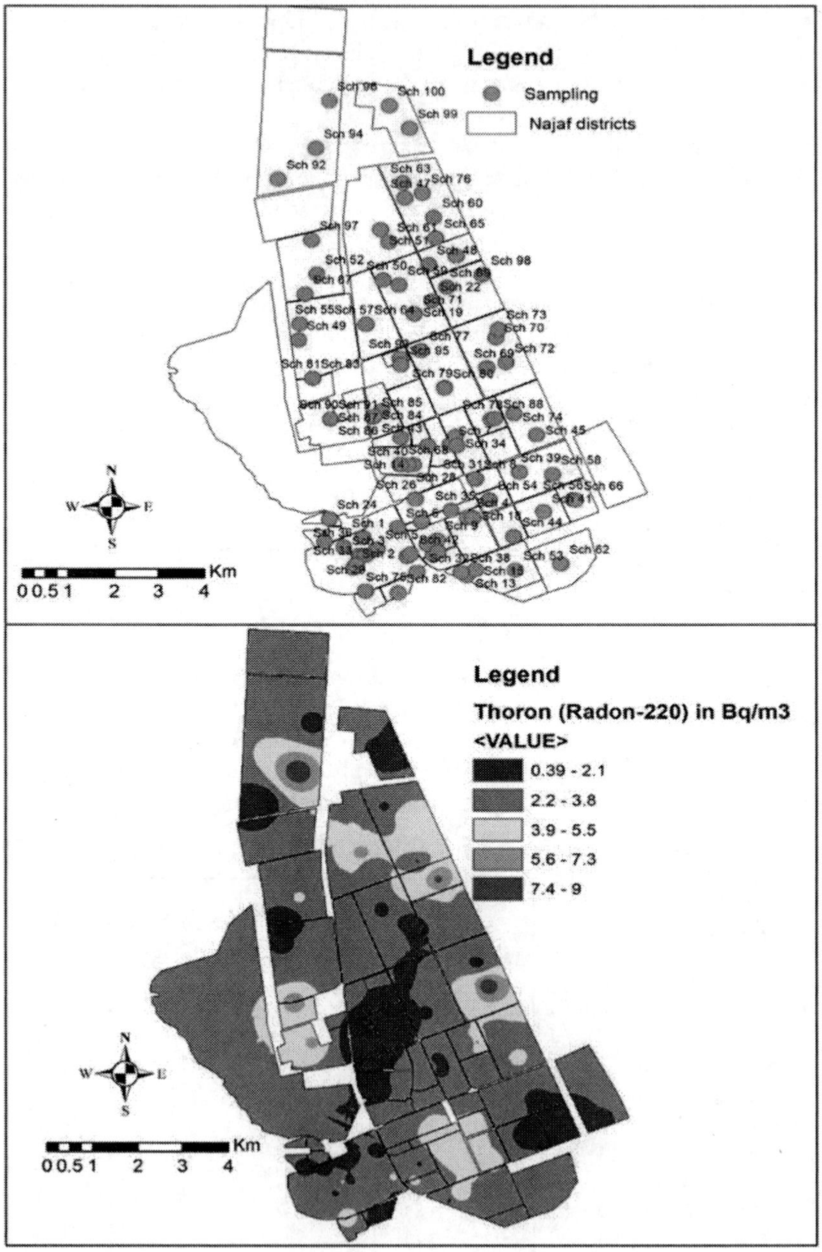

Fig. 2. Sampling of Thoron in primary School (top panel) and spatial distibution of Thoron using IDW Interpolation (bottom panel).

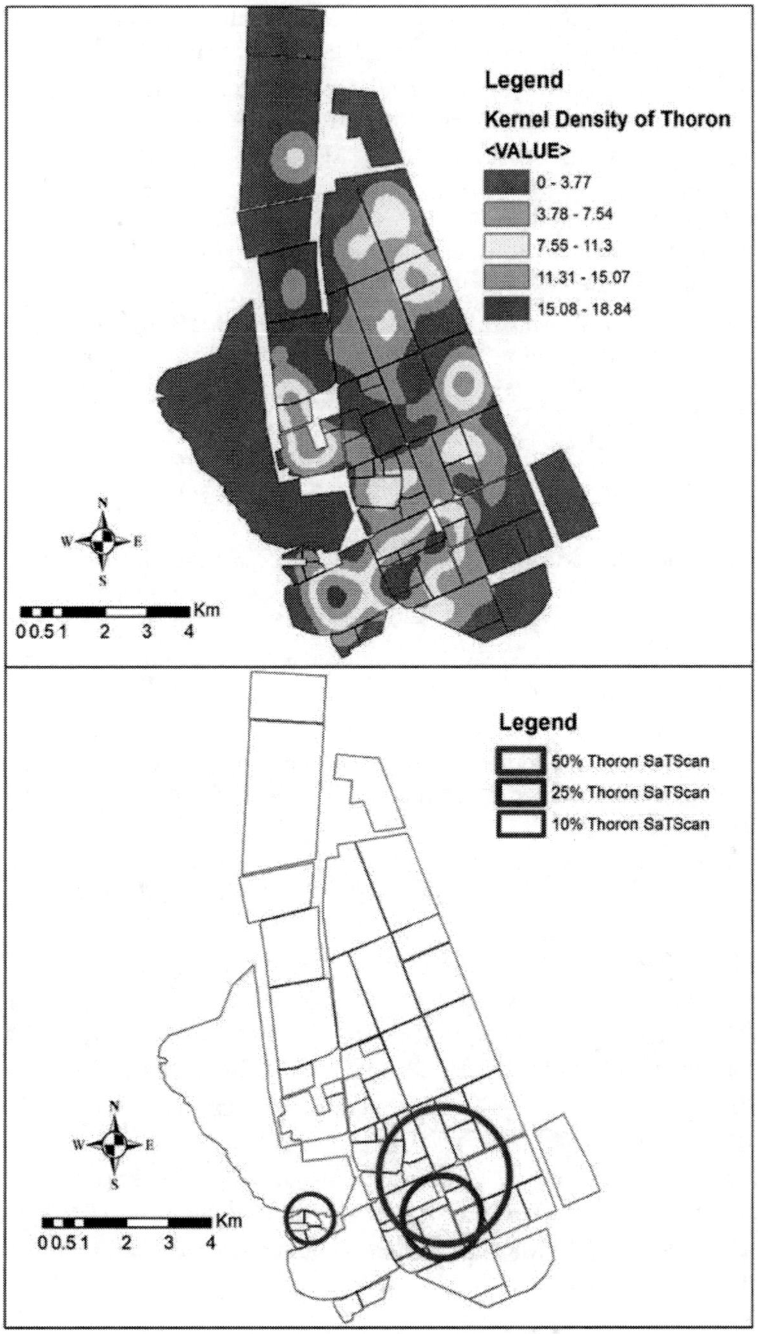

Fig. 3. Map of the estimated density of Thoron (top panel) and SaTScan (bottom panel).

References

1. United Nations Scientific Committee on the Effects of Atomic Radiation, UNSCEAR 2019 Report, Annex B-Lung Cancer from Exposure to Radon (United Nations, New York, 2019)

2. J. Temuujin, E. Surenjav, C. H. Ruescher, J. Vahlbruch, "Processing and uses of fly ash addressing radioactivity (critical review)," Chemosphere **216**, 866–882 (2019)
3. R. S. Mohammed, R. S. Ahmed, "Estimation of excess lifetime cancer risk and radiation hazard indices in southern Iraq," Environ. Earth Sci. **76**, 303 (2019)
4. M. L. Gonzalez, A. Ruano-Ravina, M. Torres-Duran, T. Kelsey, M. Provencio, I. Parente-Lamelas, V. Leiro-Fernandez, I. Vidal-Garcia, O. Castro-Anon, C. Martinez, A. Golpe-Gomez, M. Torres-Espanol, J. Abal-Arca, C. Montero-Martinez, A. Fernandez-Villar, J. M. Barros-Dios, "Residential radon, genetic polymorphisms in DNA damage and repair related," Lung Cancer **135**, 10–15 (2019)
5. S. Kumar, R. Kumar, P. Kumar, S. Kumar, A. Kumar, M. S. Kandari, K. Vats, R. B. S. Rawat, "Estimation of radon, thoron, and their progeny concentration in selected dwelling near the sugar mills of Uttar Pradesh and Uttarakhand, India," J. Surv. Fish. Sci. **9(1)**, 528–536 (2022), DOI: 10.53555/sfs.v9i1.1680
6. N. Gupta, K. Kant, M. Garg, *Indoor assessment of radon, thoron and their progeny levels in the dwelling near the fly ash dumping sites situated in Faridabad Haryana, India* (2015)
7. Z. Li, K. Wang, H. Ma, Y. Wu, "An adjusted inverse distance weighted spatial interpolation method," in *Proc. 2018 3rd Int. Conf. Commun. Inf. Manag. Netw. Secur. (CIMNS 2018)*, Wuhan, China, 27–28 September 2018, Atlantis Press, 128–132 (2018)
8. D. Silverman, "How was it for you? The Interview Society and the irresistible rise of the (poorly analyzed) interview," Qual. Res. **17(2)**, 144–158 (2017)
9. X. Y. Cui, G. E. Tjønnfjord, S. M. Kanse, A. E. A. Dahm, N. Iversen, C. F. Myklebust, ... P. M. Sandset, "Tissue factor pathway inhibitor upregulates CXCR7 expression and enhances CXCL12-mediated migration in chronic lymphocytic leukemia," Sci. Rep. **11(1)**, 5127 (2021)
10. S. Tokonami, "Characteristics of thoron (220Rn) and its progeny in the indoor environment," Int. J. Environ. Res. Public Health **17(23)**, 8769 (2020)
11. E. Zunic, A. Besirevic, R. Skrobo, H. Hasic, K. Hodzic, A. Djedovic, "Design of optimization system for warehouse order picking in real environment," in *2017 XXVI Int. Conf. Inf. Commun. Autom. Technol. (ICAT)*, 1–6 (2017)
12. S. Prasad, R. Shankar, R. Gupta, S. Roy, "A TISM modeling of critical success factors of blockchain based cloud services," J. Adv. Manag. Res. **15(4)**, 434–456 (2018)
13. R. C. Ramola, M. Prasad, "Significance of thoron measurements in indoor environment," J. Environ. Radioact. **225**, 106453 (2020)
14. R. J. Dosh, A. A. Abojassim, "Assessment of radiological risk associated with thoron gas at primary schools in Al-Najaf city, Iraq," Appl. Radiat. Isot. **205**, 111154 (2024)
15. J. D. Harrison, J. W. Marsh, "ICRP recommendations on radon," Ann. ICRP **49(1_suppl)**, 68–76 (2020)

Urban transformation towards smart cities

Zainab Khalid*

Department of Architecture, College of Engineering, University of Baghdad, Baghdad, Iraq

Abstract. Smart cities have been considered as a focal point in many research and fields of study, and a major target for national, international policies, and strategies including the urban aspects; Is the smart city a reality or an ambition by researchers. As the world's population increases and the demands resulting from the expansion and growth of urbanization increase, architects and urban engineers face economic, political, social, and technological challenges to solve problems, Population increases that have stressed the burden on the urban structure and increased the consumption of resources. It is necessary to strengthen existing cities as they are a powerful engine for economic growth and sustainable development. Hence, the research question emerged about the possibility of transforming existing cities into smart cities. The research turned to solving the research problem by analyzing international examples of smart cities and benefiting from their experiences. In the results make, discuss and assess Baghdad City Readiness checklist for Transformation towards Smart Cities. research concluded that the city is a physical place that can be the site of digital and sustainable projects with a set of indicators to transform current cities into smart cities.

1 Introduction

Nowadays, the world is going through a transition from "classical" cities to "smart" cities. This process is supported and stimulated by the continuous development of ICT, creating a new model of municipal cooperation between administration, businesses, academia and citizens [1-3]. Since the beginning of the digital age, the smart city presents as a focal point in several research, the smart city has gained increasing importance and has become the main goal of national policies and international strategies. One of the things that most researchers and those interested in the topic of the smart city have in common is their agreement on the potential of technology and smart city data to provide insights and ideal solutions to urban and societal problems, including population growth, which has increased the burden on the urban structure, and increased consumption of materials and resources, which leads to increased carbon dioxide emissions. Transportation and globalization have moved urban societies from the local to the global [1,3]. To create a more convenient life for urban residents in all parts of the city, including the possibility of everywhere having access to many services provided by the application software wherever they are, 24 hours a day [2]. Data and information needed by residents to obtain and renew government electronic cards, private transportation users and to find an empty parking location; Find a nearby hospital,

* Corresponding author: zainab.khalid@coeng.uobaghdad.edu.iq

specialist doctor or nurse, and book available hotels at reasonable prices, Affordable real estate and apartments, fees payment and electronic money transfer services, guaranteed sites to sell what they want, buy electronics, clothing and consumer products [2]. Although many global theoretical and practical research have been conducted relating smart cities, development, trends and sustainable cities, and a few local research related to smart cities, as represent in Table 1. below, the topic of the local urban possibility transformation towards smart cities (the research problem) wasn't discussed in literature. Therefore, the research studying smart cities initiatives and other related topics for solving the research problem make and assess Baghdad City Readiness checklist for Transformation towards Smart Cities according to the smart cities initiatives indicators, that clarifying the procedure adopted in the materials and methods followed by result and discussion.

2 Material and methods

In this research, we will survey the most important literature and previous studies on the smart city, including ideas and visions, as a first step. Then, followed the method of presenting and analyzing various models of European (London, Germany) smart cities, The American (San Francisco and Chicago) and Asian (China) experiences. These proposals present various specializations in the field of smart urban transformation. Followed by preparing a checklist in which the indicators accompanying the process of transformation towards a smart city tabulated, after that sorting and analyzing the indicators of urban transformation experiences locally with checking its effectiveness in a result and discussion paragraph. ICT scholars are interested in the implementation and application of ICT in cities. Urban designers and other professional managers are making efforts to improve efficiency in city operation and management by relying on digital data. Below are the relevant topics and issues.

2.1 The meaning of smart city

For about 20 years, the meaning of a smart sustainable smart city has been on the mind, and it represents an urban challenge [4]. There are many global and local efforts [5,8], that have dealt with the sustainable and smart city through study and research for a deeper understanding of the idea of the smart city. An innovative city that meets the needs of current and future generations about urban life and all its social-cultural, economic and environmental aspects [1]. In 2007, smart cities were defined in a study by the Regional Science Center in Vienna from a completely different perspective: used to describe a city with an industry "smart in relation to jobs and/or the economy. That is, industries that use data and technology in their production and manufacturing processes [3,5]. Caraglio, Del Po, Nijkamp defines the city as "smart" when it comes to wise management of natural resources with social participation and participatory action (transport), sustainable economic development, provision of modern and high-quality communication infrastructure services. For life [2]. A smart city is an integrated system of ICT technologies in urban infrastructure. Linked to this system are multi-dimensional and multi-level terminals that include many embedded devices and sensors [1-2]. Frost and Sullivan (2019) define it as a city that has an effective and active presence in at least five of the eight smart urban indicators: building, citizen, energy, health, technology, infrastructure, transportation, and smart governance [5]. Examples of the diversity of applications are Amsterdam, Aarhus and Ghent. Amsterdam adopts technology, especially carbon dioxide reduction technologies, while Aarhus innovative technologies propose specific models of actions and policies that lead to results that everyone can benefit from. In Ghent, to enhance e-democracy and e-participation tools, technologies applied to enhance cooperation between the government and citizens. Another different case in Portugal is the city of Paredes, as it aims for urban regeneration and development through technology and participation [1,6-8]. According to the above, the meaning of varies and converges in the proposals and definitions of researchers based on their research backgrounds and professional interests. It is an expanded concept of the connected city, always continuous and spread everywhere, alive and alert throughout the day, month and year.

2.2 Types of smart city

The human, socio-cultural, digital and physical components of the smart city are organized into subsystems that are interconnected within a complex urban and digital reality [1,5,10]. Components are organized into systems called smart city subsystems. Activity groups form subsystems linked by chains of collaboration, communication and feedback in the form of a platform. Shared activity groups and groups of activities interacting with each other appear in many different forms, such as land use areas, commodity supply chains, city areas, etc. within the realms of a smart city's existence and its specific entity [5]. The "smart city" is an electronic urban physical place in which digital data and electronic services improve human skills, capabilities and decision-making, that is, it relies mainly on automation to improve the city's operations and infrastructure. “Internet City” is used in cases of digital city transformation (Dubai Internet City), and refers to any type of electronic, smart or digital city. This term may be used in scientific or promotional publications, such as (Internet City, a book by Edward Elgar Publishing). “City 2.0” related to Web 2.0 and is related to the cyber-physical city. But as a concept, City 2.0 combines sustainable urban living with the democratic, social and technological aspects that revolve around innovative democratic urban management. What inhabits the previous types of cities is the “smart society” that inhabits all existing places, regions, regions and smart cities [5]. Komeno’s states: Intelligence is an internal characteristic that lies in the digital infrastructure and tools provided to the residents of society in any place, region or city that linked to the transfer of technology when there is a problem to solve. The definition is related to the ability to solve societal problems and use information and communications technology to improve this ability and facilitate innovation processes. The aspects of urban life for smart cities include public health, infrastructure, public transportation, traffic management, urban planning and design, city management and governance.

2.3 Adoption of innovation technology in the smart city.

Innovative technology can solve all urban issues and challenges, smart city planners and designers expect digital data and technology to provide insights and evidence to uncover critical urban issues. A smart city promotes digital services and content in urban areas, where computing is integrated and deployed across all city facilities to address various urban challenges. There is a need to increase efficiency in the city's operation and management. A smart city is an innovative way to solve urban problems intertwined with many disciplines, including public health, politics, social sciences, urban design, information technology, public sector and other computer sciences related to climate issues with innovative technologies. Thus, information and communications technology define the concept of the smart city, and the implementation of the idea grows and develops throughout. The development of concurrent technologies. Recent studies have begun to explore smart city technology that examines theories in urban design. Cameras, sensors and GPS devices of mobile phones have helped monitor users in public spaces. Environmental simulations, Internet-based surveys and computer vision are helping to activate people's perceptions and improve building designs. and management of the physical urban environment. This phenomenon is called the Internet of Things (IoT) (innovation). Smart cities employ this technology to connect users with public service networks, disparate urban facilities, infrastructure, etc., while simultaneously generating real-time aggregated data [6]. Every consumer product in the digital city network or piece of infrastructure could sense surrounding stimuli and communicate with each other within subsystems under the computer cloud. According to IBM's R&D department. A smart city is a three-dimensional system (instrumented, interconnected and intelligent) in the operation of a city. Harrison and others have pointed out that a smart city is "instrumented" because meters, sensors and instruments are constantly collecting live data. It "interconnected" because services, communicate, information through a platform based on Collected data and smart platform The city recreates itself based on analytics, models and data and system optimization. A smart city is a system that reorganizes itself to meet people's needs. A smart city deals with a higher level of city operation because it includes human components in addition to sensors and devices.

2.4 The requirements of urban transformation towards smart cities

The past three decades have witnessed an increase in the population of cities in the world by 65 million people annually, and in (2014) there were 28 giant cities containing 453 million people, in addition to increasing population growth rates reaching 65%. A large part of this increase is due to increased migration from the countryside to the city, which has negatively affected cities and caused environmental crises, high levels of pollution and a severe scarcity of basic facilities, which in turn has apparently affected the fabric of the suffocating cities and their infrastructure. According to expectations, population growth rose to 66% [4]. Considering the current and expected future increases, traditional regional and neoclassical theories have become inadequate to cover rapid and large urban growth. Growth theories and smart urban development add that the knowledge-intensive structure at the social and economic level has been defined as the "creative class" and is considered a major component of economic development in post-industrial cities, the fifth era of urbanization. The process of urban transformation begins with collecting complex data, providing simple electronic services used by residents, all the way to making municipal decisions. Sustainable smart city strategy works to enrich planning and urban design scenarios in the traditional city, and formulates its policies as an interconnected, equipped, open and smart city, including the Optima conceptual reference model, which is a multi-layered approach to describing smart urban areas and their structural components. The OECD study (2006) competitive Cities in the Global Economy, is important in contributing to our understanding of the urbanization process that goes hand in hand with globalization and a clear trend toward sustainability [4,9]. We are redefining the way, planning, design, and building existing city infrastructure, providing services, engage citizens, and connecting city systems and components together in the era of the Internet of Things. Major improvements in the efficiency of cities can be achieved through horizontal interconnection between systems. The goal is to transform cities into urban living environments that are more resilient to disasters, more resistant to challenges, more sustainable, and more intelligent. Considering protection from crime and ensuring cybersecurity [4]. A city that wants to become smart, sustainable, innovative, centers of creativity and entrepreneurship must properly renovate and renovate its dilapidated areas and neighborhoods, and regenerated areas can create their own "image". It has a symbolic and design value and thus determines the name of the smart city brand, as dilapidated areas do not represent an attractive urban environment for cognitive and economic investment, and this applies to many cities that are witnessing rapid development. Plans for the transformation of urban areas, from existing and deteriorating neighborhoods, various areas of industry, commercial facilities and transportation, begin with the relocation of businesses and people, the demolition of dilapidated buildings, and in many cases the government purchases private properties for development projects and fifth urbanization initiated by the city and provides business and tax incentives. In other cases, the developers formed a public-private partnership, in a profit-making model. To integrate ambitious investments into the city's services and digital infrastructure. Additional positive impacts include regeneration and improvement of housing stock, reduction of urban sprawl, economic benefits, enhancement of economic competitiveness of the city area, cultural and social amenities, improvement, safety and enhanced surveillance [4-6,10]. Some well-known urban transformation projects that have successfully designed "smart" areas are Diagonal Mar in Barcelona, Spain and London- Docklands, in Poblenou, the western harbor area of Malmö, Sweden. 100% renewable energy wear produced locally, compact built environment, green roofs and stormwater system. Open source is an international source of inspiration for planning climate-smart cities. Another example. Yokuma City in the context of becoming smart aims to rejuvenate and revitalize the economy in three urban areas: Kannai and Kangai prefectures, Minato Mirai district and the area surrounding Yokohama [2]. Mart City is one of the smart model of cities. It adopted its transformation program [8]. The research will present three initiatives and five case studies detailing urban transformation and projects, including: European Initiative (United Kingdom/London and Germany), the American Initiative (Chicago and San Francisco), and the Asian Initiative (China).

3 Smart cities initiative

The United Nations Economic proposed models to address the challenges of sustainable development represented by urbanization and rapid growth in cities. According to the committee, these can be controlled through a smart city plan. The UNEC presented two models for smart cities, Fig. 1.

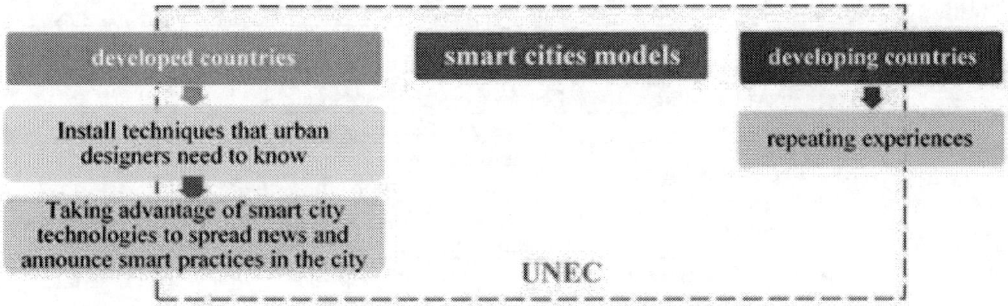

Fig. 1. UNEC smart models' researcher based on resources.

The five case studies in this research constitute only a portion of smart city implementation initiatives. Not all smart city indicators represented in each of these cities, but there are commonalities and useful differences between them. In this regard, the models of the Smart Cities Initiative have been determined below.

3.1 European Initiative:

The idea of "smart cities" began to be applied and developed at the Europe not along than (2011), "Smart Cities and Communities" was created, which explores the relationship between energy and transport and aims to reduce the impact of environmental requirements on urban life. The Partnership proposes to pool resources to support ICT, energy, transportation and urban management. The fields of communications, information technology, transport and energy are invited to work together within the Commission's initiative to combine their technologies to meet the needs of smart cities, enabling the integration of technologies and encouraging greater market penetration [2].

3.1.1 UK smart city model

In Spring of 2013 a vision of a "Smart London" that puts technology and innovation at the forefront of capital for a better place to live, work and invest. Examples of projects that London aims to address include in table 1 [2]. The European Commission (2014) has also proposed a model for developing the promotion of artificial intelligence throughout Europe. This model focuses on the main sectors in which technology will work:

Table 1. European initiative (UK) toward smart city.

	Independent indicators	**Descriptions**
European initiative (UK)	Smart Transportation	Develop transport.
	Smart Buildings	Leveraging smart technologies, such as lighthouse projects, and other replicable smart city technologies.
	Internet of Things Technology	Encouraging individuals and managers.
	Smart Platform	improve the lives by disseminating innovations in data mining and sharing. a series of smart city concepts that are replicable.
	Smart Energy	All programs geared towards energy efficiency in the built environment, and it implements and widely promotes a national smart metering campaign
	Smart Energy Saving	Promoting near-zero use renewable energy sources, or low-energy uses.
	Smart Communications	integrated infrastructure, and sustainable urban mobility for communities.
	Smart Learning	Learning develops smart cities.
	Smart traffics	Reducing traffic congestion.
	Renewable Energy	Using Renewable Energy to reduction in local air pollutants and carbon emissions.
	Smart economy	Aspects of the energy industry are well respected globally and a developing part of the UK economy.
	Smart provision of job	Using online time management and matchmaking services for job opportunities.
	Partnerships	ICT through partnerships between municipalities and industries.
	Smart Government	companies, policy consultants and city managers are developed by technology.

3.1.2 Smart City Model in Germany.

In Berlin, at the Metropolitan Technical University, a focus laboratory on smart city development currently established [2]. As for the city of Frankfurt, it has not yet promoted transformation. However, at the VDE conference, the city launched Smart Solutions for Future Living [2]. In Dortmund, it was chosen as one of the cities for IBM technologies as a show in table 2. [7].

Table 2. Germany initiative (Berlin, Frankfurt and Dortmund) toward smart city.

	Independent Indicators	Descriptions
European initiative (Germany)	Smart Transportation	It pursues the goal of making mobility easier and easier for everyone
	Internet Of Things Technology	Partnership with ASUS, Media Broadcast and Internet of Things company in Taiwan.
	Smart City Platform	It provides information and data efficiently and purposefully outside the city limits. The development of user-friendly data platforms enhances communication between citizens, city administration and other actors.
	Smart Communication	Providing new and innovative digital solutions and communication channels to their citizens.
	Smart traffics	Traffic systems and management.
	Smart lighting	Smart city lighting.
	Smart Energy	Bring together measures to implement new energy and environmental technologies.
	Smart Learning	Encourages citizens to develop in the fields of education, culture, and diverse and digital arts.
	Smart energy marketing	Named European Capital of Innovation 2021 marketing for itself.
	Smart and sustainable Energy	promoting sustainability.
	Smart Research and Development	Focus on a strong exchange of research institutions with the urban community to test and develop technological trends in a user-centered manner.
	Smart Infrastructure	Create new smart infrastructure as quickly as possible and at the same time harmoniously with the cityscape.
	Smart Privacy and Safety	Includes security monitoring and maintaining security in public places as well as cybersecurity.
	Smart Economy	It implies enhancing competitiveness, creating favorable framework conditions for innovation and investment, as well as creating a good and cooperative business climate to keep companies in place and attract new ones.
	Smart Provision of Job	The Smart City Strategy work areas lay the foundation for further development into a smart and livable city.
	Smart Waste Management	Promote recycling.
	Smart Parking	Improving parking.
	Smart Participant	Design and deliver digital engagement models that ensure civil society is better engaged and through engaged digital capabilities without excluding or separating people to development Dortmund Smart City.
	Partnerships	SIEMENS Collaboration.
	Smart Government	Adopting smart government services.
	Smart Police	detect and reduce crime rates.

3.2 United States of America initiative:

The Smart Cities Council, headquartered in the United States, launched an initiative aimed at promoting smart cities around the globe. This council consists of an organization that includes a list of key partners from the most influential global companies: Bechtel, Alstom Grid, Cisco, General Electric, Electricité de France, Qualcomm, Enel, IBM, S&C Itron, Microsoft, MasterCard, and National Grid [2, 8-10]. Focusing on policy, levels of non-governmental and governmental participation, and making data open and widely available in smart US cities such as San Francisco and Chicago, where both mayors of those cities signed orders to place an open data policy on the executive branch. In addition, most cities partner with non-governmental agencies in their development of smart city programs, including universities and companies in modern-day cities [6].

3.2.1 Smart city model in San Francisco

San Francisco presents the greenest city in the United States (2012) and known as a city of innovation and technology too. San Francisco leaders want to make the city healthier, safer, fairer, reducing carbon footprint but more affordable [10]. The city achieves zero waste and will become carbon neutral. San Francisco achieves a three-pronged approach to addresses administration, policy, and active community engagement in all available initiatives, including in table 3. [6]:

Table 3. United States of America initiative (San Francisco) toward smart city.

	Independent indicators	**Descriptions**
America initiative (San Francisco)	Smart Transportation	Implementing various intelligent transportation systems, such as contactless payment system in public transportation, using real-time data to improve traffic flow, implementing and creating a city-wide bike sharing scheme.
	Smart Buildings	Various initiatives have been implemented to improve the energy efficiency of buildings, and to rehabilitate buildings using energy-saving technology.
	Internet Of Things Technology	used IoT technology in various ways, such as using sensors to monitor air quality and improve service delivery in the city.
	Smart City Platform	developed various services.
	Smart Energy	find out the solar energy potential of their roof and save energy using the online portal and a specialized software platform.
	Smart Energy Saving	developed a smart city platform that integrates various city data and services to improve the lives of citizens and visitors.
	Smart Communications	The city has also established a modern communication system.
	Smart Monitoring	Implement intelligent monitoring systems capable of detecting potential security threats and responding to them in real time.
	Smart Street Lighting	implemented smart streetlights.
	Renewable Energy	implemented various initiatives to reduce carbon emissions.
	Electric Vehicle Charging	An extensive network of electric vehicle charging stations has been implemented to promote electric vehicles.
	Smart Waste Management	Implemented, recycling.
	Smart Parking	city's parking optimization.
	Smart Energy Marketing	Partnering with Duke Energy in this project to enable energy to market "Smart Energy in Offices" (SEiO),
	Partnerships	A strong partnership between public bodies, private and non-profit sectors with strong leadership to effective political reforms.

	Smart Government	Mandatory zero waste date, recycling and composting law, recycling, resource conservation and strengthening of the mayor's executive order to help combat global warming and reduce greenhouse gases.
	Smart Police	detect and reduce crime rates.

3.2.2 *Smart city model in Chicago*

Chicago is developing its smart city, and this decision laid the foundation for Chicago's smart city policy some time ago. solve problems by providing them with the information they need [6]. One of the initiatives that aims to achieve this true ambition is the OUT project. It aims to "measure the city and its neighborhoods in fine enough detail to provide data to help residents, planners, designers, engineers, scientists, and policymakers come together to make beautiful Chicago more livable, healthier, more efficient, more connected and open to each other." And open to the world. It will be used by everyone and accessible to everyone. Chicago collects some smart city standards, it could generate and disseminate data, but it does not collect all smart city standards. Some detailing of Chicago initiative toward smart city shows in table 4. [6].

Table 4. United States of America initiative (Chicago) toward smart city.

	Independent indicators	**Descriptions**
America initiative (Chicago)	Smart City Platform	Collect and aggregate data, make it available.
	Smart Communications	Utilizing data and Internet services for communication, Virtual maps of pipes and cables underground.
	Internet of Things Technology	used IoT technology in various ways, such as using sensors to monitor air quality and improve service delivery in the city. installed on street traffic (traffic lines), were temperature, atmospheric etc. Will be measured. Sound intensity, pedestrian and vehicle traffic, and surface temperature.
	Smart Energy	
	Smart Street Lighting	
	Smart Monitoring	One of UI Labs' endeavors is to deploy their sensors for monitoring.
	Smart Learning	Providing Digital literacy.
	Smart provision of job	Find a job and provide job opportunities in the city's low- and moderate-income neighborhoods.
	Smart privacy	Personal identification will not be public. Details, data and personal photos protected.
	Partnerships	The AoT consists of Chicago Digital city.
	Smart Government	Establishing municipal reporting and accountability requirements, i.e. establishing transparency, privacy policy for city.
	Smart Police	detect and reduce crime rates.

3.3 Asian Initiative

In the United Nations Sustainable Development Report (SDGs) by Price water house states, "The pillars of the sustainable smart city are aligned with the SDG proposal and the more integrated the initiative is, the greater the linkages with the SDGs." The Asian Development Bank has agreed in principle to allocate a budget of one billion dollars to provide loans for smart city projects in the Asian region. The World Bank can provide millions in long-term loans, which encourages support for urban transformation plans towards sustainable smart cities [6].

3.3.1 Smart city model in China

Since these first attempts to establish the SCO, many aspects of theory and practice in the field of smart cities have changed, opening the way for a better understanding of the range of terms used and proposed for the digital and urban smart city vision, their meaning and the relationships between them. And work on platforms and frameworks for high-level and more detailed domain-specific design developments, such as BFO, DOLCE and others [5]. Not surprisingly, IBM's smart city concept has been highly successful and popular in China as well [2]. For more about China's initiatives, see table 5. [5].

Table 5. Asian initiative (China) toward smart city.

	Independent Indicators	Descriptions
Asian initiative (China)	Smart Transportation	Adopting smart transportation and related applications.
	Internet Of Things Technology	Supporting the development of the Internet of Things,
	Smart City Platform	Supporting the development of Big Data, Artificial Intelligence and other emerging technology.
	Smart Building	Techniques for transform old houses to "smart" and environmentally friendly apartments in a record period and at a low cost.
	Renewable Energy	Display a model of a building that generates electricity using the LIGHT Series lightweight PV curtain wall, along with a collection of buildings that improve everyday life.
	Smart Energy	promoting the use of smart and sustainable energy.
	Smart Learning	A pioneer in the field of smart education.
	Smart marketing	Deploying digital services, data and Internet infrastructure.
	Smart Health	Notifying relevant departments of potential emergency situations, and the issue of preventing the Corona pandemic is one of the applications mentioned.
	Smart Research and Development	Technical and scientific innovation has become Beijing's new calling card, and quality controlled through modern technologies.
	Smart Infrastructure	Super magnetic environmental purification equipment, effective and broad-spectrum heavy metal sterilization for cleaner earth, purer water.
	Smart privacy and safety	Creating ideal solutions for government surveillance, cybersecurity, and privacy preservation.
	Partnerships	Alibaba, Tencent, and Baidu have set up projects.
	Smart Government	Deep application for smart city management and automate municipal functions while integrating monitoring functions.

4 Result and discuss local urban transformation

By sorting practices for cities dealing with the study, it is possible to prepare a table with indicators that can be applied locally to be considered as a measurement of local cities' readiness in the urban transformation of traditional cities into smart cities. Table 6. Baghdad City Readiness checklist, the light green color presents the effectiveness of the indicators, while the light orange color shows the ineffectiveness of the indicators. According to the testing schedule, some indicators need to be strengthened because they are not effective in the city of Baghdad. Also, some indicators, despite their effectiveness, have not been fully achieved, which requires review and support. The situation boils down to the lack of real partnerships with major companies in the field of smart technologies, and in addition to support from international organizations.

Table 6. Baghdad city readiness checklist for transformation towards smart cities.

	Independent Indicators	**Active**	**Inactive**
Baghdad initiative	Smart Transportation		
	Internet of Things Technology		
	Smart City Platform		
	Smart Communication		
	Smart Buildings	Not fully implemented	
	Smart Health	Not fully implemented	
	Smart traffics		
	Smart lighting	Not fully implemented	
	Smart Energy	Not fully implemented	
	Smart Learning		
	Smart energy marketing	Not fully implemented	
	Smart Energy Saving		
	Smart and sustainable Energy		
	Smart Research and Development	Not fully implemented	
	Smart Infrastructure		
	Smart Privacy and Safety	Not fully implemented	
	Smart Economy		
	Smart Provision of Job		
	Smart Waste Management		
	Smart Parking	Not fully implemented	
	Electric Vehicle Charging		
	Smart Participant	Not fully implemented	
	Partnerships	Not fully implemented	
	Smart Government		
	Smart Police		

Initiative has been the first shift in different areas of domestic cities in 2003, especially in major cities such as Baghdad. This change occurred because of the great openness to the world and the powerful of internet after information and data are restricted and access to a lot of information is available [2, 8]. After the Corona virus pandemic and the transmission of diseases in an era of viruses, the urgent need to activate the digital aspect and the development of the data system to maintain continuity of the city's operations and dictate. In addition, intelligent technological development in the field of urban design and its software has helped to evaluate the city's areas and its opposition between prejudices and assess their efficiency before implemented. Communication with sensors and meteorological stations and the spatial data system have helped to activate predictive design [4,7].

With the clarification of urban plans and dividing areas into digital sectors and coding them and using urban simulation programs to predict future designs through integration in communication with geographic information systems and meteorology to reduce the environmental and other challenges facing our cities and assess their reality Smartphones connected to the Internet and services launched by the central government facilitate work according to the smart transformation approach. The smart transformation process may impose a serious challenge in preserving data and confidentiality of information from hacking, and this is also what governments must do.

5 Conclusion

This research presents a study of smart city initiatives around the world, which includes within these initiatives' broad examples of case studies. What has been selected are 5 case studies within 3 basic initiatives: the European initiative, the American initiative, the Asian initiative, and the Gulf Cooperation Council initiative. The research explores the possibility of transforming existing traditional cities into smart cities, with the main indicators and performance description details of these indicators, each within its own initiative, i.e., exploring the different ways in which smart cities affect and are affected by planning, urban design and digitalization. It concluded from the urban transformation initiatives that addressed that traditional city can be transformed into smart cities within a systematic process in the era called fifth urbanization, in reference to the fifth generation of the technology and communications revolution. Cities have proven throughout the ages to be the main drivers of innovation, creativity, and urban, economic, and socio-cultural growth. Today, more than ever, cities have become a field for urban digital transformations, and this study presents several indicators at the intersection of smart city initiatives and urban design transformation. Data opens the technology way and digitization to intervene in all areas of city life. This includes the use of sensors and smart cameras to monitor and predict human behavior. Smart cities integrate different systems and services, leading to the integration of different city sectors, such as transportation, energy and healthcare, enabling the management and optimization of urban resources. Smart cities leverage technology and data to improve service delivery, reduce congestion, and enhance operational efficiency. Enhance communications through artificial intelligence, simulation, BIM and 5G connectivity to address various urban challenges. The smart city model follows a multi-layered approach to describe smart urban areas that enrich urban city design, the traditional city area (layer 0), may be topped by the green city layer (layer 1), and its smart cloud and policies formulate the connected city layers equipped, open and intelligent (layers 2,3,4 and 5), measures to mobilize the innovation ecosystem and infrastructure in the city (layer-1), Baghdad Multi-Layered Approach. According to what has been assess and discussed, it can be stated that there are measures for urban transformation towards smart cities. Although the city of Baghdad needs to provide and promote some transformation indicators, it is taking its steps towards smart urban transformation.

References

1. J. Lee, "Smart City in Urban Design," Int. J. Sustain. Build. Technol. Urban Dev. **11(4)**, 380–393 (2021), DOI: 10.22712/SUSB.20210031
2. F. Ullah, F. Al-Turjman, "A conceptual framework for blockchain smart contract adoption to manage real estate deals in smart cities," Mach. Learn. Appl. Secur. **35**, 5033–5054 (2023)
3. N. Komninos, A. Panori et al., "Smart City Ontology 2.0," URENIO Res. Discuss. Pap., Aristotle Univ. Thessaloniki (2020)
4. National League of Cities, Trends in Smart City Development-Case Studies and Recommendations (2016)
5. V. Javidroozi, C. Carter et al., "Smart, Sustainable, Green Cities: A State-of-the-Art Review," Sustainability **15**, 5353 (2023), DOI: 10.3390/su15065353
6. A. Karvonen, M. Cook, H. Haarstad, "Smart Cities: Researches, Projects and Good Practices for the City," Urban Plann. **5(1)**, 65–68 (2020), DOI: 10.17645/up.v5i1.2936
7. Z. J. Makieła, M. M. Stuss, et al., "Smart City 4.0: Sustainable Urban Development in the Metropolis GZM," Sustainability **14**, 3516 (2022), DOI: 10.3390/su14063516
8. M. Bakhit, M. Azmi, "Smart Cities Adoption in Saudi Arabia: A Comprehensive Review and Future Drivers," Br. J. Multidiscip. Adv. Stud. Eng. Technol. **4(5)**, 20–39 (2023), DOI: 10.37745/bjmas.2022.0300
9. American Planning Association, Introduction to Smart Cities, Texas Chapter State Conference & Short Course for Planning Officials, San Antonio, Texas (2016)
10. Ministry of Regional Development of the Czech Republic, Smart Cities Methodology-Methodology for the Preparation and Implementation of the Smart Cities Concept at the Level of Cities, Municipalities and Regions (2018), www.smartcities.mmr.cz

E3S Web of Conferences **633**, 07021 (2025) https://doi.org/10.1051/e3sconf/202563307021
GASES 2025

Utilization of remote sensing and GIS for land use and land cover mapping in Wasit province, Iraq

Meena K. Abdulkareem*, and *Hussein* Sabah Jaber

University of Baghdad, College of Engineering, Department of Surveying, Baghdad, Iraq

Abstract. All physical components on the Earth's surface are collectively termed land cover, such as water, vegetation, bare soil, and artificial structures made by humans. In contrast, land use refers to how humans use the land for urban growth, agricultural practices, and commerce activities. The increase in population density has resulted in changes in land use/land cover (LULC) because of urban expansion and other human activities. This study investigates the effectiveness of a supervised MLC classification algorithm to produce LULC maps of Wasit Governorate from multiple satellite images using remote sensing and GIS techniques, which is of great importance for earth observation applications, such as environmental monitoring and disaster prediction. Three Landsat 8-9/OLI satellite images were used to cover the entire study area for the year 2024. Preprocessing including radiometric and atmospheric corrections was performed using ENVI 5.3V software, and mosaicking, layer stacking, and image subsetting were performed using Arc GIS 10.8 to improve image quality. Image processing was performed using ENVI 5.3V to find LULC classes using the Maximum Likelihood (MLC) algorithm. The results indicated five LULC classifications in the study area: vegetation, water bodies, built-up areas, arid areas and saline areas. Furthermore, post-processing of the images, including accuracy assessment, was performed to assess the accuracy of the LULC mapping, yielding an overall agreement of 90.58% and a Kappa index of 88%, which measures the extent of agreement or accuracy. These results are convincing with a kappa index of more than 0.8. This study demonstrates the potential of Landsat 8-9/OLI images integrated with machine learning methods for efficient land use classification, analysis and monitoring. This study also provides broad horizons of knowledge for planners and regional authorities.

1 Introduction

LULC play an important role in many aspects, such as environmental monitoring, agricultural production, disaster prediction and urban development. Remote sensing techniques through satellite imaging have become one of the most important technologies used all over the world, and they are essential for producing LULC maps with accurate

* Corresponding author: mina.abd2412m@coeng.uobaghdad.edu.iq

classifications. Classification of LULC is critical to environmental, cultural and political research, as it is the basis for government oversight that contributes to achieving sustainable livelihoods in line with modern ecosystems. Remote sensing, in combination with GIS, provides advanced tools for managing and monitoring modern ecosystems [1]. Landsat 8 was launched in February 2013 through a collaboration between NASA and the U.S. Geological Survey (USGS), providing high-spectral coverage and spatial resolution data [2].in September 2021, Landsat 9 was launched to improve the quality of Landsat 8 data [3] .Several studies have highlighted the high capabilities of Landsat 8,9 data in LULC maps classification. A study conducted by [4] showed the effectiveness of Landsat 8,9 satellites in mapping LULC. The study's findings by demonstrated that LULC mapping is used to identify urban growth and changes in arid regions, agricultural areas, and waterways. Additionally, the Landsat imagery utilized aids in creating GIS maps that precisely depict the distributions of evaporation, precipitation, river levels, discharge, temperature, and dust storms[5]. Researchers' studies [6], [7] have highlighted the benefits of using Landsat 8 to monitor LULC, especially when related to managing archaeological and heritage sites. This study discussed the importance of using modern technologies represented by remote sensing and GIS to find maps of LULC and its changes and monitor them instead of using traditional surveying methods because they are expensive and complicated and require more effort, especially if the study area has a large area, as remote sensing technologies and GIS are considered the most efficient in terms of shortening time, effort and cost and providing more excellent coverage through satellite images and can find classifications of LULC maps for a large area such as the study area represented by Wasit Governorate, which lacks previous studies in this regard. This study aims to demonstrate the use of remote sensing and GIS in producing and analyzing LULC maps for Wasit province and assess the accuracy of classification of these maps using the Kappa coefficient.

2 Methods

2.1 Study area

The province of Wasit is situated in the eastern portion of the central area of Iraq, between latitudes 31° 52′ and 33° 36′ and longitudes 44° 30′ and 46° 40′ (Figure 1). Numerous provinces, including Baghdad, Diyala, Babil, Al-Qadisiyah, DhiQar, and Maysan, are its neighbours. It shares an eastern border with Iran State. The province of Wasit possesses abundant natural resources, including oil and water. Other significant sectors of the economy include trade and farming. The capital of the Wasit governorate, situated on the Tigris River, is Al-Kut City. With a total area of 17889.09 km2, the Wasit Governorate makes up 4% of Iraq's total land area. With limited rainfall and high temperatures, it is categorized as a coastal region with a climate between the Mediterranean and hot, arid deserts [8]

 https://doi.org/10.1051/e3sconf/202563307021

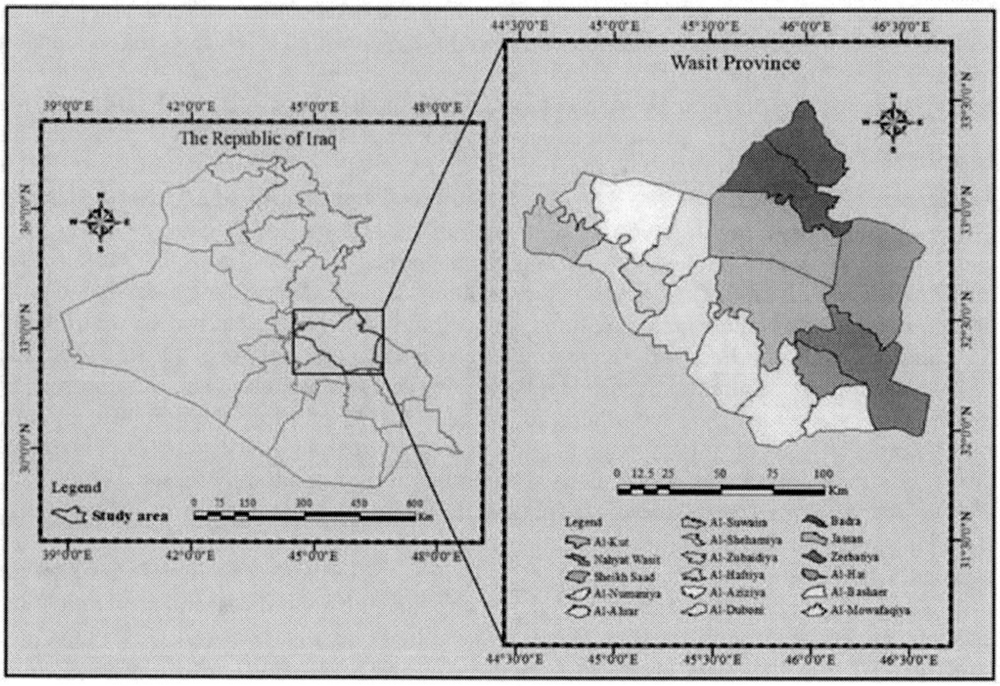

Fig. 1. Study area map.

2.2 Materials

Landsat 8-9/OLI images were used in this study since they were accessible and had a spatial resolution of 30 meters. Three images of the study area were gathered and pre-processed to correct atmospheric and radiometric errors. Figures (2,3,4, and 5) show images collected by the USGS in September 2024, which are cloud-free.

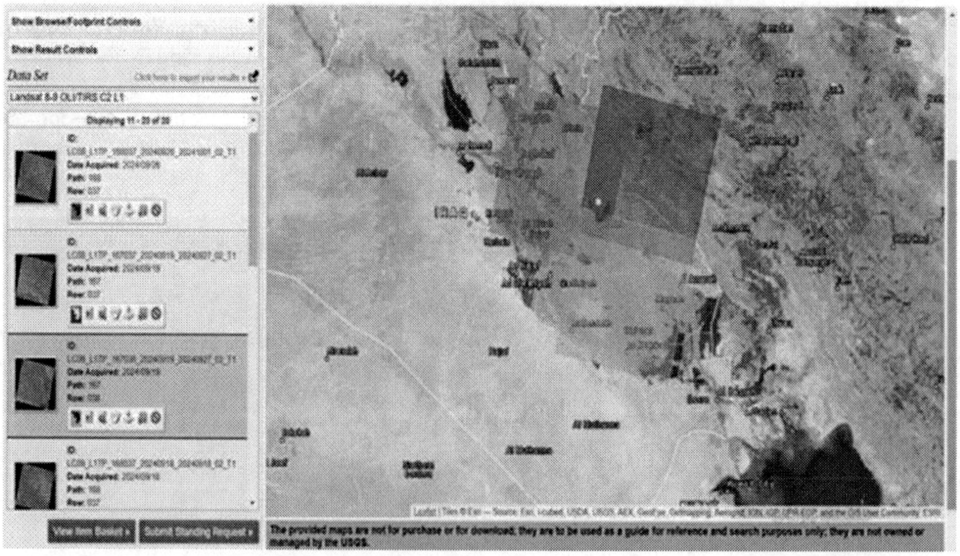

Fig. 2. USGS website.

Fig. 3. Landsat 8-9 OLI image (1) (USG).

Fig. 4. Landsat 8-9 OLI images (2)(USGS).

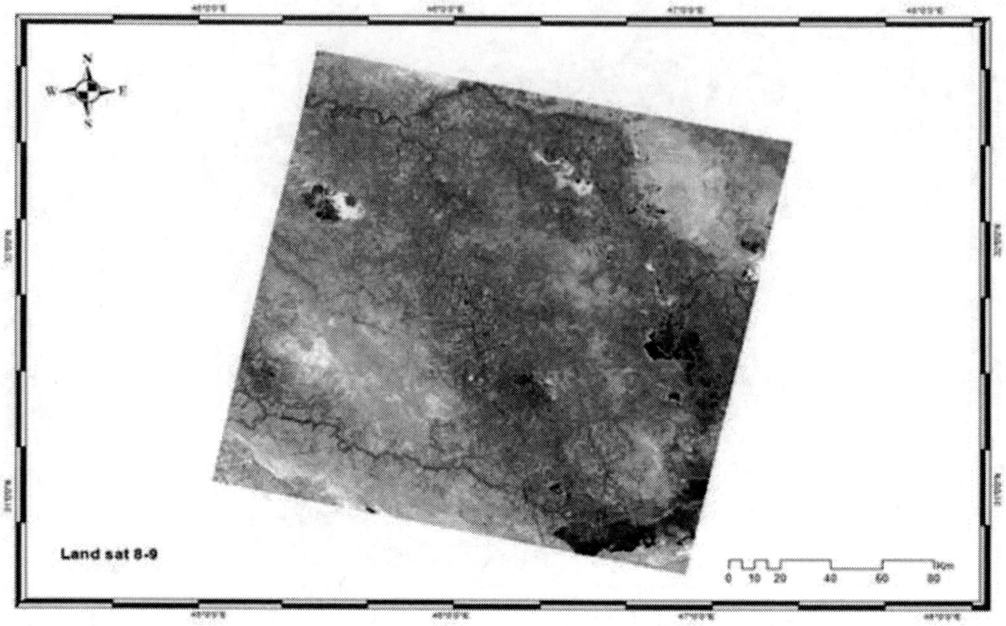

Fig. 5. Landsat 8-9 OLI images (3)(USGS).

2.3 Methodology of Research

This study employs a series of processing stages using remote sensing and GIS, as illustrated in Figure (6).

https://doi.org/10.1051/e3sconf/202563307021

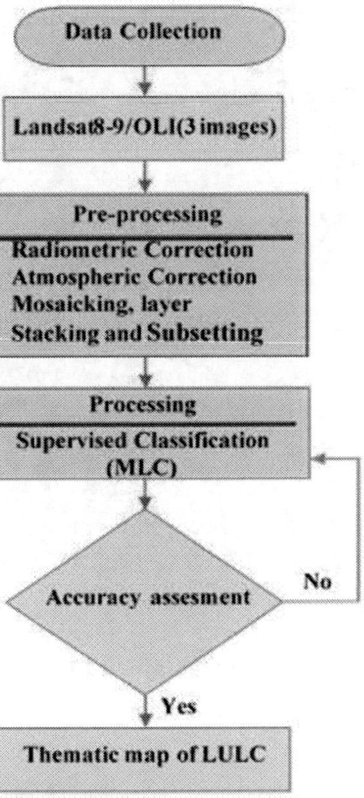

Fig. 6. The diagram of the research methodology.

2.3.1 The procedure of mosaicking satellite imagery

Upon completing the layer stacking procedure, which consolidates all bands into a singular file for enhanced accessibility inside the application, the processing is further refined using the composite band tool in ArcGIS 10.8 software [9], [10]. The metadata specifies a standard coordinate system, which is then used for image mosaicking, as illustrated in Figure (7).

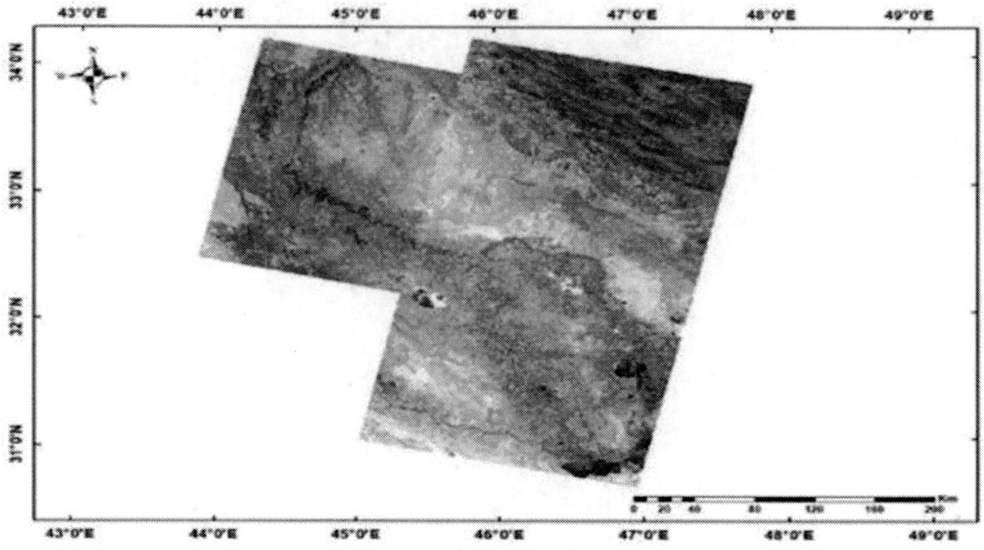

Fig. 7. Mosaiking Landsat 1-2-3 images for the study area.

2.3.2 The procedure of sub-setting satellite imagery

Fig. 8. Image after subsetting.

2.3.3 The Maximum Likelihood Classifier (MLC)

The MLC assumes that the data to be categorized follows a Gaussian or normal distribution. It uses training data to determine the categories' means, variances, and covariances, which are then used to evaluate probabilities. This method considers the mean values and the variation in brightness values within each category when assigning classifications[11]. The MLC algorithm is an effective classification method as it has the ability for a quantitative consideration of several classes and spectral bands simultaneously[12], [13]. as seen in the equation below [14] :

$$g_i(x) = \ln p(\omega i) - \tfrac{1}{2} \ln|\textstyle\sum i| - \tfrac{1}{2}(x - m_i)^T \textstyle\sum^{-1}(x - m_i) \quad (1)$$

Where

i: Category.

x: Data of n-dimensional (n: the number of bands).

p(ωi): The probability that category ωi happens in the image.

$|\Sigma_i|$: The determinant of the covariance matrix of category ωi data.

Σ_i^{-1}: The matrix inverse.

m_i: Mean vector.

Envi 5.3's Region of Interest (ROI) tool was used to gather the training data. Additionally, ROIs were randomly constructed as pixel-based validation data to assess accuracy.

2.3.4 Performance evaluation strategies

To ensure the reliability of the classification results, the last step in processing satellite images is to evaluate the accuracy of the classification. This is done by comparing the results of the classification model with reference data that accurately illustrate the groups of LULC classifications. The first stage of accuracy assessment is to collect original accurate data in specific areas which show the groups of LULC classifications. This information can be provided either through factual information obtained from field surveys or from other reliable methods, such as high-quality images and maps. A confusion matrix is then created. This tabular representation compares the classification results to the source data, and multiple accuracy measurements can then be made from the data in the confusion matrix.[15]. Producer's accuracy is calculated by dividing the total number of pixels identified in a category by the total number of pixels in that category identified by ground reference data (i.e., the column total) [14]:

$$\text{Producer's accuracy for class j} = \frac{x_{jj}}{x_{+j}} \quad (2)$$

The user's accuracy is the ratio of the accurate pixel count in a category to the overall pixel count categorized within that class[14]:

$$\text{User's accuracy for class i} = \frac{x_{ii}}{x_{i+}} \quad (3)$$

Overall accuracy (OA) is defined as the proportion of accurately categorized pixels to the total pixel count in the image. To calculate this, divide the total number of accurately categorized pixels (i.e., the diagonal sum of the error matrix) by the total number of pixels within the image(N) [14]:

$$OA = \frac{\sum_{i=1}^{k} x_{ii}}{N} \quad (4)$$

The Kappa index is an important indicator that evaluates the degree of concordance between observed and anticipated classifications while considering the likelihood of agreement occurring by chance [14]:

$$K = \frac{N\sum_{i=1}^{k} x_{ii} - \sum_{i=1}^{k}(x_{i+} \times x_{+j})}{N^2 - \sum_{i=1}^{k}(x_{i+} \times x_{+j})} \quad (5)$$

Where:

N= Total number of observations

k = The number of rows in the error matrix

$x$ii=The number of observations in row i and column i

xi+= The marginal totals for row i

x+j = The marginal totals for column j

The Kappa index value ranges from (zero to one). Zero indicates a significant disparity between the categorization outputs and the reference findings. There is strong agreement that the precision is moderate if the value is equal to 1 and ranges from 0.4 to 0.8. The lower value correlates with a stochastic categorization.

3 Results and discussion

3.1 Findings from thematic map classifications

Supervised classifications of images were conducted using the MLC algorithm in ENVI 5.3v. The classification results have been transformed into shapefiles for use in ArcGIS 10.8v, enabling the development of the LULC thematic maps, which are categorized under five principal categories depending on visual analysis and field observations of the studied area: Built-up, vegetation, saline, barren, and water bodies, as illustrated in Figure (9).

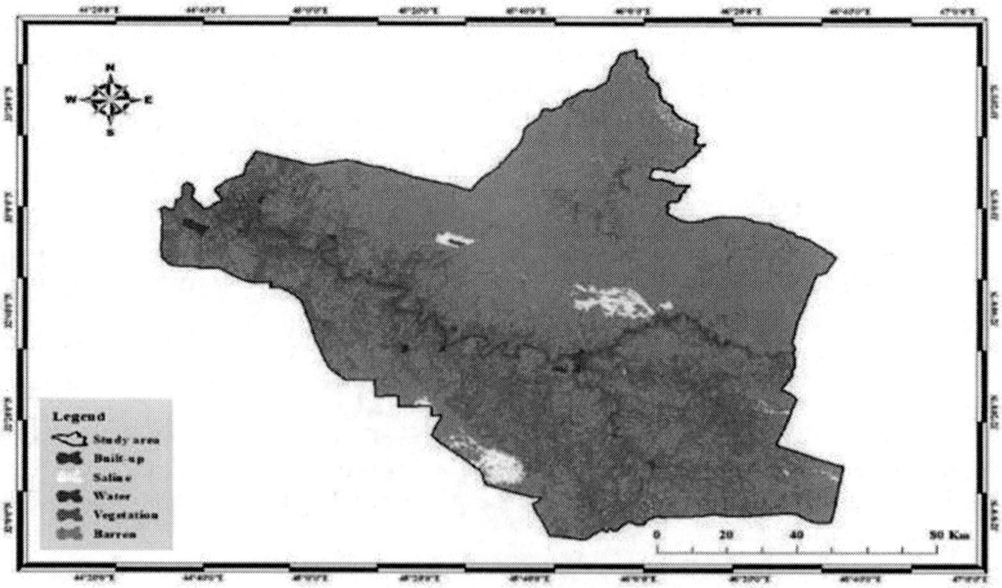

Fig. 9. LULC Thematic map utilizing MLC.

3.2 The results of area computations with percentage allocations for classes

Table (1) displays the area computation and each class's percentages. At more than 69.86%, most of the class reported a barren area. Only 1.63% of the total area came from water sources, primarily rivers and branches; another 0.82% came from built-up areas. The vegetation class had the second-highest percentage at 27.24 %. Soil area accounted for almost 0.45% of the total but covered more than 80.16 km^2 in the actual Area.

Table 1. Computation of area and percentage of classes.

Class	Area(km^2)	Percent %
Barren	12496.71	69.86%
Vegetation	4874.13	27.24%
Water	290.79	1.63%
Built-up	147.30	0.82%
Saline	80.16	0.45%
Total	17889.09	100%

3.3 Evaluation of Classification Accuracy and Verification Results

Upon concluding the classification process using ENVI 5.3v software, the accuracy of the algorithm-generated images was assessed and validated. The results showed approval, as the Kappa index was more than 0.8. Table (2) below displays the confusion matrix of five classes in the studied area. MLC classification produced an overall accuracy of 90.85% and a kappa index of 0.88 (Table 2). Table 2 indicates that around 12,000 samples were selected for each of the five classes in the research area, with a distribution of 2,400 samples per class, to methodically and thoroughly represent the region. Three classes had the highest degree of misclassification utilizing the MLC. The barren area has a deficient user accuracy of 75.04%, and approximately 25% of pixels identified as barren areas were misclassified, specifically vegetation, water bodies, built-up areas, and saline areas. The vegetation had a user accuracy of 88.09%, which is considered low. The pixels identified as vegetation were, in fact, urban areas, water bodies, saline regions, and, to a lesser extent, barren areas. The built-up class has a high misclassification rate, with a producer accuracy of only 70.39 %. This means that approximately 30% of the examined pixels selected as built-up were mistakenly classified as other categories, such as vegetation, barren areas, saline areas, and particularly water bodies. Saline regions exhibit the highest producer's accuracy among all categories. 99.38%. Conversely, water bodies exhibit superior user accuracy compared to other categories (100%).

Table 2. Confusion matrix of the MLC approach.

Actual Data (in pixels)							
Class	Built-up	Saline	Water	Vegetation	Barren	Total	User's accuracy
Built-up	7438	0	0	7	229	7674	96.92%
Saline	4	10538	85	0	116	10743	98.09%
Water	0	0	10021	0	0	10021	100%
Vegetation	133	20	397	10427	859	11836	88.09%
Barren	2991	45	6	76	9375	12493	75.04%
Total	10566	10603	10509	10510	10579	52767	
Producer's accuracy	70.39%	99.38%	95.35%	99.21%	88.6%		
Overall accuracy = (5311/5863) =90.58%							
Kappa Index for Agreement = 0.88							

The evaluation of land use map classification accuracy using the MLC algorithm is deemed satisfactory since the Kappa index exceeds 0.8.

4 Conclusions

This study adds to the existing body of knowledge by examining the efficacy of a supervised MLC classification algorithm to extract LULC maps from Wasit City satellite images by applying GIS and remote sensing. With a kappa score of 0.88 and an overall accuracy of 90.85%, the results demonstrate the efficacy of extracting LULC from satellite images using the MLC method and calculating its class areas. The study outcomes ensure the examined region's sustainability and can be utilized for environmental monitoring, water resource management, and urban planning. To determine the best, it is necessary to continue testing these methods using many categories of LULC and satellite images. It is recommended that the MLC algorithm be used to detect LULC and create a geodatabase by integrating remote sensing and GIS for all areas of Iraq. The results can be compared with different classification techniques.

References

1. M. M. Kadhim, “Monitoring land cover change using remote sensing and GIS techniques: a case study of Al-Dalmaj marsh, Iraq,” Journal of Engineering **24(9)**, 96–108 (2018), DOI: 10.31026/j.eng.2018.09.07
2. E. Chuvieco, *Fundamentals of Satellite Remote Sensing: An Environmental Approach* (CRC Press, 2020), DOI: 10.1201/9780429506482
3. J. G. Masek, J. R. Irons, A. L. M. MacDonald, C. E. Justice, S. J. Hook, "Landsat 9: Empowering open science and applications through continuity," Remote Sens. Environ. **248**, 111968 (2020), DOI: 10.1016/j.rse.2020.111968
4. Shahfahad, S. S. Ghosh, A. Ahmad, A. Rahman, "Comparative evaluation of operational land imager sensor on board Landsat 8 and Landsat 9 for land use land cover mapping over a heterogeneous landscape," Geocarto Int. **38(1)**, 2152496 (2023), DOI: 10.1080/10106049.2022.2152496
5. L. K. Abbas, "Mapping land cover/land use for change derivation using remote sensing and GIS technique," Iraqi J. Sci. **62(10)**, 3772–3778 (2021), DOI: 10.24996/ijs.2021.62.10.35
6. F. A. Zwain, T. T. Al-Samarrai, Y. I. Al-Saady, "A study of desertification using remote sensing techniques in Basra Governorate, south Iraq," Iraqi J. Sci. **62(3)**, 912–926 (2021), DOI: 10.24996/ijs.2021.62.3.22
7. H. S. Abbas, A. S. Mahdi, "Study of desertification using remote sensing imagery in south Iraq," Iraqi J. Sci. **60(4)**, 904–913 (2019), DOI: 10.24996/ijs.2019.60.4.24
8. A. Sahar, *Change detection for some land cover types of Wasit Province (Eastern Iraq) using remote sensing and GIS techniques for years 1989–2017* (2018)
9. M. H. Al-Helaly, I. A. Alwan, A. N. AL-Hameedawi, "Environmental investigation of Bahar Al-Najaf region using Sentinel-2 images," Engineering and Technology Journal **40(4)**, 732–742 (2022), DOI: 10.30684/etj.v40i5.2245
10. H. B. Ghalib, A. B. Al-Hawash, W. S. Al-Qurnaw, B. H. Sultan, A. W. Al-enzy, "Marshes waters sources hydrochemistry of the Bahr Al-Najaf at Najaf Province, Iraq," J. Phys.: Conf. Ser. **1279**, 012059 (2019), DOI: 10.1088/1742-6596/1279/1/012059
11. R. S. Dwivedi, *Remote Sensing of Soils* (Springer, 2017), DOI: 10.1007/978-3-662-53740-4
12. J. S. Alawamy, S. K. Balasundram, A. H. Mohd. Hanif, C. T. Boon Sung, "Detecting and analyzing land use and land cover changes in the region of Al-Jabal Al-Akhdar,

Libya using time-series Landsat data from 1985 to 2017," Sustainability **12(11)**, 4490 (2020), DOI: 10.3390/su12114490

13. S. Mushtaq, S. Saleem, R. Ahmed, M. S. Tass, J. A. Rather, G. M. Rather, "Spatio-temporal analysis land use land cover changes in South Kashmir region of North-western Himalayas using Landsat data," Discover Geoscience **2(1)**, 37 (2024), DOI: 10.21203/rs.3.rs-3830014/v1
14. E. Miranda, A. B. Mutiara, W. C. Wibowo, "Classification of land cover from Sentinel-2 imagery using supervised classification technique (preliminary study)," in *2018 International Conference on Information Management and Technology (ICIMTech)*, IEEE, 69–74 (2018), DOI: 10.1109/ICIMTech.2018.8528122
15. H. A. Abbas, H. S. Jaber, A. D. Salman, "Land use mapping from Sentinel satellite imagery using remote sensing and GIS techniques," in IOP Conf. Ser.: Earth Environ. Sci. **1374**, 012042 (2024), DOI: 10.1088/1755-1315/1374/1/012042

 https://doi.org/10.1051/e3sconf/202563308001

Phase transformations and physico-mechanical properties of ceramic tiles from Lower Amu Darya raw materials for sustainable construction in earth and environmental engineering

Islom Boyjanov, *Djakhangir* Djabberganov[*], *Karomat* Matyaqubova, *Erkaboy* Khudayberganov, and *Tulkinbek* Sitmuratov

Urgench State University named after Abu Rayhan Biruni, 220100 Urgench city, Uzbekistan

Abstract. This work investigates the physico-mechanical characteristics and phase transformations of ceramic tiles fabricated from local raw materials in the Lower Amu Darya region of Uzbekistan—specifically, clay from the Kulatau deposit and dune sand from the Tuprak-Kala massif. Laboratory- and industrial-scale experiments showed that an optimized composition, M-7, can be employed to produce unglazed facade and floor tiles fired at 1000°C and 1050°C. Under the same conditions, composition M-9 yields glazed facing and facade tiles with an opaque finish. The phase composition of the tiles includes mullite, anorthite, quartz, hematite, and a glassy phase, indicating that the mineral components significantly influence crystal growth and enhance physico-mechanical properties. These findings support the advancement of sustainable construction materials and resource-efficient practices within Earth and environmental engineering.

1 Introduction

After gaining independence, the production volumes of ceramic products in the Republic of Uzbekistan have been expanding year by year. However, to obtain high-quality ceramic products, including ceramic tiles, the country faces an acute shortage of low-melting fluxes, which ensure the sintering of ceramic products at low firing temperatures.

Therefore, the study of the physico-chemical properties of local mineral raw materials in the Republic, which can replace scarce low-melting fluxes such as nepheline syenite, cullet, and others, ensuring the sintering of ceramic products at low firing temperatures, is highly relevant.

Ceramic tiles, which play a crucial role in building material production, are made using clay-based raw materials. [1].

As is known, one of the ways to regulate the properties of ceramic products is by introducing fluxes into the composition of the masses. In the ceramic industry, traditional fluxes such as feldspars, pegmatites, perlites, nepheline syenites, and others are currently used [2].

[*] Corresponding author: jjs.ursu@mail.ru

Study [3] explores the varying perspectives of researchers regarding the impact of mass components on the formation of ceramic structures.

Researchers in study [4] have identified the positive influence of low-melting clays on the sintering process of ceramic masses, resembling the effect observed during the firing of low-temperature fluxes.

To date, ceramic materials science has accumulated extensive knowledge regarding mass compositions, physico-chemical properties, and technological aspects of ceramic tile production. However, in recent years, the issue of implementing resource- and energy-saving production technologies has become increasingly urgent [5].

Traditional technology involves the production of ceramic tiles from multi-component masses. However, reducing the number of raw material components in ceramic masses without compromising the quality characteristics of the ceramic body would significantly decrease material and labor costs while also reducing the production cost of finished products [6].

The authors of study [7] examined the chemical and mineralogical compositions, as well as the physico-chemical characteristics, of Yaziyavan dune sands and their suitability for use with other raw materials from Uzbekistan to develop optimal compositions for ceramic facing tiles. Studies have shown that the proportion of dune sand in ceramic facing tile compositions should be limited to a maximum of 45% in ternary mixtures and 25% in quaternary mixtures. Furthermore, the optimal technological parameters for manufacturing ceramic tiles with dune sand were determined in compliance with GOST 6141-91 standards.

During the enrichment processes feldspar quartz sand of the Khiva mine, their granulometric and chemical composition were studied [8,10].

Study [9, pp. 54-55] also explored the feasibility of utilizing dune sands from the Tuprak-Kala massif in Uzbekistan's silicate industry. The results revealed that these sands, containing approximately 30% finely ground feldspar with anorthite as the dominant component, serve as a valuable raw material for ceramic production. Due to their availability and low cost, Tuprak-Kala dune sands can successfully replace the scarce and expensive quartz-feldspar components in the production of ceramic tiles in the republic. Considering the aforementioned findings, the development of ceramic tile compositions incorporated low-melting clay from the Kulatau deposit and dune sands from the Tuprak-Kala massif.

2 Methods

GOST 9169-75 and modern physico-chemical analysis methods, such as chemical analysis, X-ray diffraction, comprehensive thermographic analysis, and others, were used.

The preparation of experimental masses and samples was carried out according to the methods described, which are adopted in the wet-process ceramic tile production technology. This included batch preparation, grinding in a ball mill, followed by drying of the resulting slip to produce press powder.

3 Results of experimental research

To investigate the impact of the ratio between Kulatau clay and Tuprak-Kala dune sand on the physical and technical properties of ceramic tiles, a series of experimental mixtures were prepared. Following initial research, 11 compositions with the most favorable performance characteristics were selected for further in-depth analysis.

The sintering processes of the experimental tile compositions were investigated by determining their physico-mechanical properties at different firing temperatures, as well as by examining phase transformations during firing. Figures 1-4 illustrate the correlation between the physico-mechanical properties of the experimental tile samples and the firing temperature. For better clarity, only the curves of compositions with odd numbers are displayed in the graphs.

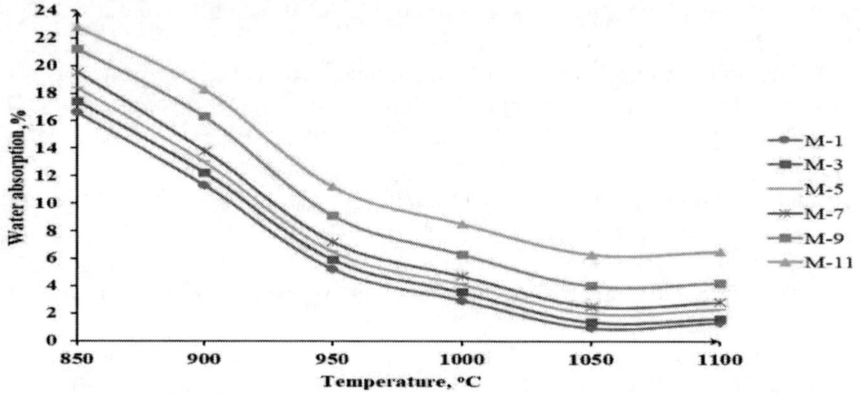

Fig. 1. Dependence of water absorption in experimental compositions on firing temperature.

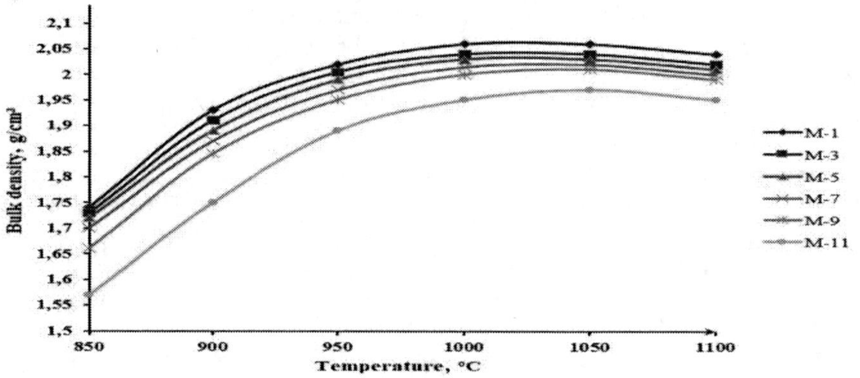

Fig. 2. Dependence of bulk density of experimental masses on firing temperature.

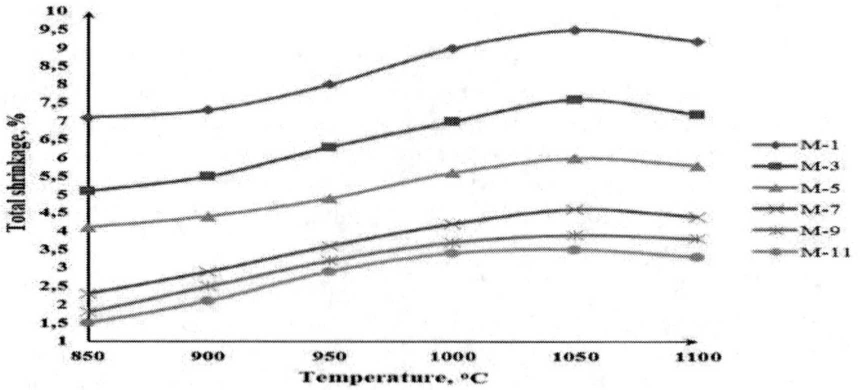

Fig. 3. Dependence of total shrinkage of experimental masses on firing temperature.

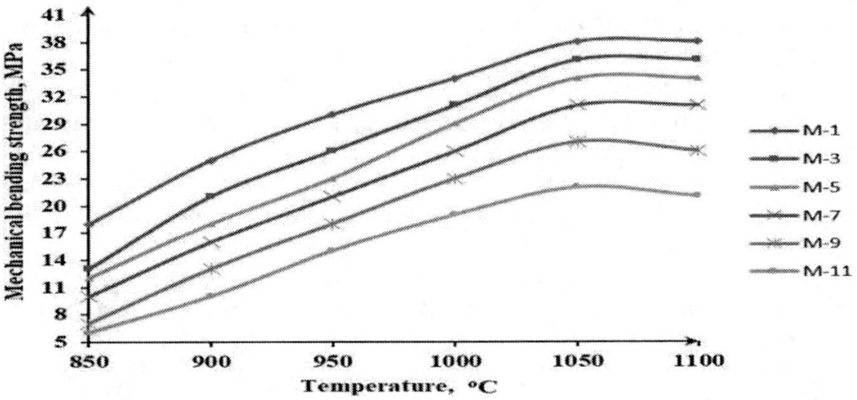

Fig. 4. Dependence of bending strength of experimental masses on firing temperature.

From Figures 1-4, it is evident that starting from 850°C, the properties of the experimental masses undergo significant changes. As the temperature increases from 1000°C to 1050°C, the overall shrinkage growth of the experimental masses decreases significantly. The curves for linear shrinkage, water absorption, bending strength, and bulk density begin to level off, forming a horizontal trend, indicating the completion of the sintering process in the tile masses.

When the firing temperature is further increased to 1110°C, a rise in water absorption is observed. Consequently, bending strength and bulk density decrease significantly. The samples exhibit signs of melting, along with deformation elements and swelling effects.

As observed in Figures 1-4, the samples from experimental compositions M-1, M-2, M-3, M-4, M-5, and M-6 exhibit higher water absorption values at 1000°C and 1050°C compared to mass M-7. However, the total shrinkage of these samples exceeds 5%, surpassing the standard limits for ceramic tiles. Meanwhile, the samples from masses M-8 and M-9 demonstrate slightly lower quality than mass M-7.

The samples from masses M-10 and M-11, fired at 1000°C and 1050°C, exhibit lower performance compared to mass M-9, with water absorption values exceeding 5.1%, rendering them unsuitable for facade tiles.

The fired experimental samples from mass M-7, at 1000°C and 1050°C, fully meet the GOST standards for facade and floor tiles, respectively.

Under industrial conditions at SP LLC "ART GLOSS GALLERY", unglazed, solid-colored tiles of a reddish-brown hue were produced from mass M-7 at the specified firing temperatures. These tiles fully comply with GOST standards for facade and floor tiles.

However, it is important to note that glazed samples made from mass M-7 developed bubbles in the glaze layer when fired under factory fast-firing conditions. This issue arises because, during rapid firing, mass densification occurs relatively early, preventing carbonate and other gases from fully escaping. At higher temperatures, residual gases trapped within the mass migrate through the glaze layer, leading to surface defects in the final product.

Experimental samples made from mass M-9, fired at 1000°C and 1050°C under laboratory conditions, fully comply with GOST requirements for cladding and facade tiles, respectively. Similarly, under industrial conditions at SP LLC "ART GLOSS GALLERY," glazed tiles produced from mass M-9 at the same firing temperatures also met GOST standards for cladding and facade tiles.

Unlike mass M-7, due to its comparatively later densification, the glaze layer of M-9 samples showed no defects during fast-firing conditions in industrial production.

Experimental data suggest that the developed compositions, incorporating Kulatau clay and Tuprak-Kala dune sand, enable a reduction in the firing temperature of ceramic tiles by

100°C and 50°C, respectively, compared to the reference factory compositions of SP LLC "ART GLOSS GALLERY."

Among the tested compositions, masses M-7 and M-9 exhibited the most favorable physico-mechanical properties.Therefore, tile samples from these masses were selected for further phase transformation studies.

To analyze the solid-phase and liquid-phase interactions within the clay–dune sand composition, differential thermal analysis (DTA) and X-ray phase analysis (XRD) were utilized.

Thermograms of samples from the optimal tile mass M-7 are presented in Figure 5. The differential thermal curve of mass M-7 exhibits multiple endothermic and exothermic effects.

On the HDSC curves of experimental mass M-7, the first endothermic effect, associated with the removal of hygroscopic and interlayer water, is observed within the temperature range of 76–125°C.

The exothermic effect within the temperature range of 125–262°C gradually increases and reaches its maximum at 449°C, which is likely associated with the combustion of organic compounds in the experimental mass.

The endothermic effects at 500°C and 580°C correspond to the dehydration of clay minerals and the polymorphic transformation of quartz, respectively. The endothermic effect resulting from the quartz polymorphic transformation is offset by the thermal decomposition of clay minerals.

The endothermic effect observed at 780°C is linked to the decomposition of carbonates within the mass. After this effect, the mass loss curve during calcination flattens out, indicating that further mass losses in the samples are minimal.

On the TG curve, a slight mass loss at 125°C is observed, which is due to the evaporation of hygroscopic and interlayer water. The mass change up to 610°C, with a sharp transition at 490°C, is associated with the release of hydroxyl groups during the decomposition of clay minerals. The sudden mass change at 1080°C is linked to the reduction of Fe_2O_3 to FeO, caused by overheating of the mass and intense gas release.

The exothermic effect in the temperature range of 850–960°C, with a maximum at 902°C, indicates the formation of a liquid phase and the onset of new crystalline phase formation. The total mass loss at 1100°C is 4.2%.

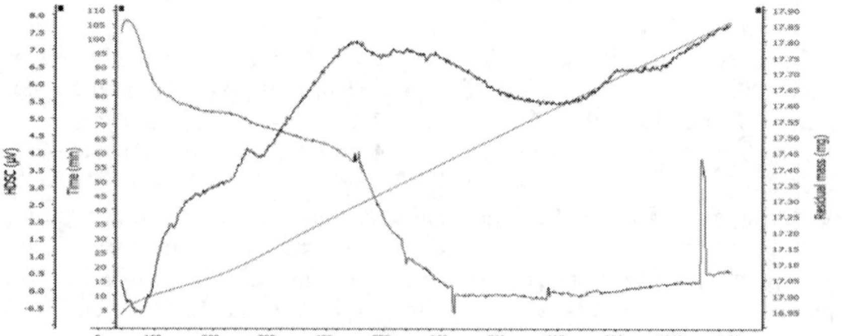

Fig. 5. Comprehensive thermogram of the tile mass M-7.

The study of phase transformations in the experimental masses M-7, fired at different temperatures using X-ray diffraction analysis, is presented in Figure 6.

From Figure 6, it can be seen that hematite lines appear on the diffractogram at 850°C, weaken at 1050°C, and disappear completely at 1100°C. In the temperature range of 850-950°C, the diffractogram also shows lines of unreacted albite, which completely disappear at 1000°C.

In the same temperature range (850-950°C), weak lines of wollastonite crystals are observed, which disappear at 1000°C and transform into anorthite. As the firing temperature increases to 1100°C, the intensity of the anorthite lines slightly increases.

At 950°C, weak mullite lines appear on the diffractogram, and with an increase in firing temperature to 1100°C, their number and intensity grow.

As the firing temperature rises from 850°C to 1100°C, quartz lines gradually weaken due to dissolution in the mass, and some of them transform into tridymite at 950°C.

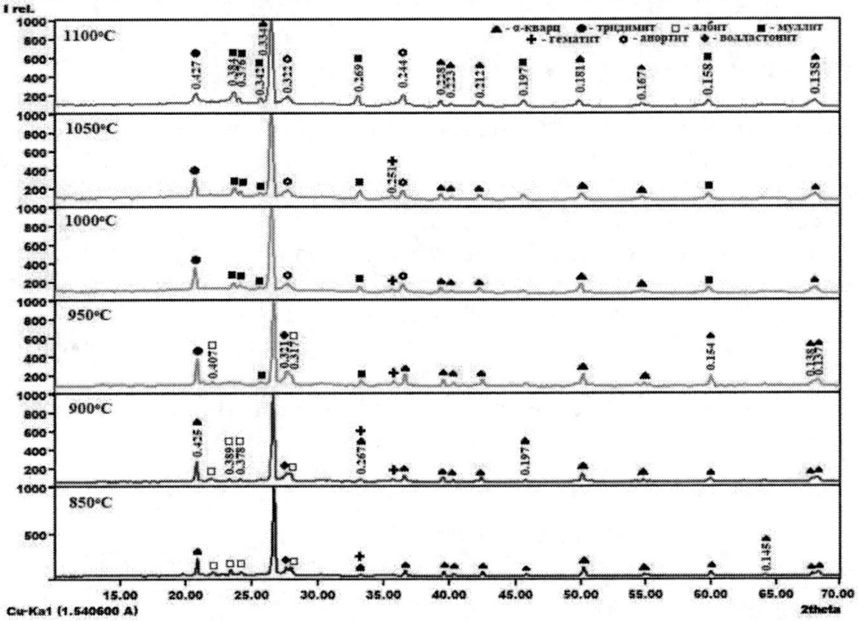

Fig. 6. X-ray diffractograms of the tile mass M-7 fired at different temperatures.

The increase in the glassy phase content in the studied samples at 1100°C is evidenced by the ratio of the integral areas of the amorphous halo and diffraction reflections in the X-ray diffractograms.

Thus, based on comprehensive analyses, it can be concluded that as the temperature rises to 950–1000°C, the viscosity of the glassy phase in the experimental mass reaches a state of fluidity. The formation of a hydromica melt from the clay component also improves the sintering conditions by wetting the surface of solid particles, including quartz. In general, favorable conditions are created for crystallization processes, leading to a quantitative increase in the crystalline phase within this temperature range. The sharp increase in strength and density of the samples during the intensive sintering temperature range is due to the reinforcing effect of the glassy phase.

Further increases in temperature to 1050–1080°C also lead to an increase in the glassy phase and a quantitative growth of the main crystalline phases, mullite and anorthite.

Firing the experimental masses above 1080°C leads to the rapid onset of deformation processes and sample bloating due to an increased liquid phase content. The intensified gas formation, resulting from the reduction of Fe_2O_3 to FeO, causes excessive bloating and structural distortion in the samples. Although crystalline phase formation continues at these temperatures, the structural integrity of the material is compromised.

It should be noted that since the experimental masses M-7 and M-9 consist of the same components, they are similar in chemical composition and positioning within the binary system "clay–dune sand." As a result, the findings of the physicochemical analysis—such as temperature intervals of endothermic and exothermic effects on the comprehensive thermogram, as well as phase transformations on the X-ray diffractogram during the firing of mass M-9—are almost identical to those of mass M-7, with very minimal differences between them.

Thus, the data obtained from differential thermal and X-ray phase analyses confirm the complex nature of solid-phase and liquid-phase reactions in the composition of Kulatau clay and Tuprak-Kala dune sand. These reactions lead to the formation of mullite, anorthite, quartz, hematite, and other phases in the experimental samples during firing, along with the development of a glassy phase that fills the gaps between these compounds.

4 Conclusion

The developed compositions, combining Kulatau clay and Tuprak-Kala dune sand, enable a reduction in the firing temperature of ceramic tiles by 100°C and 50°C, respectively, compared to the standard industrial compositions used by ART GLOSS GALLERY LLC.

Production tests confirmed that the optimized mass M-7, when fired at 1000°C and 1050°C, allows for the production of unglazed facade and floor tiles, respectively. Likewise, the optimized mass M-9, fired at the same temperatures, enables the manufacturing of glazed decorative and facade tiles with a dense glaze coating.

It was determined that the two-component composition influenced crystal formation within the experimental masses, enhancing the physico-mechanical properties of the samples at relatively low temperatures. The phase composition of the synthesized ceramic tiles is predominantly characterized by mullite, anorthite, quartz, and a glassy phase.

Thus, incorporating Kulatau clay into ceramic tile compositions, alongside Tuprak-Kala dune sands, enables the substitution of costly multicomponent mixtures currently utilized in ceramic tile manufacturing in the Republic of Uzbekistan.

References

1. Z. M. Kuryazov, Z. R. Kadyrova, A. M. Eminov, "A New Alternative Source of Clay Raw Materials for Ceramic Materials Based on the Muddy Deposits in Water-Reservoirs", Glass and Ceramics **80**, 524–528 (2024)
2. O. A. Shishakina, A. A. Palamarchuk, "Application of melts in the production of ceramic materials", International Journal of Applied and Fundamental Research **11**, 105–109 (2019)
3. A. M. Salakhov, L. R. Tagirov, "Structure formation of ceramics from clays forming different mineral phases during firing", Construction Materials **8**, 68–74 (2015)
4. Sh. M. Masharipova, Z. R. Kadyrova, D. P. Danilovich, F. A. Ataeva, Z. A. Masharipova, "Development of ceramic brick compositions for restoration of historical heritages based on loess–defecate composition", Steklo i keramika **97(5)**, 17–22 (2024). DOI: 10.14489/glc.2024.05.pp.017-022
5. B. T. Sabirov, Z. R. Kadyrova, S. S. Tairov, "Development of optimal compositions of ceramic tiles using barchan sands", Glass and Ceramics **1**, 15–17 (2018)
6. D. B. Buranova, M. Yu. Yunusov, Z. K. Babaev, Sh. K. Matchanov, "On the use of cobalt-containing metalurgical waste to obtain colored glass and glass enamel", Glass and Ceramics **79(7–8)**, 300–305 (2022)
7. A. M. Eminov, I. R. Boyjanov, Dj. S. Djabberganov, Al. A. Eminov, "Physical and mechanical properties of ceramic tile based on the clay – dune sand composition", Glass and Ceramics **81(3–4)**, 39–44 (2024)
8. M. Imen, B. Slim, F. Nathalie, M. E. Gaied, "Incorporation of Siliceous Sand and Lignite Tailings from Miocene Deposits (Zeramdine, Eastern Tunisia) in Clay Bricks and

Ceramic Tiles: Technological Feasibility", Arabian Journal of Geosciences **14**, 1–22 (2021), DOI: 10.1007/s12517-021-06582-w

9. L. İ. İsraıl, "Mechanical and Microstructural Properties of Opaque Ceramic Tiles Glazed with Some Industrial Wastes", Glass and Ceramics **79**, 340–347 (2022), DOI: 10.1007/s10717-022-00510-4
10. W. Yue, B. Wang, "Ceramic-Added Lime and Cement Mortars: A Review of Applications in Building Products", Science in Progress **107 (3)** (2024), DOI: 10.1177/00368504241266559

Kinetic study on rapid phosphating compositions for eco-efficient corrosion protection in earth and environmental engineering

Maxsud Jumaniyazov, *Sherzod* Kurambayev*, and *Surayyo* Sapayeva

Urgench State University named after Abu Rayhan Beruni, 220100 Urgench city, Uzbekistan

Abstract. This paper presents a kinetic investigation of rapid phosphating compositions designed for corrosion protection in industrial environments. The formulations incorporate extraction phosphoric acid, furfuryl alcohol, hexamethylenetetramine, and lignin-hydrolysis industry waste with the following composition (wt.% relative to absolutely dry residue): C – 17.34, H – 6.43, O – 43.50, phenol groups (–OH) – 5.06, methoxy (–OCH_3) – 3.06, carboxyl (–COOH) – 1.18, total acidic groups – 6.24, ash – 4.12–2.74. Additionally, the natural mineral vermiculite was used (bulk density 70–180 kg/m³, water absorption 400–530%, pH 6.8–7.0, Mg content 10–14%, K 3–5%, Ca 1.2–2%, Mn 0.8–1%). The kinetic parameters obtained from this study elucidate the mechanisms governing the accelerated phosphating process. These results are significant for developing environmentally safer and more durable protective coatings, thereby extending the operational life of metal structures and reducing resource consumption.

1 Introduction

The issue of anticorrosion protection for various metal products and structures remains relevant to this day. From the perspective of cost-effectiveness, availability of raw materials, and ease of processing, chemical phosphating is considered highly important for preparing the surfaces of metals and alloys for paints, coatings, or lubricants.

Rapid phosphating compositions for corrosion protection form a protective phosphate layer on the metal surface, slowing down the corrosion process and improving the adhesion of paint coatings. The phosphating process is carried out by treating the metal surface with phosphoric acid or its salts. Traditional phosphating methods often involve heavy metals that are harmful to the environment. Therefore, the development of environmentally safe, rapid phosphating compositions is an urgent issue. In this regard, the trend of using non-toxic, eco-friendly reagents in the phosphating process is gaining momentum [1].

It is known that, compared to traditional methods, electrochemical phosphating is faster and more efficient, forming uniform and dense coatings on metal surfaces [2]. The effectiveness of zinc-based phosphating compositions against corrosion has been described

* Corresponding author: sherzod.k@urdu.uz

in numerous scientific sources [3, 4]. The processes of improving the properties of phosphate coatings by adding nanoparticles are being studied. For instance, in some sources, it has been investigated that phosphate coatings enriched with graphene oxide nanoparticles enhance corrosion protection on metal surfaces. Additionally, the addition of organic substances to phosphating solutions improves the adhesion and corrosion-resistant properties of the coatings [1, 5].

2 Methods

Gravimetric Method: In the gravimetric method, metal plates of various steel samples with dimensions of 3.0 x 0.1 cm are used. Before starting the experiments, the metal samples were cleaned with sandpaper and their dimensions were accurately measured. Then, they were washed with distilled water and acetone. The experiments were conducted in 10% HCl, 10% H_2S, and 1M NaCl solutions at different time intervals. Initially, the metal sample was immersed in the test solution for 3 hours, after which its mass was measured. Then, inhibitors of various concentrations were added, and the metal sample was weighed again. Before weighing, the metal samples were washed with distilled water and then with acetone, followed by weighing on a balance. Each experiment was repeated three times, and the obtained results were recorded.

The effectiveness of the modifier is determined using the following formula:

$$\eta\% = \frac{W_0 - W_i}{W_0} \times 100$$

Here, W_0- and W_i- represent the mass loss (mg) without and with the addition of the modifier, respectively, and η% denotes the efficiency of the modifier.

The morphology of the modified metal surface was analyzed using a LaboMet-1 metallographic microscope. This microscope can magnify the metal surface up to 1000 times. The operating voltage range is between 5-15 kV. Using this microscope, images of the metal samples can be captured at various scales, accurately identifying structures up to 100 micrometers in size. Before starting the experiment, the metal samples were thoroughly washed with distilled water and acetone, then dried in a vacuum, and subsequently observed under the microscope [6-9].

3 Results and discussion

For the research, five samples were prepared with the compositions listed in Table 1 (Table 1).

Table 1. Compositions of rust modifiers.

№	Component Name	In the modifier				
		Component Amount, wt. %				
		Comp. 1	Comp. 2	Comp. 3	Comp. 4	Comp. 5
1	Hydrolyzed Lignin	14	15	16	17	18
2	Orthophosphoric Acid	20	21	22	23	24
3	Furfuryl Alcohol	1	2	3	4	5
4	Vermiculite	3,0	3,0	3,0	3,0	3,0
5	Water	62	59	56	53	50
6	Total	**100**	**100**	**100**	**100**	**100**

The proposed agent can be used on metal surfaces regardless of the presence of rust products. Its composition includes Zn^{2+}, Al, Ca, Mg, and Mn, which help to form a concentration of hydroxyl ions on the metal surface, leading to the passivation of iron. An advantage of these minerals is their low toxicity. In contrast, compositions containing chromates and lead-based substances possess toxic properties and are therefore used less frequently. The inclusion of the above-mentioned compounds in the composition contributes to the formation of micro-coatings on the metal surface.

This, in turn, reduces the penetration of aggressive substances into the surface. To enhance the ion-exchange and adsorption mechanisms, vermiculite was included in the composition. The presence of organic inhibitors, such as urotropine, strengthens chemisorption and prevents rust formation on the surface. In this composition, the formation of chelate-type compounds explains the strong adhesion and high protective properties. The presence of polar functional groups (-OH, -COOH, $-CHO_2$, -COONH) enhances cohesive energy. These groups contribute to the formation of a coating that ensures strong adhesive bonding. Additionally, the inclusion of furfuryl alcohol provides barrier-type protection. The micronized thin layer of insoluble iron phosphate salts further enhances the protection. This complex composition not only prevents rust but also facilitates the formation of insoluble organic and inorganic compounds with the rust layer itself. By introducing 3% vermiculite mineral into the composition, compounds with high adhesion and anti-corrosion properties were achieved. A simplified setup was used in the production technology, maintaining the pH of the solution within the range of 4.5-5.0.

For the research, St3 grade structural steel materials were selected. The protective properties of the obtained phosphate coatings were controlled according to GOST 9.302-88 using Akimov's drop method. The insulating properties of the obtained coatings (films) were evaluated based on their dielectric strength. Measurements were performed using the UPU-10 device with a 0.5 N load and an electrode diameter of 25 mm. Chronopotentiometric measurements were conducted using an ANION 4154 combined device in a YASE-2 electrochemical glass cell equipped with a water thermostat. A silver chloride electrode was used for comparison, and potentials were referenced against the saturated silver chloride electrode [10-12].

To measure the thickness of the phosphate coatings, an electromagnetic thickness gauge "Konstanta-K5" calibrated with an ID sensor was used. In the studies, when the coatings were subjected to friction for 5 minutes, the average protective property of the phosphate films was 172 seconds. At the same time, for the samples treated for 10 minutes, this value decreased to 155 seconds (*Table 2*).

Table 2. Results of chronopotentiometric study of the phosphating process.

Processing Time, s	Changes in Potential Values During the Phosphating of Steel Samples			
	Potential Change During Phosphating at 70°C, E, mV (AgCl/Cl)		Potential Change During Phosphating at 90°C, E, mV (AgCl/Cl)	
60	-508	-550	-510	-480
120	-475	-462	-508	-487
180	-460	-470	-500	-480
240	-465	-464	-480	-440
300	-467	-460	-485	-450
360	-470	-460	-490	-445
420	-472	-462	-492	-430
480	-475	-458	-480	-430
540	-485	-465	-500	-443

When the phosphating time was increased from 5 minutes to 10 minutes, the average roughness of the obtained coatings decreased from 2.560 μm to 1.712 μm. The effect of the temperature range of 70-90°C on the deposition process of phosphate films on St3 grade steel was studied by analyzing the chronopotentiometric values obtained during the phosphating process at different temperatures for each type of steel. As observed from Table 2 (in samples 3 and 4), the fastest shift towards positive potential values occurs during the deposition of the phosphate layer on the surface of the electrical steel at 90°C. At 240 seconds, the potential of the given sample shifts from -510 mV to -490 mV at 70°C and to -455 mV at 90°C, respectively.

Based on this, during the phosphating process at 90°C, the retention of electrical contact even after 10 minutes indicates that the process is not yet complete. Furthermore, the obtained data shows that, at 90°C, the exponential growth phase of the phosphate coating on electrical steel concludes after the first 5 minutes of the phosphating process. After this phase, the slowing phase of the process begins, during which the closure of pores associated with upward growth and the sealing of larger individual crystals in the phosphate coatings likely occur. Additionally, this phase may involve the restructuring of the coating as it stabilizes.

At the same time, increasing the temperature to 70–90°C during the phosphating of structural steel does not significantly affect the process rate. This is because, even at 70°C (as shown in samples 1 and 2 from Table 1), the active growth of the coating stops after 3 minutes from the start of the phosphating process. When the solution temperature was increased from 70°C to 90°C, the potential of the sample shifted towards positive values at 240 seconds, changing from -525 and -520 mV to -460 and -465 mV, respectively. This data indicates that at 70°C, it is possible to apply an adhesive phosphate coating on St3 grade steel within 3-4 minutes [13].

Extending the processing time to 10 minutes negatively affects the coating properties, as the potential of the sample shifts back towards negative values. The obtained results show that phosphating electrical steel with an area of up to 0.1 m^2 using 1 dm^3 of solution causes insignificant changes in the chemical composition of the solution (*Table 3*).

Table 3. Dependence of the phosphating solution composition changes on the treated area of electrical steel.

Treated Surface Area in 1 dm^3 Solution, m^2	**Total Acidity, points**	**Free Acidity, points**	**Phosphating Composition, g/dm^3**	**Mass of Sludge Formed from 1 dm^3 Solution, g**
-	61,5	8	68,13	-
0,02	60,25	7,5	67,90	0,33
0,04	59	7,25	67,63	0,14
0,06	58,25	7,2	66,83	0,13
0,08	57,25	7,1	66,17	0,09
0,10	57	7	65,15	0,13

This confirms the insignificant changes observed in the obtained phosphate coatings (*Figure 1, Curve 1*). When the surface area of electrical steel samples up to 0.1 m^2 was treated with 1 dm^3 of solution, the thickness of the resulting coating did not exceed 12-13 μm. The protective properties of the phosphate layer remained approximately at the same level (*Figure 1, Curve 2*). However, when phosphating an electrical steel surface area larger than 0.1 m^2 with the same solution volume, its protective value decreased sharply.

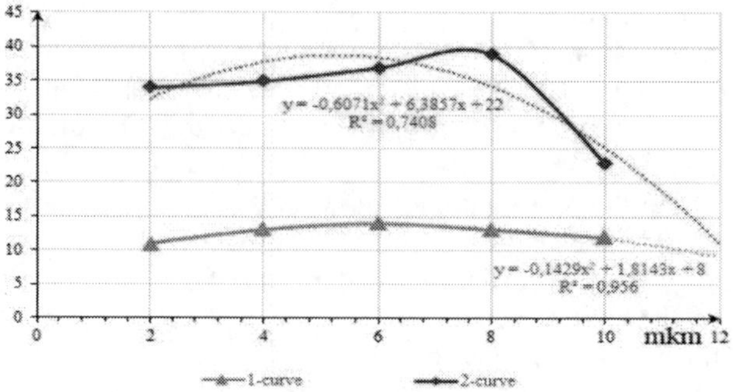

Fig. 1. Dependence of the phosphate layer thickness (1) and the protective properties of the layer (2) on the surface area of electrical steel treated with 1 dm^3 of the proposed phosphating solution.

When treating an electrical steel plate with a surface area of no more than 0.1 m^2 in 1 dm^3 of solution at a temperature of 90°C for 5 minutes, it is possible to obtain thin phosphate coatings with a thickness of 10-12 μm and a breakdown voltage of 300-500 V. Furthermore, this solution can be used for phosphating parts made of 08 KP steel at lower temperatures (70°C) following standard technological processes before applying the final lacquer-paint coatings.

The proposed technology allows for the phosphating of both electrical and structural steel in a single solution composition, reducing energy consumption and the need for additional equipment. This approach also minimizes the operational costs associated with maintaining phosphating baths. Traditionally, phosphate coatings for metals are formed in solutions at high temperatures, which leads to increased energy consumption and may limit the treatment of components susceptible to deformation. In our research, we investigated the kinetics of the phosphating process aimed at protecting metal surfaces from corrosion. For this purpose, St3 carbon steel was used. Before conducting the phosphating and electrochemical measurements, the steel electrodes were polished using fine abrasive skin, degreased with ethanol, and then washed with distilled water to ensure surface cleanliness.

The phosphating process was analyzed using a raw material containing various magnesium-based additives, including vermiculite. The process was conducted at a temperature of (25 ± 1)°C, using a solution based on 0.65M Mg^{2+} + 0.3M PO_4^{3-} with a pH of 1.95. The research began by developing a universal corrosion-resistant phosphate coating for St3 grade metals. Although the obtained phosphate coating provided sufficient protective properties at room temperature, its complex preparation process limited its technological applicability. To overcome this limitation, a phosphating composition was developed based on locally sourced raw materials.

The kinetics of phosphate coating formation on St3 steel at 25°C was studied. According to the data presented in Figure 2a, the formation of the phosphate coating begins from the first minutes of phosphating. This is evidenced by the changes in the mass of the dissolving alloy (m), the mass of the phosphate coating (p), and the shift of the potential (E) towards positive values. According to the modern theory of phosphating, initially developed for ferrous metals and later applied to some non-ferrous metals, when steel is immersed in the phosphating solution, electrochemical anodic processes occur on the metal surface. Simultaneously, the oxidizing agent present in the solution undergoes cathodic reduction, accompanied by the adsorption of hydrogen ions and an increase in pH around the electrode. This leads to reaching the solubility product necessary for the formation of sparingly soluble phosphates.

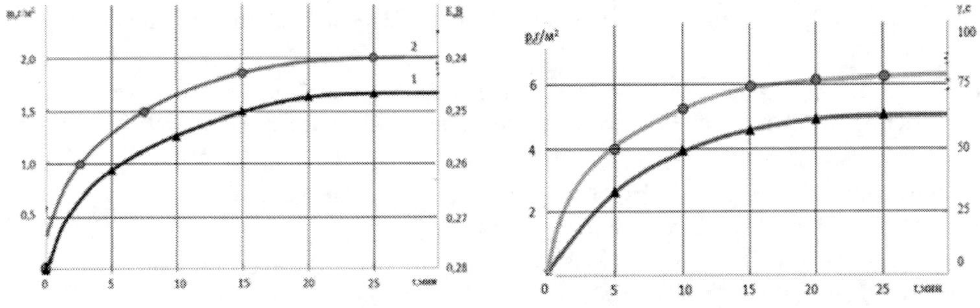

a b

Fig. 2. Formation and Transformation of Phosphate Coating: a – Dissolved metal mass (1) and potential (2). b – Phosphate coating mass (3) and corrosion resistance (4) obtained when St3 grade steel is immersed in a zinc sulfate-phosphate solution.

The initial stage of steel phosphating is characterized by the synchronous change in both the growth rate of the phosphate coating and the dissolution rate of the metal in the phosphating solution. The electrochemical processes underlying phosphating play a crucial role in the formation and development of the phosphate layer structure, particularly in facilitating the crystallization process. However, the formation of primary passivating layers can also occur through surface mechanisms, such as the specific adsorption of phosphate ions and their interaction with the base metal.

It is important to emphasize that the increased content of alloying elements in steel significantly affects both the state of the steel during phosphating and the quality of the resulting coating. As noted earlier (*Figure 2*), during the phosphating of St3 steel in a basic phosphating solution, the corrosion resistance of the phosphate coating did not exceed 90 seconds. Under these conditions, altering the concentration of salt-forming ions (Zn^{2+}, PO_4^{3-}) did not result in an improvement in the properties of the phosphate coating formed on the St3 surface.

These additives may potentially influence not only electrochemical processes but also the crystallization process. According to accelerated corrosion tests conducted in a 3% NaCl solution, steel with phosphate coatings obtained under optimal conditions meets the requirements of GOST 9.302-88 in terms of protective properties.

4 Conclusion

This solution can be considered universal, as it allows the formation of corrosion-resistant coatings at room temperature for various steel grades, including St3, 20, and 08KP. Based on the specific characteristics of the cold phosphating kinetics of St3 grade steel, it is proposed to introduce 3% vermiculite into the composition to enable the rapid formation of phosphate coatings with the required functional properties within 15 minutes. Such phosphate coatings can be recommended as a base layer under various coatings and lacquer-paint applications to protect carbon steels from corrosion under different operating conditions.

References

1. R. Kumar, A. Singh, N. Gupta, "The Environmental Significance of Phosphating Compositions Based on Plant Extracts", Green Chemistry and Technology **5(4)**, 321–328 (2019)
2. S. Wang, Q. Liu, T. Zhao, "Advantages of the Electrochemical Phosphating Process", Electrochemistry Acta **67(2)**, 234–240 (2021)

3. Y. Zhou, X. Li, J. Wang, "Effectiveness of Zinc Phosphate Coatings Against Corrosion", Corrosion Science Journal **12(3)**, 456–462 (2023)
4. H. Li, L. Zhang, Y. Chen, "Corrosion-Resistant Properties of Phosphate Coatings Enriched with Graphene Oxide", Nano Materials and Coatings **8(5)**, 789–795 (2022)
5. D. Liu, T. Guo, G. Xu et al., "Modification Mechanism of Triethanolamine on Phosphating Coating with High Corrosion Resistance", Russ. J. Phys. Chem. **97**, 3149–3155 (2023). DOI: 10.1134/S0036024423130071
6. J. Chen, P. Wu, M. Gao, "The Effect of Polymer Additives on the Efficiency of Phosphate Coatings", Journal of Coating Technologies **15(1)**, 112–118 (2020)
7. H. Du, X. Ren, D. Pan et al., "Effect of phosphating solution pH value on the formation of phosphate conversion coatings for corrosion behaviors on AZ91D", Adv Compos Hybrid Mater **4**, 401–414 (2021)
8. B. Liu, Y. Zheng, G. Xiao, C. Chen, Y. Lu, "Influence of Surface Post-Processing on Crystal Refinement and Characteristics of Hopeite Coating by Phosphating", Coatings **11**, 541 (2021)
9. L. M. Monina, X-ray diffraction. Qualitative X-ray phase analysis (Prospekt Publishing House, 2017)
10. A. A. Ayoola, O. S. I. Fayomi, I. G. Akande et al., "Inhibitive Corrosion Performance of the Eco-Friendly Aloe Vera in Acidic Media of Mild and Stainless Steels", J Bio Tribo Corros **6**, 67 (2020)
11. J. O. S. I. Akpoborie et al., "Electrochemical corrosion phenomenon and prospect of materials selection in curtailing the challenges", in *IOP Conference Series: Materials Science and Engineering* 1107(1) (2021)
12. R. Yan, X. Gao, W. He et al., "A Simple and Convenient Method to Fabricate New Type of Phytic Acid Metal Conversion Coatings with Excellent Anti Corrosion Performance on the Iron Substrate", RSC Adv. **7**, 41152–41162 (2017)
13. M. Dornbusch, Corrosion Analysis (CRC Press, Boca Raton, London, New York, 2018), p. 43

The effect of hybrid mineral additives on the genetic formation and physico-chemical processes of cement composites

Azamat Khadzhiev[1,*] *Muslimbek* Abdullaev[2], *Yusufboy* Yakubov[1], and *Jumaniyoz* Jumaniyozov

[1]Urgench State University named after Abu Rayhan Biruni, 220100 Urgench city, Uzbekistan
[2]Institute of General and Inorganic Chemistry of the Academy of Sciences of Uzbekistan, 100170 Tashkent, Uzbekistan

Abstract. The cement industry is undergoing a transformation towards sustainability, focusing on reducing CO_2 emissions and implementing energy-efficient technologies. This study investigates the impact of hybrid mineral additives, including microsilica and ashstone's slag mixture, on the genetic formation and physico-chemical properties of cement composites. The research highlights the role of hybrid additives in enhancing cement durability, optimizing raw material utilization, and contributing to the green economy. Experimental findings indicate that incorporating 20% hybrid mineral additives reduces clinker content while maintaining high mechanical properties, meeting the requirements of modern construction materials. This study underscores the importance of local raw material utilization and innovative approaches to cement production in Uzbekistan.

1 Introduction

At the United Nations Conference of the Parties 29 (UN COP29) climate summit, which was officially hosted and held in the Republic of Azerbaijan, an alarming and highly concerning global carbon budget report was formally published and made publicly available. This comprehensive and detailed report explicitly stated that worldwide emissions of carbon dioxide (CO_2), a major greenhouse gas responsible for global warming and climate change, are anticipated and projected to reach an unprecedented, record-breaking peak in the year 2024. The total volume of CO_2 emissions is expected to surge by an astonishing and concerning 1 billion metric tons in comparison to the total emissions recorded in the preceding year, ultimately reaching an estimated 41.6 billion metric tons.

This significant and troubling rise in CO_2 emissions represents a substantial and serious threat to ongoing international efforts aimed at mitigating the adverse effects of climate change, reducing global warming, and successfully achieving the long-term climate-related objectives outlined in the legally binding Paris Agreement. According to the report's findings, the vast majority of these total CO_2 emissions—approximately 37.4 billion metric tons—are directly linked to and generated by the combustion and utilization of fossil fuels,

* Corresponding author: xadjiyev2019@mail.ru

including coal, crude oil, and natural gas, as well as emissions originating from various manufacturing industries. This figure reflects a noticeable and measurable increase of approximately 0.8% in comparison to the CO_2 emissions recorded in the year 2023 [1].

In 2025, it will be ten years since the adoption of the Paris Agreement, which was developed to reduce the amount of harmful gases released into the atmosphere. This document, which came into force in 2016, has been approved by 195 countries, with 175 nations actively working within its framework. In 2021, Uzbekistan committed to reducing its greenhouse gas emissions by up to 30% by 2030 under the agreement. Additionally, 2025 has been declared the «Year of Environmental Protection and Green Economy» in the country. In line with this, efforts are being made to significantly increase the use of renewable energy as part of the transition to a green economy [2].

In the cement industry, special attention is being paid to transitioning to a green economy, energy conservation, reducing fuel and energy resource consumption, and increasing scientific research to develop and implement sustainable economic practices. Achieving these goals requires the development of innovative, cost-effective production technologies compared to existing analogs. In particular, the establishment of cement production using locally sourced active mineral additives and their hybrid derivatives, based on both natural and technogenic components, is being emphasized [3,4]. This approach not only helps increase cement production volume but also reduces production costs, enhances construction-technical properties, ensures a cleaner climate, and mitigates environmental impact.

In order to actively promote, enhance, and significantly expand the development and diversification of the production processes associated with construction materials, as well as to establish highly favorable conditions for sustainable industry growth, the government has undertaken comprehensive measures. These efforts are specifically aimed at attracting a greater volume of investments for the effective and efficient processing of locally available mineral-based raw material resources. Additionally, these initiatives are designed to substantially boost and maximize the overall export potential of various types of construction materials. As part of this strategic approach, the government has officially approved a set of forecasted parameters that serve as a roadmap for the systematic and well-planned expansion of the raw material base that underpins the entire construction industry. These forecasted indicators are carefully aligned with the target parameters established for extensive geological exploration activities, the extraction and processing of locally sourced raw materials, and the large-scale production of construction materials. Covering the period from 2019 to 2025, this comprehensive plan is structured to facilitate the diversification of construction material product types while simultaneously ensuring a consistent and substantial increase in overall production volumes. [5-7].

Modern cement production technologies require the extensive use of various additives without altering the core properties of cement while conserving essential resources [8-10]. To reduce the clinker content in cement production and form its mineral composition, many advanced countries have increasingly utilized industrial waste and replaced clinker with active mineral additives in proportions ranging from 30% to 80%, significantly cutting carbon dioxide (CO_2) emissions. The use of active mineral additives in cement production is driven not only by the need to save expensive clinker but also by the necessity to reduce CO_2 emissions into the atmosphere and enhance the physical-mechanical properties of concrete. In Uzbekistan, large-scale efforts are being undertaken to ensure the sustainable growth of competitive product manufacturing and exports, modernize the construction materials industry, and implement deep structural changes aimed at technological and technical upgrades.

In present-day Uzbekistan, the demand for locally available mineral additives in the field of cement manufacturing is experiencing significant growth, particularly among major cement production enterprises such as «Karakalpak Cement» LLC and the joint venture

«Titan Cement» LLC, both of which are strategically situated within the territorial boundaries of the Republic of Karakalpakstan. The nation's geological landscape is characterized by an extensive abundance of naturally occurring volcanic-origin rock formations, including, but not limited to, porphyry, sandstone, tuff, and other geologically similar mineralogical materials. This considerable availability of raw materials creates a pressing necessity for conducting a thorough, comprehensive, and detailed scientific investigation into their technological properties, physical characteristics, and potential applications in industrial-scale cement manufacturing. Moreover, meticulous research is required to refine and optimize the compositional structure of cement products. The integration of these mineral components into the cement formulation facilitates the production of high-quality composite cement that exhibits superior chemical stability along with enhanced physical and mechanical properties. Notably, this advancement can be achieved without necessitating any fundamental modifications to the established technological parameters and operational methodologies currently employed within the cement production process.

The primary advantage of this type of cement is the presence of special mineral additives, which enhance the resistance of construction mixtures to aggressive environments. Composite cement adapts well to low-temperature conditions and exhibits high resistance to the harmful effects of mineralized water. Due to its specialized additives, it stands out for its durability against temperature fluctuations and adverse atmospheric conditions. These mineral additives significantly improve the cement's water and frost resistance.

The utilization of local raw material resources from the Sultonuvays mountain range is of great importance for expanding and establishing cement production. The Sultonuvays range has sufficient reserves of limestone, marl, clay, and gypsum, making it a viable location for setting up cement manufacturing facilities. Establishing production in this region would eliminate the province's dependence on imported cement.

The largest cement raw material deposits—Aqtau, Jumurtau, Kuyanchiq, and Djamansoy—can serve as the foundation for expanding existing cement plants and developing new ones. In 2014, a total of 1,027,351.2 tons of gravel was produced from the Kekliktau, Qoratau, Shayxdjeyli (located in the Karauzyak district), and Jumurtau (located in the Amudarya district) porphyry deposits. The majority of this production was attributed to the North-Western SREPSADMZG and TITAN-DSZ LLC's Qoratau branch. These enterprises generate mechanically activated fine-fraction porphyry waste, an industrial by-product that cannot be used in other construction sectors.

2 Experimental methodology

As research objects, porphyry from the «Karatau-1» deposit, located 80 km southeast of Kegeyli settlement in the Karauzyak district of the Republic of Karakalpakstan, and sandstone from the Khojakul deposit were used as hybrid additives. To formulate the composition of composite Portland cements, ordinary Portland cement clinker (Cl) from the «Karakalpak Cement» LLC plant was selected as the matrix, while gypsum stone (Gs) from the Qagan deposit was used to regulate the setting time. Physico-chemical studies and physico-mechanical tests were carried out using standard methods in accordance with the requirements of relevant regulatory documents. The chemical composition analysis of Portland cement clinker from the «Karakalpak Cement» LLC plant, gypsum stone from the Qagan deposit, porphyry from the «Karatau-1» deposit, and sandstone from the Khojakul deposit was conducted following the methodology specified in GOST 5382-91 "Cements and Cement Production Materials. Methods of Chemical Analysis." The obtained results were statistically analyzed based on Student's t-test criteria and evaluated in compliance with GOST 31108-2003 "General Construction Cements. Technical Specifications." The physical and mechanical properties of hybrid additive cement samples were assessed according to

GOST 310.4-81 "Cements. Methods for Determining Flexural and Compressive Strength." The results were further evaluated in accordance with GOST 31108-2020 "General Construction Cements. Technical Specifications." The physicochemical properties of «green» cement composites were studied using the following methods. X-ray Diffraction Analysis (XRD) – conducted using a Shimadzu XRO-6100 diffractometer. Differential Thermal Analysis (DTA) – performed with a Netzsch STA 409 PG thermal analyzer. Scanning Electron Microscopy (SEM) – carried out using a JSM-6490LV scanning electron microscope equipped with INCA Energy energy-dispersive microanalysis and HKL-Basic structural analysis systems.

3 Results and discussion

The chemical composition of the clinker produced by «Karakalpak Cement» QK MCHJ complies with the requirements of Uz DST 2801 for Portland cement clinker. However, its SO_3 content of 5.5% distinguishes it from conventional Portland cement clinkers produced at other cement plants in Uzbekistan. The chemical-mineralogical composition and physicochemical properties of «Karakalpak Cement» QK MCHJ Portland cement clinker, as well as the sandstone from the Khojakul deposit and porphyrite from the «Karatau-1» section of the Karatau deposit, have been determined and presented (Table 1).

Table 1. Chemical composition of initial materials.

Material Name	Oxide Content, mass %								
	k.m.y	SiO_2	Al_2O_3	Fe_2O_3	SaO	MgO	SO_3	Pr.	∑
PS clinker	0,31	18,03	6,22	3,94	58,93	1,98	5,55	5,04	100,0
	Mineralogical composition of clinker, mass %								
	S_3S-55,04; S_2S-17,81; S_3A-5,15; S_4AF-15,47; SS-1,65								
Sandstone	2,54	76,72	8,20	4,78	0,83	1,40	0,24	5,30	100,0
Porphyrite	5,72	51,42	18,51	7,53	5,28	3,8	0,91	6,83	100,0
Gypsum stone	20,30	2,80	0,49	sl.	30,98	sl.	42,80	2,63	100,0
	$CaSO_4.2H_2O$ = 42,80x2,15 = 92,02%								

The sample of porphyry from the «Karatau-1» site of the Karatau deposit is predominantly composed of silicon oxide (SiO_2 – 76.72%) and aluminum oxide (Al_2O_3 – 8.20%). Small amounts of CaO, Fe_2O_3, and SO_3 are also present (Table 1). The sandstone from the Xo'jako'l deposit is characterized by a high content of SiO_2, Al_2O_3, CaO, MgO, and Fe_2O_3 oxides. Additionally, a small amount of SO_3 is also present.

Porphyrites are fine-grained rocks with a light gray color, and some areas appear brownish-red due to the presence of iron impurities. Due to the high silicon oxide content (51.42%) in the rock composition, it can be classified as a pyroxene-type porphyry (Table 1).

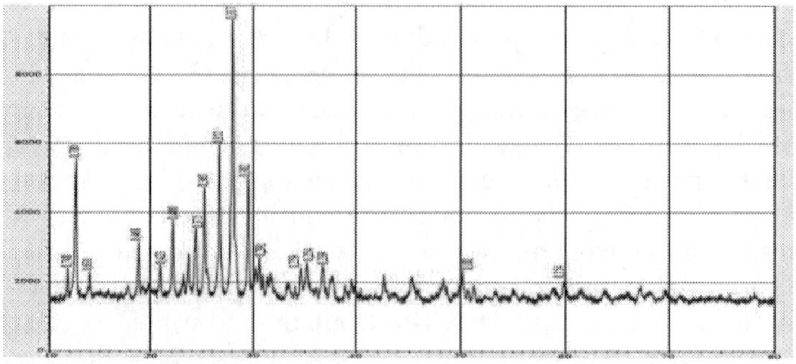

Fig. 1. X-ray diffraction pattern of porphyrite.

The mineral composition of porphyry includes quartz (d/n=0.424; 0.333; 0.244; 0.228; 0.212; 0.182...), feldspars (d/n=0.631; 0.495; 0.468; 0.400; 0.375; 0.365; 0.318; 0.291; 0.282; 0.182; 0.178), calcite (d/n=0.303; 0.249; 0.228; 0.209; 0.200; 0.188), hydrosilicates (d/n=0.495; 0.365; 0.352; 0.333; 0.303; 0.291; 0.282; 0.254), and chlorites (d/n=0.700; 0.631; 0.495; 0.468; 0.385; 0.375; 0.365; 0.291; 0.259; 0.244; 0.228; 0.212; 0.188; 0.182; 0.178; 0.156) (Figure 1).

Sandstone is a sedimentary rock with a light gray appearance, sometimes exhibiting a reddish-brown hue. Its X-ray diffraction (XRD) analysis reveals high-intensity peaks corresponding to quartz (d/n=0.421; 0.331; 0.244; 0.228; 0.212...), feldspars (d/n=0.628; 0.467; 0.443; 0.399; 0.363; 0.351; 0.319; 0.301; 0.288...; 0.181...), calcite (d/n=0.301; 0.228; 0.208; 0.188 ...), hydromicas (d/n=0.363; 0.351; 0.331; 0.301; 0.291; 0.288; 0.258; 0.254 ...), and chlorites (d/n=0.628; 0.467; 0.351; 0.291; 0.258; 0.244; 0.228; 0.208; 0.188; 0.177; 0.166 ...) (Figure 2).

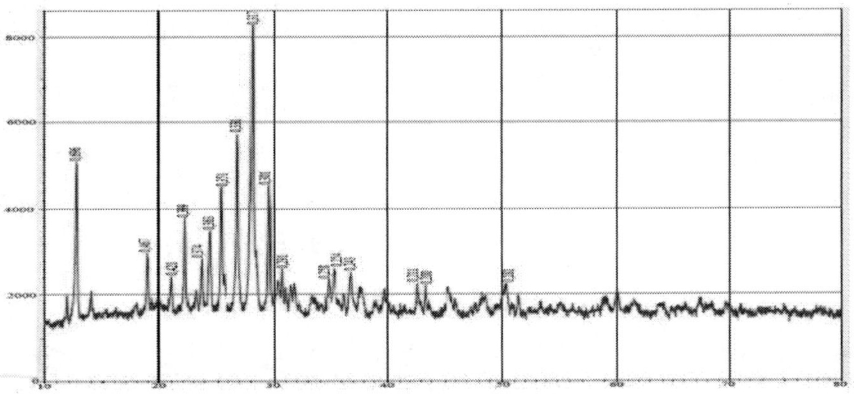

Fig. 2. X-ray diffraction (XRD) pattern of sandstone.

To assess the hydraulic activity of the hybrid mineral additive based on porphyrite and sandstone, the Student's t-test criterion (t) was determined to be t = 52.92. According to the requirements of GOST 31108-2020 «General Construction Cements. Technical Specifications,» this result indicates that the hybrid mineral additive based on porphyrite and sandstone is characterized as a highly active mineral additive.

The hydraulic activity of the hybrid mineral additive based on porphyrite and sandstone was determined to be t = 16.81 according to the Student's t-test criterion, which is higher than the minimum required value of 15.0 specified in GOST 31108-2020. Therefore, PSS falls into the category of highly active additives.

The study of the setting process of hybrid mineral additive cement showed that as the additive content increased, the strength of the cement stone gradually decreased, dropping from 55.0 MPa to 42.5 MPa over 28 days (Table 2).

When the hybrid mineral additive content ranged from 10% to 20%, the strength decreased by (19.0–22.7)% compared to PS-D0, while at 25%, the strength reduction reached 38.0 MPa, which accounts for almost one-third of the PS-D0 strength (Figure 3). In standard PS samples containing 20% hybrid mineral additive, hydraulic activity reached 35.8 MPa on the 7th day, corresponding to the 400 grade. By the 28th day, the hydraulic activity of this cement reached 42.5 MPa, which meets the requirements of GOST 10178 for PS400-D20 cement with a significant strength margin.

The results of comprehensive studies on the genetic stages of crystallohydrates during the hardening of Portland cement (PC) with hybrid mineral additives of sandstone and porphyry, as well as their correlation with the physical and mechanical properties of the cement composite, are presented.

Table 2. Physical and mechanical properties of cement with hybrid mineral additive.

№	Ratio of components, mass %		Compressive strength of 2x2x2 cm cube samples, MPa		
	Clinker	hybrid mineral additive	3 days	7 days	28 days
1	100	-	44,1	45,0	55,0
2	90	10	37,0	39,0	44,5
3	85	15	34,1	36,0	43,0
4	80	20	28,5	35,8	42,5
5	75	25	28,0	36,0	38,0
Flexural/Compressive strength of standard 4x4x16 cm samples of PS with hybrid mineral additives, MPa					
			3 days	7 days	28 days
1	Clinker 100%	-	-	-	7,2/48,6
2	Clinker 80%	20%	4,9/25,2	5,8/ 40,5	6,9/42,6

The hydration rate of PC containing 20% hybrid mineral additive is higher than that of PC-D0: after 1 day, the bound water content in hydration products is 1.51% higher. This is likely due to the rapid water absorption by the clay fraction present in the sandstone.

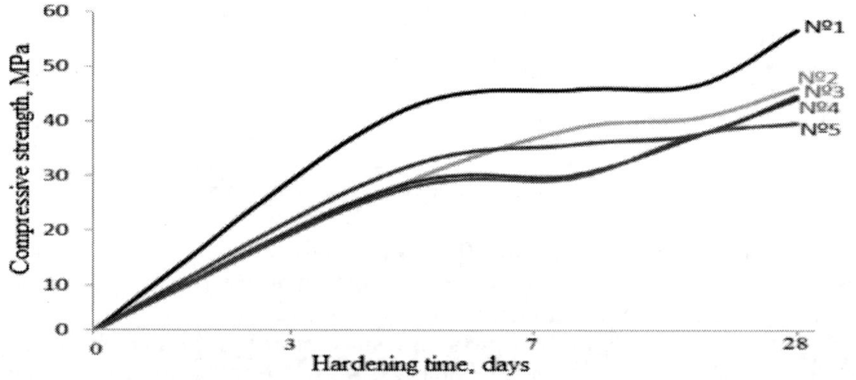

Fig. 3. The dependence of the strength gain kinetics of Portland cement (PC) on the amount of hybrid mineral additive. 1-PS-D0; 2-PS-D10; 3-PS-D15; 4-PS-D20; 5-PS-D25.

According to the X-ray phase analysis results, on the 7th and 28th days, the majority of the hydrated products in the «ground clinker-porphyrite-water» system consist of portlandite and calcium carbonate.

According to the results of this study, the hydration products formed after 3 months of water curing of Portland cement (PC) stone containing 20% porphyrite were analyzed using the X-ray diffraction (XRD) method.

The analysis results confirm the presence of portlandite ($Ca(OH)_2$), C-S-H (calcium silicate hydrates), ettringite ($C_3A \cdot 3CaSO_4 \cdot 32H_2O$), as well as carbonate compounds ($CaCO_3$) and quartz (SiO_2) as the main hydration products in the cement stone.

The detection of portlandite peaks indicates that the cement hydration process is not yet fully completed. At the same time, the formation of the C-S-H phase is considered the main factor ensuring the strength of the cement stone. The presence of ettringite suggests that it forms as a result of interactions with sulfate components during the initial hardening process.

Additionally, the formation of carbonate compounds was observed as a result of the carbonation process of the cement stone, which may affect the long-term mechanical properties of the cement. The presence of quartz in the analysis results is associated with additional mineral admixtures and may influence the mechanical and chemical stability of the cement composition.

Thus, the hydration process in the cement stone cured in water for 3 months continues, leading to the formation of the main binding phases. However, the presence of portlandite indicates that the hydration process is not yet fully completed. Therefore, it is important to conduct a more in-depth study of the effect of mineral additives in cement on the hydration process (Figure 4).

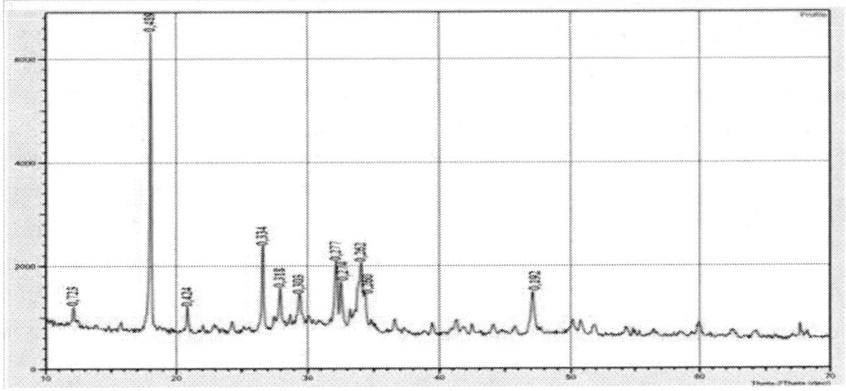

Fig. 4. Diffraction pattern of Portland cement stone containing 20% porphyrite after 3 months of curing in water.

Electron microscopy results indicate that the hydration process progresses over time, leading to significant changes in the microstructure of the cement stone. In the initial stages (3-7 days), portlandite and void structures are frequently observed, while the C-S-H phase is underdeveloped. By 28 days, the C-S-H phase develops significantly, resulting in matrix densification (Figure 5).

Fig. 5. The structure of the fracture surface of porphyrite-added PS stone cured in water for 28 days and 3 months.

As a result of 90 days of water curing, the cement stone becomes sufficiently strong, with the C-S-H phase becoming the primary binding component in the microstructure. Therefore, the hydration process of cement containing 20% hybrid mineral additive progresses gradually, reaching optimal mechanical properties at 90 days. The reduction in portlandite content and the development of the C-S-H phase contribute to increased durability and strength of the cement.

Despite the clinker content being reduced by 20%, its hydraulic activity remains consistent with the requirements for a 400-grade cement.

4 Conclusion

The article examines the impact of hybrid mineral additives on the formation process and physicochemical properties of cement composites. The study emphasizes the importance of utilizing local raw materials in Uzbekistan and Karakalpakstan's cement industry, reducing environmental impact, and transitioning to a «green» economy. It has been determined that incorporating local mineral additives such as porphyrite and sandstone into cement composition can lower production costs while enhancing strength and durability.

According to the research findings, cement with 20% hybrid mineral additives exhibits optimal mechanical properties and meets the requirements of 400-grade cement. Implementing innovative approaches in the cement industry not only reduces environmental issues but also improves production efficiency. The use of local mineral additives in cement production can contribute to sustainable development and serve as a crucial foundation for further advancements in the cement industry.

References

1. O. N. Burenina, A. V. Andreeva, M. E. Savvinova, "The Use of Local Mineral Raw Materials of Yakutia for the Manufacture of High-Strength Concrete", Polymer Science, Series D **16**, 622–626 (2023)
2. A. Zaitsev, E. Zaitseva, "Studying the Mineral Additives Effect on a Composition and Properties of a Composite Binding Agent", Oriental Journal of Chemistry **34(4)**, 1745–1750 (2018)
3. V. Makarov, T. Ivanova, "Government Policies and Investment Strategies in the Development of Construction Materials Industry", Journal of Construction Economics and Building Materials **12(3)**, 145–158 (2023)
4. Y. Zhang, H. Li, "Enhancing Export Potential of Construction Materials through Local Resource Utilization", International Journal of Construction Management **22(5)**, 345–360 (2022)
5. J. Smith, L. Brown, "Diversification Strategies in the Construction Materials Sector: A Path to Sustainable Growth", Construction Research and Innovation **12(2)**, 101–115 (2021)
6. L. Rodrigues Costa Tavares, J. F. Tavares Junior, L. M. Costa, A. C. S. Bezerra, P. R. Cetlin, M. T. Paulino Aguilar, "Influence of quartz powder and silica fume on the performance of Portland cement", in Scientific Reports (2020)
7. A. Khadzhiev, F. Atabaev, A. Jumaniyozov, Y. Yakubov, "Study on pozzolanic activity of porphyrites of the Karatau deposit", in *E3S Web of Conferences* **563**, 02029 (2024)
8. M. Yu. Yunusov, Z. K. Babaev, Sh. K. Matchanov, "Veshchestvennyy sostav i tekhnologiya obogashcheniya kvarts-kaolinovoy porody Khodzhakul'skogo mestorozhdeniya", Obogashchenie Rud **3** (2022)
9. U. Kh. Yeshimbetov, "Current state of the mining industry and mineral resource extraction in the Republic of Karakalpakstan", Economics and Innovative Technologies **6** (2016)

10. A. Chajec, "The use of granite powder waste in cementitious composites", Journal of Materials Research and Technology **25**, 4761–4783 (2023). DOI: 10.1016/j.jmrt.2023.06.253.

Improvement of the technology of obtaining mineral binders from loess-like rocks using mechanoactivation

Sharofat Masharipova[*], *Maxsud* Jumaniyazov, and *Shaxlo* Aitova

Department of Chemical Technology, Urgench State University named after Abu Rayhan Beruni, Urgench, Khorezm, Uzbekistan

Abstract. The article presents the results of a comprehensive study of the chemical, mineralogical and granulometric compositions of clay raw materials, as well as thermal properties of loess-like rocks of the Suzanlinskoye deposit using modern methods of physicochemical analysis. A technology for grinding raw materials has been developed, and a proposed process flow chart for obtaining mineral binders from loess-like rocks using mechanical activation has been presented. Using this technology to produce mineral binders ensures higher dispersion and micro-roughness of raw material particles; high concentration of surface and volumetric structural defects, and stabilization of this highly active state before sintering. The research has revealed that 20-minute grinding to 3 μm for mechanical activation of raw materials, the use of man-made waste-defecates as additional raw materials, allows to reduce the temperature to 1200□C during clinker firing. A new composition of masonry mortars was developed based on mineral binders of grade 200M (H 32.5 class), with a particle size of 3 microns, with the necessary color and chemical and mineralogical indicators for restoration materials of historical heritage.

1 Introduction

Loess-like deposits, traditionally used for the production of building ceramics for centuries, have become widely used as a building material. However, the use of natural clays is complicated by the inconsistency of their properties, which negatively affects the quality of the manufactured products. In the Central Asian republics, including Uzbekistan, the limited reserves of high-quality clay raw materials determine the use of low-plasticity loess-like rocks to produce binders. The resulting products are characterized by reduced strength, insufficient frost resistance, low decorative qualities and a tendency to form efflorescence during operation [1]. Uzbekistan has significant reserves of loess-like rocks and loams, which serve as key raw materials in the production of building ceramics. In the republic, 568 deposits and manifestations of clay rocks have been identified, represented by loess-like loams, clays of various types, argillite clays, kaolins and clay shales. Classification of clay rocks of Uzbekistan by origin includes alluvial, sedimentary (transferred) and

[*] Corresponding author: masharipova.1975@mail.ru

metamorphosed types. Formation of argillite clay deposits occurred under diagenesis conditions [2].

Production of high-quality binders requires an optimal combination of the raw materials used and the use of modern technological solutions. Regardless of the production method, the key stage in obtaining high-quality binders is the effective preparation of the raw material mass. The main goals of this technological process are the destruction of the natural structure of the clay raw material, its disaggregation and removal of unwanted impurities [3, 4].

Destruction of the natural anisotropic structure of the clay raw material can be achieved in various ways, including by dispersing clay particles or removing water-soluble salts through leaching [5].

Reducing the particle size of the raw components leads to an increase in their contact area, which has a beneficial effect on the sintering process during firing of products. The literature describes studies devoted to the use of mechanochemical activation of raw materials, which, unlike traditional preparation methods, allows changing the physicochemical state of mineral particles at the grinding stage [6].

As part of this work, a study was conducted of the chemical-mineralogical and granulometric compositions, as well as the physicochemical characteristics of loess-like rocks of the Suzanli deposit in order to develop a technology for their processing based on mechanical activation.

To achieve this goal, the following tasks were defined: assessment of the volume of loess-like rocks in Uzbekistan; selection and preparation of samples for the necessary analyses; establishment of the chemical and mineralogical composition, as well as the study of the physicomechanical properties of loess-like rocks of the Suzanli deposit; analysis and interpretation of the results of the experimental studies.

2 Research methods

This study conducted a thorough investigation into the chemical makeup, mineral content, particle size distribution, and heat-related characteristics of clay resources. The research utilized sophisticated physicochemical analytical methods in conjunction with well-established techniques from silicate materials science.

To ascertain the mineral makeup of both initial raw materials and mechanically activated samples, X-ray phase analysis was conducted using a Rigaku Corporation SmartLab3 diffractometer. The technological properties of these materials, encompassing both raw and activated forms, were assessed in accordance with GOST 9169-2021, the prevailing standard within Commonwealth of Independent States (CIS) countries. A comparative analysis of the acquired data was then performed against established findings from relevant scientific literature [7].

The granulometric composition of the initial materials was determined by employing two established techniques: dry sieving and sedimentation analysis, adhering to the guidelines outlined in GOST 21216.2-93 [8].

A sample containing raw materials suspended in water was subjected to gravitational separation. Subsequently, a Shimadzu SALD-7500 laser particle analyzer was used to assess the size distribution of the particulate matter present in the sample.

3 Results of experimental studies

The Suzanli deposit, a loess-like formation, is situated within the Khiva district of Uzbekistan's Khorezm region. A quarry operating at this site is located 4 kilometers southeast of Kasma village and borders the Karakum Desert massif. The quarry spans an area of 28 hectares and exhibits a flat geological topography.

Geologically classified as Upper Quaternary alluvial deposits (Akchadarya complex), Suzanli primarily comprises sands, sandy loams, and loess-like rocks. The productive layer consists of sheet-like loess formations overlying fine-grained quartz-feldspar sands. The upper 0.3 to 0.5 meters contain plant roots and other extraneous materials.

The northern portion of the deposit features finely dispersed aeolian sands, varying in thickness from 0.4 to 2.5 meters. Total estimated reserves amount to 829 thousand cubic meters, with 596 thousand cubic meters registered on the State Committee on Geology's balance sheet and the remaining 233 thousand cubic meters unregistered [9].

The deposit's shape resembles a lens, with a thickness ranging from 3.0 to 4.5 meters, exhibiting localized variations. The dominant color of the loess-like rocks is light brown or gray, and they are composed of minerals arranged in a complex structure [10-12].

Table 1. Results of chemical analysis of loess-like rock samples Suzanli deposit.

Oxide content, wt.%	**№ samples**										Average
	1	**2**	**3**	**4**	**5**	**6**	**7**	**8**	**9**	**10**	
SiO_2	52,54	52,55	52,53	52,49	52,56	52,60	52,57	52,51	52,65	52,50	52,55
Al_2O_3	14,72	15,60	15,59	15,27	14,56	15,72	15,39	15,57	15,54	14,51	15,36
CaO	14,04	14,09	14,32	13,31	14,14	14,27	13,48	14,11	13,65	14,55	13,99
MgO	4,70	4,65	4,52	4,62	4,53	4,49	4,48	4,61	4,43	4,42	4,55
Fe_2O_3	7,20	7,44	7,27	7,25	7,44	7,20	7,26	7,23	7,21	7,22	7,28
K_2O	2,76	2,78	2,82	2,75	2,71	2,84	2,72	2,85	2,81	2,74	2,78
Na_2O	2,12	2,08	2,08	2,11	2,02	2,15	2,05	2,14	2,28	2,07	2,11
SO_3	0,18	0,11	0,14	0,14	0,13	0,17	0,12	0,15	0,16	0,12	0,14
TiO_2	0,68	0,66	0,69	0,64	0,65	0,50	0,62	0,63	0,61	0,67	0,64
l.o.p. %	1,06	0,04	0,04	1,42	1,25	0,06	1,31	0,20	0,66	1,20	0,72

Note: Loss on ignition (L.o.p) includes: hygroscopic, constitutional, crystallized water, organic and volatile matter, and carbon(IV) oxide.

This report examines the chemical composition of loess-like rocks from the Suzanli deposit to assess their suitability for ceramic building product manufacturing.

X-ray fluorescence analysis reveals that these rocks are primarily composed of silicon dioxide (SiO_2, 52.55%), classifying them as acidic clay raw material based on aluminum oxide (Al_2O_3) content. The presence of titanium dioxide (TiO2) and iron oxide (Fe_2O_3) suggests potential for use as coloring pigment sources.

However, the relatively low alumina (15.36%) content indicates a limited presence of clay minerals like kaolinite and montmorillonite. Calcium oxide (CaO, 13.99%) and magnesium oxide (MgO, 0.17%), indicative of carbonate inclusions from limestone, could lower the firing temperature of mineral binders.

While the silicon dioxide content meets industry standards for ceramic production, the low aluminum oxide levels hinder the creation of high-quality ceramic bricks. Conversely, the elevated calcium oxide content exceeds acceptable limits, further limiting its applicability in this context.

Preliminary physical observations reveal a light gray coloration and moderate resistance to manual grinding. Foaming upon reaction with hydrochloric acid confirms the presence of carbonate compounds. Granulometric analysis using laser diffraction techniques classifies the rock as coarse clay raw material.

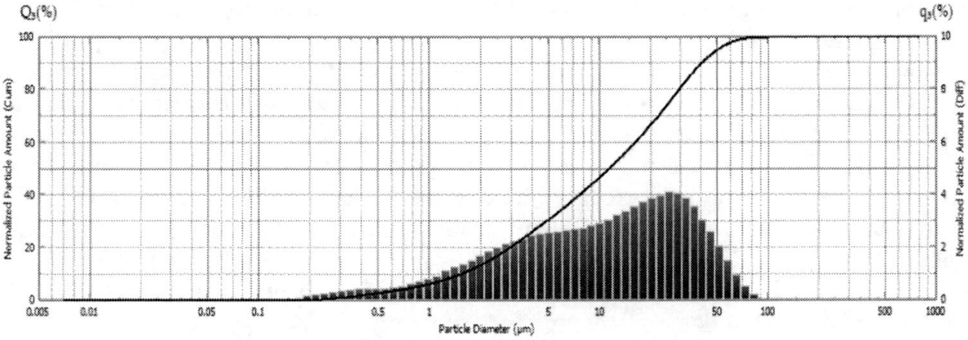

Fig. 1. Granulometric composition of a sample of loess-like rocks from the Suzanli deposit.

The analysis revealed that the raw material exhibits a bimodal particle size distribution, characterized by peaks at approximately 3.5 micrometers and 25 micrometers, resulting in an average particle size of around 9 micrometers.

Examination of X-ray diffraction patterns (Fig. 2) obtained from a sample of loess-like rocks sourced from the Suzanli deposit yielded further insights that there are diffraction maxima of the minerals quartz (SiO_2) (d/n = 0.081; 0.079; 0.102 nm), calcite ($CaCO_3$) (d/n = 0.091; 0.090; 0.102; 0.103 nm), albite (Na (Al Si_3O_8)) (d/n = 0.083; 0.089; 0.091; 0.114; 0.132; 0.079; 0.086; 0.106; 0.108; 0.090 nm).

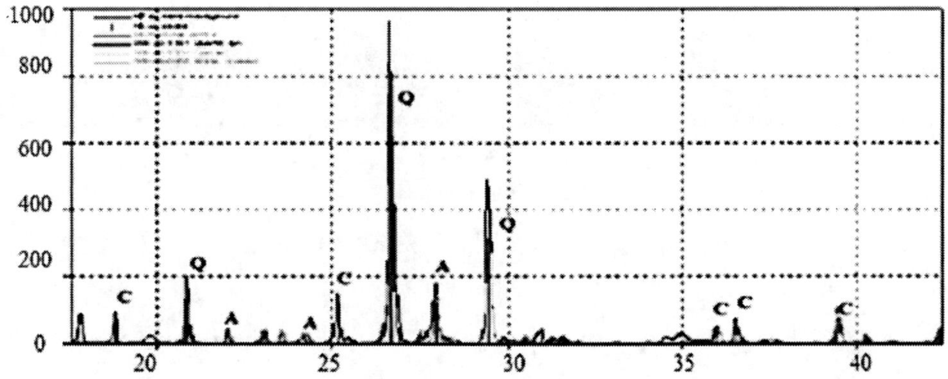

Q- quartz, A- albite, C-calcite

Fig. 2. X-ray diffraction pattern of a loess-like rock sample from the Suzanli deposit.

To improve the physicochemical characteristics of the raw mix, we used a defecation, a carbonate-containing by-product of sugar production. In the process of purifying the diffusion juice obtained after extracting sugar from beet chips and pre-purifying it by filtration from plant fibers and precipitation of proteins by heating, defecation is used. The waste material produced from sugar refining is characterized by its cream-colored, powdery texture with an oily consistency that readily absorbs moisture and becomes malleable. Analysis of this waste, collected from filters at Khorezm Shakar JSC, reveals a moisture content ranging from 20% to 24%, a density between 2135 and 2380 kg/m³, and a specific volume between 1030 and 1095 kg/m³.

Elemental composition analysis indicates the presence of various components: sugar (1.60-2.10%), pectin substances (1.40-1.65%), calcium carbonate (74.25-80.00%), nitrogen-containing organic substances (5.90-6.01%), nitrogen-free organic substances (8.04-9.50%), lime in the form of various salts (2.80-3.10%), and other mineral substances (2.78-3.91%).

Incorporation of this waste into mineral binders facilitates the formation of alite, belite, and aluminoferrite phases, thereby enhancing their mechanical and chemical properties. Moreover, the organic fraction of the waste creates a reducing environment during firing, enabling the production of clinker that aligns with the color characteristics of traditional masonry mortars.A laboratory ball mill LShM-7500 was used to grind the raw materials. Uralite balls were used as grinding media. The process of grinding the raw mixture was carried out in two stages. Loess rocks were crushed for half the entire grinding time. Then the measured defecate powder was added to the ball mill, and the process of joint grinding continued.

The results of determining the optimal parameters of mechanical activation of loess-like rocks in the presence of defecate are presented in Figure 3.

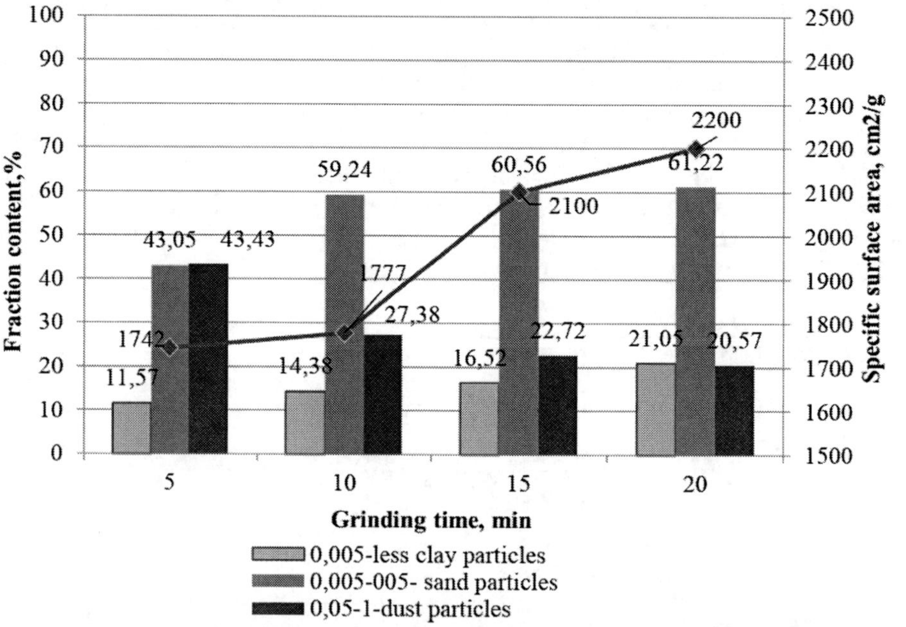

Fig. 3. Change in the granularity of the obtained powder with different grinding durations.

As can be seen from the data presented in Figure 3, grinding the raw material for 20 minutes leads to an increase in the proportion of fine clay and dust particles. To improve the rheological characteristics of the molding mixture, a joint activation of loess-like rock and defecate was carried out. To obtain a composition with binding properties, "loess-defecate" mixtures were used as a base in proportions from 70 to 95%, while the proportion of mechanically activated loess-like rock powder varied from 2 to 27%. Based on the "loess-defecate" mixture, a mass was prepared for the production of a mineral binder. Mechanically activated pressed powder of loess-like rock and defecate was ground to a specific grain surface of 2200 cm2/g, after which the defecate mass was added and dry mixing was carried out. Then the raw mix was moistened to 21.00-22.00% and pelletized to obtain spheres with a diameter of 20.00-23.00 mm. The obtained samples were dried in a laboratory drying oven ШЦ-80-02 ЦП at a temperature of 120 °C until a residual moisture content of 1.0% was achieved and fired in a muffle furnace СНОЛ 120/1200. Heat treatment was carried out at a temperature of 1200 °C with a heating rate of 245 °C/hour and holding at the maximum temperature for 2 hours. The obtained clinker was ground in a ball mill. It was found that mechanical activation of loess-like rocks in a ball mill for 20 minutes leads to an increase in the content of fine clay and dust fractions from 43.05% to 61.22%, while the proportion of coarse sand particles decreases from 11.57% to 21.05%. At the same time, the specific surface area, compared to the initial raw material, increased from 1750 cm2/g to 2200 cm2/g. The study of the dynamics of changes in the initial sample and fractions after grinding for 20 minutes was carried out using a Shimadzu SALD-7500 laser diffraction analyzer. The results of these studies are presented in the form of histograms in Figure 4.

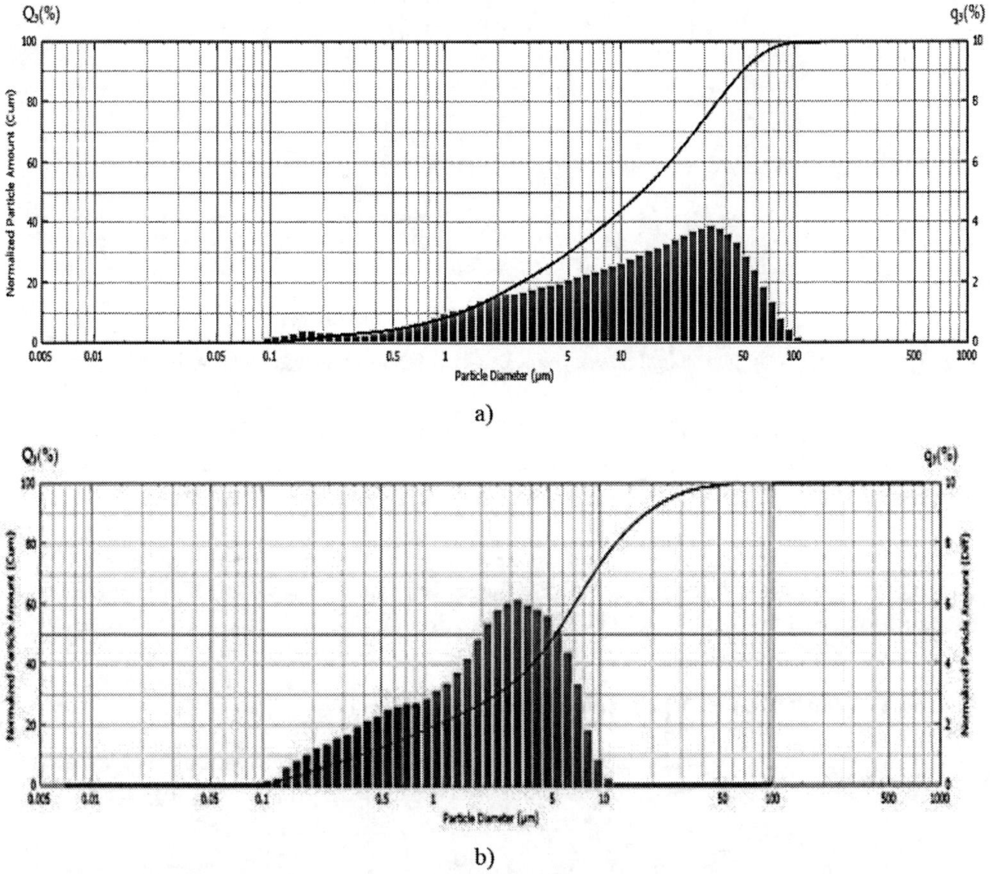

Fig. 4. Histograms illustrating changes in the ceramic mass based on the "loess-defecate" compositions and its particles after mechanical activation: a) original sample; b) sample after mechanical activation for 20 minutes.

It was found that mechanical activation of loess-like rocks in a ball mill for 20 minutes leads to an increase in the content of finely dispersed clay and dust particles from 43.05% to 61.22%, and the amount of coarse sand particles decreases from 11.57% to 21.05%. At the same time, compared to the original raw material, the specific surface area increased from 1750 cm^2/g to 2200 cm^2/g. The dynamics of changes in the initial sample and fractions after grinding for 20 minutes were studied using a Shimadzu SALD-7500 laser diffraction analyzer. The findings are visually represented in the histograms depicted in Figure 4.

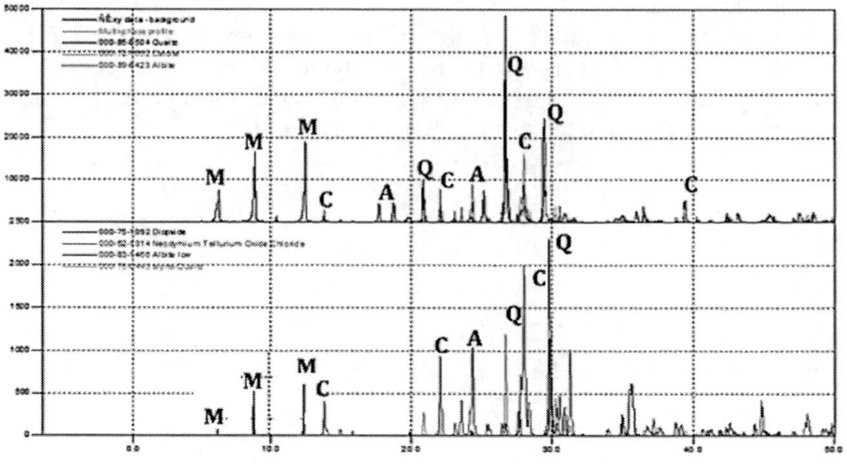

Q-quartz, C-calcite, A-albite, M-montmorillonite

Fig. 5. Results of X-ray phase analysis of non-activated (a) and mechanically activated mass (b) for 20 minutes.

Analysis of X-ray patterns of samples subjected to grinding for different periods showed that with increasing grinding duration, characteristic diffraction maxima appear on the diffraction patterns, indicating an increase in the degree of raw material amorphization. Based on this, it can be concluded that the use of this technology in the production of mineral binders provides: increased dispersion and microroughness of raw material particles; high concentration of structural defects both on the surface and in the volume of particles; as well as stabilization of this highly active state before the start of the clinker firing process.

4 Development of a process flow chart

Based on the results of tests carried out in laboratory and semi-industrial conditions, a technological scheme for obtaining a binder was developed, presented in Figure 6.

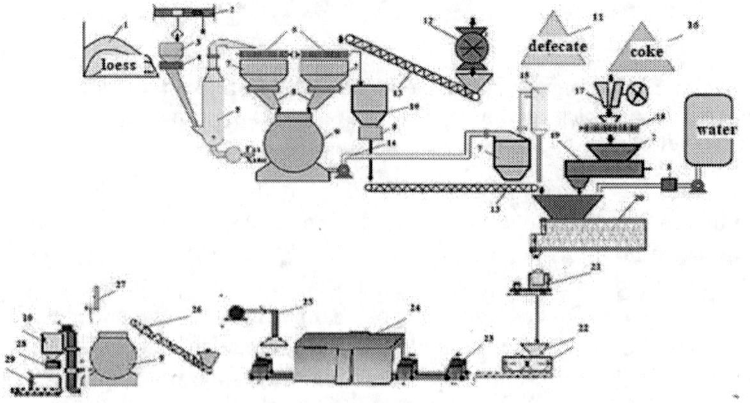

Fig. 6. Flow chart of production of mineral binder composition "loess-defecate": 1- excavation pit; 2-electric hoist-loader; 3-hopper; 4-plate feeder; 5-dryer with aerobic crusher; 6-drum sieve-borer; 7-hoppers; 8-dimensional dispenser; 9-ball mill; 10-coarse fraction hopper; 11-defecate compartment; 12-hammer crusher; 13-conveyors; 14-pneumatic conveyor; 15-cyclone; 16-coke compartment; 17-jaw crusher; 18-sieve; 19-doser hopper; 20-screw mixer; 21- measuring trolley; 22-briquette machine; 23- granule trolley; 24-chamber oven; 25-cooling system; 26-screw conveyor; 27-scales; 28- measuring dispenser; 29-packaging machine.

The conducted studies confirmed the possibility of obtaining a mineral binder using mechanical activation of loess-like rocks of the Suzanli deposit. A new composition of binder solutions with specified color characteristics was developed, taking into account the chemical and mineralogical composition of the original components. To evaluate the binding properties of the developed samples of mineral binder, their physical and mechanical characteristics were determined by the requirements established in GOST 30744-2001. The results of the tests are presented in Table 2. During the studies, it was found that the quality of binders decreases with an increase in the percentage of defecate in the "loess-defecate" mixture. Thus, for mineral binders samples MB-1 and MB-2, the setting time is 45-60 minutes, and complete hardening occurs after 120-150 minutes. While in other compositions, with an increased content of defecate, the initial setting occurs after 80-120 minutes, and complete hardening occurs after 180-210 minutes. The developed material can be used in the restoration and reconstruction of defects in the masonry of walls of historical monuments. In particular, in the presence of deep damage caused by aggressive impacts, when the depth of defects exceeds 3 cm, it is envisaged to remove defective bricks with subsequent replacement with inserts of the required size, fixed with mortar. As a result, a full restoration of defects in the masonry of historical objects is ensured.

Table 2. Physicomechanical properties of mineral binders obtained based on the "loess-defecate" composition.

Samples	Defecate, %	Loess, %	Binder grad GOST 31108 -201 (H 32,5)			km		28	14	28
MB-1	70	30	190	60	150	3,0	2,6	5,4	30,3	48,5
MB-2	75	25	200	75	120	3,5	2,8	4,5	34,6	50,4
MB-3	80	20	180	70	160	4,5	2,7	3,7	28,4	45,3
MB-4	85	15	170	80	180	5,0	2,3	3,5	25,5	43,2
MB-5	90	10	160	90	190	5,2	2,2	3,0	20,2	40,4
MB-6	95	5	150	120	210	5,4	2,1	2,5	19,6	38,6

There is a tendency for the grain size of the mineral binder to increase with the content of defecate in its composition. Determination of the bending strength of the developed mineral binders in dynamics showed a decrease in this indicator from 2.8 to 2.1 MPa after 14 days and from 5.4 to 2.5 MPa after 28 days. A similar pattern was observed when determining the compressive strength of mineral binders: the compressive strength decreased from 5.3 to 3.0 MPa after 14 days and from 20.4 to 15.4 MPa after 28 days. As a result, a mineral binder of grade 200 M (strength class H 32.5) with a particle size of 3 μm was obtained. Conclusion.

5 Conclusions

It has been established that during the mechanical activation process, the proportion of fine clay and dust particles increases by 20.01% and 61.12%, while the content of coarse sand particles decreases by 18.87% and the specific surface area increases to 2200 cm2/g, which has a positive effect on the technological properties of loess-like rocks. Changes in the phase composition of "loess-defecate" mixtures during mechanical activation have been revealed.

Based on the conducted research, a process flow chart for the production of a mineral binder based on the "loess-defecate" combination has been proposed, including mechanical activation of raw materials by 20-minute grinding in a F1830×6400 ball mill and the use of defecate (man-made waste) as an additional component, which makes it possible to reduce the clinker firing temperature to 1200 °C. In addition, the developed binder can be used in the creation of building materials intended for the restoration of cultural heritage sites.

References

1. H. Zhang, Y. Li, P. Wang, "Structural Changes in Loess During Mechanical Activation", Journal of Materials Science **54(3)**, 1125–1137 (2019), DOI: 10.1007/s10853-019-3456-7
2. R. Z. Rakhimov, R. A. Khusainova, "Optimization of High-Energy Milling Conditions for Loess-Based Materials", Construction and Building Materials **235**, 117469 (2020), DOI: 10.1016/j.conbuildmat.2020.117469
3. M. Ghorbani, S. Hashemi, "The Impact of Mechanoactivation on the Hydration Behavior of Loess-Based Blended Cements", Cement and Concrete Research **145**, 106455 (2021), DOI: 10.1016/j.cemconres.2021.106455
4. X. Liu, J. Chen, "Review of Mechanical Milling Techniques for Enhancing Pozzolanic Activity in Loess", Powder Technology **398**, 117034 (2022), DOI: 10.1016/j.powtec.2022.117034
5. L. Wang, F. Zhao, "The Role of Particle Size Reduction in Loess-Based Binders", Advanced Materials Research **1167**, 378–386 (2023), DOI: 10.4028/www.scientific.net/AMR.1167.378
6. P. Nikolaev, E. Makarov, "Influence of Milling Time on Loess Reactivity", Journal of Building Engineering **32**, 101755 (2020), DOI: 10.1016/j.jobe.2020.101755
7. S. K. Komilov, "Historical technology of production of gypsum mortar of historic monuments in Central Asia", International Journal for Innovative Research in Multidisciplinary Field, 152–157 (2019)
8. M. Jumaniyazov, S. Tadjiev, E. Atashev, "Results of X-Ray phase and IR-spectral analysis of azosuperphosphates obtained on the basis of phosphorites of central Kizilkum", Chemical Technology, Control and Management **3**, 19–25 (2021)
9. R. Ahmadov, N. Mammadov, "Loess and its use in the production of binders", Journal of Construction Materials **12(3)**, 45–52 (2020)
10. V. N. Gusev, P. A. Ivanov, "Mechanochemical activation of mineral raw materials", Materials Science and Engineering **27(2)**, 89–97 (2018)
11. T. Johnson, L. Baker, "Advances in the production of sustainable binders from loess", Green Construction Journal **9(4)**, 131–145 (2021)
12. ASTM C109/C109M-21, "Standard Test Method for Compressive Strength of Hydraulic Cement Mortars", ASTM International (2021)

Automation and modeling of copper smelting processes in Vanyukov furnace

Nazarbek Musabekov[1,*], *Irbulat* Utepbergenov[2], *Akhmetbek* Musabekov[3], and *Berik* Mirzabayev[3]

[1]S. Seifullin Kazakh Agrotechnical Research University, 010000 Astana, Republic of Kazakhstan
[2]G. Daukeev Almaty University of Power Engineering and Telecommunications, 010000 Almaty, Republic of Kazakhstan
[3]M. Auezov South Kazakhstan State University, 160010 Shymkent, Republic of Kazakhstan

Abstract. This article discusses the process of copper smelting in a Vanyukov furnace, with special attention paid to maintaining the optimal melt temperature, which is a key factor in the stable operation of the furnace. The article presents the results of experimental studies conducted at the Balkhash Copper Smelter, as well as experimental statistical data on the input and output parameters of the automated process control system (APCS). To build a mathematical model of the process, the System Identification Toolbox package in the MATLAB environment was used. As part of the work, modeling was carried out using a PID controller, its parameters were optimized to minimize the mean square error of regulation. In addition, the effectiveness of a fuzzy controller was considered, which showed a decrease in the dynamic error and a smoother process.

1 A control algorithm and parametric identification of the mathematical model of the Vanyukov process

One of the distinctive features of the Vanyukov process, which is at the same time one of the main requirements to its organization – this is continuity process. Continuously should be received in the furnace charge materials and oxygen-containing blowing, continuously must be carried out and removal of products from the oven swimming trunks [1].

For continuous, stable work of complexes Vanyukov furnace is necessary strictly observe technological discipline during operation not only the main, but also all the auxiliary equipment.

The most important parameter the melting process in the Vanyukov furnace furnace is the temperature melt. It defines copper losses with slag, stability operation of the Vanyukov furnace furnace and the most important its nodes - flows, siphons, overflow troughs, tuyeres. Low temperature swimming trunks promote over growing tuyeres; decrease in oxygen uptake; slowing down processes interactions in bubbling zone, delamination slag and matte; leads to the formation layers in the oven, build-up; increases copper losses with slag [2,3]. Unacceptable temperature decrease melting below 12000C. Excessive overheating of slag

* Corresponding author: n_mussabekov@mail.ru

can lead to melt leakage through the masonry or in the joints between caissons. Maximum recommended temperature the swimming trunks should not exceed 13000C.

Experimental studies were conducted at the Balkhash Copper Smelter and experimental statistical data were obtained for all input and output parameters of the Vanyukov Furnace APCS. Figures 1-3 show fragments of tables and trends compiled from the obtained experimental data (approximately 19,800 data). Figure 1 shows the graph of the input parameter - Cu content in the charge. Figure 2 shows the graph of the output parameter - Cu content in the matte, and Figure 3 shows the graph of the output parameter - Cu content in the slag.

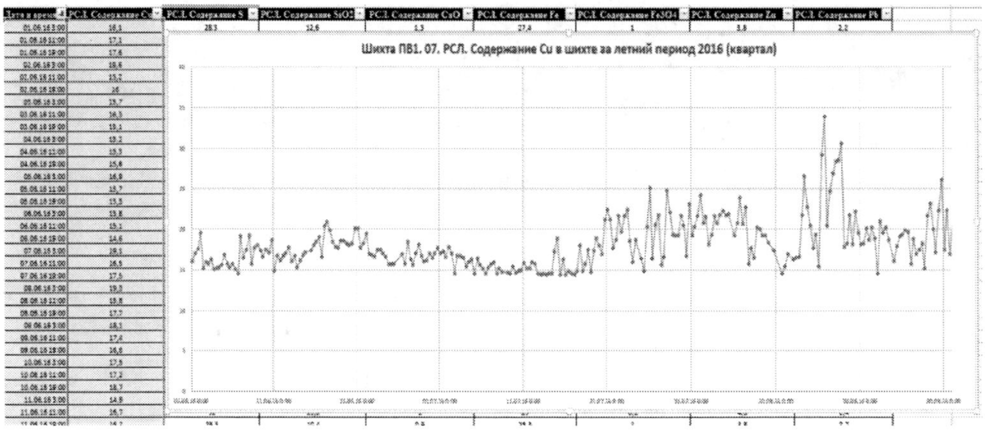

Fig. 1. Time sequence of the parameter "Cu content in the charge".

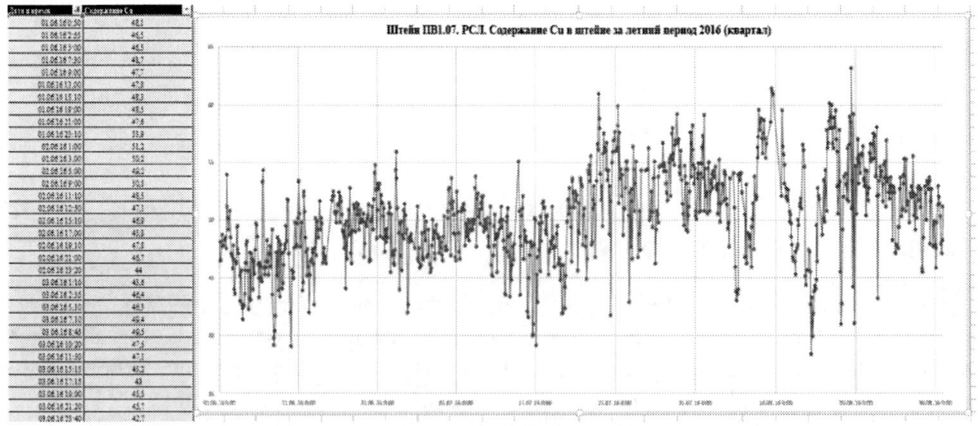

Fig. 2. Time sequence of the parameter "Cu content in matte".

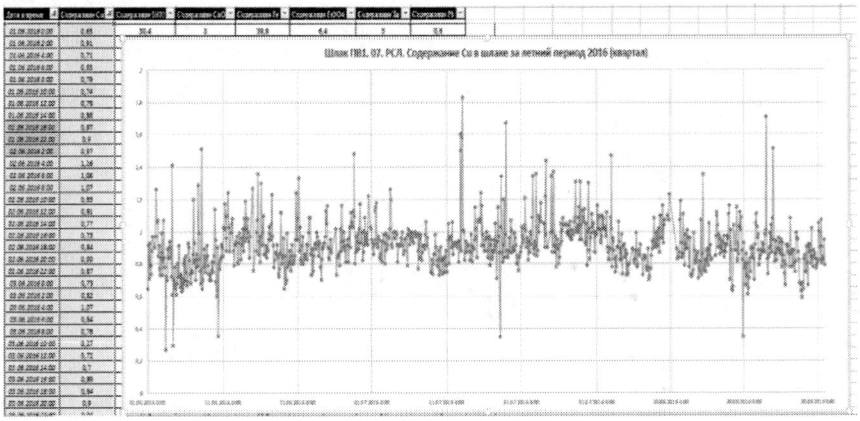

Fig. 3. Time sequence of the parameter "Cu content in slag".

In order to obtain a mathematical model of the control object – the Vanyukov furnace, data processing was performed in MATLAB in the System package Identification Tools, using the correlation method of statistical data processing. System Identification Toolbox is a specialized MATLAB package that contains tools for creating mathematical models of dynamic systems based on observed input/output data [4-7]. With the help of this package, parametric identification of the system was carried out and as a result, we received a mathematical model of the object in the form of transfer functions for each control channel.

Based on the obtained data, the channel oxygen blowing flow rate - matte temperature was selected, according to which the data processing was carried out. The choice of the channel is due to the best probability of constructing the transfer function. Data collection for both variables was carried out every 10 minutes.

The next step after data processing is to find the transfer function (Transfer Function Models), using System Identification Toolbox (Fig. 4.5) in Matlab. The interface is easy to use for working with models and data. The identification methods included in the package are applicable to a wide range of problems, from designing control systems and signal processing to time series analysis [7.8].

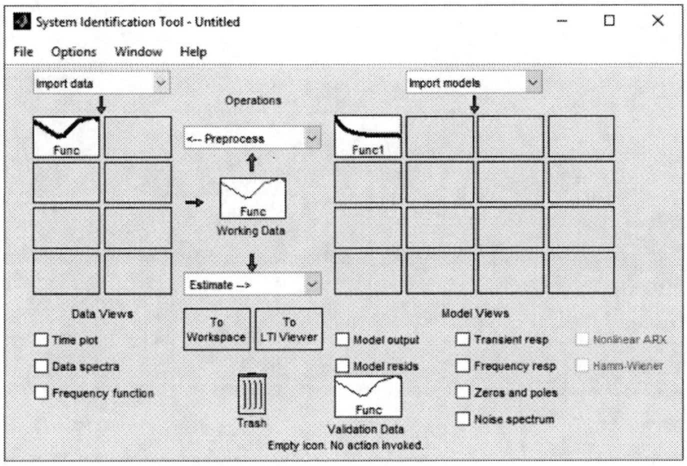

Fig. 4. Package for identification methods.

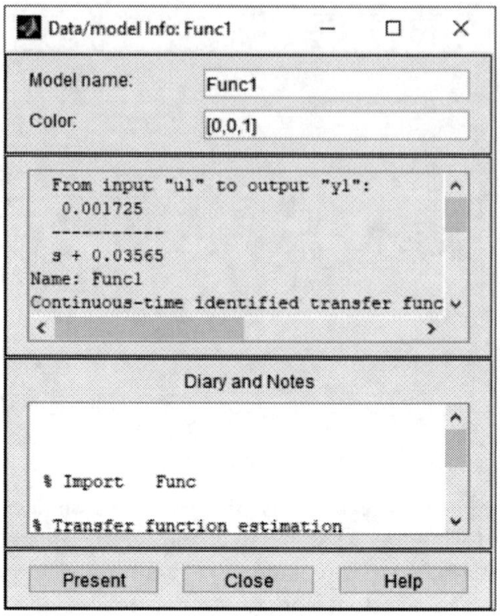

Fig. 5. First order transfer function.

As a result, using this interface, the transfer function of the first-order aperiodic link was obtained.

$$W(p) = \frac{K}{Tp+1} \tag{1}$$

Where: K is the gain factor;
T is the time constant.

$$W(p) = \frac{0.0476}{28p + 1}$$

Plotting graphs of the real model and constructed models (estimate models) is necessary for their comparison.

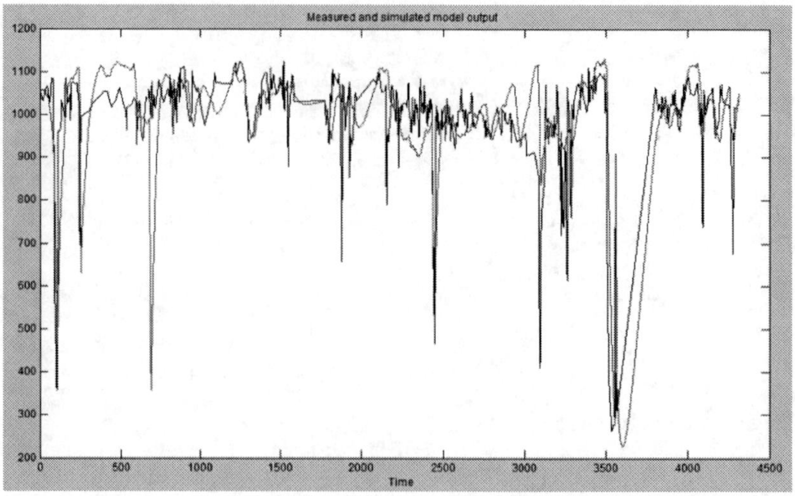

Fig. 6. Constructed models in comparison with the real model.

https://doi.org/10.1051/e3sconf/202563308005

The resulting acceleration curve for this channel as a result of the study is shown in Figure 7:

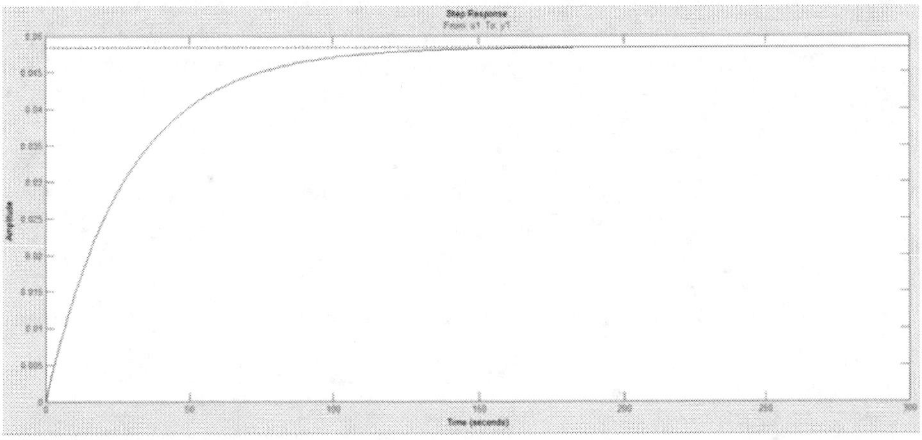

Fig. 7. Time graph of the constructed model (single jump).

The next step is to simulate the mathematical model. In this case, a structural diagram was constructed using a PID controller [9].

The PID controller is a universal controller. It is also a simple P controller with an additional integral component and a differential component. The additional I component is necessary to eliminate the static error, which is typical for the proportional component of the controller. The advantages of the PID controller prevail over the PI controller, for example, the static error of regulation is almost zero; it is quite easy to use, or rather to configure, since only two parameters are configured, which ensures control with the minimum possible mean square error of regulation; it is not sensitive to noise in the measurement channel (in comparison with the PID controller) and due to the differential component , the system response speed is ensured. Transfer function:

$$W(p) = K_p + \frac{K_i}{p} + K_d p \tag{2}$$

Where: K p – coefficient of the proportional component c;
K i – coefficient of the integral component;
K d – differential coefficient component.

To control the process, it is necessary to find the value of the PID controller coefficients. This step was carried out using the PID application Tuner in Matlab. PID Tuner uses process PID tuning, which is the process of finding the values of the proportional, integral, and derivative gains in a PID controller to achieve the desired transient response characteristics. Typically, PID controllers are tuned either manually or using specific techniques. With the help of, PID Tuner can automatically adjust the PID controller in real time, looking at the constructed graph. Then the interface itself finds the values of the coefficients, based on the constructed graph [10-12].

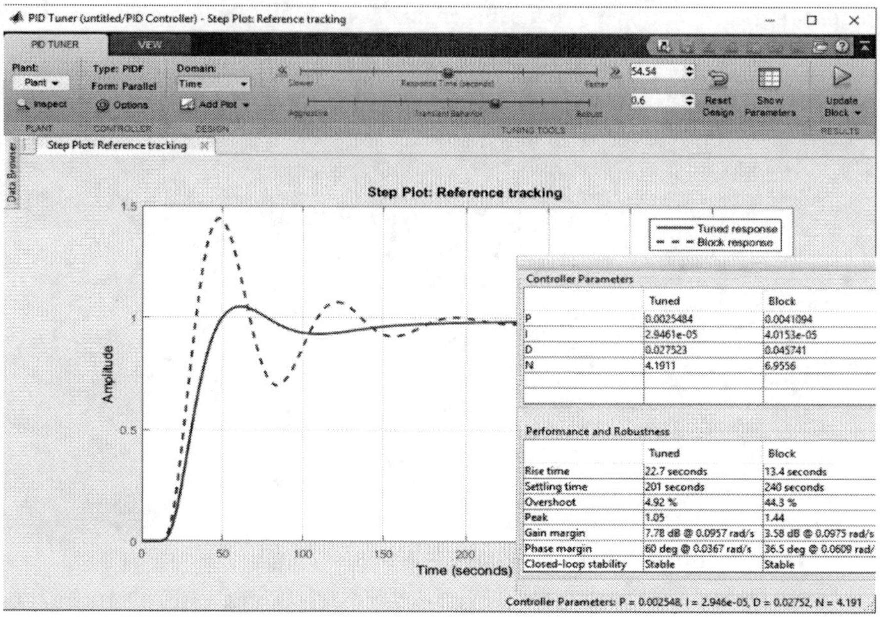

Fig. 8. Finding the values of the coefficients in PID Tuner.

As can be seen from Figure 8, the values of the coefficients are K p = 9.8, K i = 1.2.

Now it is necessary to model the system in Simulink. Figure 9 shows the structural diagram of the ACS with a PID controller.

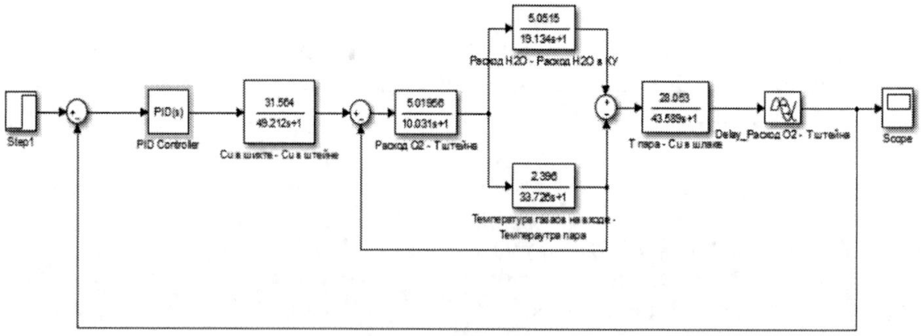

Fig. 9. Control system of the PV TC with a PID controller.

The next step is to model the system in Simulink.

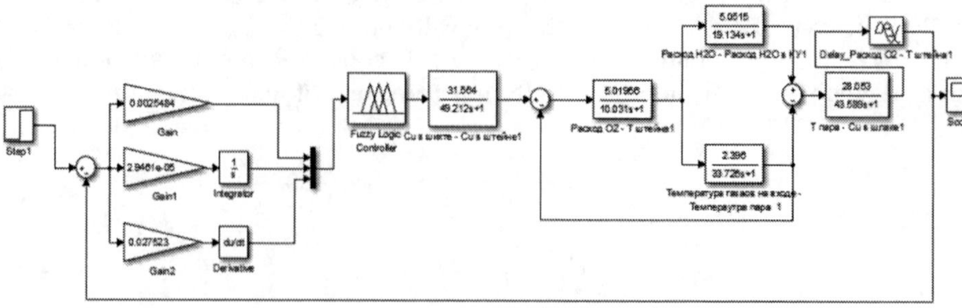

Fig. 10. Control system of the PV TC using a Fuzzy controller.

Let's plot a graph of transient processes for the PID controller and the Fuzzy controller.

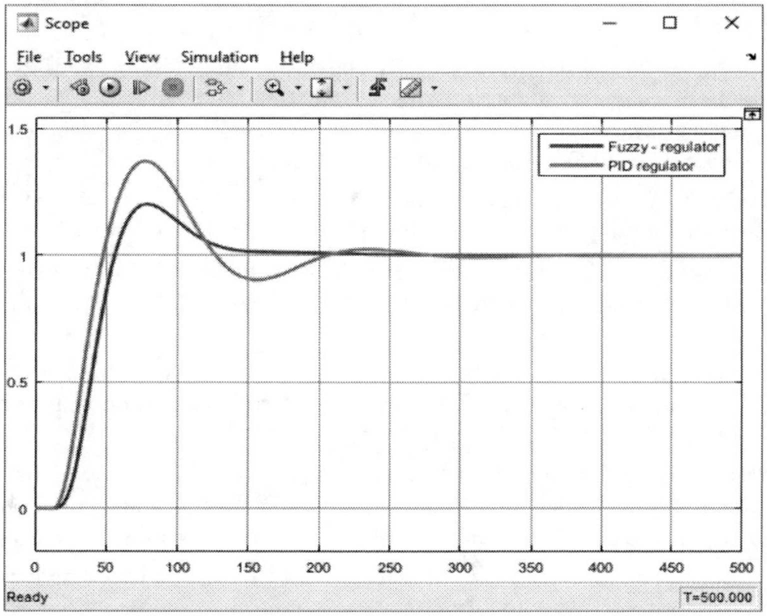

Fig. 11. Comparison of transient processes of the TC PV control system with a PID controller and a Fuzzy controller.

The use of mathematical modeling methods has already proven itself to be the best unit for implementing technological modes and designing automated control systems. System Identification Toolbox is an interface that has all the necessary tools for designing mathematical models of dynamic systems based on observed input and output data. According to the graph (Fig. 11), the regulation time and dynamic error of the Fuzzy controller have decreased and the process proceeds without strong fluctuations, i.e. there is no sharp increase in temperature, by about 20%.

2 Conclusion

Thus, maintaining a stable temperature in the Vanyukov furnace is critical to minimizing copper losses with slag and ensuring stable operation of the equipment. As a result of experimental studies and mathematical modeling using System Identification Toolbox found that the use of a PID controller provides accurate and fast regulation, eliminating the static error. At the same time, the use of a Fuzzy controller allows achieving even greater process stability by reducing the dynamic error and eliminating sharp temperature fluctuations. These control methods in combination allow optimizing the operation of the Vanyukov furnace and increasing the efficiency of the copper smelting process.

References

1. V. Osadchyy, O. Nazarova, T. Hutsol, S. Glowacki, K. Mudryk, A. Bryś, A. Rud, W. Tulej, M. Sojak, "Adjustable Vibration Exciter Based on Unbalanced Motors", Sensors **23(4)**, 2170 (2023)

2. L. Belascu, A. Horobet, G. Vrinceanu, C. Popescu, "Performance Dissimilarities in European Union Manufacturing: The Effect of Ownership and Technological Intensity", Sustainability (2021)
3. A. Bonci, G. Rizzello, S. Longhi, D. Naso, M. Pirani, S. Seelecke, E. Lorenzoni, "Simulation analysis and performance evaluation of a vibratory feeder actuated by dielectric elastomers", in *Proceedings of the 14th IEEE/ASME International Conference on Mechatronic and Embedded Systems and Applications (MESA)* (Oulu, Finland, 2018)
4. J. Zhang, X. Wang, "Design and experimental study of ultrasonic vibration feeding device with double symmetrical structure", IEEE Access **10**, 63481–63495 (2022)
5. N. Yaroshevich, I. Zabrodets, S. Shymchuk, T. Yaroshevich, "Influence of elasticity of un-balance drive in vibration machines on its oscillations", East.-Eur. J. Enterp. Technol. **5**, 62–69 (2018)
6. D. Khort, A. Kutyrev, N. Kiktev, T. Hutsol, S. Glowacki, M. Kuboń, T. Nurek, A. Rud, Z. Gródek-Szostak, "Automated mobile hot mist generator: A quest for effectiveness in fruit horticulture", Sensors **22**, 3164 (2022)
7. V. Yatsun, G. Filimonikhin, K. Dumenko, A. Nevdakha, "Equations of motion of vibration machines with a translational motion of platforms and a vibration exciter in the form of a passive auto-balancer", East.-Eur. J. Enterp. Technol. **5**, 19–25 (2017)
8. K. S. Sholov, K. Abzhaparov, B. Mirzabaev, "Justifying and choosing parameters of the wind power installation with an automatically controlled sailing working body", Journal Energy Web and Information Technologies **6**, 1–8 (2018)
9. V. Topilnytskyy, D. Rebot, M. Sokil, O. Velyka, S. Liaskovska, I. Verkhola, R. Kovalchuk, L. Dzyubyk, "Modeling the dynamics of vibratory separator of the drum type with concentric arrangement of sieves", East.-Eur. J. Enterp. Technol. **2**, 26–35 (2017)
10. M. L. Hoang, A. A. Pietrosanto, "New technique on vibration optimization of industrial inclinometer for MEMS accelerometer without sensor fusion", IEEE Access **9**, 20295–20304 (2021)
11. N. R. Mussabekov, B. K. Mukhanov, "Development of a mathematical model for a compound technological complex of vanyukov melting in order to control the material and thermal regime", Kompleksnoe Ispolzovanie Mineralnogo Syrya **4(327)**, 15–22 (2023)
12. W. Cao, N. Xing, Y. Wen, X. Chen, D. Wang, "New Adaptive Control Strategy for a Wind Turbine Permanent Magnet Synchronous Generator (PMSG)", Inventions **6(3)** (2021)

Evaluation of the natural radioactivity (^{238}U,^{232}Th,^{40}K) in building materials used in Iraqi homes

Roua Mouaid Ali[1,*], *Ansam* Qassim Gadhban[2], *Yassir* A. Baqir[3], and *Shireen* N. Alebadi[4]

[1]College of Medicine, University of Iraq, Department of Physiology, Baghdad, Iraq
[2]Middle Technical university, Baghdad, Iraq
[3]Science department, College of Basic Education, Mustansiriyah University, Baghdad, Iraq.
[4]Al-Furat Al-Awsat Technical University, Engineering Technical College of Najaf, Department of Health Physics and Radiation Therapy Technical Engineering, Najaf, Iraq

Abstract. The purpose of this study is to quantify the amounts of natural radioactivity of the building materials employed in homes throughout Iraq due to three radioactive elements present: thorium-232 (232Th), potassium-40 (40K), and uranium-238 (238U). Radiation levels and specific activity of these elements in the materials were determined by gamma-ray spectroscopy. The specific activities (average value for each element) wereUranium-238 (^{238}U): 23.3633 ± 2.035 Bq/kg, ^{232}Th: 10.1967 ± 1.482 Bq/kg, Potassium-40 (^{40}K): 217.4367 ± 25.815 Bq/kg.So, this means the rate of radiation emitted in Kg of building material in these naturally occurring radioactive elements. The research compares this data with international safety limits to look at whether the levels of natural radioactivity in the materials could have health implications for local people.

1 Introduction

Natural radioactivity and its possible effects on human health have been a topic of concern in recent years, especially in areas where there are rich natural resources of radionuclides. Here and there, the radioactive elements thorium-232 (232Th), potassium-40 (40K), and uranium-238 (238U) are found in the environment, including soils, rocks, and even in the building materials used in the homes we live in. These substances are dangerous for the environment and health when found in buildings or workplaces where humans stay for longer periods of time [1,2]. Therefore, the assessment of construction materials' inherent radioactive levels, particularly the materials that frequently used in Iraqi houses, is essential for evaluating of the possible radiological hazard and maintain of public health and safety [3]. Iraq, which is privileged with natural and geological resources, has crossed a more polluted stage due to wars, rapid urbanization and industrialization. The high levels of natural background radiation found in many areas of Iraq are due to the local soil and the rocks containing high concentrations of natural radionuclides [4]. The use of local construction materials in residential buildings, including limestone, gypsum and clay, can lead to different

* Corresponding author: roua.m.ali@aliraqia.edu.iq

levels of natural radioactivity [5]. These materials, when introduced to indoor environments, can result in cumulative radiation exposure, which could, over the long-term, pose an increased risk of health defects, including cancer [6]. Radioactivity is natural; earth crust contains trace amounts of isotopes of radioactivity. Thorium-232 (232Th), potassium-40 (40K), and uranium-238 (238U) are naturally" occurring isotopes found on earth which contribute to the background radiation. Uranium and thorium are heavy metals that occur in some rocks and soils, and potassium-40 is an ordinary occurrence isotope of potassium that is widely distributed in nature [7-9]. This study aims to assess the outdoor ionizing radiation arising from the building materials commonly used in Iraqi homes; this study accomplishes that by determining the potassium-40, thorium-232, and uranium-238 activity concentrations in a piety of building materials used in Iraqi homes, in addition to figuring out the radioactive risk indexes based upon the measured values and evaluating the possible negative health effects of prolonged radiation exposure generated from these materials. This latest research aims to present the radiological significance of building materials in Iraq, based on the data collected.

2 Materials and methods

2.1 Sample collection and preparation

The natural radio nuclides concentrations are reported in 30 Building materials Samples with an average of 15 imported and 15 local samples. These samples have been collected from various local markets of Bagdad and then sealed in plastic bags with labels and delivered to the radiation detection and measurement facility. A 1 L polyethylene Marinelli beaker serves as the sampling and measuring container for this project. Before being used, the containers are rinsed with distilled water and cleaned with diluted hydrochloric acid. Samples were placed in an oven to dry and remove moisture at 100Co for 24 hrs. for the analysis. The samples are then crushed mechanically to a degree of homogenization suitable for analysis, using a model FT102 micro soil attrition mill supplied by TAISITE; after homogenization, the samples were passed through a sieve of 0.8 mm pore size diameter. All samples were previously kept achieving secular equilibrium for four weeks before measurement. The light cap of the Marinelli breaker presses on the sample to ensure that all the air is removed. Each of the samples are then hermetically sealed and dry weighed in a standard Marinelli beaker within ±0.01% using a highly sensitive digital weighing balance, and no less than 1 kg of each sample is used. For a long-time measurement, the sample is coupled face-to-tail with the detector.

2.2 Measurements of system

In the current study, gamma-ray spectroscopy techniques were used to classify samples because of the considerable penetrating power of gamma rays in the structures. The multi-channel analyzer (MCA) (ORTEC-Digi Base) and a concern program (MAESTRO-32) are connected to the NaI(Tl) (3"×3") crystal dimension scintillation detector (purchased from Alpha Spectra, Inc.-12I12/3) in the lab PC for analysis (Fig.1). A spectrum from five common gamma radiation sources (22Na, 60Co, 54Mn, 137Cs, and 152Eu) was used to energy calibrate a NaI (Tl) detector. The energy calibration curve for these sources shows a linear relationship between energy and channel numbers. The energy resolution value for the 137Cs standard source at 662 keV was 8.6%. The secular equilibrium of 214Bi (1765 keV) and 208Tl (2614 keV) was used to determine the specific activities of each identified sample for

238U and 232Th, respectively, whereas the specific activity of 40K was directly calculated at phase position of 1460 keV.

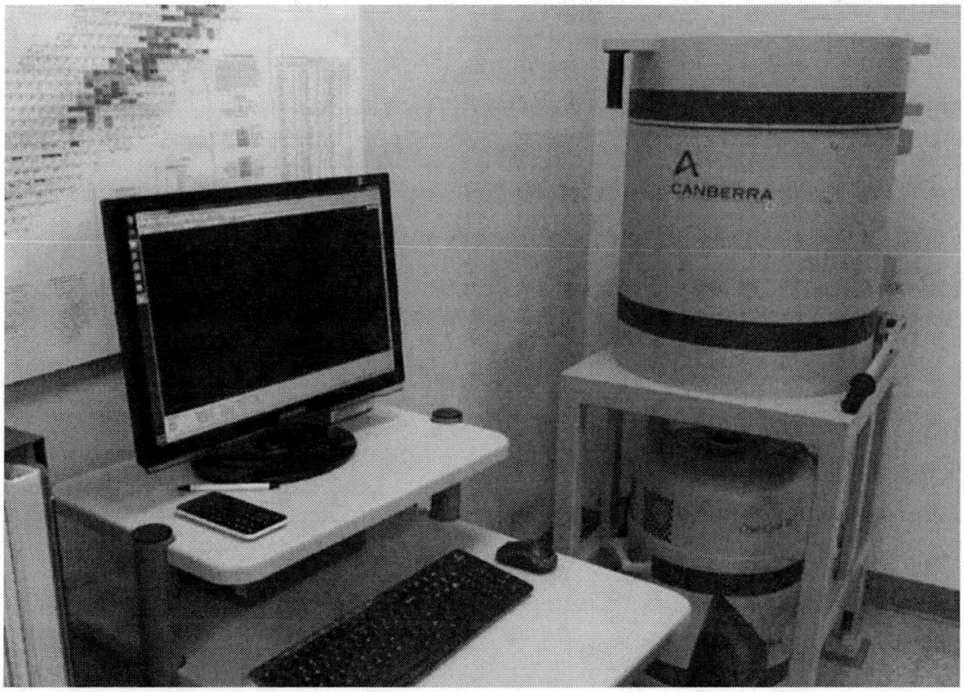

Fig. 1. Gamma-ray spectroscopy.

2.3 Calculations

Specific Activity (A)
The following formula was used to determine the specific activity of the radionuclides 238U, 232Th, and 40K (represented as AU, ATh, and Ak, respectively):

$$\mathbf{A}\left(\frac{\mathbf{Bq}}{\mathbf{kg}}\right) = \frac{\mathbf{N}}{\mathbf{I_\gamma\ \varepsilon\ M\ T}} \quad (1)$$

where N represents the net area beneath the photopeak, Iγ denotes the likelihood of gamma decay, ε signifies the detector's effectiveness, M indicates the mass of the sample, and the measurement pertains to time.

3 Results

The activities and ratios of the radionuclides 238U, 232Th, and 40K in 30 samples of construction materials are presented in Table 1. The specific activity 238U, 232Th, and 40K, were 23.3633 ±2.035 Bq/kg; 10.1967 ±1.482 Bq/kg and the average were the; 217.4367 ±25.815 Bq/kg respectively. A comparison with the average global activity is shown in Fig. 2-4.

Table 1. Results of particular activity for 40K, 238U, and 232Th in current samples.

No.	Sample code	Specific activity in Bq/kg					
		^{238}U		^{232}Th		^{40}K	
		Average	S.E	Average	S.E	Average	S.E
1	p1	12.0	0.7	25.5	0.7	476.7	4.3
2	p2	31.0	1.2	8.5	0.3	167.6	2.5
3	p3	26.8	1.4	19.6	0.6	486.1	4.3
4	p4	22.5	1.3	0.8	0.1	173.2	2.6
5	p5	30.1	1.6	12.0	0.6	141.2	2.3
6	p6	33.4	1.2	6.8	0.5	120.7	2.2
7	p7	40.6	1.3	11.3	0.4	510.9	4.4
8	p8	29.7	1.6	20.2	0.6	465.8	4.2
9	p9	25.1	1.8	27.5	0.9	352.0	3.7
10	p10	20.6	0.9	25.6	0.8	315.9	3.5
11	p11	18.8	1.1	13.3	0.5	410.8	4.0
12	p12	33.2	1.4	5.5	0.4	57.1	1.5
13	p13	26.9	1.3	8.3	0.4	161.5	2.5
14	p14	26.3	1.3	6.9	0.4	125.4	2.2
15	p15	24.0	1.4	4.0	0.3	84.4	1.8
16	p16	26.2	1.2	3.7	0.3	147.1	2.4
17	p17	29.1	1.2	6.4	0.4	125.4	2.2
18	p18	12.5	0.7	22.8	0.6	260.8	3.2
19	p19	6.3	0.6	4.2	0.2	90.6	1.9
20	p20	5.2	0.4	8.2	0.4	149.5	2.4
21	p21	10.7	0.6	2.1	0.2	91.6	1.9
22	p22	16.4	1.0	3.5	0.2	140.4	2.3
23	p23	31.3	1.3	5.3	0.3	150.3	2.4
24	p24	40.3	1.4	-1.6	0.4	100.1	2.0
25	p25	18.6	0.8	6.5	0.3	117.7	2.1
26	p26	22.3	1.1	23.7	0.6	100.6	2.0
27	p27	7.0	0.5	11.1	0.5	166.4	2.5
28	p28	12.6	0.8	2.8	0.2	189.0	2.7
29	p29	19.4	1.0	4.4	0.3	476.7	4.3
30	p30	12.0	0.7	7.0	0.4	167.6	2.5
Aveage±S.E		23.3633 ±2.035		10.1967 ±1.482		217.4367 ±25.815	
Worldwide		**33**		**45**		**420**	

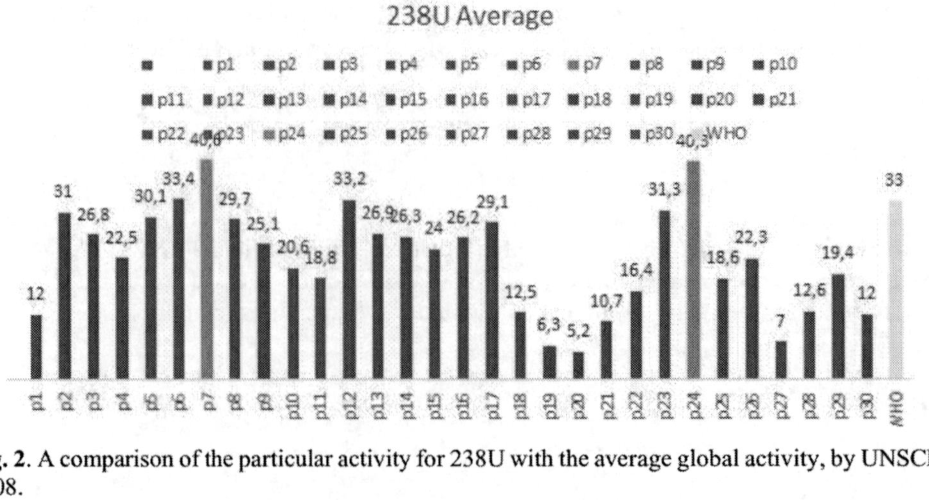

Fig. 2. A comparison of the particular activity for 238U with the average global activity, by UNSCEAR 2008.

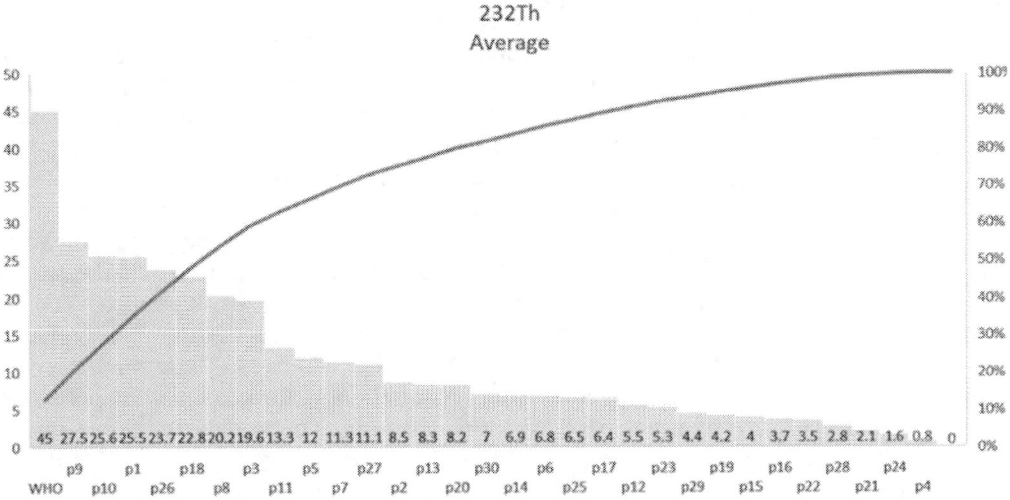

Fig. 3. A comparison of the particular activity for 232Th with the average global activity, by UNSCEAR 2008.

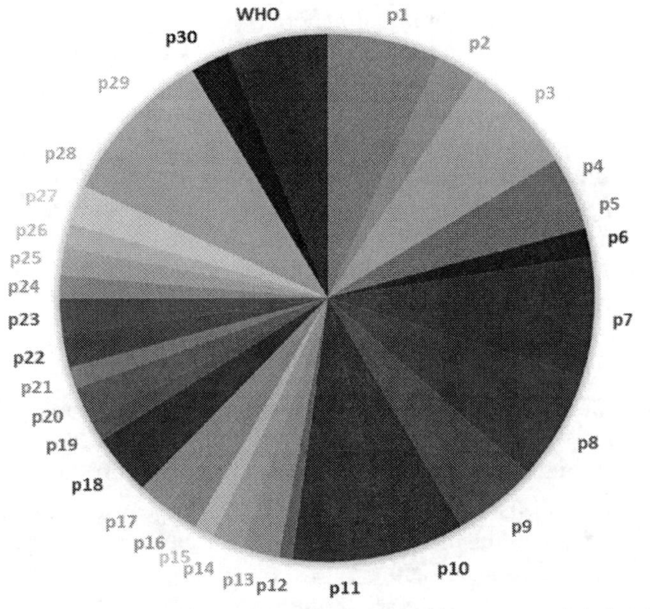

Fig. 4. A comparison of the particular activity for 40K with the average global activity, by UNSCEAR 2008.

4 Discussions

Among the radioactive elements, uranium (^{8}U), thorium (^{32}Th), and potassium (^{40}K) are among the most examples occur naturally in trace amounts in the crust of the earth. These are found naturally in a variety of materials found in the environment. These radioactive elements are often found in building materials like stone, granite, sand, concrete, and bricks but in low concentrations. Sources of Radioactivity: The three naturally occurring isotopes are Potassium-40 (^{40}K), uranium (^{238}U), and thorium (^{232}Th) will be discussed. Being in the Earth's crust, raw materials used in common construction, such as stones, sands or clays, can be contaminated with these isotopes. Variation in Concentrations: These radioactive elements may vary in concentration depending on geography and which raw materials you are using. For instance, the radioactivity of materials may be slightly greater if the earth is naturally rich in uranium or thorium. Specific Building Materials: Both of these natural stones are used for construction and countertops. Because of the geological conditions in which they are formed, they are also typically more enriched in uranium and thorium than other materials. Concrete and Bricks These building supplies are created using mixtures, including sand and aggregates, which can contain potassium. Potassium itself naturally contains the radioactive isotope potassium-40. Moreover, small amounts of uranium and thorium may also be present in concrete and bricks depending on basic ingredients that are utilized in the production of each item. Cement-like concrete and bricks are commonly produced by mixing together several raw ingredients, such as limestone, clay and other additives. These materials may contain trace amounts of uranium and thorium. In addition, if the cement contains any of these raw materials, like fly ash or slag, which may contain natural radioactive substances, the level of radioactivity in the cement varies.

5 Conclusion

This study would provide an estimate of natural radioactivity levels from (^{238}U), thorium (^{232}Th), and potassium (^{40}K) in the building materials normally used in the homes of Iraqi families. While radionuclide concentrations in the materials are within acceptable levels of safety, such radionuclides merit concern about the potential health effects they may pose on human health, particularly in poorly ventilated or high-density settlements. Ongoing monitoring of these products is important and if, based on its results, necessary regulations should be developed and implemented to prevent exposure to residential areas from ionizing radiation. Moreover, by raising knowledge of the risks of natural radioactivity through preventive action such as ensuring safer building materials are used and proper ventilation of houses, the associated risks could be minimized.

References

1. P. S. Cardoso da Silva et al., "Activity concentration of the uranium and thorium series and exhalation rates of 222Rn and 220Rn in building material derived from niobium residues," J. Build. Eng. **46**, 103657 (2022), DOI: 10.1016/j.jobe.2021.103657
2. H. Tohver, S. Salupere, "Breaking the cycle of excessive conservatism: Evaluating the γ-radiation dose from building materials," J. Environ. Radioact. **268-269**, 107250 (2023), DOI: 10.1016/j.jenvrad.2023.107250
3. D. J. Pizzarello, ed., *Radiation Biology* (CRC Press, 2020), DOI: 10.1201/9780429279034
4. C. Okogbue, M. Nweke, "The 226Ra, 232Th and 40K contents in the Abakaliki baked shale construction materials and their potential radiological risk to public health, southeastern Nigeria," J. Environ. Geol. **2(1)** (2018)
5. E. F. Khanjer, L. Awad, W. A. Hassan, B. S. Ismael, "Prepare Maps for Greenhouse Gases with Some Weather Elements for Baghdad City Using Data Observation and Arc-GIS Techniques," in IOP Conference Series: Materials Science and Engineering **757(1)**, 012031 (2020)
6. A. Abojassim, N. A. Jafer, A. J. Azeez, A. H. Abdulzahra, L. Y. Jebur, Z. R. Abdel-Al, Z. J. Ghani, "Radiation dangers in Iranian ceramic samples in local markets," in E3S Web of Conferences **583**, 04016 (2024)

7. A. A. Marzaali, M. M. Kadhim, T. A. Hussein, A. S. Jassim, A. A. Abojassim, H. A. A. B. Mraity, "Evaluation of Natural Radioactivity and Associated Hazards in Certain Ceramic Samples Commonly used in Iraq," Lat. Am. J. Pharm. **42**, 391-396 (2023)
8. K. Asaduzzaman, M. U. Khandaker, Y. M. Amin, D. A. Bradley, "Natural radioactivity levels and radiological assessment of decorative building materials in Bangladesh," Indoor Built Environ. **25(3)**, 541–550 (2016)
9. L. A. Najam, A. I. Yaseen, "Evaluation of Radioactivity in Soil Sample from Al-Hadbaa Cement Plant in Nineveh Governorate, Iraq", ARO: The Scientific Journal of Koya University **11**, 83–88 (2023), DOI: 10.14500/aro.11283
10. Ş. Turhan, A. Kurnaz, M. Karataşlı, "Evaluation of Natural Radioactivity Levels and Potential Radiological Hazards of Common Building Materials Utilized in Mediterranean Region, Turkey", Environmental Science and Pollution Research **29**, 10575–10584 (2022), DOI: 10.1007/s11356-021-16505-7

E3S Web of Conferences **633**, 08007 (2025) https://doi.org/10.1051/e3sconf/202563308007
GASES 2025

Applications of lasers in dermatology: from hair removal to skin rejuvenation

Morjan Hassan Alsobh [*]

College of Health and Medical Technologies, Department of Physiotherapy Technologies, Al-Zahraa University for Women, Karbala, Iraq

Abstract. This paper covers some of the uses in dermatology, for both cosmetic and therapeutic applications. It surveys laser technologies used in a series of common procedures in dermatology: hair removal, skin rejuvenation, acne, etc. The review highlights the mechanisms of laser interaction with skin tissue, safety concerns, efficacy, and recent developments in laser technologies. The goal is to facilitate a balanced and broad overview of laser therapies in contemporary dermatology, including who, what, and where they are used, advantages, potential hazards, along with future ramifications.

1 Introduction

Whether cosmetic or therapeutic, laser technology transformed the landscape of modern dermatology, allowing for non-invasive treatment options [1]. Over the past several decades, the evolution and enhancement of laser systems has offered practitioners ever more effective tools for the treatment of a wide range of skin conditions. Used for everything from hair removal to skin rejuvenation, the lasers have shown versatility as well as safety and resulted in various therapeutic results in the skin.Lasers in dermatology dates back to the late 1960s, with the advent of the first laser. Lasers have since also been adapted for everything from skin resurfacing to vascular lesions to help reduce pigmented spots [2]. . Today, many types of lasers, each differing in terms of wavelengths and mechanisms of action, are utilized to treat dermatological conditions.How Laser Interacts with Biological TissuesLasers Therapeutic effects are based on a light-tissue interaction. Lasers produce light at targeted wavelengths, which are absorbed by specific chromophores (substances in the skin) such as melanin, hemoglobin, and water [3]. For localized treatments, lasers are highly efficient: they can hit specific components of the skin while sparing the surrounding tissue [4].

This article reviews the most common applications of lasers in dermatologic practice: laser hair removal and skin rejuvenation. This article discusses various laser modalities, including intense pulsed light (IPL), fractional CO2 laser, erbium laser, and pulsed dye laser and - their modes of action, clinical indications, outcome, and safety. Furthermore, it will summarize the evolution of laser technology including emerging devices and the future of laser dermatology.

[*] Corresponding author: ali.alhameedawi@uokufa.edu.iq

2 Materials and methodology

This section presents a detailed account of the methods used to collect data for the research. The methodology involves qualitative and quantitative review to provide a systematic overview of studies, clinical trials, and case reports in the realm of laser applications in dermatology.

Quantitative data were analyzed to assess the clinical effectiveness of various laser modalities. Various laser treatments were compared in terms of success rate, treatment time, side effects, patient satisfaction, and long-term outcomes.
(Include t ables or graphs for summarizing the data from clinical trials or case studies. Example picture: Efficacy of different laser technologies in the treatment of specific dermatological indications.)

3 Results

This section presents the results obtained from the literature review. The data will be organized into tables and graphs to highlight key findings.

Table 1. Comparison of laser types for hair removal.

Laser Type	Wavelength (nm)	SkinType Targeted	Efficacy (%)	Side Effects
Alexandrite Laser	755	Light to Medium	85%	Hyperpigmentation
Diode Laser	810	Medium to Dark	90%	Erythema, Edema
Nd:YAG Laser	1064	Dark	80%	Scarring, Hypopigmentation

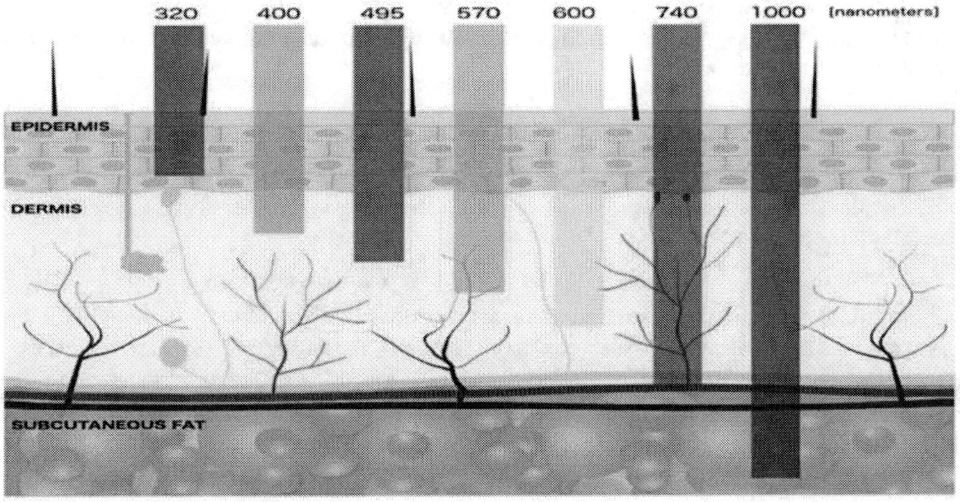

Fig. 1. Skin layers and laser absorption mechanism.

4 Discussion

Laser Hair Removal Effectiveness: Laser hair removal is becoming one of the most searched dermatological treatments. Research has demonstrated that lasers, particularly due to the Alexandrite laser and Diode laser, are extremely efficient in decreasing hair growth; patients often note that the results are long lasting or even permanent. Its effectiveness is determined by factors such as the patient's skin type, hair color, and the type of laser used [4-6].

Dermal Rejuvenation: Mechanism and Advantages: Especially, fractional CO2 and erbium laser skin rejuvenation were remarkably successful for treatment of photodamaged skin, wrinkles and scars. Collagen inducing lasers (also called regenerative lasers), work through injecting collagen to the skin. Fractional CO2 lasers, specifically, create micro-thermal zones that allow for targeted treatment with limited damage to surrounding tissue and quicker recovery times [7].

Risks and Side Effects: While effective, there are some risks to laser treatments. Common side effects consist of erythema, swelling, and temporary hyperpigmentation. In very rare cases, more serious adverse events such as scarring or changes in permanent pigmentation may occur. Implementing safety protocols, which involve the proper selection of patients and pre-treatment assessments, is essential to mitigate these risks [8].

Novel Laser Approaches: Laser technology has advanced recently, making treatments more effective, safe and accessible [9-11]. Especially, the introduction of non-ablative lasers that reach deeper skin layers without causing damage to the outer layer has able to increase the variety of lasers used in therapy in dermatology. Moreover, some other novel technologies such as picosecond laser technology or fractional laser have also been investigated, which can better rejuvenate skin and treat pigmentation.

5 Conclusions

Over the last several decades, lasers have become important tools in dermatology for combating hair growth, serving skin renewal functions, and treating other dermatological conditions. With constant advancements in technology lasers will continue to play a key role in both cosmetic and medical dermatology. Although lasers show great promise, patient safety is of utmost importance and practitioners should be familiar with best practices and current techniques. The continuing evolution of laser treatments, which will become more efficient and safer through new research, will no doubt continue to carve dermatology's place in the 21st century landscape of care.

References

1. K. Žužul, *The use of lasers in dermatology* (Master's thesis, University of Zagreb, School of Medicine, 2014)
2. S. Aleem, I. Majid, *Unconventional Uses of Laser Hair Removal: A Review* (2019)
3. P. Pezhdam, M. AliAsgari, M. Sadat Ansari, N. Mazlumi Tutunchi, A. Cheraghalian, M. Nasimi, "Safety of using laser for hair removal in patients with skin diseases: a retrospective study," J. Cosmet. Laser Ther. **26(1-4)**, 31-35 (2024)
4. C. Byalakere Shivanna, C. Shenoy, S. Lalitha, P. Jaju, A. Zorman, "Comparison of submillisecond pulse (FRAC3) and long-pulse 1064 nm Nd:YAG laser hair removal," J. Cosmet. Dermatol. **21(8)**, 3393-3397 (2022), DOI: 10.1111/jocd.1510
5. L. Bennardo, C. Patruno, E. Zappia, F. Tamburi, M. Sannino, F. Negosanti, S. Paul Nisticò, G. Cannarozzo, "Combination of Specific Vascular Lasers and Vascular Intense Pulsed Light Improves Facial Telangiectasias and Redness," Medicina **58(5)**, 651 (2022), DOI: 10.3390/medicina58050651
6. S. R. Meghe, A. Khan, S. D. Jangid, B. Sarda, N. Vangala, V. Saoji, "Shedding Light on Acne Scars: A Comprehensive Review of CO_2 vs. Erbium-Doped Yttrium Aluminium Garnet (Er:YAG) Laser Therapy" (2024)

7. S. Sky, Modern nursing practices when using lasers in dermatology and cosmetology as an effective skin treatment, Doctoral dissertation (Ternopil, 2023)
8. H. Leal-Silva, S. Leal-Delgado, "Intense Pulsed Light Hair Removal," Dermatol. Rev. **5(4)**, e252 (2024)
9. M. H. Gold, E. Weiss, J. Biron, "Novel laser hair removal in all skin types," J. Cosmet. Dermatol. **22(4)**, 1261-1265 (2023)
10. E. Sharon, A. Levi, M. Lapidoth, I. Snast, "Laser and light therapy for pediatric hair removal: a systematic review," Lasers Med. Sci. **38(1)**, 156 (2023)
11. F. Ghorbani, P. Koohsari, M. Yaseri, N. Tootoonchi, Z. Aryanian, A. Vahedi, I. Etesami, "Laser hair removal complications and the associated risk factors at a tertiary dermatology center: a case-control study," Lasers Med. Sci. **39(1)**, 218 (2024)

E3S Web of Conferences **633**, 08008 (2025) https://doi.org/10.1051/e3sconf/202563308008
GASES 2025

The role of Green Building Information Modeling (GBIM) in promoting architectural sustainability

Mays Yousif Al-Mashhadani*, and *Osamah* Abdulmunem Al-Tameemi

Department of Architecture, College of Engineering, University of Baghdad, Baghdad, Iraq

Abstract. Green Building Information Modeling (Green BIM) is a critical instrument for promoting sustainable architecture, given that it integrates environmental sustainability principles into building design and construction. By utilizing highly advanced digital technology, Green BIM optimizes energy efficiency, reduces environmental impact, and enhances resource allocation throughout a building's lifecycle. This paper analyzes Green BIM's role in promoting sustainability in new buildings as well as existing buildings. The study begins with emphasizing the importance of Green BIM in reducing the carbon footprint of the built environment. A comprehensive review of the literature and previous research is conducted to examine the benefits, challenges, levels of development, data requirements, and functionalities of Green BIM. The study then discusses selected international and regional case studies, in addition to a specific case study in Iraq, to assess the application and effectiveness of Green BIM in achieving sustainability. Finally, a theoretical Analysis of Green BIM's Environmental Impact study is presented, criticizing the environmental benefits and implications of incorporating Green BIM into design and construction practices.

1 Introduction

With climate conditions worsening, sustainability and environmental conservation have become a premise objective. The building and construction industry is responsible for more than one fifth of global emissions [1]. Sustainable building techniques have become an important response to these environmental concerns. Sustainable buildings that achieve a low environmental footprint and create healthier living environments are derived by integrating energy-saving structural designs, eco-friendly construction materials, and renewable energy sources. With this method it is possible to not only reduce carbon footprint but also natural resource consumption, thus it is an important tool for worldwide climate change combat.
Platforms such as Green Building Information Modeling (Green BIM) play a significant role in the optimization of sustainable building practices. Green BIM supports activities such as energy modeling and material choice which, consequently, allow the effective implementation of sustainable elements in construction works. Ease of use and affordability of sustainable design are facilitated by these platforms, which reduces environmental

* Corresponding author: ali.alhameedawi@uokufa.edu.iq

footprint, enhances resource utilization and building performance, eventually aiding global efforts towards sustainability.
The concept of "Green BIM" was first introduced in 2008 through the publication of Green BIM: Successful Sustainable Design with Building Information Modeling by Eddy Krygiel and Bradley Nies. This pivotal work marked the convergence of Building Information Modeling (BIM) technologies with sustainable design principles, highlighting BIM's potential in energy efficiency, waste reduction, and the use of sustainable materials. Krygiel and Nies' work highlighted the breadth of BIM applications in the building life-cycle cycle and the promotion of more eco-friendly architecture, engineering, and construction methods using digital modeling and green design principles [2,3]

Green Building Information Modeling (Green BIM) has great prospects at the entire life cycle of green projects, especially for sustainability works. However, most of these potentials are still underdeveloped and used insufficiently, especially in the construction, operation, and repair stages of these potentials. Green BIM provides key functions for sustainability analysis, including the evaluation of energy consumption, carbon emissions, natural ventilation, and solar and lighting efficiency. Green BIM is significant in various stages of a project. Despite some limitations in the phase of design, it promotes data sharing and integration, allows the visual examination of the performance of the building, and helps assess design options to improve sustainability. At the construction stage, it aids exploration of environmental footprints, waste reduction and enhanced construction performance and efficiency. During the operation phase, Green BIM provides the facility for continuous monitoring of a building's performance in terms of sustainability, in real time. In renovations and retrofits Green BIM encourages energy and capital recovery, as well as energy and waste management during building renovations and retrofits [4]. Green BIM can be fully leveraged throughout the entire lifecycle of green projects, maximizing its potential to improve sustainability outcomes in the built environment.

The history of BIM for sustainability has been marked with dramatic evolution. At the early stage of the 1990s and early 2000s, BIM was mainly applied to design coordination, and mainly relied on 3D modeling and building efficiency, while sustainability is tackled separately through outside tools. During the mid-2000s, BIM included sustainability features, like energy modeling, so that design decisions could be judged for their environmental impact at the earliest stage of the design process. By the 2010s, BIM had developed such that lifespan assessments (LCAs) are also incorporated, allowing for detailed examination of energy use, carbon footprint, and resource consumption. As Green Building Certification systems like LEED and BREEAM gained prominence, BIM platforms aligned with these standards, allowing designers to meet sustainability goals within the same tool. Green BIM currently operates with sustainability as a built-in functionality that, using data-driven approaches, maximizes building performance and reduces the environmental footprint. Moving forward, BIM is still developing to address concepts related to the circular economy as well as the net zero, facilitating reuse of materials, renewable energy generation, and maximised building performance across its life cycle [5].

Based on the examination and comparison of those studies, it seems that although they recognized the leading role of Green BIM (Building Information Modeling), it is incomplete in their methodology. Though there are numerous studies of global experiences, a noticeable lack of studies addressing local experiences exists. Certain studies were restricted in a limited geographic region or contained an isolated function.

As a result, the research question is: "What are the international and local experiences of using Green BIM, and how can they improve the application of sustainability in architecture?"

This question arises from the research problem, which highlights the lack of knowledge and local studies regarding the use of Green BIM in promoting sustainable buildings, as well as the need to reassess these experiences in comparison to successful global examples.
To address this issue, the research aims to assess international and local experiences in using Green BIM, bridging the gap in knowledge and local studies, and exploring how these experiences can enhance the application of sustainability in architecture through comparisons with successful global examples.

In line with this aim, "This research assumes that assessing and reevaluating both international and local experiences in using Green BIM is essential for bridging the knowledge gap, addressing challenges, and exploring opportunities to enhance the application of sustainability in architecture."

2 Materials and methods

During the implementation of green designs, as well as the retrofitting of existing buildings, sustainable design methods can be employed to assess the impacts of green buildings. This comprises assessment of features, e.g., lighting, energy efficiency, material sustainability, and building performance. Integrating green technology in the design and construction process and in the retrofitting, process are critical in order to achieve more effective, and optimal results. In the end, this method guarantees commonality with green building criteria and sustainability goals.
The research used desktop research to review existing literature, case studies and reports from academic journals, industry publications and reliable online sources, with a focus on green BIM implementation and sustainability challenges. Qualitative data collection involved obtaining and analyzing primary data from the CBI project, including design documents, BIM models and sustainability reports, to examine green BIM integration in detail. In addition, case study analysis was conducted on selected projects, both local and international, to evaluate green BIM implementations, identify patterns, and assess challenges and best practices.

The paper starts with the literature review to identify the research problem, focusing on the use of Building Information Modeling (BIM) in promoting sustainable building practices, while exploring the concept of Green BIM, including its benefits, challenges, and primary tools. To analyze its practical application, international and local case studies of sustainable buildings that have utilized BIM are selected, with a focus on understanding the role BIM played in achieving sustainability. Following this, an analysis of Green BIM's environmental impact is conducted to evaluate how it contributes to the sustainability performance of the selected buildings. The results highlight the effectiveness of Green BIM in promoting sustainability, emphasizing its potential to enhance environmental performance, while the discussion explores how its success is influenced by factors such as frameworks, expertise, and governmental support. Finally, the study concludes by highlighting the potential of Green BIM and provides recommendations for improving its adoption, addressing challenges, and leveraging opportunities for future advancements.

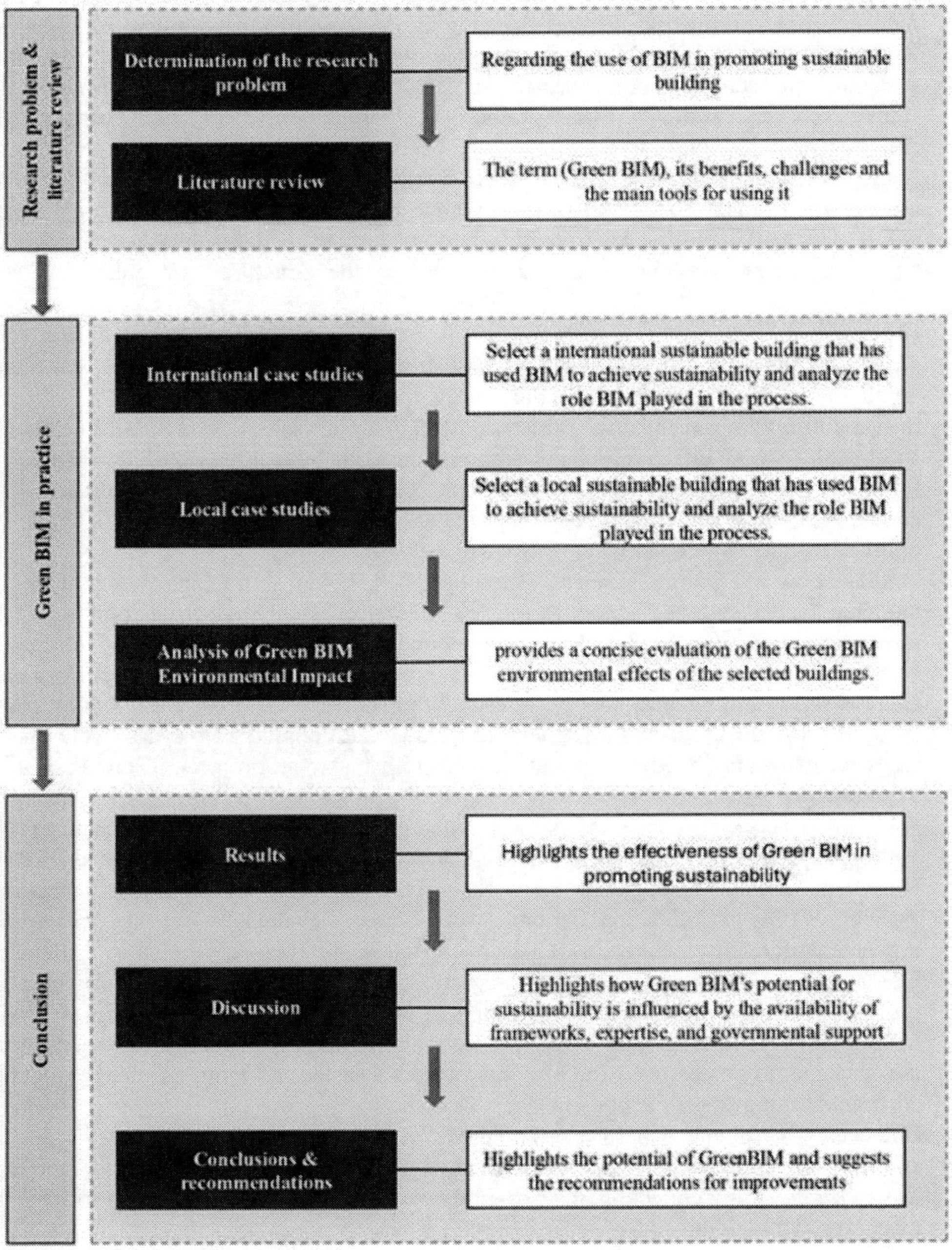

Fig. 1. Methodology of the study.

2.1 Green BIM benefits

Green BIM (Building Information Modeling) plays a pivotal role in advancing sustainable architecture by integrating environmental considerations into the design, the following are key benefits of Green BIM in promoting sustainable development.

- Team Performance and Productivity: The usage of coordinated digital models in BIM reduces errors and reworks, while enhancing effective communication and collaboration within the team. This leads to more reliable decision-making and the achievement of successful project goals. Additionally, 3D BIM models integrate rich information that supports visualization, informatization, standardization, and collaboration, all contributing to building sustainability.
- Economic Benefits:Click or tap here to enter text.Green BIM has a pivotal role in green architecture by maximizing cost savings, increasing productivity, and escalating property values and marketability. Additionally, it offers indirect benefits such as improving a building's image, reducing risks, future-proofing the structure, and enhancing self-reliance.
- Efficiency in Design and Development: Green BIM approach enforced during early design development leads to less cost and more effective work processes, especially within the engineering and sustainable energy domains.
- Sustainability Improvements: Building design optimization using BIM enhances sustainability more effectively than traditional methods by facilitating life-cycle analysis for thermal and lighting systems, allowing for more accurate results that reflect real-world conditions. Additionally, BIM reduces errors and reworks, enabling designers to achieve optimal designs more quickly and easily [6].
- Optimization of Energy-Efficient Buildings:BIM holds significant potential in the modeling, analysis, and optimization of energy-efficient industrial buildings. By integrating BIM with Building Energy Modeling (BEM), it allows for the identification of BEM requirements earlier in the design process, optimizing building performance analysis and minimizing energy consumption during the post-construction phase compared to conventional CAD methods [7,8]. Additionally, whole-life cycle energy assessments can be incorporated into BIM for refurbishment projects, further enhancing sustainability [9].
- Fabrication and assembly: BIM has the potential to enable off-site manufacturing for buildings, enhancing both sustainability and cost-effectiveness (Yin et al., 2019). However, the industry currently lacks clear standards or codes governing the various aspects of Green BIM applications, which poses challenges for its widespread implementation.
- LEED Evaluation Integration: The integration of LEED evaluation within the BIM process allows for the effective addressing of crucial environmental design issues. By linking LEED credits directly with BIM technology, this approach not only streamlines the sustainability assessment but also encourages a faster and more efficient evaluation compared to traditional methods [10]
- Life cycle assessment: A BIM-based method for green building helps experts predict the environmental effects of construction during its life cycle. The possibilities of Green BIM can be improved by collecting and providing essential information and data to facilities managers and building end-users
- Green BIM is not limited to building projects; it can also be used in sustainable infrastructures and site location planning [11]

In principle, sustainable design refers to the entire life cycle of the building or infrastructure, which includes maximizing the conservation of resources (energy, water, land and materials), protecting the environment, reducing pollution, providing people with healthy, comfortable, and efficient use of space, and establishing a harmony of nature and built environment. By using BIM it is possible to reduce waste and improve construction quality. BIM provides the simulation and analysis of scientific collaboration platforms for designers, architects, utilities engineers, developers and even end users. It helps them to take advantage of 3-D digital models in design and construction of projects and operational management [12]

2.2 The main challenge for using Green BIM

The use of Green BIM in sustainable construction offers many advantages, but its widespread adoption faces several challenges. These obstacles limit its effectiveness and slow down its integration into the industry. The main challenges associated with the implementation of Green BIM include:

1) Interoperability issues between BIM and sustainability software tools are barriers to incorporating green building principles. Effective data exchange, supported by standardized protocols and procedures, is crucial for enhancing project sustainability throughout a building's life cycle. BIM-based energy analysis can also be applied to post-occupancy studies and retrofitting, improving energy performance models and operations [12]
2) BIM's limited capability in dealing with the construction and operation stages of green projects (Hui, 2019). is shown by its inability to fully integrate real-time energy performance data during the building's operational stage. This reduces its effectiveness in optimizing long-term sustainability efforts like energy monitoring and retro-commissioning.
3) Lack of clear industry standards or codes for various aspects of Green BIM applications. A reason for the lack of definite industry-wide standards for application of Green BIM may be that there has not been any shared protocol to integrate BIM with green rating systems like LEED or BREEAM. Lack of standardized guidelines meant for utilizing green building performance in determining project requirements could render it questionable for design teams in implementing consistent sustainable evaluation using evaluation tools, enabling inconsistent determination across different projects with the objective of measuring up the performance against green buildings because of variability associated with design requirements and functionality priorities.
4) Inappropriate project delivery methods for leveraging Green BIM applications. An example is the use of traditional delivery methods such as Design-Bid-Build. In this case, the delivery methods have the tendency to separate the design from the construction phase, and Green BIM cannot be incorporated as a totality in the project life cycle. These disconnections limit cross-disciplinary collaboration, lowering the application of BIM for real-time sustainability analysis, in addition to optimization during construction [12].
5) Lack of incentives and low industrial acceptance for the adoption of Green BIM for example when construction companies avoid implementing Green BIM because of its perceived high upfront costs, such as investing in new software or training staff. Without government subsidies, tax breaks, or green certification benefits that reward the use of sustainable technologies, firms may be reluctant to adopt Green BIM, even though it can lead to long-term savings and environmental benefits [12]

2.3 Level of Detail/Development in Green BIM

Level of Development (LOD) is a significant term used in the Green Building Information Modeling (BIM) field as it defines the specific information that is needed to achieve successful model development at the various stages of building projects. LOD serves as a standard to ascertain the level of detail and accuracy needed for model elements so that communication between project stakeholders is effortless [13]

Levels of Development (LOD) represent a standardized framework used to specify the degree of detail, accuracy, and reliability that should be incorporated into a Building

Information Model (BIM) at various stages throughout a project's lifecycle. Benefits of LOD (Levels of Development) in BIM include Standardization: Provides a consistent framework for model detail and reliability, Effective Collaboration: Enhances teamwork and reduces clashes among disciplines ,Reduced Risk: Identifies issues early, minimizing costly changes during construction, Clarity: Ensures design intent is clear, reducing errors and change orders ,Optimized Resources: Allocates detail where needed, saving time and costs, Facility Management: LOD 500 models support maintenance and future modifications (Autodesk, n.d.). The Levels of Development (LOD) framework, developed by the AIA and AGC, outlines six stages of detail in a building model: LOD 100(conceptual), LOD 200 (schematic), LOD 300 (detailed design), LOD 350 (construction documentation), LOD 400 (fabrication), and LOD 500 (as-built). Each stage progressively refines the model, from basic shapes to real-world conditions, ensuring clarity and coordination throughout a project.

In a properly standardized BIM system, the environmental properties, including suppliers' Environmental Product Declarations (EPDs), are ingrained in BIM objects. The ingraining must be structured according to the project's level of development (LOD) for sustainable and integrated building practices.

Table 1. The level of development (LOD), required data and Green BIM focus.

LOD	Data required	Green BIM focus
LOD 100 – conceptual design	• Building Form and Massing: General shape, massing models, and environmental surroundings. • Site Context: Data on location, orientation, and basic climate conditions (e.g., sun path, wind). • Surface Area Estimates: Rough calculations of building surfaces, volumes, and areas for sunlight exposure and environmental impact.	• Energy and Thermal Analyses: Basic data on building orientation, massing, sun exposure, and energy consumption estimates for passive heating and cooling. • Lighting and Shading Analysis: Early daylight and shading studies using sun path and site data for shadow impacts. • Value and Cost Analysis: Rough cost estimates for implementing green technologies like solar panels and passive cooling. • Acoustic Analysis: Preliminary noise data (e.g., from highways) to plan early noise mitigation. • Water Harvesting: Initial assessment of rainwater availability and site conditions for water reuse. • Space Simulation: Early spatial layouts to explore natural ventilation and passive thermal zones. • System Simulation: High-level estimates for mechanical systems and renewable energy integration (e.g., solar PV sizing).

Continuation of Table 1.

LOD 200 – schematic design	• Building Geometry and Openings: Detailed building form, window placement, and shading devices. • Preliminary Material Properties: Data on materials affecting energy performance (e.g., thermal resistance). • Site-Specific Climate Data: Refined climate data on sun paths, wind direction, and shading from nearby buildings.	• Energy and Thermal Analyses: Refined passive energy strategies using material and geometry data, with simulations for heating, cooling, and ventilation loads. • Lighting and Shading Analysis: Initial daylight simulations to optimize window design, shading, and building orientation. • Value and Cost Analysis: Cost analysis of energy-efficient systems (e.g., insulation, windows) and payback periods. • Acoustic Analysis: Noise control strategies through window and wall placement. • Water Harvesting: Initial layouts for rainwater harvesting and water-efficient fixtures with system capacity estimates. • Space Simulation: Layouts optimized for thermal comfort, natural light, and passive ventilation. • System Simulation: Early HVAC, lighting, and renewable energy system designs for energy efficiency.
LOD 300 – detailed design	• Detailed Building Geometry: Full representation of building envelope, thermal zones, and spatial relationships. • Mechanical Systems Layout: Detailed layouts for HVAC, plumbing, and electrical systems. • Material Properties: Defined properties for insulation, windows, and other energy-related components.	• Energy and Thermal Analyses: Simulations using actual materials and systems for heating, cooling, and ventilation, with renewable energy integration. • Lighting and Shading Analysis: Advanced daylighting simulations to optimize natural and artificial lighting. • Value and Cost Analysis: Cost-benefit analysis for green technologies like geothermal and solar panels, considering life cycle costs. • Acoustic Analysis: Refined designs for soundproofing using noise barriers and sound-absorbing materials. • Water Harvesting: Detailed designs for rainwater and greywater systems, including storage and piping. • Space Simulation: Layouts optimized for energy efficiency and thermal comfort, focusing on natural ventilation and zoning. • System Simulation: Accurate simulations for mechanical and energy systems, including renewable energy.

Continuation of Table 1.

LOD 350 – construction documentation	• Construction-Ready Geometry: Building geometry information that is complete, including tolerances and installation. • Final Mechanical and Electrical Systems: Finish construction-level designs for HVAC, plumbing, and electrical systems with green features. • Final Sustainable Material Selections: Detailed material specifications with performance criteria and supplier information.	• Energy and Thermal Analyses: Conclusive energy simulations to affirm efficiency goals are being met with mechanical systems. • Lighting and Shading Analysis: Daylighting and shading strategies fully integrated into building designs. • Cost and Value Analysis: Final cost analysis with special attention to green systems and materials, and compliance with standards like LEED compliance. • Acoustic Analysis: Complete solutions for soundproofing in buildings, ensuring comfort. • Water Harvesting: Full designs for greywater and rainwater systems, including piping and cisterns. • Space Simulation: Overall space planning for comfort, ventilation, lighting, and energy efficiency. • System Simulation: Full integration of energy-efficient HVAC, lighting, and water systems for sustainability. .
LOD 400 – fabrication & assembly	• Component Fabrication: Data for the production of high-performance building components like insulated panels and low-E windows. • System Installation Data: Step-by-step guidelines for installing energy-efficient and renewable energy systems (e.g., solar panels, geothermal).	• Energy and Thermal Analyses: Delivering energy efficiency at installation to prevent thermal bridges and maintain airtightness. • Lighting and Shading Analysis: Installation of lighting system and shading devices from optimized design. • Cost and Value Analysis: Monitoring costs to ensure installation of planned sustainable materials and systems. • Acoustic Analysis: Soundproof materials are put in place in order to minimize noise. • Water Harvesting: Appropriate integration of rainwater and greywater systems for beneficial use. • Space Simulation: Adjusting final space designs to meet energy efficiency and comfort standards. • System Simulation: Ensuring the optimal performance of renewable energy, HVAC, and water systems at the time of installation.

Continuation of Table 1.

LOD 500 – as built or facility management	• As-Built Documentation: Accurate models showing the building as constructed, including mechanical, electrical, and plumbing systems. • Facilities Management Data: Real-time performance data on systems, energy use, and operational efficiency.	• Energy and Thermal Analyses: Monitoring actual energy use against simulations and making adjustments for optimization. • Lighting and Shading Analysis: Tracking the effectiveness of daylighting and artificial lighting systems and their controls. • Value and Cost Analysis: Analyzing post-construction costs of green systems and materials, focusing on ROI and operational savings. • Acoustic Analysis: Continuous evaluation of acoustic performance to ensure noise reduction standards are met. • Water Harvesting: Monitoring water systems for usage, storage, and recycling effectiveness. • Space Simulation: Adjusting layouts and system controls based on occupancy and real-time energy data. • System Simulation: Facilities management oversees the performance of energy and water systems, making adjustments to enhance efficiency and sustainability.

2.4 Green BIM techniques

Table 2. Green BIM Techniques, Geometry data, simulation data.

Green BIM Techniques / Simulations / Analysis	Geometry Data	Simulation data
Acoustic Analysis	• Building Geometry • Volume of rooms • The shape of the room • Position of the sound source • Distance from the sound source to the audience • MaterialProperties, Absorption at various frequencies Time and Sound absorption coefficient at series of octave band frequencies • Material Thickness	• External noise level • Building Occupancy, Furnitureand HVAC noise • Sound Intensity Level • Power of the sound source
Daylighting Analysis	• Building Geometry • Information about interior spatial organization and zones • Material properties • Properties of shading surfaces • Location and Orientation • Size, Shape, and Position of windows, glazing, and skylights.	• Climate Data

Continuation of Table 2.

Thermal Analysis	• Building Geometry • Information about interior spatial organization and zones • Element material constructions and their associated thermal properties -U-values (heat transfer coefficients) -R-values (measure of thermal resistance) • Location and Orientation • Size, Shape, and Position of windows, glazing, and skylights. • Shading Devices • Properties of shading surfaces	• Climate Data
Value and Cost Analysis	• Operation and maintenance costs • Maintenance, preservation and operational cost variables • Material specifications based on the design brief and the required service life • Factors influencing the deterioration and failure of components. • Net present value (NPV) indicators for the entire life cycle of assets. ❖ Initial Capital Costs (Costs of construction, procurement, and installation)	• Energy Consumption, water consumption and Utility Costs.
Energy Analysis	• material properties and their associated thermal properties ▪ U-values (heat transfer coefficients) ▪ R-values (measure of thermal resistance) • Internal heat gains from plug loads and occupancy • Natural ventilation (eg: opening and closing of windows) and Building natural infiltration rate (air leakage) • Building Geometry • Shading Devices • System types, including electrical Plug loads (such as appliances and electronic devices), lighting, heating, and HVAC systems	• Climate Data • Number of occupants • Occupant Behavior • Operating schedules

2.5 Green BIM in practice

2.5.1 International case studies

To better understand the role of Green BIM in promoting sustainable building practices, it was essential to study a selection of successful global examples that have effectively utilized BIM to enhance sustainability. The objective was to identify key indicators and derive conclusions applicable to local contexts.
Reasons for Sample Selection:

The selected case studies were chosen based on their contemporary design, innovative approaches, and strategic geographical locations. These projects exemplify the application of key sustainability indicators central to this research, providing a comprehensive reflection of these metrics in real-world scenarios. Additionally, they serve as robust, practical benchmarks that can inform and guide future research, experiments, or sustainable design practices.

The case studies include:

1. Sydney Opera House (Refurbishment)– A landmark project demonstrating sustainable retrofitting practices.
2. The Standard, London – A modern building rehabilitations integrates a comprehensive array of sustainable technologies
3. Duke Ellington School of the Arts– An educational facility refurbishment highlights a sustainable design integration.
4. The Edge, Amsterdam– A cutting-edge office building recognized as one of the greenest structures globally.
5. Museum of the Future, Dubai – A futuristic project incorporating advanced sustainability technologies and innovative design.

2.5.2 Local case studies

To understand the application of Green BIM in local contexts, local case study (Central Bank of Iraq) were analyzed to identify challenges, opportunities, and strategies relevant to the local building industry, the researchers obtained and subjected to in-depth analysis main data related to the building's design, construction, sustainability, and BIM implementation. This data provided valuable insights into the application of green BIM in a local project, providing a detailed insights of its role in achieving sustainability goals.

2.6 Analysis of Green BIM's Environmental Impact

The following analysis provides a concise evaluation of the environmental effects of selected buildings. It highlights key aspects such as energy efficiency, carbon footprint, water management, waste reduction, indoor environment quality, and biodiversity demonstrating the role of Green BIM in promoting sustainability throughout the lifecycle of these projects. In Table 3 an analysis of Green BIM's Environmental Impact will be discussed.

Table 3. Analysis of green BIM's environmental impact.

Aspect	Impact	The role of Green BIM
Energy Efficiency	**Sydney Opera House:** transitioned to 100% renewable electricity, reduced electricity consumption by 20%, incorporated natural ventilation to reduce reliance on air conditioning, and replaced incandescent bulbs with custom LED lights and that aimed to reduce electrical consumption by 75% in the Concert Hall and overall electricity use by 14%. **Duke Ellington School of the Arts:** improved energy efficiency with roof-mounted photovoltaic panels, dynamic glass that adjusts its tint with sunlight, geothermal heating and cooling systems, and a well-designed building envelope that minimizes the need for heating and cooling. **The Edge (Amsterdam):** maximized natural light and ventilation through its 15-story atrium, used over 6,000 square meters of solar panels, and produced more energy than it consumed, achieving a positive energy balance. **Museum of the Future:** generated renewable energy from an independent solar station and reduced energy consumption by 25% through a 3D energy model, passive solar design, and solar farm integration, contributing to a low-carbon building. **Central Bank of Iraq:** A potential 706m² of PV panels on louvres could offset 0.7% of emissions from regulated energy and 7% of small power load emissions.	BIM enables energy modeling, daylight analysis, and system optimization during the design phase, integrating energy-efficient systems and renewable energy technologies.
Carbon Footprint	**Sydney Opera House:** Maintained carbon-neutral status annually since 2018. **The Standard London:** Saved 5,770 tons of embodied carbon by retaining much of the original structure. **Central Bank of Iraq:** has implemented several sustainable initiatives to reduce emissions. A potential installation of 706m² of PV panels on the louvres could offset 0.7% of emissions from regulated energy and 7% of emissions from the small power load. Additionally, cooling with groundwater, which is 25% more efficient than traditional methods, is projected to reduce emissions from regulated energy loads by 13.9%. Furthermore, CHP systems running 30% of the time are expected to save 9% of emissions from regulated loads and offset 106% of emissions from the small power load, while A high-level assessment found that using structural steel, despite long-distance shipping, offers the lowest embodied carbon footprint compared to concrete.	BIM helps track and reduce embodied carbon by supporting lifecycle assessments and material reuse, guiding low-carbon material selection and building processes.

Continuation of Table 3.

Water Management	**Sydney Opera House:** Reduced water consumption by approximately 20% since 2018, Replaced chiller units connected to the seawater cooling system, achieving a 9% energy reduction and contributing to a 50% reduction in water usage. **Duke Ellington School of the Arts:** A rainwater reclamation system reduces reliance on municipal water by supplying water for toilets, irrigation, and mechanical systems. **The Edge (Amsterdam):** Rainwater is collected and reused for flushing toilets and irrigating green terraces and surrounding gardens. **Museum of the Future:** The design incorporated low-energy water solutions, reducing water use by 45%, enhancing the building's overall sustainability. **Central Bank of Iraq:** The initial rainwater harvesting plan was replaced with a greywater recycling system, meeting 70% of WC flushing needs with a $30m^3$ basement tank.	BIM supports water-efficient design through simulations, ensuring optimal integration of water recycling, reuse systems, and water-efficient fixtures.
Waste Reduction	**Sydney Opera House (Refurbishment):** Increased operational waste recycling rate from 55% to over 90%. Diverted 95% of event waste from landfills during Vivid LIVE and outdoor concerts in 2022, aiming for the same in 2023. Diverted over 90% of construction waste from major projects away from landfills.Implemented a new waste management system, increasing recycling rates from 25% to 85%, and redirected food waste to an organics facility for energy production. **The Standard London:** Approximately 94% of the primary structure, including extensive basement and piling, was able to be retained, minimising the waste and carbon release associated with demolition.	BIM facilitates waste management by optimizing material use during construction and allowing real-time tracking of waste generation and recycling efforts

Continuation of Table 3.

Indoor Environmental Quality	**Sydney Opera House (Refurbishment):** Incorporated natural ventilation in the building's design, reducing reliance on air conditioning and heating. **Duke Ellington School of the Arts:** The use of energy-efficient LED lighting, combined with occupancy, daylight, and dimming controls, minimizes energy consumption by ensuring lights are used only when needed. **The Edge, Amsterdam:** The building has nearly 30,000 sensors that track movement, lighting, humidity, temperature, and carbon dioxide levels. These sensors, paired with advanced management systems, enable efficient temperature and lighting adjustments, and optimize cleaning and maintenance based on real-time data.The building's 15-story atrium enables natural ventilation by allowing stale air to exit through the roof, while also acting as a sound buffer from nearby traffic and providing significant natural daylight. **Museum of the Future:** The exterior façade utilizing innovative technologies to optimize interior lighting and improve thermal insulation. **Central Bank of Iraq:** A double façade minimizes solar heat gain while providing excellent daylight and outdoor views.	BIM integrates ventilation, thermal comfort, and lighting simulations, enhancing indoor environmental quality while reducing energy consumption.
Biodiversity	**Sydney Opera House (Refurbishment):** Installed an artificial reef around Bennelong Point to restore habitats, clean waters, and boost marine biodiversity. **The Edge, Amsterdam:** An ecological corridor separates the building from neighboring sites,ensuring safe passage for insects across the area. **Museum of the Future:** The garden includes a smart irrigation system and over 100 native plant species adapted to Dubai's heat, with succulents covering much of the mound, similar to those in Middle Eastern coastal areas.	BIM aids in environmental planning, allowing designs to incorporate features that protect and promote biodiversity, such as green spaces and habitat preservation.

3 Results

Green BIM plays a crucial role in promoting sustainability by addressing key aspects such as energy efficiency, carbon footprint reduction, water management, waste reduction, indoor environmental quality, and biodiversity conservation. In terms of energy efficiency, Green BIM helps optimize the use of renewable energy by analyzing the most effective options for a given building and site. It also supports passive design strategies to reduce energy consumption while helping to select active systems that best suit the building conditions.

To reduce carbon footprint, Green BIM helps reduce environmental impacts in new and existing buildings. For new buildings, it helps in selecting sustainable materials and designing efficient procurement and transportation systems to reduce emissions. For existing buildings, it helps in reducing demolition waste and conserving embodied carbon. In terms of water management, Green BIM enables efficient system design by simulating water recycling and reuse while incorporating water-saving fixtures to optimize resource utilization.

Green BIM also facilitates waste management by promoting material optimization in construction, real-time waste monitoring, and recycling strategies. As for indoor environmental quality, it enhances thermal comfort, ventilation and lighting through advanced simulations, ensuring the well-being of building occupants while reducing energy consumption. In addition, Green BIM supports the preservation of biodiversity by incorporating environmentally friendly design elements, such as green spaces and habitat conservation, ensuring that construction projects are aligned with environmental conservation goals.

However, despite these benefits, the implementation of Green BIM varies significantly between international and local buildings, largely due to several challenges. One major barrier is the lack of standardized frameworks in most local environments. While international projects often rely on established environmental assessment systems such as LEED or BREEAM to guide and measure sustainability, many local projects lack similar frameworks, making it difficult to define and prove the sustainability performance of buildings.

Financial constraints also hinder the adoption of Green BIM, as the implementation of advanced software, energy-efficient technologies and expert-led processes requires significant investment. A prominent example of this is the Central Bank of Iraq, which incurred exceptionally high costs compared to local and international buildings. The heavy reliance on foreign expertise contributed significantly to these costs, as the majority of the project team consisted of international professionals. This reliance not only increased expenses, but also emphasized the shortage of local green BIM experts, leading to a long-term reliance on foreign specialists.

4 Discussion

The research outcomes reveal that the scope of Green BIM to promote sustainability in energy efficiency, carbon emission, water management, waste minimization, indoor environmental quality, and biodiversity is of huge potential. However, the adoption of Green BIM varies from international and domestic projects mainly due to variability in resources, proficiency, and policy support.

Foreign projects benefit from frameworks of organized sustainability standards like LEED and BREEAM governing the integration of Green BIM into design and construction, but availability of technology levels, extensive data-rich databases, and established professional expertise further enhance their efficiencies with palpable environmental gains in energy efficiency and consumption, waste minimization, as well as biodiversity protection.

On the other hand, local projects have several challenges that deter the complete implementation of Green BIM. The first challenge is the lack of standardized sustainability frameworks or environmental evaluation tools for the local context, which makes it challenging to evaluate and quantify the environmental performance of Green BIM projects. Moreover, the exorbitant fees for implementing Green BIM—ranging from software costs to training expenses and the necessity of hiring foreign specialists—pose financial obstacles to most local projects. The lack of Green BIM local professionals worsens these challenges by causing dependence on external experts, which raises expenses and decelerates the adoption process.

Government support is a significant factor that allows the adoption of Green BIM. Global projects are usually covered by policies, incentives, and mandates that encourage sustainability, but local projects do not have access to these supports, so they cannot use the full potential of Green BIM. Local projects also do not have access to local BIM tools and data specific to their region, which are used for creating solutions with respect to local climate, materials, and construction practices.

This concurs with present literature on Green BIM that has espoused its promising potential to revolutionize energy efficiency, waste management, and indoor environmental quality. However, identified challenges in local projects illustrate the need to fill the gap left by policy, expertise, and resources in different settings for maximum impact of Green BIM.

5 Conclusion

In conclusion, Green BIM has high potential for enhancing sustainability in various building aspects, such as energy efficiency, carbon emissions reduction, water management, minimization of waste, indoor environmental quality, and biodiversity. While international projects have the luxury of standardized frameworks, advanced technologies, and specialized human resources, local projects face challenges, including the lack of standardized frameworks, high cost of implementation, no local expertise, and insufficient governmental support. Despite these constraints, the transformative potential of Green BIM has been seen in case studies from around the world. To realize its benefits to the fullest, particularly in local contexts, policy, training, and resource gaps need to be filled.

Recommendations involve the creation of localized sustainability frameworks, greater government support and incentives, and investment in local talent and training.

One of the key limitations of this research is its reliance on high-profile projects, such as the Central Bank of Iraq, which may not be representative of typical local initiatives due to their access to greater resources, funding, and expertise. This focus makes it difficult to draw conclusions about smaller-scale or less-funded projects that face more significant constraints. Additionally, in the local context, the Central Bank of Iraq appears to be the only building that has implemented Green BIM, limiting the ability to compare or generalize findings across other local projects. Furthermore, while the research seeks in-depth and detailed information on international case studies, accessing comprehensive data proved challenging, potentially restricting the analysis and comparison of Green BIM's effectiveness in international contexts. These limitations highlight the need for broader data collection and a more diverse range of case studies in future research.

Future studies can try to develop policies for the implementation of Green BIM at the local level, in this case in Iraq, and create a Green BIM framework that is flexible enough to address the country's particular environmental, economic, and social conditions. The study can also discuss cost-efficient implementation approaches and the role played by governmental policies in promoting Green BIM uptake.

References

1. A. Andriamamonjy, D. Saelens, R. Klein, "A combined scientometric and conventional literature review to grasp the entire BIM knowledge and its integration with energy simulation," J. Build. Eng. **22**, 513–527 (2019), DOI: 10.1016/j.jobe.2018.12.021
2. Y. Chai, "Analysis and prospect of green building engineering based on BIM technology," in Proc. 2023 Int. Conf. Funct. Mater. Civ. Eng. **25(1)**, 74–82 (2023), DOI: 10.54254/2755-2721/25/20230739
3. S. Y. Chen, "Use of green building information modeling in the assessment of net zero energy building design," J. Environ. Eng. Landsc. Manag. **27(3)**, 174–186 (2019), DOI: 10.3846/jeelm.2019.10797
4. A. Ebrahim, A. S. Wayal, "Green BIM for sustainable design of buildings," in *Proc.*, 185–189 (2020)

5. R. E. Edwards, E. Lou, A. Bataw, S. N. Kamaruzzaman, C. Johnson, "Sustainability-led design: feasibility of incorporating whole-life cycle energy assessment into BIM for refurbishment projects," J. Build. Eng. **24**, 100697 (2019), DOI: 10.1016/j.jobe.2019.01.027
6. A. Ekasanti, O. C. Dewi, N. Putra, "Green BIM potential in assessing the sustainable design quality of low-income housing: a review," IOP Conf. Ser.: Mater. Sci. Eng. **1098(2)**, 022078 (2021), DOI: 10.1088/1757-899X/1098/2/022078
7. S. Eleftheriadis, D. Mumovic, P. Greening, "Life cycle energy efficiency in building structures: a review of current developments and future outlooks based on BIM capabilities," Renew. Sustain. Energy Rev. **67**, 811–825 (2017), DOI: 10.1016/j.rser.2019.03.114
8. A. Ghaffarianhoseini, J. Tookey, A. Ghaffarianhoseini, N. Naismith, S. Azhar, O. Efimova, K. Raahemifar, "Building information modelling (BIM) uptake: clear benefits, understanding its implementation, risks and challenges," Renew. Sustain. Energy Rev. **75**, 1046–1053 (2017), DOI: 10.1080/17452007.2018.1502655
9. G. Gourlis, I. Kovacic, "Building information modelling for analysis of energy efficient industrial buildings – a case study," Renew. Sustain. Energy Rev. **68**, 953–963 (2017)
10. S. C. M. Hui, "New opportunities of using building information modelling (BIM) for green buildings," in *Proc. 15th Asia Pac. Conf. Built Environ.: 5R Technol. Build. Environ.* (2019)
11. R. Jin, B. Zhong, L. Ma, A. Hashemi, L. Ding, "Integrating BIM with building performance analysis in project life-cycle," Autom. Constr. **106**, 102861 (2019), DOI: 10.1016/j.autcon.2019.102861
12. M. Khaddaj, I. Srour, "Using BIM to retrofit existing buildings," *Procedia Eng.* 1526–1533 (2016)
13. Y. Lu, Z. Wu, R. Chang, Y. Li, "Building information modeling (BIM) for green buildings: a critical review and future directions," Autom. Constr. **83**, 134–148 (2017), DOI: 10.1016/j.autcon.2017.08.024

AUTHOR INDEX

Abd Oun Al-Jamal Lina 65
Abdlwahd Jamel Alqdori Areej 220
Abduganiev Olimjon............271, 279
Abdul Hadi Mohsen Ahmed............54
Abdul-Hussein Kadhim Alaidani Abbas 377
Abdulkareem Meena K. 456
Abdulkareem Sarah K............319
Abdullaev Muslimbek............483
Abdullaeva Muborek............9
Abdullah Albakri Sundus A. 366
Abdullal Azeez Ibtihal 338
Abdulmunem Al-Tameemi Osamah 520
Abdulnabi Abbas Muna 416
Abduvaliev Khayitboy............252, 279
Abid Abojassim Ali 54, 431
Abojassim Ali Abid 46
Aghajanyan E. A. 108
Ahmed Bneen Nassr............338
Ahmed Efan Rusul............310
Ahmed Hussein Khudhur............209
Ahmed Subhi Farah............299
Aitova Shaxlo 492
Akhmadaliev Yusupjon............271, 279
Akhmadjanov Ulugbek............9
Al-Ali I. A............131
Albakry Sundus A. Abdullah 310
Alebadi Shireen N. 40, 509
Al-Gbory M. M. 181
Alhadithi Mufid............319
Al-Hadithi Mufid 173
Ali Hassan Salam 394
Ali Maryam H. 123
Ali Sami Noor............114
Alrazaq Ban A. 131, 181
Amer Wadhah 319
Amin Mustafa M............164
Arazberdieva Shemshat............9
Atashev Elyor 229, 236
Atoyants A. L............108
Avalyan R. E............108
Awad Amin K. 46
Awad Omar Qais............151
Azeez Bjwo Amtihan............338
Azez Bjoo Emtihan 59
Bakhtiyorova Dilfuza............262
Baqir Yassir A............509
Borzunova Tatyana............100
Boyjanov Islom............468
Buranova Dinara............229
Buranova Mokhigul............236
Dawoud Salman Muhammad 402
Djabberganov Djakhangir 468
Eiryan Nikolay 93
Eshmamatov Elshod 252
Falh Hassan Adnan 193
Gburi Sura I. 366
Guida Michele............46
Hade Seood Gufran............114
Hameed Mousa Elaf............423
Hamid Abbas Saif............65
Hammad Abdulateef Mohammed............408
Hashesh Tabarak S. 123
Hassan Alsobh Morjan............516
Hayder Rand A. 54, 65, 123
Hindy Tayseer............423
Hua Wang Mei............100
Husham Abdulkareem 173
Hussein Abed Alaa............287
Ibrahem Salar H. 164
Jalil Lawi Dhuha 46
Jamal Abbas Al-Badran Makram............377
Janizoqova Gulshoda 262
Jassim Abd Al-Rezzaq Adi............73
Jawad Salih Al-Haidarey Mohammed............30
Jumaniyazov Maxsud 14, 236, 476, 492
Jumaniyozov Arslon 236
Jumaniyozov Jumaniyoz............483
Jundi Rami............423
Kabulova Lola 14
Kadhim Hasan Firoz-Ali Mudhafar 30
Kadhim Hussein Ali 431
Kadhim Inaam H............164
Kadhim Zina F............366
Kambarova Galiya 23
Kamil Khaleel Hadeel 40
Khadzhiev Azamat 483
Khalid Zainab............349, 443
Khamidov Abdusamad 262
Khanjar Ebtesam F. 131, 181
Khanjer Ebtesam F............209
Khudayberganov Erkaboy 468
Khudhur Jameel Abeer............299
Khursanov Dunyobek 252
Kolchina Vera 201
Kolobov Egor............100
Komilova Nargiza............271
Kosimov Dilshodbek 271
Kumarkhan Mariam............1
Kurambayev Sherzod 236, 476
Madaminov Zafar 279
Mahdi Lateef Ishraq 287
Majeed Kareem Oras............59
Majid Abdulameer Maitham............151
Makhkamov Elyorbek 271, 279
Makki Tiba M............349
Malikov Bobur............252
Mancini Simona............46
Masharipova Sharofat 14, 492

AUTHOR INDEX

Matchanov Sherzad 244
Matchanov Sherzod 229
Matyaqubova Karomat 468
Mazhar Abdul Hamid Raghad 394
Mezaal Mahmood Rawya 151
Mirzabayev Berik 501
Mohammed Faisel G. 123, 366
Mohammed Hadi Qunoot 65
Mohammed Hala R. 73
Mohammed Mohammed A. 123
Mohammed Torki Wedyan 40, 54
Moiseenko Svetlana 201
Mouaid Ali Roua 509
Moutaz Asmael Noor 287
Mozihim Rana 173
Muhsin Israa J. 131
Mukhamedyanova Adelina 1
Musa Sarab J. 114
Musabekov Akhmetbek 501
Musabekov Nazarbek 501
Mustafina Olga 201
Nafea Yousif Saja 394
Naji Mohsion Hanan 114
Nazarbek Ulzhalgas 23
Nazarbekova Saule 23
Obayes Esraa S. 65
Pustuev Andrey 93
Qasim Ahmed Anwar 431
Qassim Gadhban Ansam 509
Qiming Sun 201
Radhi Jasim Hawraa 59
Raiymbekov Yerkebulan 23
Ravshanov Shokhrukh 252
Ruzmetova Aida 244
Rykalina Olga 93, 100
Rykov Andrey 201
Sabah Jaber Hussein 456
Sabeeh Mosleh Jehan 142
Saeed Jassim Ali 40, 54, 59, 394, 431
Saleem Noora M. 181
Salim Obayes Esraa 114
Salman Jasim M. 73
Samer Muhend K. 366
Sapayeva Surayyo 476
Sarsadskikh Anastasia 93
Sarsenova Adel 1
Saydullayeva Surayyo 262
Sitmuratov Tulkinbek 468
Smanova Zulayho 9
Srayyih Mustafa S. 181
Subhi Halboos Nagham 193
Sushchenko Maxim 86
Takhumova Oksana 93, 100
Tariq Battal Zahraa 423
Tariq Hussein Allwan Bilal 423
Turkmenbaeva Mayra 14
Utepbergenov Irbulat 501
Vlasova Irina 201
Waleed Mustafa Suha 325
Yakubov Yusufboy 244, 483
Yousif Al-Mashhadani Mays 520
Yusupova Kamola 262